Databases for Data-Centric Geotechnics

Databases for Data-Centric Geotechnics forms a definitive reference and guide to databases in geotechnical and rock engineering, to enhance decision making in geotechnical practice using data-driven methods. This first volume pertains to site characterization. The opening chapter presents an in-depth analysis of site data attributes, including the establishment of a new taxonomy of site data under "4S" (site generalizations, spatial features, sampling characteristics, and smart data) to provide a novel agenda for data-driven site characterization. Type 3 machine learning methods (disruptive value) are possible as sensors become more pervasive and more intelligent. A comprehensive overview of site characterization information is also presented with a focus on its availability, coverage, value to decision making, and challenges. The remaining 13 chapters cover databases of soil and rock properties and the application of these databases to rock socket behavior, rock classification, settlement on soft marine clays, permeability of fine-grained soils, and liquefaction among others. The databases were compiled from studies undertaken in many countries including Austria, Australia, Brazil, Canada, China, France, Finland, Germany, India, Iran, Japan, Korea, Malaysia, Mexico, New Zealand, Norway, Singapore, Sweden, Thailand, the United Kingdom, and the United States.

This volume on site characterization is a companion to the volume on geotechnical structures. *Databases for Data-Centric Geotechnics* represents the most diverse and comprehensive assembly of database research in a single publication (consisting of two volumes) to date. It follows from *Model Uncertainties for Foundation Design*, also published by CRC Press, and suits specialist geotechnical engineers, researchers, and graduate students.

Kok-Kwang Phoon is Cheng Tsang Man Chair Professor and President of the Singapore University of Technology and Design. He was awarded the ASCE Norman Medal twice in 2005 and 2020, the Humboldt Research Award in 2017, and the ASCE Alfredo Ang Award in 2024. He is the Founding Editor of *Georisk* and *Geodata and AI*. He is a Fellow of the Academy of Engineering Singapore and Singapore National Academy of Science.

Chong Tang is Professor at Dalian University of Technology in China and a former senior research fellow at the National University of Singapore. He was awarded the ASCE Norman Medal in 2020.

Challenges in Geotechnical and Rock Engineering

Challenges in Geotechnical and Rock Engineering presents advanced level books which focus on state of the art methods for handling problems across geotechnical engineering. Typical books are in include measuring and understanding material or structural behaviours, handling special environmental conditions, performing complex numerical simulation, design and construction for novel structures, life cycle and risk/reliability informed management, data-driven algorithms and AI-based decision making, resilience engineering, addressing climate change and sustainability, and digital twin and smart infrastructure.

Evolutionary Process of a Steep Rocky Reservoir Bank in a Dynamic Mechanical Environment
Luqi Wang and Wengang Zhang

Uncertainty, Modelling, and Decision Making in Geotechnics
Edited by Kok-Kwang Phoon, Takayuki Shuku and Jianye Ching

Bayesian Machine Learning in Geotechnical Site Characterization
Jianye Ching

Databases for Data-Centric Geotechnics: *Site Characterization (Volume 1)*
Edited by Kok-Kwang Phoon and Chong Tang

Databases for Data-Centric Geotechnics: *Geotechnical Structures (Volume 2)*
Edited by Chong Tang and Kok-Kwang Phoon

For more information about this series, please visit: www.routledge.com/Challenges-in-Geotechnical-and-Rock-Engineering/book-series/CGRE

Databases for Data-Centric Geotechnics

Site Characterization

Edited by

Kok-Kwang Phoon and Chong Tang

Cover image: BEST-BACKGROUNDS/Shutterstock

First edition published 2025
by CRC Press
2385 NW Executive Center Drive, Suite 320, Boca Raton FL 33431

and by CRC Press
4 Park Square, Milton Park, Abingdon, Oxon, OX14 4RN

CRC Press is an imprint of Taylor & Francis Group, LLC

ISBN: 978-1-032-57895-8 (hbk)
ISBN: 978-1-032-57988-7 (pbk)
ISBN: 978-1-003-44194-6 (ebk)

DOI: 10.1201/9781003441946

Typeset in Sabon
by Deanta Global Publishing Services, Chennai, India

Access the Support Material: www.routledge.com/9781032578958

The Open Access version of Chapter 6 was funded by University of Bristol.

Contents

Challenges in geotechnical and rock engineering

Geotechnical and rock engineering studies have made significant strides in response to different challenges and opportunities that include measuring and understanding material/structural behaviors, handling special environmental conditions, performing complex numerical simulations, design, construction methods, and circular roles for low-carbon materials, novel structures, green construction and operation, life cycle and risk/reliability informed management, data-driven algorithms and AI-based decision making, autonomous systems, digital twin and smart infrastructure, resilience engineering, climate change and sustainability, among others. These challenges are interrelated. Although some of the challenges are common to all industries, it is not meaningful to engage them in an abstract manner outside of the practice context. One example is data-centric geotechnics that takes a "data first" approach to decision making, but the data are actual "ugly" field observations (multisource, sparse, incomplete, spatially variable, corrupted, etc.) rather than ideal abstract numbers. Data-centric geotechnics should deploy digital technologies in cognizance of the context of geotechnical and rock engineering that includes physics, empirical knowledge, experience, and engineering judgment. How decision making in geotechnical and rock engineering can be revolutionized through human–machine collaboration is one grand challenge that epitomizes the motivation for launching this book series. This book series presents exciting emerging solutions in geotechnical and rock engineering that are expected to transform practice and meet fast-evolving environmental/societal trends in the twenty-first century. It is a timely response to the changing technological, environmental, and societal landscape presented in the Institution of Civil Engineers (ICE) State of the Nation Report on "Digital Transformation" and the American Society of Civil Engineers (ASCE) "Future World Vision: Infrastructure Reimagined" paper.

Book series editors
Kok-Kwang Phoon
Dongming Zhang

Foreword

Many nations have identified that the development of digital capabilities is a key strategic priority, such as the development of intelligent systems to assess the performance of infrastructure assets and inform maintenance requirements (the National Academy of Sciences 2019, Communist Party of China Central Committee (CPCCC) 2019, the Institution of Civil Engineers (ICE) 2019, and Australian Department of Infrastructure and Regional Development (DIRD) 2016). This includes big data analytics and machine learning concepts. At the same time, there is uncertainty about whether artificial intelligence will just automate repetitive tasks or make work more precarious, and how climate action and AI could create more and better jobs (World Economic Forum; weforum.org).

I do not share a fear of technological disruption. The personal computer revolutionalized engineering in the 1990s. Instead of calculating factors of safety for slope stability problems by manually drawing a circle and doing the calculations over several hours, an engineer could do thousands of calculations in 5 min. That did not mean that engineers lost their jobs, it just meant they could do many more calculations in a shorter period of time. Similarly, drafters used to create drawings by hand. They now create digital models. The role of the drafter has changed, but the number of jobs currently available in Australia is much larger than the number of people available to fill them.

Productivity in the construction industry has stagnated for 30 years (Infrastructure Australia 2022). Increasing populations demand more infrastructure but there are competing needs for investment, such as energy transition, which means less money may be available for other uses. Increasing productivity means that we can provide our communities with more of the infrastructure they desire. Automation in various forms including machine learning will increase in scope and capability to help address productivity requirements.

It is my view that in the (near) future people with a deep understanding of their subjects will survive and thrive and everything else will become automated. Hence, it is of vital importance that geotechnical professionals develop a deep understanding of machine learning as it applies to geotechnical engineering. Machine learning has two components that all engineers should understand to some extent in order to be able to effectively use emerging techniques. These are (1) the algorithms and mathematical optimization that are used to interrogate data and (2) the data. I have dabbled for some time with the use of Bayesian updating to predict the behavior of embankments constructed on soft clay. This is a physics-based application of machine learning where measurements of settlement at the surface, at various points in the ground, and pore pressures are compared to predictions from conventional consolidation theory and used to update the predictions. Typically, we have been overpredicting measured pore pressures at the end of filling and underpredicting pore pressures after a period of time. The reason for this is that real soil typically does not generate much pore pressure until the yield pressure is reached, whereas consolidation theory states that the entire applied stress increment is immediately transferred to the pore water. The likelihood function within the Bayesian algorithm tries to minimize the error between prediction and measurement, and because the consolidation model overpredicts initial pore pressure generation, the function compensates by underpredicting pore pressures later in time so that the net difference between measurement and theory is almost zero. This is both an example of the need to know how the algorithms work and an example of the need to understand what the data means relative to the consolidation model. The latter issue can potentially be eliminated using purely data-driven methods advocated by the editors of *Databases for Data-Centric Geotechnics: Site Characterization* and *Databases for Data-Centric Geotechnics: Geotechnical Structures*.

While knowledge of algorithms and data are both important, the geotechnical professional will find that they need to spend most of their time understanding, cleaning, and storing data in a format that can be interrogated by algorithms, which will mostly be written by others. Anecdotally, up to 80% of the effort is applied to data and the rest to training the algorithms and using the data. Consequently, the focus of these two books on databases for site characterization and predictive performance is of fundamental importance. All geotechnical engineers will know the challenge of collecting data, let alone managing the data, as well as the challenges posed by the MUSIC-X concept (Phoon et al, 2019). MUSIC-X means Multivariate, Uncertain and Unique, Sparse, Incomplete, and potentially Corrupted, with "X" denoting the spatial/temporal dimension. Commercial and insurance issues relating to whether an engineer can rely on data collected by other people further complicate matters. The types of data that are collected will also change in the future as methods of digital surface and subsurface measurement are developed. These changes will enhance the use of machine learning techniques and should bring improvements in productivity.

There is a view that the ability of our profession to characterize sites has regressed in recent years (DeGroot and DeJong, 2020). While it is possible that data-driven machine learning approaches can compensate for the "potentially Corrupted" element of MUSIC-X, it is also possible that the widespread use of machine learning will drive greater efforts to collect high-quality data. This has the potential to transform the practice of site characterization into the future.

Arguably, the key skill of the geotechnical professional is to make decisions, or help other people make decisions, in the face of uncertainty. I am very pleased to see the editors focus on the ability of machine learning to help make decisions in their books. In my view, decisions will be made by people who have confidence in the advice provided by geotechnical professionals who have a deep understanding of their subjects. Ideally, decisions are made as early as possible to take advantage of opportunities or efficiently manage risks, and machine learning will become a powerful tool to aid geotechnical professional in providing appropriate advice.

In focusing on site investigation and field performance, these two books provide many examples of how machine learning is being used to manage and interpret data. They provide our profession with insight into the current state-of-the-art and what the future of geotechnical practice might become. They will become invaluable reference material as the field matures. I congratulate the editors and the authors for their work and commend these books as required reading for all geotechnical professionals.

Dr Richard Kelly
General Manager of Technical Excellence & Chief Technical Principal for Geotechnical Engineering, SMEC Australia.
Conjoint Professor of Practice, University of Newcastle, Australia

REFERENCES

Australian Government Department of Infrastructure and Regional Development. 2016. Trends Transport and Australia's Development to 2040 and beyond

DeGroot, D.J. and DeJong, J.T. 2020. Best Practices for Geotechnical Site Characterization: Have we regressed from decades past? Geostrata magazine, 24(2), 50-57.

Infrastructure Australia (2023). Infrastructure market capacity 2022 report, https://www.infrastructureaustralia.gov.au/publications/2022-market-capacity-report

Institution of Civil Engineers UK 2019 What should be in the National Infrastructure Strategy? https://www.ice.org.uk/getattachment/news-and-insight/policy/what-should-national-infrastructure-strategy/What-should-be-in-the-National-Infrastructure-Strategy-ICE-July-2019.pdf.aspx

National Academies of Sciences, Engineering, and Medicine 2019. Critical Issues in Transportation 2019. Washington, DC: The National Academies Press. https://doi.org/10.17226/25314

Phoon, K. K., Ching, J., and Wang Y. 2019. Managing risk in geotechnical engineering – from data to digitalization. Proceedings, 7th International Symposium on Geotechnical Safety and Risk (ISGSR 2019), Taipei, Taiwan, 13-34. https://www.issmge.org/uploads/publications/96/97/SL.pdf

The Communist Party of China Central Committee and the State Council 2019. Program of Building National Strength in Transportation. http://www.gov.cn/xinwen/2019-09/19/content_5431432.htm (in Chinese)

Preface

An initiative to promote public sharing of databases, called 304dB, was initiated by Professor Kok-Kwang Phoon and Jianye Ching in 2017: http://140.112.12.21/issmge/tc304.htm. Despite its success (304dB has contributed to the development of data-driven methods by situating these methods in the context of real data over the past seven years), it is timely to take stock and discuss how this effort can be expanded to cover more types of databases such as multi-modal databases, more life cycle stages beyond design, more countries, and more projects. The purpose of this book which is published in two volumes covering "site characterization" and "geotechnical structures" is to expand the frontiers of 304dB, to share new data-driven methods (e.g. machine learning) and applications for data-centric geotechnics, and to identify opportunities and challenges in research and practice.

The publication of *Databases for Data-Centric Geotechnics* in 2025 is timely. The breakout of powerful generative AI such as ChatGPT in 2023 should give new impetus to collecting data, understanding data, protecting data as a critical asset, developing learning algorithms to transform data into intelligence, and reimagining infrastructure as a service to end-users through deployment of smart technologies. The vision for taking a data-centric perspective to geotechnical engineering (data-centric geotechnics) to march toward these top-level goals is more compelling than before. An initial study on large language models was conducted recently (Wu et al. 2024). Geo-everything-dB can be a hashtag for collecting, understanding, and protecting geotechnical data across the entire life cycle of the structure.

Data-centric geotechnics is distinctive from other fields in geotechnical engineering and machine learning (ML) (Phoon and Ching 2021; Phoon et al. 2022a). It is a new interdisciplinary field involving the integration of information, data, techniques, tools, perspectives, concepts, theories, and/or experiences from geotechnical engineering and machine learning (Phoon and Zhang 2023). A broader development in "data-centric engineering" (Girolami 2020) and data-driven discovery in geosciences (Chen et al. 2023) are taking place elsewhere in parallel. It is incorrect to say that geotechnical engineering has always been data-centric in the strict sense. It has been experience-centric or physics-centric, but to the knowledge of the authors, data has never supported decision making by itself. At present, data only supports physics or experience in decision making. Data centricity does not imply that experience and physics are not relevant. If data can support experience and physics, the reverse argument is also true. In machine learning, experience can be regarded as one type of "thick data" to distinguish it from the more well-known quantitative "big data." Physics-informed machine learning (PIML) methods offer the advantages of "explainability" and "interpretability" (Karniadakis et al. 2021). Large language models can be trained on the entire body of geotechnical engineering literature. One should not equate data to numerical data alone.

Many research papers on machine learning in geotechnics (MLIG) are "method first" rather than "data and practice first." Numerous foundational challenges beyond methodological ones can be identified when research is pursued under a more balanced data-centric geotechnics agenda covering data centricity, fit for (and transform) practice, and geotechnical context (Phoon and Zhang 2023; Phoon et al. 2023b). Research in the context of data-centric geotechnics must articulate its value proposition clearly and should be directed to compete within a benchmarking framework (Phoon et al. 2022b; Phoon et al. 2023a) to hasten the rate of digital transformation in the industry. Some of the databases in *Site Characterization* and *Geotechnical Structures* can provide the training and validation datasets for new benchmark examples to improve specific machine learning tasks. The value of MLIG research can be classified as follows: (1) Type 1 (incremental value) involving available data and existing conventional applications, (2) Type 2a (potentially high value) involving available data and new applications, (3) Type 2b (high value) involving new data and existing applications, and (4) Type 3 (disruptive value) involving new data and completely novel applications (e.g., precision construction, autonomous construction, human-machine teaming) (Phoon and Zhang 2023). Type 1 MLIG provides an alternate approach to an existing problem. It is only "good to have." Type 2 MLIG outperforms existing approaches (if they exist) to such an extent that it will replace them. Type 3 MLIG will make practice unrecognizable by present-day engineers. These MLIG methods are distinctive from reliability methods, although they are both data-driven. MLIG methods can improve by learning from data and may eventually even modify or propagate themselves (an AI programming a better AI). Reliability methods address uncertainty explicitly, but the classical ones do not contain a learning step. Hence, it is possible to classify reliability methods as Type 0.

Notwithstanding rapid progress in the last two or more years, there are major gaps in the current data-centric geotechnics agenda. For example, the agenda should not focus purely on maximizing the value to practice but also on minimizing the leakage of sensitive data. A proactive approach to address potential data privacy and security risks enhances the trust and confidence of users and stakeholders. Hence, Chapter 1 of "site characterization" suggested that the "data first practice central" agenda in data-centric geotechnics can be expanded to "data first, practice central, and privacy enhanced." Phoon (2025) advocated updating "data-centric geotechnics" to "trustworthy data-centric geotechnics" to manage the risk of ML/AI more systematically in response to the EU AI Act. Many exciting Type 2 MLIG research projects can be conducted. One example is data-driven site characterization (DDSC) (Phoon et al. 2022c; 2025). The construction of a quasi-site-specific ground model that includes learning cross-correlations and spatial correlations from generic databases under MUSIC-3X (Multivariate, Uncertain and Unique, Sparse, Incomplete, and potentially Corrupted with "3X" denoting 3D spatial variability) is within reach in a few years. Interest in combining site characterization information with monitoring information collected during construction to guide decision making in real time is emerging rapidly. This agenda is called the machine learning-guided observational method (MLOM). Its value is arguably higher than DDSC, and it can lead to Type 3 applications in practice. It is uncertain if a new discipline called geotechnical data science is needed to provide a stronger theoretical underpinning for data-centric geotechnics, given the large interconnected challenges.

A commonly encountered reaction to data-centric geotechnics is "data is not enough." Although this reaction is understandable, it is not helpful for two reasons: (a) there are no insights on what, where, and why data is not enough and (b) a data-poor environment is commonly seen as a showstopper rather than as a driver for data-driven innovations. It is recognized that MUSIC-3X is a basic challenge to DDSC, MLOM, and other broad application agendas in data-centric geotechnics. In fact, sparsity, uncertainty, and spatial variability are only three out of seven attributes. Current MLIG research is already handling data with much greater realism than prevailing deterministic and probabilistic methods. In fact, current MLIG research handles data "as is" with almost no idealizations. Chapter 1 of "site characterization" proposed a new taxonomy of site data under "4S" to provide a novel agenda for future MLIG research. The "4S" are site generalizations, spatial features, sampling characteristics, and smart data. The topics in the "4S" agenda are important to practice, but no work has been done testifying to the nascency of data-centric geotechnics. In addition to an in-depth analysis of geotechnical data attributes, this chapter also presents an overview of site characterization information with a focus on its availability, coverage, value to decision making, and challenges. Chapter 1 of "geotechnical structures" presents a survey of performance databases and the effectiveness of our prediction models in matching the field measurements in these databases based on three significant sources of information in the literature: (1) full-scale field tests, mainly foundation load tests, (2) prediction exercises conducted as part of international conferences, and (3) comparison between numerical analyses and in situ or field measurements.

The current book constitutes a modest step forward in data-centric geotechnics, although the database compilation efforts undertaken by the invited authors are significant. This book also represents the most diverse and most comprehensive assembly of database research in a single publication (consisting of two volumes) to date. The difference between a database assessment and a case history analysis is rarely emphasized. The current literature is replete with case history analyses, but they are primarily conducted to elucidate or improve our qualitative understanding of the governing physical mechanisms. They are not intended to quantify prediction errors in a representative way. For those studies that do compute several prediction errors, they do not attempt to characterize the prediction errors statistically with sufficient robustness for reliability-based design (sample size is too small). Reporting a prediction error for a single or a handful of case histories does not assist an engineer in decision making for a new problem. The prediction error for a new problem can be larger or smaller. It can be conservative or unconservative. In short, the engineer is none the wiser. The need to understand prediction errors statistically for a population of problems (not a single problem) is not well appreciated in a practice that remains deterministic in nature. The prediction error for a population of problems is a random variable (or it can be transformed into a random variable after removing dependencies on input parameters). Statistical characterization of the prediction error *probability distribution* requires a systematic compilation of data in the form of a database that remained a relative rarity in the past. Contrary to common perception that occasionally surfaces even in a design code, it is not possible to ensure that a calculation model is conservative with complete certainty. In the presence of a random prediction error, it is only meaningful to control the *probability* of a model being unconservative below a fractile, say 5%. The contributions in this book are already useful for design in the presence of the above model bias and imprecision. One example is the statistical characterization of the model factor to support uncertainty-informed design practice in parts of the Eurocode 7, in AASHTO's load and resistance factor design (LRFD), and in the first-order reliability-based design method permitted as an alternate design check in new design codes.

In the first volume *Site Characterization*, the properties of a wide variety of soils/rocks are covered:

- Overview – Chapter 1 (Phoon et al. 2024) presents the role of site characterization information in data-centric geotechnics regarding data availability, coverage, data attributes, value to decision making, and challenges.
- Chapter 2 (Asem and Gardoni 2024) presents the use of large databases in the development and analysis of rock socket behavior. The first database is related to full-scale field load tests on rock-socketed piles (e.g., shaft shearing, end bearing) and the other is associated with rock permeability from laboratory and in situ tests.
- Chapter 3 (Ching 2024) presents an extensive review of soil/rock property databases and their application in determining site-specific statistics for the inherent variability of soil/rock properties (e.g., mean, coefficient of variation, and scale of fluctuation) using the hierarchical Bayesian model.
- Chapter 4 (D'Ignazio and Länsivaara 2024) presents the collection and use of a high-quality database ANI-F-CLAY/7/87 to statistically establish the transformation models for estimating the anisotropic undrained shear strength of Finnish clays. They are validated against existing databases (e.g., F-CLAY/7/216 and S-CLAY/7/168) that cover a broader range of basic properties, particularly pre-consolidation stress and index properties.
- Chapter 5 (Eslami and Mo 2024) underscores the importance of soil characterization and classification in geotechnical engineering, discussing methods, uncertainties, and advancements. It presents the AUT:CPTu-Geo-Marine database that consists of 398 CPTu records from 58 different locations in 18 countries. The database is used to develop a novel triangular chart for soil behavior evaluation.
- Chapter 6 (Feng and Vardanega 2024) presents the use of a new database FG/KSAT-1358 to evaluate the variability of fine-grained soil hydraulic conductivity and study the accuracy of some established transformation models for prediction of hydraulic conductivity.
- Chapter 7 (Hov et al. 2024) presents the use of a bivariate database (Norwegian and Swedish clays) to study the correlation between normalized undrained shear strength and liquid limit for normally to slightly over-consolidated clay (OCR $\leq$ 1.5), with particular attention to uncertainties and sample quality.
- Chapter 8 (Liu and Zou 2024) presents 1,102 triaxial compression test records for 332 types of gravels with 15 parameters affecting the gravel behavior. The database is termed Geo-Gravel/15/1102, which is used to predict the shear strength and stress–strain–volume response of gravels with empirical and deep learning models.
- Chapter 9 (Löfman and Korkiala-Tanttu 2024) presents the use of an incomplete multivariate database FI-CLAY/14/856 to evaluate the compressibility of soft marine clays and its effect on settlement response.
- Chapter 10 (Ma and Li 2024) presents the use of the rock mass database SurroundingRock/8/286 in the development of an intelligent and accurate rock classification model using machine learning (e.g., convolutional neural network, random forest, Gaussian process, and support vector machine).
- Chapter 11 (Otake et al. 2024a) presents a large clay database from a single site in Japan – Tokyo-CLAY/14/67760 (2,972 boreholes covering a site area of 10 km^2) that can be used as a benchmark example for data-driven site characterization.

- Chapter 12 (Ouyang et al. 2024) presents statistical analyses of laboratory and in situ testing data to evaluate the pre-consolidation stress, over-consolidation ratio, and effective strength parameters of soil.
- Chapter 13 (Wang et al. 2024a) presents Bayesian-random field modeling of soil parameters (e.g., cohesion, friction angle, and Young's modulus) required for geotechnical analysis and design.
- Chapter 14 (Wang et al. 2024b) presents the assessment of seismic soil liquefaction with 155 CPT profiles from the New Zealand Geotechnical Database (NZGD).
- Chapter 15 (Zhou et al. 2024) presents the development and use of a large database SH-CLAY/12/4191 to evaluate the stress history (K_0), stiffness (E_s), cohesion (c), and friction angle (ϕ) of Shanghai soft clay from index parameters such as the water content, Atterberg limits, and void ratio.

In the second volume *Geotechnical Structures*, the performances of a wide variety of geotechnical structures are covered:

- Overview – Chapter 1 (Phoon and Tang 2024) presents a survey of performance databases and the effectiveness of our prediction models in matching the field measurements in these databases based on (1) full-scale field tests, (2) 39 prediction exercises organized as part of international conferences, and (3) comparison between numerical analyses and in situ or field measurements conducted by the French LCPC. The geotechnical structures include embankments, shallow and deep foundations, tunnels, and retaining walls. The focus is on the evaluation of the *statistical* degree of confidence in predicting various quantities of interest such as capacity and deformation.
- This volume covers foundations that distribute the loads from the superstructure (e.g., high-rise building, highway bridge, offshore working platform) to the bearing layer, including:
  - Spudcan foundations – Chapter 2 (Tang et al. 2024) presents an extensive review of the study (experimental, analytical, and numerical) on the punch-through behavior of spudcan in stiff-over-soft clay and clay with sand, along with the development and use of a centrifuge test database – DUT/Foundation/Punch-Through/242.
  - Deep foundations – Chapters 3–11 present the development and use of load test databases worldwide. These databases can be categorized into two types. The first is national where almost all of load tests were performed in a single country or region, e.g., France in Chapter 3 (Burlon and Frank 2024), South Africa in Chapter 5 (Dithinde and Moriasi 2024), the United Kingdom in Chapter 7 (Flynn and McCabe 2024), Malaysia in Chapter 8 (Ong et al. 2024), Japan in Chapter 9 (Otake et al. 2024b), United Kingdom in Chapter 10 (Othman et al. 2024), and Brazil in Chapter 11 (Tsuha et al. 2024). The others in Chapter 4 (Chen et al. 2024) and Chapter 6 (Eslami and Heidarie Golafzani 2024b) are generic in the sense that the databases contain load tests from many countries or regions.

These databases are widely used by researchers for the evaluation and revision of existing analysis methods that are recommended in design codes or manuals (e.g., Belgium, France, Germany, Japan, and the United States), development of more accurate and economical design methods (e.g., implementation of load and resistance factor design especially in the United States), and by engineers for the improvement of foundation design in their project (e.g., Florida and Louisiana Departments of Transportation).

- Anchors and pipelines
  - Chapter 12 (Fu et al. 2024) presents a comprehensive collection of 464 datasets obtained from centrifuge and field model tests of dynamically installed anchors in development of data-driven design methods.
  - Chapter 13 (Najjar et al. 2024) presents the collection and use of 143 uplift laboratory tests on pipes in sand and 70 field anchor pullout tests in the data-driven quantification of uncertainty and design of offshore pipelines and anchored shoring systems.
- Retaining systems and excavations
  - Chapter 14 (Bathurst and Miyata 2024) presents an extensive review of database assessment of model uncertainty in the internal stability limit states of mechanically stabilized earth walls constructed with extensible geogrids and polyester straps, and inextensible steel strip and grid soil reinforcement materials.
  - Chapter 15 (Lin 2024) presents the use of five databases to characterize the variability of predictions regarding global failure, nail bonding strength, nail load, facing tensile force, and nail displacement of soil nail walls.
  - Chapter 16 (Tan and Fan 2024) presents the use of a large database of 606 case records during 1995–2018 to address what are the main features of deep excavations in Shanghai soft clays and what are the key parameters affecting this behavior.
- Landslides – Chapter 17 (Zhang et al. 2024) presents the development of probabilistic methods to predict time to slope failure with a database of 55 landslides.

Additional supplementary resources are made available at: www.routledge.com/9781032578958 (*Site Characterization*) and www.routledge.com/9781032579108 (*Geotechnical Structures*).

The contributions of the invited authors are deeply appreciated. The editors would also like to thank Mr Tony Moore and Ms Aimee Wragg from CRC Press & Routledge for their tireless support and patient guidance.

Editors
Kok-Kwang Phoon
Chong Tang

REFERENCES

Asem, P. and Gardoni, P. 2024. Selection of rock hydromechanical parameters for rock foundation design: a database approach. Chapter 2 in *Databases for Data-Centric Geotechnics: Site Characterization*, edited by K. K. Phoon and C. Tang. Boca Raton, FL: CRC Press.

Bathurst, R. and Miyata, Y. 2024. Mechanically stabilized earth (MSE) walls. Chapter 14 in *Databases for Data-Centric Geotechnics: Geotechnical Structures*, edited by C. Tang and K. K. Phoon. Boca Raton, FL: CRC Press.

Burlon, S. and Frank, R. 2024. Development and use of LCPC pile database. Chapter 3 in *Databases for Data-Centric Geotechnics: Geotechnical Structures*, edited by C. Tang and K. K. Phoon. Boca Raton, FL: CRC Press.

Chen, G., Cheng, Q., and Puetz, S. 2023. Special issue: data-driven discovery in geosciences: opportunities and challenges. Mathematical Geosciences, 55, 287–293.

Chen, Y. J., et al. 2024. Development and use of CYCU pile load test database. Chapter 4 in *Databases for Data-Centric Geotechnics: Geotechnical Structures*, edited by C. Tang and K. K. Phoon. Boca Raton, FL: CRC Press.

Ching, J. 2024. Evaluation of soil/rock properties using databases. Chapter 3 in *Databases for Data-Centric Geotechnics: Site Characterization*, edited by K. K. Phoon and C. Tang. Boca Raton, FL: CRC Press.

D'Ignazio, M. and Länsivaara, T. 2024. Undrained shear strength of Finnish soft clays: a database perspective. Chapter 4 in *Databases for Data-Centric Geotechnics: Site Characterization*, edited by K. K. Phoon and C. Tang. Boca Raton, FL: CRC Press.

Dithinde, M. and Moriasi, H. Pile load test database for Southern Africa and evaluation of direct SPT-based pile design methods. Chapter 5 in *Databases for Data-Centric Geotechnics: Geotechnical Structures*, edited by C. Tang and K. K. Phoon. Boca Raton, FL: CRC Press.

Eslami, A. and Mo, P. Q. 2024a. Role of databases in the evaluation of soil properties. Chapter 5 in *Databases for Data-Centric Geotechnics: Site Characterization*, edited by K. K. Phoon and C. Tang. Boca Raton, FL: CRC Press.

Eslami, A., and Heidarie Golafzani, S. 2024b. Relevant data base pile design approach. Chapter 6 in *Databases for Data-Centric Geotechnics: Geotechnical Structures*, edited by C. Tang and K. K. Phoon. Boca Raton, FL: CRC Press.

Feng, S. and Vardanega, P. 2024. New laboratory database of hydraulic conductivity measurements on fine-grained soil. Chapter 6 in *Databases for Data-Centric Geotechnics: Site Characterization*, edited by K. K. Phoon and C. Tang. Boca Raton, FL: CRC Press.

Flynn, K. N. and McCabe, B. A. 2024. Piling insights from a data-centric approach. Chapter 7 in *Databases for Data-Centric Geotechnics: Geotechnical Structures*, edited by C. Tang and K. K. Phoon. Boca Raton, FL: CRC Press.

Fu, Y., Ding, K. and Han, C. C. 2024. Experimental database and data-driven design method of dynamically installed anchors. Chapter 12 in *Databases for Data-Centric Geotechnics: Geotechnical Structures*, edited by C. Tang and K. K. Phoon. Boca Raton, FL: CRC Press.

Girolami, M. 2020. Introducing data-centric engineering: an open access journal dedicated to the transformation of engineering design and practice. Data-Centric Engineering, 1:e1.

Hov, S., L'Heureux, J. S., Kahrs, K., Lundström, K. and Gaharia, D. 2024. Normalised active undrained shear strengths of soft Scandinavian clays – a data-centric and a geomechanical approach. Chapter 7 in *Databases for Data-Centric Geotechnics: Site Characterization*, edited by K. K. Phoon and C. Tang. Boca Raton, FL: CRC Press.

Karniadakis, G. E., Kevrekidis, I. G., Lu, L., Perdikaris, P., Wang, S. and Yang, L. 2021. Physics-informed machine learning. Nature Reviews Physics, 3, 422–440, DOI: 10.1038/s42254-021-00314-5.

Lin, P. Y. 2024. Statistical evaluation and calibration of model uncertainty for reliability-based design of soil nail walls. Chapter 15 in *Databases for Data-Centric Geotechnics: Geotechnical Structures*, edited by C. Tang and K. K. Phoon. Boca Raton, FL: CRC Press.

Liu, J., Zou, D. 2024. Prediction for the mechanical response of gravels. Chapter 15 in *Databases for Data-Centric Geotechnics: Site Characterization*, edited by K. K. Phoon and C. Tang. Boca Raton, FL: CRC Press.

Löfman, M. S. and Korkiala-Tanttu, L. 2024. Evaluation of compressibility properties for soft marine clays. Chapter 9 in *Databases for Data-Centric Geotechnics: Site Characterization*, edited by K. K. Phoon and C. Tang. Boca Raton, FL: CRC Press.

Ma, J. and Li, T. 2024. An engineering geological parameter database of tunnel surrounding rock and its application. Chapter 10 in *Databases for Data-Centric Geotechnics: Site Characterization*, edited by K. K. Phoon and C. Tang. Boca Raton, FL: CRC Press.

Najjar, S., Sadek, S., Ismail, S. and Chahbaz, R. 2024. Pipes and anchors. Chapter 13 in *Databases for Data-Centric Geotechnics: Geotechnical Structures*, edited by C. Tang and K. K. Phoon. Boca Raton, FL: CRC Press.

Ong, Y. H., Toh, C. T. and Chee, S. K. 2024. Bored piles in tropical soils and rocks – a database approach to design. Chapter 8 in *Databases for Data-Centric Geotechnics: Geotechnical Structures*, edited by C. Tang and K. K. Phoon. Boca Raton, FL: CRC Press.

Otake, Y., Saito, T., Wu, S., Yoshida, I. and Takano, D. 2024a. Exploring challenges via analysis of multivariate geotechnical properties: insights from large-scale sampling of Japanese marine clay. Chapter 11 in *Databases for Data-Centric Geotechnics: Site Characterization*, edited by K. K. Phoon and C. Tang. Boca Raton, FL: CRC Press.

Otake, Y., Nakamura, T., Fujita, T. and Nishida, H. 2024b. Development and Use of Databases of Pile Foundation Load Tests in Japan. Chapter 9 in *Databases for Data-Centric Geotechnics: Geotechnical Structures*, edited by C. Tang and K. K. Phoon. Boca Raton, FL: CRC Press.

Othman, M., Voyagaki, E., Crispin, J. J., Ntassiou, K., Gilder, C. E. L., De Luca, F., Mylonakis, G. E. and Vardanega, P. J. 2024. The DINGO database of axial pile load tests for the UK: determination of ultimate load. Chapter 10 in *Databases for Data-Centric Geotechnics: Geotechnical Structures*, edited by C. Tang and K. K. Phoon. Boca Raton, FL: CRC Press.

Ouyang, Z. and Agaiby, S. 2024. In situ test-based evaluation of soil effective stress strength properties and stress history. Chapter 12 in *Databases for Data-Centric Geotechnics: Site Characterization*, edited by K. K. Phoon and C. Tang. Boca Raton, FL: CRC Press.

Phoon, K. K. 2025. Trustworthy data-centric geotechnics. Geodata and AI.

Phoon, K. K. and Ching, J. 2021. Project DeepGeo – data-driven 3D subsurface mapping. Journal of GeoEngineering, 16(2), 61-74.

Phoon, K. K., and Zhang, W. G. 2023. Future of Machine Learning in Geotechnics. Georisk: Assessment and Management of Risk for Engineered Systems and Geohazards, 17(1), 7-22.

Phoon, K. K., Ching, J. and Cao, Z. J. 2022a. Unpacking data-centric geotechnics. Underground Space, 7(6), 967–989.

Phoon, K. K., Shuku, T., Ching, J. and Yoshida, I. 2022b. Benchmark examples for data-driven site characterization. Georisk: Assessment and Management of Risk for Engineered Systems and Geohazards, 16(4), 599-621.

Phoon, K. K., Ching, J. and Shuku, T. 2022c. Challenges in data-driven site characterization. Georisk: Assessment and Management of Risk for Engineered Systems and Geohazards, 16(1), 114–126.

Phoon, K. K., Shuku, T., Ching, J., and Yoshida, I. 2023a. Editorial for Special Collection on "Benchmarking data-driven site characterization methods", ASCE-ASME Journal of Risk and Uncertainty in Engineering Systems: Part A, 9(2), 02023001.

Phoon, K. K., Zhang, L. M., Cao, Z. J. 2023b. Editorial for Special Issue on "Machine learning and AI in geotechnics", Georisk: Assessment and Management of Risk for Engineered Systems and Geohazards, 17(1), 1-6.

Phoon, K. K. and Tang, C. 2024. Role of performance information in data-centric geotechnics. Chapter 1 in *Databases for Data-Centric Geotechnics: Geotechnical Structures*, edited by C. Tang and K. K. Phoon. Boca Raton, FL: CRC Press.

Phoon, K. K., Ching, J. and Tang, C. 2024. Role of site characterization information in data-centric geotechnics. Chapter 1 in *Databases for Data-Centric Geotechnics: Site Characterization*, edited by K. K. Phoon and C. Tang. Boca Raton, FL: CRC Press.

Phoon, K. K., Cai, Y. and Tang, C. 2025. Geotechnical "facial recognition" challenge. Intelligent Geoengineering.

Tan, Y. and Fan, X. Z. 2024. Statistical analyses on a database of deep excavations in Shanghai soft clays. Chapter 16 in *Databases for Data-Centric Geotechnics: Geotechnical Structures*, edited by C. Tang and K. K. Phoon. Boca Raton, FL: CRC Press.

Tang, C., Feng, X. and Phoon, K. K. 2024. Variability of predictions for punch-through of foundation in layered soils. Chapter 2 in *Databases for Data-Centric Geotechnics: Geotechnical Structures*, edited by C. Tang and K. K. Phoon. Boca Raton, FL: CRC Press.

Tsuha, C. H. C., Sousa Oliveira, C. C., Silva, B. O., Santos Filho, J. M. S. M., Schiavon, J. A. and Tang, C. 2024. Development and use of tensile loading test databases for analysis and design of helical piles. Chapter 11 in *Databases for Data-Centric Geotechnics: Geotechnical Structures*, edited by C. Tang and K. K. Phoon. Boca Raton, FL: CRC Press.

Wang, C., et al. 2024a. Mechanical-statistical evaluation of soil properties. Chapter 13 in *Databases for Data-Centric Geotechnics: Site Characterization*, edited by K. K. Phoon and C. Tang. Boca Raton, FL: CRC Press.

Wang, C., et al. 2024b. Seismic soil liquefaction assessment: approaches, data, and tools. Chapter 14 in *Databases for Data-Centric Geotechnics: Site Characterization*, edited by K. K. Phoon and C. Tang. Boca Raton, FL: CRC Press.

Wu, S., Otake, Y., Mizutani, D., Liu, C., Asano, K., Sato, N., Saito, T., Baba, H., Fukunaga, Y., Higo, Y., Kamura, A., Kodama, S., Metoki, M., Nakamura, T., Nakazato, Y., Shioi, A., Takenobu, M., Tsukioka, K. and Yoshikawa, R. 2024. Future-proofing geotechnics workflows: accelerating problem-solving with large language models. Georisk: Assessment and Management of Risk for Engineered Systems and Geohazards. https://doi.org/10.1080/17499518.2024.2381026.

Zhang, J., Yao, H., Wang, Z. and Dai, M. 2024. Probabilistic methods for slope failure time prediction. Chapter 17 in *Databases for Data-Centric Geotechnics: Geotechnical Structures*, edited by C. Tang and K. K. Phoon. Boca Raton, FL: CRC Press.

Zhou, Y., Zhang, D. and Huang, H. 2024. Prediction for soil design properties based on a multivariate database for Shanghai soft clay. Chapter 15 in *Databases for Data-Centric Geotechnics: Site Characterization*, edited by K. K. Phoon and C. Tang. Boca Raton, FL: CRC Press.

List of contributors

Shehab Agaiby
Cairo University
Giza, Egypt

Pedro Arduino
University of Washington
Seattle, WA, United States

Pouyan Asem
University of Minnesota
Minneapolis, MN, United States

Qiushi Chen
Clemson University
Clemson, SC, United States

Jianye Ching
National Taiwan University
Taipei, Taiwan

Marco D'Ignazio
Tampere University
Tampere, Finland

Abolfazl Eslami
Amirkabir University of Technology
Tehran, Iran

Shuyin Feng
Birmingham City University
Birmingham, England

David Gaharia
LabMind
Nacka, Sweden

Paolo Gardoni
University of Illinois Urbana-Champaign
Champaign, IL, United States

Sølve Hov
Norwegian Geotechnical Institute
Trondheim, Norway

Hongwei Huang
Tongji University
Shanghai, China

Katharina Kahrs
Norwegian Geotechnical Institute
Trondheim, Norway

Leena Korkiala-Tanttu
Aalto University
Espoo, Finland

Tim Länsivaara
Tampere University
Tampere, Finland

Jean-Sébastien L'Heureux
Norwegian Geotechnical Institute
Trondheim, Norway

Duo Li
Dalian University of Technology
Dalian, Liaoning, China

Tianbin Li
Chengdu University of Technology
Chengdu, Sichuan, China

Jingmao Liu
Dalian University of Technology
Dalian, Liaoning, China

Monica S. Löfman
Aalto University
Espoo, Finland

Karin Lundström
Swedish Geotechnical Institute
Linköping, Sweden

Junjie Ma
Chengdu University of Technology
Chengdu, Sichuan, China

Paul W. Mayne
Georgia Institute of Technology
Atlanta, GA, United States

Pin-Qiang Mo
China University of Mining and Technology
Xuzhou, Jiangsu, China

Fanwei Ning
Dalian University of Technology
Dalian, Liaoning, China

Yu Otake
Tohoku University
Sendai, Japan

Zhongkun Ouyang
Tsinghua Shenzhen International Graduate School
Nanshan, Shenzhen, China

Taiga Saito
Tohoku University
Sendai, Japan

Daiki Takano
Kumamoto University
Kumamoto, Japan

Daofei Tang
Shanghai University
Shanghai, China

Paul Vardanega
University of Bristol
Bristol, United Kingdom

Changhong Wang
Shanghai University
Shanghai, China

Chaofeng Wang
University of Florida
Gainesville, FL, United States

Kun Wang
Shanghai University
Shanghai, China

Stephen Wu
The Institute of Statistical Mathematics
Tokyo, Japan

Kaiyuan Xu
Xidian University
Xi'an, Shaanxi, China

Ikumasa Yoshida
Tokyo City University
Tokyo, Japan

Dongming Zhang
Tongji University
Shanghai, China

Chenguang Zhou
Dalian University of Technology
Dalian, Liaoning, China

Yelu Zhou
Tongji University
Shanghai, China

Degao Zou
Dalian University of Technology
Dalian, Liaoning, China

Chapter 1

Role of site characterization information in data-centric geotechnics

Kok-Kwang Phoon, Jianye Ching, and Chong Tang

This chapter presents an overview of site characterization information with a focus on its availability, coverage, data attributes, value to decision making, and challenges. In terms of availability, it is accurate to say that site characterization databases mainly reside in industry. However, the majority is not shared for confidentiality reasons ("dark data"). The data-centric geotechnics agenda cannot focus purely on maximizing the value of practice but also on minimizing the leakage of sensitive data. It is suggested that the "data first practice central" agenda in data-centric geotechnics can be expanded to "data first, practice central, and privacy enhanced." Data-centric geotechnics now encompasses four elements: (1) data centricity, (2) fit for (and transform) practice, (3) geotechnical context, and (4) privacy enhancement. In terms of coverage, publicly available site characterization databases are mainly geotechnical in nature containing borehole, laboratory test, and field test data. Detailed geological data such as soil stratification, rock mass features (weathering, discontinuities, etc.), and real-time data such as geoenvironmental processes (subsurface flow, contaminant transport, etc.) are lacking. This "paucity of ground truth" is a rate-limiting step in the development of machine learning algorithms for geo-related disciplines, such as soil science, geotechnical engineering, rock engineering, tunnel engineering, earthquake engineering, geoenvironmental engineering, engineering geology, mining, geo-hazard, geophysics, remote sensing, and others in geosciences. In terms of data attributes, MUSIC-3X (Multivariate, Uncertain and Unique, Sparse, Incomplete, and potentially Corrupted with "3X" denoting the spatial/temporal dimension) is a basic challenge to data-driven site characterization (DDSC) research. This chapter proposes a new taxonomy of site data under "4S" to expand the agenda for future machine learning research. The "4S" are site generalizations, spatial features, sampling characteristics, and smart data. In terms of value to design, the majority of the research is focused on the prediction of soil properties (spatial variability) and soil types (stratification) at a target site. One particularly interesting and significant challenge called "site recognition" has attracted attention in recent years. The site recognition challenge is fundamental to geotechnical engineering because it attempts to quantify the "uniqueness" (or site specificity) attribute of a target site. Recent research has demonstrated that quasi-site-specific transformation models can be developed notwithstanding the well-known limitations – site-specific data is sparse, and a generic database containing data from other sites is not directly applicable to a target site. There are three notable achievements: (1) the site recognition problem is tractable even under MUSIC, (2) inference uncertainty produced by a quasi-site-specific transformation model is smaller than that produced by a conventional generic or site-specific model, and (3) increasingly effective quasi-site-specific transformation models have been constructed, particularly those that exploit clustering. The construction of a quasi-site-specific ground model that includes learning cross-correlations and spatial auto-correlations from generic databases under MUSIC-3X is an

DOI: 10.1201/9781003441946-1

ongoing research problem. A more difficult problem is to address MUSIC-3X-G, where the attribute "G" refers to geologic uncertainty such as stratification. Research on other site characterization databases such as geophysical and remote sensing data and data fusion are limited. Interest in combining site characterization information with monitoring information collected during construction to guide decision making in real time is emerging rapidly. This new agenda is called the machine learning-guided observational method (MLOM). Its value is arguably higher than DDSC, possibly leading to a Type 3 (disruptive) outcome that can transform geotechnical practice. Other challenges such as explainability and interpretability, transferability, geo-compatibility, frugal HBM, thick data, and data protection are briefly discussed. It is uncertain whether a new discipline called geotechnical data science is needed to create novel theoretical concepts and methods for data-centric geotechnics. But trustworthiness needs to be addressed to comply with the EU AI Act and other regulatory frameworks that should be introduced soon to govern AI systems.

1.1 DATA-CENTRIC GEOTECHNICS

Foundations and underground structures have been built since antiquity. In all likelihood, the builders learnt from successful case histories. Recurring observations that are useful to practice are eventually codified as rules of thumb or passed down from masters to apprentices as traditions. Data does support decision making even when geotechnical engineering is practised as an art, except it mainly appears as a qualitative experience base. The modern form may appear as table 54.1 of Terzaghi and Peck (1967) (citing the *Kidder-Parker Architects' and Builders' Handbook* dated 1931 as the source) or table 1 of BS8004:1986 (BS 1986). These tables provide presumptive allowable bearing values that are based on local experience. The introduction of science to geotechnical engineering may be traced to Coulomb's failure wedge analysis behind a retaining wall using the laws of friction and cohesion. Data does support decision making when geotechnical engineering is practised as a science, but data is restricted to a set of inputs required by a theoretical model (e.g., friction, cohesion). It is incorrect to say that geotechnical engineering has always been data-centric in the strict sense. It has been experience-centric or physics-centric, but to the knowledge of the authors, data has never supported decision making *by itself*. Data-driven decision making is only gaining some interest because machine learning (ML) and artificial intelligence (AI) have transformed many industries.

Data-centric geotechnics is distinctive from other fields in geotechnical engineering and machine learning (Phoon and Ching 2021; Phoon et al. 2022a). It is a new interdisciplinary field involving the integration of information, data, techniques, tools, perspectives, concepts, theories, and/or experiences from geotechnical engineering and machine learning (Phoon and Zhang 2023). Broader developments in "data-centric engineering" (Girolami 2020) and data-driven discovery in geosciences (Chen et al. 2023) are taking place elsewhere in parallel. Bilal et al. (2016) and Munawar et al. (2022) surveyed the application of "big data engineering" in construction. Research in machine learning in geotechnics (MLIG) in the context of data-centric geotechnics must articulate its value proposition clearly to hasten the rate of digital transformation in the industry. For example, "ML supremacy" projects are those involving large datasets (perhaps real-time data) from multiple sources covering large spatial domains that cannot be handled effectively and efficiently using conventional methods, and there is a strong demand for

Table 1.1 Value of machine learning in geotechnics as a function of data and method

		Method	
		Available generic ML (e.g., artificial neural network, support vector machine, deep learning, transfer learning, meta learning)	*Novel geotechnical ML*
Data	Available data (e.g., site investigation data, proof test data, monitoring data)	**Type 1** Incremental value – applications of available ML to conventional geotechnical structures, e.g., foundations, slopes ("ML advantage")	**Type 2a** Potentially high value – development of new ML to address distinctive challenges in geotechnical engineering resulting in new applications, e.g., DDSC ("ML advantage/supremacy")
	Emerging data (e.g., wireless sensor network, laser scanning, drone photogrammetry, satellite imagery, etc.)	**Type 2b** High value – applications of available ML to current problems that cannot be addressed effectively using conventional methods, e.g., hazard/risk mapping ("ML advantage/supremacy")	**Type 3** Disruptive value – development of new "fit for practice" geotechnical ML to transform research and practice resulting in new products, systems, and services and may even redefine the role of geotechnical engineering

Source: table 1, Phoon and Zhang (2023).

implementation of ML methods in these projects (Shahri et al. 2019; Abubakar et al. 2021; Christensen et al. 2021; Wang et al. 2021a; Wang et al. 2021b; Lin et al. 2022; Phoon et al. 2023a). Physics cannot support decision making in these projects because it is absent or grossly incomplete. Classical statistics may entail insupportable assumptions on the attributes of the data. The value of MLIG research can be classified as follows: (1) Type 1 (incremental value) involving available data and existing conventional applications, (2) Type 2a (potentially high value) involving available data and new applications, (3) Type 2b (high value) involving new data and existing applications, and (4) Type 3 (disruptive value) involving new data and completely novel applications (e.g., precision construction, autonomous construction control) (Phoon and Zhang 2023). Details are given in Table 1.1. Type 1 MLIG provides an alternate approach to an existing problem. It is only "good to have." Type 2 MLIG outperforms existing approaches (if they exist) to such an extent that it will replace them. Type 3 MLIG will make practice unrecognizable by present-day engineers. These MLIG methods are distinctive from reliability methods (Ji and Cao 2024), although they are both data-driven. MLIG methods can be improved by learning from data and may eventually even modify or propagate themselves (e.g., an AI programming a better version of itself). Reliability methods address uncertainty explicitly, but the classical ones do not contain a learning step. Hence, it is possible to classify reliability methods as Type 0 in the context of Table 1.1.

Interest in MLIG has been gathering momentum in recent years (Faramarzi, 2020; Jaksa and Liu 2021; Morgenroth et al. 2022; Wang et al. 2022; Yuen et al. 2021; Zhang and Phoon 2022; Du et al. 2023; Phoon et al. 2023b; Phoon et al. 2023c; Zhang et al. 2023). However, the emphasis remains "method first" rather than "data and practice first." Numerous foundational challenges beyond methodological ones can be identified when research is pursued under a more balanced data-centric geotechnics agenda covering three elements (Phoon and Zhang 2023; Phoon et al. 2023b):

(1) Data centricity – The goal is to develop methods that make sense of all real-world data sets. The prevailing strategy is the opposite. The engineer screens/interprets data to fit a particular theory/paradigm. For example, only some specific constitutive model parameters are relevant to a numerical model. Another example is to reduce a dataset to a single number to fit the deterministic paradigm [e.g., characteristic value in Eurocode 7 (CEN 2004), Länsivaara et al. 2022]. The third example is to accept/reject a dataset based on its compliance with a set of conditions imposed by classical frequentist analysis (e.g., minimum sample size). These examples exhibit a "method first, data second" philosophy. The central tenet in data-centric geotechnics is that data has value if it is not fake, completely corrupted (bad data), or completely irrelevant. Hence, it may not be economical to use data selectively. Data centricity does not imply that experience and physics are not relevant. Data has supported prevailing experience-centric or physics-centric practice. In the same way, experience and/or physics can support data-centric geotechnics. In machine learning, experience can be regarded as one type of "thick data" to distinguish it from the more well-known quantitative "big data." Physics-informed machine learning (PIML) methods offer the advantages of "explainability" and "interpretability" (Karniadakis et al. 2021; Li et al. 2023; Shioi et al. 2023; Tan et al. 2023; Yan et al. 2023). Large language models such as GPT-4o are already capable of using multi-modal data.
(2) Fit for (and transform) practice – The goal is to bring significant value to critical real-world decisions (not decisions for an ideal world or decisions of minor concern to geotechnical engineers just to showcase the latest method developed in computer science). One classification of value is shown in Table 1.1. Data-centric geotechnics should, therefore, work on real-world datasets from real-world problems only and should enhance decision making where it matters most to the practitioner. Autonomous construction is one example where data can lead to intelligence that takes over routine decision making at a construction site from a human supervisor.
(3) Geotechnical context – Datasets are not abstract numbers divorced from their physical basis and their physical origin. As such, direct application of machine learning developed in other fields with cursory reference to the geotechnical context is not appropriate. The geotechnical context that gives rise to the data is important. The context can be related to statistics, physics, or experience. Statistics refer to the attributes of geotechnical data that depart significantly from the assumptions in classical statistics (large sample size, spatial/temporal/parametric independence, homogeneity, normality, etc.). Physics refers to a body of rational knowledge that associates a "number" with "meaning and casuality." An engineer distinguishes between material and state parameters, between effective and total stress parameters, and between input and output parameters from a numerical model. These distinctions exist when one approaches data from the lens of physics. Experience refers to a body of empirical knowledge accrued from deliberate practice. In contrast to statistics and physics, it is mainly subjective and qualitative. Nonetheless, many engineers regard experience as critical. Burland (2012) termed well-winnowed experience as "experience that results from a rigorous sifting of all the facts that relate to a particular empirical procedure or case history" in the geotechnical triangle. The key limitation of experience is that it cannot be transferred from one engineer to the next efficiently, certainly nothing close to the speed that one can share data and algorithms in the cloud. In addition, it is rare for an engineer to accumulate global experience. This means an engineer can only make sense of regional data that he/she is familiar with, but would struggle to do the same for the entire global database.

Given the nascent stage of development in data-centric geotechnics, it may be more realistic for this chapter to examine whether there is a plausible pathway to Table 1.1 for site investigation information. The value of performance information is discussed in the companion volume (Phoon and Tang 2025). In all likelihood, Type 3 methods will appear in the context of Industry 4.0. At the end of this chapter, it is proposed that "privacy enhancement" be added to the agenda for data-centric geotechnics.

1.2 SITE CHARACTERIZATION INFORMATION

The first step in data-centric geotechnics is not to deep dive into one source of information because it is the most convenient or most tractable. Since all data are regarded as potentially useful, the first step is to take stock of all available sources (including emerging ones related to the Internet of Things or Everything) and to lay out a strategy on how to bring the entire *geodata estate* to bear on decision support or to transform practice in novel ways. Data gaps may be identified that can motivate the development and deployment of new sensors in a more purposeful way. This requires an understanding of the traits or attributes of geotechnical information that are likely to change in tandem with the growth and convergence of digital technologies. Below are examples of classifying the attributes of site investigation information in relation to data-driven site characterization (DDSC) and the larger Industry 4.0 paradigm to push for Type 2 or 3 machine learning research. Huang et al. (2022) explored the progress, challenges, and opportunities in big earth data analytics in the same spirit as this chapter. "Big earth data" refers to data about the planet: the ocean, the land, the atmosphere, and climate.

1.2.1 Big indirect data (BID)

Phoon et al. (2019) coined the term "big indirect data" (BID) to refer to any data that are potentially useful but not directly applicable to the decision at hand. *Generic* databases containing soil/rock properties (Phoon et al. 2024), load test data (load-movement curves, load transfer curves) (Tang et al. 2024), or monitoring data from multiple sites involving different structures would belong to BID. A representative list of soil/rock property databases is shown in Table 1.2. There are several national databases (large) for public access via the Internet. The New Zealand geotechnical database (NZGD), curated by the Earthquake Commission, holds approximately 23,000 cone penetration test (CPTs), 10,000 borehole log records, 1000 piezocone (CPTu) records, and 6000 laboratory test records. The United States Community shear wave velocity profile database (VSPDB) consists of geophysical (3845 V_s/V_p profiles and 1993 spectral ratio records), in situ (1212 boreholes and 1342 CPTs), and laboratory tests (26 sieve analyses and 845 Atterberg limits) conducted at 4327 sites, where V_s and V_p are shear and compression wave velocities (Kwak et al. 2021). Another similar compilation work is the Next-Generation Liquefaction (NGL) database, which provides researchers with a substantially larger, more consistent, and more reliable source of liquefaction data than what has existed previously (Ulmer et al. 2023). The NGL database also consists of geophysical (199 V_s/V_p profiles), in situ (941 boreholes and 747 CPTs), and laboratory tests (29,023 sieve analyses and 1665 Atterberg limits). The US Geological Survey (USGS) and other governmental agencies funded the studies to compile V_{30} values for 4389 sites in the United States, where V_{30} is the time-averaged V_s in the upper 30 m (McPhillips et al. 2020). The National Geotechnical Properties Database (NGPD) maintained by the British Geological Survey (BGS) may be the most

Table 1.2 Soil/rock property databases

	Database	*Reference*	*Soil/rock parameters*	*# Data points*	*# Sites/ studies*	*% complete*
Univariate	CLAY/16	Phoon and Kulhawy (1999a)	γ, γ_d, w_n, PL, LL, PI, LI, ϕ', s_u, s_u^{FV}, q_c, q_t, SPT-N, DMT (A, B), PMT p_L		a	
	SAND/11	Phoon and Kulhawy (1999a)	ϕ', D_r, q_c, SPT-N, DMT (A, B, I_D, K_D, E_D), PMT (p_L, E_{PMT})		b	
	ROCK/8	Prakoso (2002)	γ (or γ_d), n, R, S_h, σ_{bt}, I_s, σ_c, E		c	
	ROCK/13	Aladejare and Wang (2017)	ρ, G_s, I_{d2}, n, w_c, γ, R_L, S_h, σ_{bt}, I_{s50}, σ_c, E, ν		d	
	USGS	McPhillips et al. (2020)	V_{s30}		4389 sites in the USA	Web portal
Multivariate	CLAY/5/345	Ching and Phoon (2012)	LI, s_u, s_u^{re}, σ'_p, σ'_v	345	37 sites	100%
	CLAY/7/6310	Ching and Phoon (2013)	s_u from seven different test procedures	6310	164 studies	17.7%
	CLAY/6/535	Ching et al. (2014)	s_u/σ'_v, OCR, q_{tc}, q_{tu}, $(u_2-u_0)/\sigma'_v$, B_q	535	40 sites	100%
	CLAY/10/7490	Ching and Phoon (2014a)	LL, PI, LI, σ'_v/P_a, σ'_p/P_a, s_u/σ'_v, S_t, q_{tc}, $q_{tu,}$ B_q	7490	251 studies	34.1%
	FI-CLAY/7/216	D'Ignazio et al. (2016)	s_u^{FV}, σ'_v, σ'_p, w_n, LL, PL, S_t	216	24 sites	100%
	FI-CLAY/14/856	Löfman and Korkiala-Tanttu (2021)	w_n, e_0, LL, PL, PI, γ, C_l, s_u, S_t, σ_p', OCR, C_c, C_R, C_s	856	33 sites	
	JS-CLAY/5/124	Liu et al. (2016)	M_r, q_c, f_s, w_n, γ_d	124	16 sites	100%
	JS-CLAY/7/372	Zou et al. (2017)	σ_v, σ'_v, q_{tc}, f_s/σ'_v, B_q, V_{s1}, s_u/σ'_v	372	25 sites	100%
	SAND/7/2794	Ching et al. (2017)	D_{50}, C_u, D_r, σ'_v/P_a, ϕ', q_{t1}, $(N_1)_{60}$	2794	176 studies	60.0%
	EMI-ROCK/8/26000+	Kim and Hunt (2017)	σ_c, σ_{bt}, ρ, CAI, PPI, cohesion, direction shear, triaxial confining	26,000+	–	–
	FG/5/1000	Kootahi and Moradi (2017)	e, w_n, LL, PI, C_c	1000	170 sites	100%
	ROCK/9/4069	Ching et al. (2018)	γ, n, R_L, S_h, σ_{bt}, I_{s50}, V_p, σ_{ci}, E_i	4069	184 studies	34.2%
	FG-KSAT/6/1358	Feng and Vardanega (2019)	e, k, LL, PL, PI, G_s	1358	33 studies	91.4%
	CG/KSAT/7/1278	Feng et al. (2023)	e, k, D_{10}, D_{50}, G_s, C_u, C_z	1278	> 50 studies	
	RFG/TXCU-278	Beesley and Vardanega (2020)	s_u/σ_{v0}', $\gamma_{50\ CIU}$, $\gamma_{50\ CKU}$, b_{CIU}, b_{CKU}	278	278 CU tests	100%
	SH-CLAY/11/4051	Zhang et al. (2020)	LL, PI, LI, e, K_0, σ'_v/P_a, s_u/σ'_v, S_t, q_c/σ'_v	4051	50 sites	39.5%
	CLAY/8/12225	Ching (2020)	LL, PI, w, e, σ'_v/P_a, C_c, C_{ur}, c_v	12,225	427 studies	–

(Continued)

Table 1.2 (Continued) Soil/rock property databases

	Database	*Reference*	*Soil/rock parameters*	*# Data points*	*# Sites/ studies*	*% complete*
	CLAY/12/3997	Ching (2020)	LL, PI, LI, σ'_v/P_a, σ'_p/P_a, s_u/σ'_v, K_0, E_u/σ'_v, B_q, q_{t1}, $N_{60}/(\sigma'_v/P_a)$	3997	237 studies	–
	SAND/13/4113	Ching (2020)	e, D_r, σ'_v/P_a, σ'_p/P_a, K_0, E_{dn}, q_{c1n}, B_q, $(N_1)_{60}$, K_{DMT}, E_{DMTn}, E_{PMTn}, M_{dn}	4113	172 studies	–
	ROCKMass/9/5876	Ching et al. (2021a)	RQD, RMR, *Q*, GSI, E_m, E_{em}, E_{dm}, E_i, σ_{ci}	5876	225 studies	29.3%
	CLAY-C_c/6/6203	Ching et al. (2022a)	LL, PI, w_n, e, C_c, C_{ur}	6203	429 studies	61%
	SOIL-DMT/8/7186	Ching and Kuo (2023)	LI, D_{50}, s_u^{FV}/P_a, N_{60}, $(q_t-\sigma_v)/P_a$, B_q, E_{PMT}/P_a, $(1-\nu^2) \times E_{DMT}/P_a$	7286	237 studies	28%
	Global-CPT/3/1196	Ching et al. (2023a)	q_t, f_s, u_2	1196	59 sites	100%
	Geo-Marine-CPT/3/398	Eslami et al. (2023)	q_t, f_s, u_2	398	58 sites	100%
	ROCK/10/4025	Muzamhindo and Ferentinou (2023)	γ, *n*, R_L, I_{s50}, V_p, σ_{bt}, BPI, m_i, E_i, σ_c	4025	96 studies	50%
	SAND-Small/9/939	Lo et al. (2021)	D_{50}, C_u, e_{min}, e_{max}, σ'_3, e_c, G_{max}, σ'_{1p}, ϕ'	939	15 studies	79%
National	NGL	Ulmer et al. (2023)	Geophysics – velocity profiles (199), in situ tests (941 boreholes and 747 CPTs), and laboratory tests (29,023 sieve analyses and 1665 Atterberg limits)			Web portal
	NGPD	Self et al. (2012)	Boreholes (178,436), in situ tests (3,617,186), laboratory tests (5,180,330)		7370 sites (UK)	Unavailable
	NZGD		In situ tests (23,000 CPTs, 10,000 borehole logs, 1000 CPTu records) and laboratory tests (6000)		6000 sites (New Zealand)	Web portal
	VSPDB	Kwak et al. (2021)	Geophysics – velocity profiles (3845) and spectral ratio records (1993), in situ tests (1212 boreholes and 1342 CPTs), and laboratory tests (26 sieve analyses and 845 Atterberg limits)		4327 sites	Web portal

(Continued)

Table 1.2 (Continued) Soil/rock property databases

	Database	*Reference*	*Soil/rock parameters*	*# Data points*	*# Sites/ studies*	*% complete*
SWCC	MEPDG	Zapata (2010)	20,526 soil units with enough information to compute the SWCC parameters		US	Web portal

Source: updated from Phoon et al. (2024).

Notes:

Basic – C_u = coefficient of uniformity; D_{50} = median grain size; ρ = density; G_s = specific gravity; γ = unit weight; γ_d = dry unit weight; D_r = relative density; e = void ratio; e_{min} = minimum void ratio; e_{max} = maximum void ratio; e_c = void ratio under isotropic consolidation; n = porosity; w_n (or w_c) = water content; PL = plastic limit; LL = liquid limit; PI = plasticity index; LI = liquidity index; GSI = geological strength index.

Stress – σ_v = total vertical stress; σ'_v = effective vertical stress; σ'_p = pre-consolidation stress; σ'_3 = effective confining stress under isotropic consolidation; σ'_{1p} = effective axial stress at peak state under drained triaxial compression; OCR = over-consolidation ratio; P_a = atmospheric pressure = 101.3 kPa; K_0 = at-rest lateral earth pressure coefficient.

Strength – ϕ' = effective friction angle; s_u = undrained shear strength; s_u^{FV} = field vane s_u; s_u^{re} = remoulded s_u; S_t = sensitivity; σ_{bt} = Brazilian tensile strength; σ_{ci} (or σ_c) = uniaxial compressive strength of intact rock; BPI = block punch index; m_i = Hoek–Brown constant.

Deformation – C_c = compression index; C_{ur} = unloading–reloading index; modulus; E_u = undrained modulus of clay; E_d = drained modulus of sand; $E_{dn} = (E_d/P_a)/(\sigma'_v/P_a)^{0.5}$; E_{dm} = dynamic modulus of rock mass; E_{em} = elasticity modulus of rock mass; E_i (or E) = Young's modulus of intact rock; E_m = deformation modulus of rock mass; ν = Poisson ratio; M_r = subgrade resilience modulus; M_d = effective constrained modulus determined by oedometer; M_{dn} = normalized $M_d = (M_d/P_a)/(\sigma'_v/P_a)^{0.5}$; G_{max} = small-strain shear modulus; $\gamma_{50\ CIU}$ = shear strain to mobilize $0.5s_u$ under isotropically consolidated undrained condition; $\gamma_{50\ CKU}$ = shear strain to mobilize $0.5(s_u - \tau_0)$; τ_0 = initial shear stress; b_{CIU} and b_{CKU} = coefficients to describe nonlinearity.

Permeability – k = hydraulic conductivity; c_v = coefficient of consolidation.

Dynamic – V_p = P-wave velocity; V_s = S-wave velocity; V_{s30} = time-averaged V_s to a depth of 30 m; $V_{s1} = V_s(P_a/\sigma'_v)^{0.25}$.

Field test – SPT-N = standard penetration test blow count; N_{60} = corrected SPT-N; $(N_1)_{60} = N_{60}/(\sigma'_v/P_a)^{0.5}$; q_c = cone tip resistance; q_t = corrected cone tip resistance; f_s = sleeve frictional resistance; $q_{tc} = (q_t/P_a)/(\sigma'_v/P_a)^{0.5}$; $q_{t1} = (q_t-\sigma_v)/\sigma'_v$ = normalized cone tip resistance; $q_{tu} = (q_t-u_2)/\sigma'_v$ = effective cone tip resistance; $q_{c1n} = (q_c/P_a)/(\sigma'_v/P_a)^{0.5}$; B_q = pore pressure ratio = $(u_2-u_0)/(q_t-\sigma_v)$; $(u_2-u_0)/\sigma'_v$ = normalized excess pore pressure; u_2 = pore pressure behind cone tip; u_0 = hydrostatic pore pressure; PMT (p_L, E_{PMT}) = pressuremeter limit stress, modulus; E_{PMTn} = normalized $E_{PMT} = (E_{PMT}/P_a)/(\sigma'_v/P_a)^{0.5}$; DMT ($A$, B, I_{DMT}, K_{DMT}, E_{DMT}) = dilatometer A and B readings, material index, horizontal stress index, modulus; E_{DMTn} = normalized $E_{DMT} = (E_{DMT}/P_a)/(\sigma'_v/P_a)^{0.5}$; CAI = Cerchar abrasivity index; PPI = punch penetration index; Q = Q-system; RMR = rock mass rating; RQD = rock quality designation; R = Schmidt hammer hardness (R_L = L-type Schmidt hammer hardness); S_h = Shore scleroscope hardness; I_{d2} = slake durability index; I_s = point load strength index ($I_{s50} = I_s$ for diameter 50 mm).

a – The no. of data groups varies between 2 and 42 depending on the clay parameter. Statistics are calculated at the data group level. The average no. of data points/data group varies between 16 and 564. Details given in tables 1–3, Phoon and Kulhawy (1999b).

b – The no. of data groups varies between 5 and 57 depending on the sand parameter. Statistics are calculated at the data group level. The average no. of data points/data group varies between 15 and 123. Details are given in tables 1–3, Phoon and Kulhawy (1999b).

c – The no. of data groups varies between 30 and 174 depending on the rock parameter with no differentiation of rock type [igneous (intrusive, extrusive, pyroclastic), sedimentary (clastic, chemical), metamorphic (foliated, non-foliated)]. Statistics are calculated at the data group level. The average no. of data points/data group varies between 3 and 161 for σ_c (Prakoso, 2017). Details are given in table 4.4, Prakoso (2002).

d – The no. of data groups varies between 2 and 47 depending on the rock parameter and rock type (igneous, sedimentary, or metamorphic). Statistics are calculated at the data group level. The average no. of data points/data group varies between 7 and 92. Details are given in tables 2–4, Aladejare and Wang (2017).

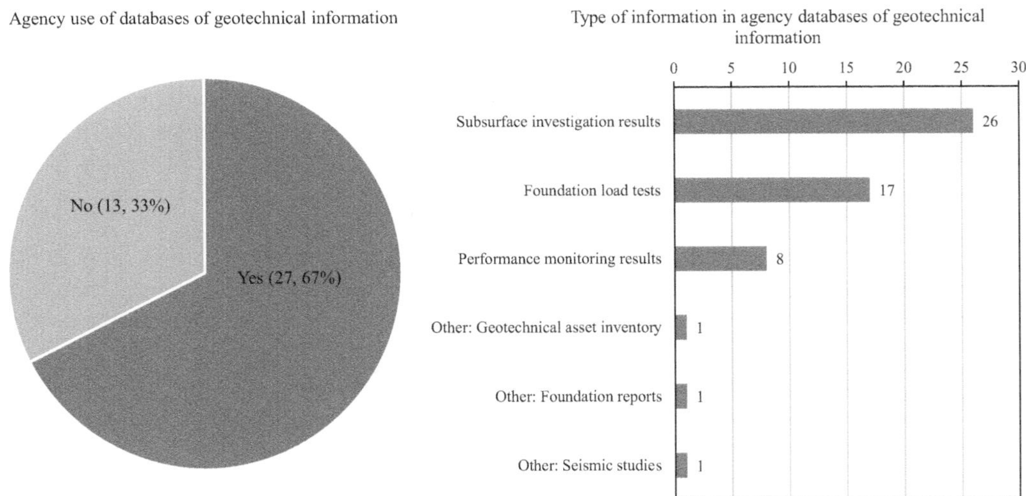

Figure 1.1 Types of information included in state DOTs' databases of geotechnical information (Source: data from Boeckmann and Loehr 2023).

comprehensive database, holding the data and information from commercial site investigations (Self et al., 2012). By 2019, the NGPD contains records for 7370 projects. Linked to the site investigation reports are 178,436 boreholes. There are a total of 3,617,186 in situ testing records and 879,293 samples linked to the boreholes, with a total of 5,180,330 laboratory testing records. Most recently, Otake et al. (2024) presented a large clay database from a single site in Japan – Tokyo-CLAY/14/67,760. It contains 14 soil parameters determined from 2972 boreholes covering a site area of approximately 10 km^2. The NCHRP Synthesis 601 indicated that 27 state Departments of Transportation (DOTs) across the United States maintain a database of geotechnical information (Boeckmann and Loehr 2023). The types of information included in the database are presented in Figure 1.1. The most common is subsurface investigation results (96%) and the next is foundation load tests (63%). Compilation, interpretation, and application of subsurface investigation results for geotechnical analysis and design constitute the main content of this volume. Development and use of performance databases for geotechnical structures (e.g., shallow/deep foundations, offshore spudcans, anchors/pipelines, slopes/retaining walls, and excavations) are the primary objectives of the companion volume (Tang and Phoon 2025), where foundation load tests are the most prevalent.

There is a widespread perception that geotechnical data is scarce and no data-driven methodologies are applicable. This is only true for site-specific data based on current site investigation technology. To insist that only site-specific data is relevant to the project at hand is not aligned to current practice. Generic transformation models that relate a field test parameter (e.g., cone tip resistance) to a design parameter (e.g., undrained shear strength) are prevalent (Kulhawy and Mayne 1990; National Academies of Sciences, Engineering, and Medicine 2019). Clause 2.4.5.2(10) of Eurocode 7 (CEN 2004) "Characteristic values of geotechnical parameters" takes a more cautious position on pooling data from different sites: "If statistical methods are employed in the selection of characteristic values for ground properties, such methods should differentiate between local and regional sampling and should allow the use of a priori knowledge of comparable ground properties." However, no data-driven method exists to differentiate between "local and regional sampling" until recently (Ching and Phoon 2020a; Ching et al. 2021a, 2021b, 2022a; Sharma et al. 2022, 2023; Cai et al. 2024a). A cursory examination of the word "local" would already reveal some complications. For example, would "local" be restricted to one site or can it encompass an adjacent site, a block, a precinct, or a city? It is also not well appreciated that drawing upon "comparable ground properties" is using some relevant information from *past* regional sampling. Vardanega and Bolton (2016) essentially raised the same concern in the selection of the coefficient of variation (COV):

> Although reliable estimates of the mean and standard deviation are easier to ascertain than the shapes of the pdfs, there remains an unjustified tendency to rely solely on published COV values from other soil deposits. Because variability in a soil deposit is a function of the processes of geological deposition and geomorphological change that have influenced the site (e.g., Hutchinson 2001), intensive efforts would be necessary to draw parallels between a new site that lacks such information and previously explored sites for which COV values have been established.

The majority of the engineers would take a more practical position that inferences drawn from BID can be useful when they are moderated by engineering judgment ("reality check"). As mentioned above, the drawback of judgment is that it cannot be applied beyond the experience base of an engineer. A global database containing sites that are unfamiliar to an engineer is not accessible to judgment. The purpose of classifying geotechnical big data as BID is to sharpen the research statement: between site-specific small data that cannot be used for machine learning training and BID that cannot be used directly without accounting for site differences, is there a quasi-site-specific machine learning solution better than both extremes? Current research shows that the concern regarding the relevance of BID to decision making at a specific site is a moot point, because it has since been addressed directly in the site recognition challenge (Phoon and Ching 2022; Phoon et al. 2025). Research motivated by BID shows that the conceptual demarcation of local and regional sampling and the assumption that only local sampling is useful for the decision at hand are overly simplistic and do not lead to fruitful research directions.

1.2.2 MUSIC-3X

Phoon et al. (2019) presented a useful mnemonic, MUSIC-X (Multivariate, Uncertain and Unique, Sparse, Incomplete, and potentially Corrupted with "X" denoting the spatial/temporal dimension) to highlight seven common ugly attributes in *site-specific* data. One example is illustrated in Table 1.3. It is common to conduct multiple tests at one site. Different test data may be cross-correlated when they are measured in close proximity (*M*). Site uniqueness is recognized as a distinctive feature of geotechnical engineering (*U*). Site-specific data are frequently sparse and incomplete in the sense that tests are conducted at limited depths and few locations (*S*), and any one test is conducted at selective depths and/or locations only (*I*). This incompleteness attribute covers both missing values in a test parameter depth profile (incomplete record) and missing tests at a particular location (entire records are missing for some test parameters) as shown in Figure 1.2. A sufficiently large database must contain potentially corrupted data (C). A manual process of screening out corrupted data will be defeated by the sheer volume of data ultimately. A yottabyte (YB) is 2^{80} bytes, or approximately a million trillion megabytes (MB). It is likely that this inconceivably large number will be breached in less than 5 years. The

Table 1.3 Data-driven methods for site characterization in Project DeepGeo

Method	No. of parameters	Spatial variability		Use generic database?	Other limitations	Reference
		Trend	Autocorrelation			
Sparse Bayesian learning (SBL)	Single	Yes	1X	No	Stationary autocorrelation	Ching and Phoon (2017)
Sparse Bayesian learning (3D SBL)	Single	Yes	3X	No	Stationary separable autocorrelation; vertically dense; lattice	Ching et al. (2020a)
Sparse Bayesian learning (3D SBL)	Single	Yes	3X	No	Stationary separable autocorrelation; vertically dense	Ching et al. (2021c)
Gaussian process regression (GPR-MUSIC)	Multiple	–	–	No	No vertical autocorrelation; perfect horizontal autocorrelation	Ching and Phoon (2019)
Gaussian process regression (GPR-MUSIC-X)	Multiple	No	1X	No	Prescribed stationary vertical autocorrelation; perfect horizontal autocorrelation	Ching and Phoon (2020b)
Gaussian process regression (GPR-MUSIC-3X)	Multiple	No	3X	No	Prescribed stationary separable autocorrelation	Ching et al. (2021d)
Gaussian process regression + hierarchical Bayesian model (HBM-MUSIC)	Multiple	–	–	Yes	No vertical autocorrelation; perfect horizontal autocorrelation	Ching et al. (2021a, 2021b)
Gaussian process regression + hierarchical Bayesian model (HBM-MUSIC-3X)	Multiple	No	3X	Yes	Prescribed stationary separable autocorrelation	Ching et al. (2021d)
Full Gaussian process regression (FGPR-MUSIC-3X)	Multiple	Yes	3X	No	Stationary separable autocorrelation	In progress

Source: table 2, Phoon and Ching (2021).

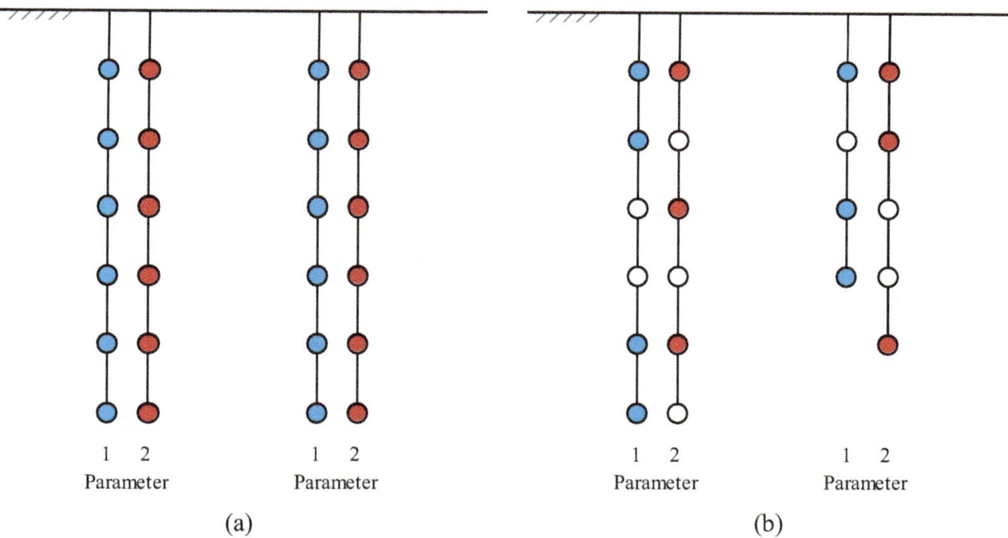

Figure 1.2 Bivariate site data: (a) complete (ideal case) and (b) incomplete (realistic case) (Source: figure 2.27, Phoon et al. 2024). Note: Open circle denotes unmeasured location.

only feasible strategy is to develop data-driven methods that can identify corrupted data automatically. Corrupted data can be bad data, which is defined as data that mislead the decision maker. In short, no data is better than bad data. Recent research on outlier analysis in data-driven site characterization takes a different perspective that inference can be improved by screening out "irrelevant" data (Ching et al. 2024; Cai et al. 2024b). Irrelevant data is not necessarily bad or corrupted data. It can be good data that so happens not to contribute to the inference of a specific property at a target site. In this chapter, the labels "outlier", "irrelevant", "bad", and "corrupted" are used interchangeably. The "outlier" label is neutral. The "irrelevant" label focuses on the impact to decision making. The "bad" and "corrupted" labels are possible explanations for outlier data. Finally, site data are spatially variable or heterogeneous (X). They can vary with time, possibly within a project duration for improved ground. Phoon (2025) has raised the ugly data challenge to MUSIC-3X-G to cover geologic uncertainty ("G"). Geologic uncertainty refers to the distribution of soil types, rock level, anomalies or other geologic features.

It is an open question whether DDSC can solve real-world 3D subsurface mapping problems based on real-world MUSIC-3X data (3X refers to 3D spatial variability as defined by Phoon et al. 2022b) with minimum ad hoc assumptions. The successes shown in the literature are based on data that addresses some attributes of MUSIC-3X only. Research in site recognition attacks the "U" attribute (Phoon and Ching 2022) (see Sections 1.3.1 and 1.4). The first attempt to address "C" in the full context of MUSIC-3X (not in the typical context of ideal data) was made only recently (Ching et al. 2024; Cai et al. 2024b) (see Section 1.4.3). The computational challenges are also very significant given the size of the 3D ground volume, but some reasonable partial solutions have been obtained recently in Project DeepGeo, which constitutes one major research effort in the emerging field of data-centric geotechnics (Phoon and Ching 2021). Phoon et al. (2022c) suggested an initial "small" ground volume of a 20 m long × 20 m wide × 10 m deep cuboid for benchmark examples. The authors suggested that "medium" and "large" ground volumes can be defined as a 40 m long × 20 m wide × 20 m deep cuboid and an 80 m long × 40 m wide × 40 m deep cuboid, respectively. A summary of the data-driven methods in Project DeepGeo is given in Table 1.3. There are other research initiatives attacking this 3D subsurface mapping challenge (Phoon et al. 2023c). Although it is easy to imagine a wider set of attributes for site investigation data such as those presented in Section 1.2.3, it is worth pointing out that MUSIC-3X has already served to drive DDSC research in a fruitful way to meet the second element of data-centric geotechnics "fit for (and transform) practice."

1.2.3 New taxonomy of site data

This section proposes a new taxonomy of site data under "4S" to expand the agenda for future MLIG research. The "4S" are site generalizations, spatial features, sampling characteristics, and smart data. This expanded agenda covers more than site investigation data and attempts to go beyond conventional applications in current practice.

1.2.3.1 Site generalizations

The concept of a "site" is fundamental in geotechnical engineering, but the full extent of its complexity is still unfolding. The current understanding of a "site" is mainly related to geology, which looks at how the earth is formed, its structure and composition, and the types of processes acting on it. The argument that every site is unique is founded on an intrinsic feature of geology, namely its natural variability. One can draw three conclusions from this line of argument: (1) differences between sites are measured by structural and/or compositional aspects of the ground, (2) the degree of difference (or similarity) between two sites is absolute in the sense that the structural and compositional aspects do not change within the lifetime of a structure for a natural ground, and (3) for a local geology that varies somewhat smoothly, an adjacent site is likely to be more similar than a distant site.

From a data-centric geotechnics perspective, one can imagine generalizing the concept of a "site" in three dimensions: (1) broaden the understanding from geology to the performance of a geotechnical structure, (2) dynamic data-driven resizing to achieve homogeneity, and (3) precision characterization and construction. For the first dimension, data on soil behaviours and soil–structure interactions can inform the similarity between two sites. The soil behaviour type index derived from the cone penetration test or the dilatometer material index is not based on compositional data, although they are used for soil classification. Monitoring, pile load tests, or other performance data have never been used for site comparison. This is surprising because an engineer is ultimately more interested in performance rather than geology. This general definition of a "site" is no longer absolute because it may depend on the soil behaviour and the limit state of interest. For example, two sites may be more similar in terms of the compression index but less similar in the small strain stiffness. The load-settlement behaviour of a shallow foundation at both sites may thus be similar. However, the deflection behaviour of a sheet pile wall at these sites may be different.

The second dimension relates to an assumption that an arbitrarily bounded "site" containing a geotechnical structure is statistically homogeneous. The volume of a "site" is currently defined by the project boundaries and the depth of its deepest foundation elements. This assumption is made when an engineer uses a single local transformation model in design to cover the whole site. However, it is possible that a "site" is governed by two or more distinct transformation models. An engineer may feel that the ground conditions are the "same" within a site because its dimensions are typically much smaller than the scale of geological structures. For linear infrastructures such as a tunnel or a levee, an engineer is certainly aware that ground conditions will change. For example, Lacasse (2015) modelled the New Orleans Levees as a series system of 560 reaches. A levee reach is defined as a portion of a levee system (usually a length of a levee) that may be considered for analysis purposes to have approximately uniform representative properties. The current definition of a reach is based on judgment rather than data-driven. A reach is a "site" in the above sense. Data-driven methods may be able to adjust site boundaries automatically to satisfy a given homogeneity requirement (e.g., foundation length for one part of a foundation layout, soil nail spacing/length for a slope section, rock bolt spacing/length for a tunnel section). Given the current data-poor environment, it goes without saying that this dynamic resizing will be probabilistic. As an insightful illustration, Otake et al. (2024) treated each borehole in the database Tokyo-CLAY/14/67,760 (from a 10 km^2 site in Japan) as a single "site" and showed that the cross-correlation scatterplots obtained are similar to those in CLAY/10/7490 (a generic database from 251 studies carried worldwide). It indicates that the prevailing physical definition of a "site" is only nominal. For DDSC, it is more meaningful to divide a physical site into statistically homogeneous sites using this dynamic resizing concept. This chapter proposes replacing the physical definition of a site boundary with a data-driven definition and terms this a "data-discovered site."

The third dimension is "precision characterization," which is implied by the "precision construction" challenge first proposed by Phoon (2018). This is unquestionably a "holy grail" pursuit in site characterization. For a critical structure such as a bridge foundation or an offshore wind turbine foundation, the current practice is to mandate sampling and testing at its location. Clearly, precision characterization is a challenge only if there is no observation at the location of interest. It is important to emphasize that precision characterization does not imply that the ground condition is known accurately at every location. "Precision" refers to spatial resolution, not an understanding of the ground. Hence, precision characterization and construction refer to taking different actions or making different decisions due to variations in ground conditions at different data-discovered "sites" that can be as small as a single pile or pile group location. These ideas are inspired by precision medicine that customizes treatment to suit the unique DNA of each individual. Precision understanding refers to a significant reduction in the range of plausible ground conditions at each data-discovered site, possibly in real time. This precision understanding may be achievable through advances in ground sensing technologies and machine learning-based data fusion (Meng et al. 2020). With this distinction in mind, precision characterization can be defined as a DDSC methodology that can produce a plausible range of ground conditions at a given location or one part of a site. The range can be narrow or broad depending on the layout of the site investigation plan, i.e., it is consistent with the available data. Kriging is the standard approach in geostatistics (Matheron 1963; Journel and Huijbregts 1978), but it does not seek to optimize its inference using data from other sites. HBM-MUSIC-3X is the first method to combine site-specific data and BID (Ching et al. 2021d). Kriging can combine data from multiple sources (Wackernagel 2003), but it does not consider performance data that may shed light on the ground condition through inverse analysis. There are other well-known limitations in cokriging (Guan et al. 2024a). The authors believe that novel machine learning or artificial intelligence solutions are available when one approaches precision characterization from a "fit for (and transform) practice" perspective rather than a method-centric perspective such as making incremental improvements to kriging.

1.2.3.2 Spatial features

The most analysed feature in geotechnical engineering is spatial variability. It was popularized by Vanmarcke (1977)'s classic paper on "Probabilistic modelling of soil profiles" and subsequently his book *Random Fields: Analysis and Synthesis* (Vanmarcke 1983). However, it is well known among engineers that uncertainty in the geological model is possibly larger (Einstein and Baecher 1982; Fookes 1997; Hoek 1999; Bárdossy and Fodor 2001; Bock 2006; Parry et al. 2014; Juang et al. 2018; Juang et al. 2019). This is denoted by "G" in MUSIC-3X-G. The spatial variability of soil/rock properties (Lumb 1966; Vanmarcke 1977, 1983) is termed "geotechnical uncertainty" to distinguish it from "geological uncertainty" (Phoon et al. 2022d). This is denoted by "3X" in MUSIC-3X-G.

Bárdossy and Fodor (2004) categorized geological uncertainty as "structured" and "unstructured." The former refers to relatively regular spatial changes that can be described by a trend-surface analysis. Examples include gradual compositional transitions of one strata/rock into another and cyclic repetitions of sedimentary features in a sequence of layers. The latter refers to features that occur unexpectedly in a stratum. The detection of anomalies would require grappling with unstructured uncertainty (Tang 1986; Tang and Halim 1988; Halim and Tang 1993). Soil stratigraphy can be structured or unstructured depending on its complexity. The characterization of soil stratigraphy has attracted some attention (Liao and Mayne 2007; Cao and Wang 2013; Li et al. 2013; Wang et al. 2014; Ching et al. 2015; Qi et al. 2016; Cao et al. 2019; Wang et al. 2019; Shuku et al. 2020; Shi and Wang 2021a, 2021b; Shuku and Phoon 2023a).

This section proposed a more general term "spatial features" to encompass geotechnical properties, geological features, ground improvement structures, and geoenvironmental processes (hydro, thermal, transport, etc.). They are basically 3D functions of space (or 4D when time is included) that influence the behaviour of a geotechnical structure constructed on or in a spatial domain. They are dominantly natural and permanent apart from ground improvement or geoenvironmental processes that can be transient. It is safe to say that data-driven site characterization focuses mainly on CPT-based soil stratigraphy, in part because richer datasets capturing more spatial features are currently unavailable and/or physics is incomplete and poorly understood.

To promote the development of DDSC methods that can infer increasingly complex spatial features (one or more), one important step is to create reference geotechnical datasets and standard performance metrics to train and measure different DDSC methods on a uniform basis. This benchmark testing or benchmarking is widely used in machine learning (ML) to support unbiased and competitive evaluation of emerging ML methods (Thiyagalingam et al. 2022). In one of the earlier attempts at establishing a benchmarking framework for DDSC, Phoon et al. (2022c) recommended the following considerations: (1) benchmark examples must be 3D, realistic in scale, stratigraphy, and properties; (2) cover a sufficient range of ground conditions; (3) restrict training datasets to only data that an engineer has at his/her disposal (MUSIC-3X or other data attributes); and (4) selection of validation datasets and performance metrics to compare methods fairly and showcase the value of DDSC to decision making in practice.

Phoon et al. (2022c) proposed the first set of eight benchmark examples for DDSC as summarized in Table 1.4. One component of a benchmark example is the virtual ground that represents the complete state of "reality." Everything is known within this virtual ground because it is generated by assuming stratigraphy and populating the properties in each layer using a random field generator (Figure 1.3). However, the rule of the game in

Table 1.4 Benchmark examples for data-driven site characterization

"Small" virtual ground	Description	Training dataset	
		T1 (3 CPT soundings)	T2 (6 CPT soundings)
S-VG1	Horizontal layers with constant property trend in each layer	S-VG1-T1	S-VG1-T2
S-VG2	Inclined layers with constant property trend in each layer	S-VG2-T1	S-VG2-T2
S-VG3	Inclined layers with linear property trend in each layer	S-VG3-T1	S-VG3-T2
S-VG4	Mixture of continuous and discontinuous layers with constant trend in each layer	S-VG4-T1	S-VG4-T2

Source: table 2, Phoon et al. (2022c).

DDSC is to assume that only the training dataset (containing CPT data from various sounding layouts) is available. Hence, the training dataset represents measured "reality" or what is known. Figure 1.3 shows two training datasets, T1 and T2. The purpose of a DDSC method, after it has been suitably trained, is to predict the soil profile and the properties at a prescribed set of validation soundings that are distinct from the set of training soundings (Figure 1.4). Note that the CPT data in the validation soundings are strictly restricted to cross-validation. They are assumed to be "unobserved" and cannot be used for training. Different DDSC methods can be compared uniformly using a set of performance metrics.

Jessell et al. (2022) generated 1 million 3D geological models for ML training. Each model is the outcome of superimposing user-defined kinematic events that represent idealized geological events such as folds, faults, unconformities, dykes, plugs, shear zones, and tilts on a base stratigraphy. An example is shown in Figure 1.5. It is effectively a "virtual ground" similar to Figure 1.3. The purpose of generating these synthetic "geologically plausible" models is to overcome challenges in applying ML to geoscience problems. Karpatne et al. (2018) presented nine commonly encountered attributes of geoscience data that should be accounted for in the design of ML algorithms: (1) spatial-temporal structure, (2) heterogeneity in space and time, (3) high dimensionality, (4) lack of concise object definitions, (5) rare classes, (6) multi-source multi-resolution data, (7) poor quality of data, (8) small sample size, and (9) paucity of ground truth. The training of a facial recognition algorithm is far simpler, because the "face" is a well-defined object, the number of facial features is small, and the training sets (photographs) are extremely large. Spatial-temporal structure and heterogeneity in space and time are two aspects of spatial features that arise from naturally occurring processes. It refers to "3X" in MUSIC-3X. Hence, some of the attributes are comparable to those highlighted in DDSC (Phoon et al. 2022b).

1.2.3.3 Sampling characteristics

MUSIC-3X is an initial attempt to describe the attributes of geotechnical data found in a typical site investigation report. However, it does not consider geotechnical data produced by a broader range of investigation methods (e.g., geophysical methods, laser scanning, drone photogrammetry, satellite imagery) and instrumentation methods (e.g., fibre optic sensors, wireless sensor networks, Internet of Things). It is useful to look at the contrasting characteristics of these data sources because they can guide the development of more general data-driven fusion methods that can tap into a wider spectrum of multi-source data multi-modal beyond those encountered in a routine site investigation.

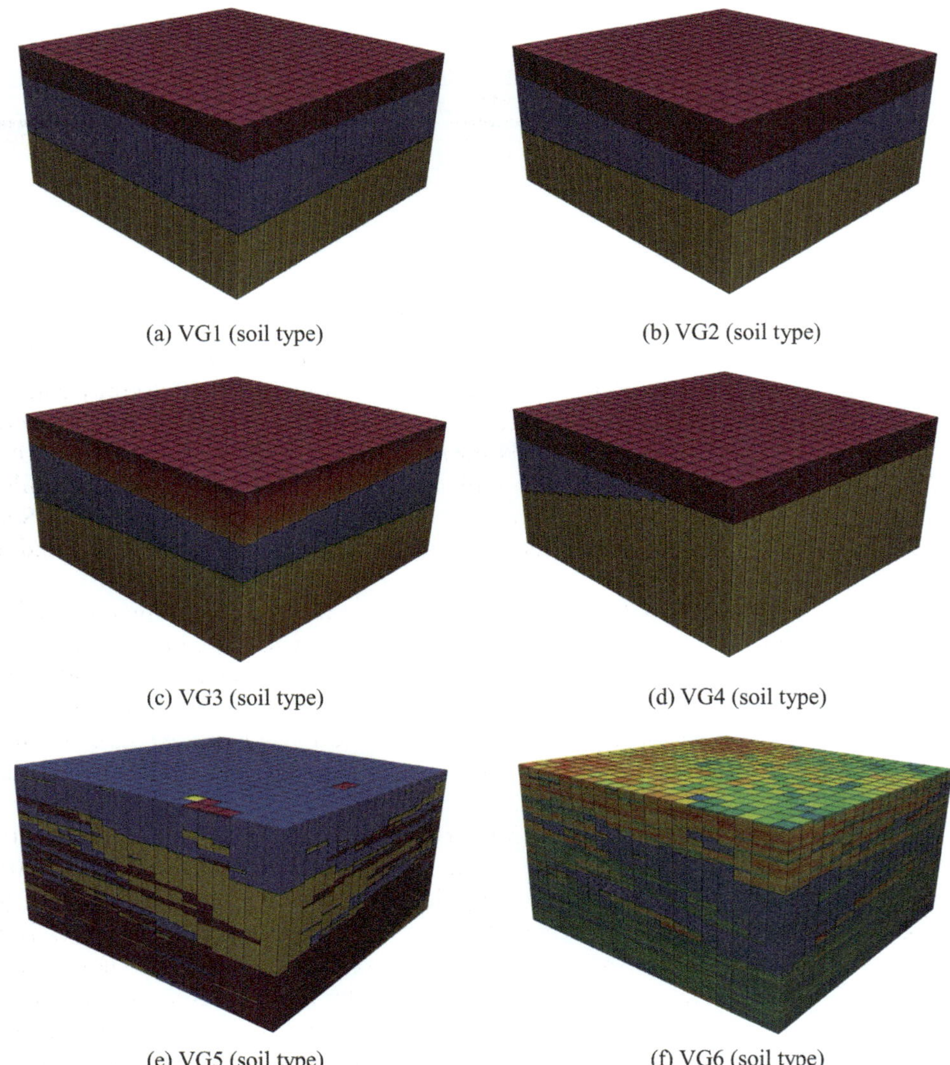

Figure 1.3 Virtual grounds (VGs) with increasing stratigraphic complexity: (a) VG1: horizontal layers with constant property trend, (b) VG2: inclined layers with constant property trend, (c) VG3: inclined layers with linear property trend, (d) VG4: mixture of continuous and discontinuous layers with constant property trend, (e) VG5: random soil type distribution based on coupled Markov chain with constant property trend, and (f) spatially varying soil property in VG5 (Source: VG1, VG2, VG3, and VG4 from figure 2, Phoon et al. 2022c; VG5 from figures 6 and 7, Shuku and Phoon 2023b).

(1) Univariate versus multivariate data – A conventional site investigation typically involves conducting different tests in a small neighbourhood such as the bivariate example shown in Figure 1.2. However, the more general multivariate data contains a spatio-temporal structure as noted by Karpatne et al. (2018). When viewed in the context of the World Wide Web, this characteristic may be a special case of networked data (Zhou and Song 2016).

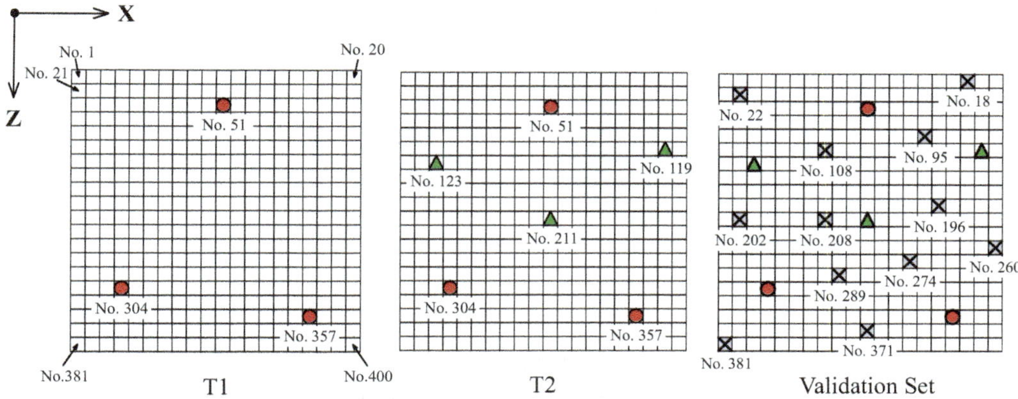

Figure 1.4 CPT soundings that produce the training and validation datasets (Source: figure 4, Phoon et al. 2022c).

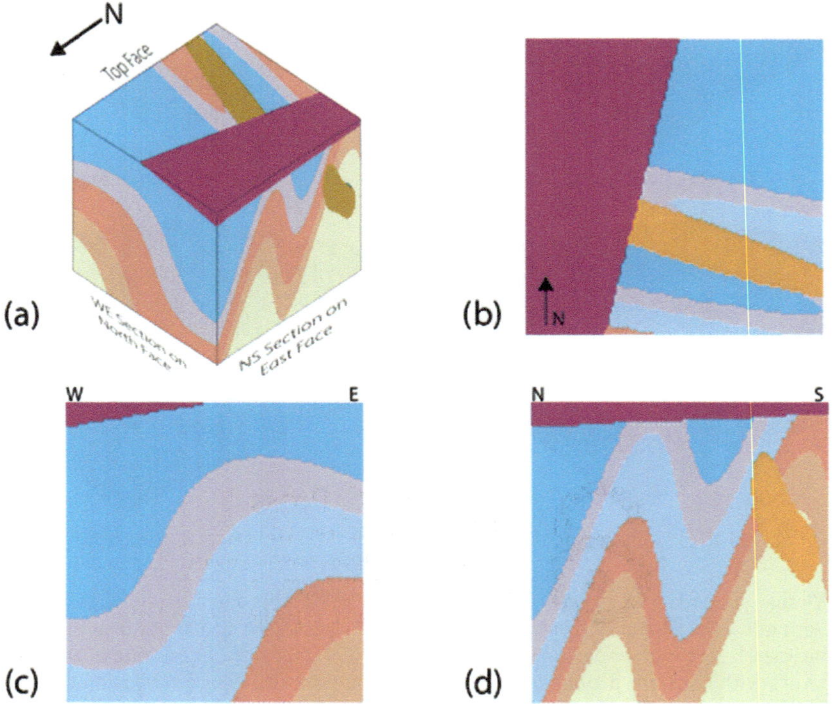

Figure 1.5 Example of simulated geological model: (a) 3D visualization, (b) top surface, (c) E–W section, and (d) N–S section (Source: figure 1, Jessell et al. 2022).

(2) Single site versus multiple sites – In the absence of the time dimension, this characteristic is addressed by the site recognition challenge (Phoon and Ching 2022). If one defines a "data-discovered site" as a spatially bounded volume of ground that is statistically homogeneous in some well-defined sense related to the limit state (e.g., governing properties, responses to a loading condition), it is possible for a

site boundary to change with time if the limit state changes with time to maintain homogeneity within a "data-discovered site" as discussed in Section 1.2.3.1. This characteristic is related to the heterogeneity in space and time attributes noted by Karpatne et al. (2018).

(3) Low-density versus high-density data – A video is high-density data in space and time, but there is no method to image a 3D subsurface continuously at low cost. It may not be necessary for the purpose at hand. A single image is high-density data in 2D space, but it does not provide any information on the time evolution of the quantity of interest. An image is also a slice of 3D space. Many field tests such as the standard penetration test (SPT) produce low-density data in 1D space, and it is usually conducted once in the lifetime of the project unless there is value in repeating these tests. For example, tests can be repeated before and after ground improvement for quality control or after a major failure in forensic engineering. It is not straightforward to combine data with different sampling densities in space and/or time (e.g., construction of a shear wave velocity profile from geophysical and borehole data). If a database is tabulated based on the source with the highest sampling density, there will be many empty cells for sources with lower sampling density. In other words, the database will be highly incomplete. If the opposite approach is taken, sources with higher sampling density will be down-sampled to match the source with the lowest sampling density (Shuku et al. 2023). This is one challenge in data fusion that has found new solutions in machine learning and artificial intelligence.

(4) Low-fidelity versus high-fidelity data – In geotechnical engineering, data produced by a field/laboratory test that can replicate in situ boundary conditions, and the tested volume is undisturbed and sufficiently representative, are regarded as high fidelity. Data produced by simple index tests on disturbed samples, such as the Atterberg limit tests, are regarded as low fidelity, even if the tests are highly repeatable (small measurement errors). Performance data produced by physical modelling could be of medium fidelity. Basically, the criterion is how closely the measurement/observation conditions match the operational context of interest.

(5) "Input" versus "output" parameters – All site investigation data are "inputs" while instrumentation data are "outputs" when viewed through the lens of physics. ML can be successful without invoking physics. In the first place, physics may not be available. In geotechnical engineering, there are problems that are poorly understood and the physics is grossly incomplete. The coefficient of variation of the model factor is very high (greater than 90%) for such problems (Tang and Phoon 2021). Hence, the distinction between inputs and outputs may not be possible. However, for problems with well-understood physics, synthetic data can be generated to supplement limited real-world data for ML training. One approach is the application of transfer learning to bridge the gap between simulation and the real world.

(6) Categorical versus numerical data – Geotechnical properties are mainly numerical. This is not the case for rock mass properties. Harrison (2019) pointed out that statistical analysis is not meaningful for some data types in rock mass properties. For example, it is not possible to determine the mean and coefficient of variation of nominal and ordinal data. Both data types are categorical, except the latter is amenable to ranking. The genetic group – sedimentary, igneous, and metamorphic – is an example of nominal data. He cited the joint set number (J_n) in the Q-system as an example of ordinal data. The reader can refer to ISO 14689-1:2003 Geotechnical investigation and testing – Identification and classification of rock – Part 1:

Identification and description for a more complete coverage of categorical data. Phoon et al. (2022b) opined that handling a mixture of categorical and numerical data of varying proportions in high random dimensional space could be a challenge. ChatGPT-4o can handle multi-modal data, which is arguably the most varied in terms of data types.

The above is a partial list. There are possible relationships between the characteristics. For example, high-fidelity data are likely to have a low sampling density due to cost.

1.2.3.4 Smart data

"Smart data" is defined as actionable data at the point of collection. In contrast, conventional data are compiled and analysed in batches. Clearly, timely decisions cannot be made in the batch processing mode. The strategy to maximize the value of data in the presence of the Internet of Things (IoT) is not likely the same as conventional monitoring instruments. The collection of all possible data and storing them in the cloud with the intent to organize and analyse big data in one batch at a later point in time is becoming less effective as the volume and complexity of data from networked sensors increase. One alternate strategy is to implement algorithms (intelligence) at every step of the data processing chain, from smart sensors with limited low-power computing resources through gateways and edge computers to large cloud computers. There are advantages to implementing algorithms as close as possible to the point of data collection (sensors). Besides enabling time-critical interventions, the refinement of data at an early stage will reduce power, storage, transmission, and subsequent analysis costs.

Smart data can be regarded as processed big data that leads to a meaningful action (or decision). The intelligence of the algorithm must be a function of the level in the data processing chain. For a single sensor collecting data in a small neighbourhood, intelligence is low. For an algorithm with access to all sensors (and connected to lower-level intelligence), trained on past data produced by similar projects, and possibly interacting continuously with human-in-the-loop, intelligence is much higher. A simple example is the action of a hand jerking away from a heat source. The intelligence resulting in this action is much lower than that of the affected person who realizes the heat source is a fire outbreak and thus takes a higher-order decision to run away or put out the fire. The value of information that defines smart data depends in part on the complexity of the data (e.g., single versus networked sensor data), an appreciation of the context (from understanding or experience), and the level of decision making (e.g., temperature exceeds threshold versus risk management considering complex trade-offs). In other words, the value of information is relative to the level of intelligence in the data processing chain. It is reasonable to speculate that there must be a balance of autonomy and interaction in intelligence at different levels because decision making is taking place at different scales (local versus global, fast versus slow) and different levels of intelligence. One example of an intelligence hierarchy is the DIKW (Data, Information, Knowledge, Wisdom) pyramid (Rowley 2007). The current direction of travel is embodied AI that combines large language models with robotics or other devices that interact with the real world in real time. Finally, communication itself is becoming smart as well. Luo et al. (2022) explained the concept of semantic communication: "The core of semantic communication is to extract the 'meanings' of sent information at a transmitter, and with the help of a matched knowledge base (KB) between a transmitter and a receiver, the semantic information can be 'interpreted' successfully at a receiver. Therefore, semantic communication essentially is a communication scheme based largely on AI."

In civil engineering, the value of information (VoI) is typically founded on Bayesian decision analysis and utility theory (Raiffa and Schlaifer 1961; Zhang et al. 2021). The purpose of evaluating VoI is to decide how to collect data to maximize its value for a given purpose before the data is measured. Hence, it is also called pre-posterior analysis by Bayesians. The underlying motivation for VoI is that additional information does not always produce sufficient benefits for the purpose at hand to compensate for the attendant costs. While the concept is useful, Straub (2014) observed that there are practical difficulties in relation to the design of monitoring systems: (a) proper probabilistic modelling of the monitored process and the monitoring itself, (b) the modelling of the action alternatives following the monitoring results, and (c) the computational efforts in evaluating the VoI. The design of monitoring systems plays a central role in the observational approach in geotechnical engineering. Hu et al. (2021) also pointed out that the application of VoI to a slope problem can be computationally prohibitive. The application of VoI in geotechnical engineering is limited at present (Yoshida et al. 2018; Hu et al. 2021). The relation between VoI and smart data is unclear. However, it is safe to say that the VoI framework needs to be revisited to guide data collection for data-driven decision making in the modern context of Industry 4.0, which consists of new hardware (IoT, 5G connectivity, cloud, edge computing, autonomous systems) and new software (AI, digital twin, blockchain, web3). The contours of a possible Value of Smart Information (VoSI) framework are not defined at this point, but a good start is to imagine how to extract more value from conventional monitoring instruments that can talk to each other by building intelligence into every stage of the data processing chain. VoSI is expected to be an integral part of machine learning-guided observational method (Section 1.5).

1.3 DATA-DRIVEN SITE CHARACTERIZATION

Site characterization is a cornerstone of geotechnical and rock engineering. It is hard to imagine a safe and economical practice in its complete absence. The classic example is the table of allowable bearing values in Terzaghi and Peck (1967) (table 54.1). No site investigation and no site-specific data are needed other than a broad qualitative appreciation of soil type and the name of the city. However, a practice based on qualitative knowledge of the site is very conservative. Note that it is possible to consider site characterization being conducted in tandem with construction. This is typical in tunnelling, say using a computer jumbo (Shuku et al. 2023). This "characterize, construct, and learn as we go" adaptive strategy is the basis of the observational method. It is arguably a more robust risk management strategy for a practice that operates in a spatially variable but data-poor environment (with occasional undetected "geologic surprises").

"Data-driven site characterization" (DDSC) refers to any site characterization methodology that relies solely on measured data, both site-specific data collected for the current project and existing data of any type collected from past stages of the same project or past projects at the same site, neighbouring sites, or beyond (Phoon et al. 2022b). This is a "data first" definition in line with data-centric geotechnics. It is silent on the role of physics and experience. As noted above, this does not mean insights from other sources are not valuable and cannot be included in DDSC. Karniadakis et al. (2021) noted that physics-informed machine learning (PIML) has two advantages: (a) it can integrate data and physics, even in partially understood, uncertain, and high-dimensional contexts, and (b) it is effective and efficient for ill-posed and inverse problems. One can speculate that PIML can operationalize the observational method in a more practical way. To the knowledge of the authors, no research has been conducted in machine learning-guided observational method (MLOM) within the agenda of data-centric geotechnics. However, some limited aspects have been studied. For example, Tan et al. (2023) mapped the spatial stress distribution over a coalmine roof by training a transfer convolutional neural network (TCNN) model using physics (finite difference method) and stress sensor data collected in the roof of the Dongtan coalmine. Tian and Wang (2023) and Tian et al. (2024) developed sparse dictionary learning methods using physics (finite element analysis) with sparse site characterization data and settlement monitoring data for improving settlement prediction in real time and continuously, particularly at locations without monitoring data and/or subsequent time steps. Section 1.5 briefly discusses the challenges of MLOM.

Thick data is qualitative and able to associate numbers with meaning and context. Hong et al. (2022) described thick data in the context of urban research:

> Thick data is descriptively rich and intensively detailed, collected from or with research participants on issues relating to their values, visions, knowledge, life experiences, and opinions, typically obtained by observing or interacting with participants in their daily lives and through in-depth interview techniques. Thick data may be collected by using and analysing a variety of formats including photographs, videos, sound recordings, sketches, and written ethnographic field notes.

Experience and other qualitative knowledge accumulated in more than half a decade of modern practice can be regarded as "thick data" in geotechnical engineering. Phoon et al. (2022b) suggested that engineers can be asked to tag a data-driven prediction with "agree," "disagree," or "not sure." These tags are valuable because an AI can learn and improve its predictions through continuous "reality checks" by human experience, but this process is expensive. With the emergence of powerful generative AI such as ChatGPT, it is reasonable to hypothesize that thick data found in the issues of the *ISSMGE International Journal of Geoengineering Case Histories* and in the proceedings of the "International Conference on Case Histories in Geotechnical Engineering" may be assimilated. In addition, the ISSMGE Online Library is open source and can serve as a training dataset for ChatGPT (https://www.issmge.org/publications/online-library). To the knowledge of the authors, there is no research on thick data or ChatGPT in geotechnical engineering although an initial study on large language models was conducted recently (Wu et al. 2023).

1.3.1 Site recognition challenge

At present, there are no physics-based models that can predict ground conditions or properties at a particular location. It is also unlikely that the geologic history at one site (which has a significant bearing on the current ground conditions and properties) can be obtained with sufficient quantitative accuracy to specify the boundary conditions of any physics-based model. There are attempts to explain transformation models more rationally, but field measurements are needed (Lim et al. 2018; 2019). Hence, DDSC is a promising direction of research (e.g., Huang et al. 2018; Cao et al. 2019; Ching and Phoon 2019; Wang et al. 2019; Ching and Phoon 2020a; Ching and Phoon 2020b; Shuku et al. 2020; Ching et al. 2021a; Ching et al. 2021b; Shi and Wang 2021a, 2021b; Wang et al. 2021; Yoshida et al. 2021; Xiao S. et al. 2021a, 2021b; Xu et al. 2021; Ching et al. 2022b; Sharma et al. 2022, 2023; Wu et al. 2022; Ching et al. 2023b, 2024; Shuku and Phoon 2023a; Guan et al. 2024a, 2024b; Hu et al. 2024). Phoon et al. (2022b) outlined three challenges in DDSC: (1) ugly data, (2) site recognition, and (3) stratification. An example of ugly data called MUSIC-3X is presented in Section 1.2.2. No data-driven research in geotechnical engineering is meaningful if it does not address real-world data. The methods proposed in many papers are data driven in taxonomy, but not in practice. The ugly data challenge is foundational and it is an integral part of all other challenges in DDSC and more broadly in data-centric geotechnics. Section 1.2.3 can be regarded as an expansion of the ugly data challenge.

The site recognition challenge is related to a fundamental feature in geotechnical practice, namely all sites are different to some extent (site specificity) (Phoon et al. 2025). For example, Clause 2.4.5.2(10) of Eurocode 7 (CEN 2004) "Characteristic values of geotechnical parameters" advises against pooling data from different sites into a single population for statistical analysis. The stratification challenge requires identification of soil type and a mapping of the spatial distribution of each soil type based on field tests typically conducted at limited discrete locations/depths in a soil mass. The stratification challenge has not been expanded to address more general spatial features discussed in Section 1.2.3.2. An example of a site investigation programme is shown in Figure 1.6. Section 1.4.3 presents an HBM-MUSIC-3X method that attempts to address the site recognition and stratification challenges together. The authors argue that the ugly data and site recognition challenges underlie all DDSC applications in geotechnical and rock engineering.

1.3.2 Data-informed decision support index (DIDI)

The site recognition and stratification challenges are not new. An engineer encounters them routinely in practice, but they are not regarded as challenges because judgment-based solutions are widely accepted as satisfactory. For instance, an engineer applies experience gained from other sites to interpret limited data at the current site of interest. However, he/she will not regard all sites as equally relevant – again, experience (occasionally complemented by some broad knowledge of local geology) is needed to recognize that some sites are more relevant or "similar." For the construction of a 2D geologic section (or fence diagram), it is not uncommon to connect the same soil type (interpreted by an engineering geologist from disturbed samples) between two adjacent boreholes using a straight line or to insert a question mark with a speculative dashed line to indicate "not sure" if the sequence of soil types does not match between two adjacent boreholes. Geotechnical failures are rare despite the adoption of a practice that is arguably more art than science.

There is a growing contrast between judgment-based site characterization that remains relatively unchanged for decades and increasingly sophisticated physics-based modelling of soil–structure interaction such as the material point method. Phoon et al. (2022d) opined that:

> Currently, numerical analysis is conducted using deterministic inputs and simple soil profiles that are inconsistent with sparse site data. This deterministic mapping (one

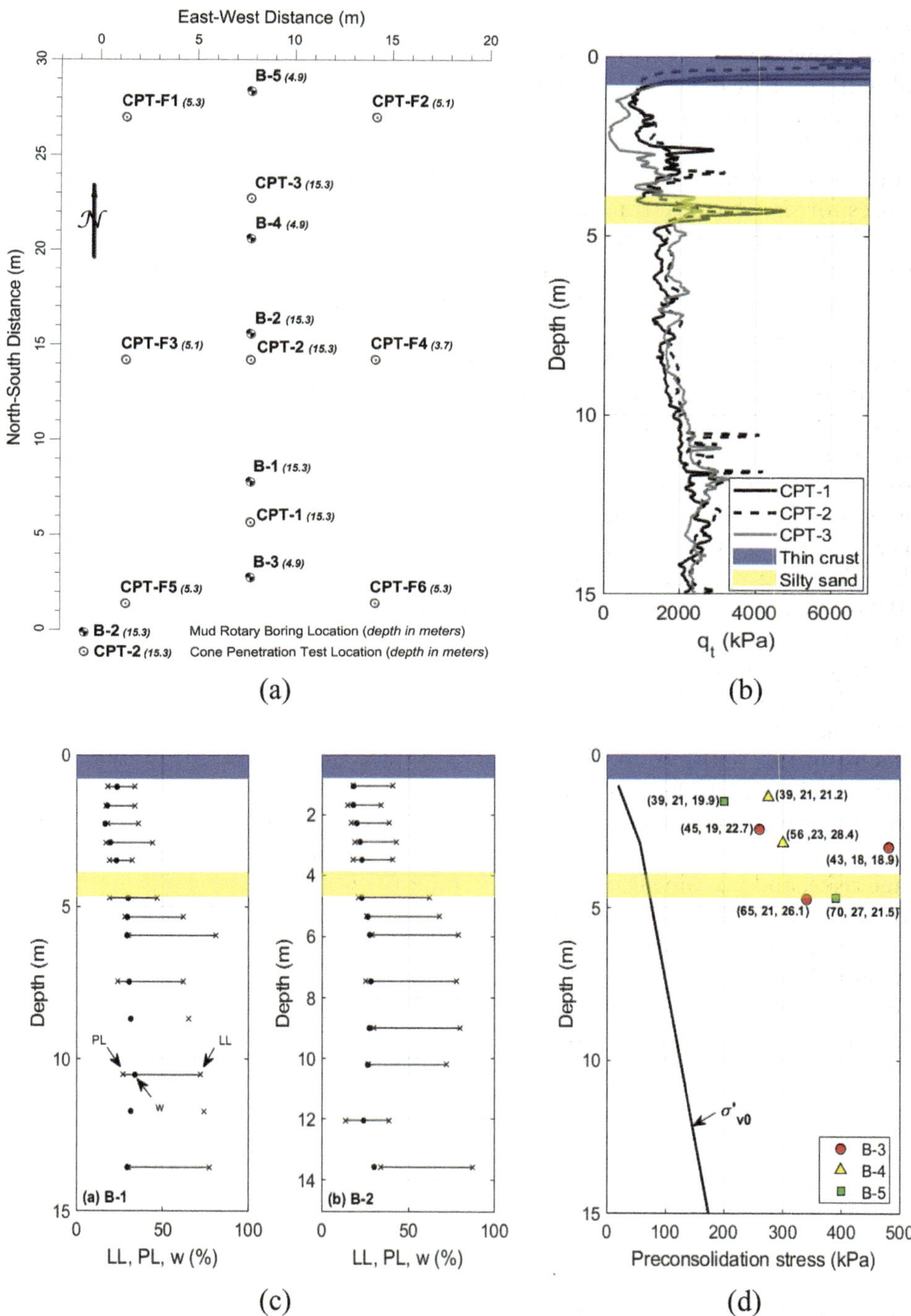

Figure 1.6 Typical site investigation programme at Baytown, Texas: (a) exploration (borings and CPT soundings), (b) some CPT test results, (c) some index test results (LL = liquid limit, PL = plastic limit, and *w* = natural water content), and (d) pre-consolidation stress results (numbers in parenthesis refer to LL, PL, *w*) (Stuedlein et al. 2012).

set of inputs to one set of outputs) does provide valuable physical insights, but cannot support risk-informed decision making on its own without appealing to engineering judgment.

In the opinion of the authors, the most important question is whether a pure judgement-based site characterization can fit Industry 4.0. Phoon (2023b) posed this question in the context of broader emerging trends: (1) rapid surge in computing power and digital technologies in general, (2) availability of more efficient and more powerful probabilistic and other data-driven methods (Phoon and Ching 2021; Phoon et al. 2022b), (3) sharing of large databases [termed big indirect data (BID) by Phoon et al. 2019], (4) incorporation of probabilistic analysis in geotechnical software (refer to table 10, Phoon 2023b), (5) increasing emphasis on risk-informed decision making in design codes (Phoon and Retief 2016; Phoon et al. 2016; Retief et al. 2016), (6) innovations should be the norm in practice to address urgent challenges such as sustainability, but current technology acceptance criteria lean towards risk-averse rather than risk-informed (ASCE 2019), and (7) the deterministic paradigm is at odds with digital transformation because it pre-dates the personal computer era (Phoon et al. 2022a). Phoon (2023b) opined that the "shift to data-informed decision making is a near certainty given the increasing power, ubiquity, and convergence of digital technologies." Phoon et al. (2022a) presented a data-informed decision support index (DIDI) (Table 1.5) to explain the expanding role of data in geotechnical engineering decisions that spans from:

(1) Almost no data (DIDI = 1) in pre-Terzaghi years.
(2) Empirical or simple closed-form solutions (DIDI = 2) found in Terzaghi (1943).
(3) Physics-centric solutions with deterministic site-specific inputs (DIDI = 3) such as those reviewed by Salgado et al. (2008) for foundation engineering.
(4) Physics-centric solutions with more attention on data representation using a frequentist interpretation of uncertainty such as those reviewed by Chwała et al. (2023) (DIDI = 3 to 4).
(5) Uncertainty-informed decision making that is more comprehensive and with increasing emphasis on the Bayesian interpretation of uncertainty (DIDI = 4).
(6) Data-centric decision making such as those reviewed by Phoon and Ching (2021), Phoon et al. (2022a), Phoon and Zhang (2023), Phoon et al. (2025) (DIDI = 5).

The arc of history in decision making can be viewed as an evolution from qualitative methods, to quantitative methods using closed-form theoretical solutions, to numerical methods, to probabilistic methods, and possibly bending to data-driven methods that include machine learning and artificial intelligence. The advent of powerful generative AI, such as GPT-4o in 2024, may lead to an elevation of DIDI beyond the current state of 3 to 4 in the near future. Human–machine teaming is now a distinct possibility. The convergence of AI, robotics, and connectivity alone is likely to disrupt many industries, including geotechnical engineering.

1.4 APPLICATIONS

Phoon and Ching (2022) and Phoon et al. (2025) presented some state-of-the-art methods in the site recognition challenge. A summary is given in Table 1.6. It is possible to

Table 1.5 Data-informed decision support index (DIDI)

Example	*Data to decision*	*Data-informed decision support index (DIDI)*
Allowable bearing values	Conservative range of design values based on qualitative soil type and name of city	1
Generic empirical correlations	Non-site-specific empirical relationship between a design value (e.g., end bearing) and an indirect field test (e.g., SPT)	2
Deterministic engineering calculations with observational approach	Design value produced by a physical model with site-specific input parameters from generic transformation models (e.g., undrained shear strength versus cone tip resistance), possibly updated by site-specific observations (e.g., load tests, monitoring)	3
Reliability analysis with Bayesian observational approach	Design value represented by a probability distribution that is related to uncertainties in input parameters, model idealizations, and possibly updated by field observations utilizing site-specific and generic data	4
Precision design and construction	Design value updated by explainable AI that: (1) processes all big and small data from generic and site-specific sources, respectively, (2) recognizes site specificity, (3) "explains" its outputs, and (4) improves from continuous interaction with engineers	–
Partially smart infrastructure lifecycle management	Data-informed management of one or more lifecycle phases using BIM, digital twin, AI, etc.	
Completely smart infrastructure that fits circular economy and delivers service to end-users and community from project conceptualization to decommission/reuse with full integration to smart city and smart society	Value engineering at system level city scale to fulfil sustainability, resilience engineering, and other societal goals through ethical human-centric deployment of digital technologies	–

Source: table 8, Phoon et al. (2022a).

organize the methods in Table 1.6 under six broad strategies with increasing emphasis on the quantification of "similarity" between sites:

I. Use the entire global database with every record assigned an equal weight independent of its similarity (at record or site level) to the target site.
II. Use the entire global database with every record assigned a weight based on its similarity (at record or site level) to the target site.
III. Use a subset of the global database ("quasi-regional" database) with every record assigned an equal weight.
IV. Use a subset of the global database ("quasi-regional" database) with every record assigned an unequal weight.

Table 1.6 Methods to construct a quasi-site-specific transformation model from MUSIC-3X data and BID

Method	*Big indirect data*	*Site-specific data*	*Quasi-site-specific model* **M**	*Site uniqueness*	*Explainable*
Probabilistic multiple regression (Ching and Phoon 2012)	G	G	$M = G$	No	
Hybrid (Ching and Phoon 2019)	G	S	$M = G \times S$	Yes, but empirical	No
Record similarity (Ching and Phoon 2020a)	$G = (G_1, G_2)$	S	Select G_2 (list of records) $\approx S$ $M = S + G_2$	Yes, at record level	Yes
Two-step Gibbs (Ching et al. 2020b)	G_2	S	G_2 (regional) $\approx S$ $M = G_2$ updated by S	Maybe	No
Hierarchical Bayesian model (Ching et al. 2021b)	$G = (G_1, G_2, \ldots, G_n)$	G_{n+1}	$M = G$ updated by G_{n+1}	Yes, at site level	No
Site similarity (Sharma et al. 2022)	$G = (G_1, G_2)$	S	Select G_2 (list of sites) $\approx S$ $M = S + G_2$	Yes, at site level	Yes
Site clustering (Sharma et al. 2023)	$G = (G_1, G_2, \ldots, G_n)$	S	Select $S \in G_i$ (cluster of sites) $M = S + G_i$	Yes at cluster level	Yes
Site clustering - tailored (Cai et al. 2024a)	$G = (G_1, G_2)$	S	Select G_2 (tailored cluster) $\approx S$ $M = S + G_2$	Yes at cluster level	Yes

Source: updated from table 1, Phoon and Ching (2022).

V. Use a subset of the global database ("quasi-regional" database) and a subset of the soil/rock parameters that is relevant to one or more governing limit states.
VI. Do not use the global database, in part or in full, to inform the target site because they are "dissimilar."

Past research studies addressing the site recognition challenge can be broadly classified under Strategies I to VI. The sub-sections below briefly review five studies to showcase the range of applications and value of DDSC for real-world decision making. Real-time decision making requires a combination of site characterization and monitoring information. Research in this machine learning-guided observational method (MLOM) is largely absent at present, although its value is arguably higher than DDSC. It is uncertain whether a new discipline called geotechnical data science is needed to provide a stronger theoretical underpinning for data-centric geotechnics, given the large interconnected challenges highlighted in the previous sections. It is accurate to say that current data-driven research in geotechnical engineering is fragmented in terms of its theoretical foundation and largely follows concepts and methods developed in the wider machine learning community.

1.4.1 Strategy 1 – Generic model for clay parameters

Ching and Phoon (2014b) compiled a generic database called CLAY/10/7490. Each record in the database constitutes a set of multivariate clay properties (e.g., liquid limit, plasticity index, water content, undrained shear strength, cone tip resistance, etc.). In Ching and Phoon (2014b), it is assumed that there is only a "generic population" for all records in the database. Furthermore, the paper assumed that the target site belongs to the same generic population and the (transformed) multivariate soil properties of each record follow a multivariate normal distribution with mean vector = μ_g and covariance matrix = C_g. The training of the model parameters (μ_g, C_g) is straightforward: all records in the database as well as all known records for the target site are considered to be equally relevant. Hence, they are adopted to update (μ_g, C_g) with equal weights. The updated parameters (μ_g, C_g) (namely, the quasi-site-specific model) are then used to predict the unknown records at the target site. This strategy of constructing a quasi-site-specific model is referred to as the probabilistic multiple regression (PMR). One limitation is already clear. If there are 7490 records in the database and only 10 records at the target site, the effect of the target-site records on (μ_g, C_g) is marginal. The quasi-site-specific is almost the same as the generic model (site-specific data not included).

For illustration, consider the $(q_t-\sigma_v)/\sigma_v'$ versus s_u/σ_v' data in CLAY/10/7490 (grey dots in Figure 1.7a) (q_{t1} denotes $(q_t-\sigma_v)/\sigma_v'$). Also, consider the q_{t1} versus s_u/σ_v' data for an Onsøy site (Norway) (Lacasse and Lunne 1982). Some target-site records have complete q_{t1} and s_u/σ_v' data (red dots), yet some records only have q_{t1} data available (yellow dots) with their s_u/σ_v' data deliberately regarded as unknown for validating the model prediction. The solid and dashed lines in the figure show the mean trend and 95% confidence interval (CI) of the quasi-site-specific model constructed by the records in the database (grey dots) and the known records of the target site (red dots). This mean trend and 95% CI can then be used to predict the s_u/σ_v' values of the known records of the target site (yellow dots) based on their q_{t1} information, as shown in Figure 1.7b.

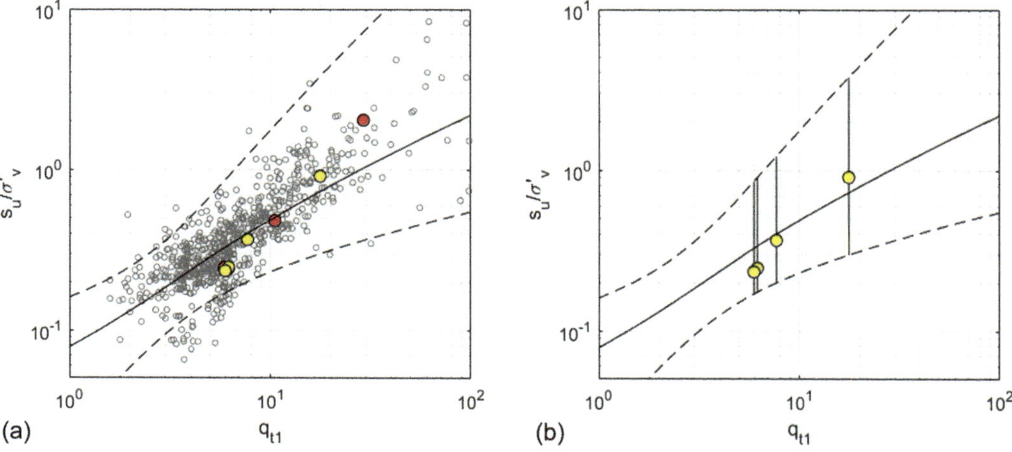

Figure 1.7 (a) The q_{t1} versus s_u/σ_v' data (grey dots are from CLAY/10/7490; red dots are from the Onsøy site; yellow dots are also from the Onsøy site, yet the s_u/σ_v' values are deliberately regarded as unknown); the dark and dashed lines are the mean trend and 95% CI of the quasi-site-specific model; (b) the prediction results for the s_u/σ_v' values of the known records of the Onsøy site.

1.4.2 Strategy II (without spatial variability) – Quasi-site-specific model for MUSIC clay parameters

Ching et al. (2021b) proposed a different strategy for modelling a generic soil database, called the hierarchical Bayesian model (HBM). In this model, the records of each site in the database are assumed to follow a distinct local population rather than a common generic population. To be more specific, the (transformed) multivariate soil properties of the records of the ith site are assumed to follow the multivariate normal distribution with mean vector = μ_i and covariance matrix = C_i. Moreover, (μ_i, C_i) and (μ_j, C_j) are different if $i \neq j$ to account for site uniqueness. More specifically, (μ_1, μ_2, …, μ_{ns}) (n_s is the number of sites in the database) are assumed to follow the multivariate normal distribution with (hyper) mean vector = μ_0 and (hyper) covariance matrix = C_0, and (C_1, C_2, …, C_{ns}) are assumed to follow the inverse-Wishart distribution with (hyper) scale matrix = $\mathbf{\Sigma}_0$ and (hyper) degree of freedom = ν_0. The parameters for the Onsøy site, denoted by (μ_s, C_s), are also assumed to follow the same (hyper) distributions. The HBM can learn the inter-site and intra-site variabilities in CLAY/10/7490. Note that the HBM is one plausible method to model site uniqueness.

Figure 1.8a shows the database data (records from the same site are in the same colour), whereas Figure 1.8b shows the behaviour of the HBM trained by CLAY/10/7490. The trained HBM can simulate (μ_s, C_s) of a hypothetical target site (each skewed ellipse represents such a hypothetical target site) whose variabilities are similar to those of the database sites. The trained HBM can be further updated by the Onsøy-site data. Many ellipses in Figure 1.9a are incompatible with the Onsøy-site data (shown as grey ellipses). A few ellipses are compatible (shown as coloured ellipses). The coloured (compatible) ellipses illustrate the quasi-site-specific model of (μ_s, C_s) after the Bayesian updating. The solid and dashed lines in Figure 1.9b show the mean trend and 95% confidence interval (CI) of the quasi-site-specific model. This mean trend and 95% CI can then be used to predict the s_u/σ_v' values of the known records of the Onsøy site (yellow dots) based on their q_{t1} information, as shown in Figure 1.9b. Figure 1.9b can be compared with Figure 1.7b. The comparison shows that the HBM (Strategy 2) is more effective than the PMR (Strategy 1)

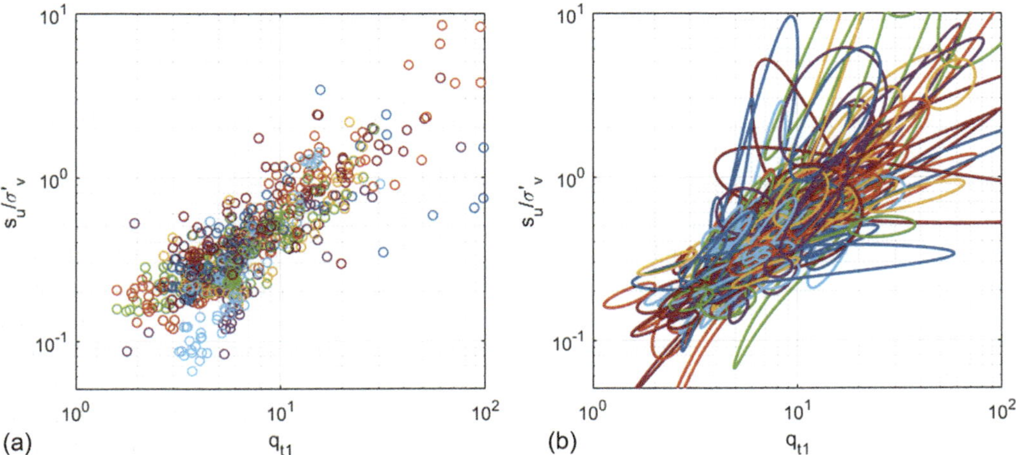

Figure 1.8 (a) The database data (records from the same site are with the same colour); (b) the behaviour of the HBM trained by CLAY/10/7490 (each skewed ellipse represents such a hypothetical target site).

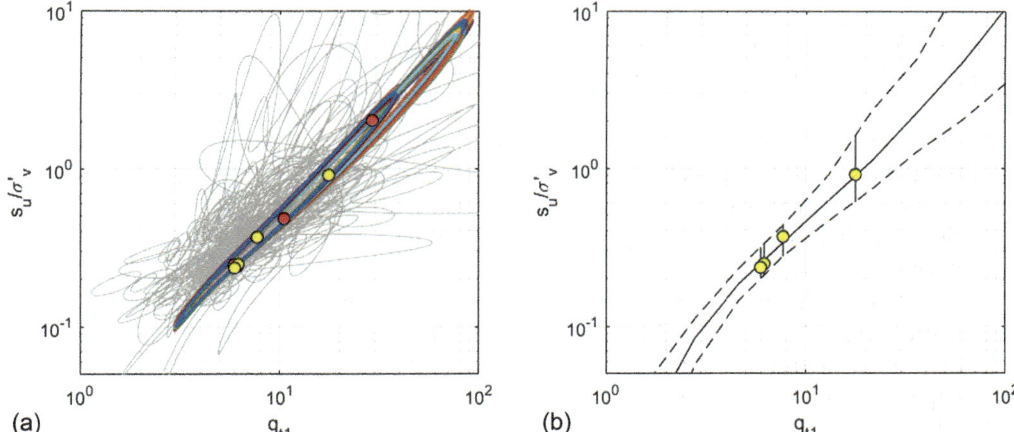

Figure 1.9 (a) Illustration of the quasi-site-specific model (grey ellipses are incompatible to the Onsøy-site data; coloured ellipses are compatible); (b) the mean trend and 95% CI of the quasi-site-specific model as well as the prediction results for the s_u/σ_v' values of the known records of the Onsøy site.

with narrower 95% CIs. Note that in the real application of HBM, the simulation of the skewed ellipses is not necessary: the quasi-site-specific model is automatically constructed by Bayes' rule. These skewed ellipses are shown herein to explain the basic idea of the HBM. The HBM has been adopted by Ching et al. (2021a) to make quasi-site-specific predictions for the deformation modulus of rock mass and by Ching et al. (2022a) to make quasi-site-specific predictions for the compressibility of clay. The key limitation of the HBM is that it is not an "explainable" method. Most engineers will feel more comfortable if the sites contributing to the quasi-site-specific inferences are real and judged by them as comparable or similar to the target site.

1.4.3 Strategy II (with spatial variability) – Quasi-site-specific model for MUSIC-3X clay parameters

In Section 1.4.2, the HBM trained by the soil database was updated by the Onsøy-site data into a quasi-site-specific model. The Onsøy-site data was assumed to be spatially independent because the locations (depths) of q_{t1}–s_u/σ'_v data are far apart. Ching et al. (2022b) showed that the HBM trained by the soil database can be updated by MUSIC-3X target-site data as well. Here, "3X" denotes spatially variable site data. The target site considered by Ching et al. (2022b) was the Baytown site (Texas, USA) (see Figure 1.6). The spatial correlation of the target site is characterized by the autocorrelation matrix **R**. By assuming that all soil properties follow the same autocorrelation structure, the total covariance matrix can be expressed as the Kronecker product between the autocorrelation matrix **R** and the cross-correlation matrix $\mathbf{C}_s$. This assumption allows the HBM trained by the soil database to be updated by MUSIC-3X site data into a quasi-site-specific model with affordable computational cost. This quasi-site-specific model, denoted by the HBM-MUSIC-3X model, can further simulate the conditional random fields for all properties because it addresses both spatial correlation and cross-correlation. Figure 1.10

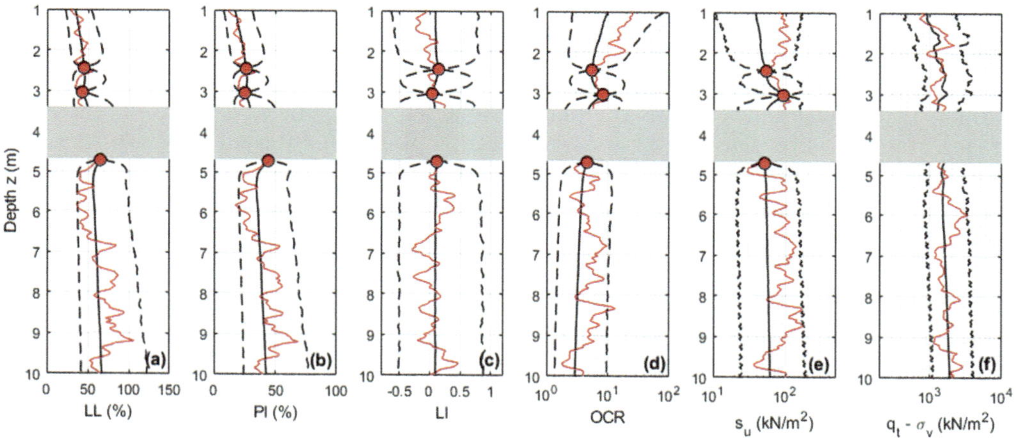

Figure 1.10 The conditional random fields simulated at the B-3 location at the Baytown site for several soil properties.

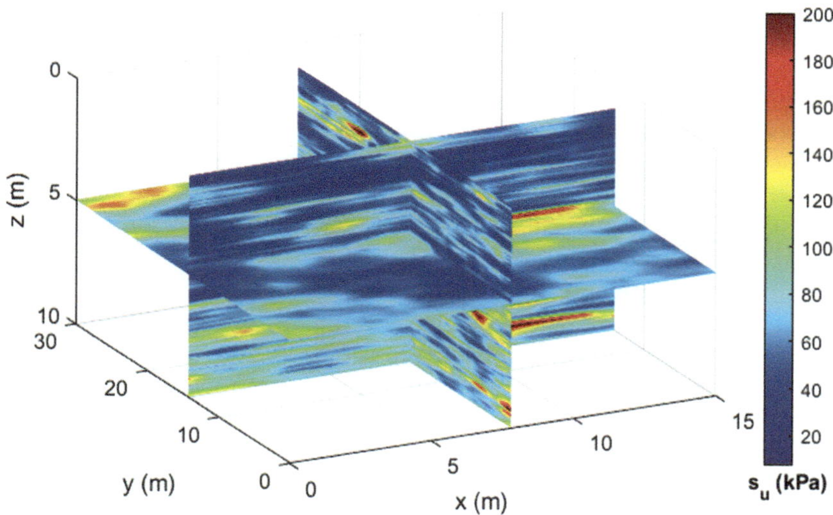

Figure 1.11 One conditional random field realization of s_u for the 3D underground space of the Baytown site.

shows the conditional random fields simulated at the B-3 location at the Baytown site (see Figure 1.6a) for several soil properties. The red lines are realizations of the conditional random fields, whereas the dark and dashed lines are the medians and 95% CIs. The red dots are the observed data at the B-3 location. Figure 1.11 shows one conditional random field realization of s_u for the 3D underground space of the Baytown site. The limitation is that only cross-correlation information in the database is updated in HBM-MUSIC-3X. Guan et al. (2024b) proposed a dictionary learning method to infer spatial variability at a target site from CPT profiles at neighbouring sites.

1.4.4 Strategy IV – Quasi-regional cluster model for MUSIC clay parameters

Ching et al. (2020b) showed that a quasi-site-specific model constructed by a Shanghai municipal database outperforms the one constructed by a generic database if the target site is a Shanghai site. Therefore, there is a question of whether the entire generic database should be adopted to train the HBM because some database sites may not be relevant to the target site. It is possible that a subset of the generic database with sites similar to the target site can train a more suitable HBM. This is known as the "regional advantage" (Phoon and Ching 2022). Sharma et al. (2022) proposed a hierarchical Bayesian similarity index that measures the similarity between (μ_i, C_i) and (μ_j, C_j) of two sites. This similarity measure can be used to construct the similarity matrix (size = $n_s \times n_s$) among the n_s sites in the generic database. Based on this similarity matrix, Sharma et al. (2023) further proposed a spectral clustering algorithm to divide the generic database into clusters. Sharma et al. (2023) showed that the HBM trained by a cluster more relevant to the target site [i.e., (μ_i, C_i) of the sites in the cluster are "closer" to (μ_s, C_s)] tends to outperform the HBM trained by the entire generic database.

1.4.5 Strategy IV – Tailored cluster model for MUSIC clay parameters

The limitation in classical clustering (CC) is that the number of clusters is small while the number of target sites can be very large. It is possible that some target sites do not fit any of the clusters well. Cai et al. (2024a) suggested a tailored clustering (TC) approach that addresses this limitation. In comparison to CC, TC is more flexible because the clustering results are dependent on the target site information. In the TC algorithm proposed by Cai et al. (2024a), a modified inter-dataset similarity measure (MIDSM) is used to calculate the similarity (*Sim*) between the target site and each database site. For each database site, MIDSM quantifies the value of *Sim* by considering the difference in geometries between the 95% confidence regions of datasets collected from the target site and the database site. The database sites that have high *Sim* values are considered to qualify as members of the desired cluster. Consequently, each target site is assigned to a cluster tailored to it.

Cai et al. (2024b) conducted a real example based on the generic database CLAY/10/7490 to show the advantage of TC in inferring soil parameters at a target site. The target site is selected from CLAY/10/7490 (#211 site) and this target site is removed from CLAY/10/7490. The two soil properties, limit liquid (LL, in %) and plasticity index (PI, in %), are considered in this example. The TC proposed by Cai et al. (2024a) and the CC proposed by Sharma et al. (2023) were adopted to identify two clusters (termed as CC cluster and TC cluster, respectively) from CLAY/10/7490 that most closely resemble the target site. Then, three inference methods [e.g., probabilistic multiple regression (PMR) (e.g., Kulhawy and Mayne 1990; Ching and Phoon 2012), classical Bayesian model (CBM) (e.g., Ching and Phoon 2019), and hierarchical Bayesian model (HBM) (e.g., Ching et al. 2021b)] were applied to the TC and CC clusters to infer clay parameters at the target site. Five PI data at the target site were assumed to be unknown and removed for checking the inference error arising from each combination of clustering method and inference method. The inference error was quantified by the root mean square error (RMSE). The value of RMSE for each combination is plotted against the corresponding computational time in Figure 1.12. Results show that the TC cluster tends to outperform the CC cluster in terms of reducing the inference error and computational

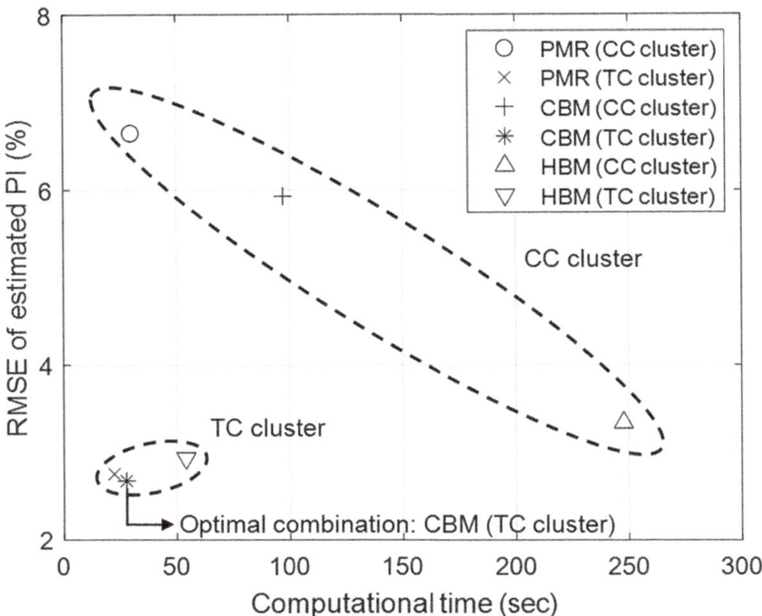

Figure 1.12 Computational time and RMSE for different combinations of cluster and inference method.

time for each of the investigated combinations. In this example, TC with CBM is considered to be the optimal approach because it can achieve the smallest inference error with the least computational cost. However, recent research showed that further improvement is possible (Cai et al. 2024c; Phoon et al. 2025).

1.5 CHALLENGES

The development of DDSC methods that are tractable and computationally efficient for large 3D spatial domains in the presence of MUSIC-3X and site uniqueness is far from mature. Research on more complex data attributes such as MUSIC-3X-G has not received the attention it deserved. For site characterization or other geotechnical applications where physics and experience are encountering difficulties in extracting useful insights from an increasing volume and complexity of data, it is clear that the role of machine learning and artificial intelligence in decision making will grow in importance. Although data science is a rapidly evolving field, research specific to geotechnical engineering is limited. The same can be said of machine learning and artificial intelligence, which are important “learning” algorithms in data science. Data-centric geotechnics is an agenda to transform geotechnical engineering practice to Geo 4.0 (geotechnical realization of Industry 4.0). DDSC is one broad application domain in data-centric geotechnics. One can easily imagine autonomous construction as another application domain, although there are complex ethical dimensions (Assaad and Boshuijzen-Van Burken 2023). This application is outside the scope of this chapter. What is missing is a geotechnical data science foundation that can guide the development of novel machine learning or artificial intelligence algorithms to support applications in data-centric geotechnics. The research questions raised in this chapter are use-inspired to meet the "fit for (and transform) practice" element in data-centric geotechnics. They may inform the agenda for geotechnical data science in the future. In the meantime, for the specific instance of DDSC, the following research questions are open:

(1) **Explainability and interpretability** – The finite element method is the most prevalent embodiment of physics in geotechnical engineering. Using finite element analysis, an engineer understands cause and effect (interpretability) and knows which input parameters affect the outputs (explainability). Decisions supported by physics-informed machine learning (PIML) are arguably more "explainable" and "interpretable." Xu et al. (2023) provide a review of the state of the art of physics-informed machine learning methods in enhancing the reliability and safety of complex systems from both technical and application perspectives. Explainability and interpretability are important in the trustworthy data-centric geotechnics agenda.

(2) **Transferability** – One can expect a generic database to expand with the addition of new sites on a regular basis. Learning is deemed "transferable" if past training outcomes can be updated without requiring brute force training from scratch. This is already implemented in ChatGPT, where responses are more relevant in the presence of few-shot learning or fine-tuning.

(3) **Geo-compatibility** – The generic databases in Table 1.2 are currently restricted to a single geomaterial type such as clay and sand. However, the target site data can contain records belonging to different geomaterial types because of the presence of different layers at different depths. All the methods listed in Table 1.7 require the target site data and the generic databases to be "geo-compatible," meaning that only layers belonging to the same geomaterial type as the records in the generic database can be analysed.

(4) **Frugal HBM** – A typical building regulation mandates a minimum three boreholes on a project site. Another typical guideline is to conduct one boring every 300 m^2. The "medium" and "large" ground volumes suggested by Phoon et al. (2022c) are a 40 m long × 20 m wide × 20 m deep cuboid and a 80 m long × 40 m wide × 40 m deep cuboid, respectively. The number of boreholes to cover these ground volumes

Table 1.7 Literature survey of site recognition research organized according to six broad strategies

Strategy	*Description*	*References*
I	Use entire global database with every record assigned an equal weight independent of its similarity to the target site.	Ching and Phoon 2012, 2013, 2014b; Ching et al. 2014, 2017, 2018, 2020b
II	Use entire global database with every record assigned a weight based on its similarity to the target site.	Ching and Phoon 2020a; Ching et al. 2021a, 2021b, 2022a; Sharma et al. 2022
III	Use a subset of the global database ("quasi-regional" database) with every record/site assigned an equal weight.	
IV	Use a subset of the global database ("quasi-regional" database) with every record/site assigned an unequal weight.	Sharma et al. 2023; Cai et al. 2025
V	Use a subset of the global database ("quasi-regional" database) and a subset of the soil/rock parameters that is relevant to one or more governing limit states.	Cai et al. 2024c
VI	Do not use the global database, in part or in full, to inform the target site because they are "dissimilar."	Ching et al. 2023b; Cai et al. 2024b

is approximately 3 and 10, respectively. In addition to borings, field tests are almost always conducted. In comparison, a full-scale performance evaluation may be conducted only once on a project site. Typical building regulations for foundation load tests are: (1) "1 number or 0.5% of the total piles, whichever is greater" for an ultimate load test on a preliminary pile (preferably instrumented) and (2) "2 numbers or 1% of working piles installed or 1 for every 50 m length of proposed building, whichever is greater" for a working load test. While site-specific soil property data are sparse, site-specific performance data are extremely sparse because they are expensive. There is no hierarchical Bayesian model (HBM) that can extract inter-site variability from a database where the site-specific data are extremely sparse. The proposed concept of a frugal HBM is to prune a conventional HBM such that it can deliver meaningful inter-site variability insights in the presence of a performance database. Even if the site-specific data are sufficient to train a conventional HBM, it is also useful to ask if it is the most frugal.

(5) **Thick data** – It is not possible to ignore the rapid pace of development in generative AIs. The ability to generate insights from multi-modal data such as unstructured text, images, and videos appears to have a direct bearing on the question of how human experience can add value to quantitative data analytics. The authors hypothesize that generative AIs such as ChatGPT offer interesting opportunities to further advance data-centric geotechnics, especially because of their strong ability to handle multi-modal data.

(6) **Machine learning-guided observational method (MLOM)** – The combination of site characterization and monitoring information can support real-time decision making during construction. This MLOM agenda is in its infancy, although its value is arguably higher than DDSC. After all, for geotechnical engineering, the purpose of understanding a site is to understand the performance of a structure. Some studies have emerged, and a representative sample is briefly presented below to illustrate the current state-of-the-art:

a. Ma et al. (2023) proposed a real-time intelligent classification method for blasting and drilling tunnels installed in rock using a machine learning algorithm. A new database is compiled from 286 case studies in China. The data include rock hardness, weathering degree, rock mass structure, structural plane integrity, rock mass integrity, in situ stress condition, groundwater condition, and surrounding rock grades. An intelligent classification method for the surrounding rock is first established using machine learning. A tunnel information management system for drilling and blasting tunnels is next built by combining the intelligent classification method and cloud technology, and it is applied to the Lexi Expressway project. The tunnel information management system can provide a timely and accurate reference for the dynamic design and construction of tunnels.

b. Yan et al. (2023) noted that developing a generalized model for axial displacement prediction for immersed tunnel joints is challenging because of the complex geo-environment and limited monitoring data. This is a typical small data problem in geotechnics since only a few features are available for training machine learning models. The one-year monitoring data in this study, including axial displacement, structure temperature, and water level, show seasonal trends and their correlations vary spatially along the tunnel. The study proposed a hybrid physical data (HPD) informed deep neural network (DNN) to improve spatial generalization for axial displacement prediction. The HPD is created based on physical analysis and contributes to the DNN as a substituting feature rather than an additional feature. The proposed model is shown to outperform conventional regression methods in terms of spatial generalization with improved physical interpretation.

c. Zhou et al. (2023) proposed a tunnel lining crack detection algorithm named improved You Only Look Once version X (YOLOX), which can detect tunnel cracks in a complex environment efficiently and accurately. The images taken during the damage detection phase of the study conducted in several tunnels are compiled and expanded to obtain a tunnel crack damage image dataset that can represent the complex physical state of the tunnel. These crack images can be used for model training and testing, and after experiments, the improved YOLOX model can identify these images and can achieve high speed, high accuracy, and real-time dynamic detection of tunnel cracks.
d. Morgenroth et al. (2023) observed that stress model updating for mine-scale finite difference models is generally time-consuming and tedious, requiring a large volume of engineering hours and computational power. Their study presents a novel machine learning approach, using a microseismic database, to update a stress model at Garson Mine in Ontario, Canada. A long-short term memory (LSTM) network is trained on the microseismic events and the geomechanical parameters from a previously manually calibrated FLAC3D model. Two LSTM networks are trained for comparison: one to predict the principal stresses and the other to predict the six-component stress tensor. The predicted values are compared to the values computed by the FLAC3D model ("ground truth").
e. Tan et al. (2023) mapped the spatial stress distribution over a coalmine roof by using a transfer convolutional neural network (TCNN) model. A convolutional neural network (CNN) model is first pre-trained based on the simulated data from the finite difference method. This CNN can be refined for a specific coalmine roof using limited structural health monitoring (SHM) data through transfer learning. The SHM data are recorded by a stress sensor installed in the roof of the Dongtan coalmine over a range of 120 m ahead of the working face. This study demonstrated that practical site-specific predictions can be made even in the presence of limited field measurements by adopting a physics-informed rather than pure data-driven approach.
f. Tian and Wang (2023) proposed a data-driven and physics-informed Bayesian learning framework that generates a project-specific training dataset by the random finite element method (reflecting the embedded physics) using sparse site investigation data. The generated training dataset is then used in Bayesian sparse dictionary learning of settlement monitoring data for improving model prediction in real time and continuously, particularly at locations without monitoring data and/or subsequent time steps. Uncertainty in settlement prediction may also be properly quantified. The proposed approach can also be applied to real-time fusion of numerical modelling results with multi-source monitoring data (Tian et al. 2024).

(7) **Data protection** – If geotechnical data is viewed as a critical asset and a repository of significant value, the data-centric geotechnics agenda cannot focus purely on maximizing the value to practice but also on minimizing the leakage of sensitive

data. Data protection is widely recognized in the machine learning community, but to the authors' knowledge, no research has been conducted in the geotechnical engineering community. This chapter proposes adding "privacy enhancement" to the agenda for data-centric geotechnics. Data protection may cover the following aspects:

a. Data minimization – The risk of data breaches, unauthorized access, or misuse of data is reduced if the minimal training set is used. This minimization principle will also encourage the use of relevant data sets only, which can improve accuracy and computational efficiency.
b. Data encryption – Data encryption is a technique that transforms data into an unreadable form using a secret key or algorithm. Homomorphic encryption is regarded as the "holy grail" in privacy-enhancing technologies (PETs). It allows operations to be performed on encrypted data such that the final results exactly match the output of the operations on unencrypted data.
c. Data protection by design – Data protection by design means implementing privacy-enhancing technologies (PETs) into the ML solutions from the earliest stages of design and throughout the entire life cycle of the system or application. PETs include homomorphic encryption differential privacy, generative adversarial networks (GANs), synthetic data generation, and federated learning. A proactive approach to address potential data privacy and security risks enhances the trust and confidence of users and stakeholders.
d. Data governance – The purpose of data governance is to define the roles, responsibilities, policies, and standards for managing and using data in an organization. This covers the rules of the game for data collection, processing, storage, sharing, and disposal. In the presence of a data governance framework, data ethics, compliance, and accountability could be enforced.

1.6 CONCLUSIONS

One of the key findings of the ISSMGE 2017 SOA/SOP Survey is that geotechnical databases, which include information on inherent spatial variability, soil and rock properties, and risk databases, should be made more readily available to the profession. Thus, compiling geotechnical databases represents one of the key missions of the ISSMGE TC304 (Engineering Practice of Risk Assessment & Management). An initiative to promote public sharing of databases, called 304dB, was initiated by Professors Kok-Kwang Phoon and Jianye Ching in 2017: http://140.112.12.21/issmge/tc304.htm. Despite its success (304dB has contributed to the development of data-driven methods by situating these methods in the context of real data over the past seven years), it is timely to take stock and discuss how this effort can be expanded to cover more types of databases, more multi-modal and real-time data, more countries, and more projects. In short, what do we need to do in research and practice to go from 304dB to Geo-everything-dB? The breakout of powerful generative AI such as ChatGPT in 2023 should give new impetus to collecting data, understanding data, protecting data as a critical asset, developing learning algorithms to transform data and devices into embodied intelligence, and reimagining infrastructure as a service to end-users through the deployment of smart technologies. The intent of taking a data-centric perspective to geotechnical engineering (data-centric geotechnics) is to march towards these top-level goals. Geo-everything-dB can be a hashtag for collecting, understanding, and protecting data across the entire life cycle of the structure. The current book and its companion volume (Tang and Phoon 2025) are edited in the spirit of expanding the boundaries of Geo-everything-dB.

This chapter presents an overview of site characterization information with a focus on its availability, coverage, data attributes, value to decision making, and challenges. In terms of availability, it is accurate to say that site characterization databases mainly reside in industry. They were compiled to support the design and construction of a particular project and as part of compliance. Their value is almost zero after the project is completed as they are shelved in some storage, physical or virtual. Phoon (2020) called this "dark data." There are several reasons for geotechnical databases remaining "dark," but one can argue that data belong to the clients and there are non-disclosure or confidentiality agreements in many projects. Nonetheless, it is worthwhile to remind ourselves that this issue is not unique to geotechnical engineering. There are different strategies in secure data sharing being developed to extract insights from databases without exposing confidential details or breaching legal agreements. They include homomorphic encryption, differential privacy, and federated learning. To the authors' knowledge, there is no systematic research on the protection of geotechnical data. The "data first practice central" agenda in data-centric geotechnics can be expanded to "data first, practice central, and privacy enhanced." Data-centric geotechnics now encompasses four elements: (1) data centricity, (2) fit for (and transform) practice, (3) geotechnical context, and (4) privacy enhancement. Phoon (2025) advocated updating "data-centric geotechnics" to "trustworthy data-centric geotechnics" to manage the risk of ML/AI more systematically in response to the EU AI Act. There is also "dirty data" (Phoon and Zhang 2023). It is very time-consuming and labour-intensive to clean, label, and organize dirty data from legacy systems. Current databases are compiled manually (Tang and Phoon 2021; Phoon and Tang 2025; Tang and Phoon 2025). There is a recent trend of using more flexible and efficient database query technology based on a large language model and vector database. Users can extract relevant and useful data from the database using natural language, which allow more interactive and meaning-based queries to the database.

In terms of coverage, publicly available site characterization databases are mainly geotechnical in nature containing borehole, laboratory test, and field test data. Detailed geological data such as soil stratification, rock mass features (weathering, discontinuities, etc.), and real-time data such as geoenvironmental processes (subsurface flow, contaminant transport, etc.) are lacking. Karpatne et al. (2018) identified the "paucity of ground truth" as one attribute of geoscience data as well. In terms of data attributes, Phoon et al. (2019) posed MUSIC-X (Multivariate, Uncertain and Unique, Sparse, Incomplete, and potentially Corrupted with "X" denoting the spatial/temporal dimension) as a basic challenge to data-driven site characterization (DDSC) research. It has since been expanded to MUSIC-3X-G. It is clear that sparsity, uncertainty, and spatial variability are only three out of seven attributes. This chapter proposed a new taxonomy of site data under "4S" to expand the agenda for future machine learning research. The "4S" are site generalizations, spatial features, sampling characteristics, and smart data. Sampling characteristics include time. No real-time picture of the subsurface is available and no one knows how to achieve this real-time understanding with available data.

In terms of value to design, the majority of the research is focused on the prediction of soil properties (spatial variability) and soil types (stratification) at a target site. Phoon et al. (2022b) argued that these predictions are intrinsically probabilistic because statistical uncertainties are significant at the current sampling density (e.g., volume fraction for logging at a Brent Field site in the North Sea is 1×10^{-6}, Chilès and Delfiner 1999). The prediction of soil types and properties is related, and the volume to be characterized is 3D. Phoon et al. (2022c) proposed the first set of eight benchmark examples to accelerate the pace of data-driven site characterization (DDSC) research. One component of a benchmark example is the virtual ground (VG) that represents the complete state of "reality." Everything is known within this virtual ground because it is generated by assuming stratigraphy and populating the properties in each layer using a random field generator. Phoon et al. (2022c) suggested an initial "small" ground volume of a 20 m long × 20 m wide × 10 m deep cuboid for benchmark examples. The authors suggested that "medium" and "large" ground volumes can be defined as a 40 m long × 20 m wide × 20 m deep cuboid and a 80 m long × 40 m wide × 40 m deep cuboid, respectively. The key point here is that there are computational challenges in terms of time and cost. In addition, there are at least two limitations to this initial set of benchmark examples. First, the soil layers are too idealized. Shuku and Phoon (2023b) proposed a more realistic VG5 model (Figure 1.3e and f) with spatially varying and discontinuous soil layers simulated using a coupled Markov chain (Qi et al; 2016). Second, real examples are needed. The database Tokyo-CLAY/14/67,760 (Otake et al. 2024) contained developed 2792 boreholes covering a site area of more than 10 km^2 from a real project in Japan. Each borehole has 29 data records on average. This database can be adopted as the basis for establishing a benchmark example for a real site.

One DDSC challenge called "site recognition" has attracted attention in recent years. The site recognition challenge is fundamental to geotechnical engineering because it attempts to quantify the "uniqueness" (or site specificity) attribute of a target site. The argument that geotechnical engineering practice is more art than science and engineering judgment is indispensable arises in large part from the recognition that site specificity is not sufficiently accounted for in our calculation models. One classical example is the use of generic transformation models to estimate design parameters from field test parameters. Recent research has demonstrated that quasi-site-specific transformation models can be developed notwithstanding the well-known limitations that site-specific data is sparse and a generic database containing data from other sites is not directly applicable to a target site (Phoon and Ching 2022). There are three notable achievements: (1) the site recognition problem is tractable even under MUSIC, (2) inference uncertainty produced by a quasi-site-specific transformation model is smaller than the conventional generic or site-specific model, and (3) increasingly effective quasi-site-specific transformation models have been constructed, particularly those that exploit clustering (Cai et al. 2024a; Phoon et al. 2025). The construction of a quasi-site-specific ground model that includes learning cross-correlations, spatial auto-correlations, and geologic uncertainties from generic databases under MUSIC-3X-G is an ongoing research problem.

Research on other site characterization databases, such as geophysical and remote sensing data and data fusion, is limited (Xie et al. 2022; Xu et al. 2022; Guan et al. 2024a). Interest in combining site characterization information with monitoring information collected during construction to guide decision making in real time is emerging rapidly (Shuku et al. 2023; Tian and Wang 2023; Tian et al. 2024). This agenda is called the machine learning-guided observational method (MLOM). Its value is arguably higher than DDSC. It may evolve to human–machine teaming, which is outside the scope of this chapter. Other challenges such as explainability and interpretability, transferability, geo-compatibility, frugal HBM, thick data, and data protection are briefly discussed. It is uncertain whether a new discipline called geotechnical data science is needed to provide a stronger theoretical underpinning for data-centric geotechnics, given the large interconnected challenges highlighted in this chapter. But trustworthiness needs to be addressed to comply with the EU AI Act and other regulatory frameworks of AI systems.

ACKNOWLEDGEMENTS

The authors gratefully acknowledge the assistance of Dr Yongmin Cai and insightful comments from Professors Zjiun Cao, Yu Otake, Yu Wang, Stephen Wu, and Tengyuan Zhao. The authors also would like to thank the members of the TC304 Committee on Engineering Practice of Risk Assessment & Management of the International Society of Soil Mechanics and Geotechnical Engineering for developing the database 304dB used in this study and making it available for scientific inquiry.

REFERENCES

Abubakar, A., Brædstrup, M. J., Di, H., Diaz, A. T., Freeman, S., Hviid, S., Karkov, K. H., Kriplani, S., Manikani, S., Salun, G., and Zhao, T. 2021. Deep learning applications for wind farms site characterization and monitoring. SEG Technical Program Expanded Abstracts, 3009–3013.

Aladejare, A. E. and Wang, Y. 2017. Evaluation of rock property variability. *Georisk: Assessment and Management of Risk for Engineered Systems and Geohazards*, 11(1), 22–41.

ASCE 2019. *Future World Vision: Infrastructure Reimagined*. American Society of Civil Engineers, Reston, VA.

Assaad, Z. and Boshuijzen-Van Burken, C. 2023. Ethics and safety of human-machine teaming. *Proceedings, First International Symposium on Trustworthy Autonomous Systems (TAS '23)*, July 11–12, 2023, Edinburgh, United Kingdom. ACM, New York.

Bárdossy, G. and Fodor, J. 2001. Traditional and new ways to handle uncertainty in geology. *Natural Resources Research*, 10(3), 179–187.

Bárdossy, G. and Fodor, J. 2004. *Evaluation of Uncertainties and Risks in Geology*. Springer, Berlin.

Beesley, M. and Vardanega, B. J. 2020. Parameter variability of undrained shear strength and strain using a database of reconstituted soil tests. *Canadian Geotechnical Journal*, 57(8), 1247–1255.

Bilal, M., Oyedele, L. O., Qadir, J., Munir, K., Ajayi, S. O., Akinade, O. O., Owolabi, H. A., Alaka, H. A., and Pasha, M. 2016. Big Data in the construction industry: A review of present status, opportunities, and future trends. *Advanced Engineering Informatics*, 30(3), 500–521.

Bock, H. 2006. Common ground in engineering geology, soil mechanics and rock mechanics: Past, present and future. *Bulletin of Engineering Geology and the Environment*, 65(2), 209–216.

Boeckmann, A. and Loehr, J. E. 2023. *Practices for local calibration of LRFD geotechnical resistance factors. NCHRP Synthesis 601*. Transportation Research Board, Washington, DC.

BS 1986. *BS8004: Code of Practice for Foundations*. British Standard, London.

Burland, J. B. 2012. The geotechnical triangle. *ICE Manual of Geotechnical Engineering: Geotechnical Engineering Principles, Problematic Soils and Site Investigation*, Eds. John B Burland, T. J. P. Chapman, H. Skinner, M. J. Brown, Vol. 1, pp. 17–26. Emerald Publishing Limited, Leeds, England.

Cai, Y., Ching, J., and Phoon, K. K. 2024a. Tailored clustering method to identify quasi-regional sites. *Engineering Geology*, 333, 107490.

Cai, Y., Pan, Q., Phoon, K. K., and Luo, W. 2024b. Modifying tailored clustering enabled regionalization (TCER) framework for outlier detection and inference efficiency. *Engineering Geology*, 335, 107537.

Cai, Y., Phoon, K. K., Ching, J., and Wang, Y. 2024c. Dimension reduction of MUSIC database for tailored clustering using Lasso. *Canadian Geotechnical Journal*, Under review.

Cai, Y., Phoon, K. K., Otake, Y., and Wang, Y. 2025. Efficient dictionary learning for constructing quasi-local transformation models. *Computers and Geotechnics*. Under review.

Cao, Z. and Wang, Y. 2013. Bayesian approach for probabilistic site characterization using cone penetration tests. *Journal of Geotechnical and Geoenvironmental Engineering*, 139(2), 267–276.

Cao, Z. J., Zheng, S., Li., D. Q., and Phoon, K. K. 2019. Bayesian identification of soil stratigraphy based on soil behavior type index. *Canadian Geotechnical Journal*, 56(4), 570–586.

CEN 2004. EN 1997-1: *Geotechnical design – Part 1: General rules*. Comité Européen de Normalisation.

Chen, G., Cheng, Q., and Puetz, S. 2023. Special issue: data-driven discovery in geosciences: opportunities and challenges. *Mathematical Geosciences*, 55, 287–293.

Chilès, J-P. and Delfiner, P. 1999. *Geostatistics: Modeling Spatial Uncertainty*. John Wiley & Sons, New York.

Ching, J. 2020. Unpublished databases.

Ching, J. and Kuo, M.C. 2023. Data-centric quasi-site-specific prediction for soil modulus. *Journal of GeoEngineering*, 18(4), 215–223.

Ching, J. and Phoon, K. K. 2012. Modeling parameters of structured clays as a multivariate normal distribution. *Canadian Geotechnical Journal*, 49(5), 522–545.

Ching, J. and Phoon, K. K. 2013. Multivariate distribution for undrained shear strengths under various test procedures. *Canadian Geotechnical Journal*, 50(9), 907–923.

Ching, J. and Phoon, K. K. 2014a. Transformations and correlations among some clay parameters – The global database. *Canadian Geotechnical Journal*, 51(6), 663–685.

Ching, J. and Phoon, K. K. 2014b. Correlations among some clay parameters – The multivariate distribution. *Canadian Geotechnical Journal*, 51(6), 686–704.

Ching, J. and Phoon, K. K. 2017. Characterizing uncertain site-specific trend function by Sparse Bayesian learning. *Journal of Engineering Mechanics, ASCE*, 143(7), 04017028.

Ching, J. and Phoon, K. K. 2019. Constructing site-specific multivariate probabilistic distribution model by Bayesian machine learning, *Journal of Engineering Mechanics, ASCE*, 145(1), 04018126.

Ching, J. and Phoon, K. K. 2020a. Measuring similarity between site-specific data and records from other sites. *ASCE-ASME Journal of Risk and Uncertainty in Engineering Systems, Part A: Civil Engineering*, 6(2), 04020011.

Ching, J. and Phoon, K. K. 2020b. Constructing a site-specific multivariate probability distribution using sparse, incomplete, and spatially variable (MUSIC-X) data. *Journal of Engineering Mechanics, ASCE*, 146(7), 04020061.

Ching, J., Phoon, K. K., and Chen, C. H. 2014. Modeling piezocone cone penetration (CPTU) parameters of clays as a multivariate normal distribution. *Canadian Geotechnical Journal*, 51(1), 77–91.

Ching, J., Wang, J. S., Juang, C. H., and Ku, C. S. 2015. Cone penetration test (CPT)-based stratigraphic profiling using the wavelet transform modulus maxima method. *Canadian Geotechnical Journal*, 52(12), 1993–2007.

Ching, J., Lin, G. H., Chen, J. R., and Phoon, K. K. 2017. Transformation models for effective friction angle and relative density calibrated based on a multivariate database of coarse-grained soils. *Canadian Geotechnical Journal*, 54(4), 481–501.

Ching, J., Li, K. H., Phoon, K. K., and Weng, M. C. 2018. Generic transformation models for some intact rock properties. *Canadian Geotechnical Journal*, 55(12), 1702–1741.

Ching, J., Huang, W. H., and Phoon, K. K. 2020a. 3D probabilistic site characterization by Sparse Bayesian Learning. *Journal of Engineering Mechanics, ASCE*, 146(12), 04020134.

Ching, J., Phoon, K. K., Khan, Z., Zhang, D. M., and Huang, H. W. 2020b. Role of municipal database in constructing site-specific multivariate probability distribution. *Computers and Geotechnics*, 124, 103623.

Ching, J. Y., Phoon, K. K., Ho, Y. H., and Weng, M. C. 2021a. Quasi-site-specific prediction for deformation modulus of rock mass. *Canadian Geotechnical Journal*, 58(7), 936–951.

Ching, J., Wu, S., and Phoon, K. K. 2021b. Constructing quasi-site-specific multivariate probability distribution using hierarchical Bayesian model. *Journal of Engineering Mechanics, ASCE*, 147(10), 04021069.

Ching, J., Yang, Z. Y., and Phoon, K. K. 2021c. Dealing with non-lattice data in three-dimensional probabilistic site characterization. *Journal of Engineering Mechanics, ASCE*, 147(5), 06021003.

Ching, J., Phoon, K. K., Yang, Z., and Stuedlein, A. W. 2021d. Constructing a quasi-site-specific multivariate probability distribution model for soil properties using sparse, incomplete, and three-dimensional (MUSIC-3X) data. *Georisk*, 16(1), 53–76.

Ching, J., Phoon, K. K., and Wu, C. T. 2022a. Data-centric quasi-site-specific prediction for compressibility of clays. *Canadian Geotechnical Journal*, 59(12), 2033–2049.

Ching, J., Phoon, K. K., Yang, Z. Y., and Stuedlein, A. W. 2022b. Quasi-site-specific multivariate probability distribution model for sparse, incomplete, and three-dimensional spatially varying soil data. *Georisk: Assessment and Management of Risk for Engineered Systems and Geohazards*, 16(1), 53–76.

Ching, J., Uzielli, M., Phoon, K. K., and Xu, X. 2023a. Characterization of autocovariance parameters of detrended cone tip resistance from a global CPT database. *Journal of Geotechnical and Geoenvironmental Engineering, ASCE*, 149(10), 04023090.

Ching, J., Yoshida, I., and Phoon, K. K. 2023b. Comparison of trend models for geotechnical spatial variability: sparse Bayesian learning vs. Gaussian process regression. *Gondwana Research*, 123, 174–183.

Ching, J., Phoon, K. K., and Huang, P. 2024. Detection of outliers with respect to a MUSIC geotechnical database. *Canadian Geotechnical Journal*, 61(7), 1275–1293.

Christensen, C. W., Harrison, E. J., Pfaffhuber, A. A., and Lund, A. K. 2021. A machine learning–based approach to regional-scale mapping of sensitive glaciomarine clay combining airborne electromagnetics and geotechnical data. *Near Surface Geophysics*, 19(5), 523–539.

Chwała, M., Phoon, K. K., Uzielli, M., Zhang, J., Zhang, L., and Ching, J. 2023. Time capsule for geotechnical risk and reliability. *Georisk: Assessment and Management of Risk for Engineered Systems and Geohazards*, 17(3), 439–466.

D'Ignazio, M., Phoon, K. K., Tan, S. A., and Lansivaara, T. 2016. Correlations for undrained shear strength of Finnish soft clays. *Canadian Geotechnical Journal*, 53(10), 1628–1645.

Du, B., Ye, J., Zhu, H., Sun, L., and Du, Y. 2023. Intelligent monitoring system based on spatio-temporal data for underground space infrastructure. *Engineering*, 25, 194–203.

Einstein, H. and Baecher, G. 1982. *Probabilistic and statistical methods in engineering geology I. Problem statement and introduction to solution. Ingenieurgeologie und Geomechanik als Grundlagen des Felsbaues/Engineering Geology and Geomechanics as Fundamentals of Rock Engineering*, Springer, pp. 47–61.

Eslami, A., Heidarie Golafzani, S., and Naghibi, M. H. 2023. Developed triangular charts; deltaic CPTu-based soil behavior classification using AUT:CPTu-Geo-Marine Database. *Probabilistic Engineering Mechanics*, 71, 103380.

Faramarzi, A. 2020. Editorial for themed issue on "Application of machine learning". *Proceedings of the Institution of Civil Engineers-Smart Infrastructure and Construction*, 173(4), 73.

Feng, S. and Vardanega, P. J. 2019. A database of saturated hydraulic conductivity of fine-grained soils: probability density functions. *Georisk: Assessment and Management of Risk for Engineered Systems and Geohazards*, 13(4), 255–261.

Feng, S., Barreto, D., Imre, E., Ibraim, E., and Vardanega, P.J. 2023. Use of hydraulic radius to estimate the permeability of coarse-grained materials using a new geodatabase. *Transportation Geotechnics*, 41, 101026.

Fookes, P. 1997. Geology for engineers: The geological model, prediction and performance. *Quarterly Journal of Engineering Geology and Hydrogeology*, 30(4), 293–424.

Girolami, M. 2020. Introducing data-centric engineering: an open access journal dedicated to the transformation of engineering design and practice. *Data-Centric Engineering*, 1, e1.

Guan, Z., Wang, Y., and Phoon, K. K. 2024a. Fusion of sparse non-co-located measurements from multiple sources for geotechnical site investigation. *Canadian Geotechnical Journal*, 61(8), 1574–1592.

Guan, Z., Wang, Y., and Phoon, K. K. 2024b. Dictionary learning of spatial variability at a specific site using data from other sites. *Journal of Geotechnical and Geoenvironmental Engineering, ASCE*, 150(9), 04024072.

Halim, I. S. and Tang, W.H. 1993. Site exploration strategy for geologic anomaly characterization. *Journal of Geotechnical Engineering*, 119(2), 195–213.

Harrison, J. 2019. Challenges in determining rock mass properties for reliability-based design. *Proceedings, 7th International Symposium on Geotechnical Safety and Risk (ISGSR 2019)*, Taipei, Taiwan, pp. 35–44.

Hoek, E. 1999. Putting numbers to geology- an engineer's viewpoint. *Quarterly Journal of Engineering Geology and Hydrogeology*, 32(1), 1–19.

Hong, A., Baker, L., Prieto Curiel, R., Duminy, J., Buswala, B., Guan, C., and Ravindranath, D. 2022. Reconciling big data and thick data to advance the new urban science and smart city governance. *Journal of Urban Affairs*, 45(10), 1737–1761.

Hu, J. Z., Zhang, J., Huang, H. W., and Zheng, J. G. 2021. Value of information analysis of site investigation program for slope design. *Computers and Geotechnics*, 131, 103938.

Hu, Y., Wang, Y., Phoon, K. K., and Beer, M. 2024. Similarity quantification of soil spatial variability between two cross-sections using auto-correlation functions. *Engineering Geology*, 331, 107445.

Huang, J., Zheng, D., Li, D., Kelly, R., and Sloan, S. 2018. Probabilistic characterization of two-dimensional soil profile by integrating CPT with MASW data. *Canadian Geotechnical Journal*, 55(8), 1168–1181.

Huang, T., Vance, T. C., and Lynnes, C. 2022. *Big data analytics in earth, atmospheric, and ocean sciences*. AGU Special Publications, American Geophysical Union.

Hutchinson, J. N. 2001. Reading the ground: Morphology and geology in site appraisal. *Quarterly Journal of Engineering Geology and Hydrogeology*, 34(1), 7–50.

Jaksa, M. B and Liu, Z. Q. 2021. Editorial for Special Issue on "Applications of artificial intelligence and machine learning in geotechnical engineering". *Geosciences*, 11(10), 399.

Jessell, M., Guo, J., Li, Y., Lindsay, M., Scalzo, R., Giraud, J., Pirot, G., Cripps, E., and Ogarko, V. 2022. Into the Noddyverse: a massive data store of 3D geological models for machine learning and inversion applications. *Earth System Science Data*, 14(1), 381–392.

Ji, J. and Cao, Z.J. 2024. Geotechnical reliability analysis for practice. Chapter 6 in *Uncertainty, Modeling, and Decision Making in Geotechnics*, edited by Kok-Kwang Phoon, Takayuki Shuku, and Jianye Ching. CRC Press, Boca Raton, FL.

Journel, A. G. and Huijbregts, C. J. 1978. *Mining Geostatistics*. Academic Press, London.

Juang, C. H., Zhang, J., Shen, M., and Hu, J. 2018. Probabilistic methods for unified treatment of geotechnical and geological uncertainties in a geotechnical analysis. *Engineering Geology*, 249, 148–161.

Juang, C. H., Ge, Y., and Zhang, J. 2019. Geological uncertainty: a missing element in geotechnical reliability analysis. *Proceedings, 7th International Symposium on Geotechnical Safety and Risk (ISGSR 2019)*, Taipei, Taiwan, pp. 1–12. http://rpsonline.com.sg/proceedings/isgsr2019/pdf/WT.pdf

Karniadakis, G. E., Kevrekidis, I. G., Lu, L., Perdikaris, P., Wang, S., and Yang, L. 2021. Physics-informed machine learning. *Nature Reviews Physics*, 3, 422–440.

Karpatne, A., Ebert-Uphoff, I., Ravela, S., Ali Babaie, H., and Kumar, V. 2018. Machine learning for the geosciences: challenges and opportunities. *IEEE Transactions on Knowledge and Data Engineering*, 31(8), 1544–1554.

Kwak, D.Y., Ahdi, S.K., Wang, P., Zimmaro, P., Brandenberg, S.J., and Stewart, J.P. 2021. Web portal for shear wave velocity and HVSR databases in support of site response research and applications. UCLAgeo, Los Angeles.

Kim, E. and Hunt, R. 2017. A public website of rock mechanics database from Earth Mechanics Institute (EMI) at Colorado School of Mines (CSM). *Rock Mechanics and Rock Engineering*, 50(12), 3245–3252.

Kootahi, K. and Moradi, G. 2017. Evaluation of compression index of marine fine-grained soils by the use of index tests. *Marine Georesources and Geotechnology*, 35(4), 548–570.

Kulhawy, F. H. and Mayne, P. W. 1990. *Manual on Estimating Soil Properties for Foundation Design*. Report EL-6800, Electric Power Research Institute, Palo Alto, CA.

Lacasse 2015. *55th Rankine Lecture: Hazard, Risk and Reliability in Geotechnical Practice.* Presentation at the Institution of Civil Engineers, London.

Lacasse, S. and Lunne, T. 1982. Penetration tests in two Norwegian clays. *Proceedings, 2nd European Symposium on Penetration Testing*, Amsterdam, pp. 661–670.

Länsivaara, T., Phoon, K. K., and Ching, J. 2022. What is a characteristic value for soils? *Georisk: Assessment & Management of Risk for Engineered Systems & Geohazards*, 16(2), 199–224.

Li, X., Li, P., and Zhu, H. 2013. Coal seam surface modeling and updating with multi-source data integration using Bayesian Geostatistics. *Engineering Geology*, 164, 208–221

Li, X., Li, H., Du, S., Jing, J., and Li, P. 2023. Cross-engineering utilization of Tunnel Boring Machines (TBM) construction data: a case study using big data from Yin-Song Diversion Project in China. *Georisk: Assessment and Management of Risk for Engineered Systems and Geohazards*, 17(1), 127–147.

Liao, T. and Mayne, P. 2007. Stratigraphic delineation by three-dimensional clustering of piezocone data. *Georisk*, 1(2), 102–119.

Lim, Y. X., Tan, S. A., and Phoon, K. K. 2018. Application of press-replace method to simulate undrained cone penetration. *International Journal of Geomechanics*, ASCE, 18(7), 04018066.

Lim, Y. X., Tan, S. A., and Phoon, K. K. 2019. Interpretation of horizontal permeability from piezocone dissipation tests in soft clays. *Computers and Geotechnics*, 107, 189–200.

Lin, Q., Ci, T., Wang, L., Mondal, S. K., Yin, H., and Wang, Y. 2022. Transfer learning for improving seismic building damage assessment. *Remote Sensing*, 14, 201.

Liu, S., Zou, H., Cai, G., Bheemasetti, T. V., Puppala, A. J., and Lin, J. 2016. Multivariate correlation among resilient modulus and cone penetration test parameters of cohesive subgrade soils. *Engineering Geology*, 209, 128–142.

Lo, M. K., Wei, X., Chian, S. C., and Ku, T. 2021. Bayesian network prediction of stiffness and shear strength of sand. *Journal of Geotechnical and Geoenvironmental Engineering*, 147(5), 04021020.

Löfman, M.S. and Korkiala-Tanttu, L.K. 2021. Transformation models for the compressibility properties of Finnish clays using a multivariate database. *Georisk: Assessment and Management of Risk for Engineered Systems and Geohazards*, 16(2), 330–346.

Lumb, P. 1966. The variability of natural soils. *Canadian Geotechnical Journal*, 3(2), 74–97.

Luo, X., Chen, H.-H., and Guo, Q. 2022. Semantic communications: overview, open Issues, and future research directions. *IEEE Wireless Communications*, 29(1), 210–219.

Ma, J., Li, T., Yang, G., Dai, K., Ma, C., Tang, H., Wang, G., Wang, J., Xiao, B., and Meng, L. 2023. A real-time intelligent classification model using machine learning for tunnel surrounding rock and its application. *Georisk: Assessment and Management of Risk for Engineered Systems and Geohazards*, 17(1), 148–168.

Morgenroth, J., Kalenchuk, K., Moreau-Verlaan, L., Perras, M., and Khan, U. T. 2023. A novel long-short term memory network approach for stress model updating for excavations in high stress environments. *Georisk: Assessment and Management of Risk for Engineered Systems and Geohazards*, 17(1), 196–216.

McPhillips, D.F., Herrick, J.A., Ahdi, S., Yong, A.K., and Haefner, S. 2020. Updated compilation of Vs30 data for the United States. U.S. Geological Survey data release.

Matheron, G. 1963. Principles of geostatistics. *Economic Geology* 58, 1246–1266.

Meng, T., Jing, X., Yan, Z., and Pedrycz, W. 2020. A survey on machine learning for data fusion. *Information Fusion*, 57, 115–129.

Morgenroth, J., Unterlaß, P. J., Sapronova, A., Khan, U. T., Perras, M. A., Erharter, G. H., and Marcher, T. 2022. Practical recommendations for machine learning in underground rock engineering – On algorithm development, data balancing, and input variable selection. *Geomechanics and Tunnelling*, 15(5), 650–657.

Munawar, H.S., Ullah, F., Qayyum, S., and Shahzad, D. 2022. Big data in construction: current applications and future opportunities. *Big Data and Cognitive Computing*, 6, 18.

Muzamhindo H. and Ferentinou, M. 2023. Generic compressive strength prediction model applicable to multiple lithologies based on a broad global database. *Probabilistic Engineering Mechanics*, 71, 103400.

National Academies of Sciences, Engineering, and Medicine. 2019. *Manual on Subsurface Investigations*. The National Academies Press, Washington, DC. https://doi.org/10.17226/25379.

Otake, Y., Saito, T., Wu, S., Yoshida, I., and Takano, D. 2025. Exploring challenges via analysis of multivariate geotechnical properties: insights from large-scale local sampling of Japanese marine clay. Chapter 11 in *Databases for Data-Centric Geotechnics: Site Characterization*, edited by K. K. Phoon and C. Tang. CRC Press, Boca Raton, FL.

Parry, S., Baynes, F.J., Culshaw, M.G., Eggers, M., Keaton, J.F., Lentfer, K., Novotny, J., and Paul, D. 2014. Engineering geological models: an introduction: IAEG commission 25. *Bulletin of Engineering Geology and the Environment*, 73(3), 689–706.

Phoon, K. K. 2018. Editorial for special collection on probabilistic site characterization. *ASCE-ASME Journal of Risk and Uncertainty in Engineering Systems, Part A: Civil Engineering*, 4(4), 02018002.

Phoon, K. K. 2020. The story of statistics in geotechnical engineering. *Georisk: Assessment and Management of Risk for Engineered Systems and Geohazards*, 14(1), 3–25.

Phoon, K. K. 2023a. Uncertainty-informed decision making in Burland Triangle. Chapter 1, *Uncertainty, Modelling, and Decision Making in Geotechnics*, CRC Press, Boca Raton, FL.

Phoon, K. K. 2023b. What geotechnical engineers want to know about reliability. *ASCE-ASME Journal of Risk and Uncertainty in Engineering Systems, Part A: Civil Engineering*, 9(2), 03123001.

Phoon, K. K. 2025. *Trustworthy data-centric geotechnics*. Geodata and AI.

Phoon, K. K. and Kulhawy, F. H. 1999a. Characterization of geotechnical variability. *Canadian Geotechnical Journal*, 36(4), 612–624.

Phoon, K. K. and Kulhawy, F. H. 1999b. Evaluation of geotechnical property variability. *Canadian Geotechnical Journal*, 36(4), 625–639.

Phoon, K. K. and Retief, J. V. 2016. *Reliability of Geotechnical Structures in ISO2394*, CRC Press/Balkema, London.

Phoon, K. K. and Ching, J. 2021. Project DeepGeo –Data-driven 3D subsurface mapping. *Journal of GeoEngineering*, 16(2), 61–74.

Phoon, K. K. and Ching, J. 2022. Additional observations on the site recognition challenge. *Journal of GeoEngineering*, 17(4), 231–247.

Phoon, K. K. and Tang, C. 2025. Role of performance information in data-centric geotechnics. Chapter 1 in *Databases for Data-Centric Geotechnics: Geotechnical Structures*, edited by Tang, C. and K. K. Phoon, CRC Press, Boca Raton, FL.

Phoon, K. K. and Zhang, W. G. 2023. Future of machine learning in geotechnics. *Georisk: Assessment and Management of Risk for Engineered Systems and Geohazards*, 17(1), 7–22.

Phoon, K. K., Retief, J. V., Ching, J., Dithinde, M., Schweckendiek, T., Wang, Y., and Zhang, L. M. 2016. Some observations on ISO2394:2015 Annex D (Reliability of Geotechnical Structures). *Structural Safety*, 62, 24–33.

Phoon, K. K., Ching, J., and Wang Y. 2019. Managing risk in geotechnical engineering – From data to digitalization. *Proceedings, 7th International Symposium on Geotechnical Safety and Risk (ISGSR 2019)*, Taipei, Taiwan, pp. 13–34. https://www.issmge.org/uploads/publications/96/97/SL.pdf

Phoon, K. K., Ching, J., and Cao, Z. J. 2022a. Unpacking data-centric geotechnics. *Underground Space*, 7(6), 967–989.

Phoon, K. K., Ching, J., and Shuku, T. 2022b. Challenges in data-driven site characterization. *Georisk: Assessment and Management of Risk for Engineered Systems and Geohazards*, 16(1), 114–126.

Phoon, K. K., Shuku, T., Ching, J., and Yoshida, I. 2022c. Benchmark examples for data-driven site characterization. *Georisk: Assessment and Management of Risk for Engineered Systems and Geohazards*, 16(4), 599–621.

Phoon, K. K., Cao, Z., Ji, J., Leung, Y. F., Najjar, S., Shuku, T., Tang, C., Yin, Z. Y., Yoshida, I., and Ching, J. 2022d. Geotechnical uncertainty, modeling, and decision making. *Soils and Foundations*, 62(5), 101189.

Phoon, K. K., Cao, Z., Liu, Z., and Ching, J. 2023a. Report for ISSMGE TC309/TC304/TC222 Third ML dialogue on "Data-Driven Site Characterization (DDSC)". *Georisk: Assessment & Management of Risk for Engineered Systems & Geohazards*, 17(1), 227–238.

Phoon, K. K., Zhang, L. M., Cao, Z. J. 2023b. Editorial for Special Issue on "Machine learning and AI in geotechnics", *Georisk: Assessment and Management of Risk for Engineered Systems and Geohazards*, 17(1), 1–6.

Phoon, K. K., Shuku, T., Ching, J., and Yoshida, I. 2023c. Editorial for special collection on "Benchmarking data-driven site characterization methods", *ASCE-ASME Journal of Risk and Uncertainty in Engineering Systems: Part A*, 9(2), 02023001.

Phoon, K. K., Ching, J., and Tao, Y. 2024. Soil and rock parametric uncertainties. Chapter 2 in *Uncertainty, Modelling, and Decision Making in Geotechnics*, CRC Press, Boca Raton.

Phoon, K. K., Cai, Y., and Tang, C. 2025. *Geotechnical "facial recognition" challenge. ASCE-ASME Journal of Risk and Uncertainty in Engineering Systems: Part A.*

Prakoso, W. A. 2017. Personal communication.

Prakoso, W. A. 2002. Reliability-based design of foundations on rock for transmission line and similar structure. PhD Thesis, Cornell University, New York.

Qi, X. H., Li, D. Q., Phoon, K. K., Cao, Z., and Tang, X. S. 2016. Simulation of geologic uncertainty using coupled Markov chain. *Engineering Geology*, 207, 129–140.

Raiffa, H. and R. Schlaifer. 1961. *Applied Statistical Decision Theory*. MIT Press, Cambridge, MA.

Retief, J. V., Dithinde, M., and Phoon, K. K. 2016. General principles on reliability according to ISO2394. Chapter 2 in *Reliability of Geotechnical Structures in ISO2394*, CRC Press, Balkema, pp. 33–48.

Rowley, J. 2007. The wisdom hierarchy: representations of the DIKW hierarchy. *Journal of Information and Communication Science*, 33(2), 163–180.

Salgado, R., Houlsby, G. T., and Cathie, D. N. 2008. Contributions to Géotechnique 1948–2008: foundation engineering. *Géotechnique*, 58(5), 369–375.

Self, S.J., Entwisle, D.C., and Northmore, K.J. 2012. The structure and operation of the BGS National Geotechnical Properties Database. Version 2. British Geological Survey, Nottingham, England.

Sharma, A., Ching, J., and Phoon, K. K. 2022. A hierarchical Bayesian similarity measure for geotechnical site retrieval. *Journal of Engineering Mechanics, ASCE*, 148(10), 04022062.

Sharma, A., Ching, J., and Phoon, K. K. 2023. A spectral algorithm for quasi-regional geotechnical site clustering. *Computers and Geotechnics*, 161, 105624.

Shahri, A. A., Spross, J., Johansson, F., and Larsson, S. 2019. Landslide susceptibility hazard map in southwest Sweden using artificial neural network. *Catena*, 183, 104225.

Shi, C. and Wang, Y. 2021a. Nonparametric and data-driven interpolation of subsurface soil stratigraphy from limited data using multiple point statistics. *Canadian Geotechnical Journal*, 58(2), 261–280.

Shi, C. and Wang, Y. 2021b. Development of subsurface geological cross-section from limited site-specific boreholes and prior geological knowledge using iterative convolution XGBoost. *Journal of Geotechnical and Geoenvironmental Engineering*, 149(9), 04021082.

Shioi, A., Otake, Y., Yoshida, I., Muramatsu, S., and Ohno S. 2023. Data-driven approximation of geotechnical dynamics to an equivalent single-degree-of-freedom vibration system based on dynamic mode decomposition. *Georisk: Assessment and Management of Risk for Engineered Systems and Geohazards*, 17(1), 77–97.

Shuku, T. and Phoon, K. K. 2023a. Data-driven subsurface modeling using a Markov random field model. *Georisk: Assessment and Management of Risk for Engineered Systems and Geohazards*, 17(1), 41–63.

Shuku, T. and Phoon, K.K. 2023b. Comparison of data-driven site characterization methods through benchmarking: methodological and application aspects. *ASCE-ASME Journal of Risk and Uncertainty in Engineering Systems, Part A: Civil Engineering*, 9(2), 04023006.

Shuku, T., Phoon, K. K., and Yoshida, I. 2020. Trend estimation and layer boundary detection in depth-dependent soil data using sparse Bayesian lasso. *Computers and Geotechnics*, 128, 103845.

Shuku, T., Phoon, K. K., Ishii, M., Kumagai, T., Yokota, Y., and Date, K. 2023. A probabilistic transformation model between two rock mass properties: specific fracture energy and p-wave velocity. *Canadian Geotechnical Journal*, 60(8), 1161–1172.

Straub, D. 2014. Value of information analysis with structural reliability methods. *Structural Safety*, 49, 75–85.

Stuedlein, A. W., Kramer, S. L., Arduino, P., and Holtz, R. D. 2012. Geotechnical characterization and random field modeling of desiccated clay. *Journal of Geotechnical and Geoenvironmental Engineering, ASCE*, 138(11), 1301–1313.

Tan, X., Chen, W., Qin, C., Zhao, W., and Ye, W. 2023. Characterisation for spatial distribution of mining-induced stress through deep learning algorithm on SHM data. *Georisk: Assessment and Management of Risk for Engineered Systems and Geohazards*, 17(1), 217–226.

Tang, C. and Phoon, K. K. 2021. *Model Uncertainties in Foundation Design*. CRC Press, Boca Raton, FL.

Tang, C. and Phoon, K. K. 2025. *Databases for Data-Centric Geotechnics: Geotechnical Structures*. CRC Press, Boca Raton, FL.

Tang, C., Phoon, K. K., and Yuan, J. 2024. Variability of predictions in geotechnics. Chapter 5 in *Uncertainty, Modelling, and Decision Making in Geotechnics*, CRC Press, Boca Raton, FL, pp. 176–216.

Tang, W. H. 1986. Updating anomaly statistics-single anomaly case. *Structural Safety*, 4(2), 151–163.

Tang, W. H. and Halim, I. 1988. Updating anomaly statistics-multiple anomaly pieces. *Journal of Engineering Mechanics*, 114(6), 1091–1096.

Terzaghi, K. 1943. *Theoretical Soil Mechanics*. John Wiley & Sons, New York.

Terzaghi, K. and Peck, R. B. 1967. *Soil Mechanics in Engineering Practice*. John Wiley, New York.

Thiyagalingam, J., Shankar, M., Fox, G., and Hey, T. 2022. Scientific machine learning benchmarks. *Nature Review Physics*, 4, 413–420.

Tian, H. and Wang, Y. 2023. Data-driven and physics-informed Bayesian learning of spatiotemporally varying consolidation settlement from sparse site investigation and settlement monitoring data. *Computers and Geotechnics*, 157, 105328.

Tian, H., Wang, Y., and Phoon, K. K. 2024. Real-time fusion of multi-source monitoring data with geotechnical numerical model results using data-driven and physics-informed sparse dictionary learning. *Canadian Geotechnical Journal*, e-First https://doi.org/10.1139/cgj-2023-0457

Vanmarcke, E. H. 1977. Probabilistic modeling of soil profiles. *Journal of the Geotechnical Engineering Division*, 103(11), 1227–1246.

Vanmarcke, E. H. 1983. *Random Fields: Analysis and Synthesis*, The MIT Press, Cambridge, MA.

Vardanega, P. J. and Bolton, M. D. 2016. Design of geostructural systems. *ASCE-ASME Journal of Risk and Uncertainty in Engineering Systems, Part A: Civil Engineering*, 2(1), 04015017.

Ulmer, K.J., Zimmaro, P., Brandenberg, S.J., Stewart, J.P., Hudson, K.S., Stuedlein, A.W., Jana, A., Dadashiserej, A., Kramer, S.L., Cetin, K.O., Can, G., Ilgac, M., Franke, K.W., Moss, R.E.S., Bartlett, S.F., Hosseinali, M., Dacayanan, H., Kwak, D.Y., Stamatakos, J., Mukherjee, J., Salman, U., Ybarra, S., and Weaver, T. 2023. *Next-generation liquefaction database*, Version 2. Next-Generation Liquefaction Consortium.

Wang, Y., Huang, K., and Cao, Z. 2014. Bayesian identification of soil strata in London clay. *Géotechnique*, 64(3), 239–246.

Wang, X., Wang, H., Liang, R. Y., and Liu, Y. 2019. A semi-supervised clustering-based approach for stratification identification using borehole and cone penetration test data. *Engineering Geology*, 248, 102–116.

Wang, H., Zhang, L. M., Wang, L., He, J., and Luo, H. 2021a. An automated snow mapper powered by machine learning. *Remote Sensing*, 13(23), 4826.

Wang, H., Zhang, L. M., Luo, H., He, J., and Cheung, R. W. M. 2021b. AI-powered landslide susceptibility assessment in Hong Kong. *Engineering Geology*, 288, 106103.

Wang, Y., Hu, Y., and Phoon, K. K. 2021c. Non-parametric modelling and simulation of spatiotemporally varying geo-data. *Georisk: Assessment and Management of Risk for Engineered Systems and Geohazards*, 16(1), 77–97.

Wang, Y., Qi, X., Zhang, W., and Ching, J. 2022. Editorial for Special Issue on "Data analytics in geotechnical and geological engineering", *Georisk: Assessment and Management of Risk for Engineered Systems and Geohazards*, 16(1), 1.

Wackernagel, H. 2003. *Multivariate Geostatistics: An Introduction with Applications*. Springer, Berlin.

Wu, S., Ching, J., and Phoon, K. K. 2022. Quasi-site-specific soil property prediction using a cluster-based hierarchical Bayesian model. *Structural Safety*, 99, 102253.

Wu, S., Otake, Y., Mizutani, D., Liu, C., Asano, K., Sato, N., Saito, T., Baba, H., Fukunaga, Y., Higo, Y., Kamura, A., Kodama, S., Metoki, M., Nakamura, T., Nakazato, Y., Shioi, A., Takenobu, M., Tsukioka, K. and Yoshikawa, R. 2024. Future-proofing geotechnics workflows: accelerating problem-solving with large language models. *Georisk: Assessment and Management of Risk for Engineered Systems and Geohazards*. https://doi.org/10.1080/17499518.2024.2381026

Xiao, S., Zhang, J., Ye, J., and Zheng, J. 2021a. Establishing region-specific N–Vs relationships through hierarchical Bayesian modeling. *Engineering Geology*, 287, 106105.

Xiao, T., Zou, H. F., Yin, K. S., Du, Y., and Zhang, L. M. 2021b. Machine learning-enhanced soil classification by integrating borehole and CPTU data with noise filtering. *Bulletin of Engineering Geology and the Environment*, 80, 9157–9171.

Xie, J., Huang, J., Lu, J., Burton, G. J., Zeng, C., and Wang, Y. 2022. Development of two-dimensional ground models by combining geotechnical and geophysical data. *Engineering Geology*, 300, 106579.

Xu, J., Wang, Y., and Zhang, L. 2021. Interpolation of extremely sparse geo-data by data fusion and collaborative Bayesian compressive sampling. *Computers and Geotechnics*, 134, 104098.

Xu, J., Wang, Y., and Zhang, L. 2022. Fusion of geotechnical and geophysical data for 2D subsurface site characterization using multi-source Bayesian compressive sampling. *Canadian Geotechnical Journal*, 59(10), 1756–1773.

Xu, Y., Kohtz, S., Boakye, J., Gardoni, P., and Wang, P. 2023. Physics-informed machine learning for reliability and systems safety applications: State of the art and challenges. *Reliability Engineering & System Safety*, 230, 108900.

Yan, W., Yan, Y., Shen, P., and Zhou, W. H. 2023. A hybrid physical data informed DNN in axial displacement prediction of immersed tunnel joint. *Georisk: Assessment and Management of Risk for Engineered Systems and Geohazards*, 17(1), 169–180.

Yoshida, I., Tasaki, Y., Otake, Y., and Wu, S. 2018. Optimal sampling placement in a Gaussian random field based on Value of Information. *ASCE-ASME Journal of Risk and Uncertainty in Engineering Systems, Part A: Civil Engineering*, 4(3), 04018018.

Yoshida, I., Tomizawa, Y., and Otake, Y. 2021. Estimation of trend and random components of conditional random field using gaussian process regression. *Computers and Geotechnics*, 136, 104179.

Yuen, K. V., Ching, J., and Phoon, K. K. 2021. Editorial for special collection on "Bayesian learning methods for geotechnical data". *ASCE-ASME Journal of Risk and Uncertainty in Engineering Systems: Part A*, 7(1), 02020002.

Zapata, G.E. 2010. A national database of subgrade soil-water characteristic curves and selected soil properties for use with the MEPDG. *NCHRP Web-Only Document 153*. Transportation Research Board, Washington, DC.

Zhang, W. G. and Phoon, K. K. 2022. Editorial for Special Issue on "Advances and applications of deep learning and soft computing in geotechnical underground engineering". *Journal of Rock Mechanics and Geotechnical Engineering*, 14(3): 671–673.

Zhang, D. M., Zhou, Y., Phoon, K. K., and Huang, H. W. 2020. Multivariate probability distribution of Shanghai clay properties. *Engineering Geology*, 273, 105675.

Zhang, W. H., Lu, D. G., Qin, J. J., Thöns, S., and Faber, M. H. 2021. Value of information analysis in civil and infrastructure engineering: A review. *Journal of Infrastructure Preservation and Resilience*, 2, 16.

Zhang, W., Gu, X., Hong, L., Han, L., and Wang, L. 2023. Comprehensive review of machine learning in geotechnical reliability analysis: Algorithms, applications and further challenges. *Applied Soft Computing*, 136, 110066.

Zhou, Y. and Song, X. K. 2016. Regression analysis of networked data. *Biometrika*, 103(2), 287–301.

Zhou, Z., Yan, L., Zhang, J., and Yang, H. 2023. Real-time tunnel lining crack detection based on an improved You Only Look Once version X algorithm. *Georisk: Assessment and Management of Risk for Engineered Systems and Geohazards*, 17(1), 181–195.

Zou, H., Liu, S., Cai, G., Puppala, A. J., and Bheemasetti, T. V. 2017. Multivariate correlation analysis of seismic piezocone penetration (SCPTU) parameters and design properties of Jiangsu quaternary cohesive soils. *Engineering Geology*, 228, 11–38.

Chapter 2

Rock hydromechanical parameters for rock foundation design

A database approach

Pouyan Asem and Paolo Gardoni

Rock sockets are often constructed in weak rock masses. Understanding the hydromechanical properties of the rock mass is required for the successful design and performance of the rock sockets. This chapter establishes separate databases for the hydromechanical properties of rock masses, with particular attention to weak shales and mudstones. The databases include information on rock socket geometry, load–displacement response, and the rock mass properties and mechanical and transport properties of the rock. The chapter then reviews the available models for the prediction of (i) the axial resistance (compression or uplift), (ii) axial deformation, and (iii) permeability. The databases are used to evaluate the existing predictive models for hydromechanical properties of rock as related to the design of rock sockets. The load test databases are also used to evaluate the effect of rock mass variability on the reliability of the foundations in weak rock. Based on the results of the analysis presented, the applicability of the existing models and approaches to the study of socket behavior in weak rocks is discussed.

2.1 INTRODUCTION

Rock sockets are being increasingly used for the support of structures that carry heavy loads (Horvath and Kenney 1979; O'Neill and Reese 1999; Seidel and Collinwood 2001; Brown *et al.* 2010; Dai *et al.* 2016; AlKhafaji *et al.* 2020), and hence determination of the rock mass properties required for their design (e.g., Asem 2018; Asem *et al.* 2018; Asem *et al.* 2019a,b; Asem and Gardoni 2019a,b,c; Asem 2019a,b; Asem 2020a,b; Asem and Labuz 2020; Guevara-Lopez *et al.* 2020; Tabandeh *et al.* 2020; Asem *et al.* 2021a,b; Asem and Gardoni 2021a,b; Asem and Gardoni 2022) is of utmost interest. Rock-socketed foundations often interact with weak rocks (i.e., a rock mass with unconfined compressive strength, q_u, from ~0.5 MPa to ~30 MPa; Geological Strength Index, GSI $\leq$ 70 after Hoek and Brown 1997; or equivalently, rock mass rating, RMR $\leq$ 40). This is because the upper zone of most rock formations hosting foundations is altered by physical and chemical weathering, especially in tropical and humid areas (Goodman 1993). A weak rock mass consists of relatively intact rock blocks that are separated by joint sets (Hoek 1983; Jaeger *et al.* 2007). The behavior of foundations in weak rock, therefore, is complex due to the existence of the rock mass secondary structures and their influence on rock mass mechanical properties.

Some researchers have based their work on the results of rock socket load tests in weak rock (e.g., Teng 1962; Coates 1967; Rosenberg and Journeaux 1976; Horvath and Kenney 1979; Meigh and Wolski 1979; Williams 1980; Gupton and Logan 1984; Rowe and Armitage 1987; Carter and Kulhawy 1988; Toh *et al.* 1989; Kulhawy and Phoon 1993; Zhang and Einstein 1998; Miller 2003; Abu-Hejleh *et al.* 2003; Paikowsky *et al.*

 DOI: 10.1201/9781003441946-2

2010; Stark *et al.* 2013; Stark *et al.* 2017; Asem 2018; Asem *et al.* 2018; Baghdady 2018; Asem *et al.* 2019a,b; Asem 2019a,b; Asem and Gardoni 2019a,b,c; Tabandeh *et al.* 2020; Asem 2020a,b). These studies have led to the development of empirical design methods where the observed side and base resistances and the deformational properties of rock sockets have been related to the rock properties (e.g., unconfined compressive strength, q_u, and rock mass deformation modulus, E_m) and to the rock socket geometry (e.g., rock socket length, L, and rock socket diameter, B). These methods are sometimes site-specific (e.g., Williams 1980; Miller 2003; Abu-Hejleh *et al.* 2003).

Other researchers (e.g., Lam 1983; Haberfield 1987; Johston and Lam 1989; Kodikara 1989; Seidel 1993; Hassan 1994; Collingwood 2000; Seidel and Collingwood 2001; Haberfield and Lochaden 2018; Johnston 2020) have used a combination of theoretical and empirical methods for the analysis of the behavior of rock sockets. These semi-empirical methods are applicable to a broader range of field conditions because they are mechanistic where the axial behavior of drilled shafts is related to the fundamental variables that characterize the weak rock mechanical and engineering properties, the rock socket geometry and properties of mobilized shear surfaces on the rock socket sidewalls, and the boundary conditions at the base of rock sockets.

Another important consideration is the selection of the correct rock parameters, and hence the degree of excess pore pressure dissipation for the rock mass during loading should be determined. Excess pore water pressure is generated in a rock mass during the application of loads (Terzaghi *et al.* 1996). The average degree of consolidation (U) of weak rock, as controlled by excess pore water pressure dissipation, dictates whether a rock mass would behave under drained or undrained conditions. Therefore, hydromechanical properties such as rock permeability are also needed for proper analysis of rock behavior during the application of loads.

This chapter will discuss the use of large databases in the development and analysis of rock socket behavior, particularly those constructed in weak clay-based rock and will present a synthesis of the work of the authors on the subject at hand (e.g., Asem 2018; Asem *et al.* 2018; Asem and Gardoni 2019; Asem *et al.* 2019; Asem and Gardoni 2019a,b; Asem 2019a,b; Asem 2020a,b; Asem and Labuz 2020; Guevara-Lopez *et al.* 2020; Tabandeh *et al.* 2020; Asem *et al.* 2021a,b; Asem and Gardoni 2021a,b,c; Asem and Gardoni 2022). The databases include (i) an in situ side-resistance database, (ii) an in situ base-resistance database, and (iii) a permeability database from laboratory and in situ tests. The databases (i) and (ii) include information on the rock socket geometry, rock socket load–displacement response, and rock mass properties. The available methods for the design of rock sockets are discussed in detail to understand the range and variability in the rock hydromechanical properties involved in design processes. In particular, the databases (i) and (ii) are used to evaluate the existing predictive models for axial resistance and settlement. Once a clear picture of the required hydromechanical rock properties is presented, we use the available data from the literature to propose models for the prediction of rock hydromechanical properties.

2.2 IN SITU LOAD TESTS FOR DETERMINATION OF MECHANICAL ROCK PROPERTIES

2.2.1 Existing databases

The existing load test databases that are relevant to this study are summarized in Table 2.1. The following should be noted:

Table 2.1 Summary of the existing load test databases

Database	*Number of load tests in rocks*	*Location of load tests*	*Rock type*	*Unconfined compressive strength range, q_u (MPa)*	*Rock mass properties reported*	*Load–displacement relationship reported*	*Load test type*
FHWA deep foundation load test database (DFLTD)	Limited	The United States			No	No	
NCHRP report 507, Paikowsky *et al.* (2004)	83					No	
NCHRP report 651, Paikowsky *et al.* (2010)	122	The United States, the United Kingdom, Italy, Russia, Taiwan, Japan, Canada, Australia, Singapore, South Africa	Claystone, mudstone, sandstone, shale, till, chalk, diabase, limestone, mudstone, gypsum, granite, tuff	0.2–103	Yes	No	Tip resistance
Rosenberg and Journeaux (1976)	8		Interbedded shale and sandstone, shale, sandstone, limestone, and andesite	0.5–55	No	No	Side and tip resistances
Horvath and Kenney (1979)	76	Canada, the United States, the United Kingdom, Australia, Czechoslovakia, Germany, South Africa, New Zealand, and Switzerland	Shale, clayshale, mudstone, siltstone, sandstone, chalk, limestone, marl, diabase, basalt, schist, and slate	0.35–110	No	No	Side resistance
Williams (1980)		Australia	Siltstone	0.2–80	No	Yes	Side and tip resistances
Williams and Pells (1981)	71	Australia, Ireland, South Africa, the United States	Mudstone, marl, claystone, clayshale, shale, and sandstone	0.48–62	No	No	Side resistance
Rowe and Armitage (1984)	158	Australia, the United States, Canada, the United Kingdom, South Africa, Czechoslovakia, Ireland	Sandstone, mudstone, shale, claystone, siltstone, chalk, andesite, schist, limestone, and slate	0.412–110	Yes	No	Side resistance

(Continued)

Table 2.1 (Continued) Summary of the existing load test databases

Database	*Number of load tests in rocks*	*Location of load tests*	*Rock type*	*Unconfined compressive strength range, q_u (MPa)*	*Rock mass properties reported*	*Load–displacement relationship reported*	*Load test type*
Hassan (1994)	139	South Africa, the United States, Singapore, Ireland, Canada, Brazil, Australia, the United Kingdom, Hong Kong	Shale, diabase, claystone, marl, till, hard clay, hardpan, basalt, siltstone, mudstone, chalk, marl, and limestone	0.14–37.23	No	No	Side and tip resistances
Zhang and Einstein (1998), Zhang (1999)	39	Australia, the United States, South Africa, Italy, Singapore	Mudstone, shale and clayshale, gypsum, till, diabase, hardpan, sandstone, siltstone, marl, and limestone	0.52–55	No	No	Tip resistance
Collingwood (2000)	162	The United States, Australia, Canada, Ireland, the United Kingdom, South Africa, Singapore	Shale, clayshale, limestone, sandstone, mica schist, mudstone, chalk, hard clay, andesite, diabase, siltstone, marl, and granite	0.375–110	No	No	Side resistance
Stark *et al.* (2013)	54 (side) and 33 (tip)	The United States, South Africa, Australia, Ireland, Canada	Shale, mudstone, claystone, and mudstone	0.12–4.45	No	No	Side and tip resistances
Baghdady (2018)	93 (side) and 62 (tip)	The United States, South Africa, Australia, Ireland, Canada	Shale, mudstone, and claystone	0.12–6.6	No	No	Side and tip resistances

1. *Number of load tests*: The number of load tests in some databases is limited. For example, the database collected by Rosenberg and Journeaux (1976) only contains eight drilled shaft load tests. The accuracy of the estimation of model parameters is dependent on the number of observations (e.g., load tests) (Gardoni *et al.* 2002). The use of small databases for the prediction of model parameters could result in inaccurate predictions and large uncertainties in the estimated parameters. Some databases only focus on side or base resistance (Zhang and Einstein 1998) and not both.
2. *Availability of complete load–displacement information*: In the majority of the current databases (see Table 2.1), the information regarding the load–displacement response (i.e., side shear stress–displacement and base contact pressure–displacement, f_s–δ and q–δ relationships, respectively) is not reported. Therefore, such databases cannot be used for a proper study of the load–displacement mechanism of drilled shafts in weak rock mass, which is a fundamental aspect of drilled shaft behavior.
3. *Lack of basic rock mass mechanical properties*: In the majority of the existing drilled shaft load test databases (e.g., Abu-Hejleh *et al.* 2003; Stark *et al.* 2013; Stark *et al.* 2017; Baghdady 2018), only the unconfined compressive strength (q_u) for weak rock is reported. Moreover, the weathering condition of weak rock is not discussed and the engineering properties of the weak rock "mass" are not reported or estimated. Therefore, the actual problem involving the interaction of a drilled shaft with the weak rock "mass" is simplified to a problem where the rock mass is represented by the properties of intact rock (i.e., unconfined compressive strength, q_u, and intact rock deformation modulus, E_i).
4. *Range of rock compressive strength*: The range of rock unconfined compressive strength (q_u) in some of the existing databases is well beyond the upper bound for the compressive strength of weak rocks (i.e., 30 MPa) (e.g., see database of Zhang and Einstein 1998).
5. *Range of rock socket size*: The socket diameter (B) reported in the existing databases covers a wide range. However, reported diameters are not distributed uniformly. Therefore, the effect of socket diameter (i.e., size) cannot be properly investigated. For example, the data in Williams (1980) consist mostly of small diameter (B) sockets and the rock socket lengths (L) were usually short.

2.2.2 New side-resistance and base-resistance databases

Two new databases are summarized in the Supplemental Data Section and in other publications (Asem 2018; Asem *et al.* 2018; Asem and Gardoni 2019; Asem *et al.* 2019; Asem and Gardoni 2019a,b; Asem 2019a,b; Asem 2020a,b; Asem and Labuz 2020; Guevara-Lopez *et al.* 2019; Tabandeh *et al.* 2020; Asem *et al.* 2021a,b; Asem and Gardoni 2021a,b,c; Asem and Gardoni 2022). One is a side-resistance database, and the other is a base-resistance database. The rock types include weak shale, siltstone, claystone, mudstone, limestone, sandstone, tuff, and granite. The case histories where q_u was larger than 30 MPa (that is, the selected upper bound for weak rock after Deere and Miller 1966; Barton *et al.* 1978; Rowe and Armitage 1987; Kanji 2014) are not used in the development of the database mainly because the aim of development of the databases in this section is to study the performance of the available design models with particular attention to weak rocks. The main contributions of the new databases include reporting: (i) the load deformatiom properties for the rock sockets and (ii) the rock mass properties (e.g., deformation modulus E_m). Additionally, the load test databases only concentrate on the axial behavior of rock sockets in weak rocks. In the following, the side- and base-resistance databases are discussed.

2.2.2.1 Side-resistance database

The side-resistance database includes 317 rock socket load tests. The peak side resistance (f_{sp}) and shear stress–shear displacement (f_s–δ) relationship for each socket sidewall is directly measured using in situ instrumentation. Therefore, it was not necessary to assume a value of base resistance to obtain the f_s–δ relationships. The side-resistance database contains the following information:

1. *Load-transfer function (f_s–δ relationship) for rock socket sidewalls*: The f_s–δ relationships are obtained based on the measured rock socket load–displacement relationship and the load distribution versus depth measurements that are gathered using strain gages or load cells (see Figure 2.1). The displacements reported in f_s–δ relationships exclude the elastic deformation $\left(\frac{PL}{AE_c}\right)$ of the test shaft to give the net rock socket deformation where P is the axial load, L is the rock socket length, A is socket cross-sectional area, and E_c is Young's modulus of concrete. A typical f_s–δ relationship is shown in Figure 2.2b.
2. *Properties of intact rock*: The unconfined compressive strength (q_u) of the intact rock specimens is reported for all cases. The internal friction angle (ϕ_i), shear strength intercept (c_i), and intact rock deformation modulus (E_i) were not available for a majority of case histories.
3. *Properties of the rock mass:* The rock mass modulus is estimated using the method of Pells and Turner (1979) – that is essentially a collection of solutions based on the theory of elasticity for the calculation of socket settlement, based on the numerical integration of Mindlin's equation – as $E_m = 2\pi LK_{si}I$, where L is the rock socket length, I is the embedment influence factor (Pells and Turner 1979), and K_{si} is the initial shear stiffness (Figure 2.2b) that is obtained from load tests. Other properties of the rock mass are estimated using the same approach explained for the base-resistance database.
4. *Side resistance*: Figure 2.2b shows a typical f_s–δ relationship that illustrates the definitions of initial shear stiffness (K_{si}), peak shear stress (f_{sp}), and ultimate shear stress (f_{su}). The initial shear stiffness (K_{si}) is the initial slope of the f_s–δ relationship.
5. *Drilled shaft geometry and material*: The diameter for each rock socket (B) and the depths of embedment from the ground surface (D_{GS}) and from the top of rock formation (D_{TOR}) to the center of the shear profile are obtained based on the idealized site stratigraphy and socket reported dimensions. The length of the rock socket (L) is also provided for each case history. The method of construction is summarized when reported. The shaft concrete compressive strength (f'_c), concrete modulus of elasticity (E_c), and concrete slump are summarized when they are available.

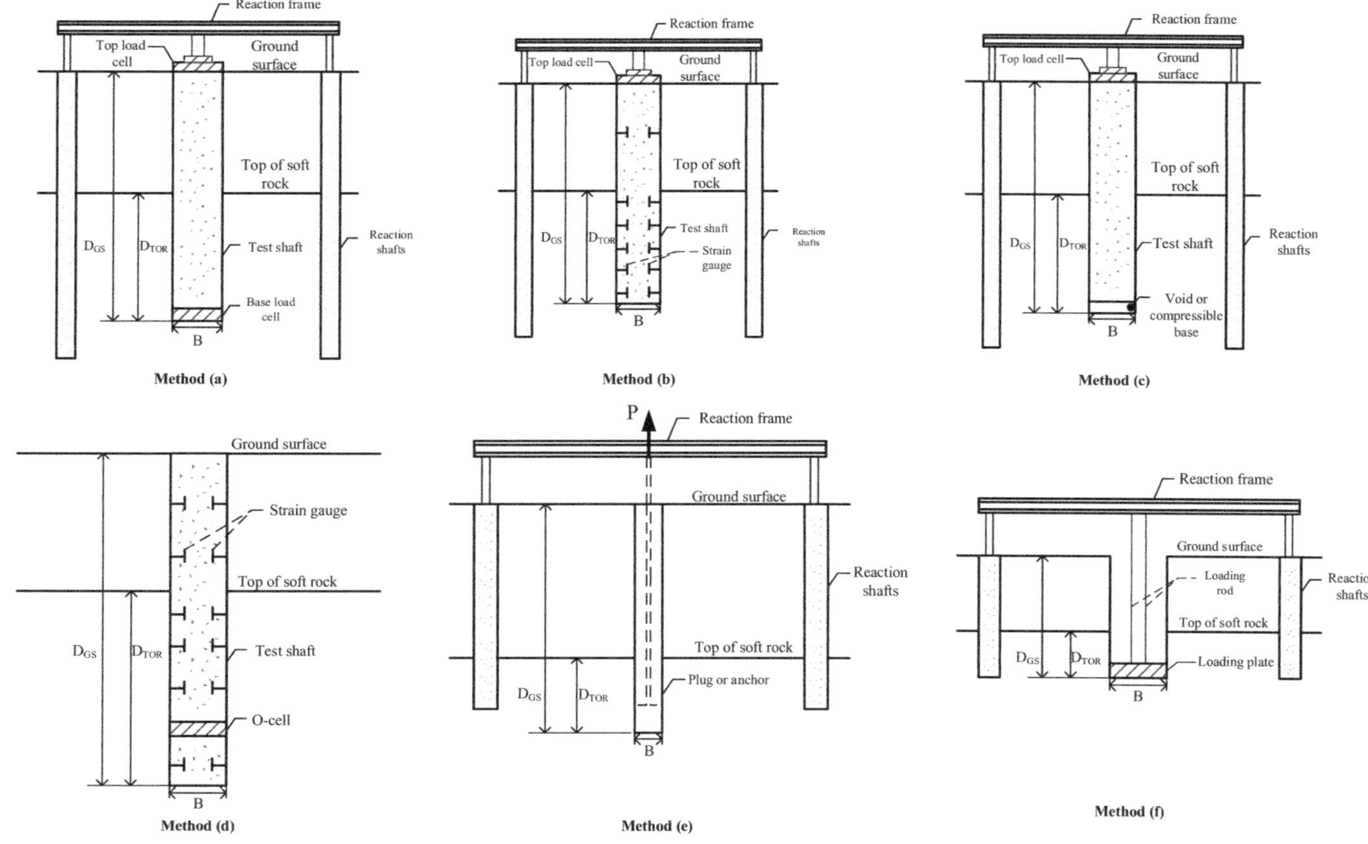

Figure 2.1 Load test methods used in the in situ test database.

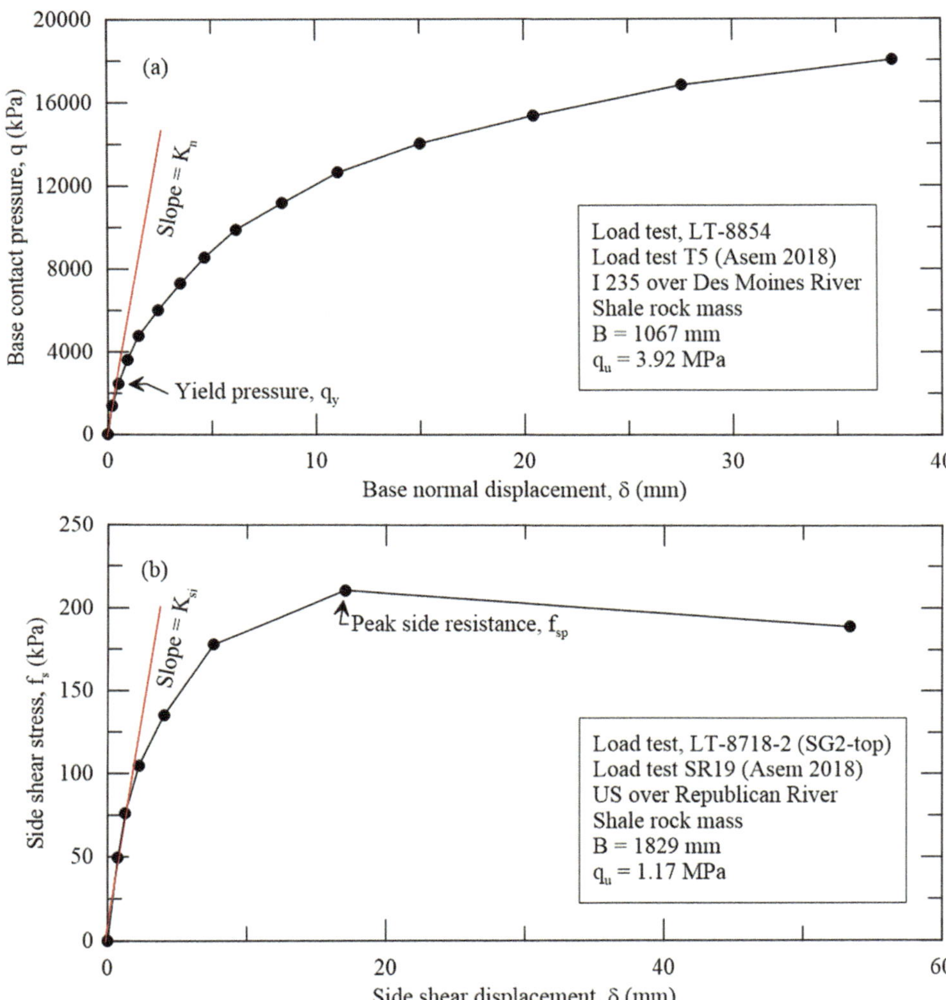

Figure 2.2 Typical side resistance and base displacement relationships for rock sockets in weak rock.

2.2.2.2 Base-resistance database

Figure 2.1 shows the load test methods that are commonly used to measure the base resistance of rock sockets. The base-resistance database includes 190 axial load tests (socket and plate load tests) and contains the following information:

1. *Load-transfer function (q–δ relationship) for socket base*: The q–δ relationships (e.g., Figure 2.2a) are developed based on the top of rock socket load–displacement relationship and load distribution measured by strain gages or tell-tales in conventional top-loaded load tests or the load–displacement relationship obtained from the O-Cell in Osterberg load tests. The elastic compression of the test shaft $\left(\frac{PL}{AE_c}\right)$ is considered and excluded in the determination of deformation of rock mass under the rock socket base in top-loaded load tests.

2. *Properties of the intact rock*: The database contains rock socket case histories that include the measured unconfined compressive strength (q_u) of weak rock specimens obtained from rock cores extracted from each load test site. The unconfined compressive strength (q_u) of the intact rock specimens is reported for all cases. The friction angle (ϕ_i), shear strength intercept (c_i), and modulus of deformation (E_i) for intact rock are not reported in the majority of cases.
3. *Properties of the rock mass*: The modulus of deformation of the rock mass (E_m) is estimated by the authors from the load test results in each case history using the method of Pells and Turner (1979) where the rock mass deformation modulus (E_m) is related to the initial normal stiffness of the rock socket base (K_n) as $E_m = 0.5\pi BK_n$ (B is the rock socket diameter and I is depth influence factor and is based on the results of finite element models Pells and Turner 1979). For design purposes, E_m can be estimated using a model of E_m as a function of q_u (Rowe and Armitage 1984; Asem 2018). This model is developed using linear regression analysis of (i) the results of in situ plate load tests reported by Chern *et al.* (2004) for weak rocks in China and Taiwan, (ii) back-analysis of q–δ relationships from 190 rock socket load tests from the United States, Puerto Rico, Canada, Australia, South Africa, Italy, the United Kingdom, and Singapore, and (iii) back-analysis of 340 f_s–δ relationships from 292 rock socket load tests in weak rock as discussed in subsequent sections. This linear model may be expressed as:

$$E_m = 150q_u^{1.1} \tag{2.1}$$

where E_m and q_u are in units of MPa. The predicted values of E_m and the method of Hoek and Diederichs (2006) – $E_m = \dfrac{100{,}000}{1+e^{[75\text{-}GSI]}}$ – are used to infer the Geological Strength Index (GSI). Rock mass Mohr–Coulomb strength parameters (Hoek and Brown 1997; Labuz and Zang 2012), namely the drained friction angle (ϕ_m) and failure envelope shear strength intercept (c_m) for rock mass are estimated based on (i) the inferred values of GSI as explained above, (ii) approximate material constant (m_i) that relates to the type of rock and frictional characteristics of rock minerals based on values reported in Hoek (1983) and Marinos and Hoek (2001) and (iii) the method of Hoek and Brown (1997) for estimation of ϕ_m and c_m. The predicted values of ϕ_m and c_m are not sensitive to m_i (Marinos and Hoek 2001), and thus estimation of m_i using qualitative description of rock will not significantly affect the predicted values of ϕ_m and c_m, especially for sedimentary and metamorphic rocks in the side-resistance database.
4. *Base resistance*: The q–δ relationships (Figure 2.2a) usually exhibit a strain hardening response and a failure point is not apparent. Therefore, the L1–L2 approach (Hirany 1988) is used to interpret two failure points, i.e., the values of q_y (i.e., contact pressure at an initial yield that is analogous to Hirany's $q1$ stress) and q_f (i.e., the interpreted failure pressure that is analogous to Hirany's $q2$ stress for Terzaghi 1943 failure pressure). When a distinct failure point was reported, it is included in the database.
5. *Initial normal stiffness* (K_n): The initial normal stiffness (K_n) is defined as the slope of the tangent line to the initial portion of q–δ relationship as shown in Figure 2.2.
6. *Rock socket geometry*: As shown in Figure 2.1, the base diameter for each rock socket (B) and the depths of embedment of the rock socket base from the ground

surface (D_{GS}) and from the top of rock formation (D_{TOR}) are obtained based on the reported site stratigraphy and rock socket reported dimensions. The method of construction and the condition of rock socket base are summarized when reported.

2.3 REVIEW OF SIDE RESISTANCE METHODS

Rock sockets carry their loads in side resistance, base resistance, or combination of both (Horvath and Kenney 1979; Rowe and Armitage 1987; Hassan *et al.* 1997; Zhang and Einstein 1998; Seidel and Collingwood 2001; CGS 2006; Turner 2006; Asem 2018; Haberfield and Lochaden 2018). Osterberg and Gill (1973) and Rowe and Armitage (1987) using finite element analysis and Goodman (1980; 1989) and Seo and Prezzi (2008) using solutions based on the theory of elasticity showed that the portion of the load transferred to the rock socket base is only a small percentage of the total applied load when the ratio of socket length to diameter, L/B, is ≥ 4. Therefore, rock sockets with $L/B \geq 4$ and especially those socketed in stiff rock masses (i.e., $E_m/E_c \geq 5$) carry most of their service load using side resistance (Seidel and Collingwood 2001; Seo and Prezzi 2008). Additionally, the design of foundations for base resistance requires inspection procedures that guarantee the cleanness of the rock socket base, which has been proven to be expensive and difficult, especially for the case of deep sockets in weathered rock mass (Seidel and Collingwood 2001; Brown *et al.* 2010). The influence of the construction procedure and the detrimental effects of soft material near the rock socket base are discussed in the literature (e.g., Holden 1984; Turner 2006; Brown *et al.* 2010; Haberfield and Lochaden 2018). Therefore, significant effort has been devoted to the development of predictive models for side resistance. Design methods for side resistance are commonly empirical and have been developed using load test databases that have a limited number of tests in a wide range of rock types with a broad range of rock compressive strengths. Exceptions include those of Rowe and Armitage (1984), Stark *et al.* (2013), and Asem (2018). Some methods, however, are based on laboratory constant normal stress and constant normal stiffness (CNS) direct shear tests on synthetic rock/concrete interfaces where bonding between rock and concrete is prevented (e.g., Seidel 1993; Collingwood 2000; Seidel and Collingwood 2001). These methods commonly relate the measured peak shear stress that is mobilized on the rock/concrete interface to the properties of the rock (e.g., the unconfined compressive strength of intact rock, q_u, and rock mass deformation modulus, E_m) and the properties of the rock/concrete interface such as interface roughness height (h).

2.3.1 Empirical side resistance models

Rosenberg and Journeaux's (1976) study is among one of the first models for peak side resistance. It is based on a total of eight rock socket load tests where intact rock compressive strength for four cases is reported. The load tests were conducted on drilled shafts in sandstone, shale, limestone, and andesite where 0.5 MPa $< q_u <$ 34 MPa. Rosenberg and Journeaux (1976) did not provide a mathematical expression for their proposed method. Kulhawy *et al.* (2005) provided a best-fit equation for the method of Rosenberg and Journeaux (1976) as:

$$\frac{f_{sp}}{\sigma_p} = 1.1\left(\frac{q_u}{\sigma_p}\right)^{0.56} \tag{2.2}$$

where f_{sp} is the peak shear stress and σ_p is the atmospheric pressure (σ_p = 0.101 MPa). In the method of Rosenberg and Journeaux (1976), (i) some of the load tests did not mobilize their peak side shear stress, (ii) the effect of rock type on side resistance is not evaluated, (iii) the effect of construction and shaft geometry on the development of shear stress is not studied, and (iv) the effect of rock mass properties on side resistance is not considered. The method of Rosenberg and Journeaux (1976) is evaluated using the new load test database in Figure 2.3a.

The method of Horvath (1978) is based on the results of six rock socket load tests in Queenston Shale Formation where (i) all test sockets are constructed at one test site near Ontario, Canada, (ii) the shale is horizontally bedded and 4.7 < q_u < 11.1 MPa, and (iii) B = 710 mm. The effect of socket geometry and embedment depth could not be investigated using the database of Horvath (1978) as all rock sockets had the same diameter and the depth of embedment was 1.37 m in all cases. Horvath (1978; 1982), based on the

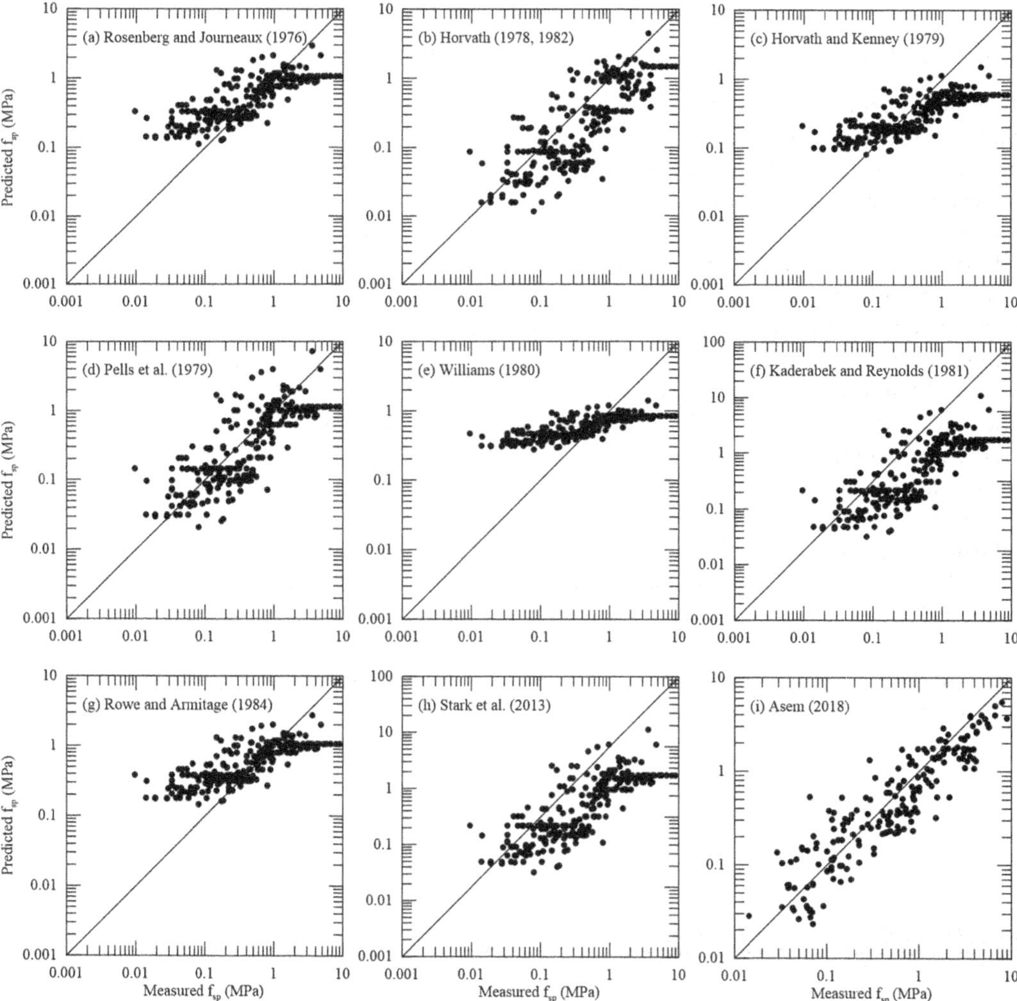

Figure 2.3 Evaluation of the empirical side resistance models using the in situ load test database (Source: data from Asem 2018).

measurement of socket wall roughness, developed a roughness factor (RF) that is used to account for the effect of the socket roughness on the peak side resistance (f_{sp}). The predictive models for RF and f_{sp} are as follows:

$$\mathrm{RF} = \frac{h}{r}\frac{l_t}{L} \tag{2.3}$$

and

$$f_{sp} = 0.8 q_u \mathrm{RF}^{0.45} \tag{2.4}$$

where h represents the average height of asperities that is measured with respect to the surface of the largest imaginary cylinder that can fit inside the rock socket, l_t is the total travel distance along the socket wall that accounts for the irregular texture of the rock socket sidewalls (e.g., see Seidel and Collingwood 2001 for additional discussion on characterization of sidewall roughness), r is the socket radius, and L is the socket length. The roughness of the interface (h) can be evaluated using the following expression (Asem 2018) that was obtained based on rock socket roughness measurements compiled by Collingwood (2000) (see Figure 2.4):

$$h = \frac{50}{q_u + 6.53} \tag{2.5}$$

where h is in mm and q_u is in MPa. The average socket sidewall roughness height (h) is inversely related to the unconfined compressive strength (q_u) because a greater q_u indicates a greater resistance to abrasion, and thus cutting tools are expected to create a smoother sidewall. The method of Horvath is evaluated in Figure 2.3. The results of recent research (e.g., Lam 1983; Haberfield 1987; Johnston and Lam 1989; Kodikara 1989; Seidel 1993; Hassan 1994; Collingwood 2000; Haberfield and Lochaden 2018; Johnston 2020) show that drilled shaft/rock "interface" roughness dominates the behavior of rock sockets as

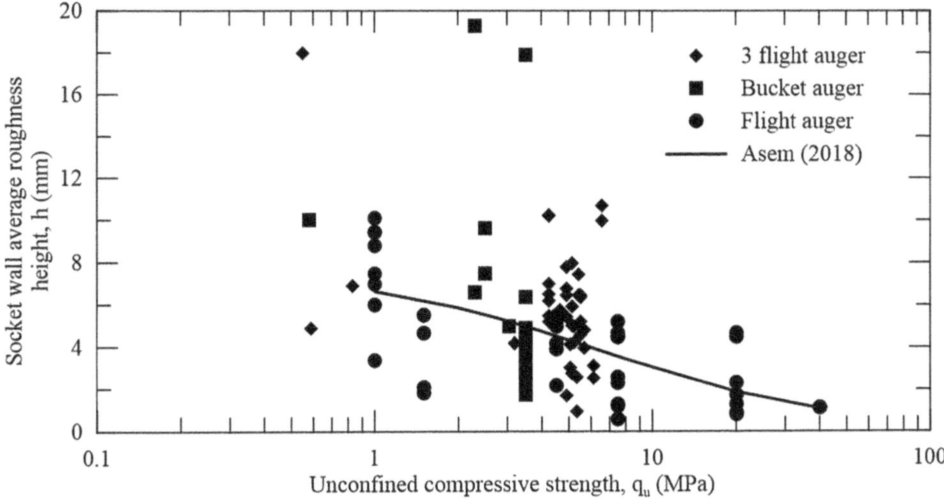

Figure 2.4 Predictive model for the drilled shaft/rock interface roughness (Source: data from Collingwood 2000; Asem 2018).

Table 2.2 Summary of the existing empirical methods for side resistance of rock sockets

Reference	*Equation*	*Comments*
Horvath and Kenney (1979)	$f_{sp} = a\sqrt{q_u}(\text{MPa})$	(i) Large- and small-scale drilled shafts and rock anchors, (ii) rocks include shale, sandstone, limestone, chalk, and igneous and metamorphic rocks, (iii) $0.35 < q_u < 110$ MPa, and (iv) $0.2 < a < 0.25$ (large diameter) and $0.25 < a < 0.33$ (small diameter), where parameter a is an empirical factor to account for the socket sidewall roughness and is obtained by fitting $f_{sp} = a\sqrt{q_u}(\text{MPa})$ to f_{sp} and q_u data.
Pells *et al.* (1979)	$f_{sp} = \alpha q_u$	(i) Load tests in Hawkesbury sandstone in the Sydney, Australia, (ii) $6 < q_u < 40$ MPa, (iii) $75 < B < 710$ mm, (iv) $240 < D_{GS} < 1370$ mm, (v) $1 < h < 10$ mm, and (vi) $0.1 < \alpha$ $(= f_{sp}/q_u) < 0.3$, where α is the ratio of f_{sp} to q_u.
Williams (1980)	$\frac{f_{sp}}{\sigma_p} = 2.69\left(\frac{q_u}{\sigma_p}\right)^{0.28}$	(i) Weak Melbourne siltstone at Stanley Avenue, Middleborough Road, and Westgate Highway test sites, (ii) $0.4 < q_u < 80$ MPa, (iii) $0.335 < B < 1.35$ m, and (iv) $D_{GS} = D_{TOR} < 2$ m.
Kaderabek and Reynolds (1981)	$f_{sp} = 0.3q_u$	(i) Miami limestone, (ii) $q_u < 8.6$ MPa, and (iii) $50 <$ rock quality designation (RQD) $< 80\%$ in the upper part of the formation and usually is near 0 in the lowest portion of the formation.
Rowe and Armitage (1984)	$f_{sp} = 0.45\sqrt{q_u}(\text{MPa})$ for smooth and $f_{sp} = 0.6\sqrt{q_u}(\text{MPa})$ for rough rock sockets	(i) Sandstone, mudstone, shale, siltstone, chalk, and andesite, (ii) $0.55 < q_u < 32$ MPa, (iii) $0.1 < B < 1.12$ m, and (iv) $0.2 < D_{GS} < 10$ m.
Stark *et al.* (2013), Stark *et al.* (2017), and Baghdady (2018)	$f_{sp} = 0.3q_u$	(i) Stark *et al.* (2013) developed a database of 54 drilled shaft load tests mostly in Intermediate Geomaterial (IGMs) (i.e., shale, mudstone, siltstone, and claystone with $0.13 < q_u < 3.2$ MPa), (ii) $0.3 < B < 2$ m, and (iii) Stark *et al.* (2017) and Baghdady (2018) updated the method of Stark *et al.* (2013) using new load tests (e.g., Vu 2013) where $f_{sp} = 0.31q_u$.

it governs the development and maintenance of the normal stresses on the sidewalls of rock sockets. Field observations (Williams 1980), however, showed that the shear surface most likely forms within the rock mass and not at the shaft/rock interface, especially in weak rocks. Therefore, the shaft/rock interface roughness may not be as important as is implied from the laboratory tests conducted by previous investigators. Other investigators have proposed similar empirical models to that of Rosenberg and Journeaux (1976) for side resistance of rock sockets in weak rocks. These models are summarized in Table 2.2 and evaluated in Figure 2.3.

Asem (2018) proposed two alternative models that account for the rock socket geometry and rock mass properties. The model in Asem (2018) is written as:

$$f_{sp} = \frac{q_u^{0.22} E_m^{0.31}}{12.8L^{0.79}} \tag{2.6}$$

where L is the length of the rock socket. Alternatively, a more refined approach in Asem (2018) is based on the Mohr–Coulomb model and is written as:

$$f_{sp} = \sigma_{np} \tan \phi_m \tag{2.7}$$

where σ_{np} is the mobilized normal stress on the rock socket sidewalls at the instant of mobilization of f_{sp} and may be obtained as:

$$\sigma_{np} = \frac{0.15 E_m^{0.37}}{L^{0.77} B^{0.01}} \tag{2.8}$$

These models (for Asem 2018; Equation 2.7) are evaluated in Figure 2.3.

2.3.2 Semi-empirical side resistance models

Hassan (1994) and Hassan *et al.* (1997) proposed that rock/concrete interface has a sinusoidal roughness pattern when the rock socket is constructed with rock auger in clayshale and distinguished between smooth and rough sockets. Hassan (1994) proposed that for the case of rough sockets, the shear plane is within the weak rock material and thus the unit side resistance (f_{sp}) is equal to the average drained shear strength of the geomaterial. Hassan (1994) and Hassan *et al.* (1997) proposed the following expression for obtaining the peak side resistance (f_{sp}) of rock sockets with rough rock/concrete interfaces:

$$f_{sp} = c_m + \sigma_{no} \tan \phi_m \tag{2.9}$$

where it is assumed that c_m and ϕ_m are the drained shear strength intercept and friction angle of rock mass, and σ_{no} is the initial normal stress on the shear surface. For smooth rock/concrete interfaces, Hassan (1994) and Hassan *et al.* (1997) proposed that the peak unit side resistance (f_{sp}) is mobilized at very small displacements (i.e., local displacements of 5–10 mm). They also assumed that the shear surface for smooth interfaces is at the rock/concrete interface and suggested that f_{sp} in smooth sockets is governed by initial normal stress on the interface and interface friction angle. For smooth rock/concrete interface, Hassan (1994) suggested the following expression:

$$f_{sp} = \alpha q_u \tag{2.10}$$

where α is the adhesion factor written as:

$$\alpha = (5 - 8.8d)\left(\frac{q_u}{\sigma_p}\right)^{d-1} \frac{\tan \phi_{int}}{\tan 30^o} \tag{2.11}$$

in which ϕ_{int} is the interface friction angle of rock/concrete interface, and

$$d = \frac{15 - \dfrac{\sigma_{no}}{\sigma_p}}{27} \tag{2.12}$$

In relation to the effect of stresses on side resistance, Vesic (1963) indicated that side resistance is a "linear function of vertical stress at failure." Vesic (1963) further maintained that "this stress is not necessarily equal nor proportional to the overburden pressure." Contrary to the proposal of Hassan (1994), rock socket load test data show no

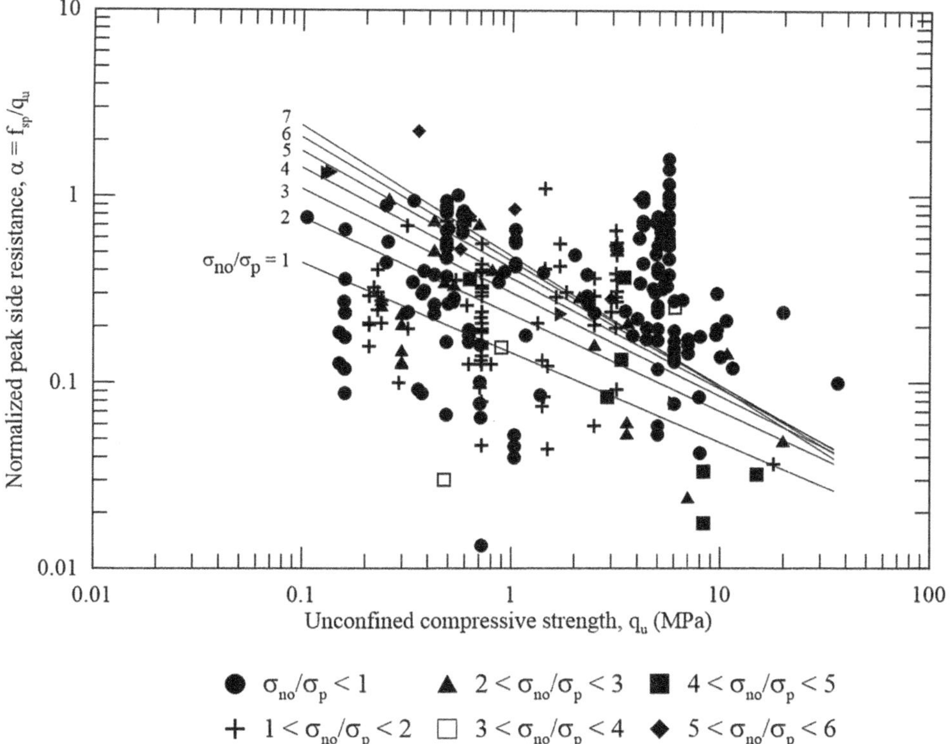

Figure 2.5 Evaluation of the model of Hassan (1994) and Hassan *et al.* (1997) using the in situ drilled shaft load test database (Source: data from Asem 2018).

relationship between peak side resistance and initial normal stress on the mobilized shear surface, especially when the rock socket sidewalls are relatively rough and large dilations are expected to take place as a result. The method of Hassan (1994) is evaluated in Figure 2.5, which shows that data do not provide a clear trend as suggested by the Hassan (1994) method.

Alternatively, the design model of Seidel and Collingwood (2001) is semi-empirical and is based on the analysis using the ROCKET computer program, which is based on the mechanistic models of Seidel (1993) and other researchers (e.g., Lam 1983; Haberfield 1987; Johnston and Lam 1989; Kodikara 1989; Seidel 1993; Hassan 1994; Collingwood 2000; Seidel and Collingwood 2001; Haberfield and Lochaden 2018). These mechanistic models were developed based on the observations of the behavior of natural and synthetic rock/concrete interfaces in drained constant normal stiffness (CNS) direct shear tests. It is noted that the bonding between the synthetic rock and the concrete was sometimes prevented in the preparation of the specimens (e.g., Seidel 1993). The method accounts for the effects of socket geometry (that is modeled using the normal stiffness, K_{ns}, Seidel and Collingwood 2001), q_u, the equivalent elastic properties of rock mass (i.e., modulus of deformation, E_m, and Poisson's ratio, ν), and the socket sidewall average roughness height (h). The method of Seidel and Collingwood (2001) is shown in Figure 2.6. The shaft resistance coefficient (SRC) in Figure 2.6 is computed as:

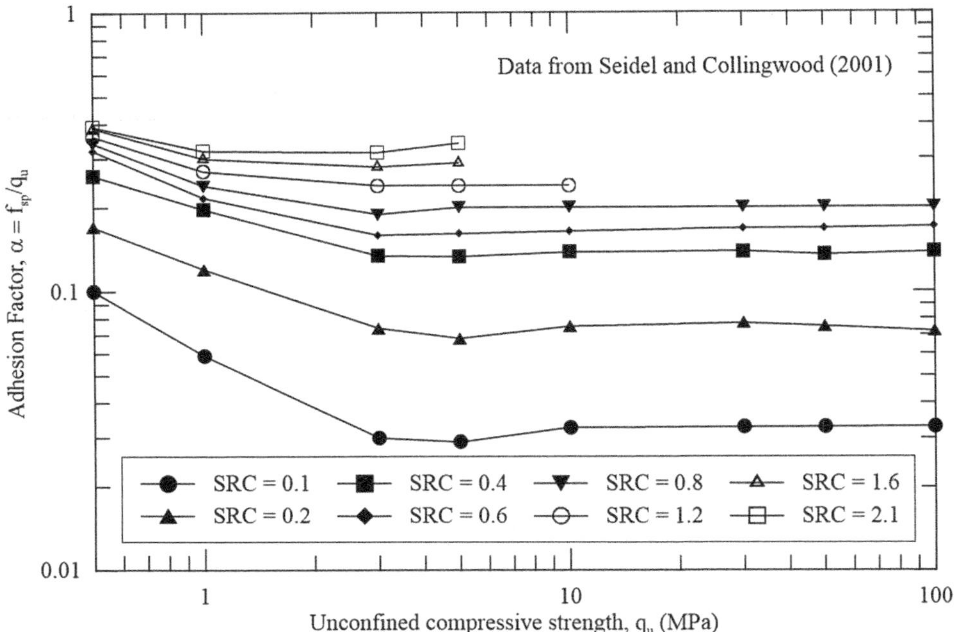

Figure 2.6 Design method of Seidel and Collingwood (2001) (Source: reproduced from Seidel and Collingwood 2001).

$$\mathrm{SRC} = \eta \frac{\frac{E_m}{q_u}}{1+\nu} \frac{h}{B} \tag{2.13}$$

where η is the construction method reduction factor. One of the important parameters that is accounted for in this model is the roughness of rock/socket interface. The roughness model is based on idealizing the interface roughness by a series of interconnected cords. The assumption is made that the distribution of the cord angle is Gaussian, and thus the standard deviation of roughness height, cord angle, and roughness angle is related as:

$$\sigma_h = l_c \sin \sigma_i \tag{2.14}$$

where σ_h is the standard deviation of roughness height, l_c is the cord length, and σ_i is the standard deviation of roughness angle. The model of Seidel and Collingwood (2001) is evaluated using the in situ load test data in Figure 2.7, which shows a reasonable degree of agreement between measured and predicted values.

2.3.3 Discussion

The previous sections presented a review of the existing literature on the predictive models for the peak side resistance (f_{sp}) of rock sockets. For each model, we compute the bias (λ) defined as:

$$\lambda = \frac{M'}{P'} \tag{2.15}$$

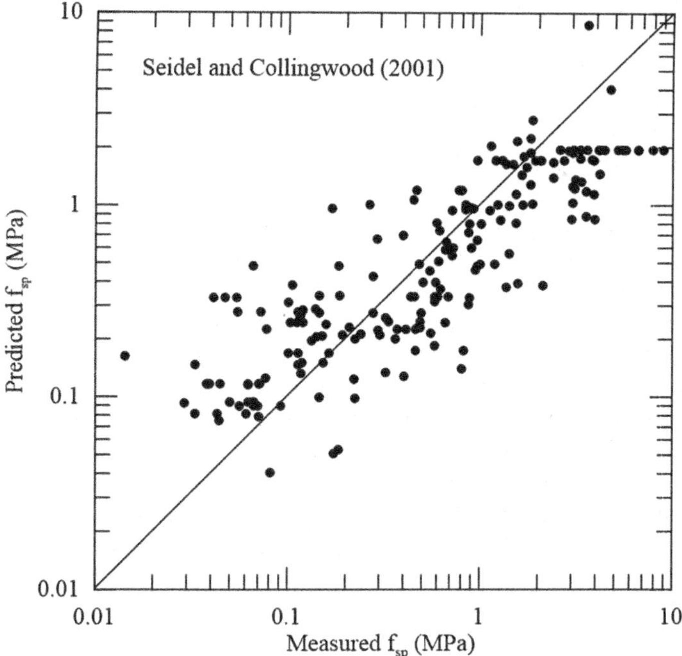

Figure 2.7 Evaluation of design method of Seidel and Collingwood (2001) using the in situ load test data (Source: data from Asem 2018).

where M' is the measured quantity of interest and P' is the predicted quantity of interest. The bias (λ) for the method of Rowe and Armitage (1984) is plotted versus various rock mass and rock socket properties in Figure 2.8 as an example that shows that λ varies with B, q_u, K_{ns}, the shear length (L) and ϕ_m that suggests $\lambda = f(B, q_u, K_{ns}, L)$. Further, because q_u is included in the method of Rowe and Armitage (1984), the variation of λ with q_u indicates that the model form is not appropriate. The existing models are also evaluated using other databases compiled by other researchers (Horvath and Kenney 1979; Rowe and Armitage 1984; Stark *et al.* 2013); the results are presented in Table 2.3. The analyses based on our new database are summarized in Table 2.4. The mean of bias (μ_λ) is:

$$\mu_\lambda = \frac{1}{n}\sum_{i=1}^{n}\left(\frac{M'}{P'}\right)_i \tag{2.16}$$

where n is the number of observations in each database. Similarly, the variance of $\frac{M'}{P'}$ (or other quantities of interest such as initial normal stiffness or yield pressure as defined later in this paper) is estimated using the following expression (Ang and Tang 2007):

$$\sigma_\lambda^2 = \frac{1}{n}\sum_{i=1}^{n}\left[\left(\frac{M'}{P'}\right)_i - \mu_\lambda\right]^2 \tag{2.17}$$

The square root of the variance is the standard deviation (σ_λ) of λ for each predictive model. The coefficient of variation (c.o.v.) for bias (δ_λ) is a measure of "dispersion" of the $\frac{M'}{P'}$ ratio about its mean which is indicative of the model accuracy:

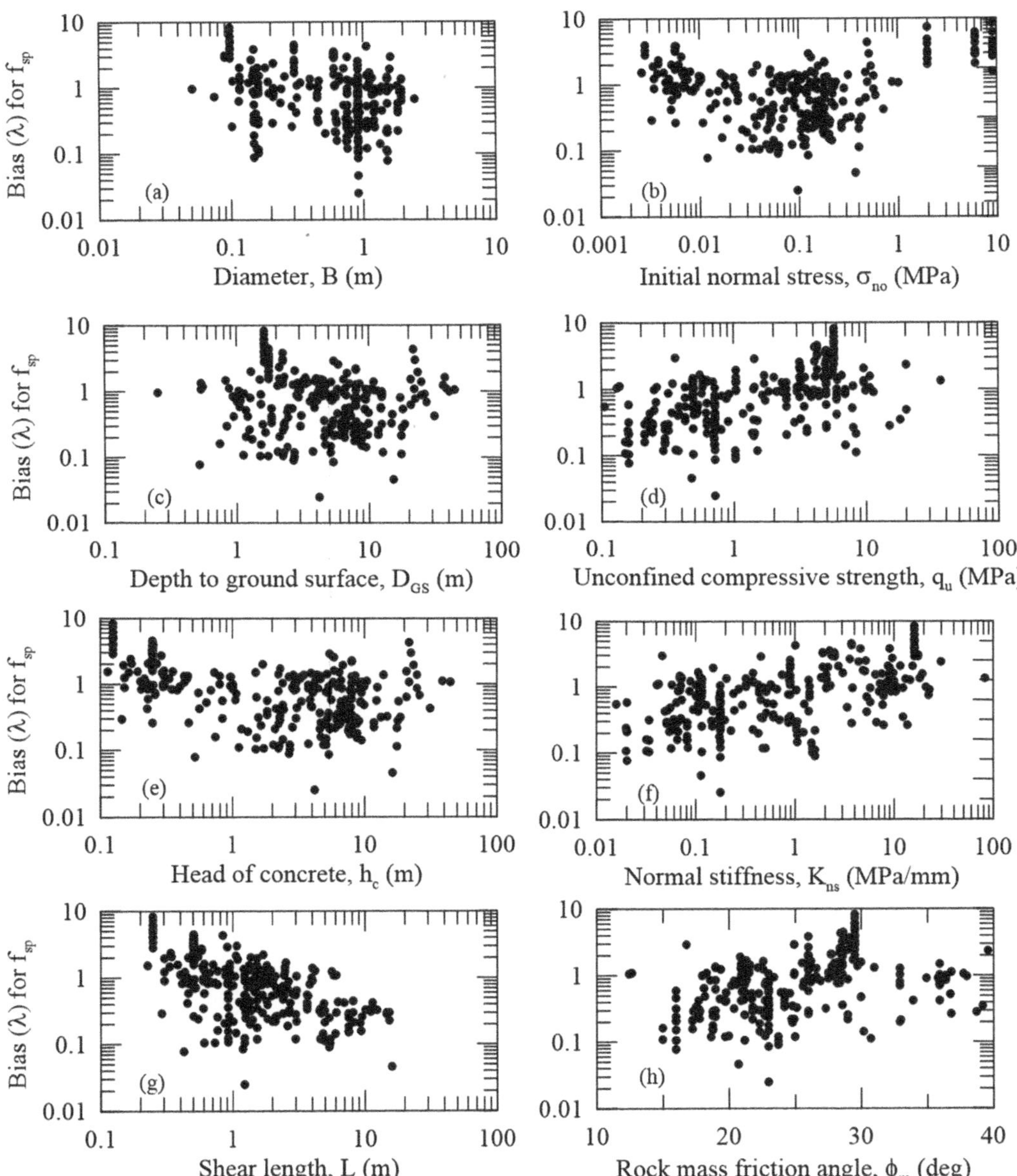

Figure 2.8 Evaluation of the variation of bias (λ) for the design method of Rowe and Armitage (1984) using the in situ load test data (Source: data from Asem 2018).

$$\delta_\lambda = \frac{\sigma_\lambda}{\mu_\lambda} \tag{2.18}$$

The data presented in Tables 2.3 and 2.4 suggest that most of the available models are biased and have high c.o.v.

Table 2.3 Evaluation of the side resistance predictive models using databases of Horvath and Kenney (1979), Rowe and Armitage (1984), and Stark *et al.* (2013)

Design methods		*Databases*		
		Horvath and Kenney (1979)	*Rowe and Armitage (1984)*	*Stark* ***et al.*** *(2013)*
Rosenberg and Journeaux (1976)	μ_λ	0.83	1.15	1.08
	σ_λ	0.45	0.58	0.44
	δ_λ	0.55	0.5	0.41
Horvath and Kenney (1979)	μ_λ	1.24	1.73	1.57
	σ_λ	0.68	0.87	0.65
	δ_λ	0.55	0.5	0.41
Pells *et al.* (1979)	μ_λ	0.89	1.07	1.78
	σ_λ	0.72	0.95	0.88
	δ_λ	0.8	0.89	0.5
Williams (1980)	μ_λ	0.89	1.3	0.96
	σ_λ	0.54	0.71	0.42
	δ_λ	0.6	0.54	0.44
Kaderabek and Reynolds (1981)	μ_λ	0.59	0.71	1.18
	σ_λ	0.48	0.63	0.59
	δ_λ	0.8	0.89	0.5
Rowe and Armitage (1984)	μ_λ	0.69	0.96	0.87
	σ_λ	0.38	0.48	0.36
	δ_λ	0.55	0.5	0.41
Carter and Kulhawy (1988)	μ_λ	1.54	2.16	1.96
	σ_λ	0.85	1.09	0.81
	δ_λ	0.55	0.5	0.41
Hassan (1994) for rough interfaces)	μ_λ	0.36	0.43	0.71
	σ_λ	0.29	0.38	0.35
	δ_λ	0.8	0.89	0.5
Miller (2003)	μ_λ	0.77	1.08	0.98
	σ_λ	0.43	0.54	0.4
	δ_λ	0.55	0.5	0.41
Kulhawy *et al.* (2005)	μ_λ	0.97	1.08	1.24
	σ_λ	0.54	0.54	0.51
	δ_λ	0.55	0.5	0.41
Stark, Long and Assem (2013)	μ_λ	0.59	0.71	1.18
	σ_λ	0.48	0.63	0.59
	δ_λ	0.8	0.89	0.5

2.4 BASE RESISTANCE METHODS

In weak rock masses, a significant portion of the axial load is carried in base resistance in short sockets (e.g., when $E_m/E_c < 0.02$; Seo and Prezzi 2008) after peak side resistance (f_{sp}) is mobilized and full-slip condition has materialized. Mobilization of the failure mechanism in base resistance requires a significant amount of displacement at the base

Table 2.4 Evaluation of the side resistance predictive models using new databases (this study) with 279 measurements at peak side shear stress

Design methods		Database (Asem 2018)			
		35 > GSI	35 < GSI <50	50 < GSI <70	Entire database
Rosenberg and Journeaux (1976)	μ_λ	0.77	2.41	1.12	1.40
	σ_λ	0.63	1.94	0.78	1.50
	δ_λ	0.82	0.81	0.69	1.07
Horvath and Kenney (1979)	μ_λ	1.11	3.60	1.71	2.07
	σ_λ	0.91	2.91	1.19	2.25
	δ_λ	0.82	0.81	0.70	1.09
Pells *et al.* (1979)	μ_λ	1.89	2.08	0.62	1.88
	σ_λ	1.55	1.54	0.42	1.54
	δ_λ	1.89	0.74	0.68	0.82
Williams (1980)	μ_λ	0.61	2.57	1.39	1.38
	σ_λ	0.52	2.12	0.99	1.65
	δ_λ	0.85	0.83	0.71	1.20
Kaderabek and Reynolds (1981)	μ_λ	1.26	1.38	0.41	1.25
	σ_λ	1.03	1.03	0.28	1.02
	δ_λ	0.82	0.74	0.68	0.82
Rowe and Armitage (1984)	μ_λ	0.62	2.00	0.95	1.15
	σ_λ	0.51	1.62	0.66	1.25
	δ_λ	0.82	0.81	0.70	1.09
Carter and Kulhawy (1988)	μ_λ	1.39	4.50	2.14	2.58
	σ_λ	1.14	3.64	1.49	2.81
	δ_λ	0.82	0.81	0.70	1.09
Hassan (1994) for rough interfaces)	μ_λ	0.75	0.83	0.25	0.75
	σ_λ	0.62	0.62	0.17	0.61
	δ_λ	0.82	0.74	0.68	0.82
Hassan (1994) for smooth interface)	μ_λ	3.35	9.47	6.47	5.55
	σ_λ	3.59	9.43	6.26	6.87
	δ_λ	1.07	1.00	0.97	1.24
Seidel and Collingwood (2001)	μ_λ	1.25	1.65	0.64	1.36
	σ_λ	1.11	1.14	0.58	1.13
	δ_λ	0.89	0.69	0.90	0.83
Kulhawy *et al.* (2005)	μ_λ	0.87	2.84	1.35	1.63
	σ_λ	0.72	2.30	0.94	1.78
	δ_λ	0.82	0.81	0.70	1.09
Stark *et al.* (2013), Stark *et al.* (2017), Baghdady (2018)	μ_λ	1.26	1.38	0.41	1.25
	σ_λ	1.03	1.03	0.28	1.02
	δ_λ	0.82	0.82	0.68	0.82

of the rock socket that is significantly greater than that required for mobilization of the peak side resistance (f_{sp}). The available design models are of empirical and semi-empirical nature.

Empirical models are based on in situ or laboratory load tests (e.g., Teng 1962; Coates 1967; Rowe and Armitage 1987; Carter and Kulhawy 1988; ARGEMA 1992; Zhang and Einstein 1998; CGS 2006; Stark *et al.* 2013; Stark *et al.* 2017; Baghdady 2018; Asem 2018; Johnston 2020). Due to the large displacement required for the development of the failure mechanism in the base resistance, most of the load tests that were used in the development of empirical models did not reach failure, and thus the maximum mobilized base contact pressure was often used to develop predictive models for base pressure (q) (e.g., Zhang and Einstein 1998). Therefore, the failure criteria vary from one predictive model to another. In many cases, the failure is based on serviceability requirements, where the resulting stresses correspond to different rock socket axial displacements. Thus, the resulting design models cannot be readily compared and the predicted base resistance using these models for one drilled shaft and a given in situ condition can vary by several orders of magnitude (Zhang and Einstein 1998). Other researchers have developed semi-empirical predictive models for the base resistance based on an assumed failure mechanism (e.g., Bishnoi 1968; Turner 2006) or by inconsistently combining an assumed failure mechanism and the results of in situ load tests (e.g., Stark *et al.* 2013; Stark *et al.* 2017; Baghdady 2018). Therefore, a clear understanding of the assumptions and failure criteria (i.e., definitions) in the existing methods is a prerequisite to their evaluation and use. Some of the suggested predictive models are shown in Figures 2.9 and 2.10 and are compared with q_y and q_f. Additionally, the assumed failure mechanism (e.g., general shear failure mode) is often not mobilized due to resistance to the amount of allowable deformation (Terzaghi 1943) or an increase in confining pressure that promotes a local type of failure. Such change in failure mode with change in confining pressure has been observed in the literature.

2.4.1 Empirical base resistance models

Teng (1962) discussed the base resistance of rock sockets in terms of allowable bearing pressure as:

$$q_{all} = \frac{1}{5} to \frac{1}{8} q_u \tag{2.19}$$

where q_{all} is the allowable unit base resistance. Teng (1962) pointed out that the evaluation of the effect of embedment on the base resistance becomes exceedingly difficult in weak rock because of the presence of joints, and bedding planes and fissures. Therefore, the effect of embedment depth is usually conservatively discounted. Additionally, it is unclear what displacement is required for mobilization of the base resistance proposed by Teng (1962). Other investigators have proposed similar models for predicting the bearing capacity of foundations on rocks. These models are reviewed in Table 2.5 and are evaluated in Figure 2.11.

2.4.2 Semi-empirical base resistance models

Coates (1967) developed a model based on Griffith's fracture theory (Griffith 1921; Hoek 1965), suggesting that microcracks in the rock mass migrate and expand when the stresses along their edges grow and become equal to the tensile strength of the material. Coates (1967) used a wedge analysis in combination with Griffith's fracture theory for the determination of the bearing capacity of strip foundations to write:

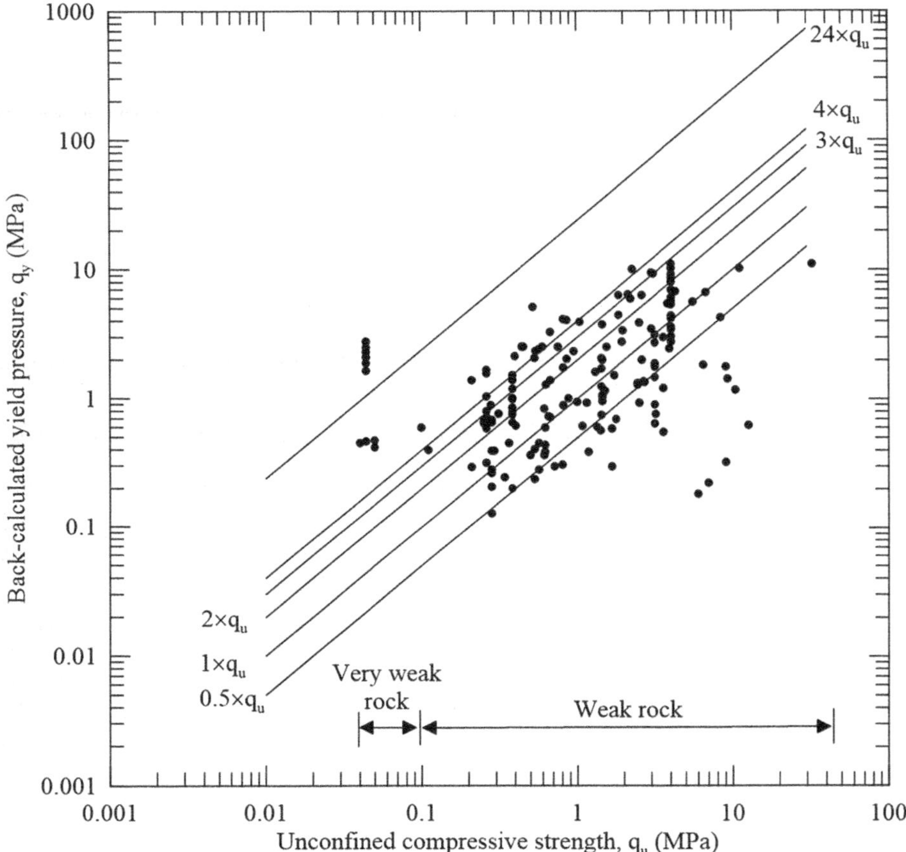

Figure 2.9 Variation of the available theories for base resistance (q) and a comparison with measured yield pressure (q_y) (Source: data from Asem 2018).

$$q = 3q_u \tag{2.20}$$

where q is the base resistance of a long foundation. The model of Coates (1967) is developed for hard rock and is based on the assumption that the strength of rock is mobilized along the entire failure surface at the same time. In reality, however, the failure in brittle materials (e.g., rock) would initiate from regions of stress concentration and subsequently propagate to less severely stressed zones (Hoek 1965; Coates 1967). The method of Coates (1967) assumes that the surcharge loads representing the embedment depth are small compared to the applied loads and thus do not contribute greatly to the bearing capacity and do not account for the effect of foundation depth on the bearing capacity of foundations. Other researchers (e.g., Zhang and Einstein 1998) have made similar conclusions about the tip resistance of sockets in weak rock and in sand (Vesic 1963).

The method of Carter and Kulhawy (1988) is based on the Hoek–Brown failure criterion for jointed rock mass and was developed for foundations placed on the surface of rock mass (Figure 2.12). It is written as:

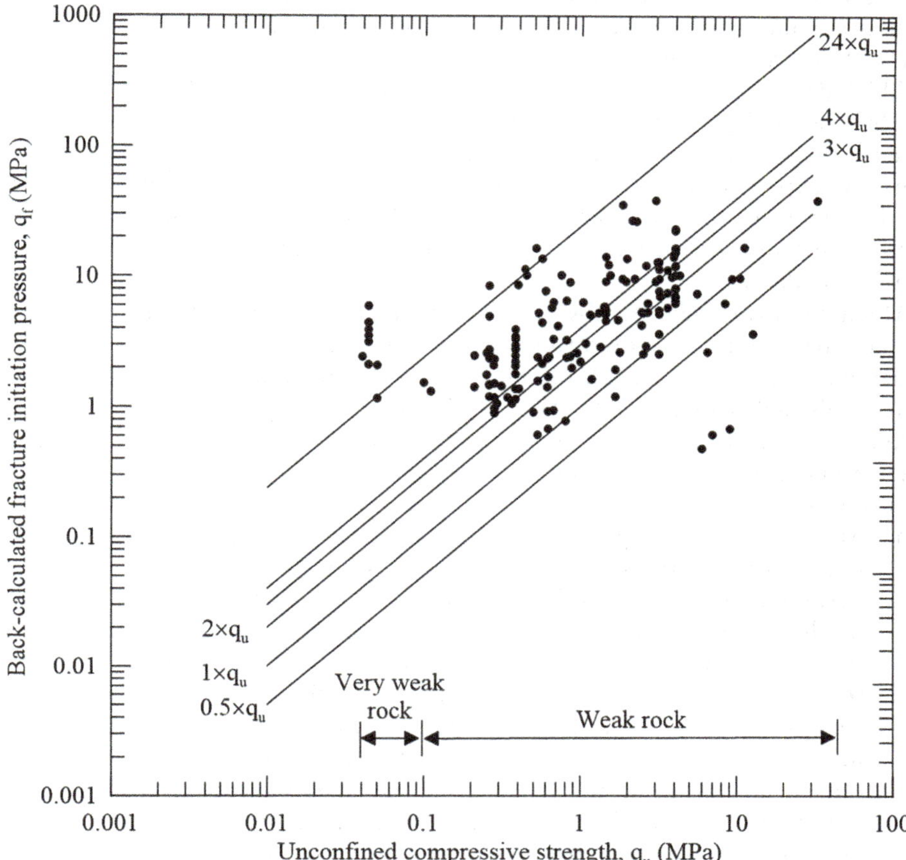

Figure 2.10 Variation of the available theories for base resistance (q) and a comparison with measured fracture initiation pressure (q_f) (Source: data from Asem 2018).

$$q = (m + \sqrt{s})q_u \tag{2.21}$$

where q is the lower bound estimate of base resistance for rock sockets, and "m" and "s" are the material constants from the Hoek–Brown (Hoek and Brown 1980, 2019) criterion. The method in Carter and Kulhawy (1988), however, is not based on rock socket load tests but may be classified as a semi-empirical method only due to material constants "m" and "s," which are determined empirically. As for the previous methods, it is unclear what displacements are required for mobilization of the base resistance proposed by Carter and Kulhawy (1988), and thus compatibility with side resistance cannot be checked. Other investigators have developed semi-empirical models which are summarized in Table 2.6. The base resistance models are evaluated in Table 2.7.

2.4.3 Relationships with rock mass properties

Peck *et al.* (1974) proposed a relationship between the allowable bearing pressure of foundations on unweathered and sound rock masses and the rock quality designation (RQD).

Table 2.5 Summary of empirical base resistance models

Reference	*Equation*	*Comments*
Rowe and Armitage (1987)	$q = 2.5q_u$	(i) Based on in situ load tests by Horvath (1980), Glos and Briggs (1983), and Williams (1980), (ii) 0.3 < *B* < 1 m, (iii) 0.65 < q_u < 9.26 MPa, and (iv) rock types in this database include siltstone, sandstone, and shalely sandstone.
Zhang and Einstein (1998)	$q = 4.8\sqrt{q_u(\text{MPa})}$	(i) Based on a database of 39 drilled shaft load tests in mudstone, shale, gypsum, till, diabase, hardpan, sandstone, siltstone, marl, and limestone, (ii) 0.3 < *B* < 1.92 m, (iii) 1 < D_{GS} < 20.7 m, (iv) 0.52 < q_u < 55 MPa, (v) 6 load tests reported are model centrifuge tests (Leung and Ko 1993) on synthetic rock, and (vi) 3 load tests from Carrubba (1997) were not instrumented and base resistance had to be back-analyzed using elastic theory.
Stark *et al.* (2013), Stark *et al.* (2017), Baghdady (2018)	$q = 2.5q_u d_c$ depth factor (d_c) after Vesic (1973)	(i) Based on 33 drilled shaft load tests where the base stress–displacement relationships (q–δ relationships) were back-calculated from the results of drilled shaft load tests, (ii) rocks include shale and clayshale, claystone and siltstone, (iii) 0.34 < q_u < 4.2 MPa, (iv) 0.3 < *B* < 1.82 m, and (v) failure mechanism in the base resistance is developed at a displacement of 5% of the drilled shaft tip diameter.

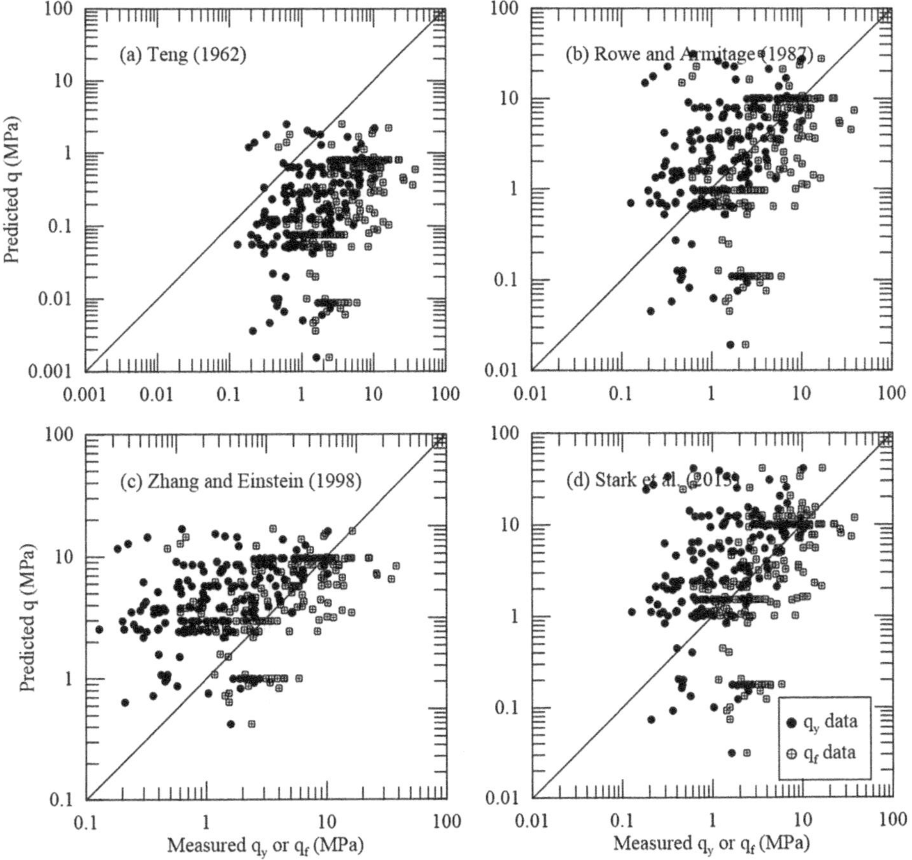

Figure 2.11 Evaluation of empirical models for base resistance with measured base resistance in in situ load tests.

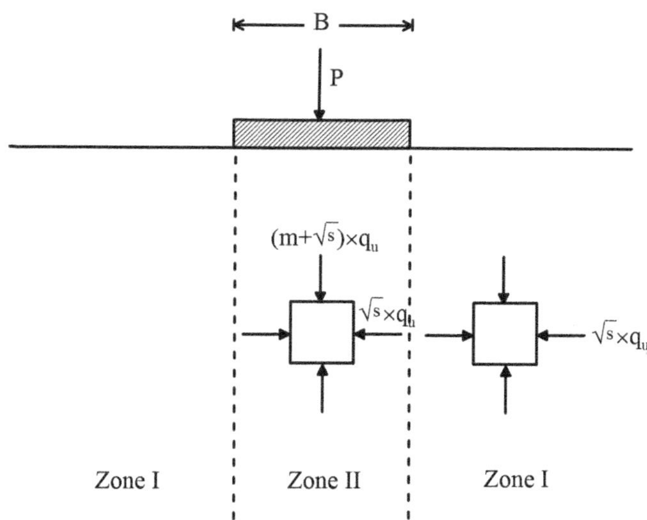

Figure 2.12 Method of Carter and Kulhawy (1988) for the end resistance of foundations on rock mass (Souce: reproduced from Carter and Kulhawy 1988).

Table 2.6 Summary of semi-empirical base resistance models

Reference	*Equation*	*Comments*
Canadian Geotechnical Society (2006)	$q = K_{sp}q_u$ $K_{sp} = \dfrac{3 + S_J / B}{10\sqrt{1 + 300\omega / S_J}}$	(i) q is the allowable bearing capacity of foundation, (ii) K_{sp} is an empirical coefficient that includes a factor of safety of 3.0 (K_{sp} that ranges from 0.1 to 0.4), and (iii) CGS defines the "favorable conditions" as conditions where "i) the rock surface is perpendicular to the foundation, ii) the foundation load has no tangential component and iii) the rock mass has no open discontinuities." $0.05 < S_J/B < 2$ and $0 < \omega/S_J < 0.02$.
Lee *et al.* (2013)	$q = 79.2\left(\dfrac{q_u S_J}{\sigma_p B}\right)^{0.315}\sigma_p$	(i) Parametric studies using numerical models and (ii) in situ load test data from South Korea

The use of this correlation (see Table 2.8 or Peck *et al.* 1974; Kulhawy and Goodman 1980) results in foundation settlements that are less than 13 mm (Peck *et al.* 1974; Kulhawy and Goodman 1980). GSI of rock mass for each load test is converted to RQD based on the recommendations of Hoek *et al.* (2013). The relationship proposed by Peck *et al.* (1974) is plotted in Figure 2.13. q_y and q_f from the base-resistance database case histories interpreted using the Terzaghi (1943) and Hirany (1988) methods and the estimated RQD values are also shown. The comparison of the data from 190 load tests (i.e., plate, O-cell, and rock socket load tests) with the proposed method of Peck *et al.* (1974) in Figure 2.13 shows that this method only provides crude estimates of q_f and q_y and that significant scatter exists around the mean correlation proposed by Peck *et al.* (1974). The data scatter in Figure 2.13 indicates that in addition to the RQD, that is mainly a measure of the joint frequency (Peck *et al.* 1974; Pells *et al.*, 2017) in the weak rock mass, other rock mass properties (i.e., rock mass friction angle ϕ_m) as well as joint and fissure alteration, socket geometry (i.e., diameter), and displacement to mobilize q_f may also be

Table 2.7 Evaluation of base resistance models using existing models

Design methods		Databases		
		Paikowsky et al. (2010)	*Stark et al. (2013)*	*Zhang and Einstein (1998)*
Teng (1962)	μ_λ	24.64	20.32	17.79
	σ_λ	24.98	10.97	10.86
	δ_λ	1.01	0.54	0.61
Coates (1967)	μ_λ	1.64	1.35	1.19
	σ_λ	1.67	0.73	0.72
	δ_λ	1.01	0.54	0.61
Rowe and Armitage (1987)	μ_λ	1.97	1.63	1.42
	σ_λ	2	0.88	0.87
	δ_λ	1.01	0.54	0.61
Carter and Kulhawy (1988)	μ_λ	23.46	19.35	16.94
	σ_λ	23.79	10.45	10.34
	δ_λ	1.01	0.54	0.61
ARGEMA (1992)	μ_λ	1.09	0.9	0.79
	σ_λ	1.11	0.49	0.48
	δ_λ	1.01	0.54	0.61
Zhang and Einstein (1998)	μ_λ	1.64	1.01	1.06
	σ_λ	2.82	0.44	0.34
	δ_λ	1.72	0.44	0.32
Canadian Foundation Engineering Manual (2006)	μ_λ	7.51		
	σ_λ	6.92		
	δ_λ	0.92		

Table 2.8 Relationship between allowable foundation pressure and rock quality designation (RQD) for unweathered rock masses

RQD (%)	*Allowable pressure (MPa)*
0	1
25	3
50	6.5
75	12
90	20
100	30

Source: After Peck *et al.*, 1974 or Peck, 1976.

important (Asem 2018; Tang *et al.* 2020). These parameters are missing in the Peck *et al.* (1974) predictive equation. Moreover, it must be noted that the method of Peck *et al.* (1974) was originally developed for "unweathered and sound rock masses" and should be used with caution with weak rocks.

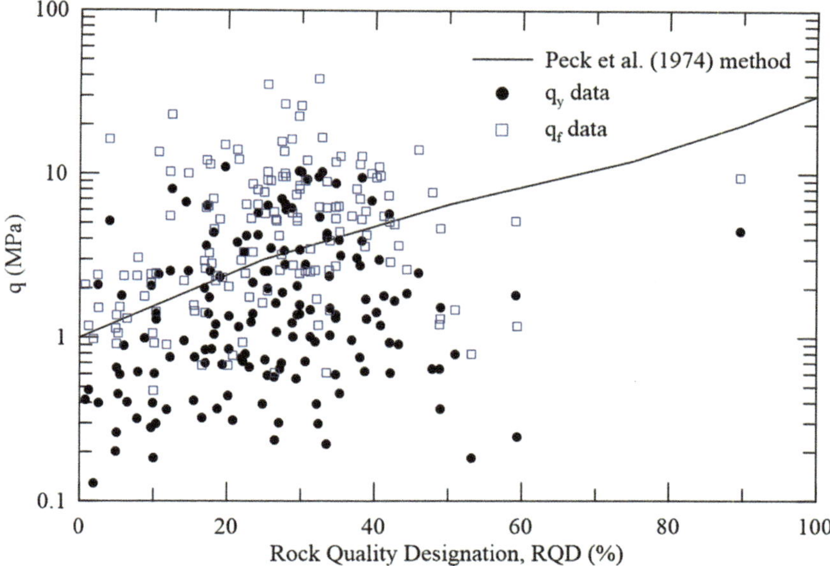

Figure 2.13 Peck *et al.* (1974) method for the prediction of base resistance based on rock quality designation (RQD)

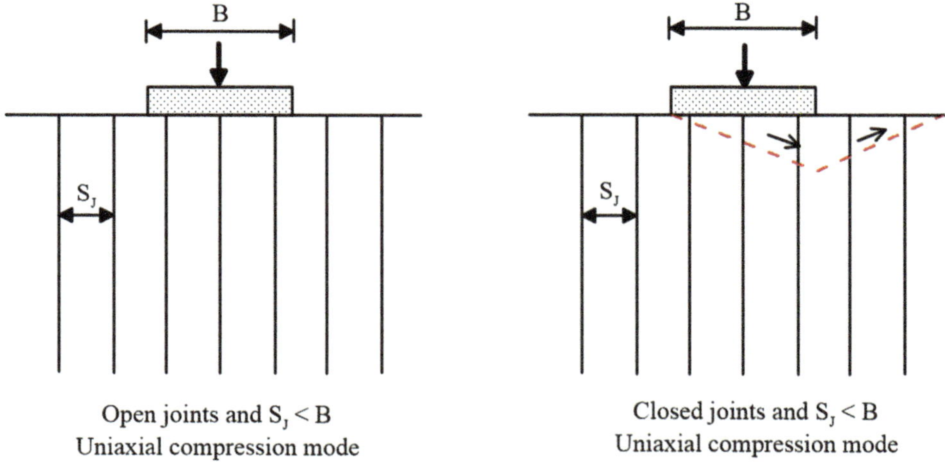

Figure 2.14 Failure modes proposed by Sowers (1979) for shallow foundations on rock masses (Source: reproduced from Kulhawy and Goodman, 1980).

Sowers (1979) solutions for the case of shallow foundations on rock masses for situations where joint spacing (S_J) is less than the foundation size (i.e., rock socket or in general foundation diameter, B) (see Figure 2.14) are discussed below. Accordingly, when rock joints are vertical, open, and $S_J < B$, the bearing capacity is obtained from the following expression:

$$q = q_u = 2c_m \tan\left(45 + \frac{\phi_m}{2}\right) \tag{2.22}$$

and when the joints are vertical, closed, and $S_J < B$, the bearing capacity is evaluated using the following expression:

$$q = c_m N_c s_c + \frac{B}{2}\gamma N_\gamma s_\gamma + \sigma_v N_q s_q \tag{2.23}$$

where all the material parameters in Equations (2.22) and (2.23) are that of the rock mass. N_c, N_γ, and N_q are the bearing capacity factors and s_c, s_γ, and s_q are the shape factors and may be obtained from Sowers (1979). The models proposed by Sowers (1979) are plotted in Figures 2.15 and 2.16 for different values of rock mass cohesion intercept (c_m) and friction angle (ϕ_m). The q_f values (obtained from q–δ relationships collected by Asem 2018) are also superposed on these figures. Examination of Figures 2.15 and 2.16 shows the following:

1. These models cannot properly represent the actual variation of q_f with the rock mass friction angle for sockets in weak rock masses.
2. The significant scatter in the plots of q_f versus friction angle indicates that other parameters in addition to friction angle and cohesion intercept must affect q_f that are missing in the above-proposed models (Asem 2018; Tang *et al.* 2020).

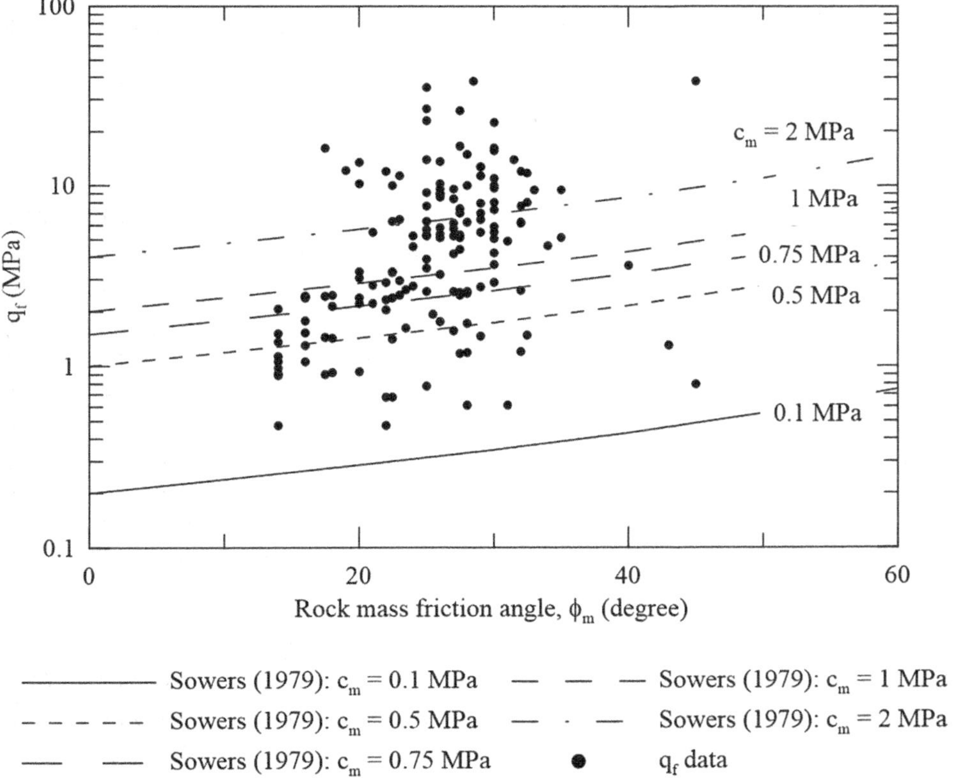

Figure 2.15 Comparison of the model of Sowers (1979) for foundations on rock masses with vertical open joints when $S_J < B$ with the *qf* data from the drilled shaft tip resistance database.

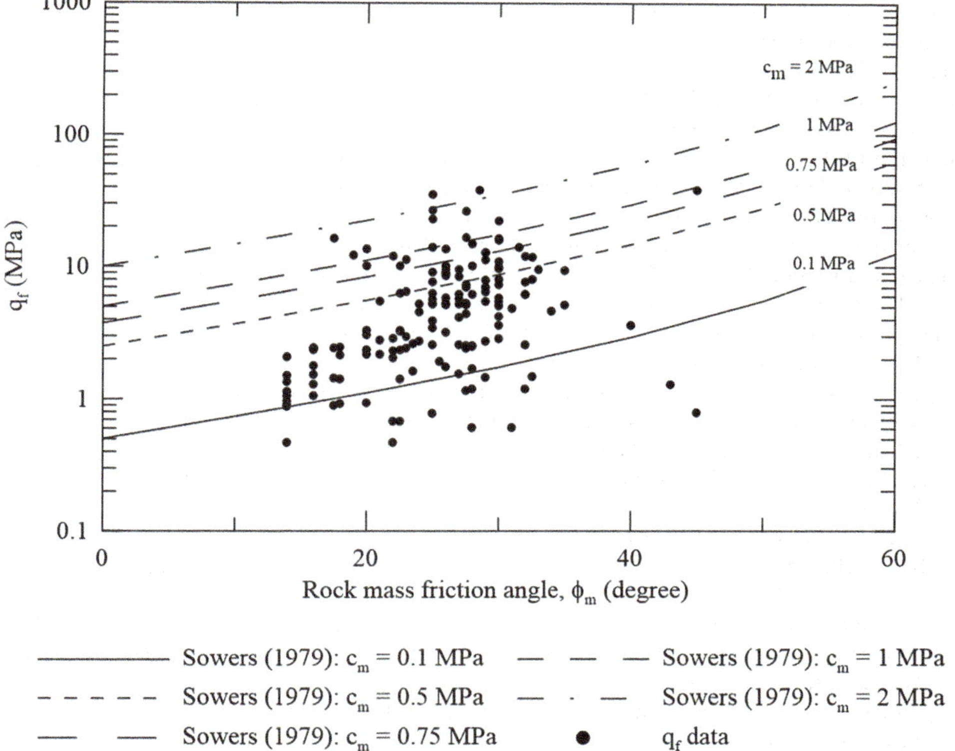

Figure 2.16 Comparison of the model of Sowers (1979) for foundations on rock masses with vertical closed joints when $S_j < B$ with the qf data from the drilled shaft tip resistance database.

3. The inability of Sowers (1979) methods to properly predict the base resistance in weak rocks can also be attributed to the assumed failure mechanism that is used to develop the above-proposed design equations, which is likely to be different from that mobilized in the field.

2.5 SETTLEMENT METHODS

2.5.1 Load-transfer methods

The serviceability limit state is important in the design of foundations on rock masses (Goodman 1980) and should be evaluated carefully. Some researchers developed predictive models based on the theory of elasticity and some used the load-transfer approach (Seed and Reese 1957) to develop nonlinear stress–displacement models for the side and the tip of drilled shafts. The models based on the theory of elasticity relate the settlement to rock socket geometry, rock mass properties, and socket concrete properties. In the load-transfer approach, the stiffness of the shear surface mobilized in the socket walls and the stiffness of the tip of rock socket may be modeled using nonlinear springs (i.e., f_s-δ and q–δ relationships). This model is robust because the effect of socket geometry, rock and concrete mechanical properties, local displacement, rock/concrete interface characteristics (e.g., socket wall roughness), and construction methods can be accounted for. The most important advantage of the load-transfer method is that it accounts for the nonlinearity of the rock behavior, which is an important aspect of the behavior of weathered and weak rocks that cannot be properly addressed when the theory of elasticity (e.g., Pells and Turner 1979; Kulhawy and Carter 1992) is used for evaluating the settlement of foundations in soft rocks unless such effects are reflected in rock elastic properties.

Williams (1980) developed a load-transfer function that is based on the results of field load tests in Melbourne siltstone where side and tip resistances were directly measured. The siltstone at the load test sites is commonly weathered to highly weathered, $0.335 < B < 1.35$ m and $D_{TOR} < 2$ m (the ground surface and the top of weak rock coincided in most of the load tests). Williams (1980) proposed a method for normalizing the f_s–δ and q–δ relationships (i.e., load-transfer functions for side and tip resistance, respectively) for sockets in weak rock. The normalization method, with particular attention to tip resistance, is shown in Figure 2.17 and is applied to 190 load tests in weak rock in Figure 2.18. The Williams (1980) method first predicts the load–displacement response of the foundation using an elastic method. Williams (1980) does not propose a method that could be specifically used for drilled shafts in weak rocks. Williams (1980), however, realizes that the behavior of actual drilled shafts in weathered rocks is highly nonlinear. To account for the effect of nonlinearity observed in the field for back-calculated f_s–δ and q–δ relationships, Williams (1980) defined plastic stress, f_p or q_p (for side and tip, respectively), at

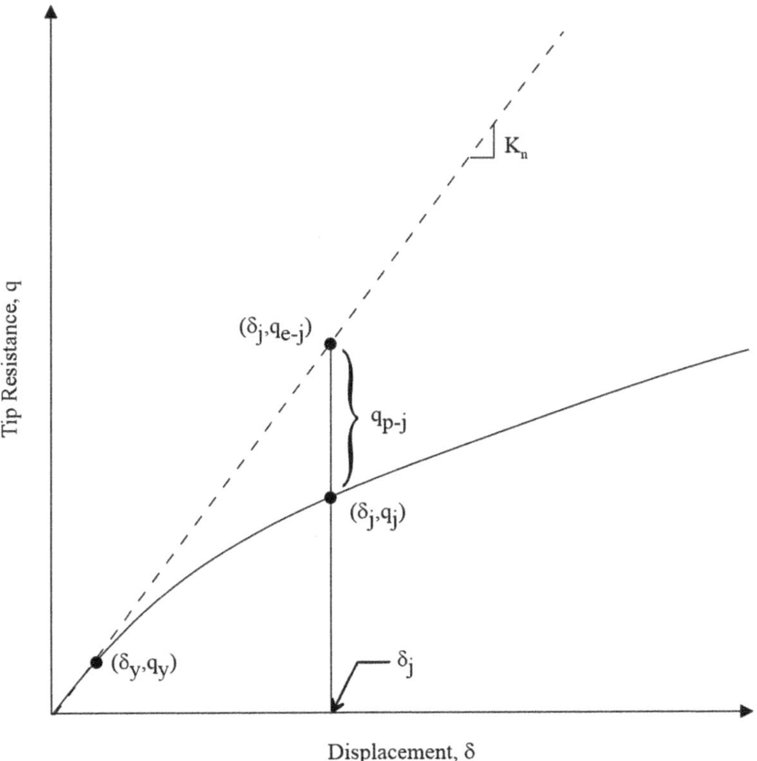

Figure 2.17 Normalization method developed by Williams (1980) (Source: after Williams, 1980, with modifications).

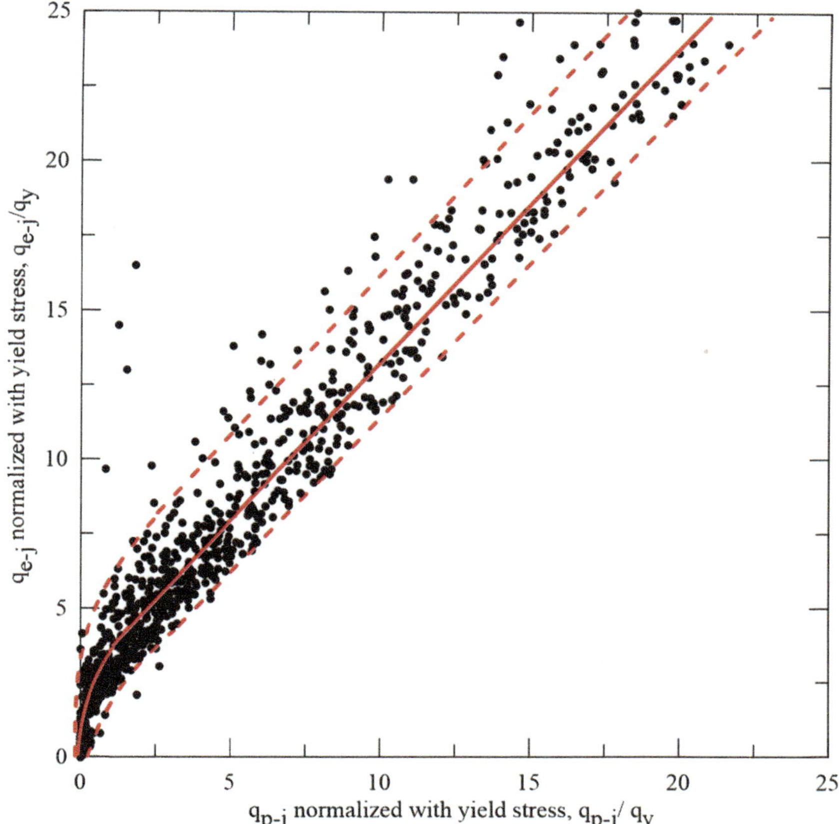

Figure 2.18 Empirical method of Williams (1980) (modified) for prediction of q_p for the purpose of determining nonlinear load-transfer functions for tip resistance (all data from the tip resistance database in Appendix A or Chapter 3 that is compiled from literature, normalization method after Williams, 1980, with modifications and updates).

each increment of axial displacement, which is described in mathematical forms for side and tip resistance in Equations (2.24) and (2.25), respectively as:

$$f_p = f_e - f_s \tag{2.24}$$

and

$$q_p = q_e - q \tag{2.25}$$

where f_e and q_e are the predicted values of side and tip stresses based on the elastic theory at any given value of local displacement along the side and near the tip of the foundation, respectively, and f and q are the corresponding values from the field "nonlinear" load-transfer functions for side and tip resistances, respectively. The q_e values in Figures 2.17 and 2.18 are normalized using the yield pressure (q_y) of the rock mass, which is an alternative to what was proposed by Williams (1980). A similar relationship may be developed for the side resistance f_s–δ relationships. The method of Williams (1980) for normalization of load-transfer functions of sockets in weak rock is a novel approach for the prediction of load-transfer functions. Although this method has been mostly applied to the investigation of load–displacement response of foundations in weak rock, it can be applied to hard rock masses as was done in some of the load tests in relatively hard sedimentary rocks by Williams (1980). Asem (2018) developed a family of equations that could be used to predict the deviation of the tip response from linear elastic behavior written as:

$$\frac{q_{p\text{-}j}}{q_y}=0.007797\left(\frac{q_{e\text{-}j}}{q_y}\right)^3+0.05209\left(\frac{q_{e\text{-}j}}{q_y}\right)^2,\frac{q_{e\text{-}j}}{q_y}\leq 5 \tag{2.26}$$

and

$$\frac{q_{p\text{-}j}}{q_y}=0.9436\left(\frac{q_{e\text{-}j}}{q_y}\right)-2.467,\frac{q_{e\text{-}j}}{q_y}>5 \tag{2.27}$$

The Hirayama (1990) method is based on the modeling of the load-transfer function for unit side resistance and tip resistance using a "two-constant" hyperbolic model written as:

$$f_s=\frac{\delta}{a_f+b_f\delta} \tag{2.28}$$

and

$$q=\frac{\delta}{a_q+b_q\delta} \tag{2.29}$$

where f_s and q are the unit side and base resistances of rock socket and a_f, b_f, a_q, and b_q are model parameters that define the shape of the hyperbolic function. Hirayama relates the model parameters to the "ultimate" side and base resistances of rock sockets. The peak side resistance of rock sockets in weak rock can be estimated using the available predictive models for the unit side resistance of sockets in weak rock. In the case of base resistance, however, the majority of the field load tests have not been carried to failure, and thus reliable methods for the prediction of failure stress in tip resistance of sockets in weak rock are not available. This is an important limitation of Hirayama's method. Hirayama (1990) does not account for the difference between peak side or ultimate base resistance predicted from the hyperbolic model and the actual measured values, and thus does not use a fitting ratio (Mesri *et al.* 1981; Terzaghi *et al.* 1996) (R_f) that has been used to account for this difference between the back-calculated and theoretical values. Other load-transfer models have been developed by other investigators that are summarized in Table 2.9.

2.5.2 Initial shear stiffness and initial normal stiffness models

2.5.2.1 *Evaluation of models for initial shear stiffness in side resistance (K_{si})*

The initial shear stiffness for rock joints has been studied extensively (e.g., Bandis 1980; Bandis *et al.* 1983; Yoshinaka and Yamabe 1986). The problem of determination of the

Table 2.9 Summary of load-transfer models

Reference	*Equation*	*Comments*
Hassan *et al.* (1997), side resistance	$f_s = \Theta_f f_{sp}$ For pre-peak and $f_s = k_f f_{sp}$ For post-peak	(i) k_f and Θ_f are correction factors that account for mobilization of unit side resistance and are functions of displacement, (ii) f_{sp} is the peak unit side resistance, and (iii) k_f and Θ_f could be obtained from Hassan *et al.* (1997)
Hassan *et al.* (1997), tip resistance	$q = \Lambda\delta^{0.67}$	Based on the results of finite element numerical models (FEM).
Gupta (2012)	$f_s = \dfrac{\frac{\delta}{B}}{\dfrac{\ln\left[\frac{5L(1-\nu)}{B}\right]}{2G_m} + \dfrac{\frac{\delta}{B}R_f}{f_{sp}}}$	Method does not provide any recommendations for the prediction of load-transfer function for tip resistance of drilled shafts.
Stark *et al.* (2013)	$q = \dfrac{3.2\frac{\delta}{B}}{\frac{\delta}{B}+1.3} q_u d_c \le 2.5 q_u d_c$	(i) Based on 33 drilled shaft load tests, and (ii) rock types in the database included shale and clayshale, claystone, and siltstone.
Lee *et al.* (2013)	$q = \dfrac{\delta}{\frac{1}{K_n} + \frac{\delta}{q_f}}$ where $K_n = 0.00037\left(\dfrac{E_m S_J}{\sigma_p B_{ref}}\right)^{0.4058}\left(\dfrac{\sigma_p}{B}\right)$ $q_f = 79.2\left(\dfrac{q_u S_J}{\sigma_p B}\right)^{0.315}\sigma_p$	(i) Based on the parametric studies that were performed using numerical models, (ii) model depends on δ, B, E_m, and the discontinuity spacing (S_J), and (iii) B_{ref} = 1.0 m and σ_p is the atmospheric pressure (0.101 MPa).

initial shear stiffness (K_{si}) in side resistance of sockets in weak rock, however, has not been explored in detail. Therefore, it is necessary that the existing models for the prediction of initial shear stiffness be evaluated against the back-calculated initial shear stiffness (K_{si}) from the results of rock socket load tests. One of the most widely used correlations is that of Randolph and Wroth (1978):

$$K_{si} = \frac{G_m}{r} \tag{2.30}$$

where K_{si} denotes the initial shear stiffness in side resistance of rock sockets, G_m is the shear modulus of geomaterial while accounting for the properties of the rock mass, and r is the radius of the rock socket. The shear modulus of rock mass (G_m) can be related to the modulus of deformation (E_m) of rock mass using the following equation (Goodman 1980):

$$G_m = \frac{E_m}{2(1+\nu)} \quad (2.31)$$

where ν is the Poisson's ratio. It must be noted that Equations (2.30) and (2.31) are based on the theory of elasticity and are only applicable to small axial displacements (i.e., displacements ≤0.5% of the socket diameter to ensure shear stresses on the socket sidewalls are significantly less than the peak shear strength of rock socket and hence to ensure socket would remain in the elastic range). Hirayama (1990) proposed an empirical design model for the estimation of initial shear stiffness (K_{si}) in side resistance of drilled shafts that is shown below:

$$K_{si} = \frac{f_{sp}}{0.0025B} \quad (2.32)$$

where f_{sp} is the peak side resistance and B is the diameter of rock socket. Hirayama (1990) recognizes that the value at which peak side resistance is developed can range from 0.1% to 0.5% of the rock socket diameter (B). The peak shear stress must be determined before this model can be used. This introduces additional uncertainty because current predictive methods for peak side resistance are not accurate as suggested by comparison of these methods with the new side resistance database.

These predictive models are evaluated herein in terms of the mean and coefficient of variation (c.o.v.) of the ratios of measured to predicted initial shear stiffness (μ_λ and δ_λ, respectively). It must be noted that the measured (i.e., back-calculated) initial shear stiffness values are obtained from the back-calculated f_s–δ relationships reported in the side-resistance database and following the procedure that is outlined in the preceeding sections of this chapter. The evaluation results are summarized in Table 2.10. The bias (i.e., ratio of the measured to predicted initial shear stiffness) is also plotted against different variables from the side-resistance database and is shown in Figures 2.19 and 2.20. The following observations can be made:

1. Table 2.10 shows the statistics of bias (λ) for initial shear stiffness (K_{si}). The statistics of bias are calculated and grouped based on the condition of soft rock mass in which the rock sockets have been constructed. These are rock masses with 50 < GSI < 70, 35 < GSI < 50, and GSI < 35.

Table 2.10 Evaluation of the predictive models for the initial shear stiffness (K_{si}) of drilled shafts in weak rock mass

Design methods		*Database (Asem 2018)*			
		35 > GSI	*35 < GSI <50*	*50 < GSI <70*	*Entire database*
Randolph and Wroth (1978)	μ_λ	2.70	0.84	0.63	1.62
	σ_λ	5.39	1.23	1.01	3.77
	δ_λ	1.99	1.46	1.61	2.33
Hirayama (1990)	μ_λ	2.14	0.92	2.13	1.6
	σ_λ	7.47	1.32	4.44	5.37
	δ_λ	3.49	1.43	2.08	3.36

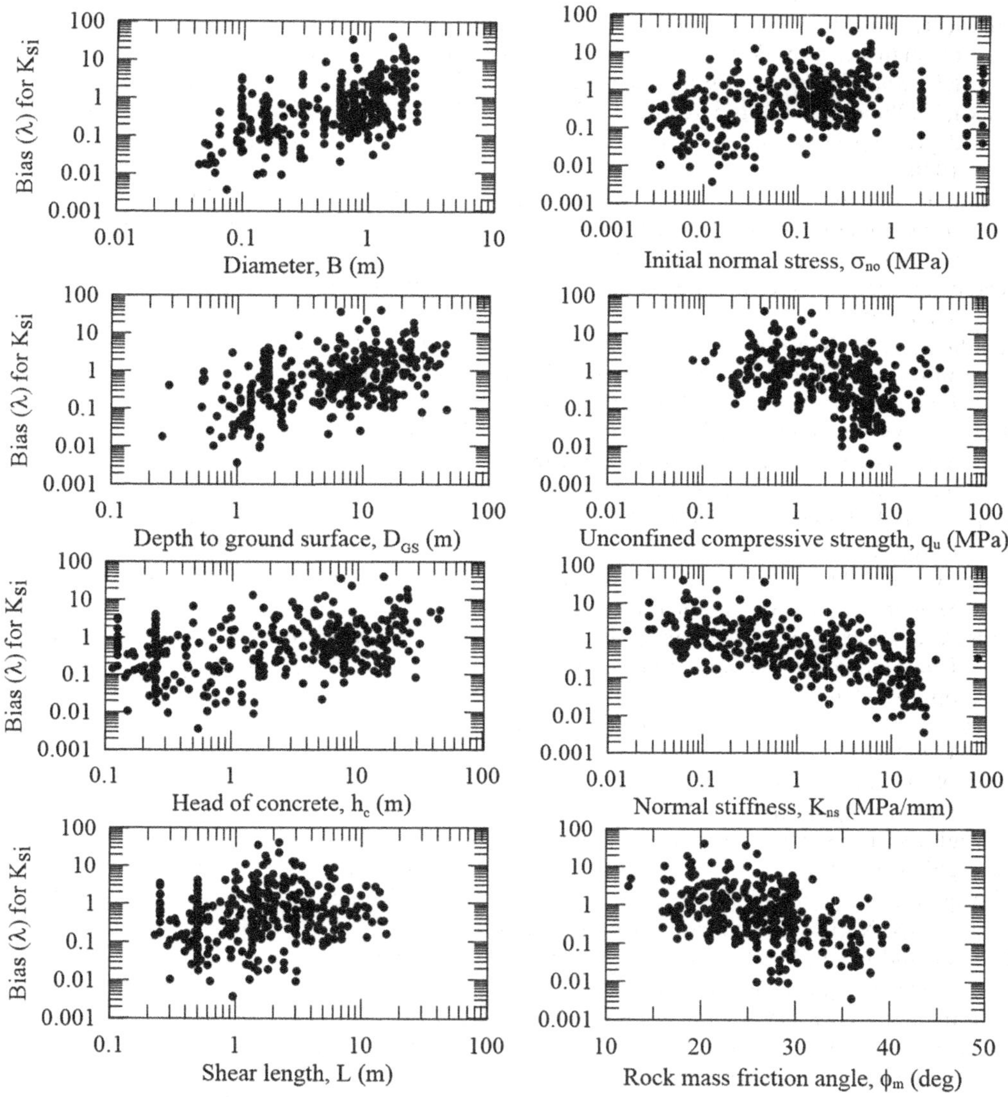

Figure 2.19 Variation of bias (λ) in the predictions of the method of Randolph and Wroth (1978) for initial shear stiffness (K_{si}) with properties of weak rock mass and drilled shaft geometry.

2. Table 2.10 shows that the method of Randolph and Wroth (1978) underestimates the initial shear stiffness (K_{si}) for the sockets in rocks with GSI smaller than 35. The analysis also shows that the variability in the predictions using the method of Randolph and Wroth (1978) is considerably larger for very weathered and jointed weak rock mass (i.e., GSI < 35) as compared to better quality rocks. The predictive method of Hirayama (1990) is unconservative for weak rock masses with GSI between 35 and 50. The variability of the predictions is also considerably larger for more weathered weak rocks as is the case for the method of Randolph and Wroth (1978).

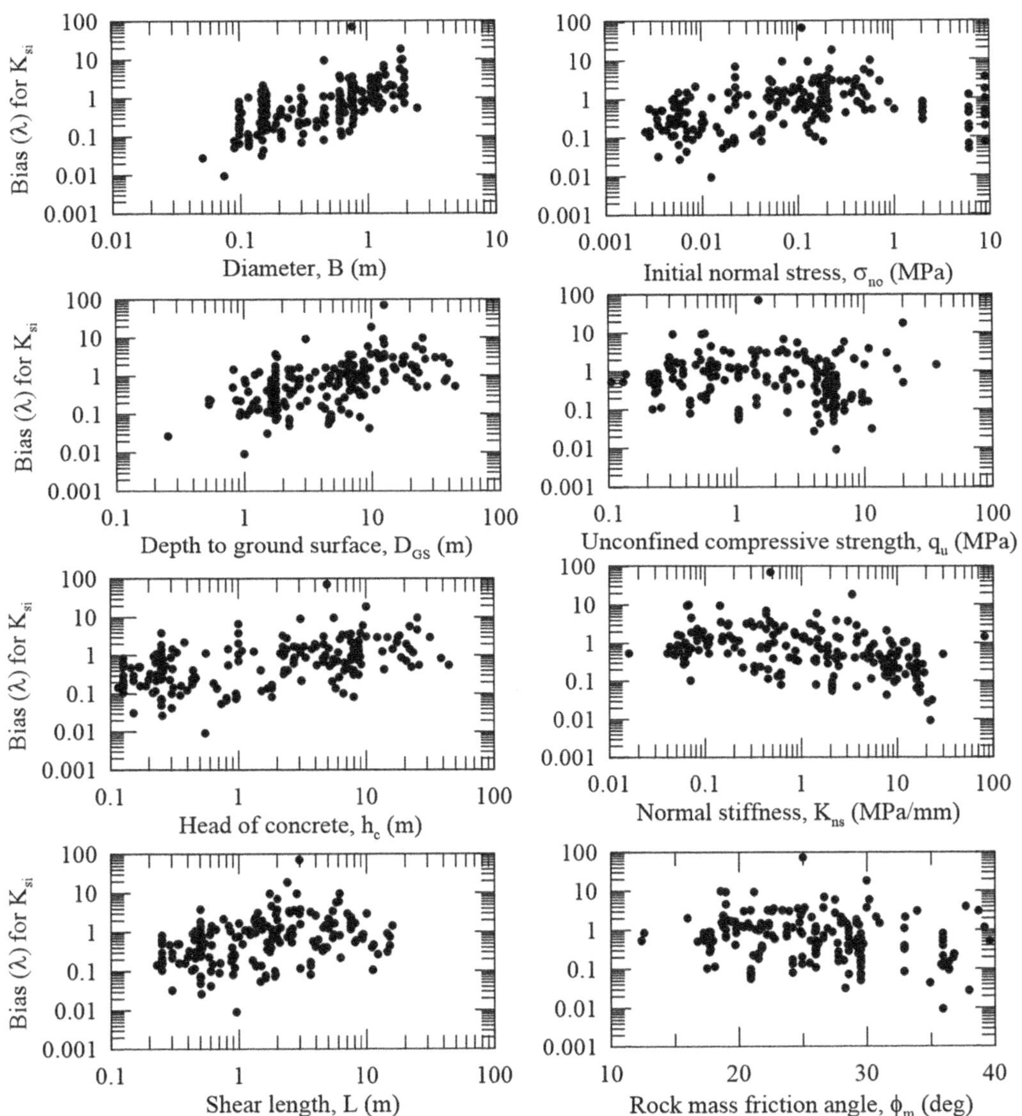

Figure 2.20 Variation of bias (λ) in the predictions of the method of Hirayama (1990) for initial shear stiffness (K_{si}) with properties of soft rock mass and drilled shaft geometry.

3. The variations of bias for the predictions of the methods of Randolph and Wroth (1978) and Hirayama (1990) with different properties of soft rock mass and socket geometry are shown in Figures 2.19 and 2.20. These figures show that the error of prediction of these methods for initial shear stiffness varies significantly with the diameter of the socket (B), depth of embedment (D_{GS}) (or equivalently the degree of confinement of the side of the drilled shaft), the normal stiffness of the surrounding rock mass (K_n), and the unconfined compressive strength of the soft rock (q_u).

4. The correlation between the bias in the predictions of the methods of Randolph and Wroth (1978) and Hirayama (1990) and weak rock mass properties and rock socket characteristics indicates that additional properties of the socket geometry and soft rock mass mechanical properties should be accounted for in the predictive models.
5. In addition to the possibility of the existence of missing parameters in the current predictive models, the strong correlation between the bias and weak rock properties and the drilled shaft geometry may result from model inexactness, meaning that the mathematical form of the current predictive models may not accurately describe the actual variation of initial shear stiffness with rock and drilled shaft properties.

2.5.2.2 Evaluation of models for initial normal stiffness in tip resistance (K_n)

Similar to the initial shear stiffness (K_{si}) of rock sockets, the initial normal stiffness (K_n) for the tip of the socket in soft rock mass has not been studied rigorously. Lee *et al.* (2013) developed a model for the prediction of the initial normal stiffness of socket tip in rock. Pells and Turner (1979), Goodman (1980), Jeong *et al.* (2010), and Lee and Jeong (2016) measured and developed models for the prediction of load–displacement behavior of shallow and socketed foundations in rocks that may be used to obtain correlations for prediction of initial normal stiffness for the tip of rock sockets.

The method of Goodman (1980) may be used to develop the following expression for the prediction of initial normal stiffness (K_n) for the tip of rock sockets:

$$K_n = \frac{E_m}{C(1-v^2)r} \tag{2.33}$$

where C is a constant that depends on the boundary conditions (Goodman 1980) and is equal to $\pi/2$ if the foundation can be shown to behave as a rigid structure compared to the soft rock or 1.7 if the foundation is a flexible structure compared to the soft rock mass (Goodman 1980), ν is the Poisson's ratio (see Gercek 2007 for typical values for rocks), and r is the radius of the foundation (i.e., $B/2$).

Pells and Turner (1979) also used a similar method to that proposed by Goodman (1980). Pells and Turner (1979), however, add a correction factor to account for the effect of embedment depth on load–displacement behavior of the tip of sockets in rock mass as:

$$K_n = \frac{E_m}{C(1-v^2)rI} \tag{2.34}$$

where I is the correction factor for embedment. This is a theoretical factor that is a function of Poisson's ratio (ν) of rock and depth of embedment in rock to diameter ratio (D_{TOR}/B). I ranges from 0.4 to 1.0 for D_{TOR}/B ratios of 0 to 10 (Pells and Turner 1979).

Lee *et al.* (2013) proposed a predictive model for the initial normal stiffness (K_n) for the tip of sockets in rock masses. The method is semi-empirical because it was formulated based on parametric studies performed using numerical models and was calibrated using a rock socket load test database that was reported by Lee *et al.* (2013) and is written as:

$$K_n = 0.00037\left(\frac{E_m S_J}{B_{ref}\sigma_p}\right)^{0.4058}\left(\frac{\sigma_p}{B}\right) \tag{2.35}$$

where K_n is in units of MPa/mm, S_J is the discontinuity spacing, B_{ref} = 1.0 m, σ_p is the atmospheric pressure (= 0.101 MPa), and B is the diameter of the rock socket. The mean discontinuity frequency per meter (i.e., $M = 1/S_J$) may be obtained from the following expression proposed by Priest and Hudson (1976):

$$RQD = 100e^{-0.1M}(0.1M + 1) \tag{2.36}$$

where RQD is the rock quality designation and M is the discontinuity frequency. RQD can be converted to GSI using the recommendations of Hoek *et al.* (2013).

These methods are evaluated herein using the mean (μ_λ) and coefficient of variation, c.o.v., (δ_λ) of ratios of back-calculated to predicted initial normal stiffness for the tip of sockets in weak rock masses. It must be emphasized that the back-calculated initial normal stiffness values are obtained from the back-calculated q–δ relationships reported in the tip resistance database. The analysis results are summarized in Table 2.11 and the diagnostic plots are shown in Figures 2.21 and 2.22. The findings and summary of observations are discussed in Tables 2.12–2.14.

1. The review of statistics of bias shows that all three initial normal stiffness models that are evaluated in this section underestimate the initial normal stiffness by a large margin.
2. The c.o.v. of the bias for predictive models for initial normal stiffness does not show a clear trend with the condition of rock mass. However, the study of the statistics of bias for the models of Pells and Turner (1979) and Lee *et al.* (2013) suggests that these models tend to become less accurate as the rock becomes more fractured (i.e., as the GSI values decrease).

Table 2.11 Evaluation of the predictive models for the initial normal stiffness (K_n) of drilled shafts in weak rock mass

		Database (Asem 2018)			
Design methods		*35 > GSI*	*35 < GSI <50*	*50 < GSI <70*	*Entire database*
Goodman (1980)	μ_λ	3.65	5.32	61.39	6.72
	σ_λ	4.43	7.18	78.92	20.54
	δ_λ	1.21	1.35	1.29	3.06
Pells and Turner (1979)	μ_λ	1.91	1.66	1.48	1.82
	σ_λ	0.93	0.44	0.66	0.83
	δ_λ	0.49	0.26	0.44	0.45
Lee *et al.* (2013)	μ_λ	2.27	5.37	30.90	4.68
	σ_λ	1.97	2.96	28.44	9.08
	δ_λ	0.87	0.55	0.92	1.94

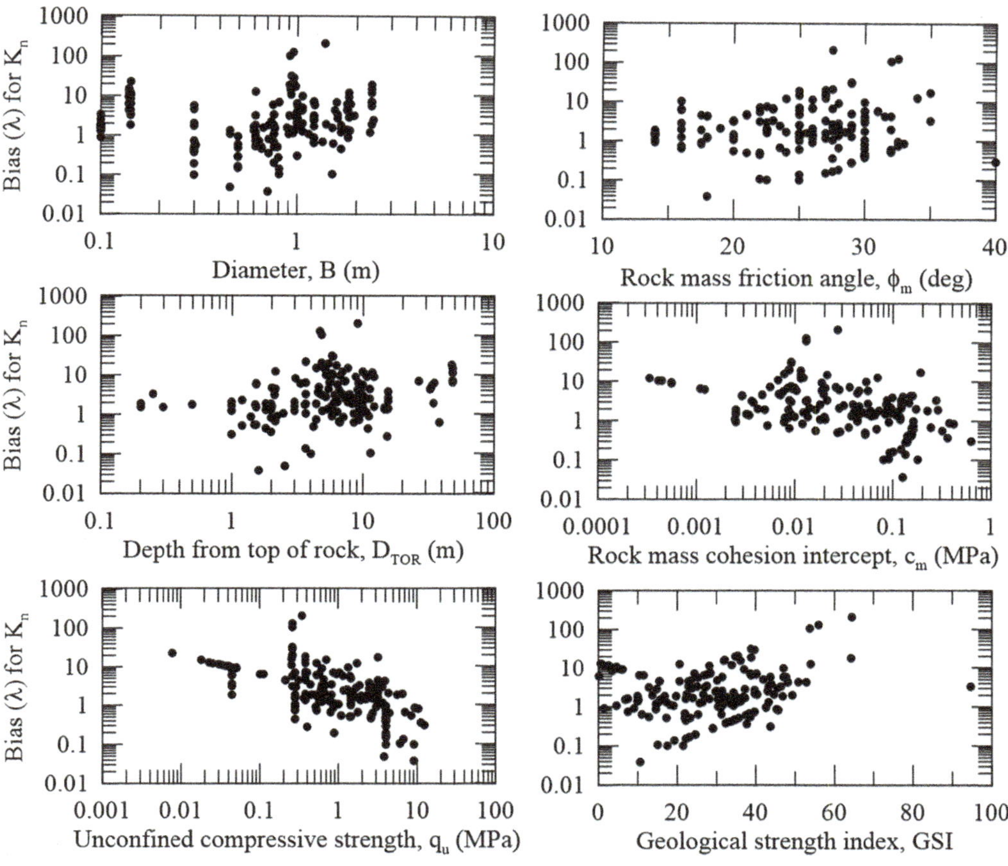

Figure 2.21 Variation of bias (λ) in the predictions of the method of Goodman (1980) for drilled shaft tip normal stiffness (K_n) with properties of weak rock mass and drilled shaft geometry using tip resistance database.

3. The examination of variation of bias (λ) with the properties of rock mass and socket geometry indicates a strong correlation between bias and socket diameter (B), unconfined compressive strength of intact rock (q_u), Geological Strength Index (GSI), and rock mass friction angle (ϕ_m).

2.6 ESTIMATION OF MODEL PARAMETERS

The rock mass deformation modulus (E_m), unconfined compressive strength (q_u), and the degree of consolidation (U) are used when deformations in rock mass are calculated (Bieniawski 1978; Hoek and Diederichs 2006). Previous investigators have used published data to develop predictive models for the strength and elastic properties of rock (e.g., Labuz *et al.* 2018; Guevara-Lopez *et al.* 2019; Asem 2020; Asem *et al.* 2021). Following this approach, we used in situ and laboratory test results to evaluate the existing methods and to recommend predictive models for E_m, q_u, and k.

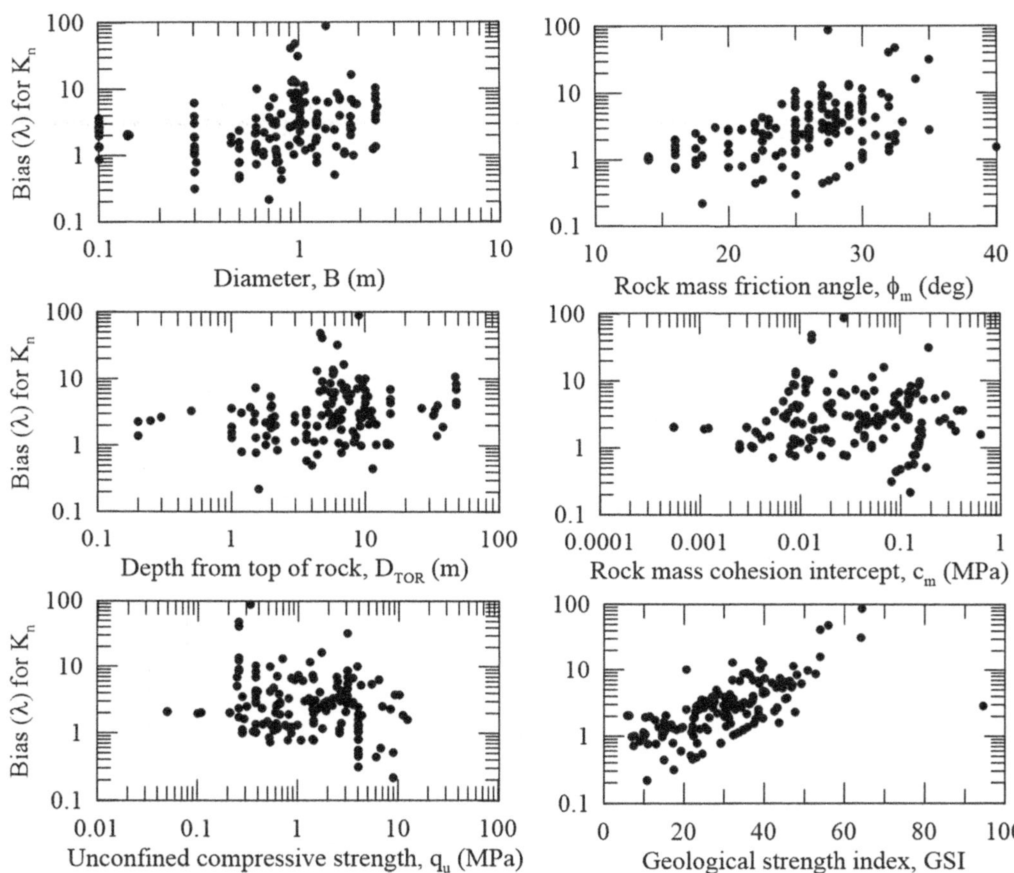

Figure 2.22 Variation of bias (λ) in the predictions of the method of Lee *et al.* (2013) for drilled shaft tip normal stiffness (K_n) with properties of soft rock mass and drilled shaft geometry using the tip resistance database.

Table 2.12 Summary of predictive methods for the modulus of deformation of rock mass

Number	*Empirical methods (E_m in units of GPa)*	*Reference*
1	$E_m = \dfrac{E_i}{100\left[0.0028\text{RMR}^2 + 0.9\exp(\dfrac{\text{RMR}}{22.82})\right]}, E_i = 50\text{GPa}$	Nicholson and Bieniawski (1990)
2	$E_m = \dfrac{E_i\left[1-\cos(\pi.\dfrac{\text{RMR}}{100})\right]}{2}, E_i = 50\text{GPa}$	Mitri *et al.* (1994)
3	$E_m = 0.1(\text{RMR}/10)^3$	Read *et al.* (1999)
4	$E_m = 10\sqrt[3]{\dfrac{Qq_u}{100}}, q_u = 100\text{MPa}$	Barton (2002)
5	$E_m = E_i\sqrt[4]{\exp\left[(\text{GSI-100})/9\right]}, E_i = 50\text{GPa}$	Carvalho (2004)
6	$E_m = 7\sqrt{10\left[\text{RMR}-44\right]/21}$	Diederichs and Kaiser (1999)

Table 2.13 Rock mass classification and rock mass deformation modulus database

Rock type	Joint infilling material	Plate diameter (cm)	Em (GPa)	RQD	RMR-76	RMR-89	GSI	Q	Reference
Fault breccia	Sand and gravel with silt	30.2	0.25		15	18	10	0.0078	Keffeler (2014)
Oxidized limestone	Silty sand with gravel	30.2	4.5		41	34	45	0.27	Keffeler (2014)
Argillized rhyolite dike	Clayey sand with gravel	30.2	0.37		16	21	15	0.017	Keffeler (2014)
Argillized rhyolite dike	Clayey sand	25.4	0.031		16	21	10	0.01	Keffeler (2014)
Decalcified limestone	Carbon and iron oxide coatings	25.4	2.5		37	47	40	0.1	Keffeler (2014)
Decalcified limestone	Carbon and iron oxide coatings	30.2	1.7		35	45	45	0.069	Keffeler (2014)
Decalcified limestone	Carbon and non-plastic clay coating	30.2	1.3		34	44	30	0.067	Keffeler (2014)
Decalcified limestone	Carbon coatings	30.2	1.1		40	49	40	1	Keffeler (2014)
Decalcified limestone	Carbon and minor soft white infillings	30.2	1.4		41	49	40	0.83	Keffeler (2014)
Decalcified limestone	Carbon and pyrite coatings	30.2	1.4		40	51	45	1.7	Keffeler (2014)
Decalcified limestone	Carbon coatings and 1-3 mm calcite	30.2	0.71		47	51	50	3.3	Keffeler (2014)
Decalcified and argillized limestone	Clayey sand	30.2	0.9		20	28	15	0.02	Keffeler (2014)
Decalcified limestone	Soft realgar coatings	30.2	4.2		45	54	50	0.67	Keffeler (2014)
Decalcified and argillized limestone	Soft realgar 1–3 mm	30.2	2.3		32	45	45	0.067	Keffeler (2014)
Siltstone and sandstone	Calcite and iron oxide		30	84	59	66			USACE (1988)
Siltstone and sandstone	Calcite and iron oxide		4.8	25	55	57			USACE (1988)
Siltstone and sandstone	Calcite and iron oxide		2.8	63	44	53			USACE (1988)
Siltstone and sandstone	Calcite and iron oxide		18	57	53	54			USACE (1988)
Siltstone and sandstone	Calcite and iron oxide		17	75	68	63			USACE (1988)

(Continued)

Table 2.13 (Continued) Rock mass classification and rock mass deformation modulus database

Rock type	*Joint infilling material*	*Plate diameter (cm)*	*Em (GPa)*	*RQD*	*RMR-76*	*RMR-89*	*GSI*	**Q**	*Reference*
Siltstone and sandstone	Calcite and iron oxide		2.1	18	42	45			USACE (1988)
Siltstone and sandstone	Calcite and iron oxide		34	73	65	66			USACE (1988)
Siltstone and sandstone	Calcite and iron oxide		8.3	26	39	40			USACE (1988)
Sarvak limestone	Calcite and iron oxide	0.971	14.1	5	44	50			Agharazi et *al.* (2012), Agharazi (2013)
Sarvak limestone	Calcite and iron oxide	0.971	16	14	43	52			Agharazi et *al.* (2012), Agharazi (2013)
Sarvak limestone	Calcite and iron oxide	0.971	19.5	3	45	52			Agharazi et *al.* (2012), Agharazi (2013)
Sarvak limestone	Calcite and iron oxide	0.971	15.6	19	44	53			Agharazi et *al.* (2012), Agharazi (2013)
Sarvak limestone	Calcite and iron oxide	0.65	5.5	6	45	52			Agharazi et *al.* (2012), Agharazi (2013)
Sarvak limestone	Calcite and iron oxide	0.65	3.7	16	45	53			Agharazi et *al.* (2012), Agharazi (2013)
Sarvak limestone	Calcite and iron oxide	0.65	3.7	15	45	54			Agharazi et *al.* (2012), Agharazi (2013)
Sarvak limestone	Calcite and iron oxide	0.65	3.2	25	44	54			Agharazi et *al.* (2012), Agharazi (2013)

Table 2.14 Evaluation of empirical models for prediction of rock mass modulus of deformation

Predictive method	*Database used for evaluation*	*Number of cases*	μ_λ	σ_λ	δ_λ
Bieniawski (1978)	Keffeler (2014)	5	1.35	1.04	0.77
Serafim and Pereira (1983)	Keffeler (2014)	30	0.81	0.74	0.91
Nicholson and Bieniawski (1990)	Keffeler (2014)	30	0.83	0.79	0.95
Mitri *et al.* (1994)	Keffeler (2014)	30	0.26	0.26	1.00
Read *et al.* (1999)	Keffeler (2014)	30	0.52	0.45	0.86
Barton (2002)	Keffeler (2014)	14	0.28	0.21	0.75
Diederichs and Kaiser (1999)	Keffeler (2014)	23	0.58	0.49	0.83
Carvalho (2004)	Keffeler (2014)	14	0.16	0.11	0.66
Palmström and Sigh (2001)	Keffeler (2014)	14	5.1	4.58	0.88
Galera *et al.* (2007)	Keffeler (2014)	30	0.65	0.59	0.92
Gokceoglu and Zorlu (2004) RMR	Keffeler (2014)	30	1.86	1.68	0.90
Gokceoglu and Zorlu (2004) GSI	Keffeler (2014)	14	0.92	0.58	0.62
This study, Equation (37)	Asem (2018)	581	0.98	1.1	1.12

Source: Based on data from Keffeler (2014) and Asem (2018).

2.6.1 Model formulation framework

Reliability analysis of structures built in rock masses requires the use of probabilistic models that can properly quantify the design uncertainties (e.g., Gardoni *et al.* 2002; Asem 2018; Asem 2019a,b; Asem *et al.* 2019; Asem and Gardoni 2019a,b,c; Asem 2020; Tan *et al.* 2020; Asem and Gardoni 2021). Following Gardoni *et al.* (2002), a probabilistic model is formulated for the estimation of permeability of rock:

$$T[f(x,\Theta)] = \gamma(x,\theta) + \sigma\varepsilon \tag{2.37}$$

where $T[.]$ is a suitable transformation of $f; \gamma(x,\theta) = \sum_{j=1}^{n} \theta_j h_j(x)$ is a function expressed in terms of observed variable, x, and the unknown parameter $\theta = (\theta_1, \theta_2, \ldots, \theta_n)$; and $\sigma\varepsilon$ is a model error term that captures the model uncertainty resulting from missing model parameters or inappropriate selected model form. In the model error term, σ is the standard deviation of the model error and ε is a random variable with zero mean and unit variance.

A suitable transformation of f should be used to justify the assumptions that $\sigma\varepsilon$ is additive to the model (additivity assumption), σ is constant and independent of the variable x (homoskedasticity assumption), and that ε follows a normal distribution (normality assumption). These assumptions are used in the estimation of the model parameters $\Theta = (\theta, \sigma)$. Here, we use the natural logarithm of f to satisfy these assumptions. The validity of this transformation may be checked using diagnostic plots as described by Rao and Toutenburg (1995).

A Bayesian approach is used to estimate the vector of unknown model parameters $\Theta = (\theta, \sigma)$ using a Markov-chain Monte-Carlo (MCMC) approach. A Bayesian framework allows different types of information to be included in the estimation of $\Theta = (\theta, \sigma)$ such as field data, laboratory test results, and engineering judgment. In a Bayesian approach, the prior probability distribution density function (PDF) for the model parameters Θ is updated following the Bayes' theorem (Box and Tiao 1992):

$$f''(\Theta) = \kappa L(\Theta) f'(\Theta) \tag{2.38}$$

where $f''(\Theta)$ is the posterior PDF of Θ that captures all available information about Θ; the prior distribution PDF, $f'(\Theta)$ is the prior PDF that represents the state of our knowledge about the model parameters Θ before new information become available; $L(\Theta)$ is the likelihood function that captures the information from the data; and κ is a constant used to ensure that $f''(\Theta)$ is a proper PDF (i.e., $\kappa = \frac{1}{\int L(\Theta) f'(\Theta) d(\Theta)}$ such that the area under $f''(\Theta)$ is equal to unity). The effect of $f'(\Theta)$ on $f''(\Theta)$ becomes smaller as the number of new measurements and observations increases. Box and Tiao (1992) suggested that when no previous information on Θ is available, a noninformative prior may be used for $f'(\Theta)$. Gardoni *et al.* (2002) showed that for a univariate model that is linear in parameter θ, the noninformative prior uniform PDF may be represented as:

$$f'(\Theta) \propto \frac{1}{\sigma} \tag{2.39}$$

The information from n sets of measured quantities (k_j, x_j) $j = 1, \ldots, n$ is included in the Bayesian estimation of the model parameters Θ using the likelihood function, $L(\Theta)$ that is the probability of the observation of a given value of Θ and its formulation depends on the type of available data. When the available observations are statistically independent, the likelihood function, $L(\Theta)$, is written as:

$$L(\Theta) \propto \prod_{j=1}^{n} \left\{ \frac{1}{\sigma} \varphi \left[\frac{r_j(\theta)}{\sigma} \right] \right\} \tag{2.40}$$

where φ [.] is the standard normal PDF and $r_j(\theta)$ is the residual term that may be given by $r_j(\theta) = k_j - \gamma(x_j, \theta)$ and is valid for equality data. Equation (2.38) can be used to update the current posterior as new data become available by using the current posterior as the new prior and using the new data to construct the likelihood function.

2.6.2 Rock mass deformation modulus (E_m)

A database of rock mass deformation modulus (E_m) and the unconfined compressive strength (q_u) of weak rock in rock socket load tests was developed by Asem (2018). These load tests were augmented by those collected by Chern *et al.* (2004), which include sedimentary, igneous, and metamorphic rocks from China and Taiwan. These data are plotted in Figure 2.23. Equation (2.41) can be used to predict E_m based on the unconfined compressive strength (q_u):

$$E_m = 150 q_u^{1.1} \tag{2.41}$$

where E_m and q_u are in units of MPa.

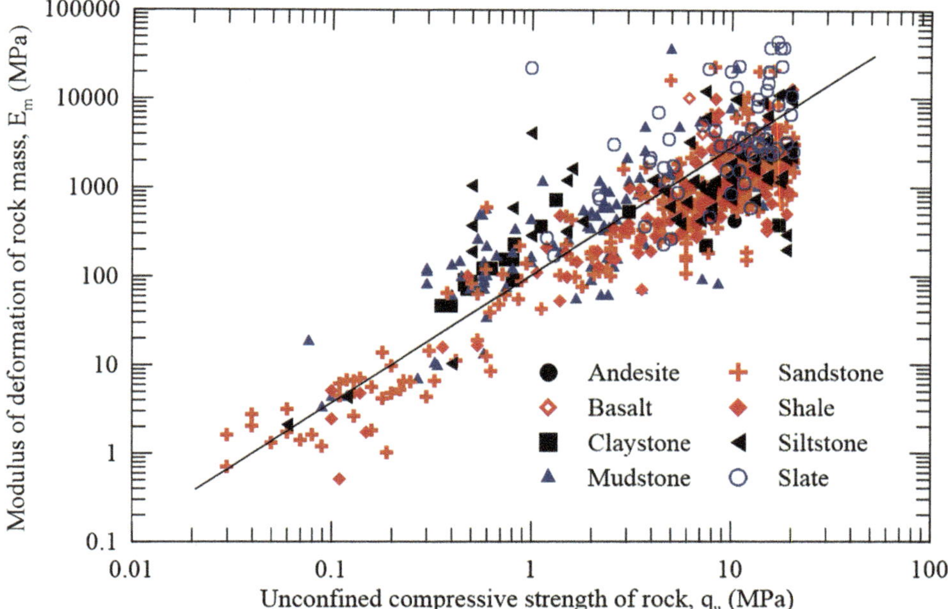

Figure 2.23 Variation of modulus of deformation of rock mass (E_m) with unconfined compressive strength for different rock types and the proposed design trend line. All data from published literature are summarized in Asem (2018).

2.6.3 Unconfined compressive strength (q_u)

The following approach is suggested for the prediction of q_u based on standard penetration test (SPT) results: (i) develop a plot of the blow counts versus penetration at each SPT depth, (ii) use the approach proposed by Stark *et al.* (2013) and Asem (2020) to estimate the penetration rate (Ψ) at each SPT depth, (iii) obtain the corrected Ψ for 85% energy transfer (Ψ_{85}), (iv) correct the corresponding value of Ψ_{85} for the effect of the mean effective stress and obtain $(\Psi_{85})_{\sigma'}$, and (v) use the corresponding values of $(\Psi_{85})_{\sigma'}$ to estimate q_u (Asem 2020):

$$q_u(\text{MPa}) = \frac{10}{4}(\psi_{85})_{\sigma'} \tag{2.42}$$

Equation (2.42) is developed by regression analysis between data for the corrected values of Ψ_{85} for the effect of σ', $(\Psi_{85})_{\sigma'}$, and the measured values of q_u and E at 21 Illinois sites. The corrected value of Ψ_{85} for the effect of the mean effective stress obtained $(\Psi_{85})_{\sigma'}$

$$(\psi_{85})_{\sigma'} = C_\psi \psi_{85} = \frac{0.23}{\sigma'}\psi_{85} \tag{2.43}$$

It must be noted that the correlation in Equation (2.43) is only intended for the range of $0 < \sigma' < 0.8$ MPa (Asem 2020). If this correlation is applied to $\sigma' > 0.8$ MPa, it will lead to unreliable results. Additional data is required to further enhance Equation (2.43) for $\sigma' > 0.8$ MPa. It is also noted that the data scatter for $0 < \sigma' < 0.8$ MPa is significant. This equation should be updated using the approach proposed by Gardoni *et al.* (2002) as more reliable data become available.

The depth correction factor in Equation (2.43), i.e., C_Ψ, was formulated for a reference effective vertical stress of $\sigma'_v = 0.1$ MPa (1 ton/ft^2 after Skempton 1986) or equivalently a mean effective stress of $\sigma' = 7/3\ (0.1) = 0.23$ MPa. C_Ψ may be obtained as (after Skempton 1986).

$$C_\Psi = \frac{\frac{a}{b} + 0.23}{\frac{a}{b} + \sigma'} \tag{2.44}$$

For $a/b = 0$ for the case of weak rocks, $C_\Psi = 0.23/\sigma'$. It must be noted that this correlation in only valid for a range of mean effective stress of $0 < \sigma' < 0.8$ MPa that is equivalent to an overall test depth of about 26 m.

2.6.4 Permeability (*k*)

The correlation between some of the parameters affecting the permeability of mudstone was examined by Gardoni and Asem (2022) who showed (i) porosity ϕ and the clay size fraction w_{CL} are not strongly related, and (ii) clay size fraction w_{CL} and mean pore radius r are more or less correlated, as w_{CL} increase, r decreases. This is expected because the size of pore spaces and the size of the particles making the porous medium are inversely related, i.e., as the percentage of small particles increases, the pore size decreases. Based on data from the literature, Asem and Gardoni (2022) presented a simple permeability model that could be formulated:

$$k = f(r, \phi) \tag{2.45}$$

Note that Equation (2.45) excludes the clay size fraction w_{CL} because $r = f\,(w_{CL})$ and hence the effect of w_{CL} on k is indirectly accounted for by including r in the predictive model. Using the method of Gardoni *et al.* (2002), a probabilistic model for the prediction of permeability of mudstone is proposed

$$k = e^{-41.6}\left(r\phi\right)^{0.6} \tag{2.46}$$

For a Terzaghi effective pressure of 10 MPa, with our formulation and results (Asem and Gardoni 2022), the effect of overall model uncertainty $\sigma\varepsilon$ as well as model parameter uncertainty $\Theta = (\theta, \sigma)$ can be readily accounted for that is an improvement over existing probabilistic models.

2.7 CONCLUSIONS

Rock sockets are often constructed in weak rock mass. Understanding the hydromechanical properties of the rock mass is required for the successful design and performance of the rock sockets. This chapter establishes separate databases for the hydromechanical properties of rock masses with particular attention to weak shales and mudstones. The databases include information on rock socket geometry, load–displacement response, and the rock mass properties and mechanical and transport properties of the rock. The chapter then reviews the available models for the prediction of (i) the axial resistance (compression or uplift), (ii) axial deformation, and permeability. The databases are used to evaluate the existing predictive models for hydromechanical properties of rock as related to the design of rock sockets. The load test databases are also used to evaluate the effect of rock mass variability on the reliability of the foundations in weak rock. Based on the results of the analysis presented, the applicability of the existing models and approaches to the study of socket behavior in weak rocks is discussed.

ACKNOWLEDGEMENTS

Professor James H. Long of the University of Illinois at Urbana-Champaign contributed to this manuscript. His suggestions and contributions are acknowledged.

REFERENCES

Abu-Hejleh, N., O'Neill, M.W., Hanneman, D., *et al.* (2003). *Improvement of the geotechnical axial design methodology for Colorado's drilled shafts socketed in weak rocks*. Denver, CO: Colorado Department of Transportation.

Agharazi, A., Tannant, D. D. and Martin, C. D. (2012). Characterizing rock mass deformation mechanisms during plate load tests at the Bakhtiary dam project. *International Journal of Rock Mechanics & Mining Sciences*, 49(2012), 1–11.

Agharazi, A. (2013). *Development of a 3D equipment continuum model for deformation analysis of systematically jointed rock masses*. Ph.D. thesis, University of Alberta.

AlKhafaji, H., Imani, M., and Fahimifar, A. (2020). Ultimate bearing capacity of rock mass foundations subjected to seepage forces using modified Hoek–Brown criterion. *Rock Mechanics and Rock Engineering*, 53, 251–268.

Ang, A. H.-S., and Tang, W. H. (2007). *Probability Concepts in Engineering: Emphasis on Applications in Civil and Environmental Engineering*, 2nd ed. Wiley, New York.

ARGEMA (1992). Design guides for offshore structures: offshore pile design. in P.L. Tirant (ed.), Paris, France: Association de recherché en geotechnique marine.

Asem, P. (2018). *Axial behavior of drilled shafts in soft rock*. Ph.D. Thesis. University of Illinois at Urbana-Champaign, Urbana, Illinois, United States.

Asem, P., Long, J. H., and Gardoni, P. (2018). Probabilistic model and LRFD resistance factors for the tip resistance of drilled shafts in soft sedimentary rock based on axial load tests. *Innovations in Geotechnical Engineering: Honoring Jean-Louis Briaud*. X. Zhang, P. J. Cosentino and M. H. Hussein. Orlando, Florida, American Society of Civil Engineers. *GSP* 299, 1–49.

Asem, P. (2019a). Load-displacement response of drilled shaft tip in soft rocks of sedimentary origin. *Soils and Foundations* 59(12), 1193–1212.

Asem, P. (2019b). Base resistance of drilled shafts in soft rock using in situ load tests: a limit state approach. *Soils and Foundations* 59(6), 1639–1658.

Asem, P. and Gardoni, P. (2019a). Evaluation of the peak side resistance for rock socketed shafts in weak sedimentary rock from an extensive database of published field load tests: a limit state approach. *Canadian Geotechnical Journal* 56(12), 1816–1831.

Asem, P. and Gardoni, P. (2019b). A load-transfer function for the side resistance of drilled shafts in soft rock. *Soils and Foundations* 59(2019), 1241–1259.

Asem, P. and Gardoni, P. (2019c). Bayesian estimation of the normal and shear stiffness for rock sockets in weak sedimentary rocks. *International Journal of Rock Mechanics and Mining Sciences* 124(2019), 104–129.

Asem, P., Jimenez, R. and Gardoni, P. (2019a). Probabilistic prediction of intact rock strength using point load tests using a Bayesian formulation. *7th International Symposium on Geotechnical Safety and Risk (ISGSR 2019)*. Taipei, Taiwan.

Asem, P., Taukoor, V., and Kane, T. (2019b). Departure of soft rock mass from undrained response in drilled shaft and plate load tests. *Soils and Foundations* 59(1), 228–233.
Asem, P. (2020a). Prediction of unconfined compressive strength and deformation modulus of weak argillaceous rocks based on the standard penetration test. *International Journal of Rock Mechanics and Mining Sciences* 133.
Asem, P. (2020b). The effect of expansive concrete on the side resistance of sockets in weak rock. *Soils and Foundations* 60(1), 274–282.
Asem, P. and J. F. Labuz (2020). *A failure mechanism around axially loaded sockets in weak rock. Geo-Congress 2020*. ASCE, Minnesota.
Asem, P., Fuselier, H., and Labuz, J. F. (2021a). On a four-parameter linear failure criterion. *Rock Mechanics and Rock Engineering* 54, 3369–3376. https://doi.org/10.1007/s00603-021-02451-w
Asem, P., X. Wang, C. Hu and J. F. Labuz (2021b). On tensile fracture of a brittle rock. *International Journal of Rock Mechanics & Mining Sciences* 144, 104823.
Asem, P. and P. Gardoni (2021a). A generalized Bayesian approach for prediction of strength and elastic properties of rock. *Engineering Geology* 289, 106187.
Asem, P. and P. Gardoni (2021b). On the use and interpretation of in situ load tests in weak rock masses. *Rock Mechanics and Rock Engineering* 54, 3663–3700.
Asem, P. and P. Gardoni (2022). A probabilistic, empirical model for permeability of mudstone. *Probabilistic Engineering Mechanics* 69, 103262.
Baghdady, A. K. (2018). *Axial behavior of drilled shafts socketed into weak Pennsylvanian shales*. Ph.D. Thesis. University of Illinois at Urbana-Champaign, Urbana, Illinois, United States.
Bandis, S. (1980). *Experimental studies of scale effects on shear strength and deformation of rock joints*. Ph.D. Thesis. University of Leeds, Leeds, United Kingdom.
Bandis, S. C., Lumsden, A. C., and Barton, N. R. (1983). Fundamentals of rock joint deformation. *International Journal of Rock Mechanics and Mining Sciences and Geomechanics Abstracts*, 20(6), 249–268.
Box, G. E. P. and Tiao, G. C. (1992). *Bayesian inference in statistical analysis*. Reading, Massachusetts, Addison–Wesley.
Barton, N., *et al.* (1978). Suggested methods for the quantitative description of discontinuities in rock masses. *International Journal of Rock Mechanics and Mining Sciences and Geomechanics Abstracts*, 15(6), 319–368.
Barton, N. (2002). Some new Q value correlations to assist in site characterisation and tunnel design. *International Journal of Rock Mechanics and Mining Sciences*, 39(2), 185–216.
Bieniawski, Z. T. (1978). Determining rock mass deformability: experience from case histories. *International Journal of Rock Mechanics and Mining Sciences and Geomechanics Abstracts*, 15(5), 237–247.
Bishnoi, B.L. (1968). *Bearing capacity of a closed jointed rock*. Ph.D. Thesis. Georgia Institute of Technology, Atlanta, Georgia, United States.
Brown, D.A., Turner, J.P., and Castelli, R.J. (2010). *Drilled shafts: construction procedures and LRFD design methods*. Washington, DC: National Highway Institute, U.S. Department of Transportation, Federal Highway Administration, 972pp.
Carrubba, P. (1997). Skin friction on large-diameter piles socketed into rock. *Canadian Geotechnical Journal*, 34(2), 230–240.
Carter, J.P. and Kulhawy, F.H. (1988). *Analysis and design of drilled shaft foundations socketed into rock*. Palo Alto: Electric Power Research Institute, p. 190.
Carvalho, J. (2004). Estimation of rock mass modulus. Pers. commun.
CGS (2006). *Canadian foundation engineering manual.*
Chern, J. C., *et al.* (2004). *Correlation study on the deformation modulus and rating of rock mass (R-GT-97-04), from Taipei*, Taiwan.
Coates, D.F. (1967). Rock mechanics principle. *Department of Energy, Mines and Resources, Canada*, 410.

Collingwood, B. (2000). *The effect of construction practices on the performance of rock socketed bored piles*. Ph.D. Thesis, Monash University, Melbourne, Australia.

Dai, G., *et al.* (2016). The effect of sidewall roughness on the shaft resistance of rock-socketed piles. *Acta Geotechnica*, 1–12, 429–440.

Deere, D.U. and Miller, R.P. (1966). *Engineering classification and index properties for intact rock*. Urbana, IL: University of Illinois at Urbana-Champaign/Air Force Weapons Laboratory, p. 327.

Diederichs, M. S. and Kaiser, P. K. (1999). Stability of large excavations in laminated hard rock-masses: the Voussoir analogue revisited. *International Journal of Rock Mechanics and Mining Sciences*, 36(1), 97–117.

Galera, J. M., Alvarez, M., and Bieniawski, Z. T. (2007). Evaluation of the deformation modulus of rock masses using RMR: Comparison with dilatometer tests. In M. Romana, A. Perucho, and C. Olalla (eds.), *Underground works under special conditions*. Taylor & Francis, London, pp. 71–77.

Gardoni, P., Kiureghian, A. D., and Mosalam, K. M. (2002). Probabilistic capacity models and fragility estimates for reinforced concrete columns based on experimental observations. *Journal of Engineering Mechanics*, 128(10), 1024–1038.

Gercek, H. (2007). Poisson's ratio values for rocks. *International Journal of Rock Mechanics and Mining Sciences*, 44(1), 1–13.

Glos, G. H. III and Briggs, O. H. Jr. (1983). Rock Sockets in Soft Rock. *Journal of Geotechnical Engineering*, 109(4), 525–535.

Gokceoglu, C. and Zorlu, K. (2004). A fuzzy model to predict the uniaxial compressive strength and the modulus of elasticity of a problematic rock. *Engineering Applications of Artificial Intelligence*, 17(1), 61–72.

Goodman, R.E. (1980). *Introduction to rock mechanics*. United States: John Wiley & Sons, Inc..

Goodman, R. E. (1989). *Introduction to rock mechanics*. United States: John Wiley & Sons, Inc..

Goodman, R. E. (1993). *Engineering geology: rock in engineering construction* (1st ed.). United States: John Wiley & Sons, Inc.

Guevara-Lopez, F., Jimenez, R., Gardoni, P., and Asem, P. (2019). Probabilistic prediction of intact rock strength using point load tests using a Bayesian formulation. *Georisk: Assessment and Management of Risk for Engineered Systems and Geohazards* 14, 206–215. https://doi.org/10.1080/17499518.2019.1634274

Gupta, R. C. (2012). Hyperbolic model for load tests on instrumented drilled Shafts in intermediate geomaterials and rock. *Journal of Geotechnical and Geoenvironmental Engineering*, 138(11), 1407–1414.

Gupton, C. and Logan, T. J. (1984). Design guidelines for drilled shafts in weak rocks of south Florida. Florida, United States: South Florida Annual ASCE Meeting.

Griffith, A. A. (1921). The phenomena of rupture and flow in solids. *Philosophical Transactions of the Royal Society of London*. Series A, Containing Papers of a Mathematical or Physical Character, 221, 582–593.

Haberfield, C. M. (1987). *The performance of the pressuremeter and socketed piles in weak rock*. Ph.D. Thesis, Monash University, Monash University, Melbourne, Australia.

Haberfield, C. M. and Lochaden, A. L. E. (2018). Analysis and design of axially loaded piles in rock. *Journal of Rock Mechanics and Geotechnical Engineering* 11, 535–548.

Hassan, Khaled M. (1994). *Analysis and design of drilled shafts socketed into soft rock*. Ph.D. Thesis, University of Houston, Houston, TX.

Hassan, K. M., *et al.* (1997). Design method for drilled shafts in soft argillaceous rock. *Journal of Geotechnical and Geoenvironmental Engineering*, 123(3), 272–80.

Hirany, A. (1988). *Test loading of drilled shafts*. Ph.D. Thesis. Cornell University, Ithaca, NY.

Hirayama, H. (1990). Load-settlement analysis for bored piles using hyperbolic transfer functions. *Journal of Soils and Foundations*, 30(1), 55–64.

Horvath, R. G. and Kenney, T. C. (1979). Shaft resistance of rock socketed drilled piers. In F.M. Fuller (ed.), *Symposium on Deep Foundations*, Atlanta, GA, 182–214.

Hoek, E. (1965). *Rock fracture under static stress conditions*. Ph.D. Thesis, University of Cape Town, Cape Town, South Africa.
Hoek, E. and Brown, E. T. (1980). Empirical strength criterion for rock masses. *Journal of the Geotechnical Engineering Division*, 106(9), 1013–1035.
Hoek, E. (1983). Strength of jointed rock masses. *Géotechnique*, 33(3), 187–223.
Hoek, E. and Brown, E. T. (1997). Practical estimates of rock mass strength. *International Journal of Rock Mechanics and Mining Sciences*, 34(8), 1165–1186.
Hoek, E. and Diederichs, M. S. (2006). Empirical estimation of rock mass modulus. *International Journal of Rock Mechanics and Mining Sciences*, 43(2), 203–215.
Hoek, E., Carter, T. G., and Diederichs, M. S. (2013). *Quantification of the geological strength index chart*. San Francisco, CA: American Rock Mechanics Association.
Hoek, E. and Brown, E. T. (2019). The Hoek-Brown failure criterion and GSI — 2018 edition. *Journal of Rock Mechanics and Geotechnical Engineering*, 11(2019), 445–463.
Horvath, R. G. (1982). *Drilled piers socketed into weak shale-methods of improving performance*. Ph.D. Thesis, University of Toronto, Toronto, Ontario, Canada.
Holden, J. C. (1984). *Construction of board piles in weathered rocks*. Victoria, Australia: Road Construction Authority of Victoria, p. 69.
Horvath, R.G. (1978). *Field load test data on concrete-to-concrete rock bond strength drilled pier foundations*. Toronto: University of Toronto, p. 46.
Horvath, R. G. and Kenney, T. C. (1979). Shaft resistance of rock socketed drilled piers. In F.M. Fuller (ed.), *Symposium on Deep Foundations*, Atlanta, GA, pp. 182–214.
Jaeger, J. C., Cook, N. G., and Zimmerman, R. W. (2007). *Fundamentals of rock mechanics*. MA: Blackwell Publishing.
Jeong, S., *et al.* (2010). Point bearing stiffness and strength of socketted drilled shafts in Korean rocks. *International Journal of Rock Mechanics and Mining Sciences*, 47(6), 983–995.
Johnston, I. W. and Lam, T. S. K. (1989). Shear behavior of regular triangular concrete/rock joints-Analysis. *Journal of Geotechnical Engineering*, 115(5), 711–727.
Johnson, I. W. (2020). Revisiting methods for the design of rock socketed piles. *Journal of Geotechnical and Geoenvironmental Engineering*, 146(12), 1–13.
Kaderabek, T. J. and Reynolds, R. T. (1981). Miami limestone foundation design and construction. *Journal of the Geotechnical Engineering Division*, 107(GT7), 859–872.
Kanji, M. A. (2014). Critical issues in soft rocks. *Journal of Rock Mechanics and Geotechnical Engineering*, 6(3), 186–195.
Keffeler, E. R. (2014). *Measurement and prediction of in-situ weak rock mass modulus: case studies from Nevada, Puerto Rico, and Iran*. Ph.D. Thesis, University of Nevada, Reno.
Kodikara, J. K. (1989). *Shear behaviour of rock-concrete joints and side resistance of piles in weak rock*. Ph.D. Thesis, Monash University, Melbourne, Australia.
Kulhawy, F. H. and Carter, J. P. (1992). Socketed foundations in rock masses. In F.G. Bell (ed.), *Engineering in Rock Masses*, Oxford: University of Oxford, pp. 509–529.
Kulhawy, F. H. and Goodman, R. E. (1980). Design of foundations on discontinuous rock. In: *International conference on structural foundations on rock*, Sydney, pp. 209–220.
Kulhawy, F. H. and Phoon, K. K. (1993). Drilled shaft side resistance in clay soil to rock. In P.P. Nelson, T.D. Smith, and E.C. Clukey (eds.), *Design and Performance of Deep Foundations: Piles and Piers in Soil and Soft Rock*, New York: American Society of Civil Engineers, pp. 172–183.
Kulhawy, F. H., Prakoso, W. A., and Akbas, S. O. (2005). Evaluation of capacity of rock foundation sockets. In *40th U.S. Symposium on Rock Mechanics*, American Rock Mechanics Association, Anchorage, AL.
Labuz, J. F., and Zang, A. (2012). Mohr–Coulomb failure criterion. *Rock Mechanics and Rock Engineering*, 45, 975–979.
Labuz, J. F., Zeng, F., Makhnenko, R., and Li, Y. (2018). Brittle failure of rock: a review and general linear criterion. *Journal of Structural Geology* 112(2018), 7–28.

Lam, T. S. K. (1983). *Shear behaviour of concrete-rock joints*. Ph.D. Thesis. Monash University, Melbourne, Australia.

Lee, J., *et al.* (2013. Proposed point bearing load transfer function in jointed rock-socketed drilled shafts. *Soils and Foundations*, 53(4), 596–606.

Lee, J. and Jeong, S. (2016). Experimental study of estimating the subgrade reaction modulus on jointed rock foundations. *Rock Mechanics and Rock Engineering*, 49, 2055–2064.

Leung, C. F. and Ko, H. Y. (1993). Centrifuge model study of piles socketed in soft rock. *Soils and Foundations*, 33(3), 80–91.

Marinos, P. and Hoek, E. (2001). Estimating the geotechnical properties of heterogeneous rock masses such as Flysch. *Bulletin of Engineering Geology and the Environment*, 60(2), 85–92.

Meigh, A.C. and Wolski, W. (1979). Design parameters for weak rock. In *7th European Conference on Soil Mechanics and Foundation Engineering*, Brighton, pp. 59–79.

Mesri, G., Febres-Cordero, E., Shields, D. R., and Castro, A. (1981). Shear stress-strain-time behaviour of clays. *Géotechnique* 31(4), 537–552.

Miller, A.D. (2003). *Prediction of ultimate side shear for drilled shafts in Missouri shales*. M.S. Thesis, University of Missouri-Columbia, Columbia, Missouri, United States.

Mitri, H. S., Edrissi, R., and Henning, J. (1994). Finite element modeling of cablebolted stopes in hard rock ground mines. In *The SME Annual Meeting*, New Mexico, Albuquerque.

Nicholson, G. A., and Bieniawski Z. T. 1990. A nonlinear deformation modulus based on rock mass classification. *International Journal of Minining and Geological Engineering*, 8, 181–202.

O'Neill, M. W. and Reese, L. C. (1999). *Drilled shafts: construction procedures and design methods*. Washington, DC: Federal Highway Administration, 758pp.

Osterberg, J. O. and Gill, S. A. (1973). Load transfer mechanism for piers socketted in hard soils or rock. In *The 9th Canadian Rock Mechanics Symposium*, Montreal, Canada, pp. 235–262.

Paikowsky, S. G., Birgisson, B., McVay, M. C., *et al.* (2004). *Load and resistance factor design (LRFD) for deep foundations (NCHRP 507)*. Washington, DC, United States: American Association of State Highway and Transportation Officials.

Paikowsky, S. G., Canniff, M. C., Lesny, K., Kisse, A., Amatya, S., and Muganga, R. (2010). *LRFD design and construction of shallow foundations for highway bridge structures*. Washington DC: American Association of State Highway and Transportation Officials.

Palmström, A. and Singh, R. (2001). The deformation modulus of rock masses - comparisons between in situ tests and indirect estimates. *Tunnelling and Underground Space Technology*, 16(2), 115–131

Peck, R. B., Hanson, W. E., and Thornburn, T. H. (1974). *Foundation engineering*. 2nd ed. New York: John Wiley & Sons.

Peck, R. B. (1976). Rock foundations for strucrures. In: *Rock Engineering for Foundations and Slopes*, Boulder, CO, pp. 1–20.

Pells, P. J. N. and Turner, R. M. (1979). Elastic solutions for the design and analysis of rock-socketed piles. *Canadian Geotechnical Journal*, 16(3), 481–487.

Pells, P. J. N., *et al.* (1979). *Socketed bored piles in Hawkesbury sandstone*. Sydney, Australia: University of Sydney, 138pp.

Pells, P., Bieniawski, Z. T. R., Hencher, S., and Pells, S. (2017). Rock quality designation (RQD): time to rest in peace. *Canadian Geotechnical Journal*, 54, 825–834.

Priest, S. D. and Hudson, J. A. (1976). Discontinuity spacings in rock. *International Journal of Rock Mechanics and Mining Sciences and Geomechanics Abstracts*, 13(5), 135–148.

Randolph, M.F. and Wroth, C. P. (1978). Analysis of deformation of vertically loaded piles. *Journal of the Geotechnical Engineering Division*, 104(GT12), 1465–1488.

Rao, C. R. and Toutenburg, H. (1995). *Linear models: least squares and alternatives*. New York, Springer.

Read, S. A. L., Richards, L. R., and Perrin, N. D. (1999). Applicability of the Hoek-Brown failure criterion to New Zealand greywacke rocks. In *The Ninth International Congress on Rock Mechanics*, Paris, France.

Rosenberg, P. and Journeaux, N. L. (1976). Friction and end bearing tests on bedrock for high capacity socket design. *Canadian Geotechnical Journal*, 13(3), 324–333.

Rowe, R. K. and Armitage, H. H. (1984). *The design of piles socketed into weak rock*. Ottawa, Ontario, Canada: National Research Council Canada, p. 380.

Rowe, R. K. and Armitage, H. H. (1987). A design method for drilled piers in soft rock. *Canadian Geotechnical Journal*, 24(1), 126–142.

Seed, H. B. and Reese, L. C. (1957). The action of soft clay along friction piles. *ASCE Transactions*, 122(2882), 731–754.

Seidel, J. P. (1993). *Analysis and design of pile shafts in weak rock*. Ph.D. Thesis, Monash University, Melbourne, Australia.

Seidel, J. P. and Collingwood, B. (2001). A new socket roughness factor for prediction of rock socket shaft resistance. *Canadian Geotechnical Journal*, 38(1), 138–153.

Seo, H. and Prezzi, M. (2008). *Use of micropiles for foundations of transportation structures (FHWA/IN/JTRP-2008/18)*. West Lafayette, IN.

Serafim, J. L. and Pereira, J. P. (1983). Consideration of the geomechanical classification of Bieniawski. In *Symposium on Engineering Geology and Underground Constructions*.

Skempton, A. W. (1986). Standard penetration test procedures and the effects in sands of overburden pressure, relative density, particle size, ageing and overconsolidation. *Geotechnique*, 36(3), 425–447.

Sowers, G. F. (1979). *Introductory soil mechanics and foundations*. New York, United States: Macmillan Publishing Co., Inc. , pp. 621.

Stark, T. D., Long, J. H., and Asem, P. (2013). *Improvements for determining the axial capacity of drilled shafts in shale in Illinois*. Urbana, IL: Illinois Department of Transportation.

Stark, T. D., Long, J. H., Baghdady, A. K., and Osouli, A. (2017). *Modified standard penetration test-based drilled shaft design method for weak rocks (Phase 2 study)*. Urbana, IL: Illinois Center for Transportation.

Tabandeh, A., Asem, P., and Gardoni, P. (2020). Physics-based probabilistic models: Integrating differential equations and observational data. *Structural Safety* 87(2020), 101981.

Tang, C., Phoon, K. K., Li, D.-Q., and Akbas, S. O. (2020). Expanded database assessment of design methods for spread foundations under axial compression and uplift loading. *Journal of Geotechnical and Geoenvironmental Engineering*, 146(11), 1–21.

Teng, W. C. (1962). *Foundation design*. New Jersey: Prentice Hall, Englewood Cliffs.

Terzaghi, K. (1943). *Theoretical soil mechanics*. New York: John Wiley & Sons.

Terzaghi, K., Peck, R. B., and Mesri, G. (1996). *Soil mechanics in engineering practice*. New York: Wiley-Interscience.

Toh, C. T., Ooi, T. A., Chiu, H. K., Chee, S. K., and Ting, W. H. (1989). Design parameters for bored piles in weathered sedimentary formation. In *12th International Conference on Soil Mechanics and Foundation Engineering*, Rio De Janeiro, pp. 1073–1078.

Turner, J. (2006). *Rock-socketed shafts for highway structure foundations*. Project 20-5, Washington, WA.

USACE. (1988). *Review of consolidation grouting of rock masses and methods for evaluation*. Report No. REMR-GT-8. Washington, DC: US Army Corps of Engineers.

Vesic, A. S. (1963). Bearing capacity of deep foundations in sand. *Highway Research Record*, (39), 112–153.

Williams, A. F. (1980). *The design and performance of piles into weak rock*. Ph.D. Thesis, Monash University, Melbourne, Australia.

Williams, A. F. and Pells, P. J. N. (1981). Side resistance rock sockets in sandstone, mudstone, and shale. *Canadian Geotechnical Journal*, 18(4), 502–513.

Yoshinaka, R. and Yamabe, T. (1986). 3. Joint stiffness and the deformation behaviour of discontinuous rock. *International Journal of Rock Mechanics and Mining Sciences and Geomechanics Abstracts*, 23(1), 19–28.

Zhang, L. and Einstein, H. H. (1998). End bearing capacity of drilled shafts in rock. *Journal of Geotechnical and Geoenvironmental Engineering*, 124(7), 574–584.

Zhang, L. (1999). *Analysis and design of drilled shafts in rock*. Ph.D. Thesis, Massachusetts Institute of Technology, Massachusetts, MA, United States.

Chapter 3

Evaluation of soil/rock properties using databases

Jianye Ching

The proper characterization of the variability of soil/rock properties for a specific site (including inherent variability and transformation uncertainty) is necessary for reliability-based design as well as for the determination of soil/rock characteristic values. This chapter reviews some useful sources that provide site-specific statistics for the inherent variability of soil/rock properties (e.g., site-specific mean, COV, and SOF). The emphasis is placed on the site-specific statistics extracted from some soil/rock property databases compiled by the author of this chapter. The soil/rock property databases can also be used to calibrate generic transformation models, but the transformation uncertainty is usually significant. The hierarchical Bayesian model (HBM) recently proposed by the author of this chapter may reduce the transformation uncertainty significantly. A real example is adopted to illustrate this reduction in the transformation uncertainty.

3.1 INTRODUCTION

The proper characterization of the variability of soil/rock properties for a specific site plays a critical role in the reliability-based design (RBD) of geotechnical structures (e.g., Phoon and Kulhawy 1999a; Baecher and Christian 2003; Fenton and Griffiths 2008). In Eurocode 7 (CEN 2004), the proper characterization of soil/rock property variability also plays an important role in the determination of the characteristic value. Clause 2.4.5.2(4) in Eurocode 7 states a list of factors for the selection of a characteristic value:

1. Geological and other background information, such as data from previous projects.
2. The variability of the measured property values and other relevant information, e.g., from existing knowledge.
3. The extent of the field and laboratory investigation.
4. The type and number of samples.
5. The extent of the zone of ground governing the behaviour of the geotechnical structure at the limit state being considered.
6. The ability of the geotechnical structure to transfer loads from weak to strong zones in the ground.

Factor #2 "the variability of the measured property values" recognizes the role of spatial variability of soil/rock property as well as measurement error. Factor #3 "the extent of the field and laboratory investigation" and Factor #4 "the number of samples" recognize the role of statistical uncertainty. Factor #5 "the extent of the zone of ground governing the behaviour of the geotechnical structure" recognizes the role of spatial average. The role of transformation uncertainty is not explicitly stated in the above list, but the draft version of

DOI: 10.1201/9781003441946-3

the revised Eurocode 7 (CEN 2020) explicitly recognizes the role of transformation uncertainty (Orr 2017; CEN 2020). The total variability involved during the evaluation of a soil/rock property can be summarized by the following equation (Phoon and Kulhawy 1999a; Schneider and Schneider 2013; Orr 2017):

$$COV_{TOT}^2 = \Gamma_S^2 \cdot COV_{inher}^2 + COV_{meas}^2 + COV_{trans}^2 + COV_{stat}^2 \quad (3.1)$$

where COV_{TOT} is the coefficient of variation (COV) for the total variability; COV_{inher} is for inherent (spatial) ground variability; Γ^2_S is a factor that considers the variance reduction due to spatial averaging over the zone of ground governing the behaviour of the geotechnical structure; COV_{meas} is for measurement uncertainty; COV_{trans} is for transformation uncertainty; and COV_{stat} is for statistical uncertainty. Schneider and Schneider (2013) adopted the following simplification for the variance reduction factor: $\Gamma_S^2 = \Gamma_x^2\Gamma_y^2\Gamma_z^2$, where Γ_x^2 and Γ_y^2 are the variance reduction factors in the two horizontal directions, x and y, and Γ_z^2 is the variance reduction factor in the vertical direction, z. The approximate value for $\Gamma_i^2 = 1$ when the size of the influence zone in the ith direction (denoted by L_i) is shorter than the scale of fluctuation in the ith direction (denoted by SOF_i). For the case of $L_i > SOF_i$, the approximate value is $\Gamma_i^2 = SOF_i/L_i$ (Vanmarcke 1977). It is customary to assume the soil property of interest follows a normal (Gaussian) population if the total COV is less than 30% (Schneider and Schneider 2013). In this case, the characteristic value can be calculated as

$$X_k = X_{mean}\left(1 - 1.645 \times COV_{TOT}\right) \quad (3.2)$$

where X_{mean} is the (sample) mean value of evaluated soil properties. If the total COV exceeds 30%, it is more reasonable to assume the soil property of interest follows a lognormal population (Schneider and Schneider 2013) to avoid a negative characteristic value:

$$X_k = X_{mean} \cdot \frac{e^{-1.645 \cdot \sqrt{\ln\left(1+COV_{TOT}^2\right)}}}{\sqrt{1+COV_{TOT}^2}} \approx X_{mean} \cdot \frac{0.2^{\sqrt{\ln\left(1+COV_{TOT}^2\right)}}}{\sqrt{1+COV_{TOT}^2}} \quad (3.3)$$

In the above equations, the evaluation of the characteristic value X_k may require site-specific statistics for inherent variability such as site-specific mean, COV_{inher}, and SOFs. However, in geotechnical site investigation, site-specific data are usually sparse and limited, particularly for small- or medium-sized projects. This leads to the difficulty in obtaining meaningful site-specific statistics of soil/rock properties. To deal with these challenges, sparse site-specific data might be integrated with prior knowledge such as typical ranges of site-specific mean, COV_{inher}, and SOFs (e.g., Phoon and Kulhawy 1999a; Wang and Cao 2013). This underlines a need to summarize the typical values of site-specific mean, COV_{inher}, and SOFs for different soil/rock properties from existing databases. The EPRI TR-105000 report (Phoon et al. 1995) and Phoon and Kulhawy (1999a) have complied the statistics for some soil parameters. Kulhawy et al. (2000) summarized the ranges of COV_{inher} for some soil/rock parameters as shown in Table 3.1, whereas Phoon et al. (1995) summarized the ranges of SOFs for some soil parameters as shown in Table 3.2.

Nonetheless, more soil/rock databases have been collected recently, and some databases are shown in Table 3.3, labelled as (soil type)/(number of parameters of interest)/(number of data points). Because these databases cover a broad range, they may be used to update the

Table 3.1 Coefficients of variation for inherent variability for some soil/rock parameters

Test type	*Property*	*Material type*	COV_{inher} (%)		
			# Groups	*Range*	*Mean*
Index	γ, γ_d (kN/m^3)	Fine-grained	14	2–20	7.8
	w_n (%)	Fine-grained	40	7–46	18.1
	PL	Fine-grained	23	6–34	15.7
	LL	Fine-grained	38	7–39	18.1
	PI – all data	Fine-grained	33	9–57	29.5
	– ≤20%	Fine-grained	13	16–57	35
	– >20%	Fine-grained	20	9–40	26
	γ, γ_d	Rock	42	0.1–3	0.9
	n (%)	Rock	25	3–71	25.9
Strength	ϕ, tan ϕ	Sand, clay	48	4–50	13.9
		Sand	32	4–15	9
		Clay	16	10–50	23.5
	s_u (kPa)	Clay	100	6–80	31.5
	q_u (kPa)	Rock	184	0.3–61	14.2
	σ_{bt} (kPa)	Rock	74	2–58	16.6
Stiffness	$E_{t\text{-}50}$ (MPa)	Rock	32	7–63	30.7
CPT	q_c (kPa)	Sand, clay	65	10–81	36.6
		Sand	54	10–81	38.2
		Clay	11	16–40	28.4
	q_t (kPa)	Clay	9	2–17	7.9
FV	s_u^{FV} (kPa)	Clay	26	13–36	25.3
SPT	N	Sand, clay	23	19–62	38
DMT	A,B	Sand, clay	56	12–59	27.9
		Sand	30	13–59	34.8
		Clay	26	12–38	19.9
	I_{DMT} (w/o outliers)	Sand	29	8–66	37.7
	K_{DMT} (w/o outliers)	Sand	29	15–67	37.6
	E_D (w/o outliers) (MPa)	Sand	30	7–69	41.1

Source: Table 1, Kulhawy et al. (2000).

Note: γ = total unit weight; n = apparent porosity; γ_d = dry unit weight; w_n = natural water content; PL = plastic limit; LL = liquid limit, PI = plasticity index; ϕ' = effective stress friction angle; s_u = undrained shear strength; q_u = uniaxial compressive strength; σ_{bt} = Brazilian indirect tensile strength; $E_{t\text{-}50}$ = tangent modulus at 0.5q_u; q_c = cone penetration test (CPT) tip resistance; q_t = corrected tip resistance; s_u^{FV} = undrained shear strength from field vane; N = standard penetration test (SPT) blow count; A, B = dilatometer test (DMT) A and B readings; I_{DMT} = dilatometer material index; K_{DMT} = dilatometer horizontal stress index; E_{DMT} = dilatometer modulus.

statistics reported in the EPRI TR-105000 report. One purpose of this chapter is to present the typical values of site-specific mean and COV_{inher} observed from the seven soil/rock databases, including CLAY/10/7490, SAND/7/2794, ROCK/9/4069, ROCKMass/9/5876, CLAY-C_c/6/6203, SOIL-DMT/8/7186, and SOIL-KSAT/7/2293 in Table 3.3. These databases are compiled by the author of this chapter. The details of these databases can be found in Chapter 1 of ISSMGE-TC304 (2021). Another purpose of this chapter is to present the typical values of site-specific SOFs observed from the literature. The

Table 3.2 Scales of fluctuation for some soil parameters

Direction	*Parameter*	*Soil type*	*# Studies*	*SOF (m)*	
				Range	*Mean*
Vertical	s_u	Clay	5	0.8–6.1	2.5
	q_c	Sand, clay	7	0.1–2.2	0.9
	q_t	Clay	10	0.2–0.5	0.3
	s_u^{FV}	Clay	6	2.0–6.2	3.8
	N	Sand	1	–	2.4
	w_n	Clay, loam	3	1.6–12.7	5.7
	LL	Clay, loam	2	1.6–8.7	5.2
	Γ	Clay, loam	2	2.4–7.9	5.2
Horizontal	q_c	Sand, clay	11	3.0–80.0	47.9
	q_t	Clay	2	23.0–66.0	44.5
	s_u^{FV}	Clay	3	46.0–60.0	50.7
	w_n	Clay	1	–	170.0

Source: Table 4-4, Phoon et al. (1995).

Table 3.3 Soil/rock databases compiled by the author of this chapter

Database	*Reference*	*Parameters of interest*	*# Records*	*# Studies*
CLAY/10/7490	Ching and Phoon (2014a)	LL, PI, LI, σ'_v/P_a, σ'_p/P_a, s_u/σ'_v, S_t, q_{t1}, q_{tu}, B_q	7490	251
SAND/7/2794	Ching et al. (2017)	D_{50}, C_u, D_r, σ'_v/P_a, ϕ', q_{c1n}, $(N_1)_{60}$	2794	176
ROCK/9/4069	Ching et al. (2018)	γ, n, R_L, S_h, σ_{bt}, I_{s50}, V_p, σ_{ci}, E_i	4069	184
ROCKMass/9/5876	Ching et al. (2021a)	RQD, RMR, Q, GSI, E_m, E_{em}, E_{dm}, E_i, σ_{ci}	5784	225
CLAY-C_c/6/6203	Ching et al. (2022)	LL, PI, w_n, e, C_c, C_{ur}	6203	427
SOIL-DMT/8/7186	Ching and Kuo (2023)	LI, D_{50}, s_u^{FV}/P_a, N_{60}, $(q_t-\sigma_v)/P_a$, B_q, E_{PMT}/P_a, $(1-\nu^2) \times E_{DMT}/P_a$	7286	237
SOIL-KSAT/7/2293	Ching et al. (2023)	k_{sat}, e, w, LL, PI, D_{10}, CC	2293	76

Note: ν = Poisson ratio; σ'_p = pre-consolidation stress; σ'_v = vertical effective stress; σ_{ci} = uniaxial compressive strength of intact rock; $(N_1)_{60} = N_{60}/(\sigma'_v/P_a)^{0.5}$; B_q = CPT pore pressure ratio = $(u_2-u_0)/(q_t-\sigma_v)$; CC = clay content (in %); C_c = compression index; C_{ur} = unload/reload index; C_u = coefficient of uniformity; D_{10} = 10% percentile grain size; D_{50} = median grain size; D_r = relative density; e = void ratio; E_{PMT} = soil modulus determined by PMT; E_{dm} = dynamic modulus of rock mass; E_{em} = elasticity modulus of rock mass; E_i = Young's modulus of intact rock; E_m = deformation modulus of rock mass; GSI = geological strength index; I_{s50} = point load strength index for diameter 50 mm; k_{sat} = saturated hydraulic conductivity; LI = liquidity index; N_{60} = corrected SPT N; P_a = atmospheric pressure = 101.3 kPa; Q = Q-system; $q_{c1n} = (q_c/P_a)/(\sigma'_v/P_a)^{0.5}$; $q_{t1} = (q_t-\sigma_v)/\sigma'_v$ = normalized cone tip resistance; $q_{tu} = (q_t-u_2)/\sigma'_v$ = effective cone tip resistance; R_L = L-type Schmidt hammer hardness; RMR = rock mass rating; RQD = rock quality designation; S_h = Shore scleroscope hardness; S_t = sensitivity; s_u^{re} = remoulded s_u; u_0 = hydrostatic pore pressure; u_2 = CPTU pore pressure; V_p = P-wave velocity; w = water content; w_n = natural water content.

SOF typical ranges are extracted from three new sources: Cami et al. (2020), Chapter 3 of ISSMGE-TC304 (2021), and Ching et al. (2023).

The COV_{trans} term in Equation (3.1) is for the transformation uncertainty. Transformation models (Phoon and Kulhway 1999b) are useful to infer soil/rock properties from indirect measurements. Although a site-specific transformation model is desirable, its calibration may require sufficient site-specific data. When site-specific data are sparse (which is usually the case), a generic transformation model calibrated by a soil/rock database may be adopted to infer soil/rock properties. The multivariate soil/rock databases in Table 3.3 can be adopted to calibrate generic transformation models. Another purpose of this chapter is to present the bias (b) and COV_{trans} for some useful generic transformation models calibrated by the databases in Table 3.3.

The main disadvantage of a generic transformation model is that it is usually subjected to significant transformation uncertainty (COV_{trans} is large). The final section (Section 3.4) presents a recent theoretical advancement called the hierarchical Bayesian model (HBM), which constructs a "quasi-site-specific" transformation model that combines the sparse target-site data with the experiences learned from a soil/rock database. The transformation uncertainty for a quasi-site-specific transformation model can be significantly less than that for a generic transformation model.

3.2 TYPICAL VALUES OF SITE-SPECIFIC MEAN AND COV_{INHER} OBSERVED IN SOIL/ROCK DATABASES

One purpose of the current chapter is to report the typical range of site-specific mean and COV_{inher} of soil/rock properties observed in the databases in Table 3.3. Another purpose is to present the bias (b) and COV_{trans} of some useful generic transformation models calibrated by these databases.

3.2.1 CLAY/10/7490

The CLAY/10/7490 database (Ching and Phoon 2014a) compiles 7490 records for ten dimensionless clay parameters, including liquid limit (LL), plasticity index (PI), liquidity index (LI), effective stress (σ'_v/P_a), pre-consolidation stress (σ'_p/P_a), normalized undrained shear strength (s_u/σ'_v), sensitivity (S_t), normalized CPT cone resistances (q_{t1} and q_{tu}), and CPT pore pressure coefficient (B_q). For s_u/σ'_v, all s_u values in CLAY/10/7490 are converted to the mobilized s_u values defined by Mesri and Huvaj (2007), denoted by s_u(mob). The clay properties cover a wide range of LL (from 18 to 515), PI (from 2 to 363), LI (from −0.75 to 6.5), OCR (from 1 to 60.2), and S_t (from 1 to 1467). Details are given by Ching and Phoon (2014a). Each record in CLAY/10/7490 is stored as one row in the Excel worksheet. There were originally 7490 rows (records) in the worksheet, but the number of records has gradually grown since the CLAY/10/7490 database was first presented by Ching and Phoon (2014a). At the time this chapter is written, there are 12,754 records. The ten parameters are not completely observed for each record, i.e., there are empty entries. If the database is complete, it contains 12,754 × 10 = 127,540 entries. The current CLAY/10/7490 contains 55,976 entries, i.e., it is 44% complete.

The basic marginal statistics of the ten parameters are listed in Table 3.4. The statistics for five derived parameters (s_u/σ'_p, OCR, w_n, q_t, s_u) are also shown. The number of records containing a given parameter (e.g., LL) is shown in the second column of Table 3.4. Note that the statistics in Table 3.4 are not for a specific site but for the entire database (generic), i.e., data from multiple sites are pooled together. In this chapter, the term "site" can be regarded as the footprint below a typical size building or building complex and its peripheral land, occupying a few football fields. Different field and laboratory tests can be conducted at different locations within a site boundary. The records obtained from these tests conducted at

Table 3.4 Summary of generic statistics for (expanded) CLAY/10/7490

Variable	*# Data*	*Mean*	*COV*	*Min*	*Max*
LL (%)	10315	62.9	0.64	18.0	612.7
PI (%)	10720	35.0	0.94	1.7	493.0
LI	8410	0.86	0.91	−2.72	8.20
σ'_v/P_a	8692	1.50	1.46	0.00	35.32
σ'_p/P_a	5608	2.84	1.81	0.01	167.82
s_u/σ'_v	3914	0.50	1.57	0.02	16.00
S_t	3228	51.7	17.08	1.0	26,679.2
B_q	1475	0.50	0.65	−0.10	3.12
q_{t1}	2083	13.29	1.53	0.68	354.50
q_{tu}	1531	7.75	1.76	−7.34	228.24
s_u/σ'_p	2238	0.21	0.46	0.072	0.94
OCR	5362	2.95	1.98	0.045	212.0
w_n (%)	9588	56.5	0.80	5.80	868.7
q_t (kPa)	2209	1474.3	1.06	66.7	15,748.0
s_u (kPa)	3468	58.6	1.48	0.47	1049.8

Table 3.5 Summary of site-specific statistics for (expanded) CLAY/10/7490

	#	*# Records/site*		*Site-specific mean*		*Site-specific COV_{inher}*	
Property	*Sites*	*Range*	*Mean*	*95% CI*	*Mean*	*95% CI*	*Mean*
LL (%)	551	5-61	12.4	27.7–144.2	62.7	0.036–0.41	0.18
PI (%)	575	5-61	12.5	7.7–88.7	35.1	0.048–0.58	0.26
LI	479	5-61	12.6	−0.16–2.29	0.89	0.068–1.63	0.40
σ'_v/P_a	526	5-64	13.2	0.17–6.57	1.40	0.12–1.01	0.44
σ'_p/P_a	326	5-61	12.0	0.27–15.74	2.60	0.058–0.91	0.38
s_u/σ'_v	234	5-63	12.2	0.14–1.80	0.52	0.078–1.62	0.45
S_t	165	5-60	11.7	1.0–175.47	24.2	0.068–1.45	0.50
B_q	98	5-46	13.9	0.028–0.99	0.52	0.065–1.46	0.37
q_{t1}	140	5-64	13.6	2.29–57.6	14.01	0.097–2.11	0.56
q_{tu}	102	5-46	13.8	1.11–27.9	7.72	0.12–2.29	0.64
s_u/σ'_p	148	5-59	10.8	0.088–0.39	0.22	0.052–0.62	0.22
OCR	315	5-60	12.2	0.88–14.67	2.90	0.068–1.54	0.50
w_n (%)	549	5-61	12.9	19.4–141.6	60.9	0.039–0.53	0.19
q_t (kPa)	149	5-64	13.6	274.8–4806.3	1513.4	0.099–1.11	0.42
s_u (kPa)	229	5-63	12.1	8.51–299.3	58.4	0.069–0.88	0.38

different depths and different locations are grouped into a "site." For CLAY/10/7490, the number of records associated with each site varies from 5 to 64. Table 3.5 shows the site-specific statistics. To obtain meaningful second-order site-specific statistics, only sites with no less than five records are considered.

Soil/rock parameters may be correlated to each other. The site-specific bivariate correlation coefficients are shown in Table 3.6. They are computed based on the multivariate

Table 3.6 Summary of site-specific correlations for (expanded) CLAY/10/7490 (Pearson, Spearman, Kendall)

	PI	LI	σ'_v/P_a	σ'_p/P_a	s_u/σ'_v	S_t	B_q
LL	0.90,0.85,0.75 (459)	−0.24,−0.22,−0.17 (423)	−0.21,−0.18,−0.13 (358)	−0.16,−0.16,−0.13 (223)	0.04,0.03,0.02 (169)	−0.10,−0.08,−0.06 (104)	0.04,0.02,0.00 (65)
PI		−0.21,−0.18,−0.14 (427)	−0.20,−0.17,−0.13 (369)	−0.13,−0.14,−0.11 (233)	0.03,0.01,0.01 (178)	−0.12,−0.10,−0.09 (107)	0.04,0.03,0.03 (68)
LI			−0.11,−0.13,−0.11 (359)	−0.31,−0.30,−0.24 (217)	−0.10,−0.05,−0.04 (180)	0.29,0.24,0.20 (104)	0..03,0.02,0.01 (68)
σ'_v/P_a				0.47,0.45,0.40 (294)	−0.57,−0.56,−0.47 (228)	0.05,0.04,0.02 (80)	0.41,0.41,0.34 (98)
σ'_p/P_a					−0.15,−0.23,−0.18 (146)	0.04,0.08,0.06 (56)	0.15,0.22,0.16 (54)
s_u/σ'_v						−0.08,0.02,0.03 (57)	−0.49,−0.41,−0.34 (43)
S_t							0.12,0.11,0.09 (22)
B_q							
q_{t1}							
q_{tu}							
s_u/σ'_p							
OCR							
W							
q_t							

Table 3.6 Continued

q_{tl}	q_{tu}	s_u/σ'_p	*OCR*	w	q_t	s_u
0.00,0.01,0.01 (86)	−0.01,0.00,0.00 (68)	−0.06,−0.09,−0.08 (103)	0.09,0.11,0.09 (204)	0.61,0.57,0.47 (428)	−0.11,−0.09,−0.07 (94)	−0.22,−0.23,−0.17 (169)
−0.01,0.00,0.00 (90)	−0.02,−0.01,−0.01 (71)	−0.06,−0.09,−0.07 (110)	0.11,0.12,0.10 (211)	0.54,0.51,0.41 (434)	−0.09,−0.09,−0.06 (98)	−0.23,−0.25,−0.19 (175)
−0.11,−0.09,−0.08 (88)	−0.07,−0.05,−0.04 (71)	0.00,0.04,0.04 (110)	−0.12,−0.09,−0.07 (199)	0.40,0.37,0.30 (421)	−0.24,−0.20,−0.17 (96)	−0.28,−0.27,−0.21 (179)
−0.45,−0.45,−0.38 (140)	−0.50,−0.52,−0.44 (102)	0.01,−0.01,0.00 (147)	−0.54,−0.51,−0.43 (296)	−0.29,−0.27,−0.22 (400)	0.63,0.62,0.55 (143)	0.61,0.56,0.49 (228)
−0.14,−0.24,−0.19 (61)	−0.13,−0.25,−0.19 (53)	−0.27,−0.28,−0.22 (146)	0.20,0.14,0.11 (296)	−0.42,−0.41,−0.33 (261)	0.70,0.66,0.57 (62)	0.67,0.63,0.54 (146)
0.69,0.57,0.46 (48)	0.66,0.52,0.42 (43)	0.18,0.28,0.25 (129)	0.73,0.62,0.53 (130)	0.00,0.03,0.01 (182)	−0.35,−0.39,−0.32 (48)	0.02,−0.03,−0.03 (228)
−0.14,−0.21,−0.17 (24)	−0.12,−0.15,−0.12 (22)	0.07,0.01,0.02 (40)	−0.07,−0.03,−0.01 (53)	0.12,0.10,0.07 (109)	−0.03,−0.08,−0.06 (24)	−0.03,−0.02,−0.02 (57)
−0.61,−0.61,−0.52 (98)	−0.76,−0.78,−0.70 (98)	0.05,0.07,0.06 (33)	−0.50,−0.50,−0.42 (54)	0.09,0.05,0.03 (71)	0.02,0.07,0.07 (98)	0.23,0.24,0.20 (43)
	0.92,0.86,0.78 (101)	−0.03,0.01,0.01 (33)	0.61,0.50,0.42 (61)	−0.05,−0.01,−0.01 (94)	0.13,0.02,0.01 (140)	−0.22,−0.27,−0.21 (48)
		−0.10,−0.06,−0.5 (32)	0.68,0.60,0.50 (53)	−0.05,−0.03,−0.02 (74)	0.03,−0.11,−0.10 (102)	−0.22,−0.27,−0.21 (43)
			−0.32,−0.27,−0.23 (130)	−0.02,−0.03,−0.4 (114)	−0.15,−0.17,−0.11 (33)	0.29,0.22,0.19 (143)
				−0.01,0.03,0.02 (235)	−0.27,−0.32,−0.27 (62)	−0.24,−0.22,−0.18 (130)
					−0.30,−0.27,−0.22 (102)	−0.39,−0.37,−0.30 (182)
						0.64,0.60,0.52 (48)

site-specific records in CLAY/10/7490. Three types of correlation coefficients are considered: (a) the Pearson product–moment correlation coefficient; (b) the Spearman rank correlation coefficient (Hotelling and Pabst 1936); and (c) the Kendall's tau rank correlation coefficient (Kendall 1938). The numbers in the parentheses indicate the numbers of sites with no less than five bivariate site-specific records. The correlations shown in the table are the mean values among these sites.

The CLAY/10/7490 database can also be used to calibrate some generic transformation models. A generic transformation model is usually developed by regression based on bivariate/multivariate records in a soil/rock database. The resulting regression equation does not provide an exact fit, and the variability is called the transformation uncertainty. When implemented to a future case that is not within the database, a generic transformation model may exhibit both bias and transformation uncertainty. Ching and Phoon (2014a) defined the bias (b) as the sample mean of the following ratio:

$$\varepsilon = \frac{\text{measured target value}}{\text{prediction made by a transformation model}} \tag{3.4}$$

The COV of the transformation uncertainty (COV_{trans} in Equation 3.1) can be estimated as the sample COV of the ε ratio. The bias (b) and COV_{trans} of a transformation model can be calibrated by a generic soil/rock database. For instance, consider the s_u-CPT transformation model in Table 3.7. The target value is s_u, and the prediction is $(q_t-\sigma_v)/[29.1 \times \exp(-0.513 \times B_q)]$. For all records in CLAY/10/7490 with simultaneous knowledge of s_u, q_t, σ_v, and B_q, their ε ratios can be computed, and b and COV_{trans} are simply the sample mean and sample COV of these ε ratios. In Table 3.7, the values of b and COV_{trans} for some transformation models are calibrated by the records in CLAY/10/7490. The numbers of records used to calibrate b and COV_{trans} are shown in the third column in the table. To implement the calibrated transformation model to a future case with known (q_t, σ_v, B_q), the unbiased prediction for its s_u is simply $b \times (q_t-\sigma_v)/[29.1 \times \exp(-0.513 \times B_q)]$, and the transformation uncertainty has a COV = COV_{trans}.

3.2.2 SAND/7/2794

The SAND/7/2794 database (Ching et al. 2017) compiles 2794 records for seven sand parameters, including the median grain size (D_{50}), coefficient of uniformity (C_u), relative density (D_r), normalized vertical effective stress (σ'_v/P_a), effective stress friction angle (ϕ'), normalized cone tip resistance $q_{c1n} = (q_c/P_a) \times C_N$ ($C_N = (\sigma'_v/P_a)^{-0.5}$ is the correction factor for overburden stress), and normalized N value $(N_1)_{60} = N_{60} \times C_N$ is recorded. The sand properties cover a wide range of D_{50} (0.1 mm to more than 100 mm), C_u (1 to more than 1000), D_r (−0.1% to 117%), and OCR (1 to 15, but mostly 1). Details are given in Ching et al. (2017). The SAND/7/2794 database is 59% complete. The SAND/7/2794 database is dominated by reconstituted sands (79%). Only 21% of the records are in situ sands. In particular, 86% of the q_{c1n} data are based on reconstituted sands (laboratory calibration chamber tests) and 93% of the ϕ' data are for reconstituted sands.

The generic statistics of the seven parameters are listed in Table 3.8. There are only limited in situ records in SAND/7/2794, so there are insufficient sites to obtain meaningful site-specific statistics. The statistics in Table 3.8 are dominated by reconstituted sands. The bivariate correlation coefficients (Pearson, Spearman, and Kendall) are computed

Table 3.7 Some transformation models calibrated by (expanded) CLAY/10/7490

Model	*Literature*	*# Data*	*Transformation model*	*b*	*COV$_{trans}$*
$s_u^{re} - LI$	Locat and Demers (1988)	915	$s_u^{re} / P_a \approx 0.0144 \times LI^{-2.44}$	6.68	3.17
$S_t - LI$	Bjerrum (1954)	2329	$S_t \approx 10^{0.8LI}$	1.48	1.27
$\sigma'_p - LI - S_t$	Ching and Phoon (2012a)	794	$\sigma'_p / P_a \approx 0.235 \times LI^{-1.319} \times S_t^{0.536}$	2.24	1.01
$s_u - \sigma'_p$	Mesri (1975, 1989)	2238	$s_u(mob) / \sigma'_p \approx 0.22$	0.97	0.46
$s_u - OCR$	Jamiolkowski et al. (1985)	2238	$s_u(mob) / \sigma'_v \approx 0.23 \times OCR^{0.8}$	1.05	0.48
$s_u - CPT$	Ching and Phoon (2012b)	658	$s_u(mob) \approx \dfrac{q_t - \sigma_v}{29.1 \times \exp(-0.513 B_q)}$	1.30	0.45
$OCR - CPT$	Kulhwawy and Mayne (1990)	856	$OCR \approx 0.32 \times (q_t - \sigma_v) / \sigma'_v$	1.04	0.67
$\sigma'_p - CPT$	Kulhwawy and Mayne (1990)	856	$\sigma'_p \approx 0.33 \times (q_t - \sigma_v)$	1.01	0.67

Table 3.8 Summary of generic statistics for SAND/7/2794

Variable	*# Data*	*Mean*	*COV*	*Min*	*Max*
D_{50} (mm)	2356	2.52	3.43	0.080	160.0
C_u	2196	14.7	5.43	1.0	1913.0
D_r (%)	2089	64.6	0.38	−0.071	117.0
σ'_v/P_a	1772	1.80	1.19	0.010	24.7
ϕ' (°)	1135	39.5	0.13	22.8	59.9
q_{c1n}	1541	157.2	0.72	0.75	536.8
$(N_1)_{60}$	611	34.1	0.76	2.15	243.5

Source: Ching et al. (2017).

based on the multivariate generic records in SAND/7/2794. The generic correlations are shown in Table 3.9. The numbers in the parentheses indicate the numbers of bivariate generic records. Table 3.10 shows the values of b and COV$_{trans}$ for some generic transformation models calibrated by the records in SAND/7/2794. It is worth mentioning that the two ϕ'-SPT transformation models are calibrated by undisturbed in situ samples (e.g., extracted by ground freezing).

3.2.3 ROCK/9/4069

The ROCK/9/4069 database (Ching et al. 2018) compiles 4069 records for nine intact rock parameters, including unit weight (γ), porosity (n), L-type Schmidt hammer hardness (R_L), Shore scleroscope hardness (S_h), Brazilian tensile strength (σ_{bt}), point load strength index (I_{s50}), uniaxial compressive strength (σ_{ci}), and Young's modulus (E_i), and P-wave velocity (V_p). Jointed rock masses are not covered by this database. The database is dominated by igneous

Table 3.9 Summary of generic correlations for SAND/7/2794 (Pearson, Spearman, Kendall)

	C_u	D_r	σ'_v/P_a	ϕ'	q_{c1n}	$(N_1)_{60}$
D_{50}	0.53,0.37,0.24 (2153)	0.10,0.13,0.08 (1844)	0.53,0.20,0.14 (1483)	0.22,0.40,0.27 (1021)	0.18,0.24,0.16 (1375)	0.29,0.24,0.17 (351)
C_u		0.08,0.09,0.06 (1763)	0.15,0.16,0.10 (1452)	0.25,0.27,0.19 (895)	−0.10,−0.04,−0.03 (1358)	0.14,0.37,0.26 (267)
D_r			0.04,−0.03,−0.02 (1465)	0.53,0.52,0.37 (716)	0.77,0.82,0.62 (1060)	0.60,0.74,0.58 (354)
σ'_v/P_a				−0.07,−0.08,−0.06 (295)	−0.07,0.10,0.00 (1255)	−0.14,−0.19,−0.13 (553)
ϕ'					0.71,0.78,0.56 (376)	0.46,0.44,0.30 (69)
q_{c1n}						0.88,0.84,0.66 (188)

Table 3.10 Some generic transformation models calibrated by SAND/7/2794

Model	*Literature*	*# Data*	*Transformation model*	*b*	COV_{trans}
D_r – SPT	Terzaghi and Peck (1967)	198	$D_r(\%) \approx 100 \times \sqrt{(N_1)_{60} / 60}$	1.05	0.23
	Kulhawy and Mayne (1990)	199	$D_r(\%) \approx 100 \times \sqrt{\frac{(N_1)_{60}}{[60 + 25\log_{10}(D_{50})] \times OCR^{0.18}}}$	1.01	0.21
D_r – CPT	Jamiolkowski et al. (1985)	681	$D_r(\%) \approx 68 \times [\log_{10}(q_{c1n}) - 1]$	0.84	0.33
ϕ'– D_r	Bolton (1986)	391	$\phi' \approx_{cv} +3 \times (D_r[10 - \ln(p'_f)] - 1)$ [b]	1.03	0.052
ϕ'– SPT	Hatanaka and Uchida (1996)	58[a]	$\phi' \approx \begin{cases} \sqrt{15.4 \times (N_1)_{60}} + 20 & (N_1)_{60} \leq 26 \\ 40 & (N_1)_{60} > 26 \end{cases}$	1.07	0.090
	Chen (2004)	59[a]	$\phi' \approx 27.5 + 9.2 \times \log_{10}[(N_1)_{60}]$	1.00	0.095
ϕ'– CPT	Robertson and Campanella (1983)	99	$\phi' \approx \tan^{-1}[0.1 + 0.38 \times \log_{10}(q_t / \sigma'_v)]$	0.93	0.056
	Kulhawy and Mayne (1990)	376	$\phi' \approx 17.6 + 11 \times \log_{10}(q_{c1n})$	0.97	0.081

Source: ISSMGE-TC304 (2021).

[a] Based on undisturbed in situ samples (e.g., extracted by ground freezing).

[b] ϕ'_{cv} = critical-state friction angle; p'_f = mean confining stress at failure in kPa.

and sedimentary rocks (27.5% and 59.4%, respectively). The remaining (about 13.1%) data points are metamorphic rocks. The records in ROCK/9/4069 cover a wide range of γ (15 to 35 kN/m^3), n (0.01 to 55%), σ_{ci} (0.7 to 380 MPa), E_i (0.03 to 120 GPa), and V_p (0.4 to 8 km/s). Details are given in Ching et al. (2018). The ROCK/9/4069 database is 34% complete.

The generic statistics of the nine parameters are listed in Table 3.11. Tables 3.12–3.14 show the site-specific mean and COV$_{\text{inher}}$ for igneous, sedimentary, and metamorphic sites, respectively. Table 3.15 shows the site-specific bivariate correlation coefficients. The

Table 3.11 Summary of generic statistics for ROCK/9/4069

Variable	*# Data*	*Mean*	*COV*	*Min*	*Max*
γ (kN/m^3)	1288	24.6	0.12	15.0	34.7
n (%)	1371	11.0	1.02	0.01	55.0
R_L	812	41.4	0.29	8.1	81.6
S_h	333	47.6	0.44	8.4	100.0
σ_{bt} (MPa)	854	8.2	0.61	0.07	34.4
I_{s50} (MPa)	1303	4.2	0.69	0.05	17.4
σ_{ci} (MPa)	3226	72.3	0.77	0.68	379.0
E_i (GPa)	1495	26.5	0.99	0.03	116.3
V_P (km/s)	1858	4.0	0.39	0.44	8.0

Source: Ching et al. (2018).

Table 3.12 Summary of site-specific statistics for ROCK/9/4069 (igneous)

		# Records/site		Site-specific mean		Site-specific COV_{inher}	
Property	# Sites	Range	Mean	95% CI	Mean	95% CI	Mean
γ (kN/m^3)	11	5-20	11.0	21.4–30.1	25.4	0.002–0.18	0.049
n (%)	7	5-47	24.3	0.22–7.78	4.03	0.15–0.73	0.46
R_L	9	5-145	30.9	35.4–62.6	48.4	0.020–0.28	0.13
S_h	1	20	20	–	54.9	–	0.16
σ_{bt} (MPa)	5	5-24	15.0	11.2–19.4	14.8	0.064–0.54	0.19
I_{s50} (MPa)	9	5-145	38.0	1.75–9.02	4.44	0.14–0.81	0.41
σ_{ci} (MPa)	24	5-145	25.4	35.4–244.1	106.8	0.016–0.87	0.35
E_i (GPa)	15	5-62	21.1	4.45–86.0	34.5	0.029–0.60	0.31
V_P (km/s)	17	5-145	27.5	2.54–7.55	4.57	0.002–0.35	0.12

Table 3.13 Summary of site-specific for ROCK/9/4069 (sedimentary)

		# Records/site		Site-specific mean		Site-specific COV_{inher}	
Property	# Sites	Range	Mean	95% CI	Mean	95% CI	Mean
γ (kN/m^3)	40	5-66	16.0	18.6–28.1	24.6	0.003–0.13	0.037
n (%)	42	5-55	17.5	1.09–37.4	13.1	0.030–0.74	0.33
R_L	14	5-44	13.6	20.4–44.4	31.5	0.026–0.32	0.14
S_h	6	8-31	20.5	13.4–76.1	43.2	0.12–0.35	0.23
σ_{bt} (MPa)	16	5-45	17.8	1.25–14.9	6.81	0.043–0.81	0.26
I_{s50} (MPa)	34	5-44	14.0	0.53–10.7	4.51	0.042–0.74	0.25
σ_{ci} (MPa)	78	5-66	14.3	7.05–134.6	61.5	0.019–0.71	0.26
E_i (GPa)	40	5-66	15.0	0.38–64.7	25.5	0.022–1.15	0.38
V_P (km/s)	58	5-66	13.3	0.89–6.30	3.67	0.004–0.37	0.10

Table 3.14 Summary of site-specific statistics for ROCK/9/4069 (metamorphic)

		# Records/site		Site-specific mean		Site-specific COV_{inher}	
Property	# Sites	Range	Mean	95% CI	Mean	95% CI	Mean
γ (kN/m^3)	11	5-20	9.5	25.7–30.7	27.1	0.010–0.058	0.025
n (%)	7	5-20	10.3	0.18–6.67	1.47	0.26–1.39	0.56
R_L	2	8	8.0	47.0–50.7	48.9	0.10–0.16	0.13
S_h	–	–	–	–	–	–	–
σ_{bt} (MPa)	7	5-20	8.7	6.67–18.1	11.9	0.13–0.54	0.30
I_{s50} (MPa)	11	5-32	13.3	1.27–6.92	3.96	0.11–0.41	0.30
σ_{ci} (MPa)	17	5-32	11.7	24.1–172.5	81.0	0.068–0.55	0.29
E_i (GPa)	6	5-20	9.0	4.49–102.3	54.6	0.081–0.82	0.25
V_P (km/s)	13	5-32	12.5	2.27–6.27	4.66	0.026–0.31	0.11

numbers in the parentheses indicate the numbers of sites with no less than 5 bivariate site-specific records. Ching et al. (2018) indicated that the correlation behaviours between rock parameters do not seem to depend on the rock class (igneous, sedimentary, and metamorphic), so Table 3.15 does not differentiate the rock class. Table 3.16 shows the values of b and COV_{trans} for some generic transformation models calibrated by the records in ROCK/9/4069.

Table 3.15 Summary of site-specific correlations for ROCK/9/4069 (Pearson, Spearman, Kendall)

	n	R_L	S_h	σ_{bt}	I_{s50}	σ_{ci}	E_i	V_P
γ	−0.42,−0.37,−0.30 (22)	0.57,0.48,0.34 (7)	0.27,0.38,0.30 (3)	0.56,0.53,0.43 (8)	0.65,0.63,0.57 (16)	0.59,0.56,0.48 (43)	0.38,0.37,0.29 (18)	0.47,0.44,0.39 (39)
n		−0.38,−0.38,−0.30 (5)	−0.68,−0.62,−0.47 (3)	−0.55,−0.54,−0.43 (15)	−0.53,−0.56,−0.47 (19)	−0.54,−0.55,−0.44 (37)	−0.56,−0.51,−0.41 (28)	−0.57,−0.53,−0.43 (29)
R_L			0.94,0.93,0.80 (1)	0.90,0.88,0.75 (2)	0.66,0.61,0.49 (8)	0.72,0.71,0.58 (16)	0.64,0.58,0.47 (7)	0.47,0.45,0.38 (15)
S_h				0.69,0.68,0.54 (2)	0.76,0.73,0.58 (3)	0.55,0.51,0.40 (6)	0.55,0.51,0.40 (5)	0.91,0.85,0.72 (2)
σ_{bt}					0.57,0.52,0.44 (15)	0.60,0.57,0.46 (23)	0.42,0.37,0.28 (12)	0.43,0.44,0.35 (14)
I_{s50}						0.72,0.63,0.52 (32)	0.27,0.50,0.39 (13)	0.75,0.62,0.51 (15)
σ_{ci}							0.60,0.57,0.45 (55)	0.69,0.62,0.52 (48)
E_i								0.63,0.67,0.55 (29)

Table 3.16 Some generic transformation models calibrated by ROCK/9/4069

Model	*Literature*	*# Data*	*Transformation model*	*b*	COV_{trans}
$\sigma_{ci} - n$	Kilic and Teymen (2008)	911	$\sigma_{ci} \approx 147.16 \times e^{-0.0835n}$	0.91	0.75
$\sigma_{ci} - R_L$	Karaman and Kesimal (2015)	664	$\sigma_{ci} \approx 0.1383 \times R_L^{1.743}$	0.78	0.53
$\sigma_{ci} - S_h$	Altindag and Guney (2010)	297	$\sigma_{ci} \approx 0.1821 \times S_h^{1.5833}$	1.15	0.65
$\sigma_{ci} - \sigma_{bt}$	Prakoso and Kulhawy (2011)	525	$\sigma_{ci} \approx 7.8 \times \sigma_{bt}$	1.31	0.50
$\sigma_{ci} - I_{s50}$	Mishra and Basu (2013)	1074	$\sigma_{ci} \approx 14.63 \times I_{s50}$	1.18	0.45
$\sigma_{ci} - V_P$	Kahraman (2001)	1247	$\sigma_{ci} \approx 9.95 \times V_P^{1.21}$	1.26	0.63
$E_i - R_L$	Katz et al. (2000)	289	$E_i \approx 0.00013 \times R_L^{3.09074}$	1.47	0.99
$E_i - S_h$	Deere and Miller (1966)	197	$E_i \approx 0.739 \times S_h + 11.51$	0.61	0.71
$E_i - \sigma_c$	Deere and Miller (1966)	1152	$E_i \approx 0.303 \times \sigma_{ci} - 0.8745$	1.23	0.94
$E_i - V_P$	Yaşar and Erdogan (2004)	192	$E_i \approx 10.67 \times V_P - 18.71$	0.90	0.72

Source: ISSMGE-TC304 (2021).

3.2.4 ROCKMass/9/5876

The ROCKMass/9/5876 database (Ching et al. 2021a) compiles 5876 records for nine rock mass parameters, including rock quality designation (RQD) (Deere 1964), rock mass rating (RMR) (Bieniawski 1973), Q-system (Barton et al. 1974), geological strength index (GSI) (Hoek and Brown 1997), deformation modulus of rock mass (E_m), elastic modulus of rock mass (E_{em}), dynamic modulus of rock mass (E_{dm}), Young's modulus of intact rock (E_i), and uniaxial compressive strength of intact rock (σ_{ci}). The database consists of 17% igneous, 37% sedimentary, and 26% metamorphic cases (the remaining 20% of the cases do not contain rock class information). The remaining (about 13.1%) data points are metamorphic rocks. The records in ROCKMass/9/5876 cover a wide range of RMR (0 to 97), Q (0.001 to 1000), GSI (8 to 100), and E_m (0.0011 to 104 GPa). Details are given by Ching et al. (2021a). The ROCKMass/9/5876 database is 29% complete.

The generic statistics of the nine parameters are listed in Table 3.17. Tables 3.18–3.20 show the site-specific mean and COV_{inher} for igneous, sedimentary, and metamorphic sites,

Table 3.17 Summary of generic statistics for ROCKMass/9/5876

Variable	*# Data*	*Mean*	*COV*	*Min*	*Max*
RQD	1380	62.5	0.48	0	100
RMR	3399	55.5	0.33	0	97
Q	1933	21.5	2.92	0.001	1000
GSI	871	47.2	0.34	8	100
E_m (GPa)	3212	9.99	1.15	0.0011	104
E_{em} (GPa)	1386	14.50	1.12	0.00014	121
E_{dm} (GPa)	466	31.78	0.60	0.0411	85
E_i (GPa)	592	29.81	0.95	0.24	278
σ_{ci} (MPa)	2071	74.07	0.78	0.15	318

Source: Ching et al. (2021a).

Table 3.18 Summary of site-specific statistics for ROCKMass/9/5876 (igneous)

		# Records/site		Site-specific mean		Site-specific COV_{inher}	
Property	*# Sites*	*Range*	*Mean*	*95% CI*	*Mean*	*95% CI*	*Mean*
RQD	13	5-70	18.2	27.7–91.7	72.5	0.022–1.21	0.31
RMR	13	5-40	20.2	34.9–81.2	61.9	0.064–0.53	0.22
Q	15	7-70	22.3	0.36–74.3	21.1	0.36–3.04	1.13
GSI	6	6-26	12.2	36.3–58.2	50.8	0.016–0.30	0.13
E_m (GPa)	5	6-21	10.4	4.90–20.7	10.6	0.13–1.02	0.62
E_{em} (GPa)	6	5-21	11.7	1.91–25.3	14.1	0.57–1.04	0.71
E_{dm} (GPa)	–	–	–	–	–	–	–
E_i (GPa)	6	5-15	8.0	12.8–77.5	32.5	0.17–1.00	0.69
σ_{ci} (MPa)	11	5-31	14.2	3.00–192.0	76.5	0.13–1.18	0.35

Table 3.19 Summary of site-specific for ROCKMass/9/5876 (sedimentary)

		# Records/site		Site-specific mean		Site-specific COV_{inher}	
Property	*# Sites*	*Range*	*Mean*	*95% CI*	*Mean*	*95% CI*	*Mean*
RQD	22	5-80	14.9	0.37–93.4	47.9	0.061–1.07	0.37
RMR	30	5-80	19.1	27.4–73.2	47.8	0.053–0.53	0.23
Q	19	5-49	16.0	0.30–18.0	5.56	0.22–1.59	0.85
GSI	20	5-93	18.1	22.2–75.6	43.9	0.054–0.57	0.23
E_m (GPa)	20	5-94	20.1	0.12–30.6	9.53	0.14–1.34	0.67
E_{em} (GPa)	8	6-27	10.5	1.02–56.9	12.6	0.36–0.99	0.63
E_{dm} (GPa)	4	5-27	13.3	8.10–54.6	24.0	0.15–0.62	0.33
E_i (GPa)	11	5-53	13.7	2.80–76.0	21.5	0.00–1.20	0.42
σ_{ci} (MPa)	35	5-80	13.0	3.24–184.0	55.4	0.077–1.06	0.40

Table 3.20 Summary of site-specific statistics for ROCKMass/9/5876 (metamorphic)

		# Records/site		Site-specific mean		Site-specific COV_{inher}	
Property	*# Sites*	*Range*	*Mean*	*95% CI*	*Mean*	*95% CI*	*Mean*
RQD	11	5-47	15.8	20.2–81.9	56.4	0.078–1.20	0.40
RMR	12	5-330	43.2	32.4–74.0	49.4	0.12–0.41	0.27
Q	11	6-330	49.1	0.13–55.2	9.75	0.50–5.26	1.54
GSI	3	10-26	16.0	23.1–53.6	39.7	0.054–0.18	0.13
E_m (GPa)	8	5-24	10.1	0.71–17.2	5.86	0.24–0.74	0.50
E_{em} (GPa)	3	5-24	11.3	2.37–9.79	5.95	0.38–0.55	0.46
E_{dm} (GPa)	2	5-5	5.0	2.94–18.7	10.8	0.30–0.45	0.37
E_i (GPa)	5	5-18	8.4	8.64–92.8	47.8	0.15–0.82	0.46
σ_{ci} (MPa)	9	5-30	13.6	30.0–99.9	55.5	0.24–0.72	0.40

respectively. Table 3.21 shows the site-specific bivariate correlation coefficients. The numbers in the parentheses indicate the number of sites with no less than 5 bivariate site-specific records. Ching et al. (2021a) indicated that the correlation behaviours between rock mass parameters do not seem to depend on the rock class (igneous, sedimentary, and metamorphic). Table 3.22 shows the values of b and COV_{trans} for some generic transformation models calibrated by the records in ROCKMass/9/5876.

Table 3.21 Summary of site-specific correlations for ROCKMass/9/5876 (Pearson, Spearman, Kendall)

	RMR	*Q*	*GSI*	E_m	E_{em}	E_{dm}	E_i	σ_{ci}
RQD	0.74,0.74,0.62 (26)	0.45,0.60,0.52 (24)	0.61,0.70,0.55 (5)	0.43,0.36,0.28 (9)	0.21,0.16,0.12 (6)	0.85,0.82,0.72 (3)	−0.03,−0.06,−0.04 (5)	0.25,0.27,0.23 (29)
RMR		0.62,0.69,0.58 (32)	0.76,0.74,0.65 (13)	0.76,0.70,0.58 (11)	0.29,0.20,0.14 (6)	0.60,0.59,0.47 (3)	0.25,0.28,0.22 (9)	0.46,0.45,0.38 (27)
Q			0.64,0.71,0.62 (9)	0.43,0.62,0.51 (4)	0.33,0.65,0.53 (2)	0.67,0.71,0.55 (1)	0.08,0.02,0.00 (7)	0.21,0.24,0.20 (13)
GSI				0.76,0.77,0.63 (4)	0.65,0.63,0.50 (1)	−,-,− (0)	0.66,0.67,0.55 (6)	0.43,0.50,0.41 (13)
E_m					0.89,0.83,0.72 (11)	0.73,0.80,0.71 (4)	0.61,0.58,0.44 (8)	0.46,0.44,0.38 (10)
E_{em}						0.69,0.73,0.65 (5)	0.60,0.62,0.52 (4)	0.34,0.38,0.35 (6)
E_{dm}							0.46,0.30,0.20 (1)	−0.34,−0.35,−0.32 (1)
E_i								0.65,0.68,0.57 (11)

Table 3.22 Some generic transformation models calibrated by ROCKMass/9/5876

Model	*Literature*	*# data*	*Transformation model*	*b*	COV_{trans}
$E_m - E_i - RQD$	Coon and Merritt (1970)	147	$E_m / E_i \approx 0.0231 \times RQD - 1.32$	1.26	1.09
	Zhang and Einstein (2004)	161	$E_m / E_i \approx 10^{0.0186 \times RQD - 1.91}$	1.54	0.89
$E_m - RMR$	Bieniawski (1978)	1091	$E_m \approx 2 \times RMR - 100$	0.57	1.48
	Gokceoglu et al. (2003)	1749	$E_m \approx 0.0736 \times e^{0.0755 \times RMR}$	1.53	1.21
	Serafim and Pereira (1983)	1749	$E_m \approx 10^{\frac{(RMR-10)}{40}}$	0.51	1.00
$E_m - Q$	Grimstad and Barton (1993)	288	$E_m \approx 25 \times \log_{10}(Q)$	0.58	0.88
$E_m - GSI$	Hoek and Diederichs (2006)	349	$E_m \approx 100 \times \frac{(1 - D/2)}{1 + e^{\frac{75+25D-GSI}{11}}}$	0.83 (D = 0)	1.00 (D = 0)

Source: ISSMGE-TC304 (2021).

Note: σ_{ci} in MPa; E_m in GPa; and D is the disturbance factor (Hoek and Diederichs 2006).

3.2.5 CLAY-C_c/6/6203

The CLAY-C_c/6/6203 database (Ching et al. 2022) compiles 6203 records for six clay parameters, including liquid limit (LL), plasticity index (PI), natural water content (w_n), void ratio (e), compression index (C_c), and unloading–reloading index (C_{ur}). The unloading–reloading index C_{ur} denotes the average of swelling index (C_s) and reloading index (C_r) (Kulhawy and Mayne 1990). 85% of the records are (natural) undisturbed clays, whereas the rest 15% are (laboratory) reconstituted clays. The records in CLAY-C_c/6/6203 cover a wide range of LL (19 to 613%), PI (5 to 493%), w_n (6 to 560%), and C_c (0.01 to 14.5). Details are given in Ching et al. (2022). The CLAY-C_c/6/6203 database is 61% complete. The generic statistics of the six parameters are listed in Table 3.23. Tables 3.24 and 3.25 show the site-specific mean, COV, and bivariate correlation coefficients. Table 3.26 shows the values of b and COV_{trans} for some generic transformation models calibrated by the records in CLAY-C_c/6/6203.

3.2.6 SOIL-DMT/8/7186

The SOIL-DMT/8/7186 database (Ching and Kuo 2023) compiles 7186 records for eight soil parameters, including liquidity index (LI), median grain size (D_{50}), normalized undrained

Table 3.23 Summary of generic statistics for CLAY-C_c/6/6203

Variable	*# Data*	*Mean*	*COV*	*Min*	*Max*
LL	3702	63.1	0.60	19.0	612.7
PI	3282	34.6	0.91	5.0	493.0
w_n (%)	3752	56.8	0.68	5.8	559.5
e	3610	1.58	0.68	0.13	14.66
C_c	6203	0.79	1.18	0.01	14.50
C_{ur}	2102	0.094	1.14	0.002	2.16

Source: Ching et al. (2022).

Table 3.24 Summary of site-specific statistics for CLAY-C_c/6/6203

		# Records/site		*Site-specific mean*		*Site-specific COV_{inher}*	
Property	*# Sites*	*Range*	*Mean*	*95% CI*	*Mean*	*95% CI*	*Mean*
LL	149	5-136	12.7	33.9–130.1	64.7	0.021–0.56	0.20
PI	149	5-60	11.9	12.5–68.8	36.1	0.035–0.66	0.28
w_n (%)	170	5-93	14.0	21.2–140.8	66.8	0.051–0.60	0.22
e	161	5-72	13.0	0.57–4.82	1.87	0.043–0.51	0.22
C_c	227	5-136	15.7	0.18–3.46	1.03	0.088–0.89	0.40
C_{ur}	118	5-115	14.2	0.024–0.41	0.11	0.083–0.88	0.40

Table 3.25 Summary of site-specific correlations for CLAY-C_c/6/6203 (Pearson, Spearman, Kendall)

	PI	w_n	e	C_c	C_{ur}
LL	0.91,0.86,0.77 (117)	0.59,0.56,0.47 (116)	0.53,0.52,0.42 (103)	0.53,0.50.0.40 (141)	0.46,0.43,0.35 (60)
PI		0.54,0.51,0.41 (118)	0.49,0.48,0.38 (100)	0.50,0.48,0.38 (142)	0.45,0.43,0.34 (60)
w_n			0.88,0.82,0.74 (102)	0.63,0.59,0.48 (161)	0.50,0.46,0.36 (84)
e				0.71,0.67,0.55 (155)	0.58,0.52,0.42 (74)
C_c					0.53,0.49,0.40 (110)

Table 3.26 Some generic transformation models calibrated by CLAY-C_c/6/6203

Model	*Literature*	*# Data*	*Transformation model*	*b*	COV_{trans}
$C_c - LL$	Skempton (1944)	3398	$C_c \approx 0.007 \times (LL - 10)$	1.59	0.90
	Terzaghi and Peck (1967)	3398	$C_c \approx 0.009 \times (LL - 10)$	1.24	0.90
$C_c - PI$	Kulhawy and Mayne (1990)	2964	$C_c \approx PI / 73$	1.31	0.91
$C_c - LL - G_s$	Nagaraj and Murty (1985)	1523	$C_c \approx 0.2343 \times (LL / 100) \times G_s$	1.06	0.81
$C_{ur} - e$	Peck and Reed (1954)	668	$C_{ur} \approx 0.208 \times e + 0.0083$	0.31	0.69
$C_{ur} - w$	Azzouz et al. (1976)	771	$C_{ur} \approx 0.003 \times (w + 7)$	0.47	0.67
$C_{ur} - PI$	Kulhawy and Mayne (1990)	847	$C_{ur} \approx PI/385$	0.96	0.63
$C_{ur} - LL$	Isik (2009)	846	$C_{ur} \approx 0.0007 \times LL + 0.0062$	1.55	0.64

Source: ISSMGE-TC304 (2021).

shear strength for field vane (s_u^{FV}/P_a), corrected SPT blow count (N_{60}), normalized CPT cone tip resistance [$(q_t-\sigma_v)/P_a$], CPT pore pressure coefficient (B_q), normalized soil modulus obtained by the pressuremeter test (E_{PMT}/P_a), and normalized soil modulus obtained by the dilatometer test [$(1-\nu^2) \times E_{DMT}/P_a$], where ν is the Poisson ratio (ν is assumed to be 0.5 for clays and 0.2 for granular soils). Most soil modulus data are from DMT. E_{DMT} is multiplied by $(1-\nu^2)$ to represent the Young's modulus (E) because according to Kulhawy and Mayne (1990), $E \approx (1-\nu^2) \times E_{DMT}$. In SOIL-DMT/8/7186, 74% of the records are in situ clays, and

Table 3.27 Summary of generic statistics for SOIL-DMT/8/7186

Variable	*# data*	*Mean*	*COV*	*Min*	*Max*
LI	3347	1.03	0.71	−2.17	7.0
D_{50} (mm)	406	0.29	3.77	0.01	11.0
s_u^{FV}/P_a	2984	0.32	0.89	0.002	3.62
N_{60}	1354	18.71	0.89	0.99	124.1
$(q_t-\sigma_v)/P_a$	3198	28.22	1.36	0.38	378.2
B_q	1408	0.50	0.51	−0.10	1.56
E_{PMT}/P_a	514	280.44	1.08	7.73	1888.7
$(1-\nu^2) \times E_{DMT}/P_a$	2966	174.06	1.09	1.12	1733.2

Source: Ching and Kuo 2023).

Table 3.28 Summary of site-specific statistics for SOIL-DMT/8/7186

		# Records/site		*Site-specific mean*		*Site-specific COV_{inher}*	
Property	*# Sites*	*Range*	*Mean*	*95% CI*	*Mean*	*95% CI*	*Mean*
LI	256	5-49	12.0	0.044–2.39	1.03	0.02–6.49	0.37
D_{50} (mm)	36	5-31	10.5	0.016–0.61	0.21	0.07–0.62	0.36
s_u^{FV}/P_a	240	5-50	12.0	0.084–1.10	0.34	0.03–1.82	0.38
N_{60}	102	5-57	12.2	1.98–46.7	17.0	0.08–1.53	0.43
$(q_t-\sigma_v)/P_a$	231	5-50	12.8	2.12–133.6	29.2	0.07–1.61	0.44
B_q	97	5-46	13.5	0.028–0.97	0.49	0.06–2.75	0.36
E_{PMT}/P_a	31	5-57	14.2	15.7–861.4	289.4	0.11–1.07	0.54
$(1-\nu^2) \times E_{DMT}/P_a$	216	5-50	12.9	7.96–529.6	167.2	0.08–1.21	0.46

the rest 26% are in situ granular soils. The records in SOIL-DMT/8/7186 cover a wide range of LI (−2.17 to 7 for clayey records), D_{50} (0.01 to 11 mm for granular soil records), N_{60} (less than 1 to 124), and soil modulus [e.g., $(1-\nu^2) \times E_{DMT}$ ranges from 0.11 to 176 MPa]. Details are given in Ching and Kuo (2023). The SOIL-DMT/8/7186 database is 28% complete. The generic statistics of the eight parameters are listed in Table 3.27. Tables 3.28 and 3.29 show the site-specific mean, COV, and bivariate correlation coefficients. Table 3.30 shows the values of b and COV_{trans} for some generic transformation models calibrated by the records in SOIL-DMT/8/7186.

3.2.7 SOIL-KSAT/7/2293

The SOIL-KSAT/7/2293 database (Ching et al. 2024) compiles 2293 records for seven soil parameters, including saturated hydraulic conductivity (k_{sat}), void ratio (e), water content (w), liquid limit (LL), plasticity index (PI), 10% percentile grain size (D_{10}), and clay content (CC). All k_{sat} data are based on laboratory tests on saturated soils. About 80% of the records are soils with clay content (CC) > 5%, and the remaining 20% are with CC < 5%. For soils with CC > 5%, D_{10} information is usually missing, and for soils with CC < 5%, (LL, PI) information is usually missing. Some records in SOIL-KSAT/7/2293 are extracted from the FG-KSAT/6/1358 and CG-KSAT/7/1278 databases developed by Feng and Vardanega (2019) and Feng et al. (2023). The 2293 records are divided into 272 groups. Each group is

Table 3.29 Summary of site-specific correlations for SOIL-DMT/8/7186 (Pearson, Spearman, Kendall)

	D_{50}	s_u^{FV}/P_a	N_{60}	$(q_t-\sigma_v)/P_a$	B_q	E_{PMT}/P_a	$(1-\nu^2) \times E_{DMT}/P_a$
LI	−,−,− (0)	−0.30,−0.29,−0.23 (199)	−0.10,−0.09,−0.07 (15)	−0.28, −0.27, −0.22 (84)	0.04,0.03,0.02 (64)	0.07,0.22,0.12 (3)	−0.29,−0.17,−0.13 (38)
D_{50}		−,−,− (0)	0.00,0.08,0.08 (8)	0.13,0.35,0.28 (17)	−,−,− (0)	−0.80,−0.63,−0.47 (1)	0.12,0.43,0.36 (19)
s_u^{FV}/P_a			0.09,0.09,0.10 (4)	0.65,0.63,0.53 (50)	0.16,0.20,0.16 (44)	−,−,− (0)	0.58,0.56,0.48 (17)
N_{60}				0.47,0.46,0.38 (59)	−0.10,−0.13,−0.11 (11)	0.52,0.50,0.40 (21)	0.43,0.40,0.32 (59)
$(q_t-\sigma_v)/P_a$					−0.12,−0.04,−0.03 (97)	0.39,0.34,0.26 (11)	0.56,0.53,0.42 (141)
B_q						−0.47,−0.46,−0.43 (1)	0.03,0.04,0.03 (39)
E_{PMT}/P_a							0.40,0.49,0.40 (4)

Table 3.30 Some generic transformation models calibrated by SOIL-DMT/8/7186

Model	*Literature*	*# Data*	*Transformation model*	*b*	COV_{trans}
$E_{PMT} - N_{60}$	Ohya et al. (1982)	94	$E_{PMT} / P_a \approx 19.3 \times N^{0.63}$ (clay)	1.99	0.60
	Ohya et al. (1982)	248	$E_{PMT} / P_a \approx 9.08 \times N^{0.66}$ (sand)	3.73	0.76
$E_{DMT} - N_{60}$	Mayne and Frost (1989)	472	$E_{DMT} / P_a \approx 22 \times N^{0.82}$ (sand)	1.40	0.57
$E_{DMT} - N_{60}$	This chapter	662	$\frac{(1-\nu^2)E_{DMT}}{P_a} \approx 14.2 \times N_{60}^{0.99}$ (all soils)	1.25	0.70
$E_{PMT} - q_t$		126	$\frac{E_{PMT}}{P_a} \approx 7.64 \times \left(\frac{q_t - \sigma_v}{P_a}\right)^{0.76}$ (all soils)	1.30	0.75
$E_{DMT} - q_t$		1796	$\frac{(1-\nu^2)E_{DMT}}{P_a} \approx 4.49 \times \left(\frac{q_t - \sigma_v}{P_a}\right)^{1.00}$ (all soils)	1.18	0.63
$E_{DMT} - s_u^{FV}$		252	$\frac{(1-\nu^2)E_{DMT}}{P_a} \approx 51.7 \times \left(\frac{s_u^{FV}}{P_a}\right)^{1.13}$ (clay)	1.08	0.38

featured with a series k_{sat} versus e (or w) data on the same type of soil (the same minerology, LL, PI, D_{10}, CC, fluid, etc.). Basically, soil specimens of the same type are consolidated to different void ratios (or water contents) in the laboratory, and their k_{sat} values are measured. LL, PI, D_{10}, and CC data for the records in the same group are constant because of the same soil type. The records in SOIL-KSAT/7/2293 cover a wide range of k_{sat} (5.2E–13 to 0.6 cm/s), e (0.19 to 4.81), w (7.1 to 178%), LL (21.7 to 675), PI (11 to 626), D10 (0.0077 to 0.62 mm), and CC (0 to 100%). Details are given in Ching et al. (2023). The SOIL-KSAT/7/2293 database is 71% complete. The generic statistics of the seven parameters are listed in Table 3.31. Tables 3.32 and 3.33 show the group-specific mean, COV, and bivariate correlation coefficients. Note that these are group-specific statistics rather than site-specific ones (there is no "site" in SOIL-KSAT/7/2293). Each group consists of a series of k_{sat} versus e (or w) laboratory data on the same type of soil (the same LL, PI, D_{10}, CC) where e (or w) is controlled by consolidation. Because (LL, PI, D_{10}, CC) within each group are controlled to be constant, their group-specific COVs and correlations are not reported in Tables 3.32 and 3.33. Table 3.34 shows the values of b and δ for some generic transformation models

Table 3.31 Summary of generic statistics for SOIL-KSAT/7/2293

Variable	*# Data*	*Mean*	*COV*	*Min*	*Max*
k_{sat} (cm/s)	2293	1.1E−02	4.62	5.2E−13	0.60
e	2293	1.37	0.59	0.19	4.81
w (%)	1625	54.5	0.60	7.12	178.0
LL	1707	98.7	0.80	21.7	675.0
PI	1256	56.9	1.39	11.0	625.9
D_{10} (mm)	166	0.22	0.78	0.0077	0.62
CC (%)	2091	50.96	0.65	0.0	100.0

Table 3.32 Summary of group-specific statistics for SOIL-KSAT/7/2293

Property	# Groups	# Records/group Range	# Records/group Mean	Group-specific mean 95% CI	Group-specific mean Mean	Group-specific COV 95% CI	Group-specific COV Mean
k_{sat} (cm/s)	227	5-49	9.4	1.9E−10–0.20	1.15E−02	0.18–2.07	0.87
e	227	5-49	9.4	0.52–2.82	1.40	0.064–0.56	0.25
w (%)	153	5-49	10.1	19.6–106.0	58.4	0.065–0.57	0.29
LL	164	5-49	9.9	31.4–346.8	108.9	–	–
PI	95	5-49	12.6	12.3–440.9	65.4	–	–
D_{10} (mm)	14	5-17	8.9	0.0077–0.62	0.27	–	–
CC (%)	209	5-49	9.3	0.0–100.0	47.7	–	–

Table 3.33 Summary of group-specific correlations for SOIL-KSAT/7/2293 (Pearson, Spearman, Kendall)

	e	w	LL	PI	D_{10}	CC
k_{sat}	0.90,0.89,0.81 (75)	0.89,0.90,0.81 (58)	–	–	–	–
e		1.00,0.99,0.96 (2)	–	–	–	–
w			–	–	–	–
LL				–	–	–
PI					–	–
D_{10}						–

Table 3.34 Some generic transformation models calibrated by SOIL-KSAT/7/2293

Model	Literature	n_{cal}	Transformation model	b	COV_{trans}
$k_{sat} - D_{10}$	Hazen (1930)	166 (coarse-grained)	$k_{sat}(cm/s) \approx 1.25 \times D_{10}^2$	0.61	1.45
$k_{sat} - e - D_{10}$	Chapuis (2004)	166 (coarse-grained)	$k_{sat}\left(\frac{cm}{s}\right) \approx 2.4622 \times \left[D_{10}^2 \times \frac{e^3}{1+e}\right]^{0.7825}$	0.81	1.41
$k_{sat} - e - PI$	Nishida and Nakagawa (1969)	1256 (fine-grained)	$k_{sat}\left(\frac{cm}{s}\right) \approx 10^{\frac{e}{0.01 \times PI + 0.05} - 10}$	2.14	5.96
$k_{sat} - e - e_L$[a]	Nagaraj et al. (1994)	1468 (fine-grained)	$k_{sat}\left(\frac{cm}{s}\right) \approx 10^{\left(\frac{e}{e_L} - 2.162\right)/0.195}$	9.13	6.51
	Sivapullaiah et al. (2000)		$k_{sat}\left(\frac{m}{s}\right) \approx \left(\frac{e}{1.97 \times e_L}\right)^{4.2}$	1.50×10^{-7}[b]	4.68

[a] e_L = void ratio at liquid limit.

[b] The large bias may be due to an error in the unit in Sivapullaiah et al. (2000). According to Sivapullaiah et al. (2000), the unit for k_{sat} is m/s, but it is likely that this unit is incorrect.

Table 3.35 Ranges of site-specific SOF_h and SOF_z for various soil types

	SOF_h (m)				SOF_z (m)			
Soil type	*# Studies*	*Min*	*Max*	*Average*	*# Studies*	*Min*	*Max*	*Average*
Alluvial	9	1.07	49	14.2	13	0.07	1.1	0.36
Ankara clay	–	–	–	–	4	1	6.2	3.63
Chicago clay	–	–	–	–	2	0.79	1.25	0.91
Clay	9	0.14	163.8	31.9	16	0.05	3.62	1.29
Clay, sand, silt mix	13	1.2	1000	201.5	28	0.06	21	1.58
Hangzhou clay	2	40.4	45.4	42.9	4	0.49	0.77	0.63
Marine clay	8	8.37	66	30.9	9	0.11	6.1	1.55
Marine sand	1	15	15	15	5	0.07	7.2	1.43
Offshore soil	1	24.6	66.5	45.6	2	0.48	1.62	1.04
Over consolidated clay	1	0.14	0.14	0.14	2	0.063	0.255	0.15
Sand	9	1.69	80	24.5	14	0.1	4	1.17
Sensitive clay	–	–	–	–	2	1.1	2.0	1.55
Silt	3	12.7	45.5	33.2	5	0.14	7.19	2.08
Silty clay	7	9.65	45.4	29.8	14	0.095	6.47	1.40
Soft clay	3	22.2	80	47.6	8	0.14	6.2	1.70
Undrained engineered soil	–	–	–	–	22	0.3	2.7	1.42
Water content	9	2.8	22.2	12.9	8	0.05	6.2	1.70

Source: Cami et al. (2020).

calibrated by the records in SOIL-KSAT/7/2293. It is clear that the transformation uncertainty (quantified by COV_{trans}) for k_{sat} of fine-grained soils is significant.

3.3 TYPICAL VALUES OF SITE-SPECIFIC SOFS

Cami et al. (2020) summarized the ranges of the site-specific horizontal SOF (SOF_h) and vertical SOF (SOF_z) reported in the literature for various soil types, which are shown in Table 3.35. In practice, a common assumption is that the two horizontal directions have the same SOF: $SOF_h = SOF_x = SOF_y$. There are cases in the literature with simultaneous information on site-specific SOF_h and SOF_z. Chapter 3 of ISSMGE-TC304 (2021) summarized the values and ranges for the site-specific SOFs of these cases as Figure 3.1. Ching et al. (2023) conducted a comprehensive analysis of a large CPT database. From the database, they extracted hundreds of homogeneous soil units from 42 sites to identify the vertical scales of fluctuation of their q_t data. Table 3.36 shows the statistics of the site-specific SOF_z for soil units of various soil behaviour types (SBT) (Robertson 2016).

3.4 QUASI-SITE-SPECIFIC TRANSFORMATION MODEL

A major challenge when evaluating soil/rock properties through generic transformation models is that the transformation uncertainty is usually significant. The significant transformation uncertainty manifests as large COV_{trans} in the tables present above. A large COV_{trans}

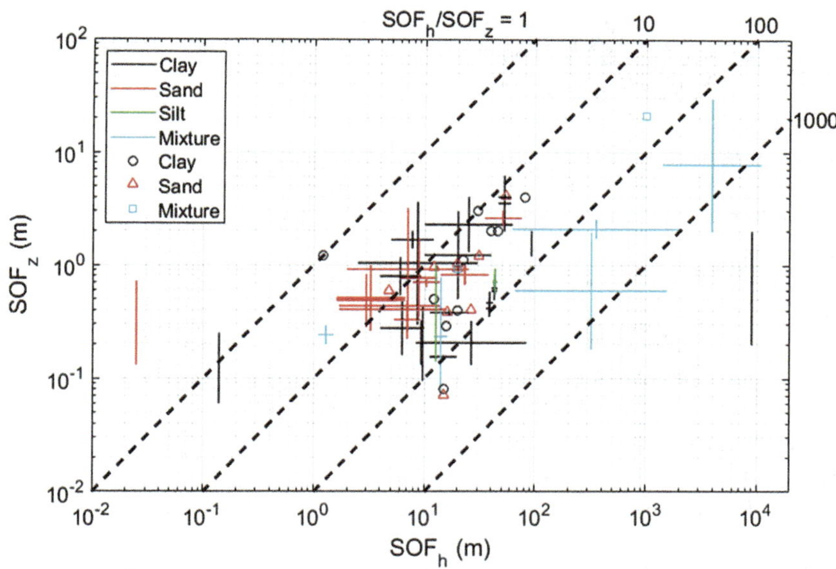

Figure 3.1 Variation of vertical scale of fluctuation (SOF_z) with horizontal scale of fluctuation (SOF_h) for various soil conditions (Source: ISSMGE-TC304 2021).

Table 3.36 Statistics for site-specific SOF_z identified from CPT cone tip resistance (q_t) for soil units of various soil behaviour types

		Site-specific SOF_z (m)	
SBT	*# Sites*	*Range*	*Mean*
2 (organic)	1	0.30	0.30
3 (clay)	7	0.10–0.46	0.26
4 (silt mixtures)	8	0.13–0.42	0.27
5 (sand mixtures)	13	0.084–0.58	0.32
6 (sand)	13	0.31–0.89	0.49
2, 3, 4 (clay-like)	16	0.10–0.46	0.27
5, 6 (sand-like)	26	0.084–0.89	0.41
All	42	0.084–0.89	0.35

Source: Ching et al. (2023b).

leads to a large COV_{TOT} in Equation (3.1), so the resulting characteristic value (X_k) is small (see Equations 3.2 and 3.3), which will further lead to a conservative design. Ching et al. (2021b) proposed the hierarchical Bayesian model (HBM) to construct a quasi-site-specific transformation model. The transformation uncertainty for the quasi-site-specific model can be significantly less than that for the generic transformation model. First, the HBM learns the site-specific statistics in the soil/rock property database. The learning outcome produces a prior model for the target site. Then, this prior model is updated into a posterior model by further conditioning on the target-site data. The posterior model can produce a "quasi-site-specific" transformation model for the target site. The resulting transformation model is quasi-site-specific because it is not entirely based on the sparse site-specific (target-site) data: it also adopts the prior knowledge learned from a database. The transformation uncertainty

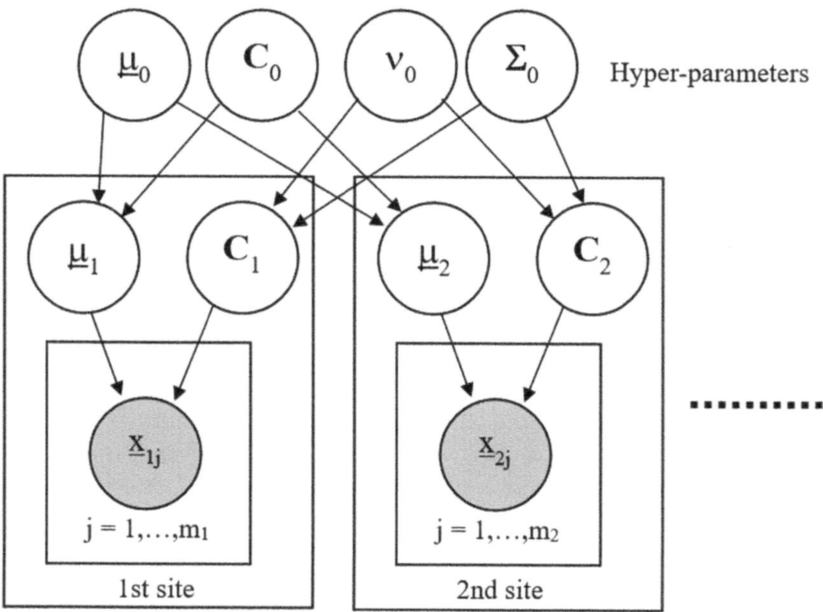

Figure 3.2 Model structure of the HBM (Source: Ching et al. 2021b).

for the quasi-site-specific model can be significantly less than that for the generic transformation model, as illustrated by an example in this section.

Figure 3.2 shows the model structure of the HBM. The (transformed) soil parameters for the jth record at the ith site in a database are denoted by the vector $\mathrm{x}_{ij} \in \mathbf{R}^{n\times1}$ (n is the dimension of the soil/rock parameters of interest, e.g., n = 10 for CLAY/10/7490). The vector x_{ij} is assumed to follow a multivariate normal probability density function (PDF) with site-specific mean vector = $\mu_i \in \mathbf{R}^{n\times1}$ and site-specific covariance matrix = $\mathrm{C}_i \in \mathbf{R}^{n\times n}$, denoted by $N(\mathrm{x};\mu_i, \mathrm{C}_i)$. The parameters (μ_i,C_i) quantify the "intra-site" variability within the ith site. Because of site uniqueness, $\mu_i \neq \mu_j$ and $\mathrm{C}_i \neq \mathrm{C}_j$ if $i \neq j$, i.e., there is also "inter-site" variability. In the HBM, $\{\mu_i: i = 1, \ldots, n_s\}$ (n_s is the number of sites) are assumed to follow the same multivariate normal PDF $N(\mu\ ;\mu_0,\mathrm{C}_0)$, where $\mu_0 \in \mathbf{R}^{n\times1}$ is the (hyper) mean vector and $\mathrm{C}_0 \in \mathbf{R}^{n\times n}$ is the (hyper) covariance matrix. Similarly, $\{\mathrm{C}_i: i = 1, \ldots, n_s\}$ are assumed to follow the same inverse-Wishart (IW) distribution (James 1964), denoted by $\mathrm{IW}(\mathrm{C};\Sigma_0,\nu_0)$, where $\Sigma_0 \in \mathbf{R}^{n\times n}$ is the (hyper) scale matrix and $\nu_0 \in \mathbf{R}$ is the (hyper) degree of freedom. The parameters $\boldsymbol{\Theta} = (\mu_0, \mathrm{C}_0,\Sigma_0,\nu_0)$, called the hyper-parameters of the HBM, quantify both inter-site and intra-site variabilities in the soil/rock property database.

The HBM contains two analysis stages: the learning and inference stages. In the learning stage, the hyper-parameters $\boldsymbol{\Theta} = (\mu_0,\mathrm{C}_0,\Sigma_0,\nu_0)$ are calibrated to learn the site-specific statistics in the soil/rock property database. The calibrated hyper-parameters can produce a prior model for the target site. In the inference stage, this prior model is updated into a posterior model by further conditioning on the target-site data. The posterior model can produce a "quasi-site-specific" transformation model for the target site. In the following two subsections, the learning and inference stages will be illustrated using a bivariate database $[(q_t - \sigma_v)/\sigma'_v$ versus $s_u/\sigma'_v]$ extracted from CLAY/10/7490 as an illustration example. It will be shown that the HBM learning stage can effectively extract site-specific statistics in the database and that the HBM inference stage can construct a quasi-site-specific transformation model that is more effective than a generic transformation model.

3.3.1 HBM learning stage: extract site-specific statistics in the database

Let us take the CLAY/10/7490 database as an example. For illustration purposes, let us consider two clay parameters $Y_1 = q_{t1} = (q_t - \sigma_v)/\sigma'_v$ and $Y_2 = s_u/\sigma'_v$. In CLAY/10/7490, there are 91 sites with Y_1–Y_2 records. These records (with site-label information) are shown in Figure 3.3a. The (Y_1, Y_2) data are transformed to (roughly) standard normal space (X_1, X_2) by the Johnson transformation (Ching and Phoon 2014b). During the HBM learning stage, the hyper-parameters $\mathbf{\Theta}$ are calibrated by the database (X_1, X_2) data. The calibration is done by performing Bayesian analysis: the posterior samples of $\mathbf{\Theta}$ are drawn by conditioning on the database data. The $\mathbf{\Theta}$ samples can simulate the site-specific mean and covariance matrix of a "hypothetical site" (denoted by μ_h and C_h) whose site-specific statistics "mimic" the 91 database sites. The red cross markers and red histogram in Figure 3.4 show the simulated site-specific statistics (means, standard deviations, and correlation) for numerous hypothetical sites. They are consistent with the actual site-specific statistics for the 91 database sites (dark open circles and dark histograms). The consistency indicates that the HBM learning stage can effectively capture the site-specific statistics in the database. The simulated hypothetical sites can be illustrated in a different way. Each (skewed) ellipse in Figure 3.3b represents the one-standard-deviation confidence region of a hypothetical site. These (skewed) ellipses mimic the site-specific records in Figure 3.3a.

3.3.2 HBM inference stage: construction of quasi-site-specific transformation model

The calibrated hyper-parameters in the HBM learning stage can produce a "prior model" for the site-specific mean and covariance matrix of the target site, denoted by μ_s and C_s. In

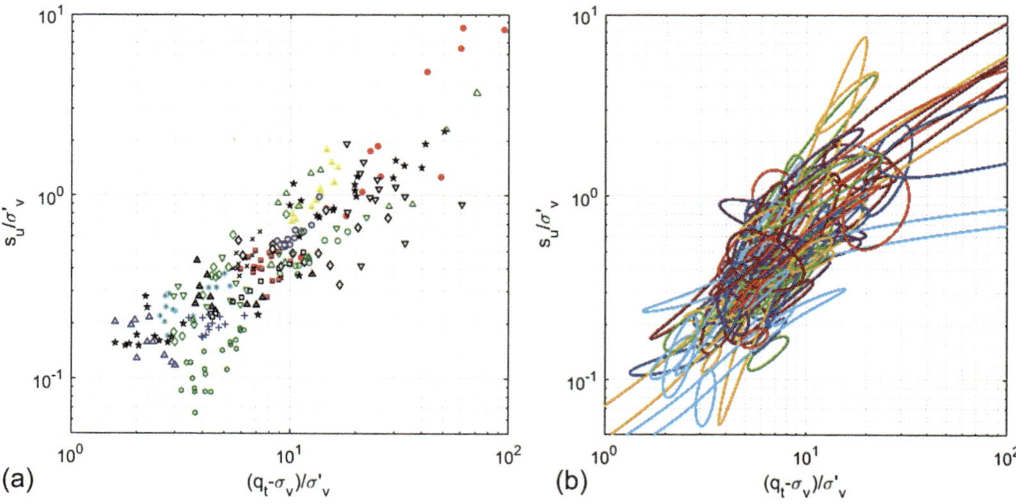

Figure 3.3 Y_1–Y_2 data points in the CLAY/10/7490 database (records from different sites are shown as distinct markers) versus simulated hypothetical sites (Source: Ching 2020).

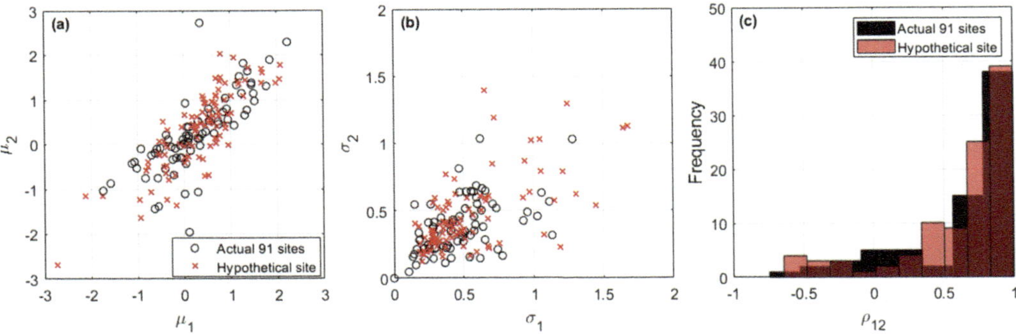

Figure 3.4 Site-specific statistics (μ and σ denote the mean and standard deviation of X, and ρ denotes the correlation coefficient), database versus hypothetical sites: (a) site-specific means; (b) site-specific standard deviations; (c) site-specific correlation (Source: Ching 2020).

the inference stage, this prior model (μ_s,C_s) is updated into a posterior model for (μ_s,C_s) by further conditioning on the target-site data. This posterior model is "quasi-site-specific" because not only the target-site data are used to construct the model but also the database is used to develop its prior.

Now consider the Onsøy (Norway) site (Lacasse and Lunne 1982) as the target site. The Onsøy site is not within the CLAY/10/7490 database. There are many in situ and laboratory tests conducted at this test site, but only those relevant to (Y_1, Y_2) are investigated in this section. The site investigation data are digitalized from Lacasse and Lunne (1982), as shown in Table 3.37. To illustrate the HBM inference stage, the s_u/σ'_v values for some depths are deliberately treated as unknowns (see the numbers in the parentheses in the table). The "unknown" s_u/σ'_v values will later serve as the validation data for the inference stage. The known s_u/σ'_v values are sparse such that a site-specific transformation model may not be reliably constructed. A possible alternative is to adopt a generic transformation model. However, the generic transformation uncertainty is usually large, e.g., COV_{trans} = 0.45 for the s_u-CPT transformation model in Table 3.7.

The prior model constructed during the HBM learning stage can be further conditioned by the sparse target-site data in Table 3.37. By conditioning on the target-site data, most ellipses in Figure 3.3b are incompatible with the target-site data. Figure 3.5a shows the incompatible

Table 3.37 Site investigation data for a site in Onsøy (Norway)

Index	*Depth (m)*	q_{t1} (Y_1)	s_u/σ'_v (Y_2)
1	1.0	29.11	2.03
2	1.9	17.69	(0.91)
3	3.5	10.52	0.48
4	5.2	7.70	(0.37)
5	7.6	5.89	0.24
6	9.5	6.19	(0.25)
7	10.8	5.93	0.25
8	13.4	5.95	(0.24)
9	16.3	6.13	0.24

Source: Lacasse and Lunne (1982).

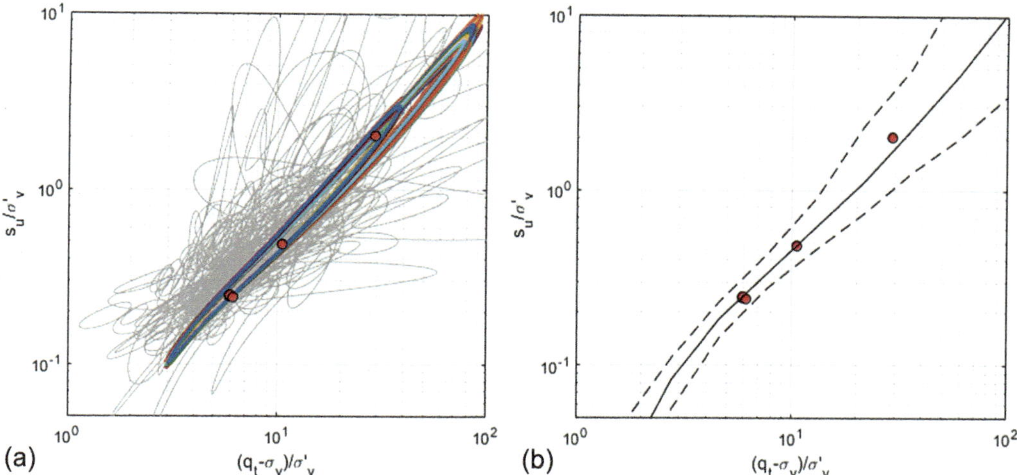

Figure 3.5 Illustration of the quasi-site-specific model.

ellipses in grey and compatible ellipses in colours. The red dots show the target-site data. The coloured (compatible) ellipses illustrate the quasi-site-specific model after the conditioning. Prediction based on the quasi-site-specific model can be carried out by reading the median values as well as the 95% confidence interval (95% CI) for Y_2 at different values of Y_1, as shown in Figure 3.5b.

Figure 3.6a shows the predictive median and 95% CI for the generic transformation model developed based on the database Y_1–Y_2 data (shown as grey dots), whereas Figure 3.6b shows the predictive median and 95% CI for the quasi-site-specific transformation model developed by the HBM inference stage. The validation data (the numbers in the parentheses in Table 3.37) are also shown as yellow dots for comparison. Although both 95% CIs contain the validation data, the transformation uncertainty for the quasi-site-specific model is much less than that for the generic model. Although the superiority of a quasi-site-specific

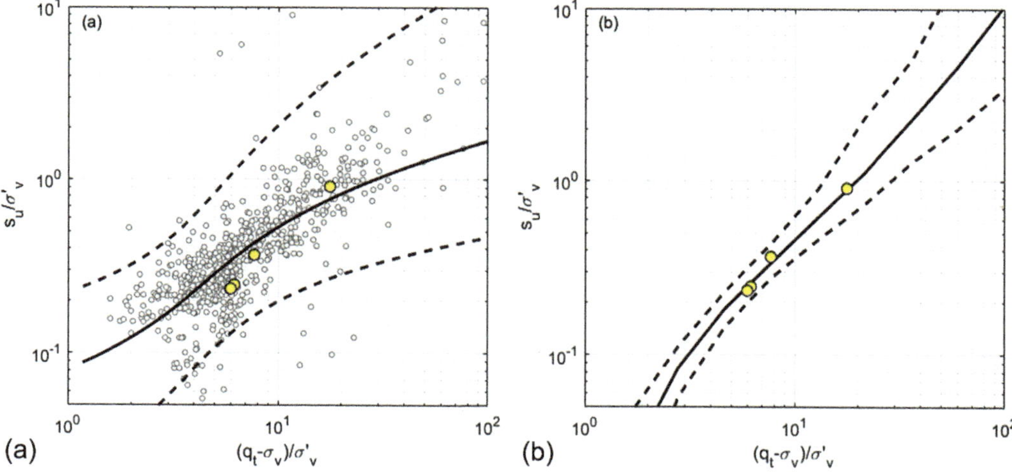

Figure 3.6 Predictive medians and 95% CIs of the generic and quasi-site-specific transformation models.

transformation model constructed by the HBM over a generic transformation model is illustrated above using a bivariate problem [$(q_t - \sigma_v)/\sigma'_v$ versus s_u/σ'_v] for the CLAY/7/7490 database, the similar superiority is found for other databases and for higher dimensions (e.g., Ching et al. 2021a, 2021b, 2022; Meng and Pei 2023).

3.5 CONCLUSION

The proper characterization of the variability of soil/rock properties for a specific site is necessary for reliability-based design as well as for the determination of soil/rock characteristic values. This chapter reviews some useful sources that provide site-specific statistics for the inherent variability of soil/rock properties (e.g., site-specific mean, COV, and SOF). Some sources are relatively old (e.g., Phoon et al. 1995; Phoon and Kulhawy 1999a; Kulhawy et al. 2000), whereas others are relatively new (e.g., Cami et al. 2020; ISSMGE-TC304 2021; Ching et al. 2023b). Some newer results in this chapter are based on the soil/rock property databases compiled by the author of this chapter. The extent of these databases is still limited, so only a limited number of soil/properties are covered. In the opinion of the author of this chapter, there is an urgent need to compile more soil/rock property databases to broaden the coverage. Although the soil/rock property databases can be used to calibrate generic transformation models, one key issue for these generic models is that the transformation uncertainty is usually significant. The large transformation uncertainty may result in a conservative design. The hierarchical Bayesian model (HBM) recently proposed by the author of this chapter may reduce the transformation uncertainty significantly by constructing a quasi-site-specific transformation model.

REFERENCES

Altindag, R. and Guney, A. (2010). Predicting the relationships between brittleness and mechanical properties (UCS, TS and SH) of rocks. *Scientific Research and Essays*, 5(16), 2107–2118.

Baecher, G.B. and Christian, J.T. (2003). *Reliability and Statistics in Geotechnical Engineering*. John Wiley & Sons, Hoboken, NJ.

Barton, N., Lien, R., and Lunde, J. (1974). Engineering classification of rock masses for the design of tunnel support. *Rock Mechanics*, 6(4), 189–236.

Bieniawski, Z.T. (1973). Engineering classification of jointed rock masses. *Transactions of the South African Institution of Civil Engineers*, 15, 335–344.

Bieniawski, Z.T. (1978). Determining rock mass deformability: Experience from case histories. *International Journal of Rock Mechanics and Mining Sciences*, 15, 237–247.

Bjerrum, L. (1954). Geotechnical properties of Norwegian marine clays. *Geotechnique*, 4(2), 49–69.

Bolton, M.D. (1986). The strength and dilatancy of sands. *Geotechnique*, 36(1), 65–78.

Cami, B., Javankhoshdel, S., Phoon, K.K., and Ching, J. (2020). Scale of fluctuation for spatially varying soils: estimation methods and values. *ASCE-ASME Journal of Risk and Uncertainty in Engineering Systems, Part A: Civil Engineering*, 6(4), 03120002.

CEN (2004). EN 1997–1: *Geotechnical Design – Part 1: General Rules*. Comité Européen de Normalisation, Brussels, Belgium.

CEN (2020). *prEN 1997–1 Geotechnical Design – General Rules (PT6) Oct 2020*. Comité Européen de Normalisation, Brussels, Belgium.

Chapuis, R.P. (2004). Predicting the saturated hydraulic conductivity of sand and gravel using effective diameter and void ratio. *Canadian Geotechnical Journal*, 41, 787–795.

Chen, J. R. (2004). *Axial Behavior of Drilled Shafts in Gravelly Soils*. Ph.D. Dissertation, Cornell University, Ithaca, NY

Ching, J. (2020). Value of geotechnical BIG DATA and its application in site-specific soil property estimation. *Journal of GeoEngineering*, 15(4), 173–182.

Ching, J. and Kuo, M.C. (2023). Data-centric quasi-site-specific prediction for soil modulus. *Journal of GeoEngineering*, 18(4), 215–223.

Ching, J. and Phoon, K. K. (2014a). Transformations and correlations among some parameters of clays – the global database. *Canadian Geotechnical Journal*, 51(6), 663–685.

Ching, J. and Phoon, K. K. (2014b). Correlations among some clay parameters – the multivariate distribution. *Canadian Geotechnical Journal*, 51(6), 686–704.

Ching, J. and Phoon, K.K. (2012a). Modeling parameters of structured clays as a multivariate normal distribution. *Canadian Geotechnical Journal*, 49(5), 522–545.

Ching, J. and Phoon, K.K. (2012b). Establishment of generic transformations for geotechnical design parameters. *Structural Safety*, 35, 52–62.

Ching, J., Lin, G.H., Chen, J.R., and Phoon, K.K. (2017). Transformation models for effective friction angle and relative density calibrated based on a multivariate database of coarse-grained soils. *Canadian Geotechnical Journal*, 54(4), 481–501.

Ching, J., Li, K.H., Phoon, K.K., and Weng, M.C. (2018). Generic transformation models for some intact rock properties. *Canadian Geotechnical Journal*, 55(12), 1702–1741.

Ching, J., Phoon, K.K., Ho, Y.H., and Weng, M.C. (2021a). Quasi-site-specific prediction for deformation modulus of rock mass. *Canadian Geotechnical Journal*, 58, 936–951.

Ching, J., Wu, S., and Phoon, K.K. (2021b). Constructing quasi-site-specific multivariate probability distribution using hierarchical Bayesian model, *ASCE Journal of Engineering Mechanics*, 147(10), 04021069.

Ching, J., Phoon, K.K., and Wu, C.T. (2022). Data-centric quasi-site-specific prediction for compressibility of clays. *Canadian Geotechnical Journal*, 59(12), 2033–2049.

Ching, J., Uzielli, M., Phoon, K.K., and Xu, X.J. (2023). Characterization of autocovariance parameters of detrended cone tip resistance from a global CPT database. *ASCE Journal of Geotechnical and Geoenvironmental Engineering*, 149(10), 04023090.

Ching, J., Phoon, K.K., and Huang, P.S. (2024). Detection of outliers with respect to a MUSIC geotechnical database. *Canadian Geotechnical Journal*, 61, 1275–1293.

Coon, R.F. and Merritt, A.H. (1970). Predicting in situ modulus of deformation using rock quality indexes. In *Determination of the In Situ Modulus of Deformation of Rock*, By Committee D-18, ASTM International, West Conshohocken, PA, pp. 154–173.

Deere, D.U. (1964). Technical description of rock cores. *Rock Mechanics Engineering Geology*, 1, 16–22.

Deere, D.U. and Miller, R.P. (1966). *Engineering Classification and Index Properties for Intact Rock*. Technical Report No. AFWL-TR-65-116, Air Force Weapons Lab, Kirtland Air Force Base, Albuquerque, NM.

Feng, S. and Vardanega, P. J. (2019). A database of saturated hydraulic conductivity of fine-grained soils: probability density functions. *Georisk: Assessment and Management of Risk for Engineered Systems and Geohazards*, 13(4): 255–261.

Feng, S., Barreto, D., Imre, E., Ibraim, E., and Vardanega, P.J. (2023). Use of hydraulic radius to estimate the permeability of coarse-grained materials using a new geodatabase. *Transportation Geotechnics*, 41, 101026.

Fenton, G. A. and Griffiths, D. V. (2008). *Risk Assessment in Geotechnical Engineering*. John Wiley & Sons, New York.

Gokceoglu, C., Sonmez, H., and Kayabasi, A. (2003). Predicting the deformation moduli of rock masses. *International Journal of Rock Mechanics and Mining Sciences*, 40(5), 701–710.

Grimstad E. and Barton N. (1993). Updating the Q-System for NMT. In: *Proceedings of the International Symposium on Sprayed Concrete-Modern Use of Wet Mix Sprayed Concrete for Underground Support*, Oslo, Norwegian Concrete Association.

Hatanaka, M. and Uchida, A. (1996), Empirical correlation between penetration resistance and internal friction angle of sandy soils. *Soils and Foundations*, 36(4), 1–9.

Hazen, A. (1930). Water supply. In: *American Civil Engineers Handbook*, eds. Merriman, T. and Wiggin, T., Wiley, New York.

Hoek, E. and Brown, E.T. (1997). Practical estimates of rock mass strength. *International Journal of Rock Mechanics and Mining Science*, 34, 1165–1186.

Hoek, E. and Diederichs, M.S. (2006). Empirical estimation of rock mass modulus. *International Journal of Rock Mechanics and Mining Sciences*, 43(2), 203–215.

Hotelling, H. and Pabst, M.R. (1936). Rank correlation and tests of significance involving no assumption of normality. *American Mathematical Statistics*, 7, 29–43.

ISSMGE-TC304 (2021). State of the art review of inherent variability and uncertainty in geotechnical properties and models. *International Society of Soil Mechanics and Geotechnical Engineering (ISSMGE) Technical Committee TC304 'Engineering Practice of Risk Assessment and Management'*, March 2nd. http://140.112.12.21/issmge/2021/SOA_Review_on_geotechnical_property_variablity_and_model_uncertainty.pdf

Jamiolkowski, M., Ladd, C.C., Germain, J.T., and Lancellotta, R. (1985). New developments in field and laboratory testing of soils. In: *Proceeding of the 11th International Conference on Soil Mechanics and Foundation Engineering*, San Francisco, CA, 1, pp. 57–153.

Kahraman, S. (2001). Evaluation of simple methods for assessing the uniaxial compressive strength of rock. *International Journal of Rock Mechanics and Mining Sciences*, 38(7), 981–994.

Karaman, K. and Kesimal, A. (2015). A comparative study of Schmidt hammer test methods for estimating the uniaxial compressive strength of rocks. *Bulletin of Engineering Geology and the Environment*, 74(2), 507–520.

Katz, O., Reches, Z., and Roegiers, J.C. (2000). Evaluation of mechanical rock properties using a Schmidt hammer. *International Journal of Rock Mechanics and Mining Sciences*, 37(4), 723–728.

Kendall, M. (1938). A new measure of rank correlation. *Biometrika*, 30(1–2), 81–89.

Kılıç, A. and Teymen, A. (2008). Determination of mechanical properties of rocks using simple methods. *Bulletin of Engineering Geology and the Environment*, 67(2), 237–244.

Kulhawy, F. H. and Mayne, P. W. (1990). Manual on estimating soil properties for foundation design. Report EL-6800, Electric Power Research Institute, Palo Alto, CA.

Kulhawy, F. H., Phoon, K. K., and Prakoso, W. A. (2000). Uncertainty in the basic properties of natural geomaterials. In: *Proceedings, First International Conference on Geotechnical Engineering Education and Training*, Sinaia, Romania, 12–14 June 2000, pp. 297–302.

Lacasse, S. and Lunne, T. (1982). Penetration tests in two Norwegian clays. In: *Proceedings, 2nd European Symposium on Penetration Testing*, Amsterdam, pp. 661–670.

Locat, J. and Demers. D. (1988). Viscosity, yield stress, remoulded strength, and liquidity index relationships for sensitive clays. *Canadian Geotechnical Journal*, 25, 799–806.

Mayne, P.W. and Frost, D.D. (1989). Dilatometer experience in Washington, D.C. and vicinity. Research Record 1169, Transportation Research Board, Washington, DC, pp. 16–23.

Meng, F. and Pei, H. (2023). Quasi-site-specific prediction of shear wave velocity from CPTu. *Soil Dynamics and Earthquake Engineering*, 172, 108005.

Mesri, G. (1975). Discussion on "New design procedure for stability of soft clays". *Journal of the Geotechnical Engineering Division, ASCE*, 101(4), 409–412.

Mesri, G. (1989). A re-evaluation of $s_u(mob) = 0.22\ \sigma'p$ using laboratory shear tests. *Canadian Geotechnical Journal*, 26(1), 162–164.

Mesri, G. and Huvaj, N. (2007). Shear strength mobilized in undrained failure of soft clay and silt deposits. ASCE Geotechnical Special Publication 173, Geo-Denver.

Mishra, D.A. and Basu, A. (2013). Estimation of uniaxial compressive strength of rock materials by index tests using regression analysis and fuzzy inference system. *Engineering Geology*, 160, 54–68.

Nagaraj, T.S., Pandian, N.S., and Narasimha Raju, P.S.R. (1994). Stress state permeability relationships for overconsolidated clays. *Geotechnique*, 44(2), 349–352.

Nisihda, Y. and Nakagawa, S. (1969). Water permeability and plastic index of soils. In: *Proceedings of the IASH-UNESCO Symposium*, Tokyo, Japan, pp. 573–578.

Ohya, S. Imai, T., Matsubara, M. (1982). Relation between N value by SPT and LLT pressuremeter results. In: Proceedings 2nd European Symposium on Penetration Testing, Amsterdam, 1, 125–130.

Orr, T.L.L. (2017). Defining and selecting characteristic values of geotechnical parameters for designs to Eurocode 7. *Georisk: Assessment and Management of Risk for Engineered Systems and Geohazards*, 11(1), 103–115.

Phoon, K.K. and Kulhawy, F.H. (1999a). Characterization of geotechnical variability. *Canadian Geotechnical Journal*, 36(4), 612–624.

Phoon, K.K. and Kulhawy, F.H. (1999b). Evaluation of geotechnical property variability. *Canadian Geotechnical Journal*, 36(4), 625–639.

Phoon, K.K., Kulhawy, F.H., and Grigoriu, M.D. (1995). *Reliability-Based Design of Foundations for Transmission Line Structure*, Report TR-105000, Palo Alto, CA, Electric Power Research Institute.

Prakoso, W.A. and Kulhawy, F.H. (2011). Effects of testing conditions on intact rock strength and variability. *Geotechnical and Geological Engineering*, 29(1), 101–111.

Robertson, P.K. (2016). Cone penetration test (CPT)-based soil behaviour type (SBT) classification system – an update. *Canadian Geotechnical Journal*, 53, 1910–1927.

Robertson, P.K. and Campanella, R.G. (1983). Interpretation of cone penetration tests: Sands and clays. *Canadian Geotechnical Journal*, 20, 719–745.

Schneider H.R. and Schneider M.A. (2013). Dealing with uncertainties in EC7 with emphasis on determination of characteristic soil properties. In: *Modern Geotechnical Design Codes of Practice*, eds. P. Arnold, G.A. Fenton, M.A. Hicks, T. Schweckendiek, and B. Simpson, IOS Press, Amsterdam, pp. 87–101.

Serafim, J.L. and Pereira, J.P. (1983). Considerations on the geomechanical classification of Bieniawski. In: *Proceedings of the Symposium on Engineering Geology and Underground Openings*, Lisboa, Portugal, pp. 1133–1144.

Sivapullaiah, P.V., Sridharan, A., and Stalin, V.K. (2000). Hydraulic conductivity of bentonite-sand mixtures. *Canadian Geotechnical Journal*, 37, 406–413.

Terzaghi, K. and Peck, R. B. (1967). *Soil Mechanics in Engineering Practice.* 2nd Ed., John Willey & Sons, New York, 729pp.

Vanmarcke, E.H. (1977). Probabilistic modeling of soil profiles. *Journal of the Geotechnical Engineering Division, ASCE*, 103(11), 1227–1246.

Wang, Y. and Cao, Z. J. (2013). Probabilistic characterization of Young's modulus of soil using equivalent samples. *Engineering Geology*, 159, 106–118.

Yaşar, E. and Erdogan, Y. (2004). Correlating sound velocity with the density, compressive strength and Young's modulus of carbonate rocks. *International Journal of Rock Mechanics and Mining Sciences*, 41(5), 871–875.

Zhang, L. and Einstein, H. H. (2004). Using RQD to estimate the deformation modulus of rock masses. *International Journal of Rock Mechanics and Mining Sciences*, 41(2), 337–341.

Chapter 4

Undrained shear strength of Finnish soft clays

A database perspective

Marco D'Ignazio and Tim Länsivaara

The undrained shear strength (s_u) of clays is a governing parameter in several geotechnical engineering problems. However, s_u is not a unique parameter; it depends on factors like stress paths, loading rates, soil composition, and stress history. This complexity results in significant uncertainty in s_u determination, impacting construction and operational safety. Further, the selection of design s_u from a given test type shall reflect the anticipated field deformation mechanism for a specific structure. To address these uncertainties, global and local databases have emerged as powerful tools. They offer statistical correlations between measured s_u values and basic clay properties, thus reducing uncertainty. In this study, high-quality data from Southern Finland is compared with extensive databases, resulting in a robust statistical correlation for anisotropic undrained shear strength of Finnish soft clays. The proposed correlations are meant for preliminary analyses, where site-specific data is limited or unavailable. Overall, databases provide invaluable resources to assess local data reliability, reduce bias, and enhance confidence in infrastructure design and construction.

4.1 INTRODUCTION

Geotechnical engineering applications such as stability of road and railway embankments, or short-term stability of excavations in soft clays, rely critically on the undrained shear strength (s_u). This parameter serves as the cornerstone for assessing safety during construction and/or under operational loads.

The direct field measurement of s_u can be achieved through various techniques, including the field vane test (FVT), which is popular in Finland (Selänpää et al. 2017), or indirect methods such as cone penetration testing (CPT/CPTU) or dilatometer testing (DMT). In the laboratory, tests like undrained triaxial (TX) and direct simple shear (DSS), as well as the Fall Cone (FC), are used to derive s_u values for unconsolidated or reconsolidated soil specimens.

However, s_u is far from a static, universal value. It is highly dependent on factors such as stress paths induced by a given test type, loading rates, in situ stress history of the soil deposit as well as reconsolidation history in the laboratory, and the specific nature and chemical composition of the soil (e.g., Bjerrum 1973; Länsivaara 1999; D'Ignazio 2016). As a result, accurately determining s_u for a particular location can be a complex endeavour. Given this inherent complexity, s_u for a specific site or soil types might be characterized by a notable degree of uncertainty (e.g., Phoon and Kulhawy 1999; Ching and Phoon 2013).

Moreover, the determination of its design value necessitates a keen consideration of the in-situ deformation mechanism relevant to a specific type of structure. For example, in

DOI: 10.1201/9781003441946-4

embankment design, the pivotal factor is the assessment of the average mobilized shear strength along a potential slip surface (Bjerrum 1973; Larsson et al. 2007; D'Ignazio et al. 2017a). In cases involving uniform soil conditions and a circular or quasi-circular slip surface, the selection of the design undrained shear strength (s_u) aligns with s_{uDSS} or $s_{u(mob)}$ values. The latter can be derived from the measured strength from FVT, s_{uFVT}, which is empirically adjusted for anisotropy and rate effects, according to the soil's plasticity characteristics (e.g., Bjerrum 1973; Liikennevirasto 2018). Notably, the literature often presumes that s_{uDSS} is approximately equal to $s_{u(mob)}$ (Jamiolkowski et al. 1985; D'Ignazio 2016; D'Ignazio et al. 2021). On the other hand, for excavations, the results of triaxial compression tests (anisotropically or isotropically consolidated undrained compression, CAUC/CIUC and extension, CAUE/CIUE) find better applicability.

To address these uncertainties, a database-driven approach emerges as a powerful tool. Such databases, like those established by D'Ignazio et al. (2016) for Finland and Scandinavia (Sweden-Norway), were used to develop statistical correlations between s_{uFVT} and $s_{u(mob)}$ and other soil properties, such as water content, plasticity index, and over-consolidation ratio (OCR), among others. These correlations can be used to estimate, e.g., s_{uFVT} of Finnish clays for local data, considering the site-specific conditions, i.e., the basic soil parameters and stress history. Furthermore, Selänpää (2021) collected data from various sites in Finland and presented regional correlations for s_u from different test types, including triaxial compression (TXC) and extension (TXE), DSS, and FVT. These comprehensive databases allow for cross-comparison and statistical analysis to derive meaningful insights and correlations, reducing the inherent uncertainty associated with local data.

In addition to the aforementioned local databases, comprehensive global databases have been compiled by Ching and Phoon (2012, 2013, 2014a,b); Ching et al (2017, 2018); D'Ignazio et al. (2019) to cover an extensive array of soils and their associated properties, including clays, sands, and rocks. Such studies demonstrated the benefits of employing databases for transformation model assessments. Furthermore, soil databases serve as indispensable resources when dealing with situations where data for a particular site is either incomplete, sparsely available, or not entirely reliable, as highlighted by Ching and Phoon (2020).

In general, databases can be used to improve the determination of a specific soil property, in our case the undrained shear strength (s_u), in a number of ways. For example, databases can be used to:

- **Identify and quantify bias in local data.** This is done by comparing the local data to a larger and more comprehensive set of data including similar soil types, from a variety of sites and conditions. For example, if the local data is consistently lower than the data in the database, this may indicate that the local data is biased low.
- **Assess the variability of s_u at a particular site.** This is done by calculating the standard deviation of the s_u data for that site. This information can be used to estimate the uncertainty associated with s_u predictions for that site, to be compared with the uncertainties from similar test sites where testing has been carried out using similar equipment..
- **Develop site-specific s_u correlations.** This is done by identifying statistical relationships between s_u and other soil properties for the local data. These correlations can then be used to estimate s_u for local data, taking into account the specific conditions at the site.

- **Validate s_u predictions from numerical models.** This is done by comparing the model predictions to the s_u data in the database. This information can be used to assess the accuracy of the numerical models and to identify any areas where the models may need to be improved (e.g. D'Ignazio et al. 2018).

In this study, data from Selänpää (2021) from six sites in Southern Finland is compared with the large F-CLAY/7/216 database by D'Ignazio et al. (2016). The dataset includes triaxial compression (CAUC/CIUC, s_{uC}), extension (CAUE/CIUE, s_{uE}), DSS, and FVT results on soft sensitive clays. The anisotropic s_u data by Selänpää (2021) is collected to derive a new database of Finnish soft clays named ANI-F-CLAY/7/87, following the nomenclature by Ching and Phoon (2014a) for multivariate database as "soil type"/"number of parameters of interest"/"number of data points." For the ANI-F-CLAY/7/87 database, the label "ANI" stands for anisotropic. The database is further compared with an extensive high-quality database of Norwegian clays presented by Karlsrud and Hernandez-Martinez (2013) and Paniagua et al. (2019). This analysis results in the establishment of a robust statistical correlation for anisotropic undrained shear strength of Finnish soft clays. By analysing Finnish and Norwegian soil data from databases exhibiting variations in water content, plasticity, and stress history, this study aims to quantify the uncertainties involved in the determination of s_u for Finnish clays in the absence of site-specific test data.

In essence, these databases serve as valuable repositories of collective knowledge and experience, enabling researchers and engineers to assess the reliability of local data, reduce bias, and gain a more comprehensive understanding of the undrained shear strength of clays in specific regions. The utilization of such databases not only aids in enhancing the precision of calculations but also fosters greater confidence in infrastructure design and construction practices.

4.2 DATABASES OF FINNISH AND SCANDINAVIAN SOFT CLAYS

4.2.1 Large regional databases

In the study conducted by D'Ignazio et al. (2016), two large regional databases were compiled, focusing on clays from Finland and Scandinavia (Sweden-Norway). These databases were denoted as F-CLAY/7/173 and S-CLAY/7/168, each containing 173 and 168 data points, respectively. These comprehensive datasets encompass multivariate information, consisting of seven fundamental parameters measured at equivalent sampling locations and depths. The parameters include in-situ undrained shear strength from FVT (s_{uFVT}), effective vertical stress (σ'_v), vertical pre-consolidation pressure (σ'_p), natural water content (w), liquid limit (LL), plastic limit (PL), and sensitivity index (S_t, calculated as the ratio of s_u to s_{ur}, where s_{ur} represents the remoulded undrained shear strength from the Fall Cone test). Additionally, ten dimensionless parameters were derived from these basic parameters, which encompass the following categories:

1. Index properties, including natural water content (w), liquid limit (LL), plasticity index (PI), and liquidity index (LI).
2. Stress and strengths, including OCR, $s_{u(mob)}/\sigma'_v$, $s_{u(mob)}/\sigma'_p$, s_{uFVT}/σ'_v, s_{uFVT}/σ'_p, $S_t = s_u/s_{ur}$.

The databases based on dimensionless parameters were renamed as F-CLAY/10/173 and S-CLAY/10/168.

The $s_{u(mob)}$ is defined as λ multiplied by s_{uFVT}, where λ is a correction factor based on the liquid limit as $\lambda = 1.5/(1 + LL/100)$ (Liikennevirasto 2018).

The clay properties within the F-CLAY/7/173 dataset exhibit a wide range of S_t values, ranging from 2 (insensitive clays) to 58 (quick clays), a broad spectrum of PI values (ranging from 2% to 95%), w values spanning from 25% to 150%, and OCR values ranging from 1 to 3.7. As for the basic clay properties within the S-CLAY/7/168 dataset, they fall within the following ranges: S_t=3.0–42.5, PI = 4–128, w=17–180, and OCR=1–6.1.

D'Ignazio et al. (2016) presented their findings in terms of normalized s_u as a function of OCR as expressed in Equation (4.1):

$$\frac{s_u}{\sigma'_v} = \left(\frac{s_u}{\sigma'_v}\right)_{NC} \mathrm{OCR}^m = \mathrm{SOCR}^m \tag{4.1}$$

Here, S represents the normalized s_u/σ'_v for the normally consolidated state, and m is an empirical material coefficient. For $s_{u(mob)}$ of Finnish soft clays, D'Ignazio et al. (2016) found $S{\approx}0.24$ and $m{\approx}0.76$ with coefficient of variation COV≈0.25 (see Figure 4.1). These coefficients align with the literature range of 0.19–0.27 as reported in previous studies by e.g., Jamiolkowski et al. (1985), Ladd and DeGroot (2003), DeGroot et al. (2019), Yang et al. (2019), and Paniagua et al. (2019). The empirical coefficient m is also consistent with a value of $m{\approx}0.8$ observed for lightly over-consolidated Scandinavian clays (Larsson et al. 2007). D'Ignazio et al. (2017b) found $m < 1$ for Finnish and Scandinavian clays from statistical analyses. Furthermore, data from Drammen clay in Norway, as presented by Andersen (2004), suggest a value of $m{\approx}0.8$ for OCR ranging from 1 to 40. Jamiolkowski et al. (1985) and Andersen et al. (2023) recommended a value of $m{\approx}0.8$ for clays.

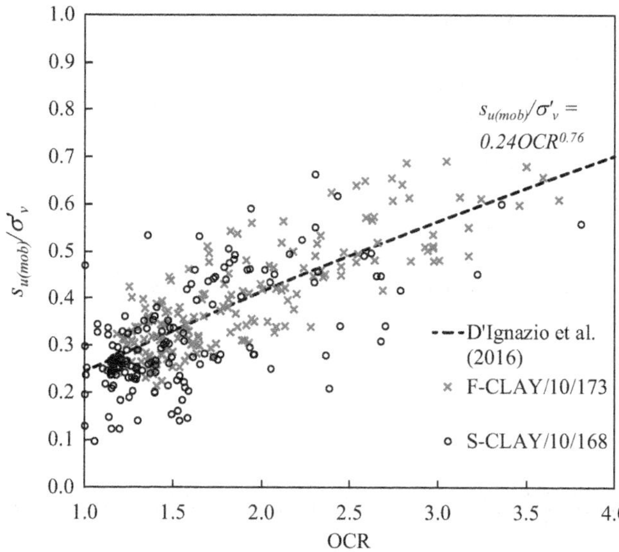

Figure 4.1 $s_{u(mob)}/\sigma'_v$ versus OCR from F-CLAY/10/173 and S-CLAY/10/168 databases and trendline by D'Ignazio et al. (2016) for Finnish clays.

Additionally, the normalized $s_{u(mob)}/\sigma'_v$ and $s_{u(mob)}/\sigma'_p$ were found to be independent of index properties, while the basic clay properties seemed to influence s_{uFVT}/σ'_v and s_{uFVT}/σ'_p. Therefore, in this study, we focus on the more fundamental relationship between s_u and σ'_p or normalized s_u and OCR=σ'_p/σ'_v.

Finally, for the purpose of this study, Equation (4.2) for the OCR-s_{uFVT}/σ'_v transformation model for Finnish clays is derived based on Equation (4.1) as follows:

$$\frac{s_{uFVT}}{\sigma'_v} = 0.27\mathrm{OCR}^{0.78} \qquad COV = 0.34 \tag{4.2}$$

The COV≈0.34 quantifies the uncertainties associated with the validation database S-CLAY/10/168.

4.2.2 Multivariate ANI-F-CLAY/7/87 database from Finland

Selänpää (2021) compiled a high-quality multivariate dataset from six sites in Southern Finland, including Perniö, Paimio, Sipoo, Masku, Lempäälä, and Kotka. This dataset was collected using a modified Laval-type sampler with a diameter of 132 mm (TUT132, Di Buò et al. 2019a), and it includes results of index tests, CAUC/CIUC and CAUE/CIUE (five sites), and DSS (four sites) tests. Specimens were reconsolidated to the in situ stress in order to preserve their original structure. Miniblock (150-mm-diameter) Sherbrook sampler (Emdal et al. 2016) data are available for the Perniö site. Such data is presented for comparison with the TUT132 sampler.

The database also incorporates FVT data points. Notably, the FVT data from Perniö is part of the F-CLAY/7/173 database by D'Ignazio et al. (2016). The clay properties within this dataset exhibit a wide range of PI values (from 17% to 63%), w values ranging from 57% to 117%, and S_t values spanning from 16 to 98. The OCR values range in the 1~2 interval. The OCR values are inferred from constant-rate-of-strain (CRS) oedometer tests carried out in the vicinity of the specimens used for triaxial and DSS tests. The data are collected from depths up to 10 m below ground level.

Sample quality was assessed according to Lunne et al. (1997). Data points fall within the sample quality category "Very good to excellent."

The multivariate TXC (CAUC/CIUC), DSS, TXE (CAUE/CIUE), and FVT data from Selänpää (2021) are presented in Appendix 4.A of this chapter. The database is named ANI-F-CLAY/7/87, as discussed in Section 4.1.

Figure 4.2 illustrates a comparison between s_{uFVT}/σ'_v and OCR from the data points in ANI-F-CLAY/7/87 and D'Ignazio et al. (2016). Figure 4.3 compares s_{uC}/σ'_v versus OCR, s_{uDSS}/σ'_v versus OCR, and s_{uE}/σ'_v versus OCR with $s_{u(mob)}/\sigma'_v$ of Finnish soft clays (D'Ignazio et al. 2016). The trend observed in the FVT data in ANI-F-CLAY/7/87 from four sites appears to align with the general trend and scatter observed in the F-CLAY/7/173 database. Furthermore, the DSS test results are in agreement with the mean trend of $s_{u(mob)}/\sigma'_v$; while TXC and TXE test data are higher and lower than $s_{u(mob)}$, respectively.

The TXC and DSS data present a clear and consistent trend, while the TXE data seem to display a lack of dependence on the OCR. However, it is worth noting that this dataset may contain outliers, as it exhibits greater variability compared to TXC and DSS. For instance, Selänpää (2021) identified n.3 outliers for the TXE dataset. Their removal results in the overall trend of the TXE data points align more closely with the TXC and

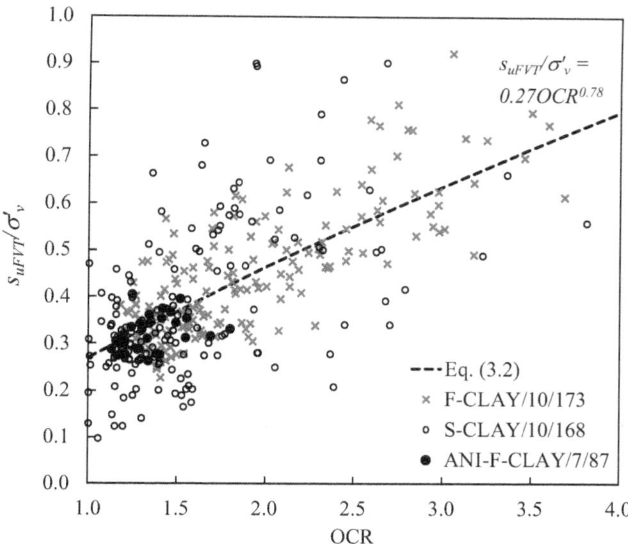

Figure 4.2 s_{uFVT}/σ'_v versus OCR from F-CLAY/10/173, S-CLAY/10/168 and ANI-F-CLAY/7/87 databases and trendline for s_{uFVT}/σ'_v of Finnish clays based on F-CLAY/10/173 database (Equation 4.2).

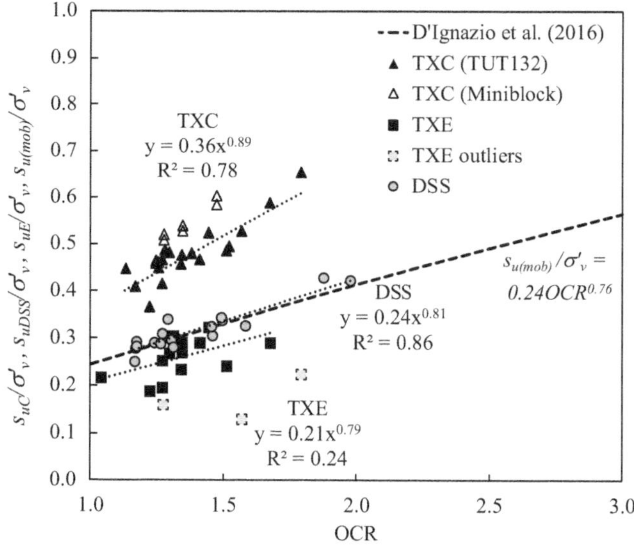

Figure 4.3 s_{uC}/σ'_v, s_{uDSS}/σ'_v, and s_{uE}/σ'_v versus OCR from ANI-F-CLAY/7/87 database and trendline for $s_{u(mob)}/\sigma'_v$ of Finnish clays based on D'Ignazio et al. (2016).

DSS data. Trendlines for TXC, DSS, and TXE of Finnish soft clays with OCR less than 2 are proposed as follows (see Figure 4.3, modified after Selänpää 2021):

$$\frac{s_{uC}}{\sigma'_v} = 0.36\ \mathrm{OCR}^{0.89} \qquad R^2 = 0.78,\ \mathrm{COV} = 0.06 \tag{4.3}$$

$$\frac{s_{uDSS}}{\sigma'_v} = 0.24\text{OCR}^{0.81} \qquad R^2 = 0.86,\ \text{COV} = 0.06 \tag{4.4}$$

$$\frac{s_{uE}}{\sigma'_v} = 0.21\text{OCR}^{0.79} \qquad R^2 = 0.24,\ \text{COV} = 0.23;\ \textit{n.3 outliers removed} \tag{4.5}$$

The coefficient of variation (COV) is defined later in Section 4.3. The S and m values calculated for the OCR-s_{uDSS}/σ'_v model for Finnish clays closely resemble those obtained by D'Ignazio et al. (2016) for the OCR-$s_{u(mob)}/\sigma'_v$ correlation. Furthermore, based on the information presented above, it is not straightforward to establish anisotropy ratios s_{uDSS}/s_{uC} and s_{uE}/s_{uC} for Finnish clays. The apparent influence of OCR on s_{uDSS}/s_{uC} and s_{uE}/s_{uC} can be attributed to two factors: (i) the limited number of data points available to establish the correlations, and (ii) the potential presence of additional outliers, although this aspect is beyond the scope of this study.

Therefore, given the restricted range of OCR, the relationships between σ'_p and s_u, as depicted in Figure 4.4, are regarded as more suitable for determining anisotropy factors. As shown in Figure 4.4:

$$s_{uC} = 0.36\sigma'_p \qquad R^2 = 0.99,\ \text{COV} = 0.07 \tag{4.6}$$

$$s_{uDSS} = 0.23\sigma'_p \qquad R^2 = 1.00,\ \text{COV} = 0.07 \tag{4.7}$$

$$s_{uE} = 0.18 - 0.19\sigma'_p \qquad R^2 = 0.95 - 0.98,\ \text{COV} = 0.24 \tag{4.8}$$

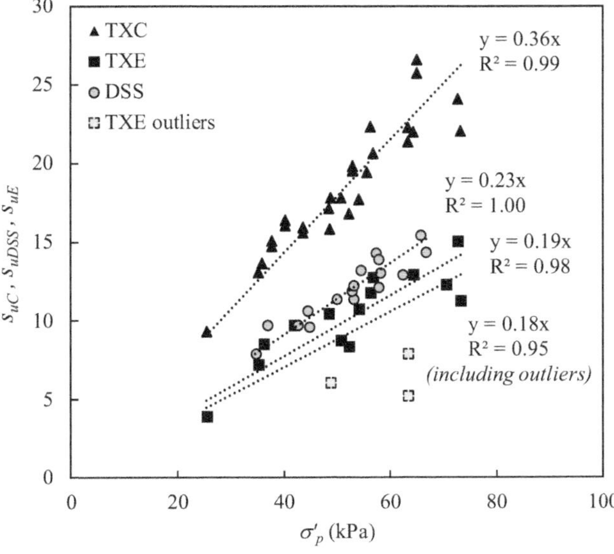

Figure 4.4 Undrained shear strengths s_{uC}, s_{uDSS}, and s_{uE} against pre-consolidation stress σ'_p from ANI-F-CLAY/7/87 database.

Consequently, for Finnish soft clays with OCR values less than 2, the anisotropy factors are estimated as s_{uDSS}/s_{uC} = 0.64 and s_{uE}/s_{uC} = 0.50–0.53. These findings align with those determined for Perniö clay based on CPTU data and finite element back analysis of the Perniö failure test, as presented in the studies by D'Ignazio et al. (2014, 2015, 2017a) and Di Buò et al. (2019b). Selänpää (2021) suggested an average s_{uE}/s_{uC} = 0.62, with a range of calculated values 0.52–0.70. The s_{uE}/s_{uC} calculated by Selänpää (2021) is based on the average of the s_{uE}/s_{uC} calculated for TXC and TXE tests at the same depths and not based on database trendlines. However, the values obtained based on Figure 4.4 are in line with typical values of soft clays from Scandinavia (Larsson et al. 2007; Karlsrud and Hernandez-Martinez 2013).

4.2.3 Comparing regional databases from Finland and Norway

Karlsrud and Hernandez-Martinez (2013) and Paniagua et al. (2019) conducted extensive data collection on block samples (250 mm diameter) of Norwegian clays from seventeen test sites spread across Norway for depths up to 23 m below ground level. Their study involves the evaluation of both TXC (n. 61) and DSS (n. 22) data. The clay properties within this dataset exhibit a wide range of PI values (ranging from 4% to 49%), w values spanning from 28% to 72%, and S_t values ranging from 2 to 240. The OCR values within this dataset range from 1 to 6, and the clay content varies between 21% and 65%. The OCR values are inferred from constant-rate-of-strain (CRS) oedometer tests carried out in proximity to the triaxial and DSS specimens.

Sample quality was assessed according to Lunne et al. (1997). Data points fall within sample quality categories "Very good to excellent" and "Good to fair."

In developing the OCR-s_{uC}/σ'_v and OCR-s_{uDSS}/σ'_v transformation models, Karlsrud and Hernandez-Martinez (2013) and Paniagua et al. (2019) observed that the normalized s_u and exponent m increase with increasing water content. It is noteworthy that, similar to the Finnish dataset, the tests on Norwegian clay samples involved reconsolidating specimens to their in situ stress levels to preserve the original soil structure. For the purpose of this study, mean trend equations for TXC and DSS, following the format of Equation (4.1), were established through linear regression analyses as follows:

$$\frac{s_{uC}}{\sigma'_v} = 0.34\ OCR^{0.58} \qquad R^2 = 0.71,\ COV = 0.16 \tag{4.9}$$

$$\frac{s_{uDSS}}{\sigma'_v} = 0.25\ OCR^{0.62} \qquad R^2 = 0.73,\ COV = 0.18 \tag{4.10}$$

Figure 4.5 provides a comparison between the TXC data in ANI-F-CLAY/7/87 and the Norwegian block-sample data. While the Finnish TXC data displays a well-defined trend with limited scatter, the Norwegian data appears to exhibit higher uncertainty.

Notably, for OCR values less than 2, the normalized s_{uC}/σ'_v for Finnish clays seems to be higher than that of Norwegian clays, with the Miniblock samples data showing slightly higher values than the TUT132 samples. This observation might be attributed to the lower OCR data points in the Norwegian database being observed at greater depths compared to the Finnish dataset, as discussed by D'Ignazio et al. (2022). This discrepancy could be an indication of the stress-dependent nature of the normalized undrained shear strength of clays. Further, both Miniblock and TUT132 triaxial samples fall into the "Very good to excellent" category by Lunne et al. (1997).

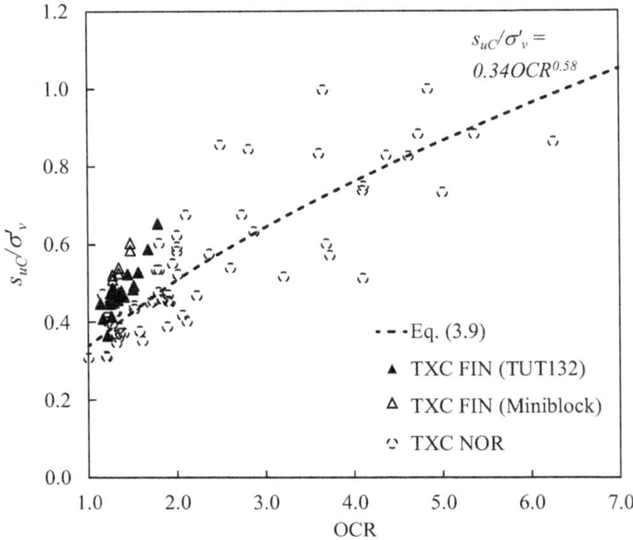

Figure 4.5 s_{uC}/σ'_v versus OCR from ANI-F-CLAY/7/87 database and trendline for s_{uC}/σ'_v of Norwegian clays based on the data in Paniagua et al. (2019) (Equation 4.9).

As discussed earlier, the variability in the Norwegian clay data appears to be linked to the natural water content, while the Finnish data, as reported by Selänpää (2021), seems to exhibit less sensitivity to basic clay properties. Nevertheless, considering that the Finnish data is characterized by higher water content, this might support the findings of Karlsrud and Hernandez-Martinez (2013) and Paniagua et al. (2019).

Furthermore, the increased scatter observed in the Norwegian data could be associated with several factors. Firstly, the Finnish data originates from 6 sites all located in Southern Finland, while the Norwegian database comprises data from seventeen sites across Norway, where regions may have experienced different depositional histories and relative sea level changes. Additionally, the Finnish data was collected using the same sampling equipment and personnel across all sites, whereas the Norwegian data was accumulated over three decades using various types of equipment.

Figure 4.6 presents a comparison between DSS data from Finland and Norway and the F-CLAY/7/173 database. The range of DSS tests on Norwegian clays appears to align with the range of DSS and FVT data from Finland, indicating that s_{uDSS} is approximately equal to $s_{u(mob)}$. As expected, results from triaxial compression tests in both datasets exhibit a generally higher trend compared to DSS and FVT data, hinting at the stress-path-dependent (or anisotropic) behaviour of soft clays.

4.3 STATISTICAL DETERMINATION OF TRANSFORMATION MODELS FOR ANISOTROPIC UNDRAINED SHEAR STRENGTH OF FINNISH CLAYS

Correlations between soil properties are frequently established through linear regression, such as the relationship between undrained shear strength and the OCR – these are known as pairwise correlations. The transformation model, represented by the regression

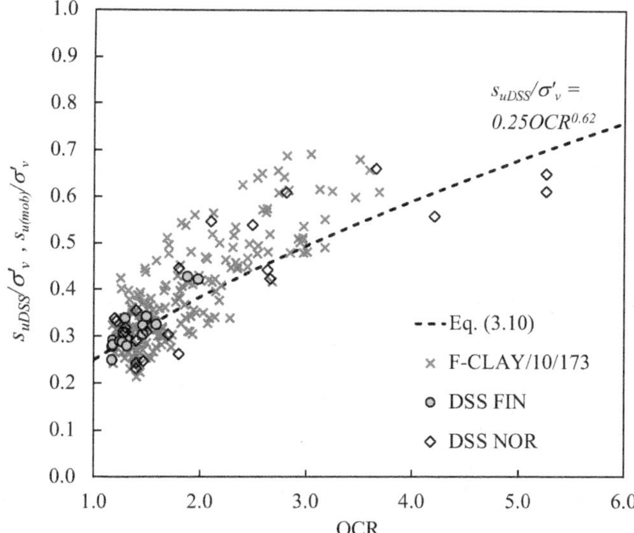

Figure 4.6 s_{uDSS}/σ'_v versus OCR from ANI-F-CLAY/7/87 database, $s_{u(mob)}/\sigma'_v$ versus OCR from F-CLAY/10/173 database, and trendline for s_{uDSS}/σ'_v of Norwegian clays based on the data in Karlsrud and Hernandez-Martinez (2013) (Equation 4.10).

line, is typically considered valid solely for the specific dataset from which it was derived. Such datasets are often tailored to a particular site or region. However, to extrapolate these correlations beyond the local context, it's necessary to validate them against a comprehensive database to quantify bias and uncertainties.

In general, the effectiveness of a transformation model concerning a database can be evaluated in terms of two key parameters: the bias factor (b') and the coefficient of variation (δ), as recommended by Ching and Phoon (2014b). These parameters, b' and δ , express the sample mean and the coefficient of variation, respectively, of the ratio (actual target value/predicted target value) and are defined as follows:

$$b' = \sum_{i=1}^{n} \frac{(\text{actual value})_i / (\text{predicted value})_i}{n} \tag{4.11}$$

$$\delta = \text{COV} = \frac{1}{b'} \sqrt{\frac{1}{n-1} \sum_{i=1}^{n} \left(\frac{(\text{actual value})_i}{(\text{predicted value})_i} - b' \right)^2} \tag{4.12}$$

When b' = 1, the model prediction is considered unbiased. For instance, for the OCR-$s_{u(mob)}/\sigma'_v$ model by D'Ignazio et al. (2016), the actual target value is s_{uDSS}/σ'_v or s_{uC}/σ'_v and the predicted target value is $0.24\text{OCR}^{0.76}$. The calibrated bias factors (b') from reference datasets or transformation models for the OCR-s_u/σ'_v models imply $s_u/\sigma'_v = b'S\ \text{OCR}^m$ with a coefficient of variation equal to δ .

The ratio of actual to predicted value is widely accepted as a model factor in geotechnical reliability literature to characterize the model uncertainty in calculated responses. For instance, in the context of the hyperbolic limit from a measured load–displacement curve compared to a calculated lateral capacity from limit equilibrium, this ratio is referred to as a model factor (Phoon and Kulhawy 2005). Equations (4.11) and (4.12) can be thought of as representing the mean and coefficient of variation of this model factor. However, in this case, they are applied at the level of soil properties rather than at the level of response. Characterizing the model factor at a soil property level offers the advantage of estimating the model factor of the response, assuming that no other uncertainties emerge from different model idealizations in the adopted calculation model that are more significant than measuring s_u from a specific laboratory test type.

Table 4.1 summarizes the bias factor and COV values for anisotropic undrained shear strength of Finnish soft clays.

Based on the data in Table 4.1, the statistically determined transformation model for s_{uDSS} is unbiased with respect to the mean trend of the $s_{u(mob)}$ of Finnish soft clays. It exhibits a relatively low coefficient of variation (COV) of approximately 0.06, which is noteworthy, considering that it finds its validity over a wider OCR range than the DSS tests alone (specifically, OCR = 1–2 as compared to OCR ≈ 1–4). In contrast, the OCR-s_{uFVT}/σ'_v model expressed in Equation (4.2) tends to overestimate the data points by 9% (with $b' = 0.91$).

When comparing the OCR-s_{uDSS}/σ'_v model for Finnish clays with the s_{uDSS} of Norwegian clays, the former appears to align well, even though it is based on a dataset with broader OCR range compared to the Finnish one (OCR = 1–2 vs OCR ≈ 1–6). Additionally,

Table 4.1 Bias factors and COV for anisotropic s_u of Finnish soft clays from ANI-F-CLAY/7/87 database.

Actual target value (ANI-F-CLAY/7/87, Selänpää 2021)	*Predicted target value (trendlines from databases)*	*Reference*	*No. of data points*	*b′*	*δ = COV*
s_{uDSS}/σ'_v	$s_{u(mob)}/\sigma'_v = 0.24OCR^{0.76}$	D'Ignazio et al. (2016)	17	1.01	0.06
s_{uDSS}/σ'_v	$s_{uFVT}/\sigma'_v = 0.27OCR^{0.78}$	Equation (4.2), derived from the data in D'Ignazio et al. (2016)	17	0.91	0.06
s_{uDSS}/σ'_v	$s_{uDSS}/\sigma'_v = 0.25OCR^{0.62}$	Equation (4.10), derived from the data in Karlsrud and Hernandez-Martinez et al (2013)	17	1.04	0.06
s_{uC}/σ'_v	$s_{uC}/\sigma'_v = 0.34OCR^{0.58}$	Equation (4.9), derived from the data in Paniagua et al (2019)	19	1.18	0.08
s_{uC}/σ'_v	$s_{u(mob)}/\sigma'_v = 0.24OCR^{0.76}$	D'Ignazio et al. (2016)	19	1.55	0.08
s_{uC}/σ'_v	$s_{uFVT}/\sigma'_v = 0.27OCR^{0.78}$	Equation (4.2), derived from the data in D'Ignazio et al (2016)	19	1.39	0.08
s_{uE}/σ'_v	$s_{uC}/\sigma'_v = 0.34OCR^{0.58}$	Equation (4.9), derived from the data in Paniagua et al (2019)	17	0.60	0.22
s_{uE}/σ'_v	$s_{u(mob)}/\sigma'_v = 0.24OCR^{0.76}$	D'Ignazio et al. (2016)	17	0.80	0.23
s_{uE}/σ'_v	$s_{uFVT}/\sigma'_v = 0.27OCR^{0.78}$	Equation (4.2), derived from the data in D'Ignazio et al (2016)	17	0.72	0.23

the OCR-s_{uC}/σ'_v model for Finnish clays tends to overestimate the mean trend of the Norwegian block-sample database by approximately 18%, as qualitatively discussed in the previous section.

Furthermore, in comparison to $s_{u(mob)}$ and s_{uFVT} of Finnish clays, s_{uC} for OCR values less than 2 is approximately 55% and 39% larger, respectively, and is characterized by a fairly low COV = 0.08.

Finally, the OCR-s_{uE}/σ'_v model underestimates the predicted target values of s_{uC}/σ'_v, s_{uFVT}/σ'_v and $s_{u(mob)}/\sigma'_v$ as anticipated. Additionally, the calculated COV is higher in comparison to the OCR-s_{uC}/σ'_v and the OCR-s_{uDSS}/σ'_v models (COV = 0.22–0.23 as opposed to 0.06–0.08). This is in line with the scatter in the original TXE dataset contained in the ANI-F-CLAY/7/87 database.

4.4 DISCUSSION

The choice of an appropriate testing setup for determining the undrained shear strength of clay depends on the expected deformation behaviour of a geotechnical structure in the field. For instance, when assessing embankment stability, it often relies on the average mobilized shear strength along a potential slip surface, which is typically approximated as $s_{uavg} \approx s_{uDSS}$ or $s_{u(mob)}$. However, it is important to note that the shear strength along a slip surface is influenced by the orientation of the major principal stress, making it stress-path dependent or anisotropic, as described by Bjerrum (1973). Therefore, for accurate modelling of shear strength in embankment stability, results from both triaxial and simple shear tests should be considered.

In the case of excavations or cuts, undrained triaxial compression and extension tests can be conducted to capture the active and passive soil resistances in undrained conditions, yielding s_{uC} and s_{uE}, respectively. For pile design, especially for axial loading, s_{uDSS} or s_{uavg} is typically adopted. However, the deformation behaviour of a pile under lateral loading involves a wide range of stress paths.

In practice, determining anisotropic undrained shear strength typically necessitates a combination of triaxial and direct simple shear tests, which are not commonly conducted in Finland, especially in medium- to low-sized projects. However, field vane tests performed in situ find practical utility in everyday geotechnical applications. To address the need for practical solutions, statistically derived transformation models, as outlined in Table 4.1, can be employed to derive equations suitable for initial assessments when specific site data is unavailable. These equations can also serve as a valuable tool for scrutinizing potentially unreliable data, given their foundation in high-quality datasets.

The following equations are proposed:

$$s_{uC} \approx 1.55 s_{u(\text{mob})} \quad \text{COV} = 0.08, \ OCR < 4 \tag{4.13}$$

$$s_{uDSS} \approx s_{u(\text{mob})} \quad \text{COV} = 0.06, \ \text{OCR} < 4 \tag{4.14}$$

$$s_{uE} \approx 0.80 s_{u(\text{mob})} \quad \text{COV} = 0.23, \ \text{OCR} < 4 \tag{4.15}$$

The equations mentioned earlier are in line with the anisotropy ratios $s_{uDSS}/s_{uC} \approx 0.64$ and $s_{uE}/s_{uC} \approx 0.50$ estimated in Section 4.2.2 for Finnish soft clays. Equations based on $s_{u(mob)}$ are favoured over those based on s_{uFVT}, primarily because of the lower COV reported for $s_{u(mob)}$ by D'Ignazio et al (2016) compared to the COV of Equation (4.2) for s_{uFVT} (COV $\approx$ 25% versus 34%).

In addition, the stress exponent m obtained from Equations (4.9)–(4.10) is notably lower than the range of m=0.79–0.89 observed in Finnish soft clays, as indicated in Equations (4.3)–(4.5). As a result, although Table 4.1 indicates δ values below 0.1 for s_{uC}/σ'_v and s_{uDSS}/σ'_v associated with the Norwegian clay database, which includes OCR values in the range 1~6, it is important to exercise caution when extrapolating Equations (4.13)–(4.15) to OCR values exceeding 4. Such extrapolation should be undertaken with due consideration for local geology and stress history.

4.5 CONCLUSIONS

This chapter presents statistical transformation models for anisotropic undrained shear strength of Finnish clays based on data from triaxial and direct simple shear tests. The correlations have been validated against comprehensive regional databases that cover a broader spectrum of soil properties.

A high-quality database collected from six distinct sites in Finland, named ANI-F-CLAY/7/87, serves as the foundation for this study. This dataset incorporates test results from laboratory triaxial compression, triaxial extension, direct simple shear tests, and in-situ field vane tests. The study focuses on establishing correlations for anisotropic undrained shear strength, validated against a large database of field vane test results from Finland and a laboratory database of high-quality Norwegian clay block samples. Both validation databases cover a broader range of basic properties, particularly pre-consolidation stress and OCR values, compared to the Finnish laboratory dataset.

The primary focus of this study is the relationship between undrained shear strength and pre-consolidation stress in clays, emphasizing normalized properties. The derived transformation models are suitable for various plasticity, water content, sensitivity, and over-consolidation ratio values in Finnish clays.

Given its relevance in the determination of undrained shear strength of clays in Finnish practice, transformation models for anisotropic shear strength based on field vane strength are proposed. These anisotropic strength formulas find relevance in a broad range of geotechnical applications, including embankment, excavation, and deep foundation design and analysis.

It is important to note that these results are most applicable for preliminary analyses, particularly when site-specific data is unavailable, and the expected in situ over-consolidation ratio is below 4. In more detailed design phases, local correlations should be verified and, if necessary, recalibrated based on specific test data.

As a final remark, for databases characterized by large scatter, using simpler equations often yields results comparable to using more complex ones, as concluded by Länsivaara et al. (2023). They noted that while adding more variables might improve a training set performance, it often worsens testing set predictions, indicating overtraining, especially with incoherent data. Hence, for scattered data, it is advisable to use simpler models with minimal input data to avoid overtraining risks.

REFERENCES

Andersen, K. H. (2004). Cyclic clay data for foundation design of structures subjected to wave loading. In *Intern. Conf. on Cyclic Behaviour of Soils and Liquefaction Phenomena, CBS04*, Bochum, Germany (pp. 371–387). A.A. Balkema, Ed. Th. Triantafyllidis. https://www.researchgate.net/publication/299810172_Cyclic_clay_data_for_foundation_design_of_structures_subjected_to_wave_loading.

Andersen, K. H., Engin, H. K., D'ignazio, M., and Yang, S. (2023). Determination of cyclic soil parameters for offshore foundation design from an existing data base. *Ocean Engineering*, 267, 113180.

Bjerrum, L. 1973. Problems of soil mechanics and construction on soft clays. State-of-the-art report. In *Proceedings, 8th ICSMFE*, Moscow. Vol. 3, pp. 111–159.

Ching J. and Phoon KK. 2012. Modeling parameters of structured clays as a multivariate normal distribution. *Canadian Geotechnical Journal* 49(5): 522–545.

Ching, J. and Phoon, K.K. 2013. Multivariate distribution for undrained shear strengths under various test procedures. *Canadian Geotechnical Journal*, 50(9): 907–923. doi:10.1139/cgj-2013–0002.

Ching, J. and Phoon, K.K. 2014a. Correlations among some clay parameters – the global database. *Canadian Geotechnical Journal*, 51(6): 663–685.

Ching, J. and Phoon, K.K. 2014b. Correlations among some clay parameters – the multivariate distribution. *Canadian Geotechnical Journal*, 51(6): 686–704.

Ching, J. and Phoon, K. K. 2020. Measuring similarity between site-specific data and records from other sites. *ASCE-ASME Journal of Risk and Uncertainty in Engineering Systems, Part A: Civil Engineering*, 6(2), 04020011.

Ching, J., Lin, G. H., Chen, J. R., and Phoon, K. K. 2017. Transformation models for effective friction angle and relative density calibrated based on generic database of coarse-grained soils. *Canadian Geotechnical Journal*, 54(4), 481–501.

Ching, J., Li, K. H., Phoon, K. K., and Weng, M. C. 2018. Generic transformation models for some intact rock properties. *Canadian Geotechnical Journal*, 55(12), 1702–1741.

Di Buò, B., Selänpää, J., Länsivaara, T. T., and D'Ignazio, M. 2019a. Evaluation of sample quality from different sampling methods in Finnish soft sensitive clays. *Canadian Geotechnical Journal*, 56(8), 1154–1168.

Di Buò, B., D'Ignazio, M., Selänpää, J., Haikola, M., Länsivaara, T., and Di Sante, M. 2019b. Investigation and geotechnical characterization of Perniö clay, Finland. *AIMS Geosciences*, 5, 591–616.

D'Ignazio, M. 2016. *Undrained shear strength of Finnish clays for stability analyses of embankments.* PhD Thesis, Tampere University of Technology.

D'Ignazio, M., Mansikkamäki, J., and Länsivaara, T. 2014. Anisotropic total and effective stress stability analysis of the Perniö failure test. In *Proceedings of Conference on Numerical Methods in Geotechnical Engineering (NUMGE)*, Delft, Netherlands. Vol. 2, pp. 609–614.

D'Ignazio, M., Di Buò, B., and Länsivaara, T. 2015. A study on the behavior of weathered clay crust in the Perniö failure test. In *Proceedings of XVI ECSMGE*, 13–17 September 2015, Edinburgh, Scotland. 7, pp. 3639–3644.

D'Ignazio, M., Phoon, K.K., Tan, S.A., and Länsivaara, T. 2016. Correlations for undrained shear strength of Finnish soft clays. *Canadian Geotechnical Journal*. DOI: 10.1139/cgj-2016–0037.

D'Ignazio, M., Länsivaara, T. T., and Jostad, H. P. 2017a. Failure in anisotropic sensitive clays: finite element study of Perniö failure test. *Canadian Geotechnical Journal*, 54(7), 1013–1033.

D'Ignazio, M., Phoon, K. K., Tan, S. A., Länsivaara, T., and Lacasse, S. 2017b. Reply to the discussion by Mesri and Wang on "Correlations for undrained shear strength of Finnish soft clays". *Canadian Geotechnical Journal*, 54(5), 749–753.

D'Ignazio, M., Phoon, K. K., Länsivaara, T., and Tan, S. A. 2018. Uncertainties in modeling undrained shear strength of sensitive clays using finite-element method. *ASCE-ASME Journal of Risk and Uncertainty in Engineering Systems, Part A: Civil Engineering*, 4(2), 04018011.

D'Ignazio, M., Lunne, T., Andersen, K. H., Yang, S., Di Buò, B., and Länsivaara, T. 2019. Estimation of preconsolidation stress of clays from piezocone by means of high-quality calibration data. *AIMS Geosciences*, 5(2), 104–116.

D'Ignazio, M., Phoon, K. K., and Länsivaara, T. T. 2021. Uncertainties in modelling undrained shear strength of clays using Critical State Soil Mechanics and SHANSEP. In *IOP Conference Series: Earth and Environmental Science* (Vol. 710, No. 1, p. 012075). IOP Publishing. https://iopscience.iop.org/article/10.1088/1755-1315/710/1/012075/pdf

D'Ignazio, M., Di Buò, B., Länsivaara, T., L'Heureux, J. S., Paniagua, P., and Selänpää, J. 2022. Piezocone testing in Nordic soft clays: Comparison of high-quality databases. In *Cone Penetration Testing* 2022 (pp. 356–362). CRC Press. https://www.taylorfrancis.com/chapters/oa-edit/10.1201/9781003308829-48/piezocone-testing-nordic-soft-clays-comparison-high-quality-databases-ignazio-di-bu%C3%B2-l%C3%A4nsivaara-heureux-paniagua-sel%C3%A4np%C3%A4%C3%A4?context=ubx&refId=eb599372-fb27-4cc0-99fc-837f960be2b8.

Emdal, A., Gylland, A., Amundsen, H.A., Kåsin, K., and Long, M. 2016. Mini-block sampler. *Canadian Geotechnical Journal*, 53(8), 1235–1245.

Jamiolkowski, M., Ladd, C.C., Germain, J.T., and Lancellotta, R. 1985. New developments in field and laboratory testing of soils. In *Proceedings of the 11th International Conference on Soil Mechanics and Foundation Engineering*, San Francisco, CA. Vol. 1, pp. 57–153.

Karlsrud, K. and Hernandez-Martinez, F.G. 2013. Strength and deformation properties of Norwegian clays from laboratory tests on high-quality block samples 1. *Canadian Geotechnical Journal*, 50(12):1273–1293.

Länsivaara, T. 1999. *A study of the mechanical behavior of soft clay*. PhD thesis, Norwegian University of Science and Technology, Trondheim.

Länsivaara, T. T., Farhadi, M. S., and Samui, P. 2023. Performance of traditional and machine learning-based transformation models for undrained shear strength. *Arabian Journal of Geosciences*, 16(3), 183.

Larsson, R., Sällfors, G., Bengtsson, P.E., Alén, C., Bergdahl, U., and Eriksson, L. 2007. *Skjuvhällfasthet: utvärdering I kohesionsjord* (2nd edition), Information 3. Swedish Geotechnical Institute (SGI), Linköping.

Liikennevirasto, 2018. Penkereiden stabiliteetin laskentaohje. *Liikenneviraston ohjeita 14/2018*, Helsinki. (In Finnish, Title in English: Guidelines for embankment stability calculations).

Lunne, T., Berre, T., and Strandvik, S. 1997. Sample disturbance effects in soft low plastic Norwegian clay. In *Proceedings of the Conference on Recent Developments in Soil and Pavement Mechanics*, Rio de Janeiro, Brazil, 25–27 June. Balkema, Rotterdam. pp. 81–102. [Also published in Norwegian Geotechnical Institute, Publication 204.]

Paniagua, P., D'Ignazio, M., L'Heureux, J. S., Lunne, T., and Karlsrud, K. 2019. CPTU correlations for Norwegian clays: an update. *AIMS Geosciences*, 5, 82–103.

Phoon, K.K., and Kulhawy, F.H. 1999. Characterization of geotechnical variability. *Canadian Geotechnical Journal*, 36(4), 612–624. doi:10.1139/t99-038.

Phoon, K. K. and Kulhawy, F. H. 2005. Characterisation of model uncertainties for laterally loaded rigid drilled shafts. *Geotechnique*, 55(1), 45–54.

Selänpää, J. 2021. *Derivation of CPTu Cone Factors for Undrained Shear Strength and OCR in Finnish Clays*. PhD Thesis, Tampere University, Tampere, Finland.

Selänpää, J., Di Buò, B., Länsivaara, T., and D'Ignazio, M. 2017. Problems related to field vane testing in soft soil conditions and improved reliability of measurements using an innovative field vane device. *Landslides in Sensitive Clays: From Research to Implementation*, 109–119.

Yang, S. L., Lunne, T., Andersen, K. H., D'lgnazio, M., and Yetginer, G. 2019. Undrained shear strength of marine clays based on CPTU data and SHANSEP parameters. Proc. XVII ECSMGE-2019, Iceland.

APPENDIX 4.A

Table 4.A1 Basic information of the ANI-F-CLAY/7/87 database (based on Selänpää, 2021)

Site	*Test type*	*Sampler type*	s_u *(kPa)*	σ'_v *(kPa)*	σ'_{pCRS} *(kPa)*	*w*	*LP*	*LL*	S_t
Kotka	DSS	TUT132	9.8	28.6	37.0	106	33	70	49
Kotka	DSS	TUT132	10.6	38.0	44.4	57	19	40	38
Kotka	DSS	TUT132	9.7	38.4	44.8	58	19	39	38
Paimio	DSS	TUT132	12.9	39.5	62.4	66	24	42	98
Paimio	DSS	TUT132	14.3	49.0	57.4	109	31	65	91
Paimio	DSS	TUT132	13.9	49.3	57.8	111	31	65	91
Paimio	DSS	TUT132	15.4	53.0	65.7	102	29	66	73
Perniö	DSS	TUT132	7.9	26.6	34.7	99	28	67	38
Perniö	DSS	TUT132	9.7	33.7	42.6	81	25	46	72
Perniö	DSS	TUT132	13.2	42.9	54.4	98	26	58	49
Sipoo	DSS	TUT132	11.4	26.6	49.9	91	30	56	45
Sipoo	DSS	TUT132	11.4	26.9	53.1	90	30	66	45
Sipoo	DSS	TUT132	11.9	35.3	52.8	109	31	85	23
Sipoo	DSS	TUT132	12.3	35.6	53.2	115	31	89	23
Sipoo	DSS	TUT132	12.1	39.7	57.8	110	31	94	22
Sipoo	DSS	TUT132	13.0	40.0	58.2	107	31	94	22
Sipoo	DSS	TUT132	14.4	51.0	66.7	98	28	79	16
Lempäälä	FV	N/A	6.7	18.8	29.2	74	27	47	25
Lempäälä	FV	N/A	12.2	30.8	46.8	69	28	54	29
Masku	FV	N/A	9.3	29.6	50.0	80	27	66	20
Masku	FV	N/A	14.6	38.9	55.2	117	36	95	18
Masku	FV	N/A	17.8	52.1	69.9	86	28	73	20
Paimio	FV	N/A	12.1	35.3	52.8	67	23	40	82
Paimio	FV	N/A	13.0	41.6	64.5	72	24	42	98
Paimio	FV	N/A	14.4	50.4	57.1	108	30	66	77
Paimio	FV	N/A	15.0	54.7	64.0	94	29	59	67
Paimio	FV	N/A	16.8	61.1	74.0	85	28	53	82
Perniö	FV	N/A	8.7	26.5	34.5	99	28	67	38
Perniö	FV	N/A	9.4	27.4	35.5	99	28	67	38
Perniö	FV	N/A	7.5	28.8	38.6	94	26	54	53
Perniö	FV	N/A	8.9	31.1	39.9	86	25	54	50
Perniö	FV	N/A	10.2	33.5	40.4	71	25	44	72
Perniö	FV	N/A	11.1	34.9	41.9	71	25	44	72
Perniö	FV	N/A	11.2	35.9	42.9	72	24	45	72
Perniö	FV	N/A	16.9	41.8	52.2	107	26	70	55
Perniö	FV	N/A	14.4	43.1	53.7	107	26	70	55
Perniö	FV	N/A	16.6	45.2	66.3	84	26	59	
Perniö	FV	N/A	17.1	47.4	63.7	95	27	64	66
Perniö	FV	N/A	15.4	49.6	66.2	95	27	64	66
Sipoo	FV	N/A	8.9	26.8	48.2	89	30	66	45
Sipoo	FV	N/A	9.8	35.5	49.9	116	31	88	23
Sipoo	FV	N/A	11.0	39.5	54.6	110	31	89	28
Sipoo	FV	N/A	18.1	51.2	72.2	99	28	79	16

Site	Test type	Sampler type	s_u (kPa)	σ'_v (kPa)	σ'_{pCRS} (kPa)	w	LP	LL	S_t
Lempäälä	CIUC	TUT132	9.4	20.1	25.5	74	27	47	25
Lempäälä	CIUC	TUT132	15.8	32.0	48.6	69	28	54	29
Masku	CIUC	TUT132	17.8	30.3	50.7	80	27	66	20
Masku	CIUC	TUT132	20.6	39.4	56.9	117	36	95	18
Paimio	CIUC	TUT132	16.8	34.7	52.3	67	23	40	82
Paimio	CIUC	TUT132	21.3	40.4	63.3	72	24	42	98
Paimio	CIUC	TUT132	22.3	49.8	56.4	108	30	66	77
Paimio	CIUC	TUT132	22.2	54.2	63.3	94	29	59	67
Paimio	CIUC	TUT132	22.0	60.0	73.3	85	28	53	82
Perniö	CAUC	Miniblock	25.7	44.0	65.0	105	30	65	55
Perniö	CAUC	Miniblock	16.4	31.5	40.3	86	25	54	50
Perniö	CAUC	Miniblock	14.7	28.0	37.7	98	26	56	54
Perniö	CAUC	Miniblock	15.1	28.0	37.7	98	26	56	54
Perniö	CAUC	Miniblock	19.8	42.6	53.0	107	26	70	55
Perniö	CAUC	TUT132	16.0	34.7	43.6	71	25	44	72
Perniö	CIUC	Miniblock	26.6	44.0	65.0	105	30	65	55
Perniö	CIUC	Miniblock	16.1	31.5	40.3	86	25	54	50
Perniö	CIUC	Miniblock	19.5	42.6	53.0	107	26	70	55
Perniö	CIUC	TUT132	15.6	34.7	43.6	71	25	44	72
Perniö	CIUC	TUT132	17.7	42.6	54.2	90	26	58	49
Perniö	CIUC	TUT132	13.7	28.1	35.9	94	26	54	53
Perniö	CIUC	TUT132	22.0	48.0	64.4	95	27	64	66
Perniö	CIUC	TUT132	13.1	27.1	35.2	99	28	67	38
Sipoo	CIUC	TUT132	17.8	27.2	48.8	90	30	73	22
Sipoo	CIUC	TUT132	17.1	36.0	48.4	116	31	88	23
Sipoo	CIUC	TUT132	19.4	40.4	55.7	110	31	89	28
Sipoo	CIUC	TUT132	24.1	51.6	72.8	80	27	71	16
Lempäälä	CIUE	TUT132	3.9	20.1	25.5	74	27	47	25
Lempäälä	CIUE	TUT132	9.7	32.0	41.9	69	28	54	29
Masku	CIUE	TUT132	12.7	39.4	56.9	117	36	95	18
Masku	CIUE	TUT132	8.8	30.3	50.7	80	27	66	20
Masku	CIUE	TUT132	12.3	52.7	70.6	86	28	73	20
Paimio	CIUE	TUT132	8.4	34.7	52.3	67	23	40	82
Paimio	CIUE	TUT132	11.8	54.2	56.4	108	30	66	77
Paimio	CIUE	TUT132	11.3	60.0	73.3	85	28	53	82
Paimio	CIUE	TUT132	5.3	40.4	63.3	72	24	42	98
Paimio	CIUE	TUT132	8.0	49.8	63.3	94	29	59	67
Perniö	CAUE	TUT132	10.8	42.6	54.2	90	26	58	49
Perniö	CIUE	TUT132	12.9	48.0	64.4	95	27	64	66
Perniö	CIUE	TUT132	8.5	28.1	36.3	94	26	54	53
Perniö	CIUE	TUT132	7.3	27.1	35.2	99	28	67	38
Sipoo	CIUE	TUT132	10.5	36.0	48.4	116	31	88	23
Sipoo	CIUE	TUT132	15.0	51.6	72.8	80	27	71	16
Sipoo	CIUE	TUT132	6.1	27.2	48.8	90	30	73	22

Chapter 5

Role of databases in the evaluation of soil properties

Abolfazl Eslami and Pin-Qiang Mo

This chapter delves into the pivotal role of databases in assessing soil properties of geotechnical engineering. It underscores the significance of site investigations, methodologies, in-situ testing, and soil variability management in advancing the field of geotechnical engineering. The chapter emphasizes the crucial function of geotechnical databases in furnishing historical data and research experiences, fostering the evolution of research methodologies, and enhancing our comprehension of the underground. The text elaborates on the pivotal importance of in-situ testing methods such as Cone Penetration Testing (CPT), Standard Penetration Testing (SPT), and Field Vane Shear Testing (FVST) in soil identification and classification. Concurrently, it explores the latest research achievements in soil correlation and parameter identification leveraging machine learning and Bayesian algorithms, emphasizing the paramountcy of accounting for uncertainties in geotechnical modeling and decision-making. Furthermore, the chapter introduces the AUT: CPTu-Geo-marine Database developed by Eslami et al. (2022), offering a comprehensive evaluation of the performance of various soil behavior classification methodologies. It presents a novel triangular classification chart aimed at enhancing the accuracy of classifying marine sediments.

5.1 INTRODUCTION

In the realm of geotechnical engineering research, soil characterization and classification stand as foundational elements supporting advancements in design and construction. This chapter emphasizes the role of databases in assessing soil properties within the context of geotechnical engineering research. Soil, as the subject of inquiry, plays a central role, in influencing the success and safety of construction projects across infrastructural development and environmental sustainability. The core of geotechnical research lies in site investigation, where researchers embark on a journey to uncover critical insights into soil characteristics and behaviors, deeply intertwined with their environmental interactions. Various methodologies are explored, including remote sensing, data review, site surveying, and continuous monitoring, thus enriching our understanding of soil mechanics. Acknowledging the inherent variability and uncertainty in soil properties, researchers draw from diverse data sources to construct robust research frameworks, mitigating the unpredictability inherent in decision making. In-situ testing emerges as a pivotal research tool, providing real-time insights into soil behavior, with a focus on tests such as the Standard Penetration Test (SPT), Field Vane Shear Test (FVST), and Cone Penetrometer Test (CPT). Geotechnical databases play a central role, providing access to historical data and research experiences, facilitating the advancement of research methodologies, and deepening our collective understanding of the complex world beneath our feet. This

DOI: 10.1201/9781003441946-5

chapter highlights the crucial role of databases in evaluating soil properties, emphasizing the significance of site investigation, methodologies, in-situ testing, and the management of soil variability, ultimately paving the way for a comprehensive exploration of the field.

5.2 DATA SOURCES IN GEOTECHNICAL ENGINEERING

Geotechnical site investigation stands as the foremost undertaking in any project, serving as the primary means to acquire critical on-site information. This comprehensive process encompasses activities ranging from drilling boreholes to collect both disturbed and undisturbed samples to conducting laboratory tests and in-situ testing, enabling the observation of soil behavior in authentic stress scenarios and under real boundary conditions.

One of the primary and fundamental phases in project design, known as geotechnical site investigation, involves the evaluation of subsurface soil and rock characteristics, which can be obtained through various approaches:

- Remote sensing by evaluating regular aerial photographs
- Reviewing existing data and published sources, such as geological and geographical maps, geotechnical reports, and geological hazard maps
- Site survey and visual observation
- Local experience and ongoing construction operations
- Geophysical and seismic evaluations
- Drilling, sampling, and groundwater assessment
- In-situ and laboratory testing
- Evaluation and analysis of environmental indices
- Instrumentation, surveillance, and monitoring

Typically, a site investigation program progresses through the following stages:

1. Collecting data and literature review
2. Site visit and conducting non-destructive tests
3. Performing operations in the field including drilling, sampling, and in-situ testing
4. Conducting laboratory testing on collected samples
5. Data synthesis and providing a geotechnical investigation report
6. Instrumentation, surveillance, and monitoring

Geotechnical engineers grapple with the inherent variability and uncertainties in natural geomaterials. To ensure the safe design of foundations and minimize uncertainties, it becomes imperative to rely on a variety of measured and collected data. This primary objective can be realized through the utilization of diverse sources of geotechnical data, which can be acquired using the following approaches:

1. Maps and technical literature review
2. Site visit
3. On-situ testing
4. Geophysical surveying
5. Drilling operations

6. In-situ penetration tests
7. Laboratory element tests
8. Physical modeling
9. Full-size field tests
10. Instrumentation and monitoring

5.3 UNCERTAINTY IN GEOTECHNICAL DESIGNS

Geotechnical design, particularly in the field of foundation engineering, encompasses the application of site investigations, design models, adherence to codes, and computational analysis. This process involves informed decision making regarding site-specific conditions, ultimately leading to an optimal design tailored to the project's significance.

In conventional design practices, engineers typically treat material properties like friction angle, soil undrained shear strength, or the uniaxial compressive resistance of concrete as fixed parameters with known values. However, when delving into borehole data or examining the results of tests like SPT or CPT, it becomes evident that there exists inherent uncertainty in soil properties. Furthermore, uncertainties extend beyond material properties; they also encompass simplified constitutive models and the estimation of material characteristics, introducing significant uncertainties into the design process. Empirical evidence in this field substantiates the notable disparities between measured values and predicted design parameters.

Uncertainty in geotechnical engineering can be broadly categorized into two main groups. The first category arises from the inherent randomness in the nature of materials and is a persistent feature that cannot be entirely eradicated. However, it can be better characterized and managed through increased testing or the use of comprehensive databases. Parameters like undrained shear strength fall into this category, as their values exhibit spatial and temporal variations. Shear strength is typically estimated from a limited number of samples and associated tests, which are then used in the design at locations different from where the samples were obtained. The disparity between the estimated resistance value at one point and its actual resistance constitutes the inherent uncertainty. This variability in soil properties is a consequence of the intricate processes governing geomaterial formation, including factors such as sedimentation, weathering, stress history, and the passage of time.

The second category of uncertainty pertains to our incomplete and constrained understanding of a system or process. In general, this type of uncertainty can be classified into three primary types: model uncertainty, transformation uncertainty, and measurement/random noise error.

Measurement errors arise from various sources, including equipment limitations, operator factors, and random fluctuations during the measurement process. Additionally, empirical methods or other models that integrate laboratory or field measurements into soil design characteristics fall into this category of uncertainty. Consequently, this form of uncertainty can be mitigated by enhancing both the quantity and precision of experiments or adopting more accurate models. Statistical uncertainty, which stems from having limited data, can be addressed by increasing the number of samples tested. However, errors attributed to the sampling or measurement procedures cannot be entirely eradicated merely by expanding the sampling effort.

Transformation error represents an additional source of uncertainty and is often associated with a limited number of case records. Every experimental correlation relies on a database that may have constraints, such as being specific to a particular location or soil type. Moreover, the development of such correlations involves assumptions and specific computational models, which means that inherent errors are implicitly introduced into the estimation of soil properties.

Model error, particularly in the context of geotechnical engineering, encompasses a multitude of uncertainties arising from factors like simplifications in equilibrium and deformation analyses, the neglect of three-dimensional modeling effects, and more. Model uncertainty is quantified as the ratio between the actual quantity and the quantity predicted by a model. The uncertainty associated with the model can be characterized by its mean and coefficient of variation. The disparity between the mean value of the unit quantity reveals a systematic error within the model, while the standard deviation denotes the variability in the model's predicted values.

Moreover, a substantial level of uncertainty surrounds the assessment of live and dynamic loads, including factors like earthquakes, winds, floods, and the like. Unlike uncertainties tied to static loads, uncertainties concerning material parameters may be less prominent. However, it is crucial to recognize that under dynamic loading conditions, material parameters may exhibit substantial variations, which might not be immediately evident. For instance, during cyclic loading, soil shear strength tends to decrease, leading to heightened uncertainty regarding the performance of this geotechnical parameter.

Recognizing the limitations of in-situ testing equipment is essential in geotechnical investigations, especially when considering the inherent variability in soil properties. Among in-situ tests, the Standard Penetration Test (SPT) is notably high in variability, while the dilatometer and electrical cone penetration tests exhibit lower levels of variability. Table 5.1 outlines the variability observed in laboratory tests. Table 5.2 illustrates the variability of in-situ tests. Soil density can be evaluated with high accuracy. The variability of shear strength test results in clays and plasticity index is high.

More recently, Phoon et al. (2022a) discussed the importance of considering uncertainties in geotechnical engineering modeling and decision making. This review paper focuses

Table 5.1 Variability from laboratory tests

Coefficient of variation (COV) (%)			
Mean	*Range*	*Property*	*Soil type*
24	5–51	Plasticity index	Fine-grained
24	7–56	Effective angle of friction, ϕ'	Clay, silt
20	19–20	Shear strength, s_u	Clay, silt
19	8–38	Shear strength, s_u	Clay, silt
14	13–14	Effective angle of friction, ϕ'	Sand
14	6–22	Effective angle of friction, ϕ'	Clay
13	3–29	Effective angle of friction, ϕ'	Clay, silt
10	7–18	Plastic limit	Fine-grained
8	2–22	Effective angle of friction, ϕ'	Clay, silt
7	3–11	Liquid limit	Fine-grained
1	1–2	Density	Fine-grained

Source: Phoon and Kulhawy (1999).

Table 5.2 Variability from in-situ tests

Coefficient of variation (COV) (%)	*Test*
15–45	SPT
15–25	Mechanical CPT
15–25	SBP
10–20	VST
10–20	PBP
5–15	Electric CPT
5–15	DMT

Source: Phoon and Kulhawy (1999).

on uncertainty quantification, uncertainty calculation, and how they enhance the role of modeling in decision making (Ching and Phoon, 2013; Phoon et al., 2016; Ching et al., 2016; Tang and Phoon, 2021). The key output from reliability analysis is the probability of failure, which takes into account both mechanics and statistics and is meaningful for both system and component failures (Phoon et al., 1995; AASHTO, 2020).

The need for system-level analysis in resilience engineering and the importance of considering non-classical failure mechanisms in spatially variable soils are also highlighted (Vessia et al., 2021; Tabarroki et al., 2022). It is emphasized that uncertainty quantification brings decision making closer to reality and supports digital transformation in geotechnical engineering practice (Phoon et al., 2022b).

Phoon et al. (2022a) provided guidance on the statistical characterization of soil properties and transformation uncertainties. Additionally, it discusses the use of machine learning and Bayesian-based algorithms for developing soil correlations and parameter identification (Zhang et al. 2020; Ching et al., 2021; Zhang et al., 2022). Overall, the recent studies can help practitioners understand and use geotechnical risk and reliability methods in conjunction with numerical modeling for improved decision making.

5.4 DESCRIPTIVE ROLE OF IN-SITU TESTS IN SOIL IDENTIFICATION

Visual inspections and a range of laboratory tests serve as standard practices for identifying and classifying geotechnical deposits. Nonetheless, visual identification is primarily suitable for preliminary investigations, while laboratory testing has its limitations, including issues related to sample recovery, size effects, and the modeling of actual stress conditions.

In the early stages of geotechnical exploration, soil type identification was often approximated, and the cone penetrometer was limited in its ability to provide detailed information about the location of soil type boundaries. Soil type confirmation typically relied on traditional borings, with empirical interpretations confined to the geological regions where they had been developed. Measured parameters such as pore pressure, tip resistance (q_c), and sleeve frictional resistance (f_s) were recognized as functions of soil type and behavior. A key application of the Cone Penetration Test (CPT) was profiling for soil type identification. Consequently, various charts were introduced in the literature to classify soil types based on CPT or CPTu data (Eslami and Fellenius, 2004; Ku et al., 2010; Cai et al., 2015).

5.4.1 Soil behavior classification based on in-situ tests

In every foundation design, understanding the pertinent parameters that govern soil behavior is imperative. Soil behavior is intricately complex, influenced by numerous factors including particle size, shape, mineral composition, pore fluid characteristics, particle packing, stress history, and various other considerations.

In recent decades, thanks to significant advancements in in-situ testing methods and the increased certainty and reliability of the obtained results, one of the primary applications of in-situ tests is now the determination of soil profiles and their behavioral classifications based on direct measurements. In cases where direct measurement of essential soil parameters is unfeasible, estimates must be derived from available data, often obtained through in-situ tests. Researchers have proposed numerous correlations between these tests and soil parameters. To effectively utilize these interpreted results, it is vital to establish correlations with data from other sources, both field and laboratory, and to draw from accumulated experience. The selection of properties for any correlation is a critical decision in this process.

5.4.2 Soil classification based on SPT blow counts

The Standard Penetration Test (SPT) is a field test carried out during borehole drilling to estimate the soil's resistance to the penetration of a split-spoon sampler at different depths beneath the ground surface. The SPT offers several advantages, including its long-standing use, which has contributed to a wealth of practical experience. Moreover, it is a relatively straightforward and cost-effective testing method. However, it is crucial to understand that the test results are influenced by a multitude of factors, and there exist various sources of potential errors, despite some correlation between the N value and soil engineering parameters (as shown in Table 5.3).

5.4.3 Soil classification based on FVST records

In fine-grained soils, the accuracy of Standard Penetration Test (SPT) records is compromised by elevated pore water pressure. In contrast, the Field Vane Shear Test (FVST) and small-scale Vane Shear Test (VST) are more reliable and suitable for these soil types. A general correlation between the soil consistency and undrained shear strength is presented in Table 5.4, highlighting the usefulness of these alternative tests in assessing fine-grained soils.

5.4.4 Soil classification based on CPT data

One of the primary objectives of conducting a Cone Penetrometer Test (CPT) in a site investigation project is to delineate the subsoil profile based on the test results. Early attempts at classifying soils using CPT data were made by Begemann (1965), who established a relationship between soil type and the ratio of cone resistance to sleeve friction. Subsequently, researchers like Schmertmann (1978), Robertson and Campanella (1983), and Olsen and Mitchel (1995) have utilized parameters such as cone resistance (q_c) or normalized cone resistance and the friction ratio (Fr) in developing the Soil Classification Chart, as depicted in Figure 5.1a, c, offering valuable tools for soil classification based on CPT results.

Table 5.3 Strength from SPT on different types of soils

Material	*Description*	*SPT (N value)*	*Corrected SPT $(N_1)_{60}$*
Clay	Very soft	≤2	–
	Soft	2–5	–
	Firm	5–10	–
	Stiff	10–20	–
	Very stiff	20–40	–
	Hard	>40	–
Clean sand	Very loose	≤4	≤3
	Loose	4–10	3–8
	Medium dense	10–30	8–25
	Dense	30–50	25–42
	Very dense	>50	>42
Fine sand	Very loose	–	≤3
	Loose	–	3–7
	Medium dense	–	7–23
	Dense	–	23–40
	Very dense	–	>40
Medium sand	Very loose	–	≤3
	Loose	–	3–8
	Medium dense	–	8–25
	Dense	–	25–43
	Very dense	–	>43
Coarse sand	Very loose	–	≤3
	Loose	–	3–8
	Medium dense	–	8–27
	Dense	–	27–47
	Very dense	–	>47

Source: Adapted from Look, (2007).

Table 5.4 General correlation between consistency and undrained shear strength

Consistency	*s_u (kPa)*
Very soft	0–25
Soft	25–50
Medium	50–100
Hard	100–200
Very hard	200–400

Source: Das (2006).

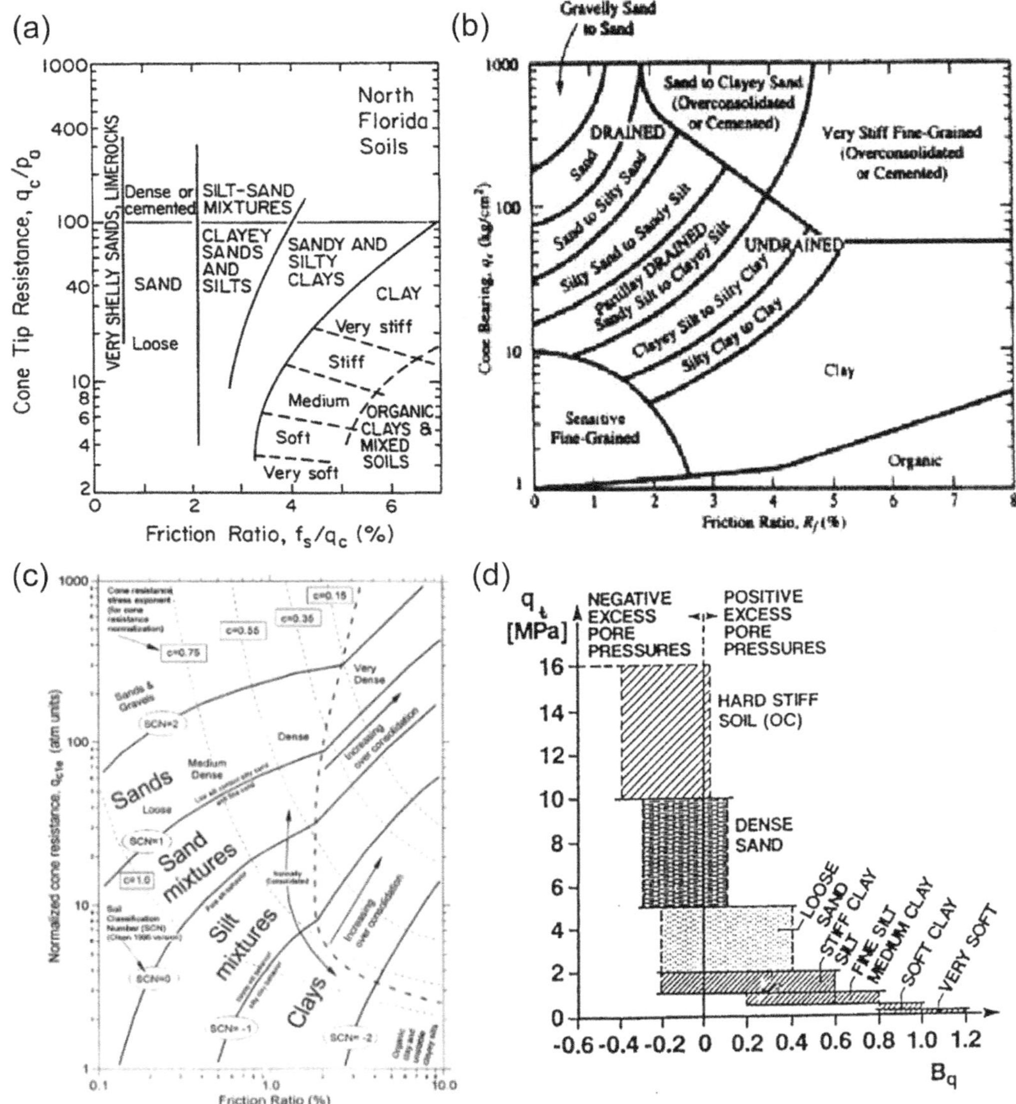

Figure 5.1 Some of the pioneer soil profiling charts: (a) Schmertmann (1978); (b) Campanella (1986); (c) Senneset et al. (1989); and (d) Olsen and Mitchel (1995).

The introduction of the piezocone brought about the capability to measure pore pressure during the test, significantly improving soil classification. The pore pressure ratio B_q, a critical parameter for soil classification, can be calculated using Equation (5.1). Senneset et al. (1989) presented a graph for utilizing this parameter in soil classification, as depicted in Figure 5.1d, enhancing the accuracy and effectiveness of classification based on CPT results.

$$B_q = \frac{u_2 - u_0}{q_t - \sigma'_{v0}} \tag{5.1}$$

where u_0 = pore pressure due to groundwater table, u_2 = pore pressure measured at the shoulder, σ'_0 = total overburden stress, and q_t = corrected tip resistance.

Corrected tip resistance (q_t) can be calculated by using Equation (5.2):

$$q_t = q_c - u_2(1 - a_n) \tag{5.2}$$

where q_c = cone resistance and a_n = net area ratio, approximately equal to the ratio of shaft cross-section and cone cross-section areas (usually measured in a calibration cell, Lunne et al., 1997).

Over the years, numerous researchers have put forth various soil classification methods and charts based on either Cone Penetration Test (CPT) or Cone Penetration Test with pore pressure measurements (CPTu) results. These approaches encompass works by Jones and Rust (1982), Jefferies and Davies (1991), and Ramsey (2002).

Jefferies and Davies (1993) introduced the soil behavior index (I_{SBT}), which delineates the demarcation between distinct soil types and can be computed using Equation (5.3), offering an innovative perspective on soil classification.

$$I_{\mathrm{SBT}} = \left[\left(3.47 - \log\left(q_c/P_a\right)\right)^2 + \left(\log R_f + 1.22\right)^2\right]^{0.5} \tag{5.3}$$

where R_f = friction ratio and q_c = cone resistance.

The values of I_C defined on the boundaries of various soil types on the Robertson (1990) classification chart are depicted in Figure 5.2 (Table 5.5).

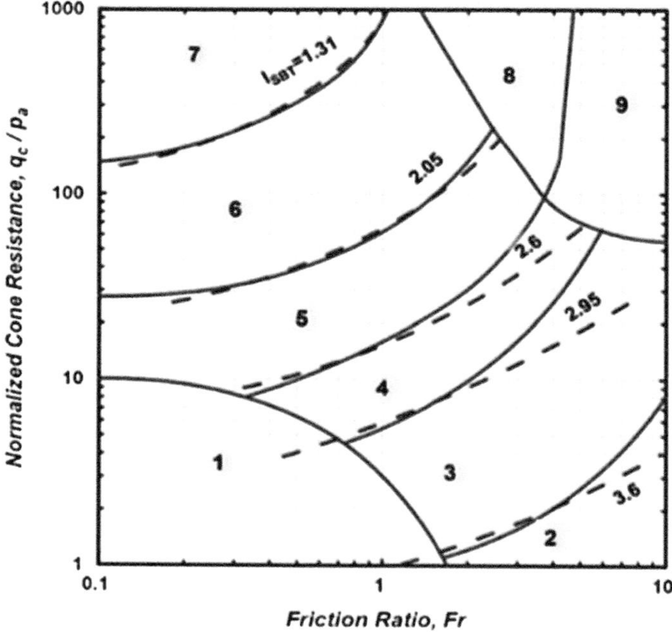

Figure 5.2 Values of I_C on the Robertson (1990) classification chart (Source: Jefferies and Davies, 1991).

Table 5.5 Soil classification based on I_{SBC}

Soil classification	*Zone*	*CPTu index* (I_{SBC})
Gravelly sands	7	$I_{SBC} < 1.25$
Sands – clean sand to silty sand	6	$1.25 < I_{SBC} < 1.90$
Sand mixture – silty sand to sandy silt	5	$1.90 < I_{SBC} < 2.54$
Silt mixture – clayey silt to silty clay	4	$2.54 < I_{SBC} < 2.82$
Clays	3	$2.82 < I_{SBC} < 3.22$
Organic clays	2	$I_{SBC} < 3.22$

Source: Li et al. (2015).

Robertson (1990) introduced an advancement of the profiling chart developed by Robertson et al. (1986). This chart, as depicted in Figure 5.3, features a normalized cone resistance plotted against a normalized friction ratio, providing valuable insights into soil behavior and classification. Additionally, an accompanying pore pressure ratio chart represents the relationship between the normalized cone resistance and pore pressure ratio, further contributing to the understanding of soil properties.

Robertson (2016) presented an update of some popular CPT-based SBT classification systems, since textural-based descriptions, such as sand and clay are used in providing them. Moreover, it includes a method for identifying the presence of microstructures in soils. Figure 5.4 illustrates examples of these updated classification systems, offering a more nuanced and behavior-driven approach to characterizing soil properties.

Eslami and Fellenius (1997) compiled a comprehensive database comprising CPT and CPTu data, which were linked with results from borehole investigations, sampling, laboratory testing, and soil properties. These data points were utilized to create a profiling chart, and distinct envelopes were delineated, encapsulating five soil types, as illustrated in Figure 5.5a. Notably, the plotting of an "effective" cone resistance, defined by Equation (5.4), was observed to offer a more consistent demarcation of these envelopes compared to using solely the cone resistance.

$$q_E = q_t - u_2 \tag{5.4}$$

where q_E = effective cone resistance, q_t = cone resistance corrected for pore water pressure at cone shoulder (Equation 5.2), and u_2 = pore pressure measured at cone shoulder.

A new format of classification chart, i.e., triangular form, is proposed by Eslami et al. (2016) containing cone tip resistance (q_c), sleeve friction (f_s), and pore pressure (u_2). The proposed chart, as shown in Figure 5.5b with more accuracy and less scatter of data than the previous charts, is able to identify soil types, particularly for deltaic soils. Moreover, Eslami et al. (2022) developed a triangular classification chart for deltaic soils by compiling a marine database of the CPTu soundings and their soil profiling in their vicinity. The proposed chart is presented in Figure 5.5c. Overall a qualitative chart is suggested for the general behavior classification of soils, dividing into five generic zones (types) as illustrated in Figure 5.5d.

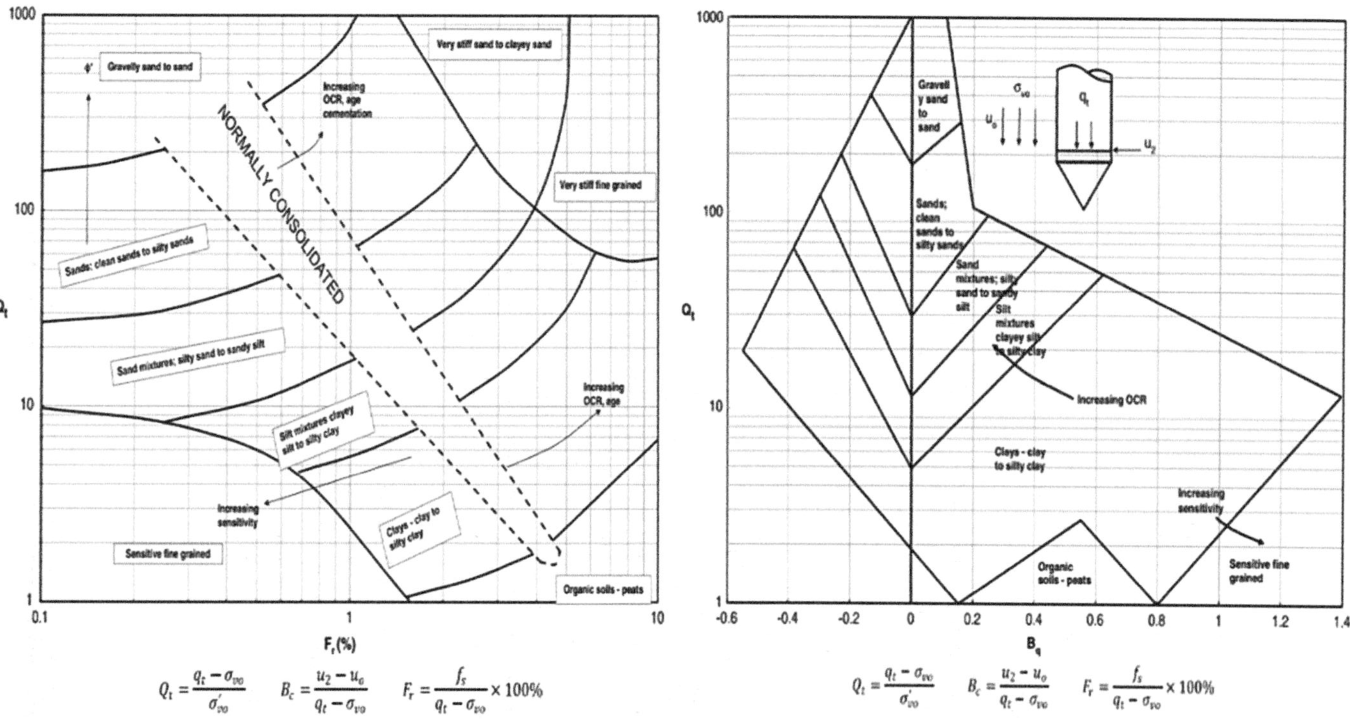

Figure 5.3 Proposed soil behavior type classification system from CPT/CPTu data (Source: adapted from Robertson, 1990).

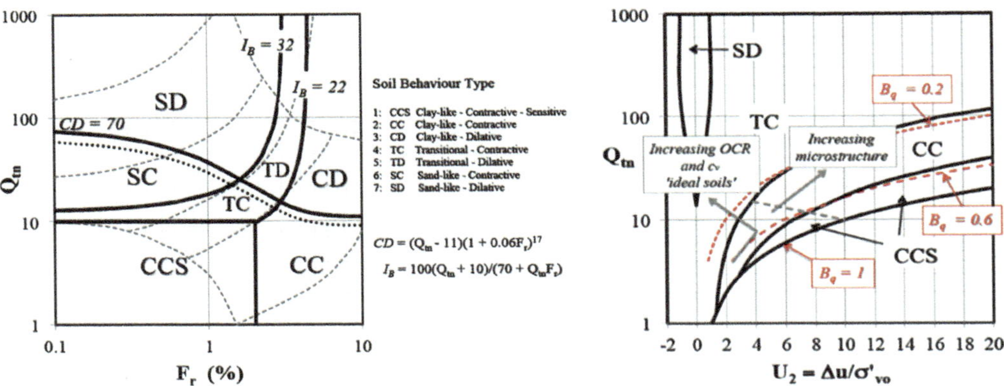

Figure 5.4 Examples of updated classification charts (Source: Robertson, 2016).

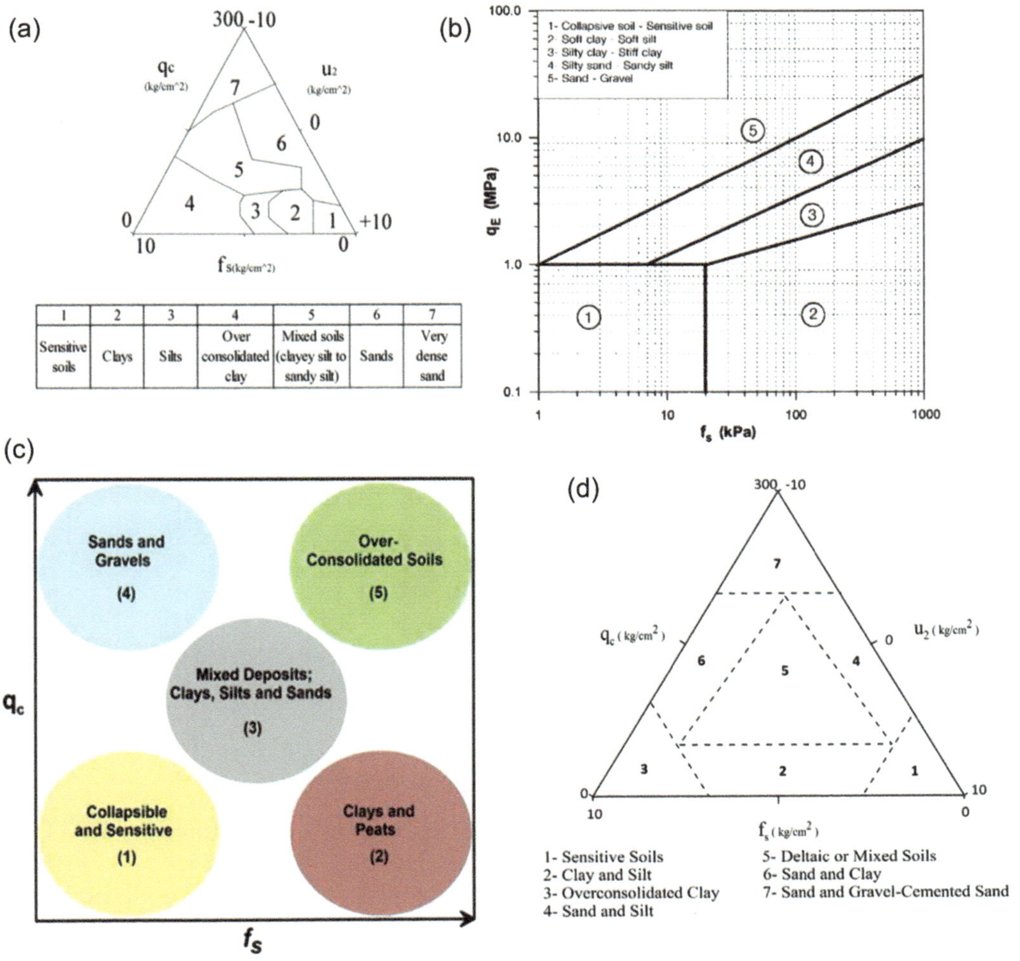

(a)

300 -10

qc (kg/cm^2)

u2 (kg/cm^2)

fs (kg/cm^2)

1	2	3	4	5	6	7
Sensitive soils	Clays	Silts	Over consolidated clay	Mixed soils (clayey silt to sandy silt)	Sands	Very dense sand

(b)

1- Collapsive soil - Sensitive soil
2- Soft clay - Soft silt
3- Silty clay - Stiff clay
4- Silty sand - Sandy silt
5- Sand - Gravel

q_E (MPa)

f_s (kPa)

(c)

Sands and Gravels (4)

Over-Consolidated Soils (5)

Mixed Deposits; Clays, Silts and Sands (3)

Collapsible and Sensitive (1)

Clays and Peats (2)

q_c

f_s

(d)

300 -10

q_c (kg/cm^2)

u_2 (kg/cm^2)

f_s (kg/cm^2)

1- Sensitive Soils
2- Clay and Silt
3- Overconsolidated Clay
4- Sand and Silt
5- Deltaic or Mixed Soils
6- Sand and Clay
7- Sand and Gravel-Cemented Sand

Figure 5.5 (a) Classification chart based on effective cone resistance and shaft resistance (Eslami and Fellenius, 1997), (b) first generation of triangular charts (Eslami et al., 2016), (c) enhanced triangular charts for deltaic soils (Eslami et al., 2022), and (d) general qualitative classification chart.

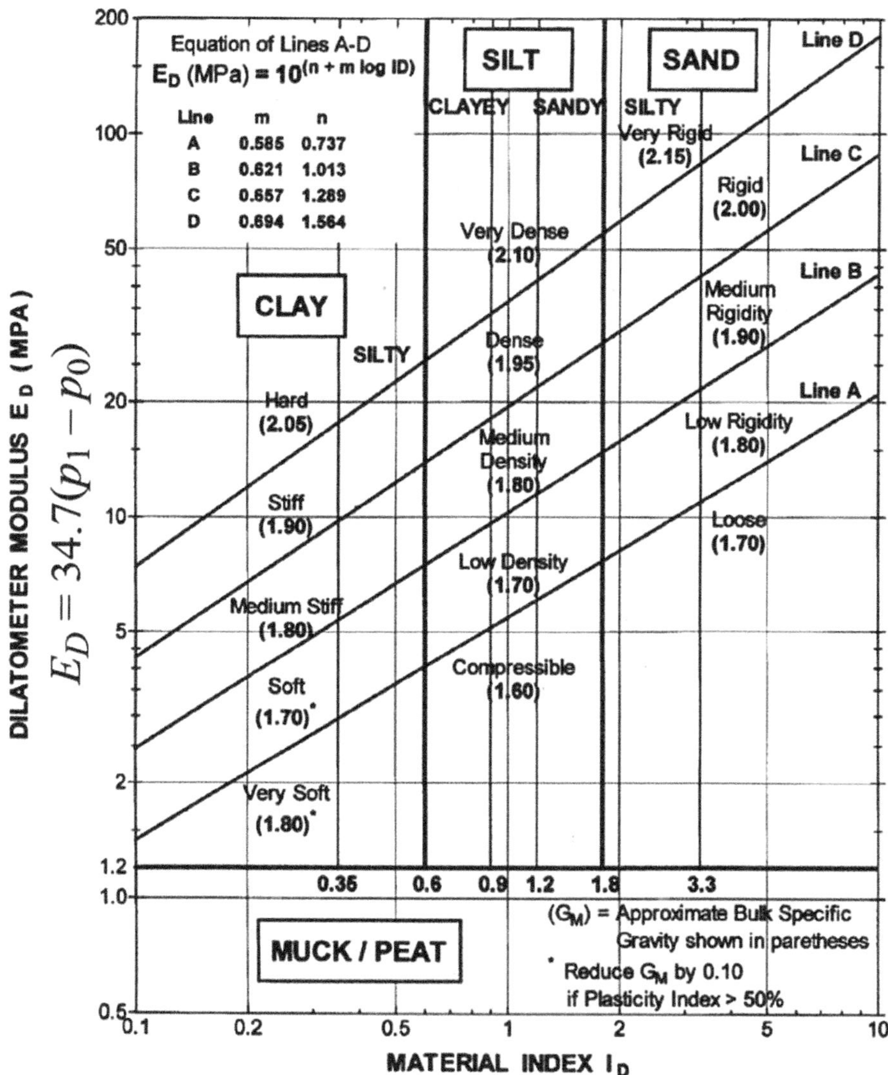

Figure 5.6 Chart for estimating soil type and unit weight for soil based on DMT result (Source: Marchetti, 1980).

5.4.5 Soil classification based on DMT records

In recent years, the static probing method using dilatometer penetration tests (DMT) has become a widely employed in-situ investigation technique, providing essential information for civil engineers involved in design, construction, permitting, and operational control. Diagram charts are commonly used for soil classification based on the dilatometer test (DMT). Marchetti (1980) introduced a diagram chart (Figure 5.6 and Table 5.6) that relies on the dilatometer modulus (E_D) and the material index (ID). This chart serves as a valuable tool for identifying various soil types, including minerals (clay, silt, and sand), as well as organic soils, based on the results obtained from DMT.

Table 5.6 Soil classification based on material index

Soil type		*Material index (I_D)*
Organic soils	Peat/sensitive clays	<0.10
Cohesive soils	Clay	0.10–0.35
	Silty clay	0.35–060
	Clayey silt	0.60–0.90
	Silt	0.90–1.20
	Sandy silt	1.20–1.80
Noncohesive soils	Silty sand	1.80–3.30
	Sand	>3.30

Source: Marchetti (1980).

Table 5.7 Typical Menard pressuremeter values

Type of soil	*Limit pressure (kPa)*	E_m/p_l
Soft clay	50–300	10
Firm clay	300–800	10
Stiff clay	600–2500	15
Loose silty sand	100–500	5
Silt	200–1500	8
Sand and gravel	1200–5000	7
Till	1000–5000	8
Old fill	400–1000	12
Recent fill	50–300	12

Source: Canadian Foundation Engineering Manual (2006).

5.4.6 Soil classification based on PMT measurements

The Borehole Pressuremeter Test (PMT), which was initially developed by Menard in 1956, has gained extensive use in France. While PMT is not commonly employed for soil classification, as indicated in Table 5.7, the typical Menard pressuremeter values can be utilized to help identify soil types, offering valuable insights into the soil's characteristics.

5.5 IN-SITU TESTS AND PREDICTION OF GEO-PARAMETERS

5.5.1 Geotechnical parameters

Every geotechnical design entails the determination of subsoil stratification and the estimation of geotechnical parameter values, often referred to as soil engineering parameters, for each distinct soil layer. Yu (2007) reviews the use of fundamental mechanics in developing rational interpretation methods for deriving soil properties from in-situ test results. The interpretation of in-situ tests is challenging due to complex boundary value problems, poorly controlled drainage conditions, differences in effective stress paths, and the highly nonlinear behavior of soils. Simplifying assumptions need to be made in the interpretation of in-situ tests, including assumptions about geometry and boundary conditions, soil behavior, and water drainage conditions. The cavity expansion methods can be used to analyze the stress–strain relationship and deformation characteristics of the soil and are typically adopted for interpretation of in-situ testing data (Yu, 2000; Mo and Yu, 2017, 2018; Chen and Mo, 2022; Mo et al., 2022). Cavity expansion testing provides valuable information for geotechnical engineers to assess the stability and behavior of soils in different engineering applications (Mo et al., 2017, 2020, 2024).

Soil engineering parameters in the context of foundation engineering can be broadly categorized into two main groups: those directly applied in foundation engineering, such as unit weight, internal friction angle, and relative density, and a few additional parameters, including sensitivity and permeability, which will be elaborated upon further.

5.5.2 Index parameters

5.5.2.1 Unit weight

In geotechnical engineering, a critical property of soil is its unit weight (γ), a parameter essential for various earthwork applications. Unit weight refers to the density of soil, defined as the ratio of the weight to the volume of an undisturbed soil sample, and it is typically determined in laboratory settings. For in-situ measurements of unit weight, methods such as sand cones, rubber balloons, or nuclear densitometer tests are commonly employed.

Table 5.8 shows that there is no disparity between the unit weights of the bulk and saturated weights in cohesive soils. However, in granular soils, the difference is somewhat more pronounced in loose materials compared to dense ones.

There have been several studies linking the unit weight to CPT measurements. The Robertson and Cabal (2010) correlations for calculating γ / γ_w for different soils using CPT records are shown in Figure 5.7.

Table 5.8 Typical values for bulk and saturated unit weights

		Bulk unit weight (kN/m³)		*Saturated unit weight (kN/m³)*	
	Soil type	*Loose*	*Dense*	*Loose*	*Dense*
Granular soils	Gravel	16	18	20	21
	Well-graded sand and gravel	19	21	21.5	23
	Coarse or medium sand	16.5	18.5	20	21.5
	Well-graded sand	18	21	20.5	22.5
	Fine or silty sand	17	19	20	21.5
	Rock fill	15	17.5	19.5	21
	Brick hardcore	13	17.5	16.5	19
	Slag fill	12	15	18	20
	Ash fill	6.5	10	13	15
Cohesive soils	Peat	12		12	
	Organic clay	15		15	
	Soft clay	17		17	
	Firm clay	18		18	
	Stiff clay	19		19	
	Hard clay	20		20	
	Stiff or hard glacial clay	21		21	

Source: Das (2016).

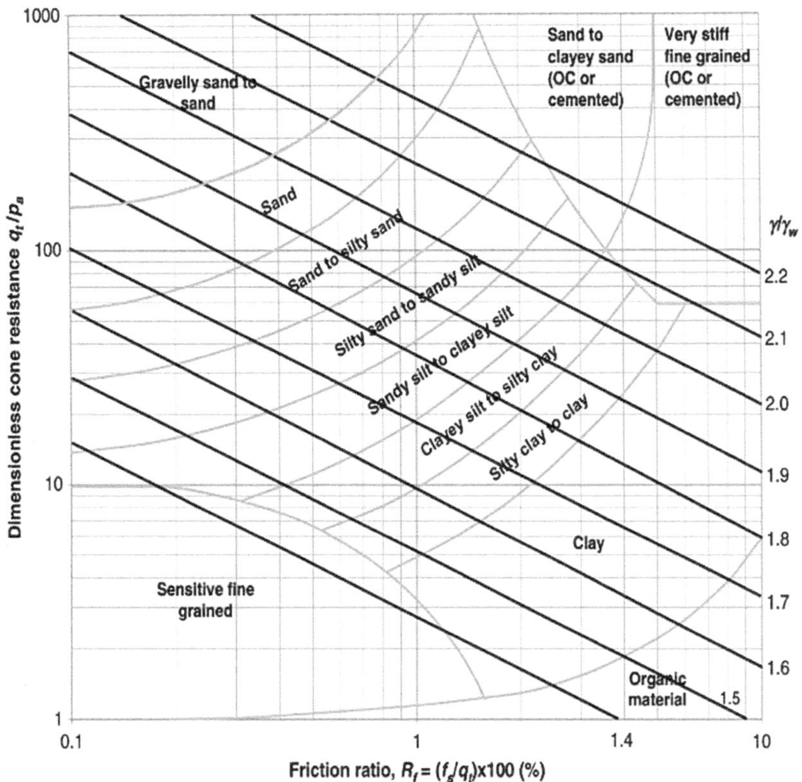

Figure 5.7 Normalized unit weight and R_f (%) (Source: adapted from Robertson and Cabal, 2010).

5.5.2.2 Relative density

As shown in Figure 5.8, Holtz and Gibbs (1979) presented the correlation of N and D_r in a more usable form. Some correlations suggested by researchers are listed in Table 5.9.

The application of in-situ density tests is constrained, as they can only be effectively measured at shallow depths. Therefore, the utilization of Cone Penetration Test (CPT) records to evaluate relative density (D_r) through empirical correlations has gained prominence. Several researchers have observed the influence of parameters like q_c and σ_v' on D_r, as illustrated in Figure 5.9. In this context, a selection of essential correlations has been compiled in Table 5.10, offering valuable tools for estimating relative density using CPT data.

5.5.3 Strength parameters

5.5.3.1 Internal friction angle

The internal friction angle is a fundamental physical property of earth materials, representing the slope of a linear representation of the shear strength of these materials. Typical values for the friction angle of sand and silt can be found in Table 5.11, providing important reference points for these materials.

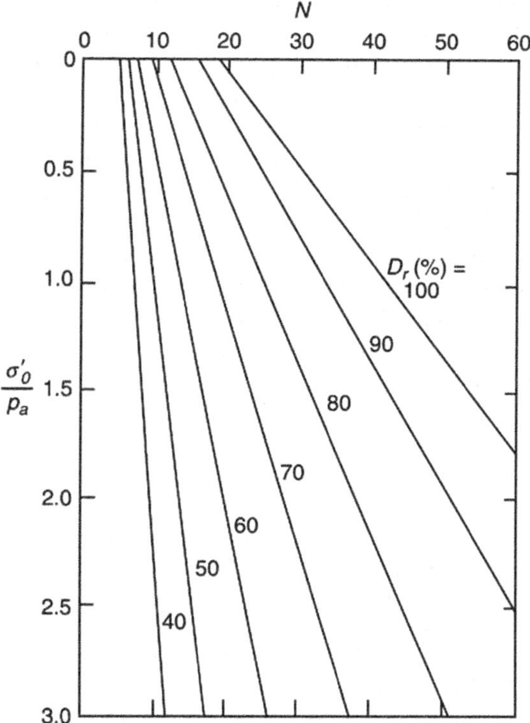

Figure 5.8 Variation of N with σ_0' / p_a and D_r (Source: adapted from Gibbs and Holtz, 1979).

Table 5.9 Proposed correlation for D_r based on SPT number

Reference	*Proposed correlation*	*Remarks*
Meyerhof (1957)	$D_r = 20.4\left(\frac{N}{0.7+\left(\sigma_o' / P_a\right)}\right)^{0.5}$	σ_o' = effective overburden pressure P_a = atmospheric pressure
Kulhawy and Mayne (1990)	$D_r = \left(\frac{(N_1)_{60}}{(60+25\log D_{50})C_A C_{OCR}}\right)^{0.5}$	C_A = correction factor for aging $C_A = 1.2 + 0.05\log(t/100)$ t = time, in years, since deposition C_{OCR} = correction for OCR $C_{OCR} = OCR^{0.18}$
Yoshida et al. (1988)	$D_r = C_0\left(\sigma_o'\right)^{-C_1} N^{C_2}$	The unit of σ_o' is kN/m² $C_0 = (18-25)$ (best fit value = 25) $C_1 = (0.12-0.14)$ (best fit value = 0.12) $C_2 = (0.44-0.57)$ (best fit value = 0.46)

Source: Adapted from Gibbs and Holtz (1957).

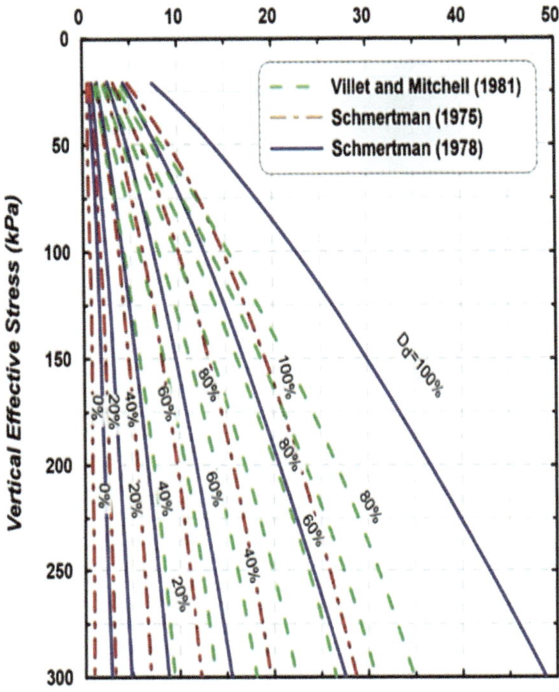

Figure 5.9 Comparison of $D_r - q_c$ relationships.

Table 5.10 Correlations predicting D_r from CPT records

Reference	*Proposed correlation*	*Remarks*
Baldi et al. (1986)	$D_r = \frac{1}{C_2}\ln(\frac{q_c}{C_0(\sigma_v^{'})^{0.55}})$	C_0 and C_2: soil constants, C_0=157 and C_2=2.41 normally consolidated sand. q_c and σ_v' are in the kPa unit
Jamiolkowski et al. (2001)	$D_r = 26.8\ln\frac{q_c / p_a}{(\sigma_v^{'} / p_a)^{0.5}} - b_x$	b_x = 52.5 for high compressibility sands b_x = 67.5 for medium compressibility sands b_x = 82.5 for low compressibility sands
Kulhawy and Mayne (1990)	$D_r = \frac{Q_{cn}}{305 Q_c Q_{OCR}}$	$Q_{cn} = q_t / p_a / (\sigma_v^{'} / p_a)^{0.5}$ Q_c = Compressibility factor (0.91 for high, 1.0 for medium, and 1.09 for low). Q_{OCR} = Over-consolidation factor, $OCR^{0.18}$
Mayne (2007)	$D_r = 100\left(0.268\ln(\frac{q_t / p_a}{\sqrt{\sigma_v^{'} / p_a}}\right) - 0.675)$	q_c and σ_v' are in kPa unit

Table 5.11 Common amount of friction angle for sand and silt

Soil type		ϕ (°)
Angular sand	Loose	27–30
	Medium	30–35
	Dense	35–38
Rounded sand	Loose	30–35
	Medium	35–40
	Dense	40–45
Sandy gravel		34–48
Silt		26–35

Source: Das (2006).

Shear strength in coarse-grained soils is typically characterized by the friction angle, denoted as ϕ. Several researchers have proposed methods for determining the friction angle in sands, based on one of the following theories:

- Cavity expansion
- Bearing capacity
- Empirical approaches often derived from calibration chamber tests

The friction angle (ϕ) of granular soils exhibits an increase with factors such as grain angularity, surface roughness, and relative density. It is important to note that poorly graded granular soils generally have lower ϕ values than well-graded ones. Figures 5.10

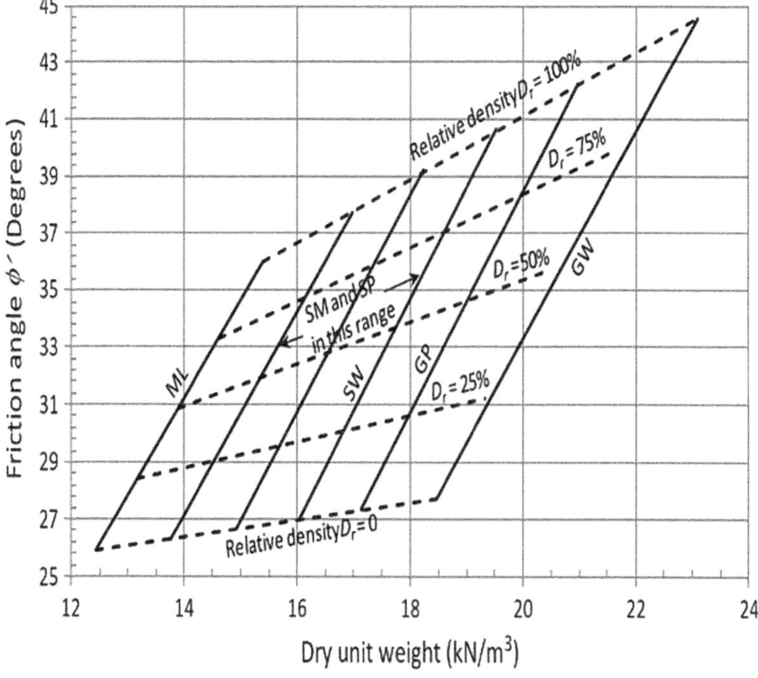

Figure 5.10 Friction angles of granular soils (Source: U.S. Navy, 1982).

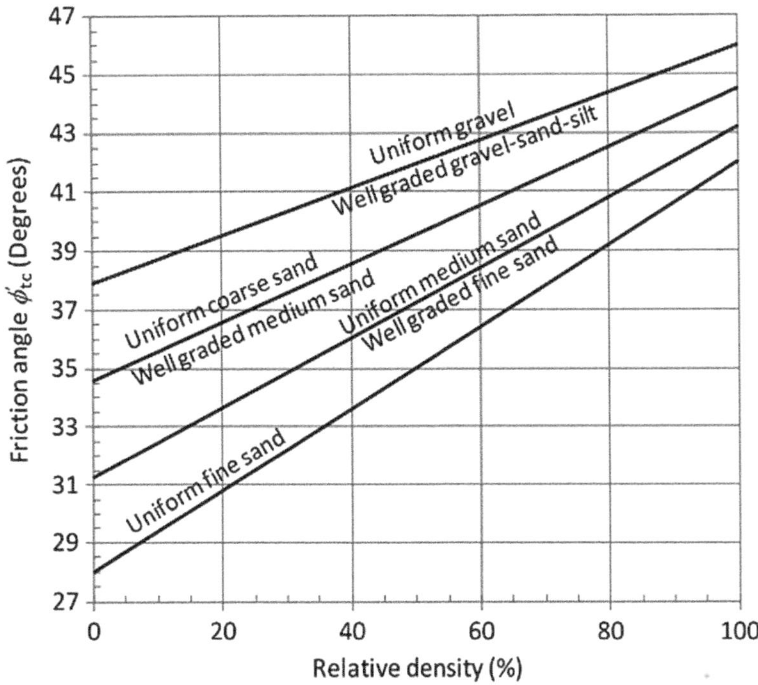

Figure 5.11 $\phi'_{tc} - D_r$ relations (Source: adapted from Schmertmann, 1978).

and 5.11 illustrate the friction angles determined from triaxial compression tests for various granular soils with the absence of plastic fines, providing a visual representation of the relationship between these factors and friction angle values.

Schmertmann (1975) proposed a correlation between effective overburden pressure σ_0, N, and ϕ, and this correlation is shown in Figure 5.12. Some other correlations are listed in Table 5.12.

Bergdahl et al. (1993) proposed a correlation of q_c and relative density with friction angle for cohesionless or mixed soils, as shown in Table 5.13. Mitchell and Durgunoglu (1983) suggested a correlation between cone tip resistance and internal friction angle as depicted in Figure 5.13.

Other researchers (e.g., Meyerhof, 1974; Robertson et al., 1986; Kulhawy and Mayne, 1990; and Uzielli et al., 2013) introduced a couple of equations to predict ϕ, as summarized in Table 5.14. Moreover, Mayne (2014) proposed correlations to predict the friction angle based on empirical correlation and case histories.

By employing two fundamental equations for assessing the bearing capacity of deep foundations, one for the corrected tip resistance q_t (q_E) and the other for the sleeve friction (f_s), and adopting the effective bearing capacity approach instead of the total stress method, it is possible to establish a dual equation system with two unknowns for static loading conditions, as outlined below:

$$CN_C + qN_q + 0.5\gamma BN_\gamma = q_E = q_t - u_2 \tag{5.5}$$

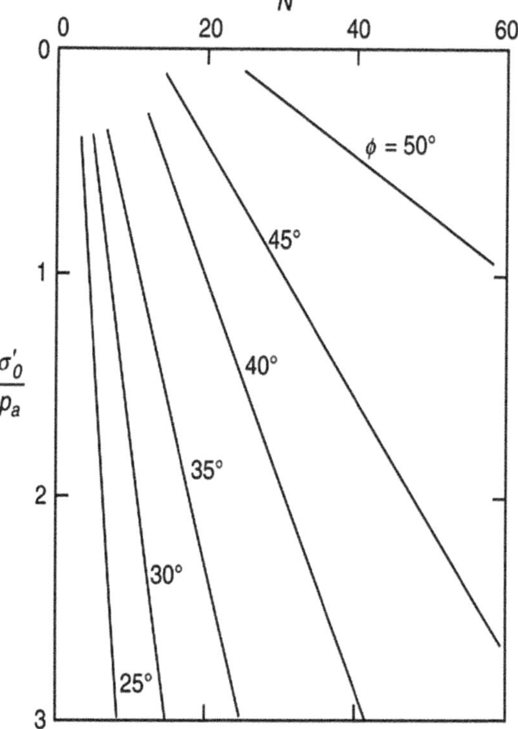

Figure 5.12 Variation of ϕ with SPT number and σ_0' / P_a (Source: adapted from Schmertmann, 1975).

Table 5.12 Variation of ϕ with N

Reference	*Proposed correlation*
Meyerhof (1959)	$\phi = 28 + 0.15\left[20.4\left(\frac{N}{0.7 + (\sigma_o' / Pa)}\right)^{0.5}\right]$
Wolff (1989)	$\phi = 27.1 + 0.3N_1 - 0.00054N_1^2$
Kulhawy and Mayne (1990)	$\phi = tan^{-1}\left[\frac{N}{12.2 + 20.3(\sigma_o' / Pa)}\right]^{0.34}$
Hatanaka and Uchida (1996)	$\phi = \sqrt{20(N_1)_{60}} + 20$

Table 5.13 Correlation of q_c and relative density with friction angle for cohesionless or mixed soils

Relative density	q_c *(MPa)*	ϕ *(°)*
Very weak	0.0–2.5	29–32
Weak	2.5–5.0	32–35
Medium	5.0–10.0	35–37
Large	10.0–20.0	37–40
Very large	>20.0	40–42

Source: Adapted from Bergdahl et al. (1993).

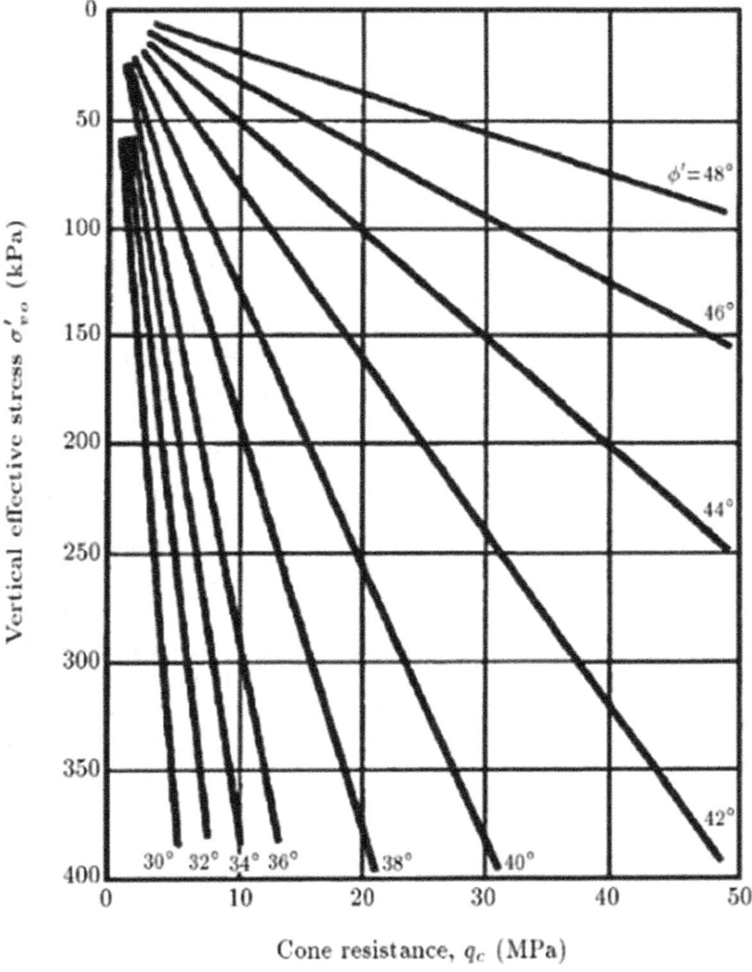

Figure 5.13 Friction angle changes and cone tip resistance (Source: Mitchell and Durgunoglu, 1983).

Table 5.14 Proposed correlations for friction angle based on CPT result

Reference	*Correlations*	*Soil type and Remarks*
Meyerhof (1974)	$\phi = \tan^{-1}(q_c / 0.5N_q)$	Sand, q_c (MPa)
Robertson et al. (1986)	$\phi = \tan^{-1}[0.1 + 0.38\log(q_c / \sigma'_v)]$	Sand
Kulhawy and Mayne (1990)	$\phi = 17.6 + 11\log(q_c / \sqrt{100\sigma'_v})$	Sand, σ_v' and q_c are in kPa unit
Uzielli et al. (2013)	$\phi = 25(q_c / \sqrt{100\sigma'_v})^{0.1}$	Sand, σ_v' and q_c are in kPa unit
Robertson and Cabal (2012)	$\tan\phi' = \frac{1}{2.68}(\log(q_c / \sigma'_v) + 0.29)$	Uncemented, unaged, moderately compressible quartz sands
Mayne (2014)	$\phi = 29.5B_q^{0.121}[0.256 + 0.33B_q + \log Q_t]$	Cohesive soils $Q_t = \frac{q_t - \sigma_v}{\sigma'_v}$ $B_q = \frac{u_2 - u_0}{q_t - \sigma_v}$

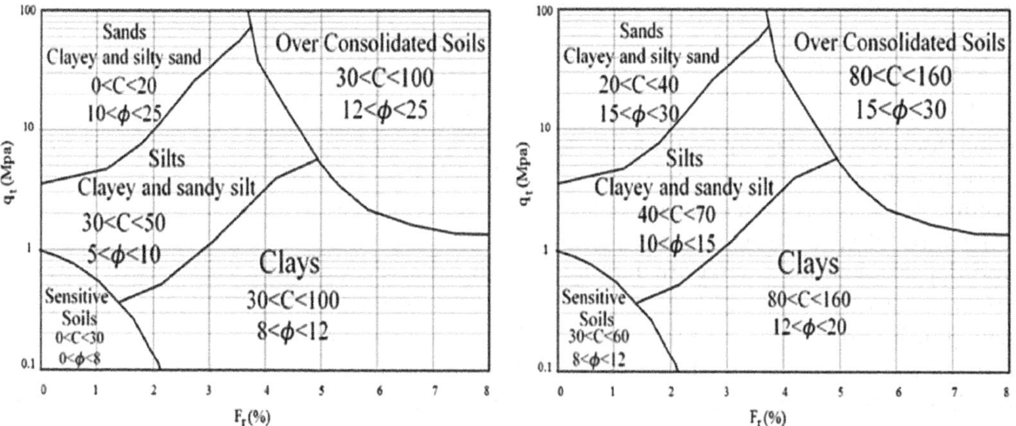

Figure 5.14 Variation range for C (kPa) and ϕ (degree), in soil classification chart: (a) 200 kPa < σ_v' < 300 kPa and (b) 300 kPa < σ_v' < 400 kPa (Source: Eslami and Mohammadi, 2016).

$$C + \sigma' \tan(\delta) = f_s \tag{5.6}$$

Eslami and Mohammadi (2016) have presented approaches to predict ϕ and C of marine deposits from CPTu records based on bearing capacity and cavity expansion theories. They summarized the Soil Classification Chart in five categorized areas based on CPTu test results. For specified q_t and F_r, i.e., friction ratio, the soil type can be determined in every depth interval. However, the variation range of C and ϕ using the proposed method, at 200 kPa < σ_v' < 300 kPa into five categorized soil types, is presented and also for 300 kPa < σ_v' < 400 kPain Figure 5.14.

Combining bearing capacity theories and direct shear modes of failure considering the pore water pressure, a set of equations is derived. By inputting CPTu at a certain

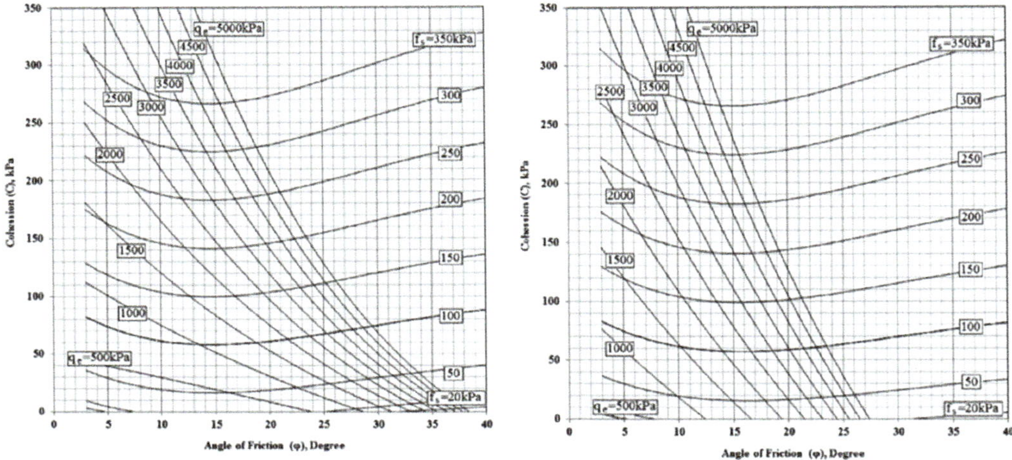

Figure 5.15 Graphic estimation of C and φ : (a) $\sigma_v^{'} = 50$ kPa, (b) $\sigma_v^{'} = 200$ kPa (Source: Eslami and Mohammadi, 2016).

depth, soil shear strength parameters can be calculated simultaneously as presented in Figure 5.15.

5.5.3.2 Undrained shear strength

The shear strength of soil represents the maximum internal resistance to applied shearing forces, and undrained shear strength (s_u) specifically refers to the shear strength of soil when it is subjected to shearing at a constant volume, typically under conditions where the pore water cannot drain or dissipate.

Numerous correlations have been proposed to relate Standard Penetration Test (SPT) results to the shear strength of soil. Terzaghi and Peck (1967) offer approximate relationships between consistency, corresponding N values, and undrained cohesion (s_u) for clay soils, as presented in Table 5.15. However, it is essential to exercise caution when using these values. Furthermore, Figure 5.16 illustrates the relationship between the standard penetration blow count (N) and undrained shear strength (s_u), as suggested by Sowers (1979), providing an alternative perspective on this correlation.

Table 5.15 Approximate variation of consistency, N, and undrained cohesion of clay

Consistency	*SPT (N value)*	*s_u (kPa)*
Very soft	0 – 2	$<$ 15
Soft	2 – 4	12 – 25
Medium	4 – 8	25 – 50
Stiff	8 – 15	50 – 100
Very Stiff	15 – 30	100 – 200
Hard	$>$ 30	$>$ 200

Source: Terzaghi and Peck (1967).

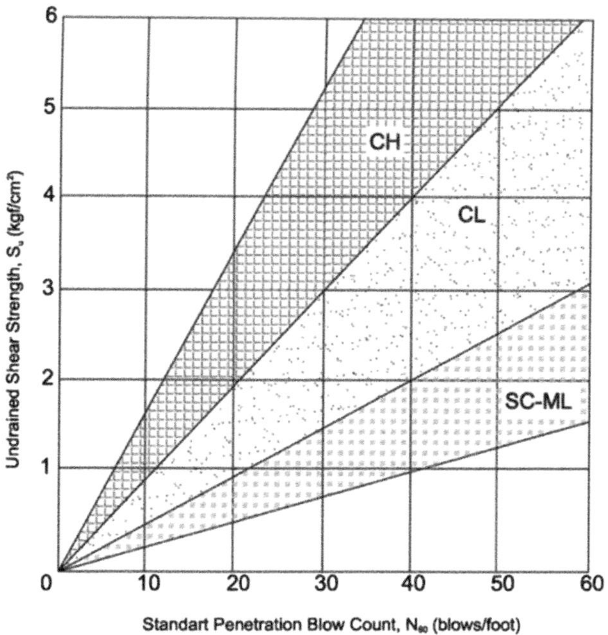

Figure 5.16 Relationship between *N*, and s_u (Source: adapted from Sowers, 1979).

In addition to soil profile classification, one of the primary engineering parameters derived from Cone Penetration Tests (CPT) for fine-grained soils is the undrained shear strength (s_u) of clayey soil. It is important to note that s_u cannot be directly measured by the CPT; however, empirical correlations can be employed to estimate it, and these estimates should be calibrated using laboratory tests such as triaxial compression or direct shear tests. Numerous researchers have conducted investigations into the relationship between s_u and CPT parameters. Some of the methods used for interpreting s_u from in-situ tests are detailed in Table 5.16, offering various approaches for estimating undrained shear strength using CPT data.

5.5.4 Stiffness and compressibility parameters

5.5.4.1 Stiffness

Soil modulus is a complex parameter in geotechnical engineering due to the various conditions it can be involved in. The Young's modulus (E) is a parameter associated with the deformability of soil, representing its linear elastic behavior. However, it is important to note that soil cannot be treated as an ideal linear elastic material, and as such, the soil modulus is not equivalent to the slope of the stress–strain curve. Instead, it is a stress-dependent parameter, meaning that the soil modulus changes with varying stress levels and must be considered as such in geotechnical analyses and design.

In predicting settlement for a foundation placed on a normally consolidated deposit under its initial loading, the secant modulus (E_s) is a valuable parameter. To calculate E_s , a slope is drawn from the origin to a selected point on the stress–strain curve. In addition, the tangent slope (E_t), associated with the tangent modulus, can be used to estimate

Table 5.16 Methods in s_u interpretation of in-situ tests

References	*Reviews*	*In-Situ Test*
Chandler (1988)	$\mu \cong 2.5(PI)^{-0.3} \leq 1.1$	VST: $s_{uv} = 6T/(7\pi D^3)$, $H/D = 2$
Windle and Wroth (1977) Baguelin et al. (1972)	$N_c = 5.5$	PMT: $s_{uPMT} = dp/d(Ln\varepsilon_v)$ $s_{uPMT} = (P_L - P_O)/N_c$
Stroud (1974) Stroud (1989)	$f_1 = 4.5, PI = 50$ $f_1 = 5.5, PI = 15$	SPT: $s_{u(N60)} = f_1 N_{60} P_a / 100$
Meyerhof (1951) Vesic (1977) Aas et al. (1986) Eslami et al. (2020) and Anagnostopolous et al. (2003)	$N_{kt} = 10(TC)$ $N_{kt} = 15(DSS)$ $N_{kt} = 20(TE)$	CPT & CPTu : $s_{uCPT} = (q_c - \sigma_{vo})/N_c$ $s_{uCPT} = (q_t - \sigma_{vo})/N_{kt}$ $s_{uCPT} = (0.8 - 1.2) f_s$
Marchetti (1980) Schmertmann (1991) Lacasse and Lunne (1988)	TC: $d_s = 0.20$ VST: $d_s = 0.19$ DSS: $d_s = 0.14$	DMT: $s_{uDMT} = 0.22\sigma'_{vo}(0.5K_D)^{1.25}$ $s_{uDMT} = (P_o - u_o)/10$ $s_{uDMT} = d_s\sigma'_{vo}(0.5K_D)^{1.25}$
Meyerhof (1951)	-	PLT: $s_{uPLT} = q_{ult}/6.18$

Source: FHWA (2002).

Table 5.17 Drained elastic modulus for soils using CPT records

Soil		q_c *(MPa)*	*Drained elastic modulus (MPa)*	E_t *from CPT*
Sand	Very loose	<2.5	<10	(2~4)q_c
	Loose	2.5–5	10–20	
	Medium dense	5–10	20–30	(3~7)q_c
	Dense	10–20	30–60	(6~10)q_c
	Very dense	>20	>60	
Clay	Very Soft	<2	<5	(1~2)q_c
	Soft	2–5	5–15	(1~2)q_c
	medium	5–10	15–25	(2~4)q_c
	Hard	10–20	>25	(3~8)q_c

Source: Look (2007).

progressive settlement when incremental loads are applied. Table 5.17 provides proposed relationships between elastic modulus and CPT results, offering a means to estimate these moduli based on CPT data, which can be crucial for settlement predictions in geotechnical engineering.

Table 5.18 Elastic modulus parameter of in-situ tests

SPT	*CPT*	*Soil type*
$E_s = 500(N+15)$ $= 7000\sqrt{N}$ $= 6000N$	$E_s = (2-4)q_u$ $= 8000q_u$	Sand
$E_s = (15000-22000)\cdot \ln N$	$E_s = 1.2(3D_r^2+2)\cdot q_c$ $E_s = (1+D_r^2)\cdot q_c$	
$E_s = 250(N+15)$	$E_s = F\cdot q_c$ $e = 1.0$ $F = 3.5$ $e = 0.6$ $F = 7.0$	Saturated sand
$E_s = 40000 + 1050\times N$ $E_{s(OCR)} \approx E_{s(NC)}\sqrt{OCR}$	$E_s = (6-30)q_c$	OCR sand
$E_s = 1200(N+6)$ $= 600(N+6)$ $N \le 15$ $= 600(N+6)+2000$ $N > 15$	-	Sandy gravel
$E_s = 320(N+15)$	$E_s = (3-6)q_c$	Clay sand
$E_s = 300(N+6)$	$E_s = (1-2)q_c$ $q_c < 2500kPa$ $E_s' = 2.5q_c$ $2500 < q_c < 5000kPa$ $E_s' = 4q_c + 5000$	Silty sand
-	$E_s = (3-8)q_c$	Soft clay

Source: Bowles (1996).)

As Baldi et al. (1989) suggested, the stiffness of fine-grained soil is affected by different factors, including mineralogy, grain size, particle shape, grain fabric, drainage conditions, mean effective stress, and stress–strain behavior. In a given soil, cone tip resistance is mainly controlled by the void ratio and the effective stress condition. Some methods in E_s interpretation of in-situ tests are listed in Table 5.18.

5.5.4.2 Constrained modulus

The constrained modulus of clays can be calculated in the laboratory using the consolidation test:

Table 5.19 Constrained modulus coefficient for cohesive soils

Soil type	q_c *(MPa)*	α
Low plasticity clay (CL)	$q_c < 0.7$	$3 < \alpha < 8$
	$0.7 < q_c < 2$	$1 < \alpha < 5$
	$q_c > 2$	$1 < \alpha < 2.5$
Silt, low liquid limit (ML)	$q_c < 2$	$3 < \alpha < 6$
	$q_c > 2$	$1 < \alpha < 2$
High plasticity clay (CH)/ silt, high liquid limit (MH)	$q_c < 2$	$2 < \alpha < 6$
	$q_c > 2$	$1 < \alpha < 2$
Organic silt (OL)	$q_c < 1.2$	$2 < \alpha < 8$
Organic clay (OH)/peat (Pt) with $50 < w < 100$	$q_c < 0.7$	$1.5 < \alpha < 4$
Organic clay (OH)/peat (Pt) with $100 < w < 200$		$1 < \alpha < 1.5$
Organic clay (OH)/peat (Pt) $w > 200$		$\alpha < 4$

Source: Adapted from Sanglerat (1972); adapted from Frank and Magnan (1995).

$$M = {}^{1}\!/_{m_v} \tag{5.7}$$

where m_v = the coefficient of volume compressibility.

Lunne et al. (1997) pointed out that most correlations between penetrometer test results and drained constrained modulus refer to the tangent modulus as found from oedometer tests. The correlations are typically represented as:

$$M = \alpha q_c \tag{5.8}$$

where M = the constrained modulus and α = the constrained modulus factor. Relations between α and q_c for cohesive soils are summarized in Table 5.19.

5.5.4.3 Over-consolidation ratio

In geotechnical engineering, in order to describe the stress history condition of a soil deposit, the over-consolidation ratio (OCR) is introduced as in Equation (5.9):

$$\text{OCR} = \frac{\text{maximum past effective vertical stress}}{\text{current effective vertical stress}} = \frac{\sigma'_p}{\sigma'_v} \tag{5.9}$$

The value of over-consolidation ratio (OCR), denoted as σ'_p, can be determined through oedometer tests. However, oedometer tests, especially those involving complex stress path approaches, can be time-consuming. Moreover, various factors, such as sample disturbance and the methods used to identify the maximum past consolidation pressure, can influence the test results. Therefore, it is recommended to conduct

Table 5.20 Proposed correlations for OCR

Reference	*Correlations*	*Remarks*
Mayne and Kemper (1988)	$OCR = 0.37\left(\frac{(q_c - \sigma_v)}{\sigma'_v}\right)^{1.01}$	–
Mayne and Kemper (1988)	$OCR = 0.193\left(\frac{N}{\sigma'_o}\right)^{0.689}$	σ'_o is in MPa
Trevor and Mayne (2004)	$OCR = 2(0.029 + 0.409M)\left[\frac{1}{1.95M+1}\left(\frac{q_t - u_2}{\sigma'_v}\right)\right]^{\frac{1}{\theta}}$	$\theta = 0.8 - 0.9$ $M = \frac{6\sin\varphi'}{3-\sin\varphi'}$
Mayne (2007)	$\sigma'_p = k(q_t - \sigma_v)$	*k*: Pre-consolidation cone factor with an expected range of 0.2–0.5
Robertson (2009)	$OCR = 0.25Q_t^{1.25}$	–
Robertson (2012)	$OCR = (2.625 + 1.75\log Fr)^{-1.25} Q_t^{1.25}$	–
Chanmee et al. (2017)	$OCR = k\left(\frac{q_t - \sigma_v}{OCR\sigma'_v}\right)$	Under consolidated deposits $k = 0.14–0.4$

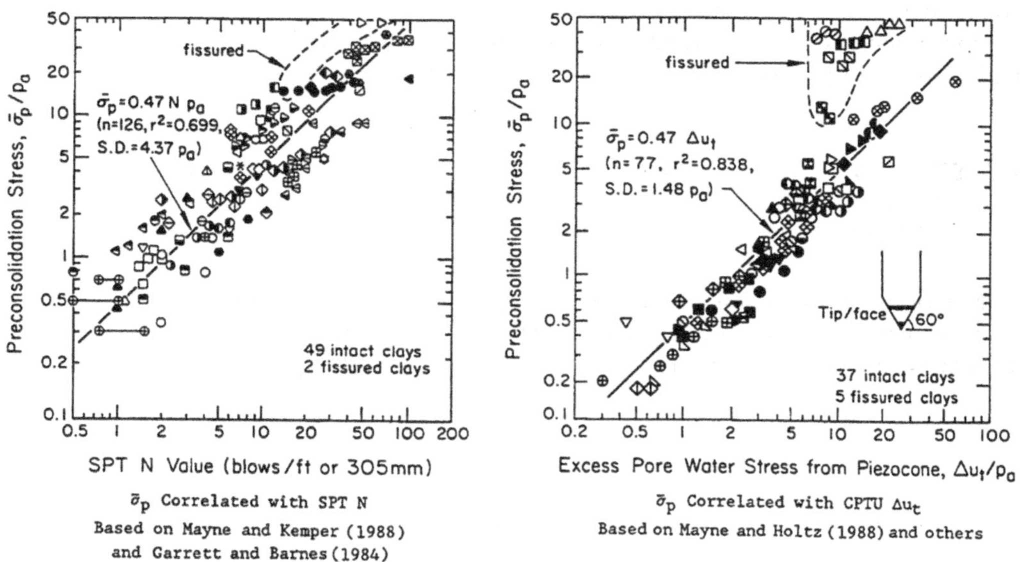

Figure 5.17 SPT and CPT correlation with OCR (Source: Kulhawy and Mayne, 1990).

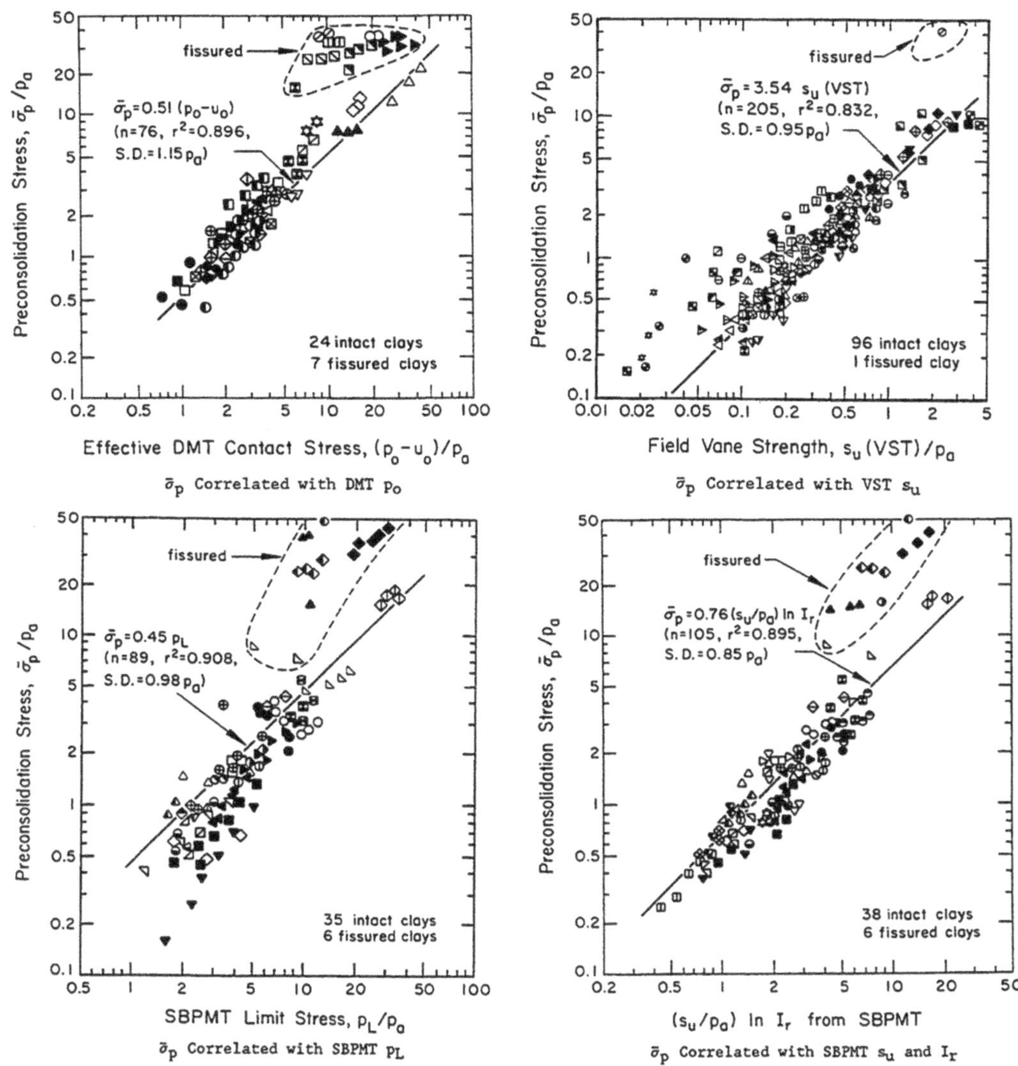

Figure 5.18 DMT, VST, and SBPMT correlation with OCR (Source: Kulhawy and Mayne, 1990).

in-situ tests as a more efficient and cost-effective means to estimate OCR using field data, which can provide valuable insights without the complications associated with oedometer testing.

There are several existing methods based on in-situ tests' data for assessing the OCR. Table 5.20 and Figure 5.17 show some proposed SPT and CPT correlations for OCR. Additionally, Figure 5.18 shows some other correlations for other in-situ tests, such as DMT and VST, based on data collected by Kulhawy and Mayne (1990).

Table 5.21 Empirical correlation for the compression ratio

Soil type	*Correlation for C_r*	*Reference*
Marine clays of Southeast Asia	$0.0043\ \omega_n$	Azzouz et al. (1976)
	0.0045 *LL*	Balasubramaniam and Brenner (1981)
Bangkok clays	0.00.463 *LL* − 0.013	Balasubramaniam and Brenner (1981)
	$0.00566\ \omega_n - 0.037$	
French clays	$0.0039\ \omega_n - 0.013$	Balasubramaniam and Brenner (1981)
Indiana clays	$0.0249 + 0.003\ \omega_n$	Lo and Lovell (1982)
	0.0294 + 0.00238 *LL*	
	$0.0125 + 0.152\ e_0$	
	$0.2037(e_0 - 0.2465)$	Goldberg et al. (1979)
Clays from Greece and parts of the United States	0.002(*LL* + 9)	Azzouz et al. (1976)
	$0.14(e_0 + 0.007)$	
	$0.003(\omega_n = 7)$	
	$0.126(e_0 + 0.003\ LL - 0.06)$	
Chicago clays	$0.208\ e_0 + 0.0083$	Azzouz et al. (1976)
Inorganic and organic clays and silty soils	$0.156\ e_0 + 0.0107$	Elnaggar and Krizek (1970)

Source: Ameratunga et al. (2016).

5.5.4.4 Consolidation indexes

From the comprehensive compilation presented by Djoenaidi (1985), several empirical correlations for the compression ratio have been selected. These correlations offer valuable tools for estimating compression ratios in geotechnical engineering analyses and design, as listed in Table 5.21.

The slope of the consolidation line in the $e - \log\sigma_v^{'}$ space is known as the compression index (C_c). When the clay is normally consolidated, the $e - \sigma_v^{'}$ values will be located on this line, irrespective of the stress level. When clay is normally consolidated, C_c is a measure of its stiffness, and it is a key parameter in evaluating the final settlements due to consolidation. It is usually related to the initial void ratio (e_0), in-situ natural water content (w_n), liquid limit (LL), or plasticity index (PI). Some of the suggested empirical correlations for C_c are listed in Table 5.22.

5.5.5 Related parameters

5.5.5.1 Shear wave velocity and shear modulus at small strain

A shear wave velocity (V_s) induced shear modulus is a critical geotechnical parameter corresponding to a small strain, which is essential in earthquake engineering design. Due to difficulties in collecting undisturbed samples, especially in granular soils, in-situ seismic

Table 5.22 Empirical correlations for C_c

Correlation for C_c	*Comment*	*Reference*
$0.009(LL - 10)$	Undisturbed clay of sensitivity <4. Reliability ± 30%	Terzaghi and Peck (1948)
$0.007(LL - 10)$	Remolded clay	Skempton (1944)
$0.0046(LL - 9)$	Sao Paulo, Brazil clays	Cozzolino (1961)
$0.0186(LL - 30)$	Soft silty Brazilian clays	Cozzolino (1961)
$0.01(LL - 13)$	All clays	USACE (1990)
$0.008(LL - 8.2)$	Indiana soils	Lo and Lovell (1982)
$0.21 + 0.008LL$	Weathered and soft Bangkok clays	Balasubramaniam and Brenner (1981)
$0.3(e_0 - 0.27)$	Inorganic silty clay	Hough (1957)
$1.15(e_0 - 0.35)$	All clays	Azzouz et al. (1976)
$0.75(e_0 - 0.50)$	Soils of very low plasticity	Azzouz et al. (1976)
$0.4(e_0 - 0.25)$	Clays from Greece and parts of the United States	Azzouz et al. (1976)
$0.256 + 0.43(e_0 - 0.84)$	Brazilian clays	Cozzolino (1961)
$0.54(e_0 - 0.35)$	All clays	Nishida (1956)
$0.22 + 0.29e_0$	Weathered and soft Bangkok clays	Balasubramaniam and Brenner (1981)
$0.575e_0 - 0.241$	French clays	Balasubramaniam and Brenner (1981)
$0.5363(e_0 - 0.411)$	Indiana soils	Goldberg et al. (1979)
$0.496e_0 - 0.195$	Indiana soils	Lo and Lovell (1982)
$0.40(e_0 - 0.25)$	Clays from Greece and parts of the United States	Azzouz et al. (1976)
$0.01\omega_n$	Chicago clays	Azzouz et al. (1976)
$0.01\omega_n$	Canada clays	Koppula (1981)
$0.0115\omega_n$	Organic soils, peat	USACE (1990) and Azzouz et al. (1976)
$0.012\omega_n$	All clays	USACE (1990)
$0.01(\omega_n - 5)$	Clays from Greece and parts of the United States	Azzouz et al. (1976)
$0.0126\omega_n - 0.162$	Indiana soils	Lo and Lovell (1982)
$0.008\omega_n + 0.20$	Weathered and soft Bangkok clays	Balasubramaniam and Brenner (1981)
$0.0147\omega_n - 0.213$	French clays	Balasubramaniam and Brenner (1981)
$(1 + e_0)[0.1 + 0.006(\omega_n - 25)]$	Varved clays	USACE (1990)

Source: Ameratunga et al. (2016).

Table 5.23 Some correlations between V_s (m/s) and SPT-N value

Source	Soil type	V_s (m/s)
Imai (1977)	All soils	$91N^{0.337}$
	Sand	$80.6N^{0.331}$
	Clay	$80.2N^{0.292}$
Ohta and Goto (1978)	All soils	$85.35N^{0.348}$
Seed and Idriss (1981)	All soils	$61.4N^{0.5}$
Sykora and Stokoe (1983)	Sand	$100.5N^{0.29}$
Okamoto et al. (1989)	Sand	$125N^{0.3}$
Pitilakis et al. (1999)	Sand	$145N^{0.178}$
	Clay	$132N^{0.271}$
Kiku et al. (2001)	All soils	$68.3N^{0.292}$
Jafari et al. (2002)	Sand	$22N^{0.77}$
	Clay	$27N^{0.73}$
Hasancebi and Ulusay (2007)	All soils	$99N^{0.309}$
	Sand	$90.82N^{0.319}$
	Clay	$97.89N^{0.269}$
Dikmen (2009)	All soils	$58N^{0.39}$
	Sand	$73N^{0.33}$
	Silt	$60N^{0.36}$
	Clay	$44N^{0.48}$

Source: Ameratunga et al. (2016).

tests are preferred to laboratory tests for achieving the V_s. By using down-hole and cross-hole techniques, as well as surface wave velocity determining techniques, a shear wave velocity profile with depth can be established.

Several correlations have been proposed to estimate the shear wave velocity based on SPT and CPT results for different soil types. The basic variables as input parameters can be q_c, f_s, N, and σ_v' in the correlation functions linear, logarithm, and power form. Some of these correlations are listed in Tables 5.23 and 5.24.

In general, it is accepted that the best way of assessing small strain (<0.001%) shear modulus (G_0) is to determine it using shear wave velocity ($G_0 = \rho V_s^2$). However, it is mainly the practice to carry out in-situ penetration tests to assess G_0 using empirical correlations, except that project-specific detailed investigations are performed. Some of the correlations for G_0 prediction based on CPT records are presented in Table 5.25.

Table 5.24 Empirical correlations between V_s and CPT data

Reference	*Proposed correlation (m/s)*	*Soil type*	*Units of parameters*	
			q_c	f_s
Hegazy and Mayne (1995)	$V_s = 12.02(q_c)0.319(f_s) - 0.0466$	Sand	kPa	kPa
	$V_s = 3.18(q_c)0.549(f_s)\ 0.025$	Clay	kPa	kPa
Mayne and Rix (1995)	$V_s = 1.75(q_c)0.627$	Clay	kPa	kPa
Iyisan (1996)	$V_s = 55.3(q_c)0.377$	Clay	kgf/cm²	–
	$V_s = 218 + 0.70\ q_c$	Sand	kgf/cm²	–
Piratheepan (2002)	$V_s = 25.3(q_c)0.163(f_s)\ 0.029D\ 0.115$	Sand	kPa	kPa
	$V_s = 11.9(q_c)0.269(f_s)\ 0.108D\ 0.127$	Clay	kPa	kPa
Tun (2003)	$V_s = 109.29 + 52.674\mathrm{Ln}(q_c)$	All	MPa	MPa
Madiai and Simoni (2004)	(1) $V_s = 211(q_c)0.23$	All	MPa	MPa
	(2) $V_s = 155(q_c)0.29(f_s) - 0.10$			
Mayne (2006)	$V_s = 18.5 + 118.8\log(f_s)$	All	–	kPa
Paoletti et al. (2010)	$V_s = 50[(q_c/P_a)0.43 - 3]$	Sand	kPa	–
MolaAbasi et al. (2015)	$V_s = 100[1.36 - 0.35f_s + 0.15q_c - 0.05f_s^2 - 0.018q_c^2 + 0.39f_sq_c]$	Clay	MPa	MPa
	$V_s = 100[1.73 + 2.74f_s + 0.03q_c - 4.015f_s^2 - 0.00026q_c^2 + 0.007f_sq_c]$	Sand	MPa	MPa
	$V_s = 100[1.47 + 2.07f_s + 0.10q_c + 9.50f_s^2 - 0.0023q_c^2 - 0.034f_sq_c]$	Mixed	MPa	MPa
	$V_s = 100[1.40 + 1.59f_s + 0.09q_c - 1.33f_s^2 - 0.002q_c^2 + 0.05f_sq_c]$	All	MPa	MPa

Source: Ameratunga et al. (2016).

Table 5.25 Empirical correlations for G_0 based on q_c

Reference	*Correlations*	*Remarks*
Mayne and Rix (1995)	$G_{max} = 99.5pa^{0.305}(qc)^{\frac{0.695}{e_0^{1.13}}}$	Cohesive soils, e_0 = initial void ratio
Eslaamizaad and Robertson (1997)	$\left(\frac{G_0}{q_c}\right) = 1634\left(\frac{q_c}{\sqrt{\sigma'_v}}\right)^{-0.75}$	Cohesionless soils
Schnaid (2009)	$G_0 = b(q_c\sigma'_v Pa)^{0.3}$	Cohesionless soils, b = 280 and 110 for an upper and lower bond

Source: Ameratunga et al. (2016).

Table 5.26 Permeability from CPT results

Soil behavior type	*Soil permeability (m/s)*
Sensitive fine-grained	3×10^{-9} to 3×10^{-8}
Organic soils – peats	1×10^{-8} to 1×10^{-6}
Clays – clay to silty clay	1×10^{-10} to 1×10^{-7}
Silt mixtures clayey silt to silty clay	3×10^{-9} to 1×10^{-7}
Sand mixtures; silty sand to sandy silt	1×10^{-7} to 1×10^{-5}
Sands; clean sands to silty sands	1×10^{-5} to 1×10^{-3}
Gravelly sand to sand	1×10^{-3} to 1
Very stiff sand to clayey sand	1×10^{-8} to 1×10^{-6}
Very stiff fine-grained	1×10^{-9} to 1×10^{-7}

Source: Robertson (1990).

5.5.5.2 Sensitivity

Sensitivity (S_t) is defined as the ratio of peak undisturbed shear strength to remolded shear strength (Equation 5.10). Many researchers have reported that the shear strength of remolded samples is affected by skin friction, f_s.

$$S_t = s_u / s_{u(\text{remolded})} = (q_t - \sigma_v) / N_{kt} \left(1/fs \right) \tag{5.10}$$

Schmertmann's (1978) suggested formula is based on this and can be simply written as follows:

$$S_t = N_s / R_f \tag{5.11}$$

where R_f = the friction ratio and N_s = a constant.

5.5.5.3 Permeability

Robertson (1990) proposed a relationship between permeability (k) and the soil behavior type for CPT, as shown in Table 5.26. Robertson and Cabal (2010) suggested Equation (5.12) to evaluate the average permeability (k_{ave}):

$$\begin{cases} k = 10^{(0.952 - 3.04 I_c)} & 1 < I_c \le 3.27 \\ k = 10^{(-4.52 - 1.37 I_c)} & 3.27 < I_c \le 4 \end{cases} \tag{5.12}$$

where $I_c = \sqrt{(3.47 - \log Q_t)^2 + (\log F_r + 1.22)^2}$.

5.6 GEO-MARINE DATABASE AND TRIANGULAR CHARTS

5.6.1 AUT: CPTu-Geo-marine Database

Eslami et al. (2022) proposed a marine database that was meticulously assembled through an extensive examination of approximately 350 geotechnical reports, articles, research

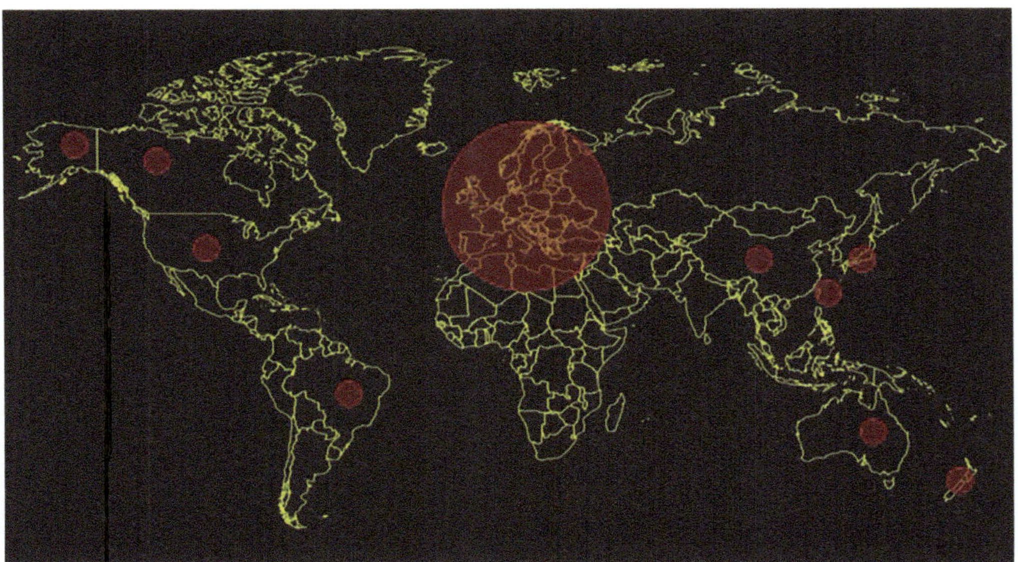

Figure 5.19 The scatter of the AUT: CPTu-GMD case studies.

programs, and doctoral theses. This comprehensive database comprises soil profiles extracted from boreholes and CPTu (Cone Penetration Testing) results obtained in their proximity. It encompasses a total of 398 cases derived from 58 different locations spanning 18 countries, with the majority of data originating from the United States, Europe, China, and New Zealand. The distribution of CPTu-GMD (Geotechnical and Marine Data) cases across these countries is illustrated in Figure 5.19, while Figure 5.20 provides a visual representation of a selection of typical CPTu logs.

The marine database consists of a diverse range of site types, with 21% located onshore along the coastline, 26% in near-shore areas within seas less than 30 m deep, 15% in offshore locations within waters exceeding 30 m, and 38% situated in riverine settings, as reported by the Federal Geographic Data Committee in 2012.

For each penetrometer test conducted in proximity to a borehole, CPT records were digitized at 10 cm intervals, capturing essential parameters such as cone tip resistance (q_c), sleeve friction (f), and excess pore water pressure (u_2). Soil profiles were meticulously characterized based on the available borehole data. Additionally, various other parameters, including soil density, effective stress, and eight additional factors, were deduced from CPT records utilizing relevant correlations as outlined by Robertson and Cabal (2010). This extensive database encompasses approximately 10,000 m of recorded soil profiles, with primary soil types comprising clay, sensitive soils, sand and gravel, overconsolidated clay, as well as mixed or deltaic soils. Figure 5.21 provides detailed insights into the AUT: CPTu-GMD dataset.

5.6.2 Soil Behavioral Classification methods

Soil Behavioral Classification diagrams can be categorized into three distinct groups, each based on the incorporation of CPTu records. The first group, often referred to as "basic diagrams," includes diagrams developed by Douglas and Olsen (1981), Campanella et al.

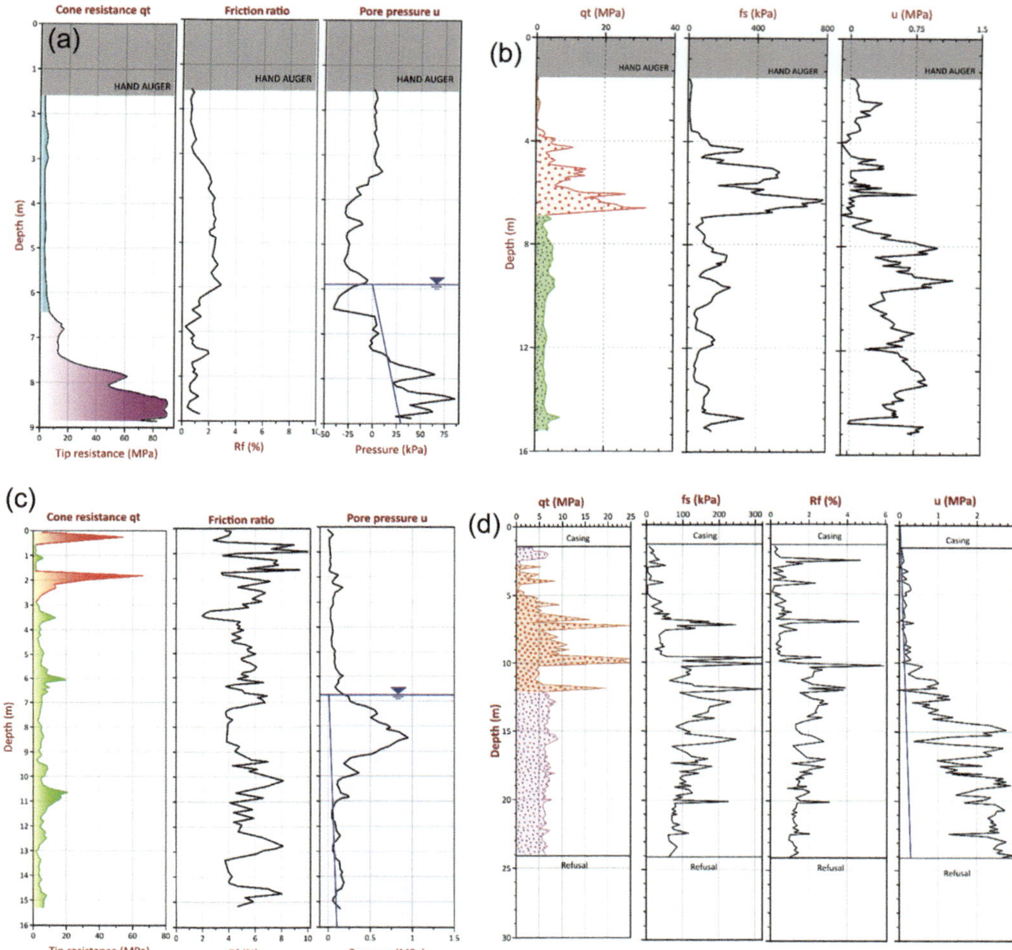

Figure 5.20 Typical CPTu logs: (a) onshore project, California, USA (Geotechnical Report, 2015); (b) near-shore project, New Zealand (Geotechnical Report, 2018); (c) offshore project, Australia (Geotechnical Report, 2019); and (d) riverine project, Canada (Geotechnical Report, 2020).

(1985), and Robertson (2010). These diagrams rely on only two of the three piezocone test records and employ mathematical equations. The most commonly utilized parameters in this group are the cone tip resistance (q_c) and the sleeve friction (f).

The second group, known as "indirect diagrams," encompasses diagrams created by Robertson (1990), Jefferies and Davies (1993), Eslami and Fellenius (1997), and Eslami et al. (2019). These diagrams utilize all three piezocone test parameters, in conjunction with mathematical equations, to provide a comprehensive description of subsurface layers. It is worth noting that the parameters considered in these second-group diagrams may vary, setting them apart from the basic diagrams.

The first set of charts includes those developed by Robertson (1990) and Jefferies and Davies (1993). These charts account for the depth-related variations in piezocone parameters, such as cone tip resistance, sleeve friction, and excess pore water pressure, by specifying the parameters Q_t, B_q, and F_r. In contrast, the second group, which comprises the

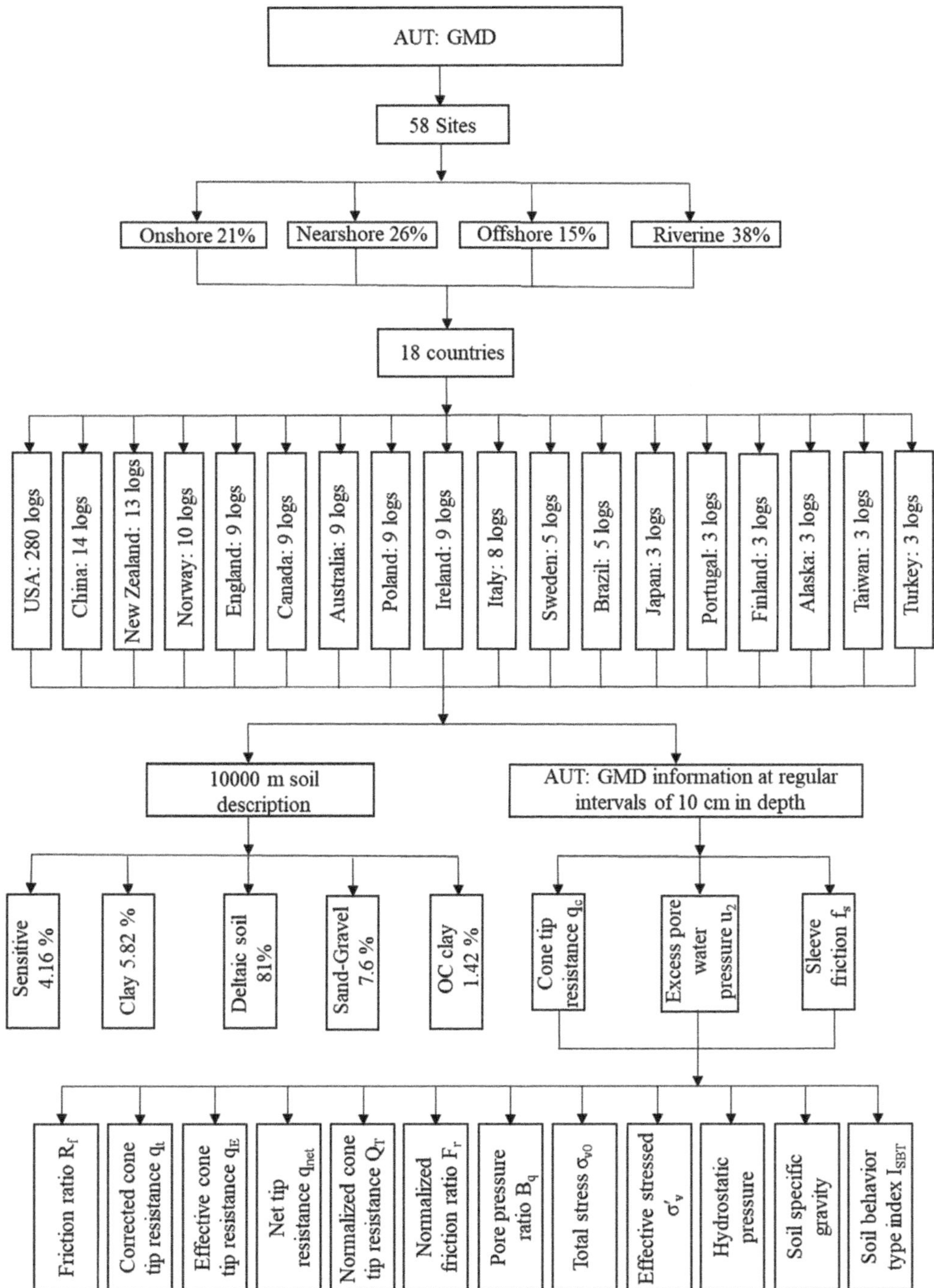

Figure 5.21 The details of AUT: CPTu-GMD.

diagrams created by Eslami and Fellenius (1997) and Eslami et al. (2019), does not address the depth effect separately through mathematical calculations for piezocone parameters.

These diagrams in the second group offer a broader classification of soils, resulting in fewer soil identification zones compared to the first group. As a consequence, each zone covers a larger area and accommodates a wider range of variations. Notably, even at considerable depths, the effect of depth on CPTu records does not prompt the relocation of the subsurface layer to a different zone in the Eslami and Fellenius (1997) and Eslami et al. (2019) diagrams.

Conversely, the diagrams presented by Robertson (1990) and Jefferies and Davies (1993) feature a greater number of zones, resulting in a smaller area allocated to each soil type. These diagrams incorporate normalized parameters, such as Q_t and B_q, to account for depth-related variations. Neglecting these equations leads to inaccuracies in identifying deep-seated soil layers, often resulting in these layers being incorrectly categorized within neighboring zones.

In contrast, the third group, referred to as "direct diagrams," directly employs the three specified parameters (q_c, f_s, and u_2) without the use of mathematical equations to identify subsurface layers. Prominent examples include the work of Eslami et al. (2015). This approach reduces uncertainties and minimizes the reliance on mathematical correlations. The key distinction between the second group of diagrams and the direct diagrams, such as those by Eslami et al. (2015), lies in the application of mathematical relationships.

5.6.3 Results and discussions

The study conducted by Eslami et al. (2022) evaluated the performance of CPT-based Soil Behavioral Classification (SBC) diagrams using 57 case studies sourced from the Marine Database (AUT: CPTu-GMD). These case studies encompassed a selection of widely recognized and efficient methods for soil behavior classification, including Douglas and Olsen (1981), Campanella et al. (1985), Robertson (1990), Robertson (2010), Jefferies and Davies (1993), Eslami and Fellenius (1997), Eslami et al. (2015), and Eslami et al. (2019).

These methods were chosen for their popularity and reliability in classifying soil behavior. Figure 5.21 provides a geographical overview of the analyzed cases, with a distribution as follows: the United States (10 logs), China (5 logs), Canada (3 logs), New Zealand (3 logs), Italy (3 logs), Norway (3 logs), Poland (3 logs), Finland (3 logs), Sweden (3 logs), Ireland (3 logs), Brazil (3 logs), England (3 logs), Japan (3 logs), Portugal (3 logs), Australia (3 logs), and Turkey (3 logs).

Furthermore, the study encompassed an assessment of all four site types found in AUT: CPTu-GMD, which includes 15 onshore locations, 20 near-shore areas, 4 offshore sites, and 18 riverine regions.

The 57 cases under investigation include a total of 1420 m of CPTu testing and soil descriptions obtained from borehole drilling. These data are distributed as follows: 117 m of sensitive soil, 98 m of clay, 1135 m of mixed or deltaic soil, 54 m of sand–gravel, and 16 m of over-consolidated clay.

Figures 5.22–5.25 provide visual representations of the distribution of these four soil types, namely sensitive, clay, mixed, and sand–gravel, on various charts.

To evaluate the accuracy of each diagram, Equation (5.13) is employed. This assessment method involves calculating the ratio of the accurately identified length of each soil

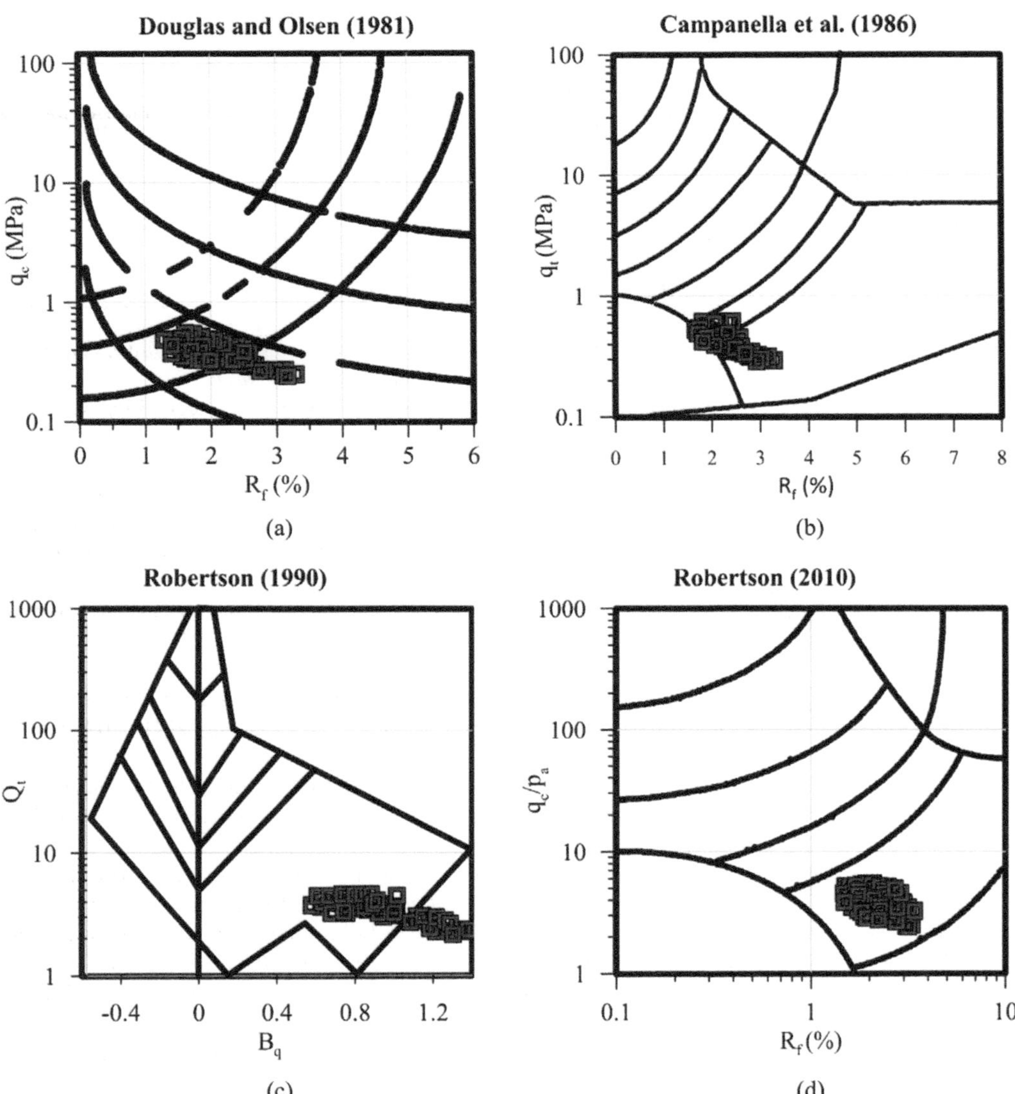

Figure 5.22 The scatter of sensitive soils in different SBC charts for the compiled database: (a) Douglas and Olsen (1981); (b) Campanella et al. (1985); (c) Robertson (1990); (d) Robertson (2010); (e) Jefferies and Davies (1993); (f) Eslami and Fellenius (1997); (g) Eslami et al. (2015); and (h) Eslami et al. (2019).

type to the total length investigated, based on the available borehole data. This ratio is defined as the success rate (SR) and is expressed as a percentage.

$$SR = \left(L'/L\right) \times 100 \tag{5.13}$$

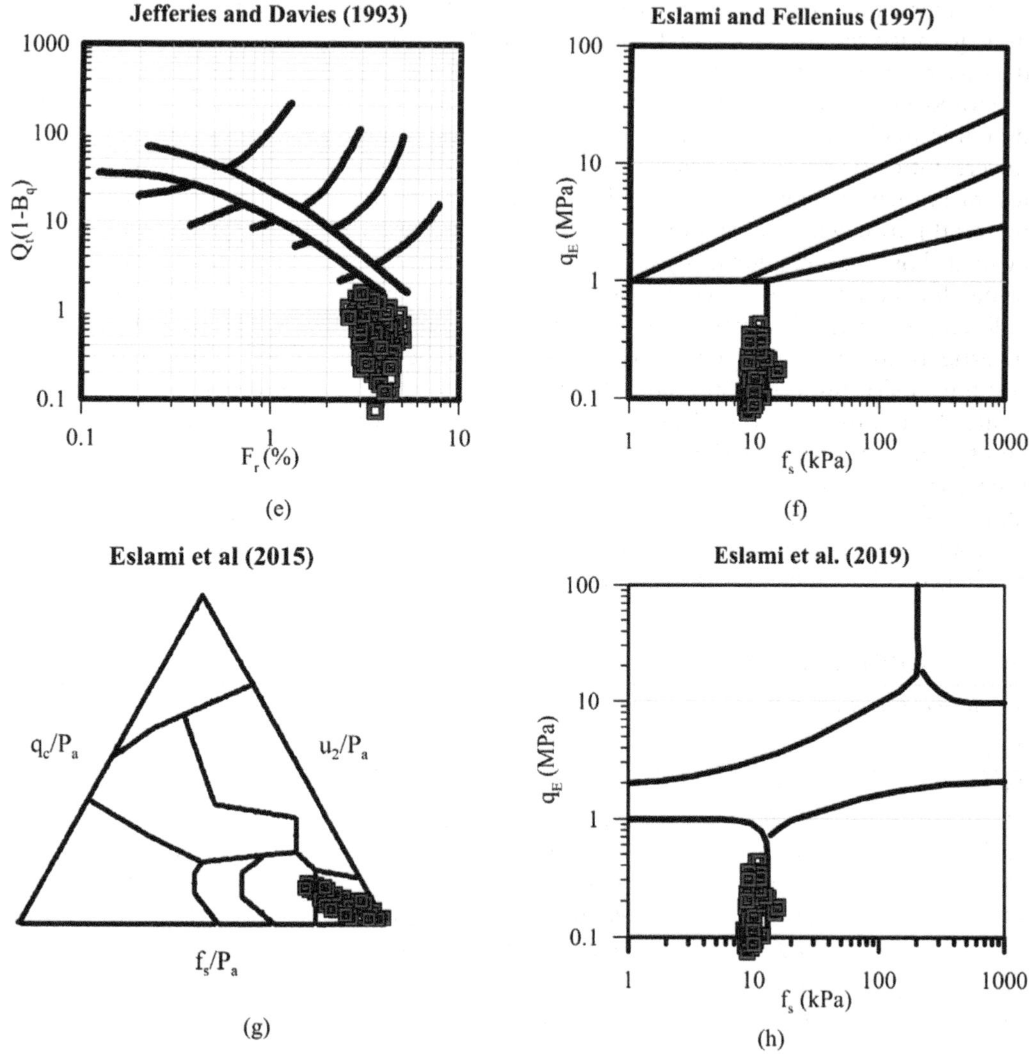

Figure 5.22 (Continued).

where SR = success rate; L' = the correct predicted length of a specific soil type in the investigated borehole; L = the total length of borehole with a particular soil type that exists in the investigated case.

The statistical assessment of each diagram is prominently displayed on a radar chart in Figure 5.26. Each radar chart is equipped with five axes, representing sensitive soils, clay, deltaic soils, sand–gravel, and over-consolidated clay, and these axes are scaled from 0% to 100%. These values effectively quantify the precision with which each diagram identifies soil types associated with their respective axes.

As indicated in Figure 5.26, the capacity to accurately identify sensitive soils has notably improved with the adoption of the second group of diagrams, which consider all

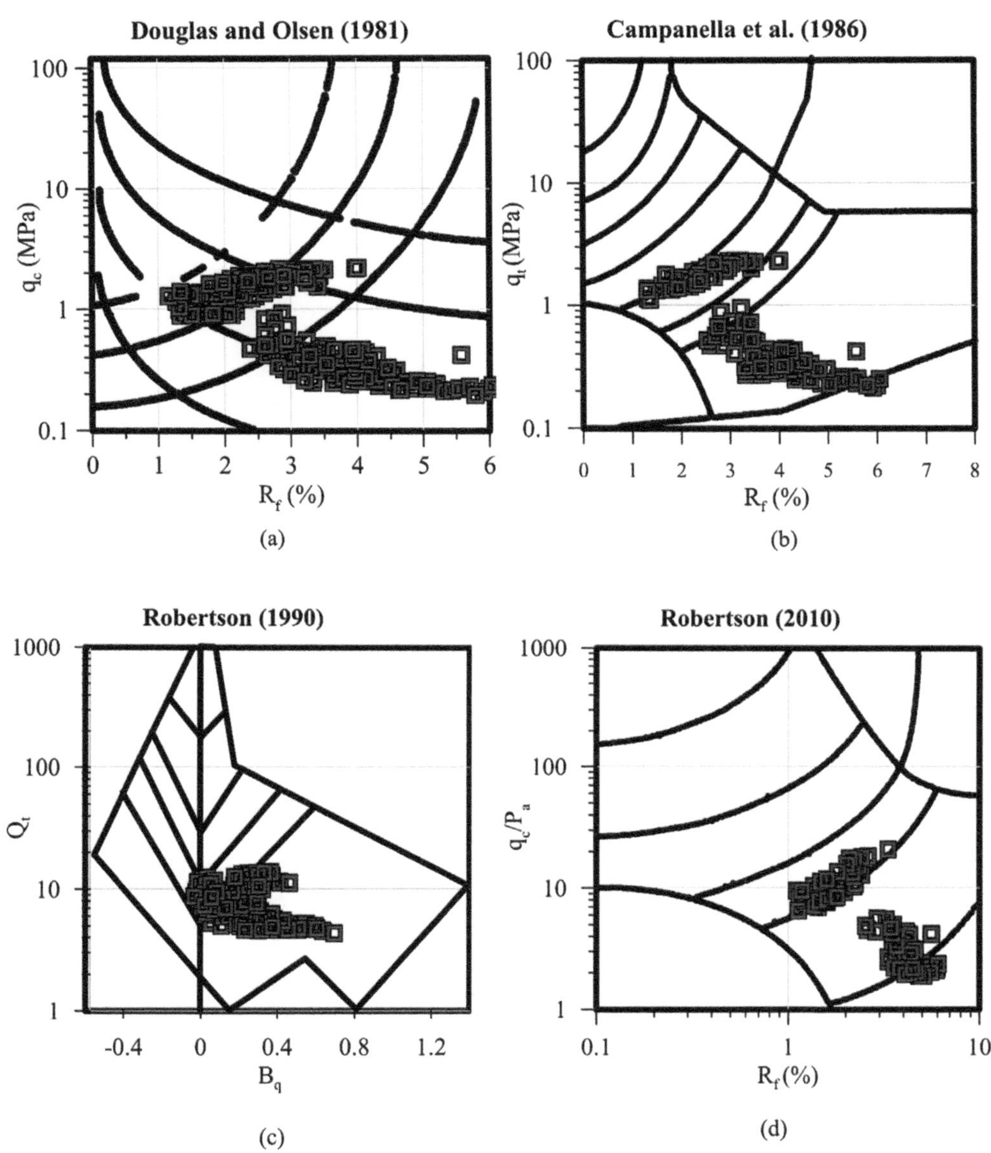

Figure 5.23 The scatter of clayey soils in different SBC charts for the compiled database: (a) Douglas and Olsen (1981); (b) Campanella et al. (1985); (c) Robertson (1990); (d) Robertson (2010); (e) Jefferies and Davies (1993); (f) Eslami and Fellenius (1997); (g) Eslami et al. (2015); and (h) Eslami et al. (2019).

three factors – q_c, f_s, and u_2 – along with mathematical correlations for soil classification. Basic charts, such as those developed by Douglas and Olsen (1981) and Campanella et al. (1985), exhibit less than 40% accuracy in detecting sensitive soils. In contrast, the implementation of indirect charts has led to an enhanced reliability in characterizing sensitive soils. This trend continues until the Eslami and Fellenius (1997) diagram, which achieves a remarkable success rate of over 90% in identifying sensitive soils.

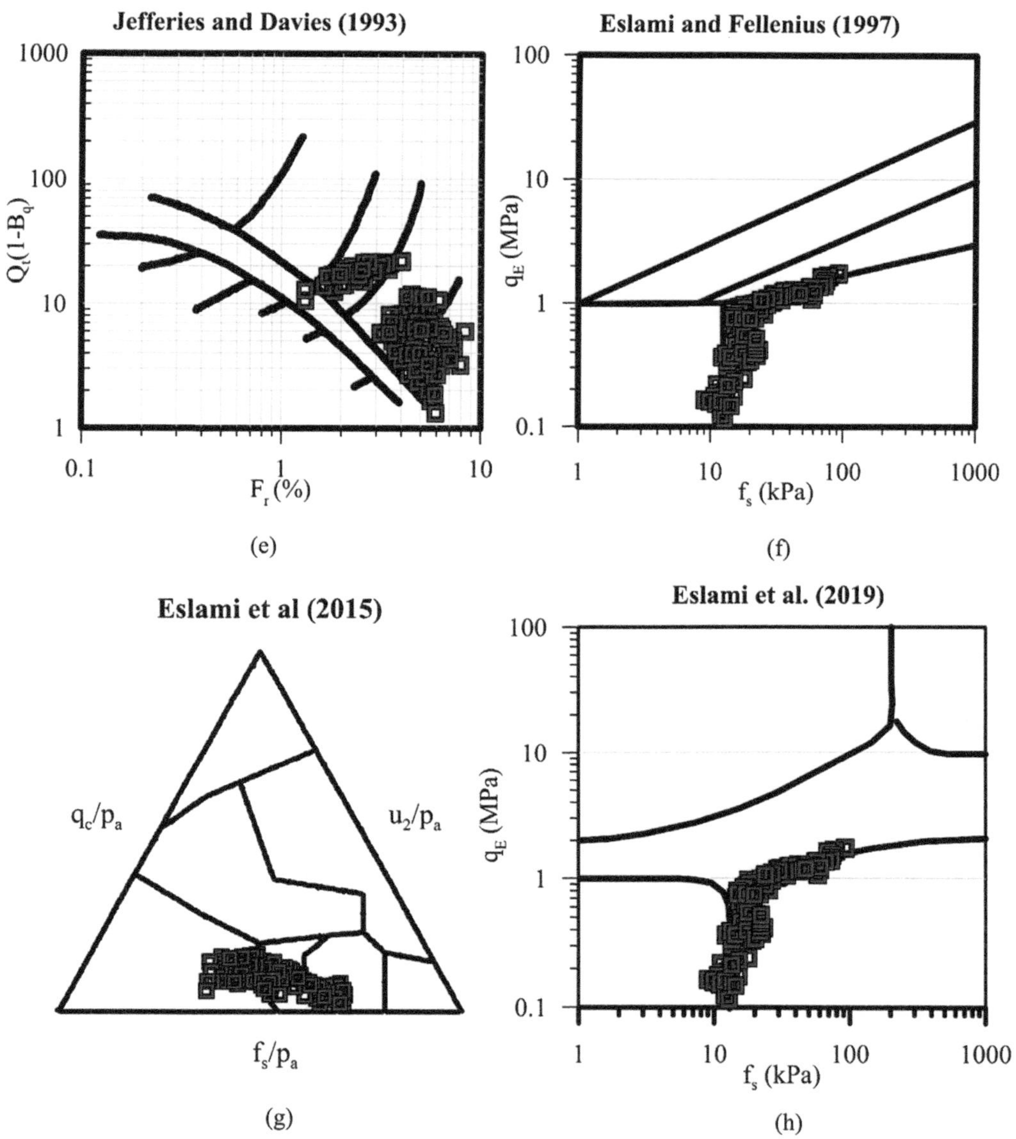

Figure 5.23 (Continued).

However, the Robertson (2010) chart falls short in effectively identifying sensitive soils due to its alignment with the first group of diagrams and the omission of the excess pore water pressure component. This underscores the pivotal role of measuring excess pore water pressure in accurately identifying sensitive soils. The Eslami and Fellenius (1997), Eslami et al. (2015), and Eslami et al. (2019) diagrams, on the other hand, exhibit a remarkable ability to identify sensitive soils with more than 90% accuracy.

Moreover, the first and second groups of diagrams exhibited challenges in accurately identifying normally consolidated clays, achieving less than 60% accuracy. This is attributed to the fact that the cone tip resistance, q_c, tends to be low for silts and

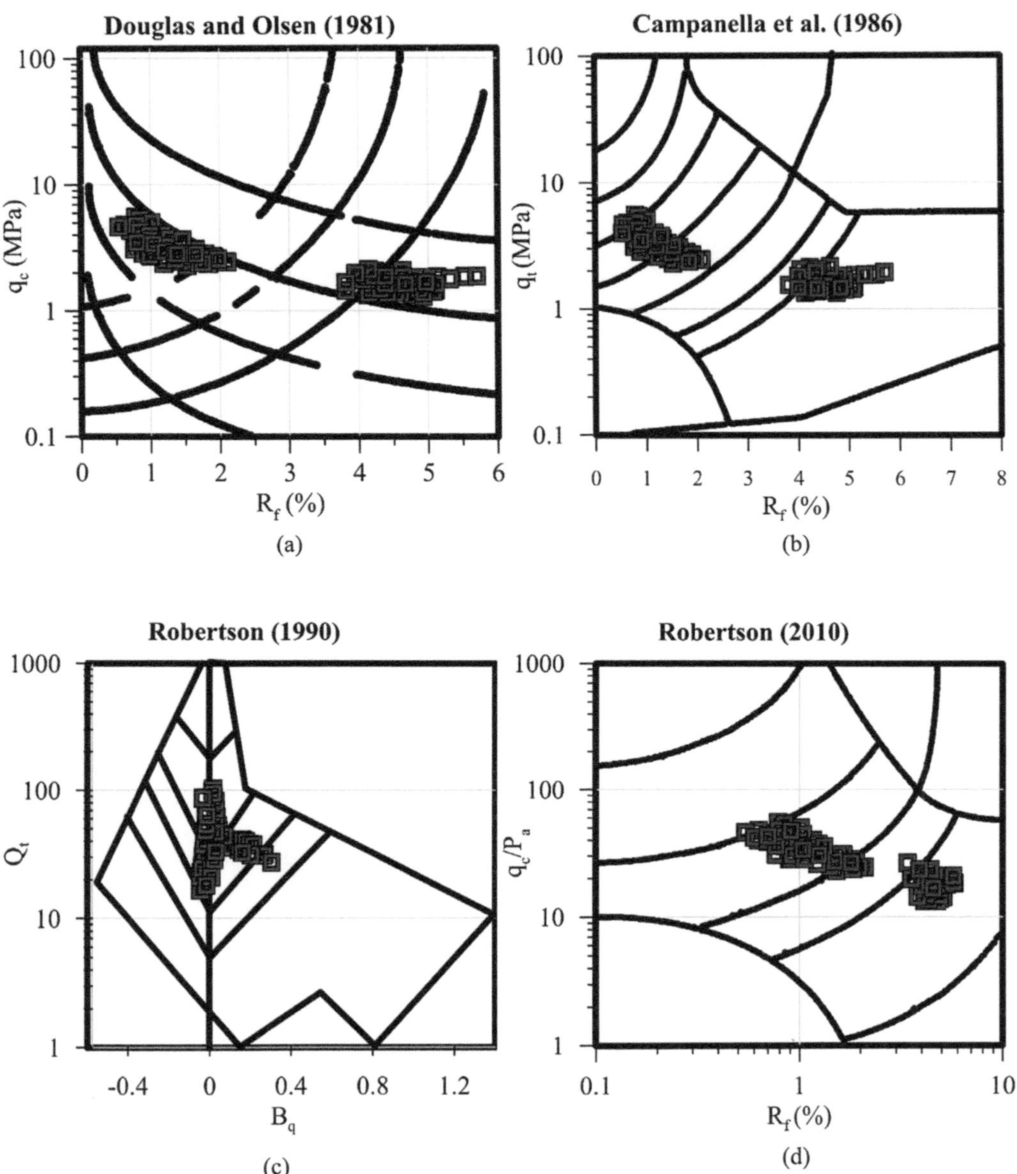

Figure 5.24 The scatter of deltaic soils in different SBC charts for the compiled database: (a) Douglas and Olsen (1981); (b) Campanella et al. (1985); (c) Robertson (1990); (d) Robertson (2010); (e) Jefferies and Davies (1993); (f) Eslami and Fellenius (1997); (g) Eslami et al. (2015); and (h) Eslami et al. (2019).

normally consolidated clays, which limits its effectiveness in distinguishing between these soil types. Consequently, these diagrams now primarily rely on parameters such as sleeve friction and pore pressure for the Soil Behavioral Classification (SBC) of these specific soil types.

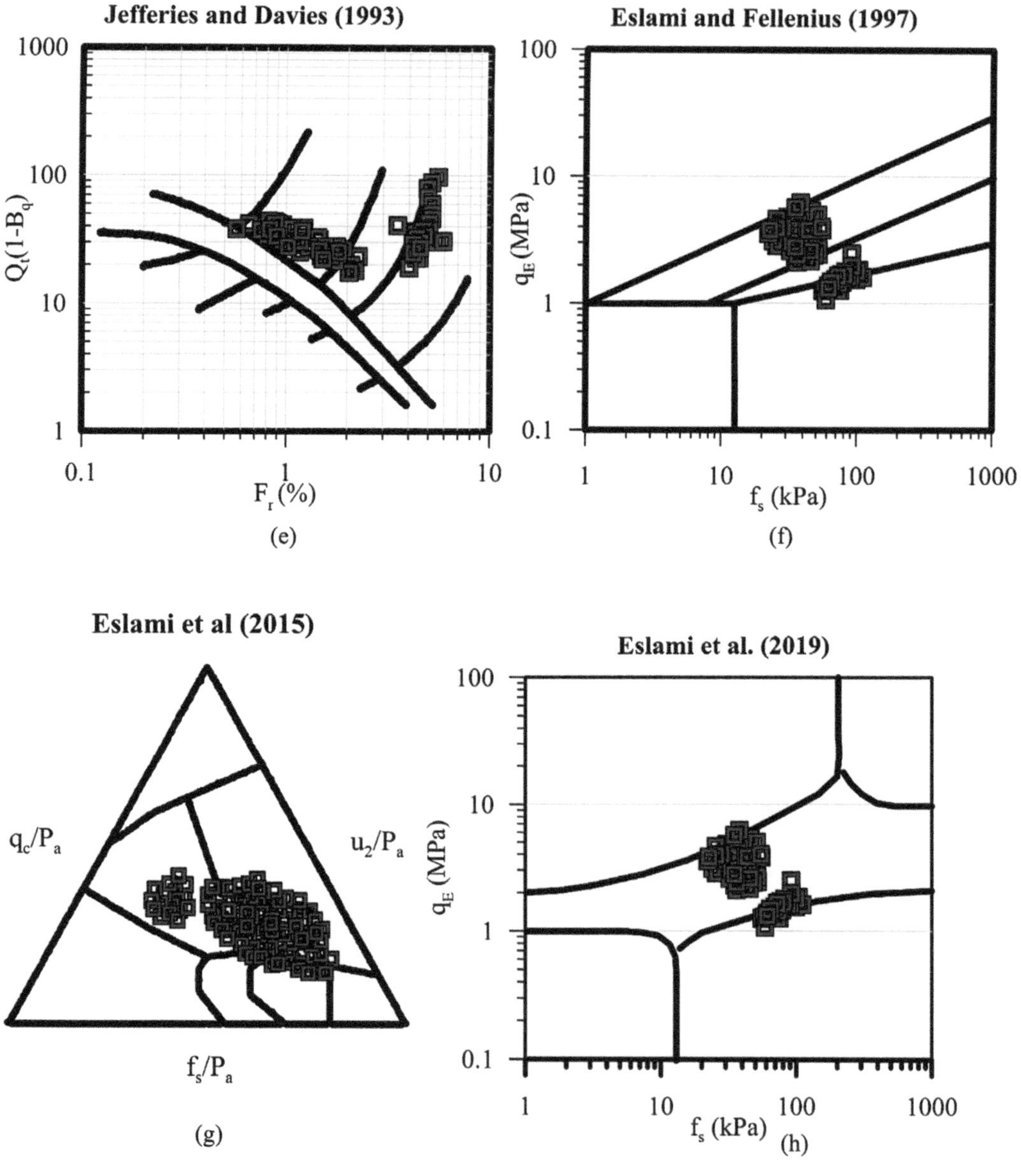
Jefferies and Davies (1993)
1000
100
10
1
Q_t(1-B_q)
0.1
1
10
F_r (%)
(e)
Eslami and Fellenius (1997)
100
10
1
0.1
q_E (MPa)
1
10
100
1000
f_s (kPa)
(f)
Eslami et al (2015)
q_c/P_a
u_2/P_a
f_s/P_a
(g)
Eslami et al. (2019)
100
10
1
0.1
q_E (MPa)
1
10
100
1000
f_s (kPa)
(h)

Figure 5.24 (Continued).

With the exception of the Eslami et al. (2019) diagram, other diagrams in the first and second groups displayed an accuracy rate of less than 60% when detecting over-consolidated clays. The Eslami et al. (2019) diagram, however, showed a considerably higher accuracy of approximately 80%. It is worth noting that due to the high cone tip resistance observed in over-consolidated clay, the other diagrams in the first and second groups often misclassify it as sandy soil.

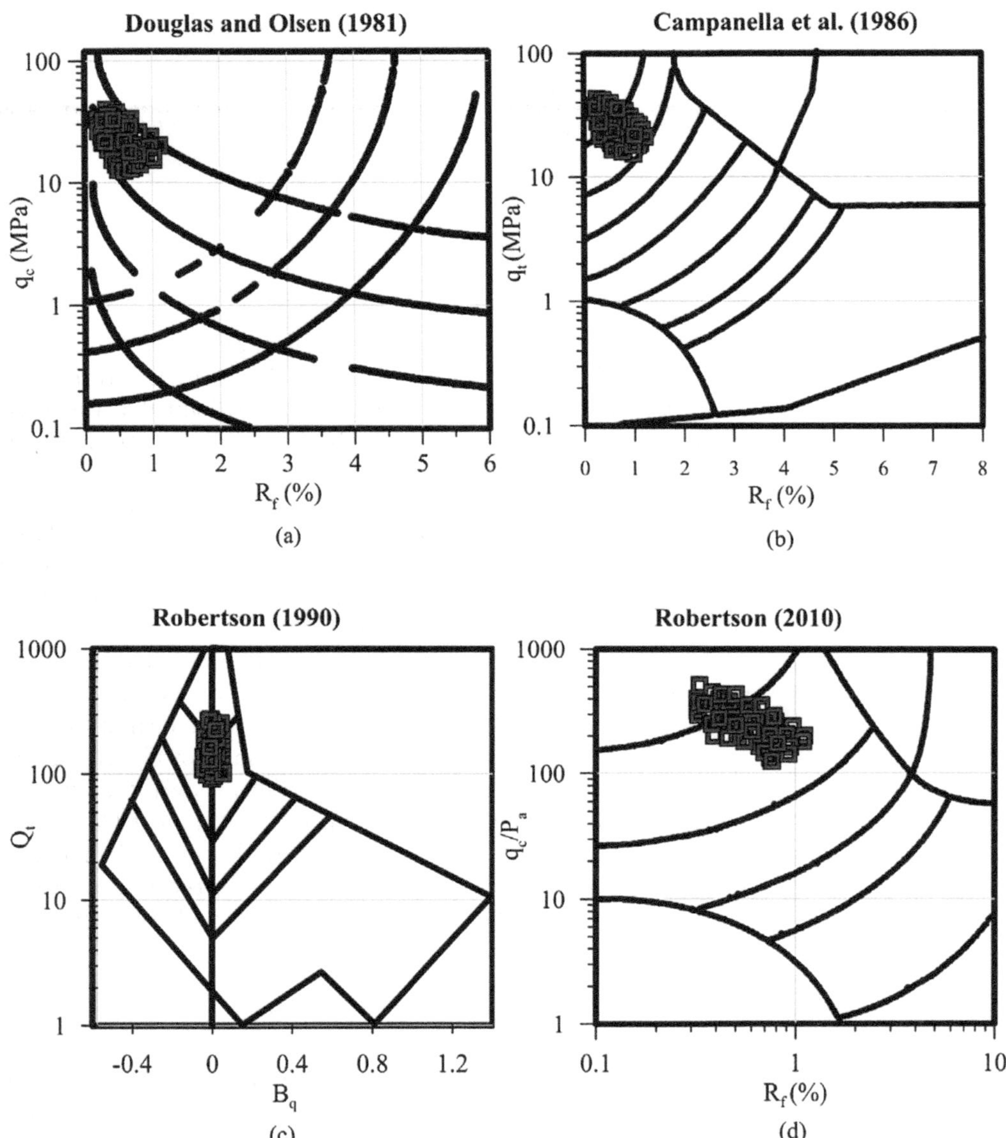

Figure 5.25 The scatter of sand and gravel in different SBC charts for the compiled database: (a) Douglas and Olsen (1981); (b) Campanella et al. (1985); (c) Robertson (1990); (d) Robertson (2010); (e) Jefferies and Davies (1993); (f) Eslami and Fellenius (1997); (g) Eslami et al. (2015); and (h) Eslami et al. (2019).

As depicted in Figure 5.26, the first two types of diagrams demonstrated a good level of accuracy, correctly identifying deltaic soils and sand–gravel mixtures with rates ranging from 80% to 90%. In contrast, the Eslami et al. (2015) triangle diagram, which directly employs all three CPTu records rather than relying on mathematical correlations, exhibited the highest accuracy in predicting sensitive soils, over-consolidated clays, and sand–gravel, with success rates of approximately 96%, 85%, and 94%, respectively.

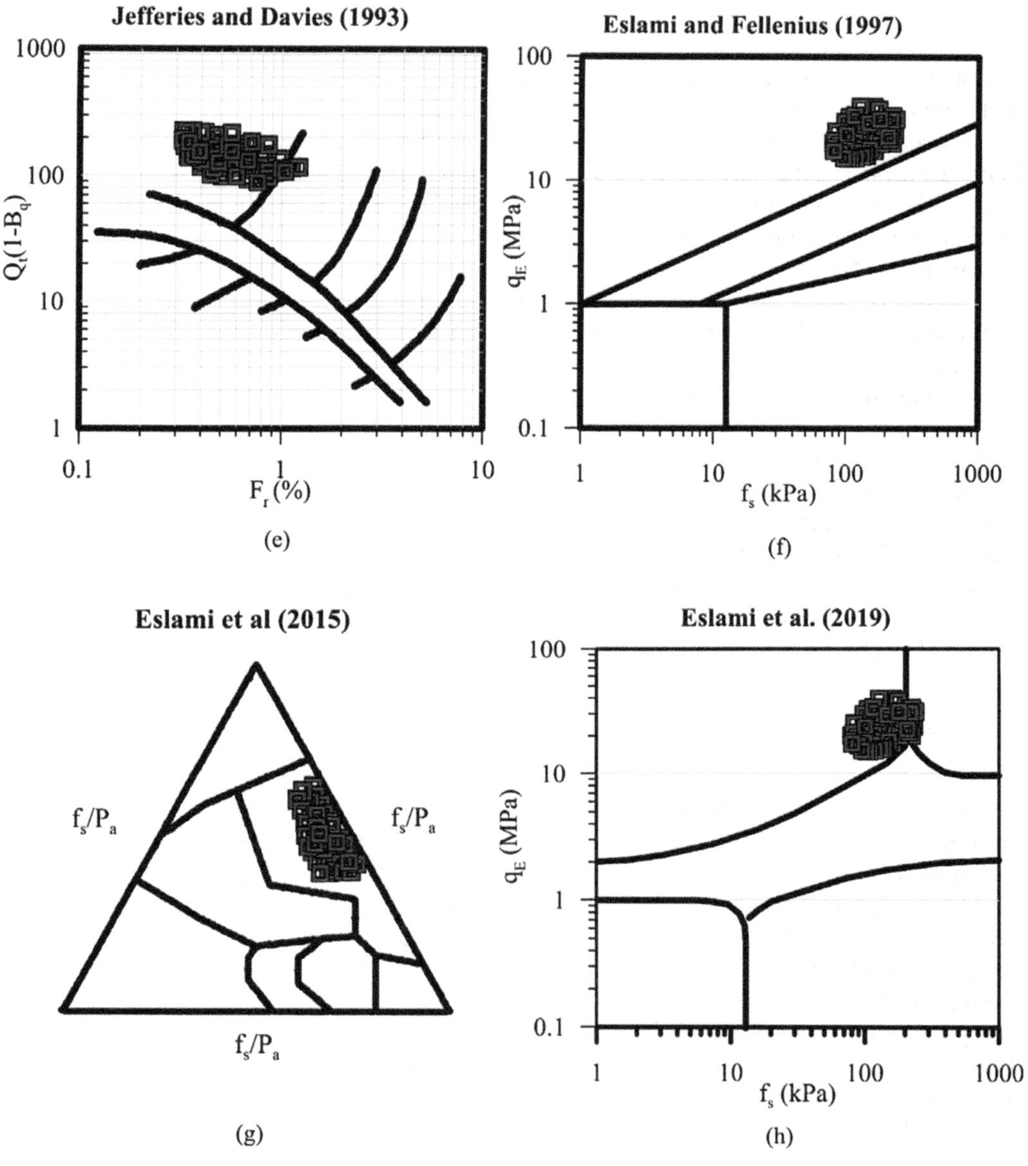

Figure 5.25 (Continued).

Nevertheless, it should be noted that this diagram tended to inaccurately classify deltaic soils and normally consolidated clays. This suggests that by refining the applied model assumptions and minimizing uncertainties introduced by mathematical relationships, the accuracy of SBC for deltaic soils and normally consolidated clay can be significantly improved with minor adjustments.

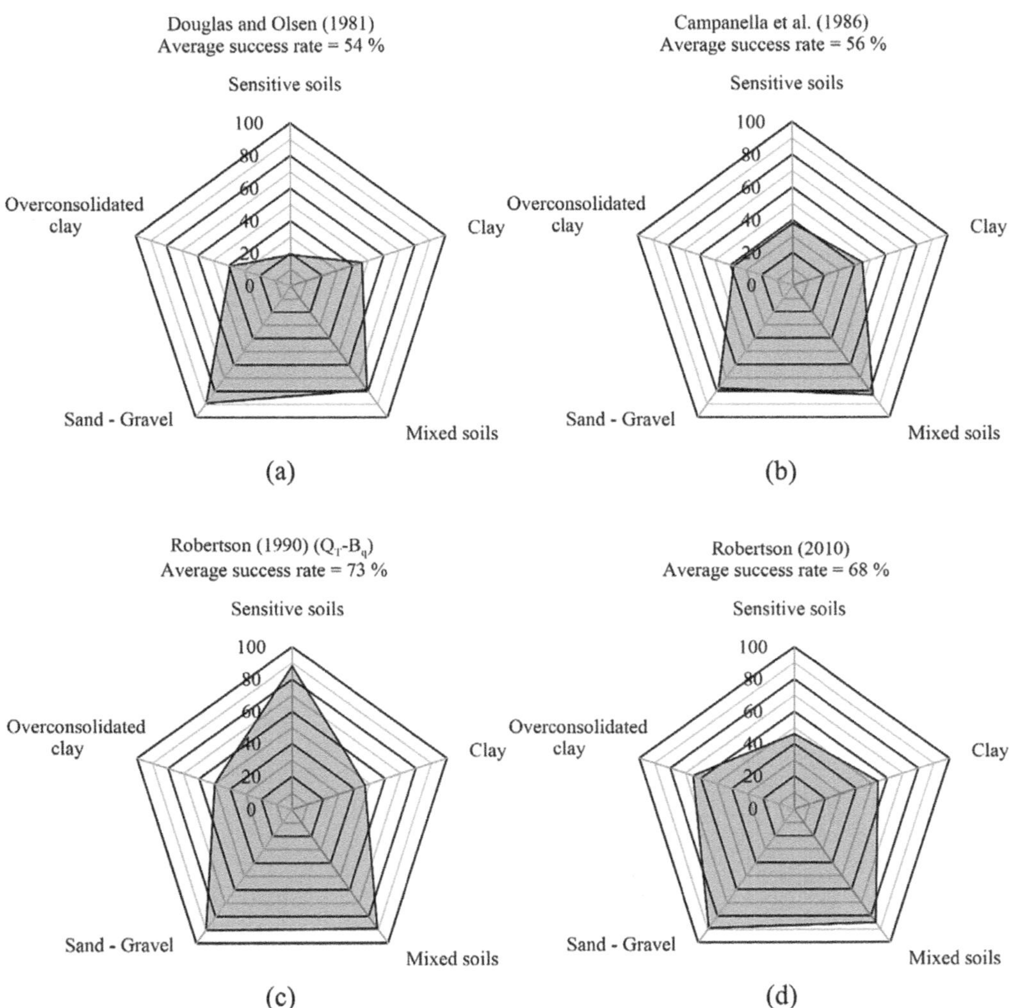

Figure 5.26 The statistical assessment of SBC diagrams via radar charts.

5.6.4 Developed triangular chart

The performance evaluation of the Soil Behavioral Classification (SBC) diagrams in accurately identifying all five deltaic soil types within the examined database highlights the critical role of input parameters, mathematical correlations, and the database source from which these diagrams are derived. This underscores the need for the development of a chart that can effectively classify marine deposits with precision.

The third group of SBC diagrams, exemplified by the Eslami et al. (2015) diagram, demonstrates significant potential in accurately identifying all soil types. In the Eslami et al. (2015) diagram, all three parameters – cone tip resistance, sleeve friction, and pore water pressure — are directly incorporated, eliminating the need for mathematical correlations to classify subsurface layers. This results in a reduction of uncertainties associated with subsurface layer identification, as mathematical formulas, are not relied upon.

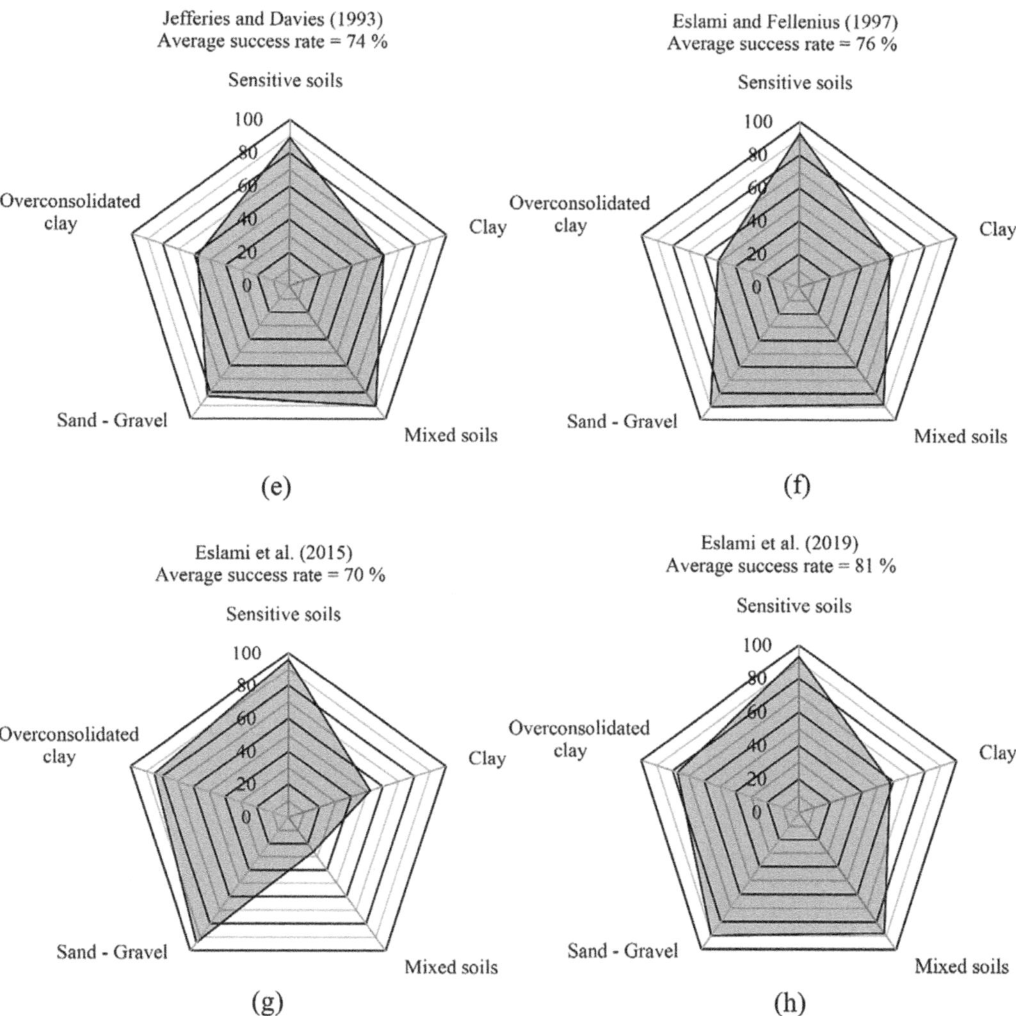

Figure 5.26 (Continued).

In traditional triangular charts, the sum of parameters on the three sides is typically a constant value, usually set at 100. For instance, in the Soil Classification Chart recommended by the United States Department of Agriculture (USDA) for deltaic soil classification, the sum of the three parameters (percentage of silt, clay, and sand) must equal 100. However, the output parameters of Cone Penetration Testing (CPTu) are not interrelated, and their sum cannot be constrained to equal 100. Consequently, the Eslami et al. (2015) diagram introduces a novel approach to triangular diagrams, integrating all three CPTu records within a single chart.

In the case of the Eslami et al. (2015) diagram, the developed triangular chart consists of three axes that correspond to the CPTu parameters, namely cone tip resistance, sleeve friction, and pore water pressure, as depicted in Figure 5.27. The application of this developed triangular chart closely mirrors that of the Eslami et al. (2015) diagram and involves the following steps.

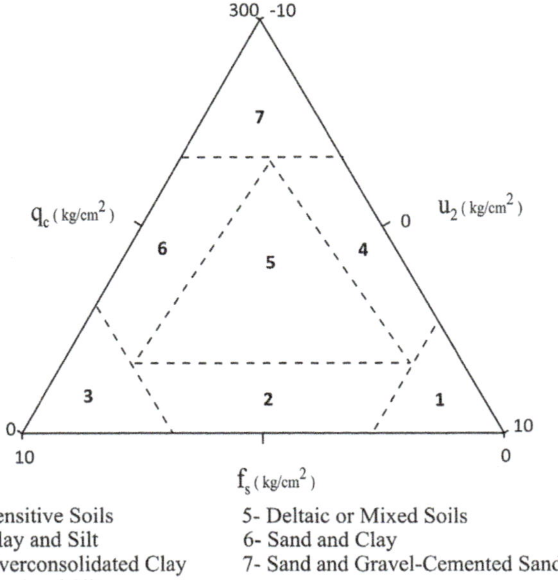

Figure 5.27 The delineated zones for the developed triangular chart using a deltaic deposit.

1. Convert the CPTu records, including q_t, f_s, and u_2, into normalized values by dividing them by a reference stress. In this case, the reference stress is equivalent to atmospheric pressure, which is 1 atm = 1 kg/cm² = 100 kPa = 1 bar.
2. Locate the q_t value on the chart and draw a line parallel to the f_s axis.
3. Locate the u_2 value on the chart and draw a line parallel to the q_t axis.
4. Locate the f_s value on the chart and draw a line parallel to the u_2 axis.
5. The point of intersection of these three lines forms an isosceles triangle.
6. The center of this triangular region on the chart accurately represents the soil condition.

As shown in Figure 5.27, the developed triangular diagram consists of seven zones. The larger middle triangle is the location of mixed or deltaic soils, i.e., clay, silt, and sand, in varying proportions. As the center of the isosceles triangle moves upward, i.e., toward the middle triangle vertex, the clay ratio decreases and the proportion of sand increases. If the ratio of one of the three soil types mentioned is negligible, it will be located in one of the three trapezoidal areas depending on the other two soil types. The three smaller triangles at the edges of the developed triangular chart are used to identify sensitive soil, over-consolidated clay, sand and gravel, and cemented sand. Figure 5.28 shows the mentioned soil types on the developed triangular chart.

5.7 SUMMARY

In the realm of geotechnical engineering, the accurate characterization and classification of soils stand as pivotal factors influencing the success of design and construction

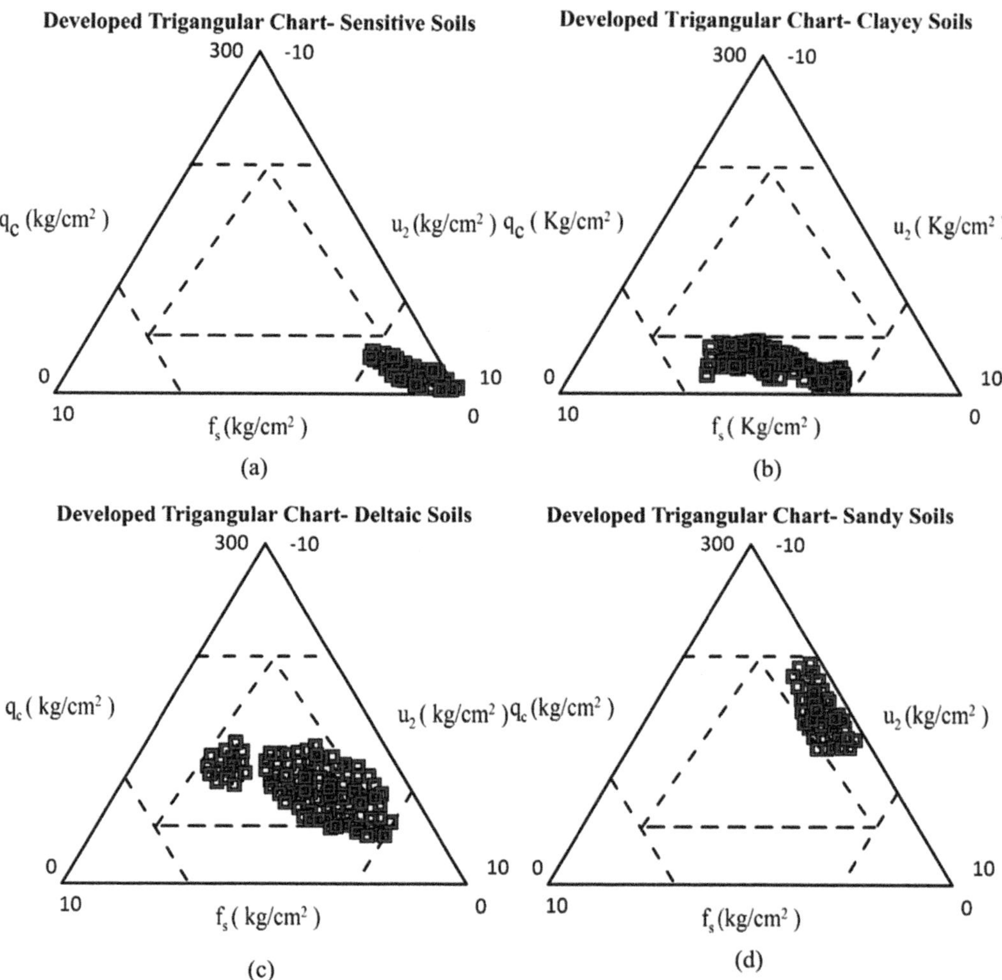

Figure 5.28 Various soil types on developed triangular chart: (a) sensitive soils, (b) clay, (c) deltaic soil, and (d) sand.

projects. This chapter delves into the profound significance of leveraging databases to assess soil properties in geotechnical engineering, highlighting the crucial role of geotechnical site investigation in acquiring vital insights into soil characteristics and behavior. The various methods explored encompass remote sensing, data review, site survey, testing, and monitoring.

The narrative emphasizes the inherent uncertainty and variability intrinsic to soil properties, underscoring the imperative to source geotechnical data from diverse channels. These uncertainties are categorized into natural variations in material properties and an incomplete understanding of the system. Addressing the former requires more testing and robust databases, while the latter involves tackling measurement errors, transformation uncertainties, and model errors through enhanced experiments and accurate models. The narrative underscores recent advancements in uncertainty quantification, machine learning, Bayesian algorithms, and system-level analysis as key elements in enhancing geotechnical decision making, supporting resilience engineering, and advancing digital transformation.

Turning attention to in-situ tests, the chapter elucidates their significant role in soil identification and classification within geotechnical engineering. Notably, the Cone Penetration Test (CPT), Standard Penetration Test (SPT), Field Vane Shear Test (FVST), and Dilatometer Penetration Test (DMT) are underscored as pivotal contributors to understanding soil behavior. The narrative accentuates the development of correlations and charts that aid in classifying soils based on test results, thus elevating the accuracy of soil type identification. Additionally, it sheds light on recent advancements, including triangular classification charts and behavior-driven approaches, providing more precise methodologies for characterizing soil properties and thereby enhancing the reliability of geotechnical investigations and foundation design.

The exploration extends to the prediction of geotechnical parameters through in-situ tests, encompassing soil engineering parameters, index parameters, strength parameters, stiffness, and compressibility parameters. The narrative navigates the intricate interpretation of in-situ test results, elucidating the use of empirical correlations to estimate geotechnical properties. Key parameters such as unit weight, relative density, internal friction angle, undrained shear strength, stiffness, constrained modulus, over-consolidation ratio, and consolidation indexes are meticulously covered. Additionally, the chapter addresses shear wave velocity and its correlation with shear modulus, sensitivity, and permeability. A wealth of empirical correlations and insights are provided to facilitate parameter estimation in geotechnical engineering applications.

Finally, the chapter delves into the AUT: CPTu-Geo-marine Database, meticulously crafted by Eslami et al. (2022). This comprehensive repository comprises 398 cases from 58 locations, offering invaluable insights into geotechnical and marine data. The chapter introduces Soil Behavioral Classification methods, categorizing diagrams into basic, indirect, and direct types. The performance of these diagrams is rigorously evaluated through 57 case studies, with statistical assessments and radar charts providing a comprehensive overview. A notable addition is the introduction of a novel triangular chart, enhancing accuracy in classifying marine deposits based on Cone Penetration Testing parameters. This chapter serves as a thorough exploration of geotechnical data analysis and classification techniques, marking a significant contribution to the field.

REFERENCES

Aas, G., Lacasse, S., Lunne, T., Hoeg, K. (1986). Use of in situ tests for foundation design on clay. In: *Proceedings of In Situ '86: Use of In Situ Tests in Geotechnical Engineering*, Virginia, pp. 1–30.

AASHTO. (2020). *LRFD Bridge Design Specifications*, 9th ed. American Association of State Highway and Transportation Officials, Washington, DC.

Ameratunga, J., Nagaratnam Sivakugan, N., Das, B. M. (2016). *Correlations of Soil and Rock Properties in Geotechnical Engineering*. Springer, India, New Delhi.

Anagnostopoulos, A., Koukis, G., Sabatakakis, N., Tsiambaos, G. (2003). Empirical correlation of soil parameters based on cone penetration tests (CPT) for Greek soils. *Geotechn Geol Eng*, 21(4), 373–387.

Azzous, A. S., Krizek, R. J., Corotis, R. B. (1976). Regression Analysis of Soil Compressibility. *Soils Found*, 16(2), 19–29.

Baguelin, F., Jezequel, J. F., Le Mee, E., Le Mehaute, A. (1972). Expansion of cylindrical probes in cohesive soils. J Soil Mech Found Eng Div ASCE, 98(SM11), 1129–1142.
Balasubramaniam, A. S., Brenner, R. P. (1981). Chapter 7: Consolidation and settlement of soft clay. In: Brand EW, Brenner RP (eds) *Soft Clay Engineering*. Elsevier, Amsterdam, pp. 481–566.
Baldi, G., Bellotti, R., Ghionna, V., Jamiolkowski, M., Pasqualini, E. (1986). Interpretation of CPT's and CPTU's. 2nd Part: drained penetration. In: *Proceedings 4th International Geotechnical Seminar*, Singapore, pp. 143–156.
Baldi, G., Jamiolkowski, M., Lo Presti, D.C.F., Manfredini, G., Rix, G.J. (1989). Italian experience in assessing shear wave velocity from CPT and SPT. In: *Proceedings of Discussion Session on Influence of Local Conditions on Seismic Response*, vol. XII. ICSMFE, Rio De Janerio, pp. 157–168.
Begemann, H. K. S. (1965). The friction jacket cone as an aid in determining the soil profile. In: *Proceedings of the 6th International Conference on Soil Mechanics and Foundation Engineering*, ICSMFE, Montreal, September 8–15, vol. 2, pp. 17–20.
Bergdahl, U., Ottosson, E., Malmborg, B. S. (1993). *Plattgrundlaggning. AB Svensk Byggtjanst*, Stockholm.
Bowles, J. E. (1996). *Foundation analysis and design*, 5th ed. McGraw-Hill, New York.
Cai, G. J., Liu, S. Y., Puppala, A. J., Tong, L. Y. (2015). Identification of soil strata based on general regression neural network model from CPTU data. *Marine Georesources & Geotechnology*, 33(3), 229–238.
Campanella, R. G., Robertson, P. K., Gillespie, D., Grieg, J. (1985). Recent development in in-situ testing of soils. In: *Proceedings of 11th International Conference on Soil Mechanics and Foundation Engineering*, vol. 2, ICSMFE, San Francisco, CA, pp. 849–854.
Canadian Foundation Engineering Manual, CFEM, 4th ed., 2006. Canadian Geotechnical Society, BiTech Publishers, Vancouver, p. 488.
Chandler, R. J. (1988). *The in-situ measurement of the undrained shear strength of clays using the field vane. Vane shear strength testing of soils: field & lab studies*, STP1014, ASTM, West Conshohocken, pp. 13–44.
Chanmee, N., Chai, J., Hino, T., Wang, J. (2017). Methods for evaluating overconsolidation ratio from piezocone sounding results. *Underground Space*, 2(3), 182–194.
Chen, H., Mo, P. Q. (2022). An undrained expansion solution of cylindrical cavity in SANICLAY for K0 consolidated clays. *J Rock Mech Geotechn Eng*, 14, 922–935.
Ching, J. and Phoon, K.K. (2013). Probability distribution for mobilized shear strengths of spatially variable soils under uniform stress states. *Georisk*, 7(3), 209–224.
Ching, J., Li, D.Q., and Phoon, K.K. (2016). *Statistical Characterization of Multivariate Geotechnical Data*. Chapter 4, Reliability of Geotechnical Structures in ISO2394, CRC Press, Balkema, pp. 89–126.
Ching, J., Wu, S., Phoon, K.K. (2021). Constructing quasi-site-specific multivariate probability distribution using hierarchical Bayesian model. *J Eng Mech ASCE* 147(10), 04021069.
Cozzolino, V. M. (1961). Statistical forecasting of compression index. In: *Proceedings of the 5th ICSMFE*, Paris, pp. 51–53.
Das, M. B. (2006). *Principles of Geotechnical Engineering*, 7th ed. Cenghage Learning, United States.
Das, M. B. (2016). *Fundamentals of Geotechnical Engineering*, 5th ed. Cenghage Learning, United States.
Dikmen, U. (2009). Statistical correlations of shear wave velocity and penetration resistance for soils. *J Geophys Eng*, 6(1), 61–72.
Djoenaidi, W. J. (1985). *A compendium of soil properties and correlations*. MEngSc thesis, University of Sydney, Australia.
Douglas, B. J., Olsen, R. S. (1981). Soil classification using electric cone penetrometer. In: American Society of Civil Engineers, ASCE, *Proceedings of Conference on Cone Penetration Testing and Experience*, St. Louis, 1981, October 26–30, pp. 209–227.

Elnaggar, M. A., Krizek, R. J. (1970). Statistical approximation for consolidation settlement. *Highway Research Record*, no. 323, HRB, pp. 87–96.

Eslaamizaad, S., Robertson, P. K. (1997). Evaluation of settlement of footings on sand from seismic in-situ tests. In: *Proceedings of the 50th Canadian Geotechnical Conference*, Ottawa, Ontario, vol. 2, pp. 755–764.

Eslami A., Heidarie Golafzani S., Moshfeghi S. (2020). CPT and pile database for performance-based design of pile axial bearing capacity. In: *Proceedings of the 45th Annual Conference on Deep Foundations, Deep Foundation Institute (DFI)*, Oxon Hill, MD, p. 12.

Eslami, A., Alimirzaei, M., Aflaki, E., Molaabasi, H. (2016). Deltaic soil behavior classification using CPTu records-proposed approach and applied to fifty-four case histories. *Marine Georesour Geotechnol* 35(1), 62–79.

Eslami, A., Fellenius, B. H. (1997). Pile capacity by direct CPT and CPTU data. In: *Year 2000 Geotechnics, Geotechnical Engineering Conference*, AIT Bangkok, 18pp.

Eslami, A., Fellenius, B. H. (2004). *CPT and CPTu Data for Soil Profile Interpretation: Review of Methods and a Proposed New Approach.*

Eslami, A., Heidarie Golafzani, S., Naghibi, M. H. (2022). Developed triangular charts; deltaic CPTu-based soil behavior classification using AUT: CPTu-Geo-Marine Database. *Probabilistic Engineering Mechanics*, 71, 103380.

Eslami, A., Mohammadi, A. (2016). Drained soil shear strength parameters from CPTu data for marine deposits by analytical model. *Ships Offshore Struct*, 11(8), 913–925.

Eslami, A., Moshfeghi, S., Molaabasi, H., Eslami, M.M. (2019). *Piezocone and Cone Penetration Test (CPTu and CPT) Applications in Foundation Engineering*, Butterworth-Heinemann.

Federal Geographic Data Committee, *Coastal and Marine Ecological Classification Standard*, Publication# FGDC-STD-018-2012, 2012.

Frank, R., Magnan, J.P. (1995). Cone penetration testing in France: national report. In: *Proceedings CPT'95, Linkoping, Swedish Geotechnical Society*, vol. 3, pp. 147–156.

Gibbs, H. J., Holtz, W. G. (1957). Research on determining the density of sand by spoon penetration testing. In: *Proceedings, 4th International Conference on Soil Mechanics and Foundation Engineering*, London, I, pp. 35–39.

Goldberg, G. D., Lovell, C. W., Miles, R. D. (1979). Use of geotechnical data bank, *Transportation Research Record*, No. 702, TRB, pp. 140–146.

Hasancebi, N., Ulusay, R. (2007). Empirical correlations between shear wave velocity and penetration resistance for ground shaking assessments. *Bull Eng Geol Environ*, 66(2), 203–213.

Hatanaka, M., Uchida, A. (1996). Empirical correlations between penetration resistance and internal friction angle of sandy soils. *Soils Found*, 36(4), 1–10.

Hegazy, Y. A., Mayne, P. W. (1995). Statistical correlations between vs and cone penetration data for different soil types. In: *Proceedings of the International Symposium on Cone Penetration Testing (CPT'95)*, Linkoping, Sweden, 4–5 October 1995, vol. 2. Swedish Geotechnical Society, pp. 173–178.

Holtz, W. G., Gibbs, H. J. (1979). Discussion of "SPT and relative density in coarse sand". *J Geotech Eng Div ASCE*, 105(3), 439–441.

Hough, B. K. (1957). *Basic Soils Engineering.* The Ronald Press Co., New York.

Imai, T. (1977). P- and S-wave velocities of the ground in Japan. In: *Proceedings, 9th International Conference on Soil Mechanics and Foundation Engineering*, Tokyo, Japan, 2, pp. 257–260.

Iyisan, R. (1996). Correlations between shear wave velocity and in-situ penetration test results. Chamber of Civil Engineers of Turkey, *Teknik Dergi*, 7(2), 1187–1199.

Jafari, M. K., Shafiee, A., Razmkhah, A. (2002). Dynamic properties of the fine grained soils in south of Tehran. *J Seismol Earthquake Eng*, 4(1), 25–35.

Jamiolkowski, M., Lo Presti, D. C. F., Manassero, M. (2001). Evaluation of relative density and shear strength of sands from CPT and DMT. In: *CC Ladd Symposium.* MIT, Cambridge, MA. ASCE Geotechnical Special Publication No. 119: 201–238.

Jefferies, M. G., Davies, M. P. (1991). Soil classification using the cone penetration test. *Discussion Canadian Geotechnical Journal*, 28(1), 173–176.

Jefferies, M. G., Davies, M. P. (1993). Use of CPTu to estimate equivalent SPT N60. American Society for Testing and Materials, *Geotechnical Testing Journal* 16(4), 458–468.

Jones, G. A., Rust, E. (1982). Piezometer penetration testing, CPTU. In: *Proceedings of the 2nd European Symposium on Penetration Testing*, ESOPT-2, Amsterdam, May 24-27, vol. 2, pp. 607–614.

Kiku, H., Yoshida, N., Yasuda, S., Irisawa, T., Nakazawa, H., Shimizu, Y., Ansal, A., Erkan, A. (2001). In-situ penetration tests and soil profiling in Adapazari, Turkey. In: *Proceedings, 15th International Conference on Soil Mechanics and Geotechnical Engineering*, TC4 Satellite Conference on Lessons Learned from Recent Strong Earthquakes, Istanbul, Turkey, pp. 259–269.

Koppula, S. D. (1981). Statistical evaluation of compression index. *Geotech Test J ASTM* 4(2), 68–73.

Ku, C. S., Juang, C. H., Ou, C. Y. (2010). Reliability of CPT Ic as an index for mechanical behavior classification of soils. *Géotechnique*, 60(11), 861–875.

Kulhawy, F. H., Mayne, P. W. (1990). *Manual on estimating soil properties for foundation design*, Report EL-6800. Electric Power Research Institute, Palo Alto.

Lacasse, S., Lunne, T. (1988). Calibration of dilatometer correlations. *Proc. ISOPT-1, Orlando*, Vol. I, pp. 539–548.

Li, J., Cassidy, M. J., Huang, J., Zhang, L., Kelly, R. (2015). Probabilistic identification of soil stratification. *Géotechnique*, 66(1), 16–26.

Lo, Y. K. T, Lovell, C. W. (1982). Prediction of soil properties from simple indices, *Transportation Research record*, No. 873, Overconsolidated clays: Shales, TRB, pp. 43–49.

Look, B. G. (2007). *Handbook of Geotechnical Investigation and Design Tables*. 1st ed., Taylor & Francis, London.

Lunne, T., Robertson, P. K., Powell, J. J. M. (1997). *Cone Penetration Testing in Geotechnical Practice*. Blackie Academic & Professional/Chapman-Hall Publishers, London, 312pp.

Madiai, C., Simoni, G. (2004). Shear wave velocity-penetration resistance correlation for Holocene and Pleistocene soils of an area in central Italy. In: Viana da Fonseca, Mayne (Eds.), *Proceedings ISC-2 on Geotechnical and Geophysical Site Characterization*. Mill Press, *Rotterdam*, pp. 1687–1694.

Marchetti, S. (1980). In situ tests by flat dilatometer. *J Geotech Eng Div ASCE*, 106(GT3), 299–321.

Mayne, P. W. (2006). Undisturbed sand strength from seismic cone tests. *Geomech Geoeng* 1(4), 239–257.

Mayne, P. W. (2007). Cone penetration testing State-of-Practice. In: *NCHRP Synthesis. Transportation Research Board Report Project 20-05*, 118pp.

Mayne, P. W. (2014). Interpretation of geotechnical parameters from seismic piezocone tests. In: *3rd International Symposium on Cone Penetration Testing*. Las Vegas, NV.

Mayne, P. W., Kemper, J. B. (1988). Profiling OCR in stiff clays by CPT and SPT. *Geotechn Test J*, 11(2), 139–147.

Mayne, P. W., Rix, G. J. (1995). Correlations between shear wave velocity and cone tip Resistance in natural clays. *Soils Found*, 35(2), 107–110.

Menard, L. (1956). Mesures in situ des proprietes physiques des sols. *Ann Ponts Chaussees Paris*, No. 14, pp. 357–377.

Meyehof, G. G. (1957). Discussion on research on determining the density of sand by spoon penetration testing. In: *Proceedings, 4th International Conference on Soil Mechanics and Foundation Engineering*, London, 3, pp. 110–114.

Meyerhof, G. G. (1951). The Ultimate Bearing Capacity of Foundations. *Géotechnique*, 2, 301.

Meyerhof, G. G. (1959). Compaction of sands and the bearing capacity of piles. *J Soil Mech Found Div ASCE*, 85(6), 1–29.

Meyerhof, G. G. (1974). Ultimate bearing capacity of footings on sand layer overlying clay. *Canad Geotech J*, 11(2), 223–229.

Mitchell, J. K., Durgunoglu, H. T. (1983). Cone resistance as measure of sand strength. *J Geotechn Eng Div ASCE*. 104(GT7), 995–1012.

Mo, P. Q., Yu, H. S. (2017). Undrained cavity expansion analysis in a unified state parameter model for clay and sand. *Géotechnique* 67(6), 503–515.

Mo, P. Q. and Yu, H. S. (2018). Drained cavity expansion analysis with a unified state parameter model for clay and sand. *Canad Geotechn J* 55(7), 1029–1040.

Mo, P. Q., Cai, G. J., Wang, K. J., Eslami, A., Yu, H. S. (2024). Cavity expansion-based interpretation of piezocone penetration test (CPTu) data in clays. *Géotechnique*. Ahead of Print. https://doi.org/10.1680/jgeot.23.00045

Mo, P. Q., Chen, H., Yu, H. S. (2022). Undrained cavity expansion in anisotropic soils with isotropic and frictional destructuration. *Acta Geotechn*, 17, 2325–2346.

Mo, P. Q., Gao, X. W., Yang, W. B. and Yu, H. S. (2020). A cavity expansion–based solution for interpretation of CPTu data in soils under partially drained conditions. *Int J Numer Anal Methods Geomech* 44(7), 1053–1076.

Mo, P. Q., Marshall, A. M. and Yu, H. S. (2017). Interpretation of cone penetration test data in layered soils using cavity expansion analysis. *J Geotechn Geoenviron Eng ASCE* 143(1), 04016084.

Mola-Abasi, H., Dikmen, U., Shooshpasha, I. (2015). Prediction of shear-wave velocity from CPT data at Eskisehir (Turkey), using a polynomial model. *Near Surface Geophys* 13(2), 155–167.

Nishida, Y. (1956). A brief note on compression index of soils. *J Soil Mechan Found Div* ASCE, 82, SM3, 1027.

Ohta, Y., Goto, N. (1978). Empirical shear wave velocity equations in terms of characteristic soil indexes. *Earthquake Eng Struct Dyn*, 6, 167–187.

Okamoto, T., Kokusho, T., Yoshida, Y., Kusuonoki, K. (1989). Comparison of surface versus subsurface wave source for P-S logging in sand layer. In: *Proceedings, 44th Annual Conference*, Japan Society of Civil Engineers, 3, pp. 996–997 (in Japanese).

Olsen, R. S., Mitchell, J. K. (1995). CPT stress normalization and prediction of soil classification. In: *Proceedings of International Symposium on Cone Penetration Testing*, pp. 257–262. CPT95, Link oping, Sweden, SGI Report 3:95, 2.

Paoletti, L., Hegazy, Y., Monaco, S., Piva, R. (2010). Prediction of shear wave velocity for offshore sands using CPT data e Adriatic Sea. In: Second International Symposium on Cone Penetration Testing, Huntington Beach, CA, pp. 1–8.

Phoon, K. K., Kulhawy, F. H. (1999). Characterization of geotechnical variability. *Canad Geotechn J*, 36, 612–624.

Phoon, K. K., Kulhawy, F. H., and Grigoriu, M. D. (1995). *Reliability based design of foundations for transmission line structures*. Report TR-105000, Electric Power Research Institute, Palo Alto, CA.

Phoon, K.K., Cao, Z., Ji, J., Leung, Y.F., Najjar, S., Shuku, T., Tang, C., Yin, Z.Y., Yoshida, I., Ching, J. (2022a). Geotechnical uncertainty, modeling, and decision making. *Soils Found*, 62, 101189.

Phoon, K.K., Prakoso, W.A., Wang, Y., and Ching, J. (2016). *Uncertainty representation of geotechnical design parameters*. Chapter 3, Reliability of Geotechnical Structures in ISO2394, CRC Press: Balkema, pp. 49–87.

Piratheepan, P. (2002). *Estimating shear-wave velocity from SPT and CPT Data*, Master of Science thesis, Clemson University.

Pitilakis, K., Raptakis, D., Lontzetidis, K. T., Vassilikou, T., Jongmans, D. (1999). Geotechnical and geophysical description of euro-seistests using field and laboratory tests, and moderate strong ground motions. *J Earthquake Eng*, 3, 381–409.

Ramsey, N. (2002). A calibrated model for the interpretation of cone penetration tests (CPT's) in North Sea quaternary soils. In: *Proceedings International Conference Offshore Site Investigation and Geotechnics*, SUT 2002, London, pp. 341–356.

Robertson, P. K. (1990). Soil classification using the cone penetration test. *Canadian Geotechnical Journal* 27(1), 151–158.

Robertson, P. K. (2009). Interpretation of cone penetration tests e a unified approach. *Canadian Geotechnical Journal* 46(11), 1337–1355.

Robertson, P. K. (2010). Soil behavior type from CPT. In: *2nd International Symposium on Cone Penetration Testing*, Huntington Beach, CA, pp. 575–583.

Robertson, P. K. (2012). Interpretation of in-situ tests-some insights. Mitchell Lecture-ISC 4, 1–22.

Robertson, P. K. (2016). Cone penetration test (CPT)-based soil behaviour type (SBT) classification system — an update. *Canadian Geotechnical Journal* 53, 1910–1927.

Robertson, P. K., Cabal, K. L. (2012). *Guide to Cone Penetration Testing for Geotechnical Engineering*, 6th edn. Gregg Drilling & Testing, Inc., Signal Hill, CA.

Robertson, P. K., Campanella, R. G. (1983). Interpretation of Cone Penetration Tests. Part I: Sand, *Canadian Geotechnical Journal*, 20(4), 718–733.

Robertson, P. K., Campanella, R. G., Gillespie, D., Grieg, J. (1986). Use of piezometer cone data. In: Clemence, S. (Ed.), *Proceedings of American Society of Civil Engineers, ASCE, In-Situ 86 Specialty* Conference, pp. 1263–1280. Blacksburg, June 23–25, Geotechnical Special Publication GSP No. 6.

Robertson, P.K., Cabal, K. (2010). Estimating soil unit weight from CPT. In: *2nd International Symposium on Cone Penetration Testing*, pp. 2–40.

Robertson, P.K., Campanella, R.G., Gillespie, D., Greig, J. (1986). Use of piezometer cone data. In: *Use of in Situ Tests in Geotechnical Engineering*, ASCE, pp. 1263–1280.

Sanglerat, G. (1972). *The Penetrometer and Soil Exploration*. Elsevier Pub, Amsterdam, 488pp.

Schmertmann, J. H. (1975). Measurement of in-situ shear strength. In: *Proceedings of ASCE Specialty Conference on In-Situ Measurements of Soil Properties*, Raleigh, NC, vol. 2, pp. 57–138.

Schmertmann, J. H. (1978). *Guidelines for Cone Test, Performance, and Design*. Report FHWATS-78209. Federal Highway Administration, Washington, 145pp.

Schmertmann, J. H. (1991). The mechanical aging of soils. *J Geotech Eng ASCE*, 117(9), 1288–1330.

Schnaid, F. (2009). In: *Situ Testing in Geomechanics*. Taylor & Francis, London, 329pp.

Seed, H. B., Idriss, I. M. (1981) Evaluation of liquefaction potential of sand deposits based on observations and performance in previous earthquakes, Preprint No. 81-544, *In situ testing to evaluate liquefaction susceptibility, Session No. 24, ASCE Annual Conference*, St. Louis, MO.

Senneset, K., Sandven, R., Janbu, N. (1989). The evaluation of soil parameters from piezocone tests. In: *Proceedings In Situ Testing of Soil Properties for Transportation Facilities, Research Council*, Trans. Research Board, Washington, DC.

Skempton, A. W. (1944). Notes on the Compressibility of Clays. *Quarterly Journal of Geological Society of London*, 100, 119–135.

Sowers, G. F. (1979). *Introductory Soil Mechanics and Foundations: Geotechnical Engineering*. 4th ed. Macmillan, New York.

Stroud, M. A. (1974). The Standard Penetration test in insensitive clays and soft rocks. *Proc. Eur Semin Penet Test*. Vol. 2:2, pp. 366–375. Stockholm.

Stroud, M. A. (1989). The Standard Penetration Test – its application and interpretation. In: *Proceedings of the ICE Conference on Penetration Testing in the UK*, Thomas Telford, London.

Sykora, D. E., Stokoe, K. H. (1983). Correlations of in-situ measurements in sands of shear wave velocity. *Soil Dyn Earthquake Eng*, 20, 125–136.

Tabarroki, M., Ching, J., Phoon, K.K., Chen, Y.Z. (2022). Mobilisationbased characteristic value of shear strength for ultimate limit states. *Georisk: Assessment and Management of Risk for Engineered Systems and Geohazards* , 16, 413–434. https://doi.org/10.1080/17499518.2020.1859121.

Tang, C., Phoon, K.K. (2021). *Model Uncertainties in Foundation Design.* CRC Press, Boca Raton, Florida, US.

Terzaghi, K., Peck, R. (1948). *Soil Mechanics in Engineering Practice.* Wiley, New York.

Terzaghi, K., Peck, R. (1967). *Soil Mechanics in Engineering Practice*, 2nd edn. Wiley, New York.

Trevor, F. A., Mayne, P. W. (2004). Undrained shear strength and OCR of marine clays from piezocone test results. In: Viana da Fonseca, M. (Eds.), *Proceeding ISC-2 on Geotechnical and Geophysical Site characterization.* Mill Press, Rotterdam, pp. 391–398.

Tun, M. (2003). *Investigation of the Characteristics of Eskisehir Soils Due to Shear Wave Velocity and Determination of Their Fundamental Vibration Periods.* Master Thesis. Anadolu University Institute of Science and Technology Department of Physics (in Turkish).

U.S. Navy. (1982). *Soil mechanics* – design manual 7.1, Department of the Navy, Naval Facilities Engineering Command, U.S. Government Printing Office, Washington, DC.

USACE. (1990). *Engineering and design – settlement analysis*, EM1110-1-1904, Department of the Army, US Army Corps of Engineers.

Uzielli, M., Mayne, P. W., Cassidy, M. J. (2013). Probabilistic assessment of design strengths for sands from in-situ testing data. In: *Modern Geotechnical Design Codes of Practice*, eds. Patrick Arnold, Gordon A. Fenton, Michael A. Hicks, Timo Schweckendiek, Brian Simpson, IOS Press, Amsterdam, Netherlands. pp. 214–227.

Vesic, A. S. (1977). *Design of Pile Foundations, National Cooperative Highway Research Program*, Transportation Research Board, National Research Council, National Academy of Sciences, Washington, Synthesis of Highway Practice No. 42.

Vessia, G., Zhou, Y. G., Leung, Y. F., Pula, W., Di Curzio, D., Tabarroki, M., and Ching, J. (2021). *Numerical evidences for worstcase scale of fluctuation. TC304 State-of-the-art Review of Inherent Variability and Uncertainty*, pp. 204–215.

Windle, D., Wroth, C. P. (1977). The use of a self-boring pressuremeter to determine the undrained properties of clays. *Ground Eng*, 10(6), 37–46.

Wolff, T. F. (1989). Pile capacity prediction using parameter functions. In: *Predicted and Observed Axial Behavior of Piles, Results of a Pile Prediction Symposium*, ASCE Geotechnical Special Publication 23, pp. 96–106.

Yoshida, Y., Ikemi, M., Kokusho, T. (1988). Empirical formulas of SPT blow-counts for gravelly soils. In: *Proceedings*, 1st International Symposium on Penetration Testing, Orlando, Florida, 1, pp. 381–387.

Yu, H. S. (2000). *Cavity Expansion Methods in Geomechanics*, Kluwer Academic, Dordrecht.

Zhang, P., Yin, Z.Y., Jin, Y.F. (2022). Bayesian neural network-based uncertainty modelling: application to soil compressibility and undrained shear strength prediction. *Canad Geotechn J* 59(4), 546–557.

Zhang, P., Yin, Z.Y., Jin, Y.F., Chan, T.H.T. (2020). A novel hybrid surrogate intelligent model for creep index prediction based on particle swarm optimization and random forest. *Eng Geol* 265, 105328.

Chapter 6

New laboratory database of hydraulic conductivity measurements on fine-grained soils

Shuyin Feng and Paul J. Vardanega

Geotechnical databases are vital for engineers wishing to make well-informed, empirical estimations of key soil mechanics parameters. Transformation models linking more difficult-to-measure parameters from readily attainable ones allow engineers to make low-cost a-priori estimations of the more difficult-to-measure parameters. Databases and associated transformation models allow for new test data to be benchmarked against prior knowledge. Other smaller databases (which may represent a specific data subset) can also be benchmarked against larger ones. The use of databases is especially valuable for assessing permeability (or hydraulic conductivity) – an important parameter that exhibits a very large range of variation across different soil types and within datasets obtained from individual laboratory testing campaigns. The issue of parameter uncertainty is often difficult to discount when hydraulic conductivity is a key parameter in a geotechnical model due to said variation. This chapter provides a comprehensive review of the variability and uncertainty of soil hydraulic conductivity with a focus on fine-grained materials, followed by a comparative study on the compilation of fine-grained soil hydraulic conductivity databases and transformation models for hydraulic conductivity making use of the recently established soil hydraulic conductivity database, FG/KSAT-1358. This chapter concludes with recommendations for future development and analysis of soil hydraulic conductivity databases.

6.1 PERMEABILITY OF SOILS

6.1.1 Definitions

Water, as one of the three phases of soil, governs both the physical and mechanical properties of the material. Understanding the distribution and percolation of water (or other permeants) in the soil is of crucial importance in geotechnical engineering design work since it is associated with many geotechnical phenomena (e.g., Taylor 1948, Lambe & Whitman 1969). The percolation of water through soil is often characterised by the hydraulic conductivity k (Length, L. Time, T^{-1}), described in Darcy's empirical law (e.g., Darcy 1856, Taylor 1948, Craig 2004):

$$\frac{Q}{A} = v = ki \tag{6.1}$$

where Q is the flow rate ($L^3.T^{-1}$), A is the area of cross section of the soil (L^2), v is the superficial velocity ($L.T^{-1}$), and i is the hydraulic gradient in the flow direction. The hydraulic conductivity k is governed by both the soil and the permeating fluid (i.e., water), and can be expressed as:

DOI: 10.1201/9781003441946-6

$$k = K\frac{\gamma}{\mu} \tag{6.2}$$

where K is the intrinsic permeability (L^2) which is independent of the permeating pore fluid, γ is the unit weight of the pore fluid (Mass, $M.L^{-2}.T^{-2}$), and μ is the dynamic viscosity ($M.L^{-1}.T^{-1}$). The reader is also directed to the doctoral thesis of Feng (2022) for preliminary work on some of the topics presented in this chapter.

6.1.2 Variation of hydraulic conductivity

Compared to the other commonly used geotechnical parameters, one of the most notable characteristics of k is its extensive range of variation (e.g., Carrier & Beckman 1984, Mbonimpa *et al.* 2002). The k value reported in soil mechanics textbooks generally ranges from 1×10^{-11} m/s to 1×10^{0} m/s, i.e., over 11 orders of magnitude (cf., Terzaghi & Peck 1948, Lambe & Whitman 1969, Craig 2004), and ranges up to 14 orders of magnitude in some compiled hydraulic conductivity datasets (databases) (cf., Ren & Santamarina 2018, Feng 2022, Feng *et al.* 2023).

Table 6.1 summarises the parameter variation range of four global multivariate databases focusing on fine-grained/clay soils in 304dB (the TC304 database of databases that can be found at: http://140.112.12.21/issmge/tc304.htm). The collected soil parameters in the databases encompass basic physical properties (e.g., void ratio (e), specific gravity (G_S), water content (w)), soil classification properties (e.g., liquid limit w_L, plasticity index I_P, liquidity index I_L), mechanical properties (e.g., compression index C_c, unload–reload index C_{ur}, over-consolidation ratio OCR, sensitivity S_t, undrained shear strength s_u), and hydraulic conductivity k. The variation range of k for fine-grained soil (FG/KSAT-1358) spans over 7 orders of magnitude, while the corresponding physical and intrinsic soil parameters within the same database show variations within 2 orders of magnitude. The variation of the soil properties listed in Table 6.1 is mostly within 3 orders of magnitude across four global fine-grained soil/clay databases available from the TC304 website mentioned above (with the exception of k).

6.1.3 Factors influencing hydraulic conductivity

As both γ and μ are temperature-dependent parameters, a specific soil with a specific permeating fluid may exhibit different k levels under different **testing temperatures**, i.e., k measured at 0°C is 1.8 times the value measured at 20°C, and k at 40°C is 0.58 times the value measured at 20°C (data from BSI 1990). The hydraulic conductivity measured at a specific test temperature can be converted to a corresponding permeability at different temperatures based on Equation (6.2), using a correction factor (e.g., BSI 2019). A test temperature of 20°C is often deemed as the standard or reference temperature condition for laboratory measurement (e.g., ASTM 2022, BSI 1990).

The permeant percolation mechanism in most geomaterials exhibits an anisotropic nature. The **flow direction** during testing is another factor that influences the measured level of k. The horizontally measured hydraulic conductivity k_h is generally greater than the vertically measured hydraulic conductivity k_v as reported in various studies (e.g., Al-Tabbaa & Wood 1987, Dewhurst *et al.* 1996, Clennell *et al.* 1999). The reported anisotropy ratio ($r_k = k_h/k_v$) typically falls below 4, and the maximum r_k observed in most engineering soil is around 2.5 (Liu, 2015). The anisotropic behaviour in soil k is often less

Table 6.1 Soil properties variation in some global multivariate fine-grained soil database (see: http://140.112.12.21/issmge/tc304.htm[a] for the downloadable database files, available at the time of writing)

Database	*Reference*	*Soil type*	*Data points*	*No. studies/sites*		*Range of parameters*
FG/KSAT-1358	Feng and Vardanega (2019a, b)	Fine-grained soil	1358	33	k	1.44×10^{-13} to 7.5×10^{-6} m/s
					e	0.19 to 8.57
					w_L	22 to 675%
					I_P	5 to 625.9%
					G_S	2.09 to 2.9
CLAY-C_C/6/6203	Ching *et al.* (2022)	Clay	6203	429	e	0.13 to 14.66
					w_L	19 to 612.69%
					I_P	5 to 493%
					w	5.8 to 559.5%
					C_c	0.01 to 14.50
					C_{ur}	0.002 to 2.16
CLAY/10/7490	Ching and Phoon (2014)	Clay	7490	251	w_L	18.1 to 515 %
					I_P	1.9 to 363%
					I_L	−0.75 to 6.45
					q_{t1}	0.48 to 95.98
					q_{tu}	0.61 to 108.20
					B_q	0.01 to 1.17
					S_t	1 to 1467
					OCR	1.0 to 60.23
CLAY/5/345	Ching and Phoon (2012)	Clay	345	37	I_L	0.14 to 4.85
					s_u^{re}	0.06 to 33.69 kPa
					$s_{u(test)}$	1.97 to 162.95 kPa
					$s_{u(mob)}$	1.19 to 162.52 kPa
					σ_v'	3.70 to 364.09 kPa
					σ_p'	7.94 to 712.71 kPa

Definitions: e = void ratio; w_L = liquid limit; I_P = plasticity index; G_S = Specific gravity; w = water content; C_c = compression index; C_{ur} = unload–reload index; I_L = liquidity index; q_{t1} = $(q_t - s_v)/s_v'$ = normalised cone tip resistance; q_{tu} = $(q_t - u_2)/s_v'$ = effective cone tip resistance; B_q = pore pressure ratio = $(u_2 - u_0)/(q_t - s_v)$; S_t = sensitivity; OCR = overconsolidation ratio; s_u = undrained shear strength; s_u^{re} = remoulded s_u; σ_v' = vertical total stress; and σ_p' = preconsolidation stress.

[a]This website was last accessed on 10 July 2023.

pronounced at a higher void ratios, e.g., for static compacted sand, $r_k = 1.8$ for $e = 0.4$ and $r_k = 1$ for $e = 0.8$ (Chapuis *et al.* 1989).

The permeability of soil, when tested under different **saturation levels**, is closely dependent on the air–water interface (Baver 1948). The k value measured under partially saturated conditions is generally lower than under saturated conditions (e.g., Corey 1957, Lambe & Whitman 1969, Bear 1972). The unsaturated k may be estimated from the saturated k value in conjunction with the soil water retention curve (correlation between soil suction and its degree of saturation) and hydraulic conductivity function (relationship between k and soil suction) (e.g., Mualem 1976, Van Genuchten 1980, Zhai *et al.* 2021).

In addition to the aforementioned factors, the **chemical constituent**(s) of the permeant can also have a substantial influence on the k measured in soil with high clay content (see also the discussion in Feng 2022). The presence of certain chemical constituents in the permeant may cause the soil particles to either flocculate or disperse, thus leading to changes in hydraulic conductivity (e.g., Lambe & Whitman 1969, Mesri & Olson 1971). The ratio of hydraulic conductivity for the same soil, measured using permeants with different chemical constituents at a constant void ratio, can vary significantly by several orders of magnitude, e.g., measured k of smectite at e = 2 is around 10^{-11} cm/s in water with NaCl, 10^{-9} cm/s in water with $CaCl_2$, 10^{-7}cm/s in ethyl alcohol, and 10^{-5} cm/s (Mesri & Olson 1971).

6.1.4 Uncertainty in permeability test results

Soil permeability can be measured with various laboratory and field tests (e.g., falling head tests, constant head tests, oedometer test, and well pumping tests; see the detailed discussion in Feng 2022). The application of different permeability testing methods has been discussed in various publications (e.g., Tavenas *et al.* 1983, Chapuis 2012, Nagy *et al.* 2013, Daniel *et al.* 1985). While each permeability testing approach has its merits (and limitations), they are all established based on specific assumptions, e.g., Darcian flow, homogeneous soil, and soil saturation. The results obtained from the tests are dependent on the selected testing methodology (Chapuis 2012). The inherent "inaccuracy" of permeability measurement approaches leads to a degree of variation close to an order of magnitude from the same test method (e.g., Pane *et al.* 1983, Nagy *et al.* 2013), and up to three orders of magnitude higher when comparing results from different test methods (cf. Moulton & Seals 1979, Chapuis 2012, Nagy *et al.* 2013). The range of variation of measured permeability data coupled with laborious and time-consuming testing procedures (see Feng 2022 for further discussion of soil testing methodologies) make the determination of soil permeability challenging – this challenge can be partially mitigated by the availability of extensive geotechnical databases.

6.2 TRANSFORMATION MODELS

6.2.1 Predictors of hydraulic conductivity

Transformation models can be used to estimate soil parameters using more easily accessible ones. Similar to all the other geotechnical parameters, the k of soil can also be assessed with more basic soil parameters using empirical correlations established from past database studies. The main soil characteristics that influence the hydraulic conductivity k are summarised as follows:

(a) **Void ratio/structure-related parameters:** Void ratio e, porosity n, water content w, void ratio function $e^3(1+e)$ (e.g., Kozeny 1927, Carman 1937, Carman 1939, Taylor 1948, Terzaghi & Peck 1948, Baver 1948, Lambe & Whitman 1969, Carrier 2003, Chapuis 2004, Liu 2015, Ren & Santamarina 2018, Nagaraj *et al.* 1993, Nagaraj *et al.* 1994, Mbonimpa *et al.* 2002);

(b) **Particle size distribution-related parameters:** Effective particle size D_{10} D_{50} (e.g., Hazen 1895, Taylor 1948, Shahabi *et al.* 1984, Shepherd 1989, Alyamani & Sen 1993, Chapuis 2004), specific surface by volume S_A (Feng 2022, Feng *et al.* 2023), specific surface by mass S_S (Ren *et al.* 2016, Ren & Santamarina 2018, Feng and

Vardanega 2019a), coefficient of uniformity C_U (Shahabi *et al.* 1984), mean constriction size D_c^m (Jaafar & Likos 2014), and normalised grading entropy coordinates A, B (Feng *et al.* 2019, Feng 2022);

(c) **Atterberg Limits:** Liquid limit (w_L) (Feng & Vardanega 2019a, Feng & Vardanega 2019b, Nagaraj *et al.* 1993, Nagaraj *et al.* 1994, Chapuis 2012, Ren & Santamarina 2018), plastic limit (w_p) (Carrier & Beckman 1984, Mbonimpa *et al.* 2002), and shrinkage limit (w_s) (Sridharan & Nagaraj 2005);

(d) **Soil fabric, shape, and composition:** Soil composition (Jones 1955, Shahabi *et al.* 1984, Indrawan *et al.* 2006, Shafiee 2008, Gao *et al.* 2019), soil fabric (Lambe & Whitman 1969, Qiu & Wang 2015), and soil shape (Cabalar & Akbulut 2016, Taiba *et al.* 2019, Katagiri *et al.* 2020, Nguyen & Indraratna 2020).

Burnham and Anderson (2004) suggested that the selection of predictors should always follow the principle of "simplicity and parsimony." The complexity of the prediction model is controlled based on whether the selection of parameters can be assessed from a relatively small number of standard tests (Whittle 1993). There is always a trade-off between model accuracy and the number of constituent parameters – the more the parameters in a model, the more difficult the calibration is.

6.2.2 Model structure

A prediction model can be established using one or multiple predictors, with a generalised form of prediction model allowing further determination of the empirical factors in the model. The generalised form may be structured directly between the predictors and k using simple linear correlation, famous examples include Hazen's formula (Hazen 1985) and the Kozeny–Carman equation (Kozeny 1927, Carman 1937, Carman 1939). More recently developed prediction models often involve the data transformation of k, such as the works of Carrier and Beckman (1984), Nagaraj *et al.* (1994), Mbonimpa *et al.* (2002), and Chapuis (2012).

Feng *et al.* (2023) reviewed some commonly used prediction models for assessing the permeability of coarse-grained soil, with a compiled permeability database (CG/KSAT/7/1278) consisting of 1278 test data sourced from over 50 studies. The model adequacy analysis of the prediction models revealed that all linear-correlating models exhibited a "funnel" pattern in the residuals versus predicted plots, indicating anomalies in prediction (see Figure 6.1, full results are given in the recent paper of Feng *et al.* 2023) when calibrated using CG/KSAT/7/1278. This observation indicates the need for data transformation in the predicted value k (Montgomery *et al.* 2007, p. 303). The need for such data transformation may be attributed to the difference in the range of variation between k and corresponding predictors, as detailed in Table 6.1.

6.2.3 Evaluation of regression strength

Kulhaway and Mayne (1990) suggested that for any correlation for estimating soil properties it is essential to supplement the coefficient of determination (R^2) with other statistical measures, particularly the number of data points used in the regression (n), to properly assess the strength of fitted models (see also e.g. Paradine and Rivett 1953). Phoon and Kulhawy (1999a, 1999b) emphasised the significance of reporting the standard error (SE) to provide a measure of "transformation uncertainty" in correlations. The *p-value* of the

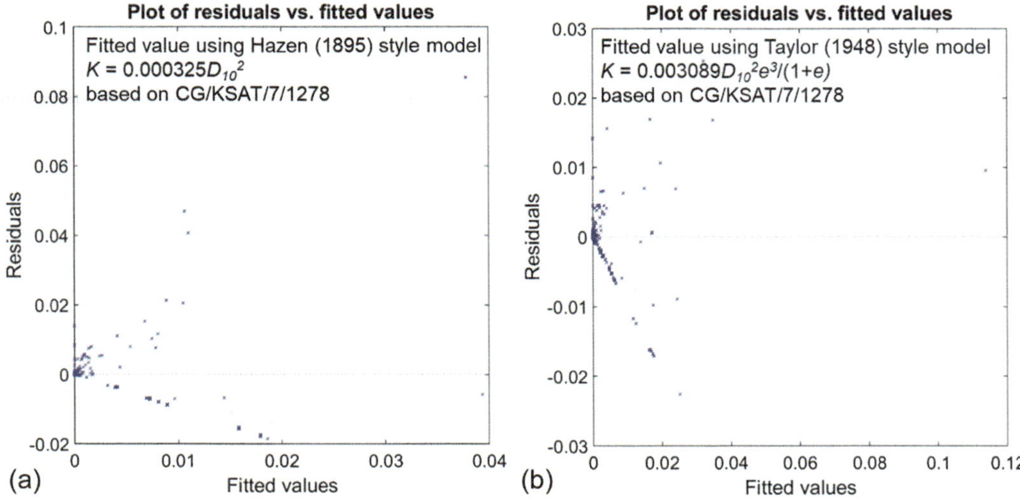

Figure 6.1 Examples of "funnel" pattern residuals versus predicted plot: (a) residual versus predicted plot using Hazen (1895) style model using CG/KSAT/7/1278; (b) residual versus predicted plot using Taylor (1948) style model using CG/KSAT/7/1278 (Source: adapted from Feng *et al.* 2023, used under the terms of the cc-by 4.0 licence).

correlation, which explains the probability of the absence of correlation (i.e., rejecting the null hypothesis, see Montgomery *et al.* 2007), should also be considered. For details on the statistical measures employed in this work, see Feng (2022).

6.2.4 Database subdivision analyses

Further subdivision analysis is often required in model development, especially when dealing with data that varies over an extensive range (e.g., k) to evaluate how generally applicable a developed transformation model is. Correlation may exhibit distinct patterns among data over different value ranges or under various test conditions, as illustrated in Figure 6.2. The subdivision analysis helps to examine the potential influencing

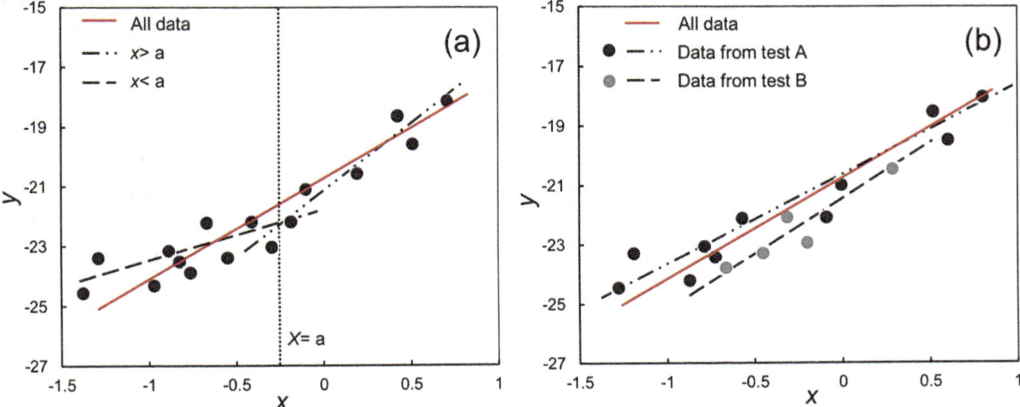

Figure 6.2 Divergency in subdivision analysis: (a) subdivision over different value ranges and (b) subdivision over different test conditions (Source: adapted from Feng 2022).

factors (e.g., permeability testing methods) and uncover underlying trends or correlations that may exist within specific subgroups which may be rather different from the transformation model calibrated with the whole dataset (see the detailed discussion in Feng 2022).

6.3 PERMEABILITY DATABASES FOR FINE-GRAINED SOIL

6.3.1 Database summary

As discussed previously, databases play a crucial role in geotechnical engineering (e.g., Phoon *et al.* 2020). Geodatabases with sufficient and reliable data are often used to improve understanding of geotechnical phenomena and further calibrate proposed models (Phoon & Tang 2019). Geotechnical databases (or datasets) can be populated through direct laboratory/on-site testing (e.g., FI-CLAY/7/216 from D'Ignazio *et al.* 2016) or sourced from geotechnical literature (e.g., RFG/TXCU-278 from Beesley & Vardanega 2020). Some databases are compiled using a combination of literature and industry sources (e.g., the DINGO database on piled foundations, Vardanega *et al.* 2024).

Table 6.2 presents a summary of eight fine-grained soil permeability database/sets used for constructing permeability prediction models, including the recently compiled database FG/KSAT-1358 (Feng & Vardanega 2019a, 2019b). These databases include datasets from individual laboratory campaigns, regional compiled databases, and global compiled databases. Each database has been supplemented with information on the size of database, soil type, property range, number of studies, and details regarding the permeability influencing factors, including the test type, test temperature, flow direction, saturation level, and chemical constituent. The property ranges are given for the soil properties that have been reported or can be digitised from the listed sources.

6.3.2 Database evaluation

All database/set sources reviewed in this chapter have offered detailed information on the soil type, reported/digitisable size, *k* range, and the number of studies. As discussed earlier in this chapter, different test types can introduce significant uncertainty and lead to variations up to orders of magnitude in the measured *k* value. However, there is a noticeable information gap concerning the permeability test type, particularly in the compiled databases, i.e., the test type is either missing or partially reported in 3 out of 5 compiled database sources. Similarly, there appear to be gaps in the reported data pertaining to other factors influencing *k*, including test temperatures, saturation levels, and permeant chemical constituents. Half of the reviewed databases (4 out of 8) provide information on one factor influencing *k*, and only two databases (Sridharan & Nagaraj 2005, Feng & Vardanega 2019a, 2019b) report relevant information on all four influencing factors.

It was noticed that more recently compiled databases tend to encompass data from earlier databases. For example, Mbonimpa *et al.* (2002) incorporated data from Nagaraj *et al.* (1994), Sivapullaiah *et al.* (2000), while Ren & Santamarina (2018) included data from Sridharan and Nagaraj (2005) and Sivapullaiah *et al.* (2000). The data sources collected to develop FG/KSAT-1258 detailed in Feng and Vardanega (2019a, 2019b) also overlap with the compiled databases in Mbonimpa *et al.* (2002) and Ren and Santamarina (2018).

Table 6.2 Summary of permeability database/set for fine-grained soil

Sources	Size	Soil type	Property range	No. of studies	Test type	Test temp.	Flow direction	Saturation level	Chemical constituent
Dataset from individual laboratory test campaign									
Nagaraj *et al.* (1994)	54	Four local residual soils: red soil, brown soil, black cotton soil, marine soil	k (9.19×10^{-12} to 1.82×10^{-9} m/s); w_L (50% to 106%); G_s (2.68 to 2.9)	1	Falling head test	n/a	Vertical, one dimensional	n/a	n/a
Sivapullaiah *et al.* (2000)	291 (no. digitised, include sand and sand mixture)	Bentonite–sand mixture (includes mixture with over 50% sand)	k (1.21×10^{-12} to 3.76×10^{-8} m/s); w_L (35% to 344%); G_s (2.60 to 2.75)	1	Consolidation test	n/a	One dimensional	n/a	n/a
Sridharan and Nagaraj (2005)	63 (no. digitised)	Red earth, silty soil, kaolinite, Cochin clay, brown soil, illitic soil, B C soil	k (1.20×10^{-11} to 6.89×10^{-8} m/s); e (0.53 to 1.84); w_L (37.0% to 73.5%); w_P (18.0% to 51.9%); w_S (11.9% to 46.4%); G_s (2.58 to 2.70)	1	Falling head test	20°C	One dimensional	Saturated	Distilled water
Regional compiled database									
Chapuis (2012)	76 (no. digitised)	Quebec Champlain Sea clay	k (2×10^{-11} to 5×10^{-9} m/s)	4	More than two types of tests involved (constant head, falling head)	n/a	n/a	Saturated	n/a
Multi-source (global) compiled database									
Carrier and Beckman (1984)	66	22 phosphatic; 13 dredged materials; 26 remoulded natural clays	k (10^{-10} to 10^{-5} m/s)	8	Constant head; consolidation test (stress-controlled and constant rate of deformation)	n/a	One dimensional	n/a	Permeating fluid with same chemical composition as the porewater

(*Continued*)

Table 6.2 (Continued) Summary of permeability database/set for fine-grained soil

Sources	*Size*	*Soil type*	*Property range*	*No. of studies*	*Test type*	*Test temp.*	*Flow direction*	*Saturation level*	*Chemical constituent*
Mbonimpa *et al.* (2002)	342	Bentonite silts mixture, Kaolin, marine soil, black cotton soil, clay, Louiseville soil, St-Esprit soil, Bächebol soil, Matagami soil, New Liskeard soil, Don Valley soil, and Leda soil	k (1.3×10^{-12} to 6.79×10^{-9} m/s); w_L (33.4% to 436.3%); e (0.51 to 5.96); G_s (2.61 to 2.78)	8	n/a; author's test conducted under constant head or falling head condition.	20°C	n/a	Saturated	n/a
Ren and Santamarina (2018)	1440 (includes sandy soil data)	Natural and remoulded sediments, from coarse sands to fine-grained clays	k (2.55×10^{-14} to 9.33×10^{-4} m/s, range for silty and clayey soil, data digitised from plot); e (0.24 to 9.1, data digitised from plot)	23	n/a	n/a	Most likely to be vertical	n/a	n/a
Feng and Vardanega (2019a)	1358	Natural or laboratory fine-grained soil with less than 50% coarse particles ($d >$ 75 mm) and a measured w_L	k (1.44×10^{-13} to 7.5×10^{-6} m/s); e (0.19 to 8.57); w_L (22% to 675%); I_P (5% to 625.9%); G_S (2.09 to 2.9)	33	Constant head test, falling head test, consolidation test, flow pump test	20 ± 1°C to 26 ± 0.1°C	Vertical	Saturated	Water

There has been an obvious increase in the size of databases for soil k due to the aforementioned research efforts. Among the compiled databases, FG/KSAT-1358 by Feng and Vardanega (2019a, 2019b) contains the largest number of k test data on fine-grained soil (over 1300), along with comprehensive details on the key factors influencing k.

6.3.3 Model calibration

Table 6.3 provides a summary of the calibrated transformation models to assess the permeability of fine-grained soil. The majority of the established models are based on fundamental soil parameters such as e, w, w_L, void ratio at liquid limit e_L, w_L, I_P, and G_S. Sridharan and Nagaraj (2005) used the shrinkage limit (w_s) as a predictor, which is widely used especially in pavement engineering in dry regions. Most of the calibrated prediction models, such as the transformation model developed by Nagaraj *et al.* (1994), Sivapullaiah *et al.* (2000), Chapuis (2012), Ren and Santamarina 2018, and Feng and Vardanega 2019a, 2019b, involve some form of data transformation for k.

Statistical measures have been reported in prediction models calibrated using individual laboratory test datasets and regional compiled databases (Nagaraj *et al.* 1994, Sivapullaiah *et al.* 2000, Sridharan & Nagaraj 2005, Chapuis 2012). However, the sample size (n) used in model calibration is missing in some cases (e.g., Sivapullaiah *et al.* 2000, Sridharan & Nagaraj 2005, Chapuis 2012). Statistical measures are largely missing (i.e., unreported) from the model calibration when using compiled databases with relatively larger sample sizes, such as Carrier and Beckman (1984) with $n = 66$, Mbonimpa *et al.* (2002) with $n = 342$, and Ren and Santamarina (2018) with $n = 1440$ (including data on sandy soils).

Transformation models calibrated with larger sample sizes, encompassing a range of soil types and sourced from multiple studies, tend to exhibit a wider range of prediction accuracy up to an order of magnitude, which is to be expected as the model has to account for a wider range of materials and testing conditions. Therefore, this transformation uncertainty may be attributed to the inherent variability among soil types and sources within the database (Phoon & Kulhawy 1999a, 1999b). Subdivision analysis has been conducted in Ren and Santamarina (2018) and Feng and Vardanega (2019a) to further analyse and verify the accuracy and applicability of the established transformation models.

6.3.4 Benchmarking databases

In order to validate and compare the predictive performance of transformation models for k calibrated using databases of varying source type, size, and data range, a benchmarking analysis is conducted using the largest fine-grained soil database – FG/KSAT-1358. Figure 6.3 summarises the analysis results for the hydraulic conductivity prediction models of fine-grained soil calibrated using database/sets with a smaller variation range of k compared to FG/KSAT-1358. The measured k values from FG/KSAT-1358 are plotted against the predicted k values based on the examined model with the corresponding calibration dataset range following the method described in Piñeiro *et al.* (2008). It was observed that models established using the global compiled database generally demonstrate better predictive performance for FG/KSAT-1358, with a higher percentage of the data points falling within the one-order-of-magnitude prediction range. Transformation models developed based on regional compiled databases and individual laboratory test campaigns tend to exhibit an overall biased prediction trend for the global compiled database FG/KSAT-1358.

Table 6.3 Summary of calibrated prediction models (data sources as shown in Table 6.2)

Sources	*Database/set*	*Calibrated transformation models*	*Reported statistical measures*	*Prediction accuracy*	*Subdivision analysis*
Dataset from individual laboratory test campaign					
Nagaraj *et al.* (1994)	Regional soil database from laboratory testing dataset on 54 samples of four local residual soils	$\frac{e}{e_L} = 2.162 + 0.195 \log k$	$r = 0.98, n = 54$	n/a	n/a
Sivapullaiah *et al.* (2000)	Laboratory testing dataset on bentonite–sand mixtures, 291 samples (no. of data digitised from plot, includes mixture with over 50% sand)	$\log_{10} k = \frac{e - 0.0535 w_L - 5.286}{0.0063 w_L + 0.2516}$; k in m/s for $w_L > 50\%$	$r = 0.94$, SE = 1.486	n/a	n/a
Sridharan and Nagaraj (2005)	Laboratory testing dataset on 10 soil types, 70 samples (no. of data digitised from plot)	$k_p = Ce^x / (1+e)$; $C = 2.5 \times 10^{-4} (I_S)^{-3.69}$ $x = 4$	$SEE = 24.34 \times 10^{-10} m/s$	Usually within 0.4 to 2.5 times range	n/a
Regional compiled database					
Chapuis (2012)	Regional clay database from 4 studies on 76 Quebec Champlain Sea clays (no. of data digitised from plot); test type not available	$k_{sat}(m/s) = 6.68 \times 10^{-6} \left[\frac{e^3}{(1+e)(w_L^{-1} + z)^2} \right]^{1.339}$; $z = 0.00836$	$R^2 = 0.81; n = 76$ (number digitised)	Within 1/3 to 3 times range	n/a
Global compiled database					
Carrier and Beckman (1984)	Compiled database from 8 studies on 66 clay samples; cover data from 2 test types (constant head, consolidation test)	$k(m/s) = \mu \frac{(e-\delta)^{\nu}}{1+e}$; $\mu(\text{m/s}) = \left[\frac{0.389}{PI} \right]^{4.29}$ $\delta = 0.027 [w_P - 0.242 I_P]$ $\nu = 4.29$	n/a	Within an order of magnitude	n/a

(Continued)

Table 6.3 (Continued) Summary of calibrated prediction models (data sources as shown in Table 6.2)

Sources	*Database/set*	*Calibrated transformation models*	*Reported statistical measures*	*Prediction accuracy*	*Subdivision analysis*
Mbonimpa *et al.* (2002)	Compiled database from 6 studies on 342 plastic/cohesive soils sample	$k_{sat}(cm/s) = C_P \frac{\gamma_w}{\mu_w} \frac{e^{3+x}}{1+e} \frac{1}{\rho_s^2 w_L^{2\chi}}$; $C_P = 5.6g^2/m^4$ $x = 7.7w_L^{-0.15} - 3$ $\chi = 1.5$ $\gamma_w \approx 9.8kN/m^3$ $\mu_w \approx 10^{-3} Pa.s$	n/a	Usually within half of one order of magnitude	n/a
Ren and Santamarina (2018)	Compiled database from 23 studies on 1440 natural and remoulded sediment sample, from coarse sand to fine-grained soil	$k(cm/s) = 10^{-5}\left(\frac{S_s}{m^2/g}\right)e^{\beta}$; $S_S = 1.8w_L - 34$ β may be derived from the power relation between k and e for the specific soil types: $\beta = 3 \pm 1$ for coarse-grained soils $\beta = 5 \pm 1$ for fine-grained soils	n/a	Mostly within 0.2 to 5 times, significant scatter below $k < 10^{-5}$ cm/s (silty/clayey soil region).	Prediction model with varying β for each soil type
Feng and Vardanega (2019a)	Compiled database from 33 studies on 1358 fine-grained soil sample	$k_{sat}(m/s) = 1.91\times10^{-9}\left(\frac{w}{w_L}\right)^{4.083}$	$R^2 = 0.62, n = 1352$, SE = 1.58, $p < 0.0001$	Mostly within an order of magnitude	Sub database analysis on Atterberg limits, test types, sample states
Feng *et al.* (2022)	Compiled database from Feng and Vardanega (2019a) with identified outliers removed	$k_{sat}(m/s) = 1.86\times10^{-9}\left(\frac{w}{w_L}\right)^{4.226}$	$R^2 = 0.65, n = 1272$, SE = 1.303, $p < 0.0001$	Mostly within an order of magnitude	

Datasets from individual laboratory test campaign

(a)

OOut of the 0.1 to 10 times range
●Within the 0.1 to 10 times range
Nagaraj et al. (1994)
72% within prediction range
42% overpredicted, 58% underpredicted
Calibration dataset range
Line of Equality
k-measured=10×k-predicted
k-measured=0.1×k-predicted
ksat-measured (m/s)
ksat-predicted (m/s)

(b)

OOut of the 0.1 to 10 times range
●Within the 0.1 to 10 times range
Sivapullaiah et al. (2000)
14% within prediction range
18% overpredicted, 82% underpredicted
Calibration dataset range
Line of Equality
ksat-measured (m/s)
ksat-predicted (m/s)

Regional compiled database

(c)

OOut of the 0.1 to 10 times range
●Within the 0.1 to 10 times range
Chapuis (2012)
6% within prediction range
98% overpredicted, 2% underpredicted
Line of Equality
ksat-measured (m/s)
ksat-predicted (m/s)

Global compiled database

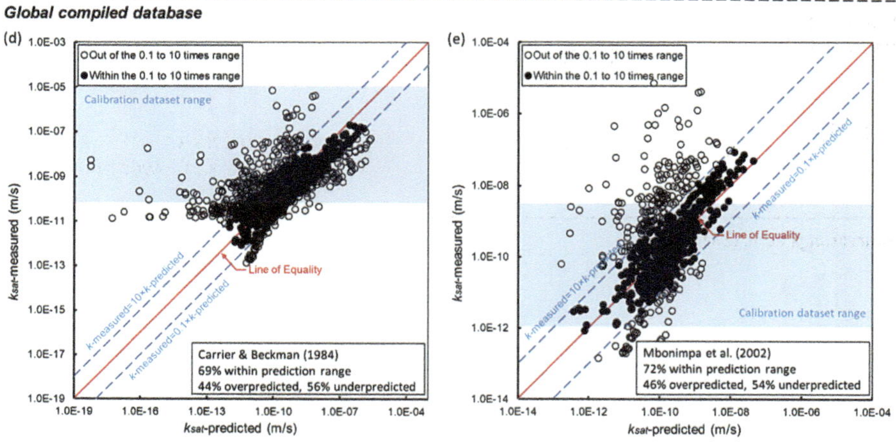

Figure 6.3 Benchmarking analysis result for models with smaller calibration dataset range using FG/KSAT-1358 as the measured values on the plots: (a) predicted values using transformation model from Nagaraj *et al.* (1994); (b) predicted values using transformation model from Sivapullaiah *et al.* (2000); (c) predicted values using transformation model from Chapuis (2012); (d) predicted values using transformation model from Carrier and Beckman (1984); (e) predicted values using transformation model from Mbonimpa *et al.* (2002).

This is to be expected as smaller datasets are unlikely to yield models representing the behaviour of a wide range of soil types/testing approaches, and this observation serves as a reminder that the use of empirical models based only on small datasets should be used with caution (even if they have a high coefficient of determination, for instance) unless calibration of the model form is undertaken with a larger and more extensive dataset.

Among the reviewed fine-grained soil database studies, the database compiled by Ren and Santamarina (2018) stands out due to its extensive variation in k value and a database size comparable to FG/KSAT-1358. Figure 6.4 presents the benchmarking analysis results using the model presented in Ren and Santamarina (2018) with $\beta = 4$, 5, and 6 (selection of β based on the range reported by Ren and Santamarina 2018: $\beta = 5 \pm 1$ for fine-grained soils), in comparison with the model developed by Feng and Vardanega using FG/KSAT-1358 (Feng and Vardanega 2019a). The prediction models by Ren and Santamarina (2018) generally underpredict k for FG/KSAT-1358, with approximately 50% of the data points within the one order of magnitude prediction range, while the model by Feng and Vardanega (2019a) yields slightly overpredicted results, with 89% of the prediction within the same prediction range. It should also be noted that while Ren and Santamarina (2018) report a β range of 5 ± 1 for fine-grained soil, individual soil types in FG/KSAT-1358 give a variation of β mostly ranging from 2 to 7 (full results in online supplement, Feng & Vardanega 2019a). This discrepancy in the β value range used in the analysis may partially account for the reduced prediction accuracy of k.

6.4 SUMMARY AND CONCLUSIONS

This chapter has explored the variability and uncertainty associated with fine-grained soil k. This was followed by a comparative analysis of the compilation of fine-grained soil k databases and the development of transformation models, using the recently established fine-grained laboratory k database, FG/KSAT-1358. Information gaps regarding the factors influencing k and transformation model prediction strength have been identified for the reviewed databases and datasets.

6.4.1 Database compilation

It is important to stress the variability and uncertainty in k measurements during the database compilation process. The measured k value of the soil is highly dependent on factors such as the test type, test condition, and sample state. Therefore, information on the relevant influencing factors, such as k testing type, test temperature, sample states, and permeating fluid, should be included in the assembled test database, together with the other corresponding soil properties where possible – this assists the development of new transformation models and also future database expansion and harmonisation efforts.

6.4.2 Transformation models

There remain information gaps in the published literature regarding the prediction strength (i.e., sample size, means of statistical measures) and accuracy (i.e., prediction bandwidth) of transformation models used to predict k. Considering the inherent uncertainty and variability of soil k assessment, adequate model evaluations that include

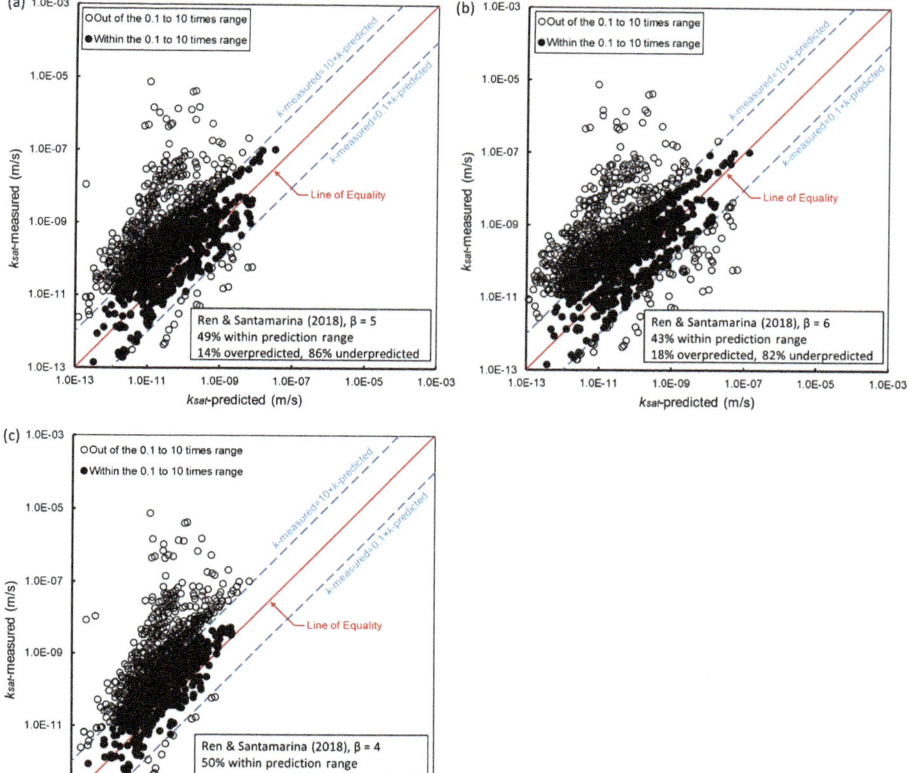

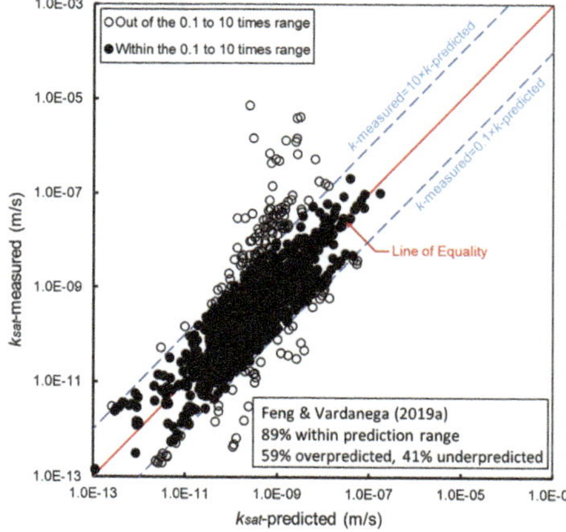

Figure 6.4 Benchmarking analysis results (using data from FG/KSAT-1358 as the measured data on the plots) for Ren and Santamarina (2018) model with varying β: (a) predicted values using transformation model from Ren and Santamarina (2018) with β = 5; (b) predicted values using transformation model from Ren and Santamarina (2018) with β = 6; (c) predicted values using transformation model from Ren and Santamarina (2018) with β = 4; (d) Predicted values using transformation model from Feng and Vardanega (2019a), data from FG/KSAT-1358, plot based on Feng *et al.* (2022).

statistical measures, prediction accuracy, and model adequacy should be provided to measure the validity and reliability of the calibrated model. For compiled databases, further subdivision analysis should also be performed to examine the variations in model prediction performance across different soil types, testing conditions, and sample states to further evaluate the uncertainty in the calibrated transformation model.

With the digitisation of geotechnical resources, significant efforts have been made to compile and publish studies on databases of k measurements. However, even with the availability of such studies, the predictions obtained using existing transformation models still exhibit a considerable range of variation, often spanning up to an order of magnitude, particularly when considering globally compiled databases. High-quality soil sampling and testing programs are still essential to reduce the uncertainty and variation associated with permeability prediction. Complete database compilation and classification are also required to enhance the accuracy of future assessments of soil hydraulic conductivity.

REFRENCES

Al-Tabbaa, A. & Wood, D.M. (1987). Some measurements of the permeability of kaolin. *Géotechnique*, **37**(4): 499–514. https://doi.org/10.1680/geot.1987.37.4.499

Alyamani, M.S. & Sen, Z. (1993). Determination of Hydraulic Conductivity from Complete Grain-Size Distribution Curves. *Groundwater*, **31**(4): 551–555. https://doi.org/10.1111/j.1745-6584.1993.tb00587.x

ASTM International (2022). *Standard test methods for measurement of hydraulic conductivity of coarse-grained soils.* D2434-22, West Conshohocken, PA.

Baver, L. (1948). *Soil Physics.* John Wiley Sons, New York, NY.

Bear, J. (1972). *Dynamics of Fluids in Porous Media.* American Elsevier, New York, NY.

Beesley, M.E.W. & Vardanega, P.J. (2020). Parameter variability of undrained shear strength and strain using a database of reconstituted soil tests. *Canadian Geotechnical Journal*, **57**(8): 1247–1255. https://doi.org/10.1139/cgj-2019-0424

British Standards Institution (BSI) (1990). *Methods of test for soils for civil engineering purposes – Part 5: Compressibility, permeability and durability tests.* BS 1377-5:1990, London, UK.

BSI (2019). *Geotechnical investigation and testing. Laboratory testing of soil. Permeability tests.* BS EN ISO 17892-11:2019, London, UK.

Burnham, K.P. & Anderson, D.R. (2004). Multimodel inference: Understanding AIC and BIC in model selection. *Sociological Methods and Research* **33**(2): 261–304. https://doi.org/10.1177/0049124104268644

Cabalar, A.F. & Akbulut, N. (2016). Evaluation of actual and estimated hydraulic conductivity of sands with different gradation and shape. *SpringerPlus*, **5**(1): 820. https://doi.org/10.1186/s40064-016-2472-2

Carman, P.C. (1937). Fluid flow through granular beds. *Transaction Institution of Chemical Engineers*, **15**: 150–166.

Carman, P.C. (1939). Permeability of saturated sands, soils and clays. *The Journal of Agricultural Science*, **29**(2): 262–273.

Carrier, W.D. & Beckman, J.F. (1984). Correlations between index tests and the properties of remoulded clays. *Géotechnique*, **34**(2): 211–228. https://doi.org/10.1680/geot.1984.34.2.211

Carrier, W.D. (2003). Goodbye, Hazen; Hello, Kozeny-Carman. *Journal of Geotechnical and Geoenvironmental Engineering*, **129**(11): 1054–1056. https://doi.org/10.1061/(ASCE)1090-0241(2003)129:11(1054)

Chapuis, R. P., Gill, D. E. & Baass, K. (1989). Laboratory permeability tests on sand: Influence of the compaction method on anisotropy. *Canadian Geotechnical Journal*, **28**(1): 172–173. https://doi.org/10.1139/t89-074

Chapuis, R.P. (2004). Predicting the saturated hydraulic conductivity of sand and gravel using effective diameter and void ratio. *Canadian Geotechnical Journal*, **41**(5): 787–795. https://doi.org/10.1139/t04-022

Chapuis, R.P. (2012). Predicting the saturated hydraulic conductivity of soils: A review. *Bulletin of Engineering Geology and the Environment*, **71**(3): 401–434. https://doi.org/10.1007/s10064-012-0418-7

Ching, J. & Phoon, K-K. (2012). Modeling parameters of structured clays as a multivariate normal distribution. *Canadian Geotechnical Journal*, **49**(5): 522–545. https://doi.org/10.1139/t2012-015

Ching, J. & Phoon, K-K. (2014). Transformations and correlations among some clay parameters – the global database. *Canadian Geotechnical Journal*, **51**(6): 663–685. https://doi.org/10.1139/cgj-2013-0262

Ching, J., Phoon, K-K. & Wu, C.T. (2022). Data-centric quasi-site-specific prediction for compressibility of clays. *Canadian Geotechnical Journal*, **59**(12): 2033–2049. https://doi.org/10.1139/cgj-2021-0658

Clennell, M.B., Dewhurst, D.N., Brown, K.M. & Westbrook, G.K. (1999). Permeability anisotropy of consolidated clays. In: *Muds and mudstones: Physical and fluid flow properties*, vol. 158, Aplin, A.C., Fleet, A.J. & MacQuaker, J.H.S. (eds.). Geological Society of London, London, UK, pp. 79–96. http://doi.org/10.1144/GSL.SP.1999.158.01.07

Corey, A.T. (1957). Measurement of water and air permeability in unsaturated soil. *Soil Science Society of America Journal*, 21(1): 7–10. https://doi.org/10.2136/sssaj1957.03615995002100010003x

Craig, R.F. (2004). *Craig's soil mechanics*. 7th Edition. Taylor & Francis, Abingdon, UK.

D'Ignazio, M., Phoon, K-K., Tan, S.A. & Länsivaara, T.T. (2016). Correlations for undrained shear strength of finnish soft clays. *Canadian Geotechnical Journal*, **53**(10): 1628–1645. https://doi.org/10.1139/cgj-2016-0037

Daniel, D., Anderson, D., & Boynton, S. (1985). Fixed-wall versus flexible-wall permeameters. In *Hydraulic barriers in soil and rock*. Johnson, A. I., Frobel, R. K., Cavalli, N. J., & Pettersson, C. B., (eds.), American Society for Testing and Materials, Philadelphia, PA, pp. 107–126.

Darcy, H. (1856). *Les fontaines publiques de la ville de Dijon. Victor Dalmont.*

Feng, S. & Vardanega, P.J. (2019a). Correlation of the hydraulic conductivity of fine-grained soils with water content ratio using a database. *Environmental Geotechnics*, **6**(5): 253–268. https://doi.org/10.1680/jenge.18.00166

Feng, S. & Vardanega, P.J. (2019b). A database of saturated hydraulic conductivity of fine-grained soils: probability density functions. *Georisk: Assessment and Management of Risk for Engineered Systems and Geohazards*, **13**(4): 255–261. https://doi.org/10.1080/17499518.2019.1652919

Feng, S., Vardanega, P.J., Ibraim, E., Widyatmoko, I., & Ojum, C. (2019). Permeability assessment of some granular mixtures. *Géotechnique*, **69**(7): 646–654. https://doi.org/10.1680/jgeot.17.T.039

Feng, S. (2022). *Hydraulic conductivity of road construction materials: With a focus on freeze-thaw effects*. Ph.D. thesis, University of Bristol, Bristol, UK.

Feng, S., Ibraim, E. & Vardanega, P.J. (2022). Databases to assess the hydraulic conductivity of road construction materials. In: *Proceedings of twentieth international conference on soil mechanics and geotechnical engineering (ICSMGE 2022), A Geotechnical Discovery Down Under, Sydney, New South Wales, Australia, 1–5 May 2022*, vol. 2, Rahman, M.M. & Jaksa, M. (eds.), Australian Geomechanics Society, Sydney, Australia, pp. 4547–4552. <https://www.issmge.org/uploads/publications/1/120/ICSMGE_2022-777.pdf> [19/10/2024].

Feng, S., Barreto, D., Imre, E., Ibraim, E. & Vardanega, P.J. (2023). Use of hydraulic radius to estimate the permeability of coarse-grained materials using a new geodatabase. *Transportation Geotechnics*, **41**: [101026]. https://doi.org/10.1016/j.trgeo.2023.101026

Gao, Q.F., Dong, H., Huang, R. & Li, Z.F. (2019). Structural characteristics and hydraulic conductivity of an eluvial-colluvial gravelly soil. *Bulletin of Engineering Geology and the Environment*, **78**(7): 5011–5028. https://doi.org/10.1007/s10064-018-01455-1

Hazen, A. (1895). *The filtration of public water supplies*. John Wiley & Sons, New York, NY.

Indrawan, I.G.B., Rahardjo, H. & Leong, E.C. (2006). Effects of coarse-grained materials on properties of residual soil. *Engineering Geology*, **82**(3): 154–164. https://doi.org/10.1016/j.enggeo.2005.10.003

Jaafar, R. & Likos, W.J. (2014). Pore-scale model for estimating saturated and unsaturated hydraulic conductivity from grain size distribution. *Journal of Geotechnical and Geoenvironmental Engineering*, **140**(2): 04013012. https://doi.org/10.1061/(ASCE)GT.1943-5606.0001031

Jones, C.W. (1955). The Permeability and Settlement of Laboratory Specimens of Sand and Sand-Gravel Mixtures. *Symposium on Permeability of Soils*, ASTM, Philadelphia, PA, pp. 68–79.

Kozeny, J. (1927). *Über kapillare leitung des wassers im boden:(aufstieg, versickerung und anwendung auf die bewässerung). Hölder-Pichler-Tempsky* (in German).

Kulhawy, F.H. & Mayne, P.W. (1990). *Manual on estimating soil properties for foundation design*, Electric Power Research Institute, Report No. EL-6800, Palo Alto, CA.

Lambe, T.W. & Whitman, R.V. (1969). *Soil mechanics*. John Wiley Sons, New York, NY.

Liu, Y. (2015). *Permeability and pore structure of granular materials*. Ph.D. thesis. McMaster University, Hamilton. Available from: <https://macsphere.mcmaster.ca/bitstream/11375/18104/2/Dissertation%20final.pdf> [10/07/2023]

Mbonimpa, M., Aubertin, M., Chapuis, R. & Busssière, B. (2002). Practical pedotransfer functions for estimating the saturated hydraulic conductivity. *Geotechnical and Geological Engineering*, **20**(3): 235–259. https://doi.org/10.1023/A:1016046214724

Mesri, G. & Olson, R. E. (1971). Mechanisms controlling the permeability of clays. *Clays and Clay Minerals*, **19**(3): 151–158. https://doi.org/10.1346/CCMN.1971.0190303

Montgomery, D. C., Runger, G. C., & Hubele, N. F. (2007). *Engineering statistics* (4th ed.). John Wiley & Sons Inc, New York, NY.

Moulton, L. & Seals, D. (1979). *Determination of the in situ permeability of base and subbase courses*. Federal Highway Administration, Washington, D.C.

Mualem, Y. (1976). A new model for predicting the hydraulic conductivity of unsaturated porous media. *Water Resources Research*, **12**(3): 513–522. https://doi.org/10.1029/WR012i003p00513

Nagaraj, T.S., Pandian, N.S. & Narashimha Raju, P.S.R. (1993). Stress state–permeability relationships for fine-grained soils. *Géotechnique*, **43**(2): 333–336. https://doi.org/10.1680/geot.1993.43.2.333

Nagaraj, T.S., Pandian, N.S. & Narashimha Raju, P.S.R. (1994). Stress state–permeability relations for overconsolidated clays. *Géotechnique*, **44**(2): 349–352. https://doi.org/10.1680/geot.1994.44.2.349

Nagy, L., Tabácks, A., Huszák, T., Mahler, A. & Varga, G. (2013). Comparison of permeability testing methods. In *18th International Conference on Soil Mechanics and Geotechnical Engineering, Paris, France*, Presses des Ponts, Paris, France, pp. 399–402. Available from: <https://www.issmge.org/uploads/publications/1/2/399-402.pdf> [10/07/2023].

Nguyen, T.T. & Indraratna, B. (2020). The role of particle shape on the hydraulic conductivity of granular soils captured through Kozeny-Carman approach. *Géotechnique Letters*, **10**(3): 398–403. https://doi.org/10.1680/jgele.20.00032

Pane, V., Croce, P., Znidarcic, D., Ko, H.-Y., Olsen, H.W. & Schiffman, R.L. (1983). Effects of consolidation on permeability measurements for soft clay. *Géotechnique*, **33**(1): 67–72. https://doi.org/10.1680/geot.1983.33.1.67

Paradine, C.G. & Rivett, B.H. (1953). Statistics for technologists. English Universities Press, London, UK.

Phoon, K-K. & Kulhawy, F.H. (1999a). Characterization of geotechnical variability. *Canadian Geotechnical Journal*, **36**(4): 612–624. https://doi.org/10.1139/t99-038

Phoon, K-K. & Kulhawy, F.H. (1999b). Evaluation of geotechnical property variability. *Canadian Geotechnical Journal*, **36**(4): 625–639. https://doi.org/10.1139/t99-039

Phoon, K-K. & Tang, C. (2019). Characterisation of geotechnical model uncertainty. *Georisk: Assessment and Management of Risk for Engineered Systems and Geohazards*, **13**(2): 101–130. https://doi.org/10.1080/17499518.2019.1585545

Phoon, K-K., Ching, J. & Wang, Y. (2020). Managing risk in geotechnical engineering – From data to digitalization. In: *Proceedings of the 7th international symposium on geotechnical safety and risk (ISGSR)*, Research Publishing, Taipei, Taiwan, pp. 13–34. Available from: <http://rpsonline.com.sg/proceedings/isgsr2019/index.html> [10/07/2023]

Piñeiro, G., Perelman, S., Guerschman, J. P., & Paruelo, J. M. (2008). How to evaluate models: Observed vs. predicted or predicted vs. observed? *Ecological Modelling*, **216**(3–4), 316–322. https://doi.org/10.1016/j.ecolmodel.2008.05.006

Qiu, Z.F. & Wang, J.J. (2015). Experimental study on the anisotropic hydraulic conductivity of a sandstone-mudstone particle mixture. *Journal of Hydrologic Engineering*, **20**(11): 04015029. https://doi.org/10.1061/(ASCE)HE.1943-5584.0001220

Ren, X., Zhao, Y., Deng, Q., Kang, J., Li, D. & Wang, D. (2016). A relation of hydraulic conductivity — void ratio for soils based on Kozeny-Carman equation. *Engineering Geology*, **213**: 89–97. https://doi.org/10.1016/j.enggeo.2016.08.017

Ren, X.W. & Santamarina, J.C. (2018). The hydraulic conductivity of sediments: A pore size perspective. *Engineering Geology*, **233**: 48–54. https://doi.org/10.1016/j.enggeo.2017.11.022

Shahabi, A., Das, B., and Tarquin, A. (1984). An Emperical Relation for Coefficient of Permeability of Sand. In *Fourth Australia-New Zealand Conference on Geomechanics*, , vol. 1, The Institution of Engineers, Australia, Barton, ACT, pp. 54–57. Available from: <https://www.issmge.org/uploads/publications/89/91/4ANZ_008.pdf> [19/10/2024]

Shepherd, R.G. (1989). Correlations of permeability and grain size. *Groundwater*, **27**(5): 633–638. https://doi.org/10.1111/j.1745-6584.1989.tb00476.x

Sivapullaiah, P.V, Sridharan, A. & Stalin, V.K. (2000). Hydraulic conductivity of bentonite-sand mixtures. *Canadian Geotechnical Journal*, **37**(2): 406–413. https://doi.org/10.1139/t99-120

Sridharan, A. & Nagaraj, H.B. (2005). Hydraulic conductivity of remolded fine-grained soils versus index properties. *Geotechnical and Geological Engineering*, **23**(1): 43–60. https://doi.org/10.1007/s10706-003-5396-x

Taiba, C.A., Mahmoudi, Y., Hazout, L., Belkhatir, M. & Baille, W. (2019). Evaluation of hydraulic conductivity through particle shape and packing density characteristics of sand–silt mixtures. *Marine Georesources and Geotechnology*, **37**(10): 1175–1187. https://doi.org/10.1080/1064119X.2018.1539891

Tavenas, F., Leblond, P., Jean, P. & Leroueil, S. (1983). The permeability of natural soft clays. Part I: Methods of laboratory measurement. *Canadian Geotechnical Journal*, **20**(4): 629–644. https://doi.org/10.1139/t83-072

Taylor, D.W. (1948). *Fundamentals of soil mechanics*. John Wiley & Son Inc. New York, NY.

Terzaghi, K. & Peck, R.B. (1948). *Soil mechanics in engineering practice*. John Wiley & Sons, New York, NY.

Van Genuchten, M.T. (1980). A closed-form equation for predicting the hydraulic conductivity of unsaturated soils. *Soil Science Society of America Journal*, **44**(5): 892–898. https://doi.org/10.2136/sssaj1980.03615995004400050002x

Vardanega, P.J., Othman, M., Voyagaki, E., Crispin, J.J., Gilder, C.E.L. & Ntassiou, K. (2024). *The DINGO database: Summary report: March 2024, v1.2*. University of Bristol, Bristol, UK. https://doi.org/10.5523/bris.1jraem68g7ara21p2oi6hv4z22

Whittle, A. (1993). Evaluation of a constitutive model for overconsolidated clays. *Géotechnique*, **43**(2): 289–313. https://doi.org/10.1680/geot.1993.43.2.289

Zhai, Q., Rahardjo, H., Satyanaga, A., Zhu, Y., Dai, G. & Zhao, X. (2021). Estimation of wetting hydraulic conductivity function for unsaturated sandy soil. *Engineering Geology*, **285**: 106034. https://doi.org/10.1016/j.enggeo.2021.106034

Chapter 7

Normalised active undrained shear strengths of soft Scandinavian clays

A data-centric and geomechanical approach

Sølve Hov, Jean-Sébastien L'Heureux, Katharina Kahrs, Karin Lundström, and David Gaharia

The undrained shear strength is one of the most important engineering parameters of soft clays. This chapter specifically deals with active undrained strengths from triaxial tests and their normalisation with the preconsolidation pressure from oedometer tests (c_{uA}/σ'_p) on normally to slightly overconsolidated clays (OCR ≤ 1.5). A bivariate database is presented containing data from Norwegian and Swedish clays with a large range in plasticity quantified by the liquid limit (LL). Linear regression analyses are performed to show that c_{uA}/σ'_p and LL are poorly correlated; however, there is a clear trend of increasing c_{uA}/σ'_p with increasing LL with respect to the 95% confidence interval. On average, c_{uA}/σ'_p ranges from just below 0.3 at $LL \approx 20$–30% to around 0.36–0.38 at $LL \approx 80\%$. The results are explained within a geomechanical framework where a simplified yield surface, defined by a drained failure criterion (φ') and the yield stress σ'_p, is used to show that C_{uA}/σ'_p is dependent on φ'. An increase in organic content results in an increase in both φ' and LL, and by pure geometrical constructions of the simplified yield surface it is shown that $\varphi' = 30°$ results in $c_{uA}/\sigma'_p = 0.33$ for low-to-medium plastic clays and $\varphi' = 55°$ results in $c_{uA}/\sigma'_p = 0.45$ for highly plastic organic clays. This agrees with the bivariate database. Specific attention is given to uncertainties and sample quality.

7.1 INTRODUCTION

The undrained shear strength (c_u) of clays is one of the most important parameters in geotechnical engineering and has over the years been researched extensively. The parameter is vital to the assessment of slope stability, embankments, and excavations in clay as well as the bearing capacity of structures founded on clayey soils. The magnitude of c_u depends on several factors, e.g., direction of loading, rate of loading, and temperature; however, it is also found to be inherently dependent on the magnitude and orientation of effective stresses in the clay before and during loading. In particular, it depends on the effective stress at yielding, i.e., the preconsolidation pressure (σ'_p) in oedometer boundary conditions. Normalised undrained shear strengths (c_u / σ'_p), i.e., the ratio between the undrained strength and the drained yield stress, are thus of great value in clay characterisation – both in research and in engineering practice.

This chapter specifically deals with normalised active undrained shear strengths (c_{uA} / σ'_p) of soft Scandinavian clays in a normally to slightly overconsolidated state, i.e., overconsolidation ratio $OCR \leq 1.5$. The c_{uA} is obtained from active undrained triaxial tests. This normalised strength has also been the topic for several previous studies, which have shown that c_{uA} / σ'_p exhibits a fairly constant value of around 0.3–0.35 for

DOI: 10.1201/9781003441946-7

$OCR \leq 1.5$ (Ladd et al. 1977; Larsson 1980; Jamiolkowski et al. 1985; Tavenas and Leroueil 1987; Ladd and DeGroot 2003; Larsson et al. 2007; Karlsrud and Hernandez-Martinez 2013; Sällfors et al. 2017; Lundström and Dehlbom 2019).

Normalised undrained shear strengths obtained through direct simple shear tests (DSS), field vane tests (FV), and fall cone (FC) tests have also previously been dealt with by several researchers. Although c_{uDSS} / σ'_p clearly depends on the OCR, varying results have been found on its dependence of the clay's plasticity. Some studies have for example shown that c_{uFV} / σ'_p is around 0.22 independently of plasticity, typically quantified by the plasticity index (IP) or liquid limit (LL) (Mesri 1975, 1989; D'Ignazio et al. 2016). Other studies have shown the opposite, i.e., values of c_{uDSS} / σ'_p increase with increasing plasticity (Jamiolkowski et al. 1985; Jardine and Hight 1987; Mayne and Mitchell 2011; Hov et al. 2021).

The reason why some data tend to show a dependence of plasticity on c_{uDSS} / σ'_p and others do not is not clear. In fact, the causes may be attributed to various factors such as the intrinsic clay properties or systematic errors in the database due to, e.g., sample disturbance or measurement errors. These are important uncertainties, and with the fast-growing amount of data available within geotechnical engineering, the topic of representative results is more important than ever. Myriads of databases with both simple and advanced statistical analyses are available to the geotechnical community. These works and databases are highly valuable and give new insights into clay behaviour (Ching and Phoon 2013, 2014a, 2014b; D'Ignazio et al. 2016; Beesley and Vardanega 2020; Hov et al. 2021), and it is transforming the understanding of soil mechanics in engineering practice, and provide a huge opportunity for further research (Tang and Phoon 2021; Phoon 2020; Phoon et al. 2022).

However, if care is not taken, a risk of database contamination with non-representative values is imminent. Consequences of such can lead to unrepresentative machine learning models and unreliable geotechnical designs. One could perhaps argue that historically published data likely had a higher degree of quality control and assurance since there were fewer test results to process. As the number of tests and data over time increases, the sheer volume of data makes quality control and assurance more difficult, simply by practical limitations. In addition, data from studies are used exchangeably throughout the world where data is being re-analysed by other researchers, often located on different continents, using different laboratory tools and different interpretation methods. This is most often advantageous, since it avoids any dogmatism that researchers can fall into. On the contrary, there is also an increased likelihood that important preconditions are not addressed since they are unknown.

In the light of all uncertainties linked to the assessment of geotechnical parameters, it could be advantageous to use a geomechanical framework to assess data quality when comparing data from different sources. This is however not always possible. The FC test, as an example, is interpreted by the cone penetration depth which is correlated to c_{uFC} . This is easily modelled with numerical tools; however, the correlation is too dependent on the (unknown) adhesion between the cone and the clay and, subsequently, a correlation cannot be established numerically. As a few more examples, the principal stress rotation in a DSS sample is unknown, and for the FV test, the failure pattern and soil structure disturbance prior to rotation are too uncertain. Similarly, the piezocone (CPTu) test can be challenging to analyse geomechanically since the effective stress paths around the cone are uncertain and depend on many variables. One test that however can be analysed within a geomechanical framework is the triaxial test. Here, the principal total stresses and pore pressure measurements enable the calculation of all effective stresses, and it is known how these change over time and with strain. This is vastly advantageous since its data sets can be analysed not only by statistical analyses but also by assessing credibility from a geomechanical point of view.

The study of normalised active undrained shear strengths (c_{uA} / σ'_p) outlined in this chapter is based on experimental results – the data-centric approach – and by analysing the results by a simplified yield surface in an effective stress plane – the geomechanical approach. A bivariate database is presented which contains the variables c_{uA} / σ'_p and liquid limit, LL. The c_{uA} is obtained from triaxial tests consolidated anisotropically and sheared undrained in an active (compression) loading (CAUC), and the σ'_p is obtained from the constant rate of strain (CRS) oedometer tests. The LL is used as a measure of plasticity, instead ofplasticity index (IP). The database is "only" bivariate since all data points originate from normally to slightly overconsolidated clays ($OCR \leq 1.5$), and hence OCR is not a variable and no other parameter has been shown to affect normalised strengths from CAUC tests. Further, the strength is normalised with σ'_p instead of the in situ effective stress (σ'_{v0}) since σ'_p is interlinked with shear strength, as will be shown.

The database contains test results from both Norwegian and Swedish clays. These clays are similar from a mineralogical and stress history point of view; however, they differ in other aspects. Norwegian clays typically have a lower clay content and a lower organic content compared to Swedish clays, resulting in lower natural water content and lower plasticity. The data thus provides a great opportunity to analyse the impact of clay plasticity on the normalised undrained shear strength of clays.

One however needs to be careful when comparing data from different areas and countries, even from Norway and Sweden which have had extensive collaboration and have had many similarities within geotechnical engineering throughout history. In fact, it turns out that not all parameters are interpreted using the same methods. In this case, although the CRS oedometer tests are performed in a similar way, the interpretation of σ'_p differs, as will be explained. This difference has been in place at least since the beginning of the 1980s when researchers in Sweden introduced the CRS test and implemented an interpretation which still today has a very limited international use. In Norway, the σ'_p interpretation has also varied with time (Janbu 1963, 1989; Karlsrud and Hernandez-Martinez 2013). The difference in interpretation method still exists since there is no ISO or EN standard for CRS oedometer tests. This fact might have resulted in erroneous analyses that have been made when including Norwegian and Swedish data containing σ'_p as a variable.

These differences in interpretation techniques will firstly be highlighted, followed by a presentation of the bivariate database. Afterwards, the results are analysed through a geomechanical framework to attempt a critical validation of the empirical data. The aim is to analyse if and how c_{uA} / σ'_p is affected by the LL. The two approaches result in similar conclusions.

7.2 DATA AND METHODS

7.2.1 Laboratory tests

All laboratory tests were performed according to their respective EN ISO standard if available. The LL was determined by the fall cone test according to ISO 17892-12:2018 and is defined as the water content at 10 mm penetration with a 60 g steel fall cone with an apex angle of 60°. The LL is used since there is a very strong correlation between LL and IP as seen in most databases (Ching and Phoon 2014b; NIFS 2014; D'Ignazio et al. 2016; J. DeGroot et al. 2019), and any of these will thus work to quantify the clay's plasticity. Another reason is that values of *IP* are often missing in Swedish data since the plastic limit (*PL*) is not measured in engineering practice in Sweden. One also needs to remember that the plasticity limits, in total 11 of them originally developed by Atterberg in the early 1910s for agricultural purposes (Atterberg 1911, 1912) and introduced to soil mechanics by Casagrande in the 1930s (Casagrande 1932), are not determined based on any scientific principle. It is instead an arbitrary property of a gradual shift in the consistency of a clay in a remoulded state. Hence, the *LL* alone is a sufficient measure to describe a clay's plasticity and the word "plasticity" is used hereafter as a common term for *LL* and *IP*.

The CAUC tests were performed according to ISO 17892-9:2018. The sample dimensions were typically 50 or 54 mm in diameter with 100 mm in height. The sample is consolidated to its calculated in situ effective stress state, which for these normally consolidated clays is anisotropic with earth pressure at rest coefficients (K_0) values around 0.5–0.7. Samples were sheared actively under undrained conditions with a strain rate of around 0.6–0.7% per hour. The active undrained shear strength (c_{uA}) is interpreted as the peak strength that typically occurs within 0.5–3% axial strain for inorganic clays with relatively low values of *LL*. For highly plastic organic clays, the peak strength typically occurs at higher strains, but the interpretation of c_{uA} is limited to 15% axial strain.

As previously mentioned, the OCR highly impacts the clay behaviour and, in particular, the effective stress path to failure. To illustrate the effects of OCR on effective stress path type, Figure 7.1 plots a compilation of effective stress paths from the Norwegian Geotechnical Institutes (NGI) database (the database is presented in the following section).

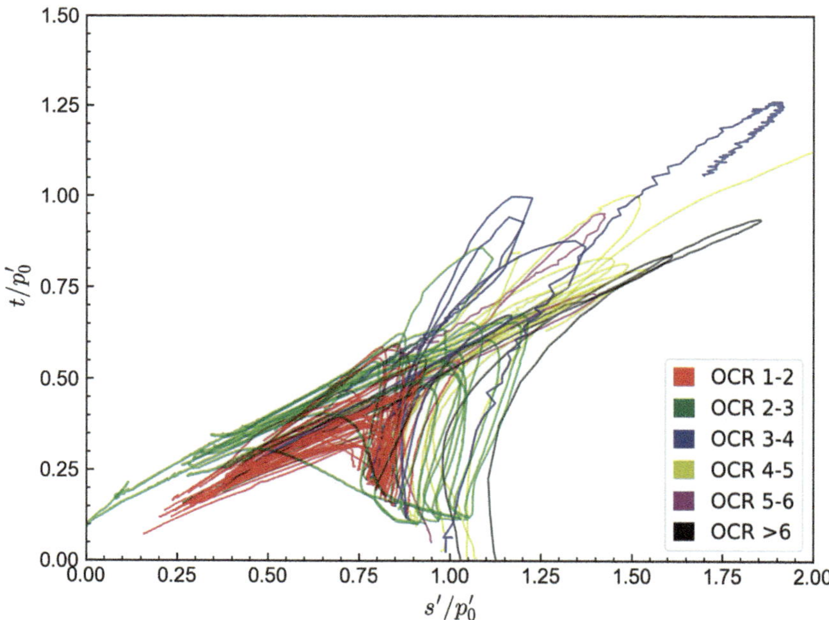

Figure 7.1 Compilation of effective stress paths in the $s' = (\sigma_v' + \sigma_h')/2$ vs. $t = (\sigma_v - \sigma_h)/2$ stress plane, i.e., the MIT plot. The axes in this plot are normalised with respect to in situ effective vertical stress (p_0') for comparison. Internal NGI data (NGI 2017).

It is seen that clays in a normally to slightly overconsolidated state ($OCR \leq 1.5$) tend to share a common type of effective stress path marked in red. From a nearly elastic initial response, stress path inclination 1:3 in this plot, yielding occurs whereupon a contractive behaviour leads to an increased pore pressure development and the effective stress path turning left and towards the origin. Samples with $OCR \gg 1.5$ exhibit a dilatant stress path which turns right and follows a drained failure envelope, e.g., Mohr–Coulomb, before complete failure occurs. The results presented herein only originate from clays in a normally to slightly overconsolidated state ($OCR \leq 1.5$).

CRS oedometer tests are executed similarly in Norway and Sweden. Samples of 50 or 54 mm in diameter with 20 mm height are compressed with a strain rate of around 0.6–0.8% per hour. An undrained boundary at the bottom enables pore pressure measurements, whilst the top boundary is drained. The interpretation of oedometer moduli is also similar in both countries; however, the interpretation of σ'_p differs. In Norway, the most commonly used method today is referred to as the "Karlsrud" method (Karlsrud and Hernandez-Martinez 2013) and is used on all Norwegian test results presented in this chapter. In Sweden, the "Sällfors method" (Sällfors 1975) has been used in Sweden since the early 1980s, and again, all Swedish test results presented in this chapter are based on the Sällfors method.

Figure 7.2 shows two CRS oedometer test results to illustrate the differences between the Norwegian and Swedish methods of interpreting σ'_p . The Karlsrud method, shown in red, interprets the σ'_p value based on the modulus curve, the first derivative of the oedometer curve. A curve-fitting procedure matches linear parts of the modulus curve in the overconsolidated range (M_i), the normally consolidated range (M_L), and after a threshold is reached where the modulus increases linearly after M_L . The σ'_p is interpreted as the average between this threshold and the end of the M_i line. The Sällfors method, shown in blue in Figure 7.2, uses a curve-fitting procedure of the oedometer

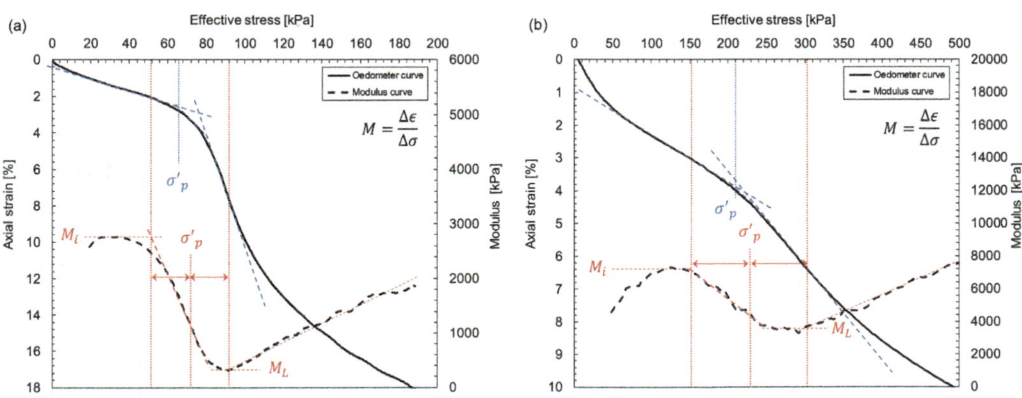

Figure 7.2 Examples of interpretation of preconsolidation pressure (σ'_p) from CRS oedometer tests with the Karlsrud (Norwegian) and Sällfors (Swedish) methods marked in red and blue, respectively: (a) test result on a Swedish clay with $M_i / M_L \approx 8$ giving $\sigma'_{p,NO} \approx 72$ kPa and $\sigma'_{p,SE} \approx 66$ kPa, (b) test result on a Norwegian clay with $M_i / M_L \approx 2$ giving $\sigma'_{p,NO} \approx 230$ kPa and $\sigma'_{p,SE} \approx 210$ kPa.

curve instead of the modulus curve. A similar procedure is performed to find linear parts of the curve representing M_i and M_L . An isosceles triangle is then drawn with its base tangential to the oedometer curve as shown in the figure. The σ'_p is defined as the intersection of the baseline and the M_i line. It has been shown that this approach gives similar values as incremental oedometer tests and field measurements (Sällfors 1975).

Another distinction is that the threshold stress, i.e., the onset of modulus increase after M_i, is directly interpreted by the apparent modulus curve according to the Karlsrud method, whilst a strain rate correction is made in the Sällfors method to take into account the considerably higher strain rate in CRS tests compared to incremental oedometer tests or field behaviour (Sällfors 1975; Tavenas and Leroueil 1987).

The interpretation method is thus different between the two countries, affecting values of c_{uA} / σ'_p. There has been no systematic study of this difference between these methods, although it is known that the Karlsrud method yields roughly similar results to other commonly used methods (Paniagua et al. 2016). For the two examples in Figure 7.2, the Karlsrud method shows σ'_p of 72 and 230 kPa, respectively, whilst the Sällfors method correspondingly shows σ'_p of 66 and 210 kPa. The difference is around 9–10% in both cases, i.e., the normalised strength c_{uA} / σ'_p will be around 9–10% lower if a Norwegian interpretation practice is used compared to Swedish practice. Based on the authors' experience and judgement, this difference of ~10% between the two methods seems reasonable for typical CRS oedometer test results.

7.2.2 Bivariate database

The database, labelled CLAY/2/213 (Ching et al. 2016), is summarised in Table 7.1. In total, it contains 213 bivariate data points of c_{uA} / σ'_p and LL, of which 46 originated from Norway and the remaining from Sweden. The reason that Swedish clays are dominant is because a large proportion of the Norwegian clays are overconsolidated ($OCR > 1.5$) and have thus been removed.

Most of the Norwegian data originates from the NGI database of onshore piston samples (72–75 mm diameter samples) and high-quality block samples (160–250 mm samples). Some of this data has previously been presented but has been supplemented with more recent test data (Karlsrud and Hernandez-Martinez 2013; Paniagua et al. 2019). The data originates from sites distributed in the central and south-eastern parts of Norway. The Swedish data originates from several sources, both from research and commercial projects on organic clays and inorganic clays. The Swedish data includes piston samples (50 mm diameter) and high-quality block samples (160 mm diameter). The data originates from sites distributed around southern Sweden.

The range of LL in the database is large; between 16 and 349, reflecting the range from silty clays, or perhaps clayey silts, to highly organic clays and gyttja (defined as clay with organic content >6%). The range of c_{uA} / σ'_p is between 0.22 and 0.55.

Common for all data is that the test results have been subjected to a thorough quality control by engineers during evaluation and processing of the data. Even though a large proportion of the data originate from commercial projects, e.g., Sällfors and Larsson (2016) and the NGI and LabMind (LM) databases, subsequent analyses have been performed and when needed, data has been removed if found not representative due to, e.g., sample disturbance. The database does not include a sample quality designation since this is not available for all test results, e.g., (Lunne et al. 1997).

Table 7.1 Overview of data ($OCR \leq 1.5$)

Reference	n	Country	c_{uA} / σ'_p		LL		Remarks
			min	max	min	max	
NGI database	43	Norway	0.23	0.48	16	107	Data from Karlsrud and Hernandez-Martinez (2013), L'Heureux et al. (2018), and Paniagua et al. (2019) supplemented with new data not previously published. Data from 11 sites in Norway.
LM database	18	Sweden	0.30	0.46	44	83	Data from Hov and Garcia de Herreros (2020) supplemented with new data not previously published. Data from seven sites in South-eastern Sweden.
Lundström and Dehlbom (2019)	16	Sweden	0.28	0.43	36	104	Data from three sites in Sweden
Sällfors and Larsson (2016)	55	Sweden	0.25	0.44	25	107	Data from 15 sites across southern Sweden
Larsson et al. (2007)	51	Sweden	0.26	0.36	25	85	The number of sites uncertain
Westerberg (1999)	19	Sweden	0.26	0.37	25	84	Data from one specific site
Holstad (2016); Holstad and Degago (2021)	3	Norway	0.30	0.55	90	230	Data included to illustrate normalised strengths of organic clays and gyttja
Larsson (1990)	8	Sweden	0.33	0.51	20	349	Data included to illustrate normalised strengths of organic clays and gyttja
All compiled data (CLAY/2/213)	213	–	0.23	0.55	16	349	–

7.3 A DATA-CENTRIC APPROACH

Figure 7.3(a) plots all data of c_{uA} / σ'_p vs. LL available in the database (Table 7.1). The values of c_{uA} / σ'_p tend to increase with increasing LL until a value of $LL \approx 100\%$ is reached. Such a value typically corresponds to a rough threshold used to distinguish inorganic clay from organic clay (gyttjig clay or gyttja). A few test results obtained from organic clay and gyttja with $LL > 100\%$ show that values of c_{uA} / σ'_p are fairly constant at around 0.5.

Looking only at results from inorganic clays, Figure 7.3(b) plots values for $LL < 100\%$. This figure also shows a trend line from a linear regression analysis with a 95% confidence interval of the mean value. The linear trend shows values of c_{uA} / σ'_p ranging from just below 0.3 at $LL \approx 20 - 30\%$ to around 0.36–0.38 at $LL \approx 80 - 100\%$. Despite a large scatter and a very weak correlation ($R^2 = 0.24$), there is clearly a trend of increasing c_{uA} / σ'_p with increasing LL of the 95% confidence interval.

To account for the difference in interpretation method, the c_{uA} / σ'_p for the Norwegian data in Figure 7.3(c) is increased by 10%. It is chosen to correct the Norwegian data

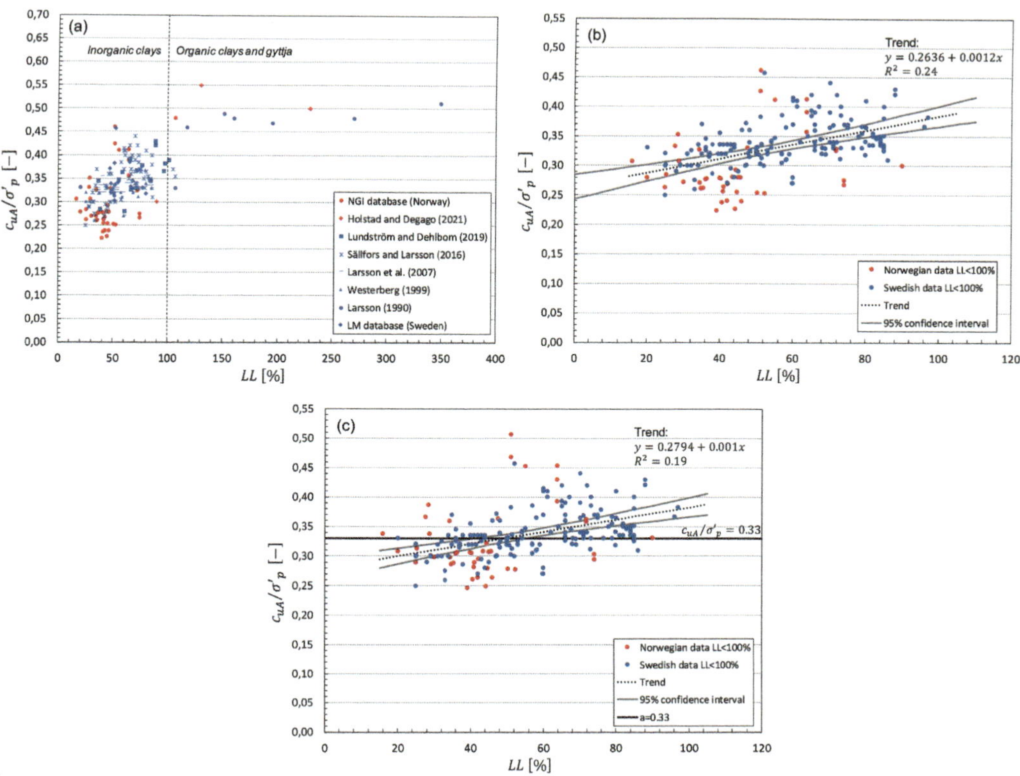

Figure 7.3 Normalised active undrained shear strengths (c_{uA} / σ'_p) vs. liquid limit (LL): (a) all data and all LL , (b) data with $LL < 100\%$, and (c) data $LL < 100\%$ with Norwegian data corrected for σ'_p interpretation (+10%). The line in (a) serves as an approximate transition between inorganic and organic clays. Linear regression analysis and 95% confidence interval are visualised in (b) and (c).

instead of the Swedish because of the reasonable strain rate effects that should be considered in CRS oedometer testing. Again, the linear regression analysis shows a weak correlation with an $R^2 = 0.19$; however, the trend of increasing c_{uA} / σ'_p with increasing LL is still seen.

This trend is contradictory with most other studies of normalised strengths from CAUC (c_{uA} / σ'_p) that show a constant c_{uA} / σ'_p independent of IP or LL (Jamiolkowski et al. 1985; Tavenas and Leroueil 1987; Ladd et al. 1977; Larsson 1980; Larsson et al. 2007; Karlsrud and Hernandez-Martinez 2013; Sällfors et al. 2017; Lundström and Dehlbom 2019). There are however a few published studies, or compilations of data, that show similar trends of increasing c_{uA} / σ'_p with increasing IP or LL, e.g., (Westerberg 1999; Lundström and Dehlbom 2019; Vesterberg and Andersson 2022). Other studies show a very slight increase with plasticity (DeGroot et al. 2019).

Why do we see these differences in data? Is it due to inherent behaviour of different clays, effects of sample disturbance, or simply by chance? From a pure data-centric approach, these effects are difficult to investigate, and hence it is difficult to validate if the data demonstrates real representative soil behaviour or not.

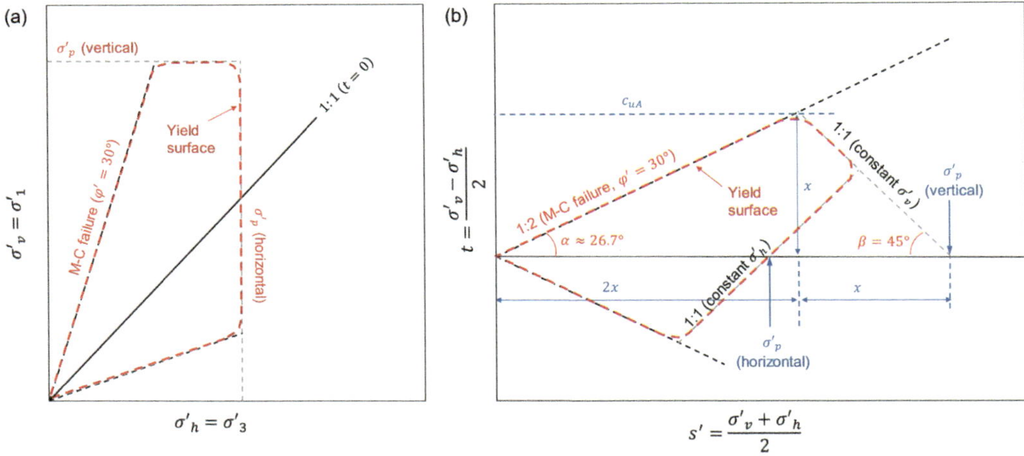

Figure 7.4 Simplified yield surface in (a) σ'_v vs. σ'_h plot and (b) s' vs. t plot (MIT). A friction angle of 30° is used to illustrate typical values for low-to-medium plastic clays.

7.4 A GEOMECHANICAL APPROACH

Active triaxial tests and oedometer tests differ in terms of boundary conditions such as drainage and radial constraints; however, they also share the principal stress orientation and the active (compressive) loading. Thus, yielding in both tests can be assumed to reflect the same vertical soil properties. This is illustrated in Figure 7.4 where (a) shows a σ'_v vs. σ'_h stress plane with a simplified yield surface consisting of the Mohr–Coulomb failure criterion, and the σ'_p in the vertical and horizontal directions. For CRS oedometer tests, the vertical effective stress is continuously increased, resulting in yield around σ'_p. In undrained triaxial shear, however, the effective stress path cannot go beyond this yield surface since the yielding results in an excess pore pressure response and hence a change in direction. This yield surface is illustrated in red in Figure 7.4. It has great value as a basis for explaining and analysing effective stress paths for clays with $OCR \lesssim 1.5$, and has previously been used to analyse both Norwegian and Swedish clays (Larsson and Sällfors 1981; Sällfors et al. 2017).

Figure 7.4(b) illustrates the same yield surface in the $s' = (\sigma'_v + \sigma'_h)/2$ vs. $t = (\sigma_v - \sigma_h)/2$ stress plane, i.e., the MIT plot. Here, constant values of vertical stress (σ'_v) are at a −45° angle, and constant values of horizontal stress (σ'_h) are at a +45° angle, and consequently, the respective σ'_p are also at 45° angles. If the cohesive intercept is negligible, as it often is for these clays, the yield surface is simply defined as a function of the friction angle (φ') and σ'_p in the vertical and horizontal directions. It should be noted that this concept of the simplified yield surface only applies to normally and slightly overconsolidated clays ($OCR \lesssim 1.5$), i.e., where the effective stress paths during triaxial shearing reach the "top" of the yield surface in the MIT plot. Examples of such stress paths were shown in Figure 7.1.

This simplified yield surface allows for a simple geometrical calculation of c_{uA} / σ'_p as a function of φ' (Equation 5.1):

$$\frac{c_{uA}}{\sigma'_p} = \frac{\sin\varphi'}{1+\sin\varphi'}$$

If the friction angle increases, the "top" of the yield surface will go up, whilst the σ'_p remains constant, and therefore, c_{uA} / σ'_p will increase. An increase in c' instead of φ' also result in an increase in c_{uA} / σ'_p.

Experimental results on low-to-medium plastic clays ($LL \approx 30-50\%$) typically show values of $\varphi' \approx 28-32°$ (Larsson 1977; Tavenas and Leroueil 1987; Karlsrud and Hernandez-Martinez 2013; L'Heureux et al. 2019). If one assumes $\varphi' = 30°$, then $c_{uA} / \sigma'_p \approx 0.33$. In the MIT stress plane, a $\varphi' = 30°$ failure criterion equals a line with $\alpha \approx 26.7°$, and the geometrical construction can then be simplified, Equation (5.2) (see Figure 7.4):

$$\frac{c_{uA}}{\sigma'_p} = \frac{x}{2x + x} = \frac{1}{3}$$

Experimental results on organic clays and gyttja, however, indicates much higher friction angles, up to 55–60° (Larsson 1990; Cheng et al. 2007; Holstad 2016; O'Kelly 2017; Mięḑlarz et al. 2019). For example, $\varphi' = 55°$ results in $c_{uA} / \sigma'_p \approx 0.45$. These higher friction angles are possibly due to embedded fibres in the organic clay (Larsson 1990). Also, due to large strains to failure in organic clays and the undrained (constant volume) shearing, the samples are typically considerably elongated in the radial direct, i.e., in the minor principal stress direction. This can cause cracking even if the minor principal stress is compressive (Janbu 1979). One can thus question if a Mohr–Coulomb failure criterion is meaningful for these types of soils; however, it still reflects real stress paths in triaxial tests, and the measured c_{uA} / σ'_p are as a result higher than those for clays with a lower plasticity and hence a lower (and more meaningful) friction angle.

Figure 7.5 show two examples of effective stress paths in the s' vs. t plot: (a) from a site with low-to-medium plastic clay, and (b) from a site with highly plastic clay, both Swedish clays. All samples are consolidated to around 85% of σ'_p, i.e., $OCR \approx 1.15$. The clay with low-to-medium plasticity, where LL ranges between 40% and 68%, shows a friction angle of around 33°. Simplified yield surfaces can be drawn for each of these stress paths, however, in the figure only that for the sample from 8.0 m depth are shown. This sample seems slightly disturbed resulting a lower σ'_p than expected (giving $OCR = 1.0$), and the effective stress path immediately yields and follows a constant

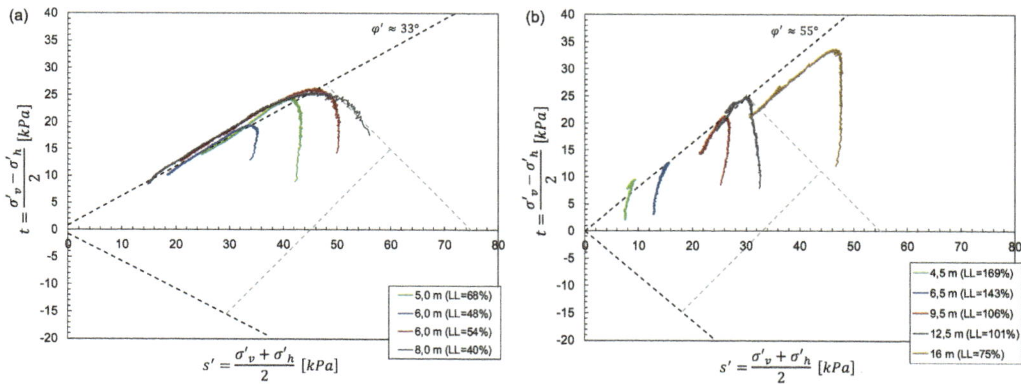

Figure 7.5 Examples of effective stress paths: (a) inorganic clay and (b) organic clay/gyttja.

vertical stress, i.e., along the σ'_p line. Notably, this disturbance does not affect the normalised strength.

Figure 7.5(b) shows the example of high plastic clays with LL ranging from 75% to 169% and where the plasticity decreases with depth. Four of the samples have considerable organic content with LL around or above 100%. For illustration, a simplified yield surface is shown for the 12.5 m sample ($LL = 101\%$). The φ' for the organic clays is interpreted to ~55°. The sample from 16 m ($LL = 75\%$) indicates a lower φ'.

The simplified yield surface and these two examples illustrate an important indirect effect of plasticity: c_{uA} / σ'_p appears to increase with increasing plasticity since the organic content affects both LL and φ'.

Figure 7.6 combines the data-centric and the geomechanical approach. Here, all data points in the database are plotted together with two c_{uA} / σ'_p points calculated based on the simplified yield surface: $c_{uA} / \sigma'_p = 0.33$ for a medium plastic clay ($\varphi' = 30°$) and $c_{uA} / \sigma'_p = 0.45$ for a high plastic organic clay ($\varphi' = 55°$). The magnitudes and trends are similar, serving as illustrative examples of how c_{uA} / σ'_p apparently depends on the apparent φ', in effect on the plasticity. Further, it seems that there is no significant threshold of organic content with a subsequent shift in values of c_{uA} / σ'_p. The gradually higher φ' values with increasing plasticity thus seem to result in a gradual increase of c_{uA} / σ'_p.

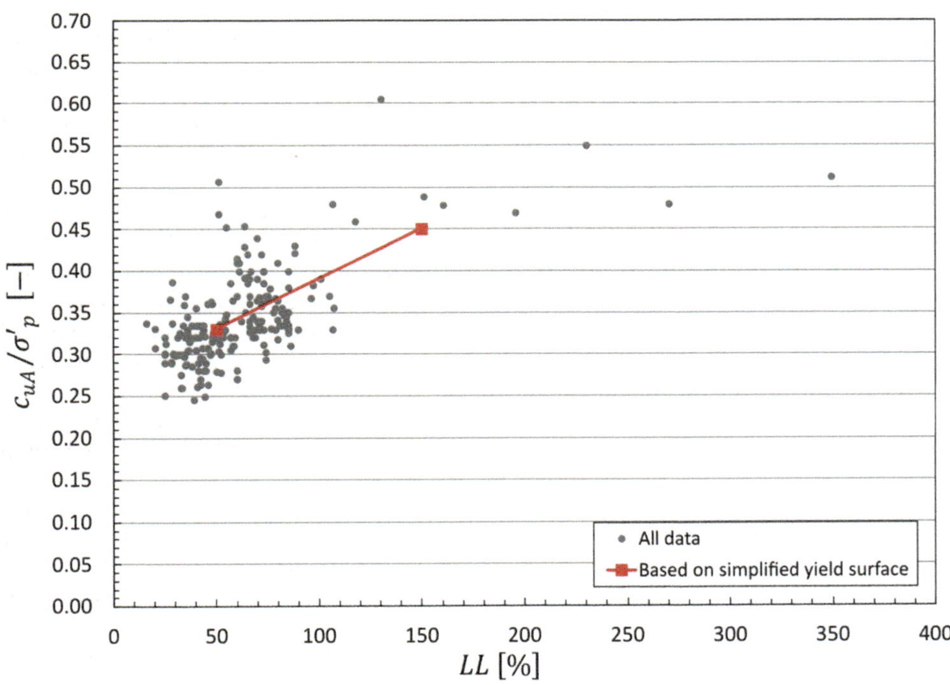

Figure 7.6 Normalised strength c_{uA} / σ'_p vs. plasticity LL. All bivariate data along with c_{uA} / σ'_p calculated from the simplified yield surface (two points $c_{uA} / \sigma'_p = 0.33$ at $LL = 50\%$ and $c_{uA} / \sigma'_p = 0.45$ at $LL = 150\%$ with an interpolation). The figure illustrates the data-centric and geomechanical approach combined.

7.5 UNCERTAINTIES AND SAMPLE QUALITY

The uncertain factors affecting normalised strengths (c_u / σ'_p) are, of course, the same as those affecting the included parameters: shear strength (c_u) and preconsolidation pressure (σ'_p). Examples of these factors are direction of loading, rate of loading, temperature, and sample quality. Naturally, this also applies to normalised strengths from CAUC tests (c_{uA} / σ'_p). The direction of loading can be neglected since it is always the same for oedometer and active triaxial tests. The rate of loading, on the contrary, is shown to affect both c_u (Lacasse 1995; Ladd and DeGroot 2003; Lunne and Andersen 2007) and σ'_p (Larsson and Sällfors 1986; Tavenas and Leroueil 1987). The combined effect is however more difficult to assess, but Ladd and DeGroot (2003) have shown a slight increase in c_{uA} / σ'_p with increased rate of loading, meaning that the effect on c_{uA} is higher than the effect on σ'_p. Furthermore, Lunne and Andersen (2007) concluded that the rate of loading does not impact the relation between strength and plasticity, indicating that the effect of plasticity on rate of loading vs. c_{uA} / σ'_p can be neglected.

Similarly, the effect of temperature on c_{uA} / σ'_p is barely studied; however, recently performed tests on a Swedish clay indicate that c_{uA} / σ'_p could increase with decreasing temperature (Vesterberg and Andersson 2022). If this is reflected in a plasticity effect is unknown, and further, the temperature at the laboratory tests is unknown for many of the test results available in the database. It is therefore not possible to analyse a possible temperature effect on the relationship between c_{uA} / σ'_p and LL.

The effect of sample quality on c_{uA} and σ'_p is extensively researched, and it is shown that low plastic clays are more sensitive to disturbance than high plastic clays (Lacasse et al. 1985; Lunne et al. 2006; Karlsrud and Hernandez-Martinez 2013). But again, the combined effect on c_{uA} / σ'_p is more difficult to assess and has not been studied in detail to the authors' knowledge. The few stress paths in Figure 7.5 indicate that there is no effect of sample quality on c_{uA} / σ'_p, and the simplified yield surface framework can give further indications since sample quality has not been shown to affect φ'. When a clay sample is disturbed, causing a destructuration of the clay fabric, the yield stresses (σ'_p in both vertical and horizontal directions) will typically decrease. In the simplified yield surface, this results in a shift in the 45° σ'_p line affecting also the peak strength in undrained loading (c_{uA}); however, the geometrical construction and the resulting value of c_{uA} / σ'_p remain constant as it is only dependent on φ'. The only effect of sample disturbance would be if the stress path does not reach the "top" of the yield surface, but there are too few studies on this to conclude. Further, there is no apparent effect of different samplers on c_{uA} / σ'_p in the NGI or LM data (Karlsrud and Hernandez-Martinez 2013; Hov and Garcia de Herreros 2020). This together indicates that there is probably no significant effect of sample quality on normalised active strengths c_{uA} / σ'_p.

7.6 CONCLUSIONS

This chapter has presented a data-centric and geomechanical approach to investigate the effect of plasticity on normalised strengths (c_{uA} / σ'_p) of normally to slightly overconsolidated soft clays ($OCR < 1.5$). A few initial remarks were made on the data origins of which one of the most important is knowing any execution or interpretation differences. In this case, the σ'_p interpretation is believed to give around 10% higher values using the

Norwegian practice compared to the Swedish practice. It was also noted that LL alone is a sufficient parameter for describing plasticity.

Linear regression analyses of the bivariate database showed that c_{uA} / σ'_p is weakly correlated to LL with a coefficient of determination (R^2) of 0.19 and 0.24, depending on if the difference in interpretation method of σ'_p was accounted for. Despite the very weak correlation, there was a clear trend of increasing c_{uA} / σ'_p with increasing LL with respect to the 95% confidence interval. On average, $c_{uA} / \sigma'_p \approx 0.30$ at $LL = 20\%$ and $c_{uA} / \sigma'_p \approx 0.38$ at $LL = 100\%$. A few data points show that organic clays, where $LL > 150\%$, have $c_{uA} / \sigma'_p \approx 0.5$. A visual trend over a large LL range is evident.

Furthermore, the data trend was analysed using a geomechanical approach where the drained Mohr–Coulomb failure criterion (c' and φ') and the vertical yield stress, i.e., σ'_p, were combined to form a simplified yield surface. A geometrical construction shows an increase in c_{uA} / σ'_p with increasing apparent φ'. Notably, $\varphi' = 30°$ results in $c_{uA} / \sigma'_p = 0.33$ and $\varphi' = 55°$ results in $c_{uA} / \sigma'_p = 0.45$. As the organic content increases, both LL and φ' increase, linking plasticity and normalised active strengths. The simplified yield surface thus provides a logical explanation of the data.

Having addressed the uncertainties and effects of sample quality, it is worth noting that the geomechanical framework is simplified. However, a useful maxim in science and engineering is "as simple as possible, but no simpler," adhered to in geotechnical engineering by, e.g., (Jardine 2020). Here, although the yield surface is simplified, it nonetheless can explain the behaviour resulting in the normalised strength seen from experimental data. And this is important, since data should be attempted to be explained, not only described, giving more confidence to the data-centric approach which includes real data along with its uncertainties.

REFERENCES

Atterberg, A. 1911. Clay and Its Relation to Water Content and Plasticity Limits [In Swedish]. *Kungliga Lantbruksakademiens Handlingar och Tidskrift* 50(2):132–158.

Atterberg, A. 1912. Soil Consistencies and Stiffness [In Swedish]. *Kungliga Lantbruksakademiens Handlingar och Tidskrift* 51(1):93–123.

Beesley, M.E.W., and Vardanega, P.J. 2020. Parameter Variability of Undrained Shear Strength and Strain Using a Database of Reconstituted Soil Tests. *Canadian Geotechnical Journal* 57(8):1247–1255.

Casagrande, A. 1932. Research on the Atterberg Limits of Soils. *Public Roads* 13(8):121–136.

Cheng, X.H., Ngan-Tillard, D.J.M., and den Haan, E.J. 2007. The Causes of the High Friction Angle of Dutch Organic Soils. *Engineering Geology* 93 (1–2):31–44.

Ching, J., and Phoon, K.K. 2013. Multivariate Distribution for Undrained Shear Strengths under Various Test Procedures. *Canadian Geotechnical Journal* 50(9):907–923.

Ching, J., and Phoon, K.K. 2014a. Correlations among Some Clay Parameters - The Multivariate Distribution. *Canadian Geotechnical Journal* 51(6):686–704.

Ching, J., and Phoon, K.K. 2014b. Transformations and Correlations among Some Clay Parameters - The Global Database. *Canadian Geotechnical Journal* 51(6):663–685.

Ching, J., Li, D.Q., and Phoon, K.K. 2016. Statistical Characterization of Multivariate Geotechnical Data. In: *Reliability of Geotechnical Structures*, CRC Press, Balkema, pp. 89–126.

D'Ignazio, M., Phoon, K.K., Tan, S.A., and Länsivaara, T. 2016. Correlations for Undrained Shear Strength of Finnish Soft Clays. *Canadian Geotechnical Journal* 53(10):1628–1645.

Holstad, Ø.B. 2016. *Shear Strength and Deformation Parameters on Gyttja* [In Norwegian]. Norwegian University of Science and Technology, Trondheim.

Holstad, Ø.B., and Degago, S. 2021. Strength and Deformation Characterization of Norwegian Organic Cohesive Soil (Gyttja). In: *18th Nordic Geotechnical Meeting. IOP Conference Series: Earth and Environmental Science*, Helsinki, 170:012018.

Hov, S., and Garcia de Herreros, C. 2020. Comparison of the Mini Block and Standard Piston Sampler in a Varved East Swedish Clay [In Swedish]. *LabMind*, A2018-A2026.

Hov, S., Prästings, A., Persson, E., and Larsson, S. 2021. On Empirical Correlations for Normalised Shear Strengths from Fall Cone and Direct Simple Shear Tests in Soft Swedish Clays. *Geotechnical and Geological Engineering* 39(7):4843–4854.

Jamiolkowski, M., Ladd, C.C., Germaine, J.T., and Lancellotta, R. 1985. New Developments in Field and Laboratory Testing of Soils. In: *11th International Conference on Soil Mechanics and Foundation Engineering*, 57–153. Balkema.

Janbu, N. 1963. Soil Compressibility as Determined by Oedometer and Triaxial Tests. In: *European Conference on Soil Mechanics and Foundation Engineering*, Wiesbaden. 19–25.

Janbu, N. 1979. Mechanics of Failure in Natural and Artificial Soil Structures. In: *International Symposium on Soil Mechanics*. Oaxaca, Mexico. pp. 95–124.

Janbu, N. 1989. *Fundamentals of Soil Mechanics* [In Norwegian]. Tapir, Trondheim.

Jardine, R.J. 2020. Geotechnics, Energy and Climate Change. *The 56th Rankine Lecture. Géotechnique* 70(1):3–59.

Jardine, R.J., and Hight, D.W. 1987. The Behaviour and Analysis of Embankments on Soft Clay. *Bulletin of the Public Works Research Center*, 159–244, Athens.

DeGroot, D., Lunne, T., Ghanekar, R., Knudsen, S., Jones, C.D., and Yetginer-Tjelta, T.I.. 2019. Engineering Properties of Low to Medium Overconsolidation Ratio Offshore Clays. *AIMS Geosciences* 5(3):535–567.

Karlsrud, K., and Hernandez-Martinez, F.G. 2013. Strength and Deformation Properties of Norwegian Clays from Laboratory Tests on High-Quality Block Samples. *Canadian Geotechnical Journal* 50(12):1273–1293.

Lacasse, S. 1995. Stress-Strain Behaviour: Importance of Mode and Rate of Load Application for Engineering Problems. In: *International Symposium on Pre-Failure Deformation Characteristics of Geo-Materials*. 2:887–907. Sapporo, Japan.

Lacasse, S., Berre, T., and Lefevbre, G. 1985. Block Sampling of Sensitive Clays. In: *11th International Conference on Soil Mechanics and Foundation Engineering*, 887–892. San Francisco.

Ladd, C.C., and DeGroot, D.J. 2003. Recommended Practice for Soft Ground Site Characterization: Arthur Casagrande Lecture. In: *12th Panamerican Conference on Soil Mechanics and Geotechnical Engineering*, 1–60. Cambridge, MA.

Ladd, C.C., Foott, R., Ishihara, K., Schlosser, F., and Poulos, H.G. 1977. Stress-Deformation and Strength Characteristics. In: *9th International Conference on Soil Mechanics and Foundation Engineering*, 421–494. Tokyo.

Larsson, R. 1977. Basic Behaviour of Scandinavian Soft Clays. *Swedish Geotechnical Institute*, Report 4, Linköping.

Larsson, R. 1980. Undrained Shear Strength in Stability Calculation of Embankments and Foundations on Soft Clay. *Canadian Geotechnical Journal* 17: 591–602.

Larsson, R. 1990. Behaviour of Organic Clay and Gyttja. *Swedish Geotechnical Institute*, Report 38, Linköping.

Larsson, R., and Sällfors, G. 1981. Hypothetical Yield Envelope at Stress Rotation. In: *International Conference on Soil Mechanics and Foundation Engineering*, 693–696. Stockholm.

Larsson, R., and Sällfors, G. 1986. Automatic Continuous Consolidation Testing in Sweden. *Consolidation of Soils: Testing and Evaluation*, ASTM STP 892, 299–328.

Larsson, R., Sällfors, G., Bengtsson P.E., Alén, C., Bergdahl, U., and Eriksson, L. 2007. Undrained Shear Strength [In Swedish]. *Swedish Geotechnical Institute, Information* 3: 1–68.

L'Heureux, J.S., Lindgård, A., and Emdal, A. 2019. The Tiller-Flotten Research Site: Geotechnical Characterization of a Very Sensitive Clay Deposit. *AIMS Geosciences* 5(4):831–867.

L'Heureux, J.S., Gundersen, A.S., D'Ignazio, M., Smaavik, T., Kleven, A., Rømoen, M., Karlsrud, K., Paniagua, P., and Hermann, S. 2018. Impact of Sample Quality on CPTU Correlations in Clay—Example from the Rakkestad Clay. In: *Cone Penetration Testing*, 395–400. CRC Press, London.

Lundström, K., and Dehlbom, B. 2019. Change in Properties over Time under Existing Embankments [In Swedish]. *Swedish Transport Authority and Swedish Geotechnical Institute*, Report A2016-10, Linköping.

Lunne, T., and Andersen, K. 2007. Soft Clay Shear Strength Parameters for Deep-Water Geotechnical Design. In: *6th International Site Investigation and Geotechnics Conference, Confronting New Challenges and Sharing Knowledge*, 151–176. London.

Lunne, T., Berre, T., and Strandvik, S. 1997. Sample Disturbance Effects in Soft Low Plastic Norwegian Clay. In: *Recent Developments in Soil and Pavement Mechanics*, 81–102. Rotterdam, Balkema.

Lunne, T., Berre, T., Andersen, K.H., Strandvik, S., and Sjursen, M. 2006. Effects of Sample Disturbance and Consolidation Procedures on Measured Shear Strength of Soft Marine Norwegian Clays. *Canadian Geotechnical Journal* 43(7):726–750.

Mayne, P.W., and Mitchell, J.K. 2011. Profiling of Overconsolidation Ratio in Clays by Field Vane. *Canadian Geotechnical Journal* 25(1):150–157.

Mesri, G. 1975. Discussion on 'New Design Procedure for Stability of Soft Clays.' *Journal of Geotechnical Engineering* 101: 409–412.

Mesri, G. 1989. A Reevaluation of Su(Mob) = 0.22σ 'p Using Laboratory Shear Tests. *Canadian Geotechnical Journal* 26(1):162–164.

Mię
dlarz, K., Konkol, J., and Bałachowski, L. 2019. Effective Friction Angle of Deltaic Soils in the Vistula Marshlands. *Studia Geotechnica et Mechanica* 41(3):143–150.

NGI., 2017. *OCR and Type of Stress Path* [In Norwegian]. Internal Report, Norwegian Geotechnical Institute, Trondheim.

NIFS, 2014. *Choice of Characteristic SuA-Profile Based on Field and Laboratory Investigations* [In Norwegian]. Norwegian Water Resources and Energy Directorate, Oslo.

O'Kelly, B.C. 2017. Measurement, Interpretation and Recommended Use of Laboratory Strength Properties of Fibrous Peat. *Geotechnical Research* 4(3):136–171.

Paniagua, P., L'heureux, J.S., Yang, S.Y., and Lunne, T. 2016. Study on the Practices for Preconsolidation Stress Evaluation from Oedometer Tests. In: *Proceedings 17th Nordic Geotechnical Meeting*, 547–557, Reykjavik.

Paniagua, P., D'Ignazio, M., L'Heureux, J.S., Lunne, T., and Karlsrud, K. 2019. CPTU Correlations for Norwegian Clays: An Update. *AIMS Geosciences* 5(2):82–103.

Phoon, K.K. 2020. The Story of Statistics in Geotechnical Engineering. *Georisk* 14(1):3–25.

Phoon, K.K., Ching, J., and Cao, Z. 2022. Unpacking Data-Centric Geotechnics. *Underground Space* 7(6):967–989.

Sällfors, G. 1975. *Preconsolidation Pressure of Soft, High-Plastic Clays*. Doctoral thesis, Chalmers University of Technology, Gothenburg.

Sällfors, G., and Larsson, R. 2016. *Compilation of Case Records* [In Swedish]. Swedish Transport Authority, Publication 2017:078. Gothenburg.

Sällfors, G., Larsson, R., and Bengtsson, P.E. 2017. *Recommendations for Undrained Shear Strength* [In Swedish]. Swedish Transport Authority, Publication 2017:078. Gothenburg.

Tang, C., and Phoon, K.K. 2021. Model Uncertainties in Foundation Design. Model Uncertainties in Foundation Design. In: *Model Uncertainties in Foundation Design*, CRC Press, Boca Raton, FL.

Tavenas, F., and Leroueil, S. 1987. Laboratory and in Situ Stress-Strain-Time Behavior of Soft Clays I: Laboratory Tests. In: *Simposio Internazionale de Ingegneria Geotecnica*, Mexico City. 7–47.

Vesterberg, B., and Andersson, M. 2022. *Settlement and Strength Properties in Sulphide Soils* [In Swedish]. Swedish Geotechnical Institute, Linköping.

Westerberg, B. 1999. *Behaviour and Modelling of a Natural Soft Clay*. Doctoral thesis, Luleå University of Technology, Luleå.

Chapter 8

Prediction for the mechanical response of gravels

Jingmao Liu, Degao Zou, Duo Li, Fanwei Ning, Chenguang Zhou, and Kaiyuan Xu

This chapter presents an extensive study on predictions for the mechanical response of gravels. Large amounts of laboratory testing data for gravels are not sufficiently utilized to support the prediction of the mechanical properties of gravels. A dataset of gravels from a triaxial test, termed Geo-Gravel, was established to provide the basis for estimating the mechanical response using physical properties and external conditions in the absence of test results. The dataset involved 332 types of gravels with 1102 triaxial compression test records. Fifteen indicators affecting mechanical behaviors were collected, including particle properties (e.g., particle hardness, σr, and roundness ρ), soil mass properties (e.g., initial void ratio e0, typical particle sizes, and gradation parameters), and external condition (e.g., confining pressure $\sigma 3$). The dataset was employed to predict the shear strength and stress–strain–volume responses. Based on this dataset, empirical formulations and deep learning models were developed to predict shear strength and stress–strain–volume response, respectively. Firstly, the detail of establishing the dataset is presented; subsequently, the methods of predicting shear strength (including empirical formulas and models) are presented; finally, the methods of predicting curves (including empirical formulas and models) are presented. The dataset can supply the possibility for data-driven geotechnical design.

8.1 INTRODUCTION

Gravels are widely used in hydraulic construction, road, and railway projects due to their high strength, low compressibility, and widespread availability (Indraratna et al. 1998, Sevi and Ge 2012, Xiao et al. 2014). It is important to accurately assess the mechanical behavior of gravels for the stability analysis and safety evaluation of engineering. Triaxial test is a commonly used approach to evaluate the mechanical properties of soils, primarily employed to determine the shear strength and deformation characteristics of gravels (Xu et al. 2012; Chu 2014; Huang et al. 2019). However, performing experimental investigations involves high costs and long period, including processes from quarry blasting for extraction, material transportation, to laboratory tests. Moreover, many projects are unable to conduct relevant experiments at the preliminary design phase. Experimental parameters are often obtained through simple analogies with similar projects, resulting in a lack of a reasonable basis for parameter determination. Additionally, the absence of test curves poses a challenge in determining the parameters for advanced constitutive models. To address this, efficiently collecting and leveraging existing experimental data can offer a practical solution. Establishing a reliable dataset is crucial for this purpose, and integrating these data with data-driven approaches supports the evaluation of gravel's mechanical properties and informs engineering decisions.

DOI: 10.1201/9781003441946-8

Recent years have seen the establishment of valuable datasets focusing on geotechnical materials (Phoon and Retief 2016; Phoon 2020). Notably, Ching and Phoon et al. provided useful multivariate datasets on clay (Ching and Phoon 2012; Ching and Phoon 2014; Ching and Phoon 2019), sand (Ching et al. 2017), and rock (Ching et al. 2018) properties. Tang and Phoon (2018; 2019a; 2019b) further provide a comprehensive survey of datasets on geotechnical structures. These studies provide important support for the assessment of geotechnical variability and uncertainty in site data. Regarding gravel materials, Kaunda (2015) gathered data on 47 types of rockfills with 165 test records, utilizing machine learning for shear strength prediction. Ovalle et al. (2020) collected data on 33 types of gravels with 158 test records.

Further, there are also extensive datasets that have been used to model and estimate the mechanical behaviors of the soils. Datasets regarding the mechanical behavior of soils primarily come from three sources: constitutive models (Javadi et al. 2012; Zhao et al. 2014; Zhang et al. 2020; 2021b; Wu et al. 2023b), discrete element simulations (Li et al. 2017; Wang and Sun 2019; Wu and Wang 2022; Wang et al. 2022; Guan et al. 2023), and laboratory tests (Habibagahi and Bamdad 2003; Nassr et al. 2018; Rashidian and Hassanlourad 2014; Banimahd et al. 2005; Araei 2014). Constitutive models and discrete element simulations can generate up to thousands of data records, through calculations based on specific requirements. These datasets are mainly used to validate and improve the effectiveness of machine learning (ML) methods, as well as to fit and model the mechanical behavior of specific materials, serving as a substitute for complex constitutive models. However, their accuracy is generally limited by the rationality of model assumptions and parameter selections. The data struggle to reliably reflect the true patterns of soil characteristics still depends on test data for validation.

Phoon and Zhang (2023) emphasize placing data at the core and underscore the significance of a data-centric approach in geotechnics. Unlike areas like computer technology and the Internet of Things, where data is readily accessible through automated means, collecting primary gravel test data is challenging due to diverse, non-digital sources. This necessitates extensive manual effort for collection, hindering dataset development, and constraining the advancement of data-driven approaches like machine learning. Thus, establishing a reliable dataset of laboratory tests is crucial.

With the rapid development of artificial intelligence (AI) technology, data-driven methods, particularly machine learning (ML), have demonstrated significant potential in the field of geotechnical engineering due to their powerful capabilities in handling intricate problems (Reich 1997, Vadyala et al. 2022, Aydin et al. 2023). ML techniques efficiently harness available data, directly learning and recognizing patterns without making any assumptions. They discover hidden patterns and capture complex nonlinear relationships (Janiesch et al. 2021). The cornerstone for the success of ML methods in civil engineering lies in the available datasets and efficient modeling algorithms (Schmidhuber 2015).

Machine learning (ML) is an important approach to achieving AI and has been widely applied in civil engineering practice and shown great promise (Lazarevska et al., 2014; Adeli, 2001; Zhu et al., 2021; Wei et al., 2021). Based on these datasets, various ML algorithms were applied to predict the mechanical properties using index properties, such as rocks (Mahmoodzadeh et al., 2021; Liu et al., 2015), soils (Zhang et al., 2021; Hao and Pabst, 2022; Zhang et al., 2023; Zhang et al., 2021), and concretes (Patil et al., 2022; Ly et al., 2019; Ling et al., 2019; Chen et al., 2018). Recently, Kaunda (2015) established a dataset with 165 sets of data for rockfill materials and used index properties to predict the shear strengths of rockfill materials using an artificial neural network (ANN).

Based on this dataset (Kaunda, 2015); Ahmad et al. (2021, 2022) and Zhou et al. (2019) applied support vector machine (SVM), Random forest (RF), AdaBoost, k-nearest neighbor (KNN), Gaussian process regression (GPR), and Cubist models for shear strength prediction.

Additionally, the feasibility of various ML algorithms in capturing the mechanical behavior of soils has been explored and validated (Zhang et al. 2020, 2022b, Qu et al. 2023). Shallow ML algorithms dominate, such as evolutionary polynomial regression (EPR) (Javadi et al. 2012), genetic programming (GP) (Javadi and Rezania 2009), support vector machine (SVM) (Zhao et al. 2014), radial basis function (RBF) (Peng et al. 2008), artificial neural network (ANN) (Wu et al. 2023b), etc., which can effectively utilize simple parameters (such as void ratio) and computation results obtained from constitutive models and discrete element simulation datasets to model the mechanical response of specific soils. However, these algorithms often exhibit limitations when predicting the mechanical behavior of unknown materials through test datasets containing multiple physical parameters and a large amount of data (LeCun et al. 2015). For instance, the ANN model proposed by Penumadu and Zhao (1999) encounters challenges in simultaneously capturing the mechanics behavior of soil under varying confining pressures, prompting the need for distinct models for different pressure stages. This is due to the small capacity of shallow models with limited representational capabilities and inadequate ability to capture complex features and relationships. Furthermore, their performance relies heavily on manual feature engineering, presenting a challenge in handling intricate and extensive test data. There exists a risk of overfitting on large datasets, which may result in a reduction of the model's generalization ability (Schmidhuber 2015; LeCun et al. 2015).

In comparison to shallow ML models, deep learning (DL) models possess the capability to automatically capture essential features and facilitate end-to-end predictions using original data as input. DL, characterized by its flexible structure, robust nonlinear mapping capacity, and powerful generalization, surpasses other shallow ML algorithms in various domains such as computer vision, speech recognition, and bioinformatics (Szegedy et al. 2015; Chauhan and Singh 2018). In the domain of large-scale visual recognition, with the increase of images in the ImageNet dataset and the development from conventional MLs to advanced DLs, the recognition error rate reduced from the initial at least 25% to less than 5% (Russakovsky et al., 2015; Deng et al., 2009), which is lower than that of humans (5.1%). Sun et al. (2017) attribute this significant success to the advanced algorithmic model based on the availability of large-scale data and the benefit of high-performance computing power.

Recurrent neural networks (RNNs), especially a variant Long Short-Term Memory (LSTM) network, demonstrate significant advantages in handling sequential data, such as stress–strain–volume responses. This has proven effective in capturing the mechanics behavior of soils (Zhang et al. 2020; Qu et al. 2021; Benabou 2021; Wu and Wang 2022). However, the challenge lies in fully utilizing complex spatial relationships among multiple physical properties, potentially limiting the model's performance in managing spatiotemporal data. For instance, the study of Zhang et al. (2021b) with LSTM reported unsatisfactory prediction results under low-stress levels. Recognizing the potential of convolutional neural networks (CNNs) in extracting spatial features, hybrid models combining CNNs and RNNs have gained popularity in various applications, including air quality prediction (Zhang et al. 2022a, Wu et al. 2023a), motor fault detection (Mohammad-Alikhani et al. 2023), energy consumption forecasting (Albelwi 2022), and agricultural farming monitoring (Du et al. 2022). Inspired by this approach, the implementation of such a hybrid structure to capture the mechanical behavior of gravels, leveraging both spatial features in physical parameters and sequential information in stress–strain–volume curves, holds promise to significantly enhance predictions of mechanics behavior for gravels. Notably, this approach has yet to be explored in applications for modeling the mechanical behavior of soils.

8.2 ESTABLISHMENT OF GEO-GRAVEL DATASET

8.2.1 Establishment process

This study introduces "Geo-Gravel," a gravel dataset using particle and soil mass properties to estimate peak deviatoric stress. To create a robust dataset, diverse experiments from 63 public papers and 22 practical lab reports, including triaxial compression tests on 332 gravels with 1102 records, were collected. 90.2% of the material is associated with rockfill used in dam construction. The data was sourced from the Web of Science and the China National Knowledge Infrastructure. The triaxial compression tests were performed all under drained conditions. Generally, the peak point of the deviatoric stress is considered as q_{peak} (equal to $\sigma_{1peak} - \sigma_3$, σ_{1peak} is the peak major principal stress, σ_3 is minor principal stress or confining pressure) in triaxial compression tests. If there is no peak point of the deviatoric stress as for strain hardening soils, the corresponding value of deviatoric stress at the strain of 15% is used as q_{peak} (GB/T 50123. 2019; ASTM D7181-11). q_{peak} is the key parameter for assessing soil shear strength. Based on multiple q_{peak}, other strength parameters such as the cohesion (c) and angle of internal friction (φ) in the Mohr–Coulomb criterion (Labuz and Zang, 2012) can be easily converted by fitting. q_{peak} obtained directly from the experiments and avoids any assumptions and simplifications in the fitting process as compared to calculating c and φ. Therefore, the inclusion of q_{peak} contributes to more accurate assessment results.

Previous research (Xiao et al., 2014; Liu and Chen, 2019; Varadarajan et al., 2003) highlights the influence of factors like lithology, density, gradation, particle shape, and stress conditions on gravel mechanics. Based on previous studies (Sadrekarimi and Olson, 2010), the factors affecting q_{peak} are divided into three categories: particle properties, soil mass properties, and external conditions (specifically, confining pressure in our dataset). The 12 indicators for gravels collected in the dataset were as many as possible.

(1) Particle properties

Particle shape. Generally, gravels are primarily obtained by artificially blasting parent rock or collected from natural river alluvium with different shapes. The regularity factor ρ is selected as the indicator of particle shapes, which has been adopted in previous studies (Cho et al., 2007; Krumbein, 1941).

Particle hardness. The uniaxial compressive strength of saturated parent rock σ_r is selected as the indicator of particle hardness.

For the missing data, the value was selected based on descriptions of materials as Table 8.1 (Bareither et al., 2008) and Table 8.2 (GB50021-2001. 2001), and the average value within each category was chosen as the empirical value.

Table 8.1 Roundness categories including the range of roundness and average roundness

Roundness	*Range*	*Average*
Angular	0.10–0.26	0.18
Sub-angular	0.26–0.42	0.34
Sub-rounded	0.42–0.58	0.50
Rounded	0.58–0.74	0.66
Well-rounded	0.74–0.90	0.82

Table 8.2 Hardness categories including the range of hardness, average hardness, and description

Hardness	*Range (MPa)*	*Average (MPa)*	*Description*
Highly hard	>60	75	Fresh granite, diorite, diabase, basalt, andesite, gneiss, quartzite, quartz sandstone, siliceous conglomerate, siliceous limestone, etc.
Sub-hard	30–60	45	Weak-weathered hard rock; fresh marble, slate, limestone, dolomite, calcareous sandstone, etc.
Sub-soft	15–30	22.5	Medium-weathered hard rock or sub-hard rock; fresh tuff, phyllite, marl, sandy mudstone, etc.
Soft	5–15	10	Highly weathered hard rock or sub-hard rock; medium-weathered sub-soft rock; fresh to weak-weathered shale, mudstone, argillaceous sandstone, etc.
Highly soft	≤5	2.5	Completely weathered rock; hypabyssal rock.

(2) Soil mass properties

Initial void ratio. The initial void ratio e_0 is the important control index, especially for rockfill dams. To maintain data integrity, initial void ratio e_0, dry density ρ_d, and specific gravity G_s are collected to facilitate conversion and supplementation.

Particle gradation. The coefficient of uniformity coefficient C_u and curvature coefficient C_c, as well as the typical particle size D_{10}, D_{30}, D_{50}, D_{60}, D_{80}, and D_{max} corresponding to the passing percentage of 10%, 30%, 50%, 60%, 80%, and 100%, respectively, are collected as the comprehensive indicators of particle gradation.

(3) External condition

The confining pressure σ_3 is selected as an indicator of the external conditions for estimating q_{peak}.

Overall, a total of 12 indicators of gravels were collected, including ρ, σ_r, e_0, D_{max}, D_{80}, D_{60}, D_{50}, D_{30}, D_{10}, C_u, C_c, and σ_3.

(4) Stress–strain–volume curves

The equal-length sequences (axial strain values of 10%) of stress–strain–volume curves were intercepted from the test curves that were loaded with different axial strains.

8.2.2 Data statistical analysis

Figure 8.1 visually displays the data's statistical properties using a scatter matrix. The lower triangular section exhibits scatter plots, the diagonal presents data distribution histograms, and the upper triangular section showcases the correlation matrix.

The statistical characteristics of the indicators are as follows:

Regularity factor. The range of ρ is from 0.18 to 0.82, with a mean value of 0.48 and a standard deviation of 0.19. The 25% percentile (Q_1), 50% percentile (Q_2 or median), and 75% percentile (Q_3) values are 0.34, 0.34, and 0.66, respectively. ρ is distributed centrally at both ends with a bimodal distribution, indicating the presence of two main material types (blasting and river alluvium gravels). Sub-angular and angular particles ($\rho < 0.5$) constitute 66.1%, mainly from parent rock blasting. Sub-rounded, rounded, and well-rounded particles ($\rho \geq 0.5$) from river alluvium make up 33.9%.

Particle hardness. σ_r is in a wide range from 2.5 to 131.1MPa. The mean value is 68.1 MPa, with a standard deviation of 16.9 MPa, and Q_1, Q_2, and Q_3 values of 60, 75, and 75 MPa, respectively. The majority (95.3%) has σ_r >30 MPa, indicating highly hard and sub-hard rock gravels. The minority (4.7%) with σr ≤30 MPa represents sub-soft, soft, and highly soft materials. Thus, the dataset is primarily suited to evaluate the materials for highly hard and sub-hard rock gravels.

Initial void ratio. The range of e_0 is from 0.142 to 0.890. The mean value is 0.297, with a standard deviation of 0.136. Q_1, Q_2, and Q_3 values are 0.230, 0.250, and 0.300, respectively. e_0 shows a positive skew distribution and a medium negative correlation with q_{peak}.

Gradation. C_u is in a wide range from 1.0 to 2140.0, with a mean value of 48.67 and a standard deviation of 196.06. Q_1, Q_2, and Q_3 values are 7.0, 14.0, and 34.5, respectively. 13.5% falls under $C_u < 5$ (uniformly graded soil), and 86.5% falls under $C_u \geq 5$. A small portion with clay has $C_u > 500$ values, making up 1.5%. C_c is found in a range from 0.10 to 149.60, with a mean value of 4.60 and a standard deviation of 14.93. Q_1, Q_2, and Q_3 values are 1.1, 1.6, and 2.5, respectively. Continuous graded soil ($1 < C_c < 3$) constitutes 64.4%, while $C_u \leq 1$ and ≥3 make up 35.6%. D_{max} ranges from 10 to 160 mm, with a significant concentration at 60 mm (74.5%), influenced by the commonly employed specimen diameter limit. The typical particle size distributions are concentrated, and share a similar discrete degree with negative skew and leptokurtic distribution (i.e., the shape of the distribution curve is more pointed and steeper). D_{max} and D_{80} have a weak positive correlation with q_{peak}, while D_{30}, D_{10}, and C_u have a weak negative correlation.

External condition. σ_3 spans 0.001–4.0 MPa, encompassing diverse engineering project scenarios. The mean value is 0.985 MPa, respectively. Q_1, Q_2, and Q_3 values are 0.4, 0.8, and 1.4 MPa, respectively. There is a strong positive correlation observed between σ_3 and q_{peak}.

8.3 PREDICTION OF SHEAR STRENGTH

8.3.1 Machine learning model

Three algorithms such as artificial neural network (ANN), support vector machine regression (SVR), and random forest (RF) are used. For a more in-depth understanding

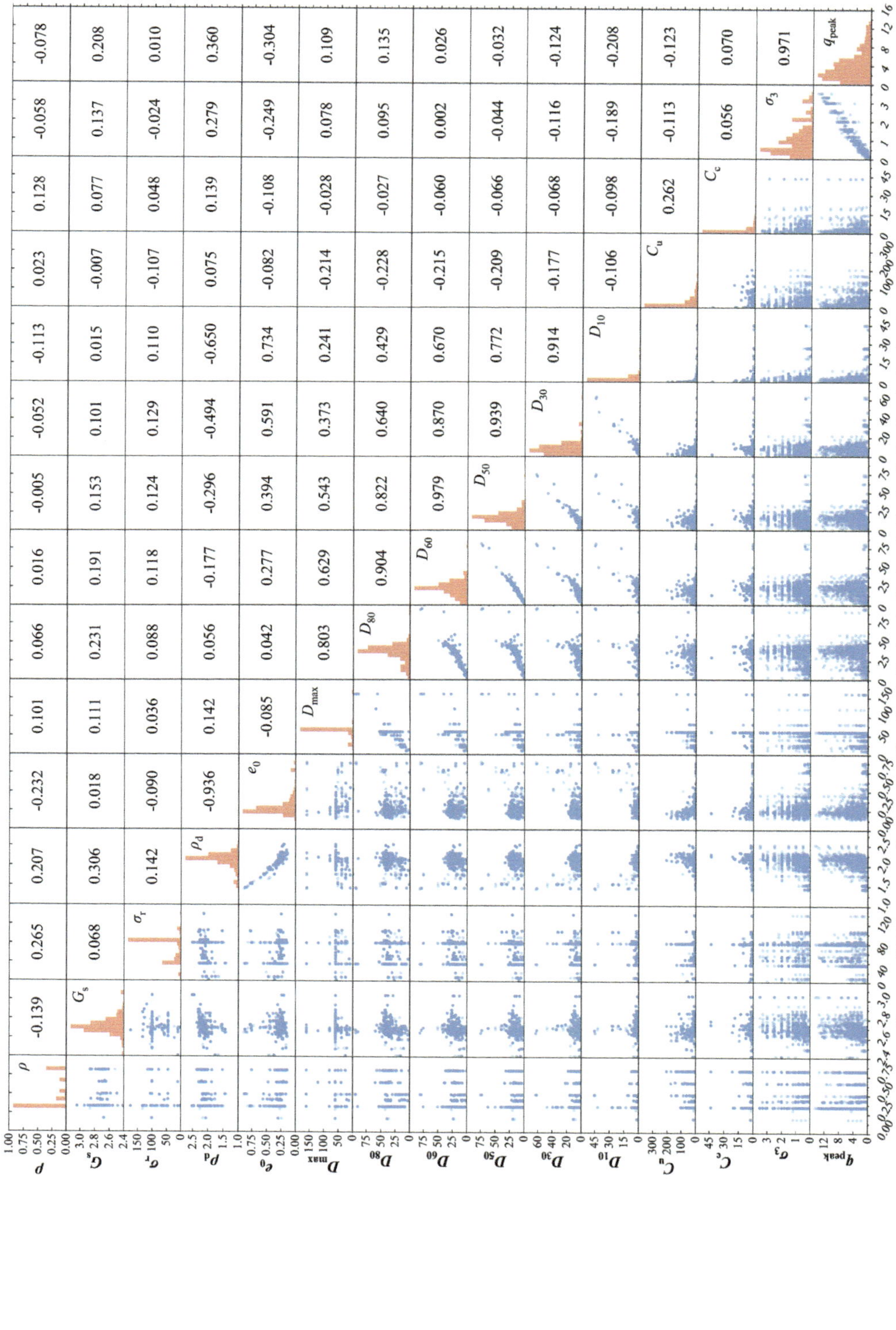

Figure 8.1 Scatter matrix

of the basic theory behind these models, readers are referred to previous studies, specifically those conducted by Phoon and Zhang (2023), Aydin et al. (2023), and Janiesch et al. (2021). Shear strength prediction was approached as a nonlinear regression optimization task, using 12 input indicators (e_0, C_u, C_c, D_{10}, D_{30}, D_{50}, D_{60}, D_{80}, D_{max}, σ_r, ρ , and σ_3). The single output variable was shear strength (i.e., σ_{1peak}, corresponding to the empirical equations in the later section). The dataset was split into 5:1 training and testing sets. The training set was used to learn patterns, with model parameters fine-tuned through optimization and fivefold cross-validation. This involved dividing the dataset into five subsets, using four for training and one for validation in each iteration. The test set evaluated the model's proficiency on unseen data, gauging its generalization capabilities. Crucial hyperparameters for each model underwent iterative optimization using the trial and error method. The ANN has a hidden layer with 50 neurons and a sigmoid activation function. For the SVR, the regularization parameter (C) is set to 10, the regression loss parameter ε is 0.1, and the kernel type is radial basis function (RBF). The RF consists of 150 trees, and each limits to a depth of five levels. Models for the shear strength of gravels are implemented using the scikit-learn library in Python.

Three performance metrics, i.e., coefficient of determination (R^2), root mean square error (RMSE), and mean estimation error (MAE), were employed to evaluate the predictive results of the models. The calculations for these metrics are as follows:

$$R^2 = 1 - \frac{\Sigma_{i=1}^{n}(A_i - P_i)^2}{\Sigma_{i=1}^{n}(A_i - \bar{A})^2} \tag{8.1}$$

$$MAE = \frac{1}{N}\Sigma_{i=1}^{n}\left|A_i - P_i\right| \tag{8.2}$$

$$RMSE = \sqrt{\frac{1}{N}\Sigma_{i=1}^{n}(A_i - P_i)^2} \tag{8.3}$$

where A_i and P_i are the actual value and predicted value, respectively, $\bar{A}$ is the average of the actual values, and n is the amount of the data.

Table 8.3 presents the evaluation results for the models, highlighting RF's superior performance in both the training set (R^2 = 0.972, MAE = 0.379 MPa, RMSE = 0.516 MPa) and the test set (R^2 = 0.917, MAE = 0.575 MPa, RMSE = 0.698 MPa) compared to ANN and SVR. Based on the optimal prediction performance of RF, the importance of 12 input indicators was assessed by analyzing their contributions to decision trees (Babar et al., 2020). Each indicator's contribution is gauged through the variation in node Gini

Table 8.3 Performance metrics of ANN, SVR, and RF models

Model	*Phase*	R^2	*MAE (MPa)*	*RMSE (MPa)*
ANN	Training	0.948	0.514	0.709
	Testing	0.835	0.633	0.998
SVR	Training	0.956	0.420	0.584
	Testing	0.881	0.625	0.861
RF	Training	0.972	0.379	0.516
	Testing	0.917	0.575	0.698

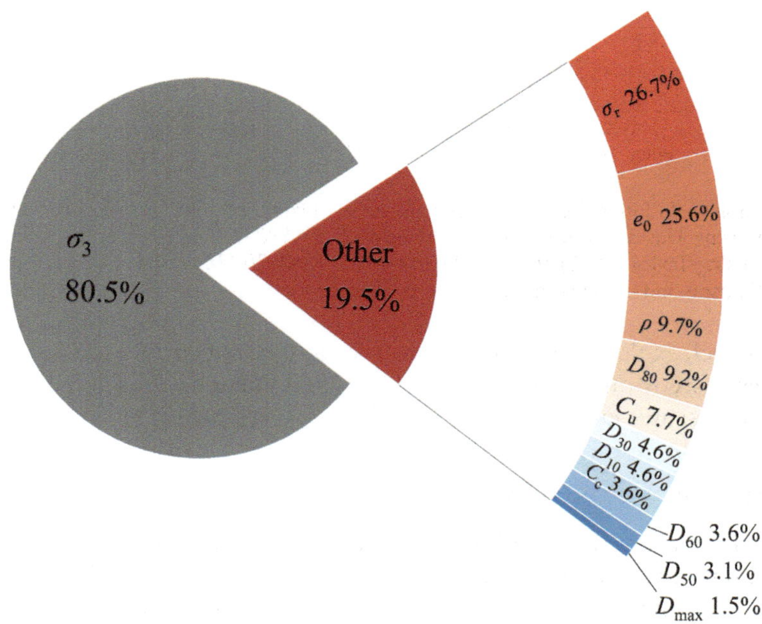

Figure 8.2 Relative importance for 12 indicators to evaluate q_{peak}.

impurity before and after branching in decision trees, with total contribution obtained by summing these variations across all trees. Normalization of contributions yields relative importance, allowing for the ranking of indicators (Archer and Kimes, 2008; Gómez-Ramírez et al., 2020). The importance is depicted in Figure 8.2.

The importance of ranking reveals that the external condition σ_3 holds the highest importance (80.5%) in estimating gravel q_{peak}, followed by σ_r, e_0, ρ , D_{80}, C_u, D_{30}, D_{10}, C_c, D_{60}, D_{50}, and D_{max}, totaling 19.5%. Among particle and soil mass properties, σ_r, e_0, ρ , D_{80}, and C_u stand out as the top five indicators, constituting 28.0%, 26.9%, 10.2%, 9.7%, and 8.1%, respectively. These key indicators formed the basis for a statistical model used to create an empirical formula for predicting shear strength.

8.3.2 Deep learning model

8.3.2.1 Seft-Net model architecture

In this chapter, the prediction of q_{peak} using index properties was regarded as a nonlinear regression optimization problem that can be presented as follows:

$$q_{peak} = f^{DL}(\sigma_r, \rho, e_0, D_{\max}, D_{80}, D_{60}, D_{50}, D_{30}, D_{10}, C_u, C_c, \sigma_3) \tag{8.4}$$

where q_{peak} is peak deviatoric stress; f^{DL} denotes the relation between the input variables and q_{peak} from the model; a total of 12 input variables of gravels are collected as much as possible, including uniaxial compressive strength of saturated parent rock σ_r, regularity factor ρ , initial void ratio e_0, D_{max}, D_{80}, D_{60}, D_{50}, D_{30}, D_{10}, uniformity coefficient C_u, curvature coefficient C_c, and confining pressure σ_3.

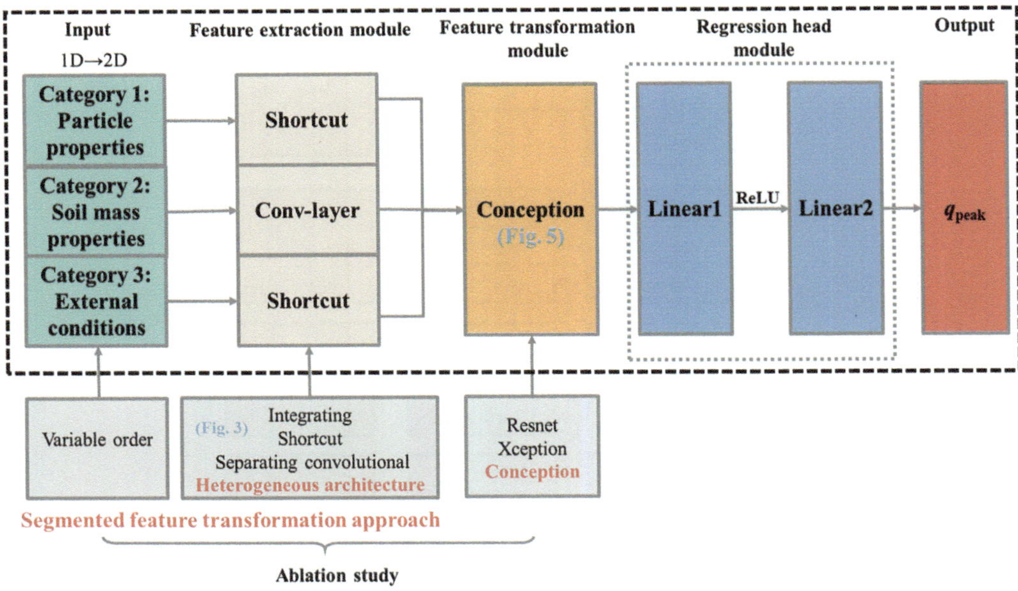

Figure 8.3 Seft-Net (Segmented feature transformation-Network) model architecture.

Thus, a deep learning model to achieve this complex task is proposed, which can abstract deep features and learn the potential mapping relationship of the data with excellent nonlinear transformation capability. The proposed methodology for q_{peak} prediction is based on the Segmented Feature Transform Network model (Seft-Net), which incorporates depthwise separable convolutions and a segmented feature transformation approach adopted for the gravels' physical properties.

The Seft-Net model architecture executes a three-step process: feature extraction module, feature transform module, and regression head module, as shown in Figure 8.3. First, the feature extraction module extracts rough features from the segmented input properties of gravels. Then, the feature transform module takes over and utilizes multi-scale receptive fields and multi-branch architectures to abstract high-level features from the gravel properties. Finally, the regression head module with fully connected layers is responsible for reducing the dimension of learned high-level features and providing the ultimate output as q_{peak} prediction. The utility of the proposed modules in Seft-Net is validated by an ablation study. The implementation process details of the proposed module are as follows.

8.3.2.1.1 Proposed feature extraction module

The Seft-Net model employed 12 input variables in the dataset. In pre-training, it was found that the ordering had a certain impact on the model performance, and the model performance is better when similar attribute variables are ordered together. Inspired by it, a novel segmented feature transformation approach is proposed. Generally, these properties influencing q_{peak} could be segmented into three different categories (Sadrekarimi and Olson, 2010): category 1 is about the particle properties (i.e., parent rock strength σ_{r}, regularity factor ρ), category 2 is about the soil mass properties (i.e., initial void ratio e_0, typical particle sizes D_{max}, D_{80}, D_{60}, D_{50}, D_{30}, D_{10}, and gradation parameters C_{u} and

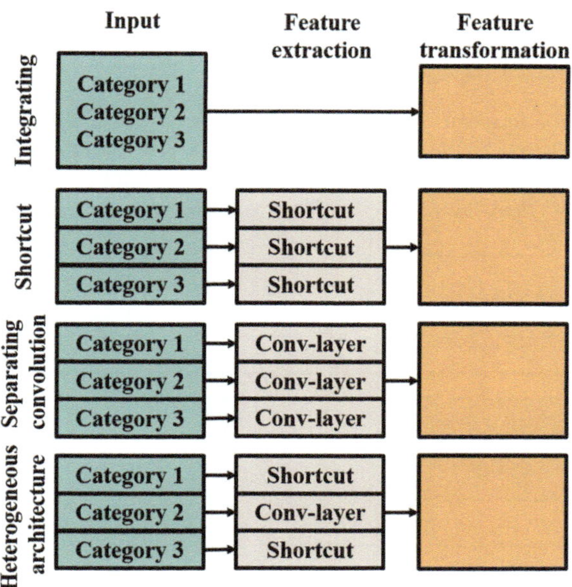

Figure 8.4 Architecture of different feature extraction modules.

C_c), and category 3 is about the external conditions (i.e., confining pressure σ_3). It is the segmenting of different variables based on their intrinsic physical properties, using the experience of long-term human research as a priori knowledge.

The input variables are first arranged in a 1 × 12 1D matrix. This initial arrangement is then subjected to a dimensionality expansion process, resulting in a 3 × 12 2D matrix that effectively expands the input space to optimize the model's efficiency in capturing spatial features associated with gravel properties. Subsequently, the variables within the 2D matrix are segmented into the aforementioned three categories and calculated by the feature extraction module. Each category is processed, respectively, by transmitting them through the heterogeneous architecture of neural networks. The utility of heterogeneous architecture of feature extraction in Seft-Net is validated by an ablation study, which compares the heterogeneous architecture module with three different modules, including the integrating module, shortcut module, and separating convolutional module, whose architectures are shown in Figure 8.4.

8.3.2.1.2 Proposed feature transform module

Depthwise separable convolution

Depthwise separable convolution has been proven to be successful in neural image classification compared to standard convolution. Standard convolution extracts feature both the spatial (width and height) and channel dimensions of the input with a high degree of coupling, where the convolutional kernel needs to describe spatial and cross-channel correlations simultaneously (Shang et al., 2020). In contrast to standard convolution, depthwise separable convolution firstly performs depthwise convolution independently on each channel of the input, followed by a pointwise convolution which processes the output of depthwise convolution with a 1 × 1 kernel to a new channel space, as shown in Figure 8.5.

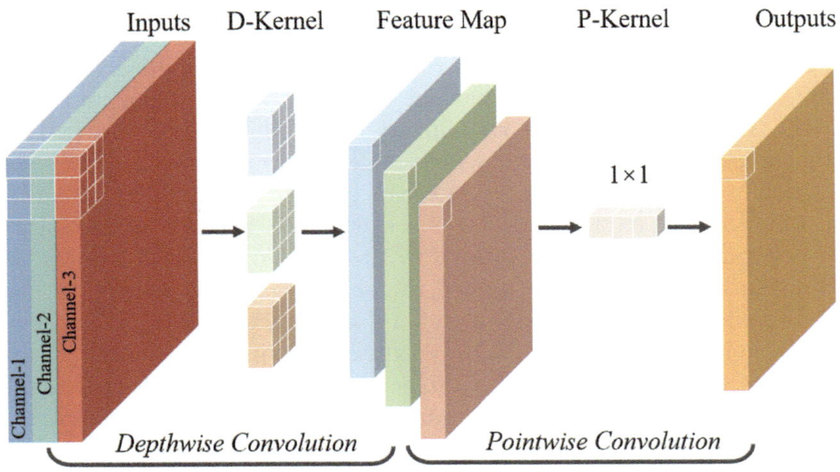

Figure 8.5 Depthwise separable convolution.

In depthwise convolution, a single kernel is applied to each input channel so that each channel can output one feature map, where the number of channels does not change after depthwise convolution. This process ($D\text{-}Conv$) can be expressed as Equation (8.2):

$$D\text{-}Conv(W, x)_{(i,j)} = \sum_{m,n}^{M,N} W_{(m,n)} \cdot x_{(i+m,j+m)} \tag{8.2}$$

In pointwise convolution, a 1 × 1 convolution is applied to combine the outputs of the depthwise convolution to extract spatial features. This does not change the spatial size of feature maps but can change the channel number. This process ($P\text{-}Conv$) can be expressed as Equation (8.3):

$$P\text{-}Conv(W, x)_{(i,j)} = \sum_{m,n}^{M,N} W_{(k)} \cdot x_{(i,j)} \tag{8.3}$$

where W is the weight matrix of convolutional kernels; x is the input of the convolutional layer, and (i, j) is the coordinate point of output feature maps; m, n, and k are the three dimensions of the convolutional kernel.

In further, the overall process ($DP\text{-}Conv$) of depthwise separable convolution can be expressed as Equation (8.4):

$$DP\text{-}Conv(W_p, W_d, x)_{(i,j)} = P\text{-}Conv(W_p, x)_{(i,j)}(W_p, D\text{-}Conv(W_d, x)_{(i,j)}) \tag{8.4}$$

Depthwise separable convolution can considerably reduce computational cost, by applying a separate filter to each input channel which reduces redundancy in the parameters. This facilitates better generalization because it reduces the risk of overfitting, and the efficiency of depthwise separable convolution is significantly higher than standard convolution (Chollet, 2017).

Conception module

Convolutional neural network (CNN)-based networks have achieved remarkable success in various image processing domains (Gu et al., 2018). Deep CNNs, in particular, have made noteworthy contributions to tasks like image classification and segmentation (Yamashita et al., 2018). They become widely known standards (Dhillon and Verma, 2020; Ajit et al., 2020). Increasing the depth and width of deep neural networks (DNNs) is a common approach to improving network performance (Khan et al., 2020; Tan and Le, 2019). Residual connections in models like ResNet (He et al., 2016) enable effective training even in deeper networks. Architectures inspired by the Inception module, such as GoogLeNet (Szegedy et al., 2015), have demonstrated the ability to extract multi-scale features. By concatenating different convolutional layers and stitching the resulting matrices together, the depth and width of the network can be efficiently expanded. This approach prevents overfitting and enhances model accuracy (Szegedy et al., 2016; Szegedy et al., 2017). Taking a step further, Xception (Chollet, 2017) is an improved architecture based on InceptionV3 (Szegedy et al., 2016) that separates channel relationships from spatial relationships through depthwise separable convolutions to efficiently extract features and improve performance.

Inspired by Xception (Chollet, 2017) and Densenet (Huang et al., 2017), a modified Xception architecture was used as the backbone of the feature transform module, called Conception. Similar to Xception (Chollet, 2017), the Conception architecture could be divided into three parts: entry flow, middle flow, and exit flow. The architecture of Conception is shown in Figure 8.6. The entry flow, which contains three residual blocks, utilizes two separable convolution layers and one normal convolution layer in parallel and then concretes the results as a bigger matrix. The middle flow simply shortcuts from three repeated separable convolutional layers and finally concatenates them in one matrix. This part is repeated eight times, as a higher count of duplication did not show great improvement in performance while increasing model complexity and calculating cost (Chollet, 2017). The upper part of the exit flow is similar to the residual blocks in the entry flow, then the exit flow uses two separable convolutional layers to adjust the output into vectors. Rectified Linear Unit (ReLU) is used as the activation function to perform nonlinear transforms. The successful part could be modified as a linear layer or convolutional layer to adapt to the demand of downstream tasks.

Unlike the conventional Xception (Chollet, 2017) architecture, Conception incorporates residual blocks (He et al., 2016) by directly adding the shortcut to the residual. The residual and shortcut vectors are concatenated into a two-row matrix, intending to retain more information about the original features from the output of the previous block. This change does show an improvement in final performance while replacing those blocks.

In the Conception module, as the input from the feature extraction module passes through subsequent convolutional layers, the network learns to extract increasingly abstract and high-level features. Lower layers tend to capture low-level features, while higher layers capture more complex features. Fusing hierarchically extracted features through multi-scale receptive fields and multi-branching architectures to abstract high-level features of gravel properties, which refer to abstract representations of the input gravel properties that capture complex and meaningful patterns relevant to the q_{peak} prediction. However, understanding these high-level features is difficult. In deep networks with multiple layers, the relationship between the input data and the high-level features becomes highly nonlinear and intricate. The models learn a huge number of parameters, and understanding the precise meaning and relevance of each parameter or connection becomes extremely difficult, impeding interpretability, as black boxes (Samek et al., 2021). We use an indirect approach to improve the interpretability of the model in a later section.

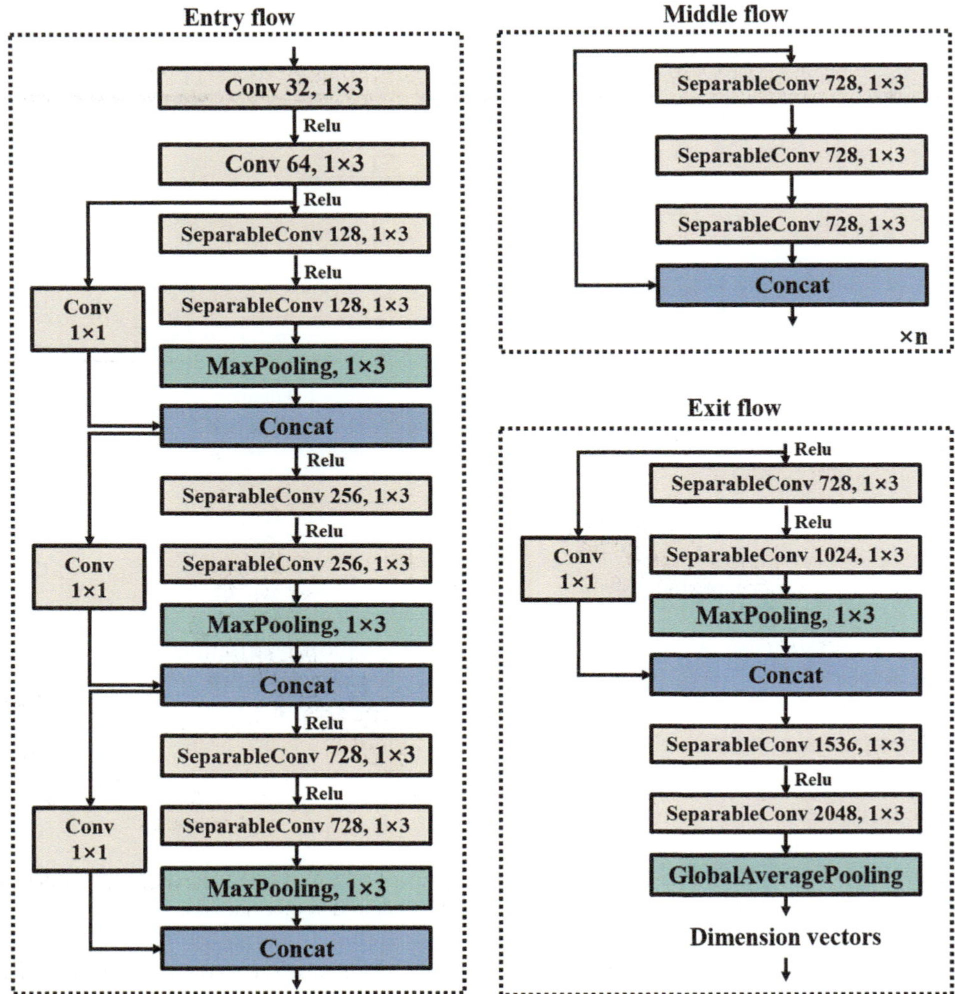

Figure 8.6 Conception module architecture.

Once the Conception module has extracted these high-level features of 12 gravel properties, they are fed into a few fully connected layers as the regression head module to reduce the dimension of features extracted from above and map to learned high-level features of gravel properties, providing the ultimate q_{peak} prediction in regression tasks.

8.3.2.2 Training strategy

In this section, the performance metrics of the evaluation and training strategy of the model are provided. The Seft-Net model can be deployed as below: First, the Seft-Net model is established and initialized. Then, the entire dataset was separated into training data and testing data with a ratio of 5:1. The ratio may vary depending on the specific requirements of the model and the dataset. The training data was used to decide the parameters of the model, and testing data was used to evaluate the model performance.

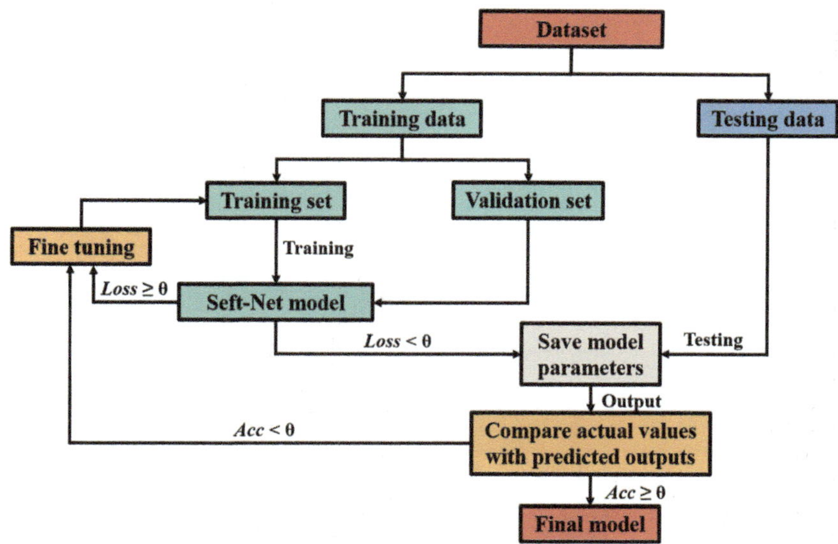

Figure 8.7 Calculation process of model.

During the experiments, Python is utilized as the programming language, and the PyTorch framework is used to perform the neural network. The model has a total of 22,756,959 parameters and was deployed on a workstation equipped with NVIDIA Tesla A100 GPU for training. The complete calculation process is shown in Figure 8.7.

(1) Performance metrics

In this study, the coefficient of determination (R^2), mean absolute error (MAE), root mean square error (RMSE), and mean absolute percentage error (MAPE) between the predicted and actual values were chosen as the criterion for defining the model's accuracy. Additionally, mean square error (MSE) and MAE are used as loss functions. These parameters can be calculated as follows:

$$R^2 = 1 - \frac{\Sigma_{i=1}^{n}(A_i - P_i)^2}{\Sigma_{i=1}^{n}(A_i - \bar{A})^2} \tag{8.5}$$

$$MAE = \frac{1}{N}\Sigma_{i=1}^{n}\left|A_i - P_i\right| \tag{8.6}$$

$$MSE = \frac{1}{N}\Sigma_{i=1}^{n}(A_i - P_i)^2 \tag{8.7}$$

$$RMSE = \sqrt{\frac{1}{N}\Sigma_{i=1}^{n}(A_i - P_i)^2} \tag{8.8}$$

$$MAPE = \frac{1}{N}\Sigma_{i=1}^{n}\frac{|A_i - P_i|}{A_i} \quad (8.9)$$

where A_i and P_i are the actual value and predicted value of q_{peak}, respectively, $\bar{A}$ is the average of the actual values, and n is the amount of data.

(2) Model training

During the training phase, the network learns the optimal weights for the convolutional filters and regression mapping layers by minimizing a loss function that measures the discrepancy between the predicted regression outputs and the actual values of q_{peak}. The network's parameters are adjusted through optimization techniques like gradient descent and backpropagation (Lecun et al., 2015) to improve the accuracy of the regression predictions.

During the training phase, the training set is iteratively fed into the Seft-Net model to optimize its parameters. The model parameters are updated and adjusted through the training process until the loss function, which measures the difference between the predicted values and the true values, decreases to a threshold of 1e−6. At this point, the model parameters are considered optimized, and they are saved for further evaluation and testing.

The training data is randomly divided into the training set and validation set. The ratio is set as 9:1, although other rational ratios can also be considered based on the specific requirements of the task. The training set is iteratively fed into the Seft-Net model to optimize its parameters. The model parameters are updated and adjusted through the training process making the loss function decrease. The validation set is used to fine-tune the hyperparameters and helps monitor the model's performance during training and detect signs of overfitting. The initial learning rate η is 0.001 with a rate decay of 0.98 every 10 epochs, and weight decay is 0.99 every 10 epochs. The specific values for these hyperparameters may vary depending on the model, dataset, and training objectives (Bergstra and Bengio, 2012). The settings used in this case are based on our former experiments and experience of the task domain. Two loss functions (MAE and MSE) were chosen for comparison.

(3) Model testing

In the testing phase, the testing data is used to evaluate the model performance with the parameters trained above. Once the accuracy achieved the goal pre-set, the model can be used as the final model; otherwise, the architecture of the model needs to be adjusted and trained again.

8.3.2.3 Experimental evaluation and results

8.3.2.3.1 Ablation study

The ablation study is used to identify the most important features of a model by adding, removing, or altering individual components of the model and observing the effect on the model's performance. It is used to identify which frameworks or features are most useful for the model's accuracy and to identify potential areas of improvement (Wang et al., 2021). According to the ablation study, the usefulness and effectiveness of the proposed modules to improve the accuracy of the model were verified. The evaluation metrics use the performance of the testing phase.

8.3.2.3.2 *Input variable order*

In the variable order experiments, the effects of different orders of input variables were compared. Due to the numerous ways of arranging and combining the 12 variables, five typical orderings (Table 8.4) were selected for experimentation based on the pre-training: (1) the variables were arranged in the order of the three categories; (2) the order of the variables was changed within the category; (3) the order of the variables was interchanged between category 1 and category 2; (4) the order of the variables was interchanged between category 2 and category 3; (5) the variables were interchanged in the order of all three categories.

As shown in Table 8.5(a), the result indicated that the ordering had a certain impact on the model performance. The model performance is better when similar attribute variables are ordered together as No. (1). No. (2) shows that changes in the order of variables within the category have little effect on model performance. No. (3)–No. (5) show that the interchange of variables between different categories has an effect on model performance, especially the ordering of σ_3 in category 3, which is related to the variable importance discussed later.

Different ordering of the variables may cause changes in the response pattern of the convolution kernel, and a reasonable ordering of the data facilitates the convolution network to capture useful local patterns and features more easily (Zhang et al. 2020, Dosovitskiy et al. 2021). By arranging related variables together, the model is more likely to find correlations and dependencies between these features. As for this task, it makes more sense to incorporate a priori knowledge indicating a certain way of ordering the input variables to help the network converge faster or extract relevant features better.

Table 8.4 Input variable order

No.	Variables											
(1)	σ_r	ρ	e_0	D_{max}	D_{80}	D_{60}	D_{50}	D_{30}	D_{10}	C_u	C_c	σ_3
(2)	ρ	σ_r	e_0	D_{60}	D_{50}	D_{max}	D_{30}	C_c	D_{10}	C_u	D_{80}	σ_3
(3)	D_{50}	ρ	e_0	D_{max}	D_{80}	D_{60}	σ_r	D_{30}	D_{10}	C_u	C_c	σ_3
(4)	σ_r	ρ	e_0	D_{max}	D_{80}	D_{60}	σ_3	D_{30}	D_{10}	C_u	C_c	D_{50}
(5)	D_{80}	C_u	e_0	D_{max}	σ_3	D_{60}	σ_r	D_{30}	D_{10}	D_{50}	C_c	ρ

Table 8.5 Experimental results of input variable order

	(a) Integrating				*(b) Heterogeneous architecture*			
No.	R^2	*MAE (MPa)*	*RMSE (MPa)*	*MAPE (%)*	R^2	*MAE (MPa)*	*RMSE (MPa)*	*MAPE (%)*
(1)	0.909	0.591	0.840	18.551	0.959	0.336	0.445	9.127
(2)	0.905	0.593	0.848	18.922	0.958	0.342	0.450	9.201
(3)	0.902	0.599	0.852	19.556	0.957	0.356	0.464	9.842
(4)	0.888	0.609	0.857	21.438	0.931	0.395	0.576	13.502
(5)	0.878	0.615	0.865	21.929	0.924	0.426	0.591	14.204

Inspired by it, we further propose a segmented feature transformation approach using a heterogeneous architecture in the feature extraction module. We also investigate the effects of variable order in heterogeneous architecture. The results in Table 8.5(b) indicated that the segmentation of variables into the wrong category is detrimental to the network decision making, resulting in some degradation of the model performance, and the degree of influence of the sort is related to the importance of the variable. It is speculated that correct categorization guides the model to enhance the learning of important features and improve the overall performance of the task through a priori knowledge, whereas incorrect categorization can somewhat misguide the model learning. In addition, it is noted that most deep convolutional networks usually automatically learn to adapt to feature representations of different ordering through convolutional operations. Thus, for non-sequentially related tasks, the order of the data usually has a limited impact on performance, and the impacts need to be assessed based on task requirements and data characteristics.

8.3.2.3.3 Feature extraction module

In the feature extraction module experiment, the utility of heterogeneous architecture within the segmented feature transformation approach is validated by comparing the heterogeneous architecture module with three different modules. Figure 8.8 illustrates the performance variance of different feature extraction modules.

The integrating module shows a decaying result due to its integrated architecture. This issue arises from the absence of segmentation in the input, preventing the fully connected (FC) layers from converging efficiently. The shortcut module shows a significant improvement in accuracy, and separating the convolutional module shows a close result, from which our idea comes from. By combining the shortcut module and separating the convolutional module, the heterogeneous architecture achieves optimal performance. The segmentation of variables is the grouping of similar indicator attributes, where a certain correlation or multicollinearity in the same categories may make the model become unstable and the predictive power reduced. By feature extraction from heterogeneous modules, this effect may be alleviated by segmenting to feature extraction, thus facilitating the convergence of the model and providing better feature abstraction for subsequent processing.

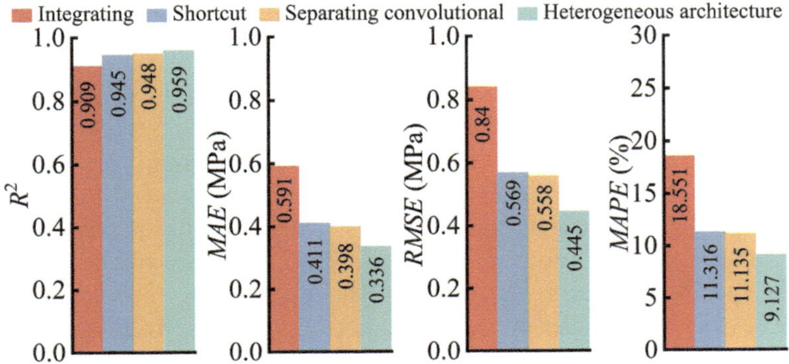

Figure 8.8 Ablation study results of different feature extraction modules.

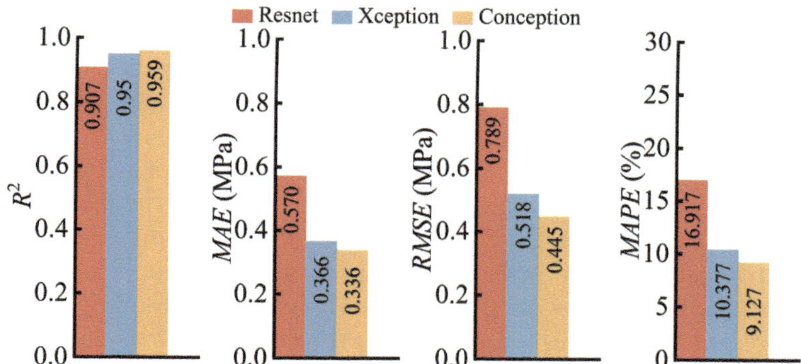

Figure 8.9 Ablation study results of different feature transformation modules.

8.3.2.3.4 Feature transformation module

To verify the utility of the proposed Conception module, compare the Conception with the original Xception module (Chollet, 2017) and ResNet module (He et al., 2016) in the Seft-Net. Figure 8.9 illustrates the performance variance of different feature extraction modules.

ResNet increases the depth of the network, while Xception and Conception expand the depth and width of the network efficiently. Additionally, Xception and Conception are able to extract features at different scales through convolutional operations with depthwise separable convolution and multi-scale fusion, using model parameters more efficiently and allowing the model to learn richer features. Therefore, Xception and Conception modules exhibit significantly superior performance to ResNet. By concatenating the residual and shortcut rather than adding both, the Conception module squeezes more performance by keeping more useful information as compared to Xception.

8.3.2.4 Seft-Net model evaluation

Figure 8.10 shows the progress of Seft-Net model training using the loss function of MAE and MSE for 3000 epochs. The loss value decreases gradually and the loss tends to converge with the increase in the epoch. The model parameters in the minimum loss are saved as the best model. It is indicated that the model using the loss function of MSE presents a faster convergence after about 300 epochs, while the model using MAE tends to converge after more than 1500 epochs.

Additionally, the accuracy of the model using the loss of MSE is higher than MAE, with MAPE of 9.127% and 9.901%, respectively. MSE gives higher weights to larger errors because of the squaring operation, making it more sensitive to outliers or large deviations between the predicted and actual values in this model. On the other hand, MAE treats all errors equally, regardless of their magnitude, but may provide a better understanding of the average prediction error (Goodfellow et al., 2016). Consequently, MSE was chosen as the loss function for the subsequent experiments. It applies to the present model and considers the specific characteristics of the problem.

The accuracy of the Seft-Net model in predicting q_{peak} of gravels is displayed in Figure 8.11, with the scatter plot of the actual and the predicted values for the training phase shown in Figure 8.11(a) and the testing phase in Figure 8.11(b), respectively. The closer

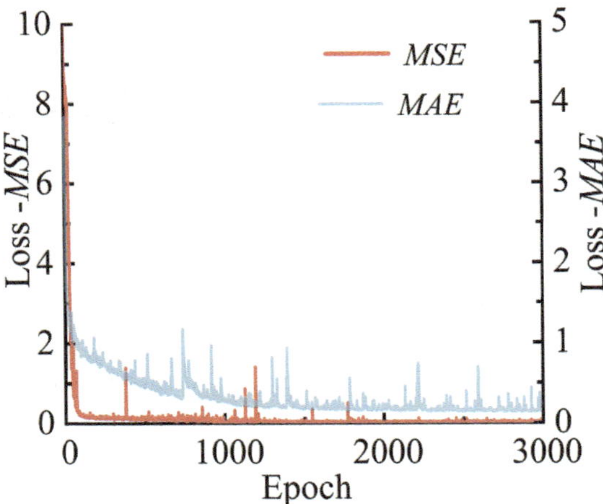

Figure 8.10. Progress of model training for 3000 epochs.

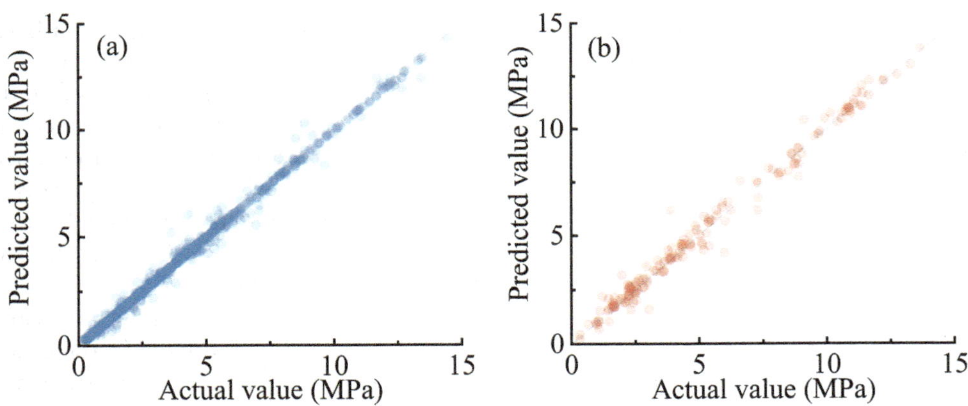

Figure 8.11 Scatter plots of actual vs. predicted q_{peak}: (a) training phase and (b) testing phase.

the points are to the diagonal line, the higher the accuracy. It is indicated that the higher aggregation of the results around the diagonal line both in the training and testing set, except for a few noise points.

The evaluation results in the training and testing phase are summarized in Table 8.6. As expected, the error in the testing phase is higher than that in the training phase, which is in line with the typical behavior of deep learning models (Zhang et al., 2021).

Figure 8.12 illustrates the percentage distribution of MAE and MAPE in both the training and testing phases. Notably, samples with high errors (MAE ≥ 1MPa, MAPE ≥ 50%) have a low occurrence percentage in both phases. The majority of predictions and fits

Table 8.6 Performance metrics of Seft-Net

Phase	R^2	*MAE (MPa)*	*RMSE (MPa)*	*MAPE (%)*
Training	0.994	0.111	0.232	4.102
Testing	0.959	0.336	0.445	9.127

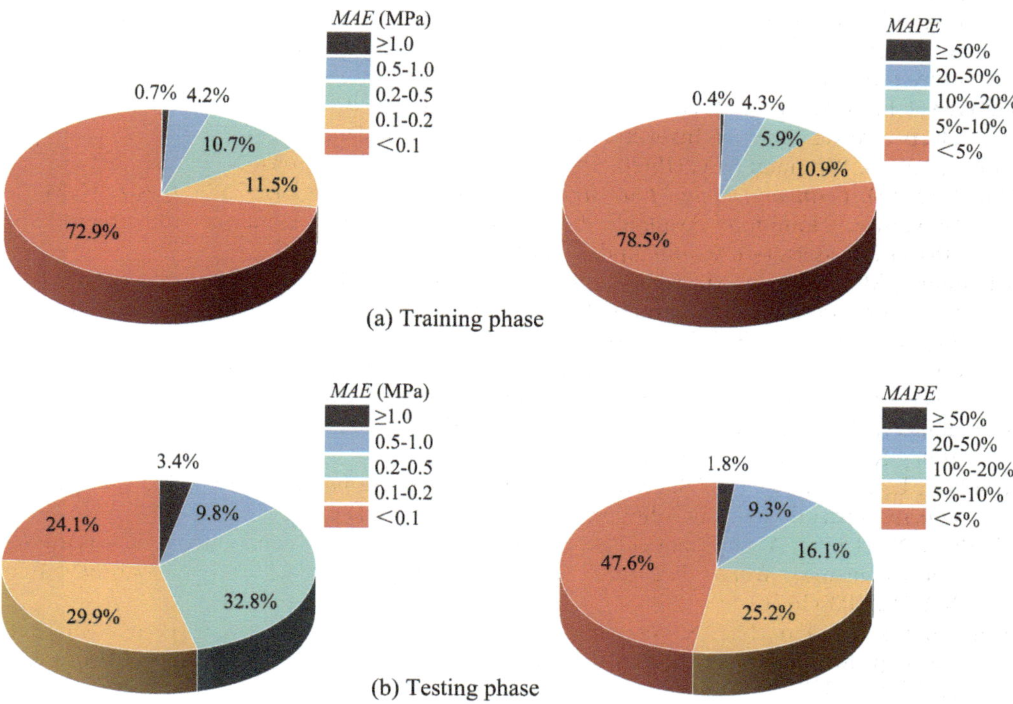

Figure 8.12 Percentage distribution of errors (MAE, MAPE): (a) training phase and (b) testing phase

were satisfactory, indicating that the model was able to accurately capture the intricate nonlinear relationships underlying patterns between gravel properties and q_{peak}. Upon analysis and examination, it was found that the samples with larger errors primarily stemmed from ballast materials and soft rock materials. The dataset included relatively few samples of these materials, resulting in the model being poorly fitted and struggling to accurately predict their properties. To enhance the accuracy of the model, the next step is to collect more data on the ballast materials and soft rock materials and include a more representative sample of them in the dataset.

In addition, three conventional ML models, i.e., naive ANN (Kaunda, 2015), RF (Zhou et al., 2019), and SVM (Ahmad et al., 2021), are used for comparison. The comparison of model performance is presented in Figure 8.13. In the training phase, Seft-Net achieved the best prediction performance. Although the R^2 is closer, other error metrics (i.e., MAE, RMSE, MAPE) of Seft-Net are reduced by more than 50% compared to ANN, RF, and SVM. As well, the Seft-Net achieved the best prediction performance in the testing phase, demonstrating the highest R^2 values, and MAE, RMSE, and MAPE of Seft-Net are reduced by more than 30% compared to ANN, RF, and SVM.

DL models like Seft-Net are superior to traditional ML methods in terms of scalable architecture (Dean et al., 2012) and advanced feature extraction (Schmidhuber, 2015; Deng and Yu, 2014), but are less interpretable (Carvalho et al., 2019; Lecun et al., 2015). In the cases of the established dataset, the developed Seft-Net model provided better performance. Additionally, the division of data in training and testing sets may influence the model performance indicators. Overall, based on the larger dataset and advanced deep learning algorithm structure, the proposed Seft-Net model has better performance.

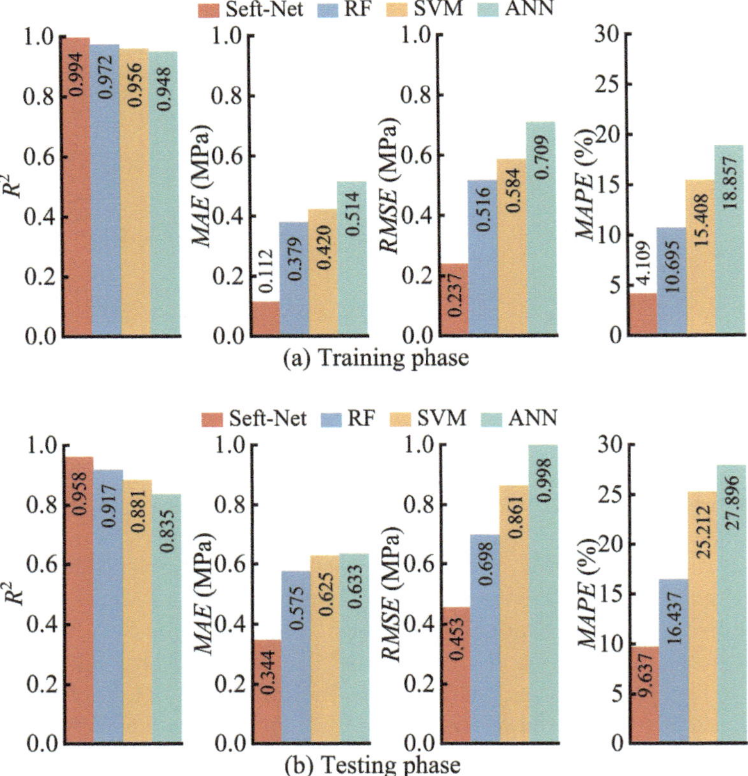

Figure 8.13 Comparison of performance metrics for Seft-Net, RF, SVM, and ANN: (a) training phase and (b) testing phase

8.3.2.5 Variable importance analysis

Deep neural networks are often considered difficult to interrogate as black box (Shrikumar et al., 2017). However, explanations for models play an important role in researching, developing, and using predictive models as information on what features were important for a given output enabling us to better understand, validate, and interpret model decisions (Doshi-Velez and Kim, 2017). To address this apparent dichotomy between performance and interpretability, perturbation-based methods are used for feature importance estimation (Li et al., 2016; Fong and Vedaldi, 2017; Dabkowski and Gal, 2017). It is mainly to mask or replace a feature of all samples and then compare the change in model performance before and after the masking or replacement. For example, a large drop in accuracy indicates that the features are more important.

Specifically, a model is trained in a training set with a specific variable masked, then predicted with the testing set with the same masked variable. The importance of each variable can be individually evaluated by the performance metrics MAE and MAPE of each masking model. The Seft-Net model in the database employed 12 input variables. By comparing and ranking (Figure 8.14), σ_3 is the most influential variable for predicting q_{peak}, where masking σ_3 results in a sharp increase in the mean absolute percentage error (MAPE) up to 69.552%, followed by σ_r, e_0, D_{80}, C_u, ρ , D_{30}, D_{10}, C_c, D_{max}, D_{60}, and D_{50}, respectively.

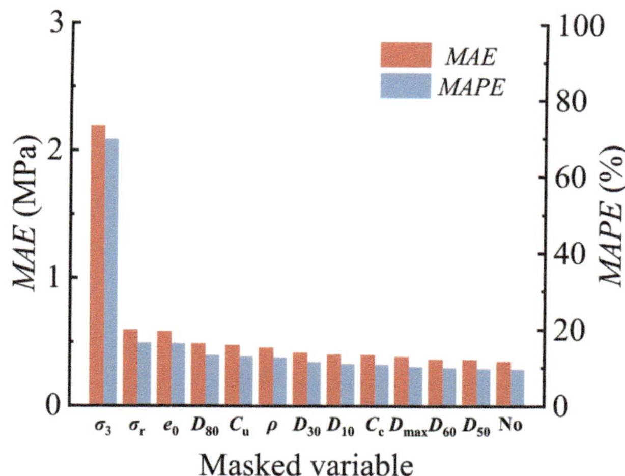

Figure 8.14 Performance metrics of different masking models and ranking.

After analyzing the variable importance, the top six variables that exhibit a MAPE variation greater than 2% have been chosen as inputs to train a simplified model, aiming to enhance convenience in practical application. The simplified Seft-Net model achieved a prediction performance with R^2 = 0.949, MAE = 0.407 MPa, RMSE = 0.562 MPa, and MAPE = 11.03% in testing, which was superior to RF (R^2 = 0.915, MAE = 0.678 MPa, RMSE = 0.816 MPa, and MAPE = 18.41%), SVM (R^2 = 0.877, MAE = 0.737 MPa, RMSE = 0.982 MPa, and MAPE = 27.49%), and ANN (R^2 = 0.831, MAE = 0.753 MPa, RMSE = 1.147 MPa, and MAPE = 29.68%). Although the prediction of the simplified model is also relatively effective as well, its prediction performance is lower than that of the original Seft-Net model due to the discarding of information on some variables. These observations and findings are dependent on the data obtained from the tests, and the amount and quality of data may affect the evaluation results.

8.3.3 Empirical formula for predicting shear strength

The dataset employed ML and DL regression models aimed at predicting shear strength based on particle and soil mass indicators. However, a significant drawback of these methods is their "black box" nature, as they do not provide explicit and straightforward formulas for interpretation. To overcome this limitation, key indicators were identified by ranking their importance. These key indicators were then utilized in a statistical modeling approach to formulate a multivariate nonlinear empirical formula for predicting shear strength.

The relationship between gravel shear strength and various factors like particle sizes (e.g., Dmax), parent rock strength, and particle characteristics has been extensively studied (Abbas, 2003; De Mello, 1977; Indraratna et al., 1993; Honkanadavar and Gupta, 2010; Varadarajan et al., 2003; Varadarajan et al., 2006). However, these studies often lack comprehensive data on particle and soil properties, and certain crucial aspects like particle shape are not widely available or easily measured empirically. Utilizing extensive data from the Geo-Gravel dataset, particularly focusing on key particle and soil mass properties (σ_3, σ_r, e_0, ρ , D_{80}, and C_u), an empirical regression formula is proposed based on Equation (8.4) (Indraratna et al., 1993). In general, confining pressure σ_3 is the controlled experimental condition in performing the triaxial compression tests, and σ_{1peak} and q_{peak} can be derived from each other. The empirical regression formula for σ_{1peak}, derived through multiple nonlinear regression using the Levenberg–Marquardt algorithm (Gavin, 2019), adopts a power exponential form as presented in Equation (8.5).

$$\frac{\sigma_{1peak}}{\sigma_r} = \alpha\left(\frac{\sigma_3}{\sigma_r}\right)^{\beta} \tag{8.4}$$

$$\sigma_{1peak} = 2.372\sigma_3^{0.826}\sigma_r^{0.089}e_0^{-0.170}\rho^{-0.084}\underbrace{D_{80}^{0.032}C_u^{0.004}}_{\alpha_g} \tag{8.5}$$

where α and β are considered to be characteristic parameters obtained by curve fitting in Equation (8.4); σ_{1peak}, σ_3, and σ_r in units of MPa, D_{80} in units of mm, and the other indicator are dimensionless. In the application of Equation (8.5), σ_3 and e_0 can be taken according to the project design. The use of measured values for other indicators is beneficial for obtaining more accurate predictions. In the absence of measured values, empirical values of ρ and σ_r in Tables 8.1 and 8.2 can be used. For missing data of gradation, the gradation parameter α_g is taken as 1.101.

In Equation (8.5), the training set achieved R^2 = 0.968, MAE = 0.452 MPa, RMSE = 0.721 MPa, and for the test set: R^2 = 0.902, MAE = 0.687 MPa, RMSE = 0.896 MPa. Figure 8.15 illustrates scatter plots of actual versus predicted values. Despite the empirical formula's performance being less than ML and DL models, its simplicity makes it suitable for shear strength estimates.

Subsequently, to further evaluate the prediction performance of the formula, the five top important indicators σ_r, e_0, ρ , D_{80}, and C_u affecting the value of σ_{1peak} under four ranges of σ_3 divided by quartiles (σ_3=0.001–0.4 MPa, 0.4–0.8 MPa, 0.8–1.4 MPa, and 1.4–4.0 MPa) are analyzed.

Clear comparisons are made by categorizing σ_r (>30 MPa: highly hard and sub-hard, ≤30 MPa: sub-soft, soft, highly soft), ρ (≥0.5: sub-rounded, rounded, well-rounded; <0.5: sub-angular, angular), and dividing e_0, D_{80}, and C_u based on median values of 0.250, 37.5 mm, 14.0, respectively.

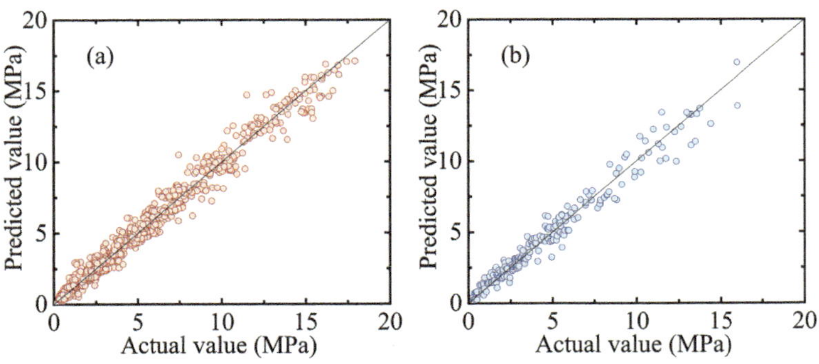

Figure 8.15 Scatter plots of actual value vs. predicted value of Equation (8.5): (a) training set and (b) testing set.

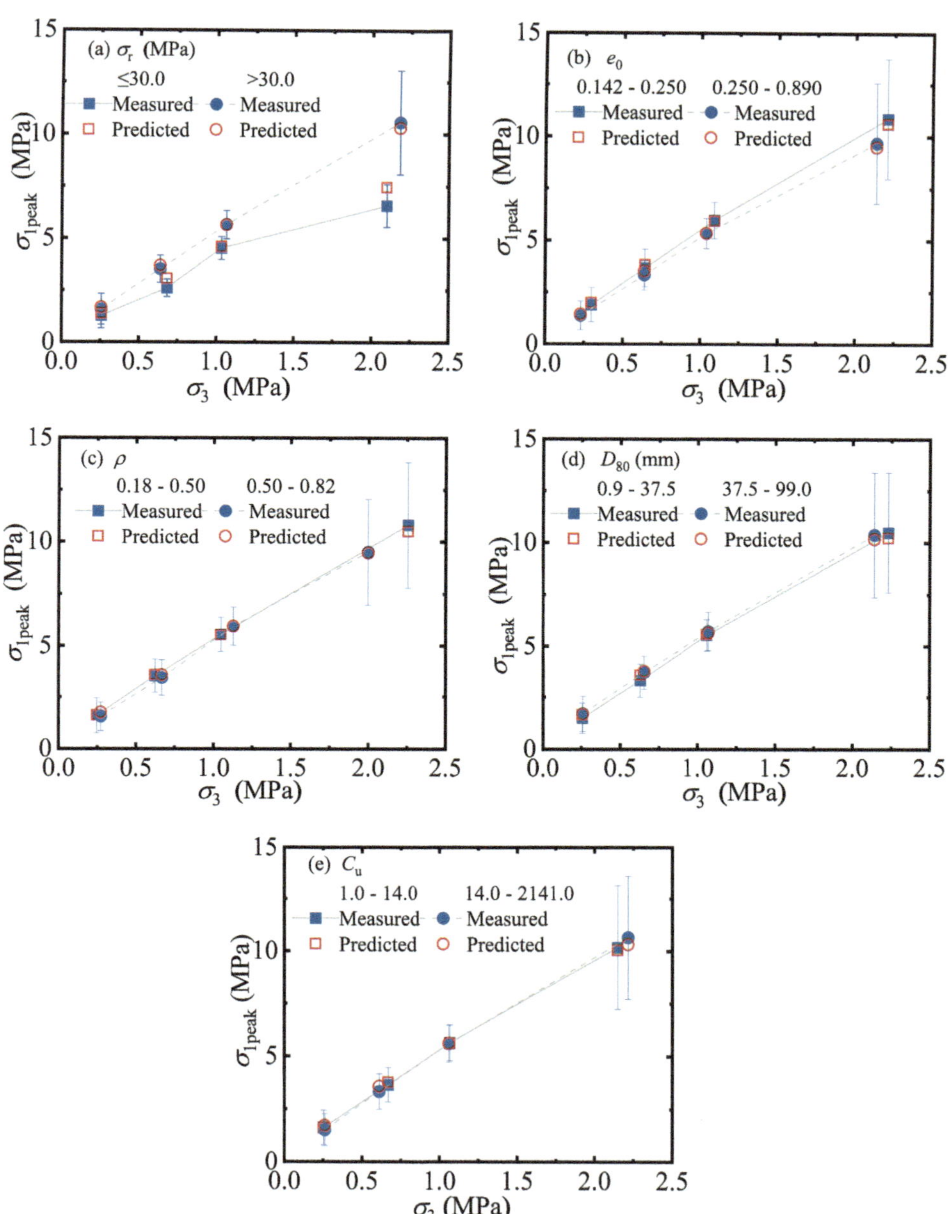

Figure 8.16 Variation of σ_{1peak} with σ_3 in different indicators ranges: (a) σ_r, (b) e_0, (c) ρ, (d) D_{80}, and (e) C_u.

In Figure 8.16(a), Equation (8.5) effectively captures the influence of σ_r at various σ_3 levels. Notably, the σ_{1peak} with $\sigma_r > 30$ MPa exceeds that of $\sigma_r \leq 30$ MPa, with the difference amplifying at higher σ_3, as weaker particles are more prone to break under increased confining pressure (Alonso et al., 2016; Dong et al., 2021). Predictions align better with $\sigma_r > 30$ MPa, reflecting the empirical formula's statistical representativeness for this subset, constituting 95.3% of the total data. Additionally, in Equation (8.4), the characteristic parameters α and β are 2.593 and 0.841 with $R^2 = 0.952$, MAE = 0.574

MPa, RMSE = 0.883 MPa in training set, and R^2 = 0.885, MAE = 0.715 MPa, RMSE = 1.094 MPa in test set. Comparing Equations (8.4) and (8.5), it shows that the best-fitting value of the power exponent of σ_r is 0.089, which is different from that of 0.159 (=$1-\beta$) in Equation (8.4).

Figure 8.16(b) reveals that the measured and predicted σ_{1peak} values of e_0 in 0.142–0.250 are higher than that of 0.250–0.890, and the difference becomes more pronounced with the increase in σ_3. The closer particle contact and the stronger interlocking forces among particles contribute to the loading transmission in the interaggregate structure at high confining pressure and low e_0 (Xiao et al., 2014).

Figure 8.16(c) shows that the measured and predicted σ_{1peak} of $\rho < 0.5$ (i.e., sub-angular and angular) gravels is slightly higher than that of $\rho \geq 0.5$ (i.e., sub-rounded, rounded, and well-rounded) gravels because the angular particle interlocking could frustrate the rotation, while this difference decreases with the increase in σ_3 (Giang et al., 2017).

Figure 8.16(d) shows that the measured and predicted σ_{1peak} of D_{80} in 37.5–99.0 mm gravels is slightly higher than that of D_{80} in 0.9–37.5 mm gravels, which may be related to the large-size particles having larger contact force and strong interlocking force (Huang et al., 2017).

Figure 8.16(e) indicates that the measured and predicted dots in different C_u ranges are relatively close, owing to the gradations of the gravels being mostly well-graded by design, which may make the differences relatively insignificant (Liu et al., 2021).

8.4 PREDICTION OF STRAIN–STRESS BEHAVIOR

Based on Geo-Gravel, this section develops a simulator for predicting triaxial compression test curves of gravels directly using physical indicators. We developed a DL model, Prior information-based ResNet-LSTM network (PRL-Net), as the core deep learning model in the simulator Figure 8.17.

It employs deep residual networks to extract high-dimensional spatial features of physical indicators, and combined with LSTM (Long Short-Term Memory) networks to extract the time-series information from test curves, as well as introducing prior information constraints in the loss function. Through ablation studies, the optimal architecture of the model was determined, with training and validation predictive accuracy significantly surpassing the baseline. This model effectively captures the intricate nonlinear relationships in the mechanical behavior of gravels. The simulator has been demonstrated to effectively predict the strain–stress and strain–volumetric behaviors of gravels under different physical indicators and confining pressure (Table 8.7), including fundamental characteristics such as softening/hardening and shrinkage/dilatancy (Figure 8.18).

This study provides efficient and accurate support for predicting mechanical behaviors the prediction of stress–strain curves and their mechanical properties of gravels through a data-driven approach.

8.5 SUMMARY AND CONCLUSION

In this chapter, a novel dataset Geo-Gravel for triaxial compression test of gravels was established. Utilizing this dataset, deep learning, machine learning, and empirical formulas were applied to predict the shear strength and strain–stress behavior of gravels. The main contributions and conclusions are summarized as follows:

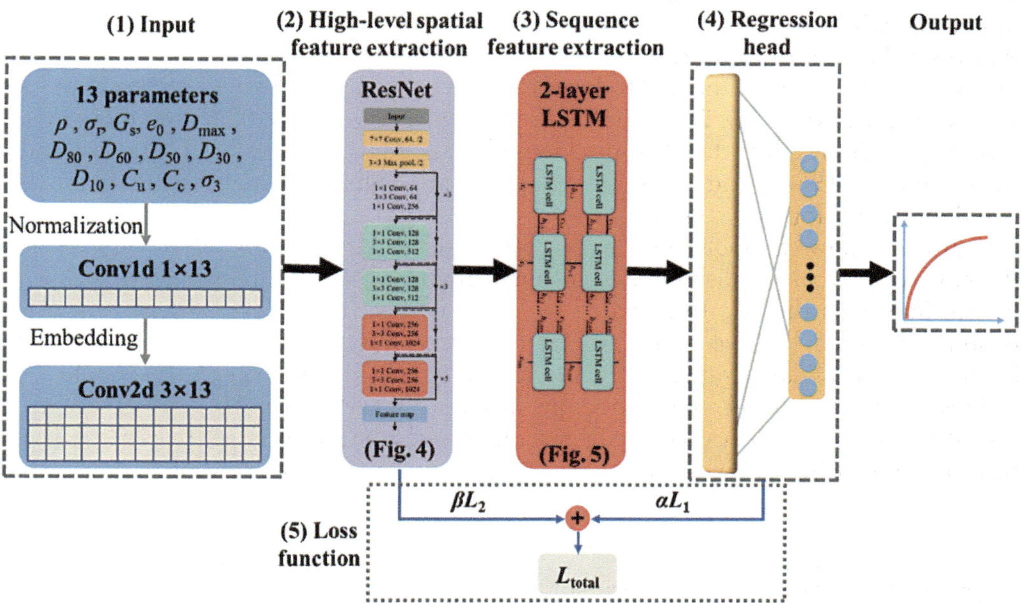

Figure 8.17 PRL-Net architecture.

(a) The Geo-Gravel dataset includes triaxial compression test data for 332 types of gravels with 1102 test records. A total of 12 attributes affecting shear strength and strain–stress behavior were collected, including particle properties (i.e., σ_r, ρ), soil mass properties (i.e., e_0, D_{10}, D_{30}, D_{50}, D_{60}, D_{80}, D_{max}, C_u, and C_c), and external condition (i.e., σ_3). The dataset is useful for stability evaluation and project design involving gravels in the absence of test results.

(b) Three machine learning regression models, i.e., artificial neural network (ANN), support vector regression (SVR), and random forest (RF), were employed to predict shear strength based on particle and soil mass indicators. RF exhibited superior performance among these models. The analysis of the importance of 12 indicators on shear strength revealed that the external condition σ_3 is the most important indicator for the estimation of shear strength and accounts for 80.5% of relative importance, while particle and soil mass indicators make up the remaining 19.5%. For particle and soil mass properties, σ_r, e_0, ρ , D_{80}, and C_u are the top five important indicators, which account for 28.0%, 26.9%, 10.2%, 9.7%, and 8.1%, respectively.

(c) A novel deep learning-based model Seft-Net was proposed to directly predict the peak deviatoric stress (q_{peak}) of gravels using gravel properties. The innovative network architecture incorporates advanced techniques, including a tailored segmented feature transformation approach and depthwise separable convolutions. Ablation studies confirmed the superiority of the proposed modules and approaches, and comparative evaluations with ANN, SVM, and RF further validated the good performance of Seft-Net. It achieved an optimal prediction performance with R^2 = 0.994, MAE = 0.111 MPa, RMSE = 0.232 MPa, and MAPE = 4.102% in the training part, and with R^2 = 0.959, MAE = 0.336 MPa, RMSE = 0.445 MPa, and MAPE = 9.127% in testing part.

Table 8.7 Test materials

No.	*Lithology*	σ_r (*MPa*)	ρ	e_0	D_{max} (*mm*)	D_{50} (*mm*)	C_u	C_c	*Reference*
(a)	Micaceous, sandstone, quartzite	77.3	0.82 Well-rounded	0.25	50	10.5	3.52	10.34	Varadarajan et al. (2006)
(b)	Marble, limestone, quartz sandstone	106.5	0.82 Well-rounded	0.189	60	26.4	5.54	1.57	Xiao et al. (2016)
(c)	Weakly weathered fresh tuff	66.1	0.34 Angular	0.290	60	17.8	14.02	1.93	Liu and Chen (2019)
(d)	Weakly weathered fresh tuff	66.1	0.34 Angular	0.205	60	17.8	14.02	1.93	Liu and Chen (2019)

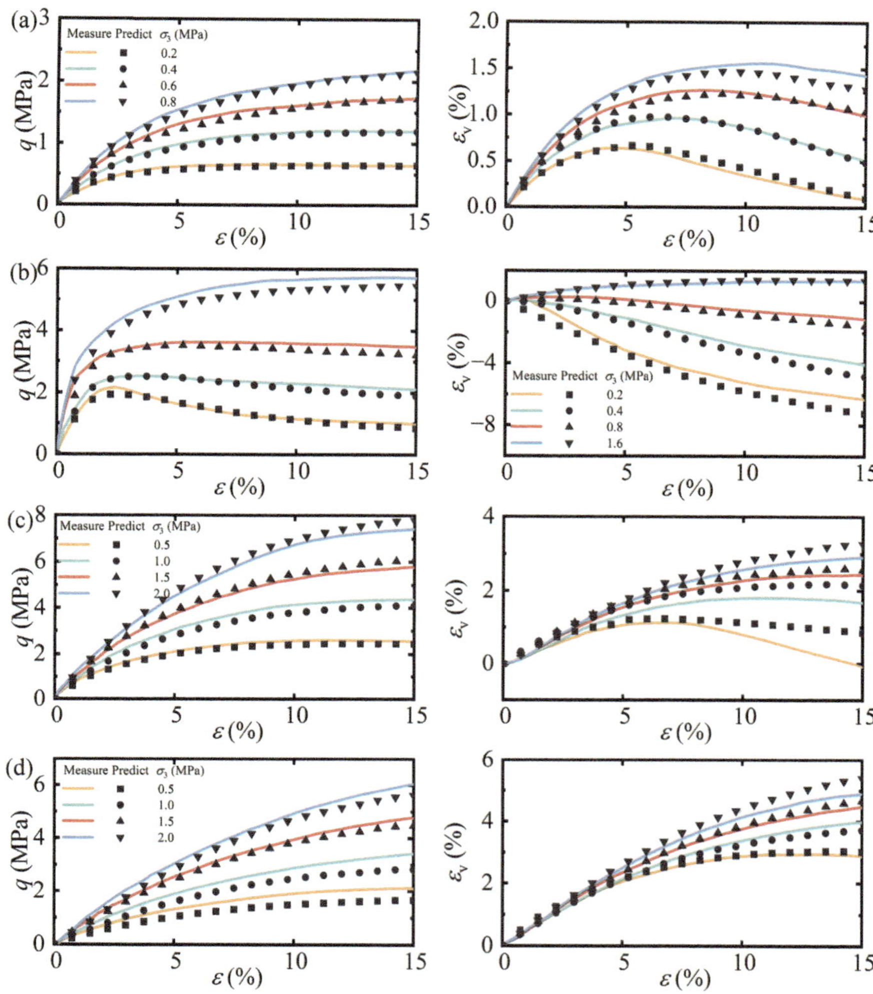

Figure 8.18 Actual curves vs. predicted curves.

(d) The modified empirical formula is proposed for estimating the strength of the gravels by considering the influence of five important particle and soil mass indicators (σ_{r}, e_0, ρ , D_{80}, and C_u) and external condition σ_{3}, as well as its simplified form for missing data of particle or soil mass indicators is also given. It also showed that the best-fitting value of the power exponent of σ_{r} is 0.089, which is different from that of 0.159 in the original empirical formula. The modified empirical formula is more statistically representative for $\sigma_{r} > 30$ MPa (highly hard and sub-hard) gravels due to its proportion is 95.3% of the total data.

(e) The DL model PRL-Net is proposed by leveraging deep residual networks for extracting spatial features from the physical parameters and LSTM networks for capturing sequential features from curves, as well as introducing prior information constraints in the loss function. The model can efficiently capture the hidden associations and patterns of mechanical behavior of gravels and reasonably predict them. The optimal model architecture is obtained through ablation studies, and its accuracy is significantly superior to the baseline model (LSTM), with the mean absolute percentage error (MAPE) reduced by 46.84%. It exhibits effectiveness in characterizing and predicting the mechanical behaviors of gravels with diverse physical properties, spanning different confining pressures, lithologies, particle shapes, gradations, and void ratios. This capability allows the model to reflect the softening and dilatancy characteristics inherent in various gravel types. Offering a novel approach to data-driven geotechnical engineering design provides a reasonable reference and support for evaluation even in the absence of test data.

REFERENCES

Abbas, S.M., 2003. *Testing and Modeling the Behaviour of Riverbed and Quarried Rockfill Materials*. IIT Delhi.

Adeli, H., 2001. Neural networks in civil engineering: 1989–2000. *Computer-Aided Civil and Infrastructure Engineering* 16 (2), 126–142.

Ahmad, M., Al-Mansob, R.A., Jamil, I., Al-Zubi, M.A., Sabri, M.M.S., Alguno, A.C., 2022. Prediction of Rockfill Materials' Shear Strength Using Various Kernel Function-Based Regression Models—A Comparative Perspective. *Materials* 15 (5), 1739.

Ahmad, M., Kamiński, P., Olczak, P., Alam, M., Iqbal, M.J., Ahmad, F., Sasui, S., Khan, B.J., 2021. Development of prediction models for shear strength of rockfill material using machine learning techniques. *Applied Sciences* 11 (13), 6167.

Ajit, A., Acharya, K., Samanta, A., 2020. A review of convolutional neural networks. In: *2020 International Conference on Emerging Trends in Information Technology and Engineering (ic-ETITE)*. IEEE. pp. 1–5.

Albelwi, S. 2022. A Robust Energy Consumption Forecasting Model using ResNet-LSTM with Huber Loss. International Journal of Computer Science and Network Security, 22(7): 301–307.

Alonso, E.E., Romero, E.E., and Ortega, E. 2016. Yielding of rockfill in relative humidity-controlled triaxial experiments. Acta Geotechnica, 11: 455–477.

Araei, A.A. 2014. Artificial Neural Networks for Modeling Drained Monotonic Behavior of Rockfill Materials. International Journal of Geomechanics, 14(3): 04014005.

Archer, K.J., and Kimes, R.V. 2008. Empirical characterization of random forest variable importance measures. Computational Statistics & Data Analysis, 52(4): 2249–2260.

Aydin, Y., Bekdaş, G., Işıkdağ, Ü., and Nigdeli, S.M. 2023. The State of Art in Machine Learning Applications in Civil Engineering. *In* Hybrid Metaheuristics in Structural Engineering: Including Machine Learning Applications. Edited by G. Bekdaş and S.M. Nigdeli. Springer Nature Switzerland, Cham. pp. 147–177.

Babar, B., Luppino, L.T., Boström, T., and Anfinsen, S.N. 2020. Random forest regression for improved mapping of solar irradiance at high latitudes. Solar Energy, 198: 81–92.

Banimahd, M., Yasrobi, S.S., and P.K.Woodward. 2005. Artificial neural network for stress–strain behavior of sandy soils: Knowledge based verification. Computers and Geotechnics, 32(5): 377–386.

Bareither, C.A., Edil, T.B., Benson, C.H., Mickelson, D.M., 2008. Geological and physical factors affecting the friction angle of compacted sands. *The Journal of Geotechical and Geoenvironmental Engineering* 134 (10), 1476–1489.

Benabou, L. 2021. Development of LSTM Networks for Predicting Viscoplasticity With Effects of Deformation, Strain Rate, and Temperature History. Journal of Applied Mechanics, 88(7): 071008.

Bergstra, J., Bengio, Y., 2012. Random search for hyper-parameter optimization. *Journal of Machine Learning Research* 13 (2), 281–305.

Carvalho, D.V., Pereira, E.M., Cardoso, J.S., 2019. Machine learning interpretability: A survey on methods and metrics. *Electronics* 8 (8), 832.

Chauhan, N.K., Singh, K., 2018. A review on conventional machine learning vs deep learning. In: *2018 International Conference on Computing, Power and Communication Technologies (GUCON)*. Greater Noida, India, IEEE. pp. 347–352.

Chen, H., Qian, C., Liang, C., Kang, W., 2018. An approach for predicting the compressive strength of cement-based materials exposed to sulfate attack. *PLoS One* 13 (1), e0191370.

Ching, J., Li, K.-H., Phoon, K.-K., and Weng, M.-C. 2018. Generic transformation models for some intact rock properties. Canadian Geotechnical Journal, 55(12): 1702–1741.

Ching, J., Lin, G.-H., Chen, J.-R., and Phoon, K.-K. 2017. Transformation models for effective friction angle and relative density calibrated based on generic database of coarse-grained soils. Canadian Geotechnical Journal, 54(4): 481–501.

Ching, J., and Phoon, K.-K. 2012. Modeling parameters of structured clays as a multivariate normal distribution. Canadian Geotechnical Journal, 49(5): 522–545.

Ching, J., and Phoon, K.-K. 2014. Transformations and correlations among some clay parameters — the global database. Canadian Geotechnical Journal, 51(6): 663–685.

Ching, J., and Phoon, K.-K. 2019. Constructing Site-Specific Multivariate Probability Distribution Model Using Bayesian Machine Learning. Journal of Engineering Mechanics, 145(1): 04018126.

Cho, G., Dodds, J., Santamarina, J.C., 2007. Closure to "particle shape effects on packing density, stiffness, and strength: natural and crushed sands" by Gye-Chun Cho, Jake Dodds, and J. Carlos Santamarina. *The Journal of Geotechical and Geoenvironmental Engineering* 133 (11), 1474–1474.

Chollet, F., 2017. Xception: Deep learning with depthwise separable convolutions. In: *Proceedings of the IEEE Conference on Computer Vision and Pattern Recognition*. Honolulu, HI, pp. 1251–1258.

Chollet, F., 2021. *Deep Learning with Python*. Simon and Schuster. Shelter Island, NY.

Chu, F.Y. 2014. Influence Factor on Dilatancy of Coarse-Grained Soil Based on Large-Scale Triaxial Test. Advanced Materials Research, 919–921: 687–692.

Dabkowski, P., Gal, Y., 2017. Real time image saliency for black box classifiers. *Advances in Neural Information Processing Systems* 30, 1–10.

De Mello, V.F., 1977. Reflections on design decisions of practical significance to embankment dams. *Géotechnique* 27 (3), 281–355.

Dean, J., Corrado, G., Monga, R., Chen, K., Devin, M., Mao, M., Ranzato, M., Senior, A., Tucker, P., Yang, K., 2012. Large scale distributed deep networks. *Advances in Neural Information Processing Systems* 25, 1–9.

Deng, J., Dong, W., Socher, R., Li, L., Li, K., Fei-Fei, L., 2009. Imagenet: A large-scale hierarchical image database. In: *2009 IEEE Conference on Computer Vision and Pattern Recognition*. Miami, FL, IEEE. pp. 248–255.

Deng, L., Yu, D., 2014. Deep learning: methods and applications. SpringerBriefs in Computer Science 7 (3–4), 197–387.

Dhillon, A., Verma, G.K., 2020. Convolutional neural network: a review of models, methodologies and applications to object detection. *Progress in Artificial Intelligence* 9 (2), 85–112.

Dong, H., Peng, B., Gao, Q.-F., Hu, Y., and Jiang, X. 2021. Study of hidden factors affecting the mechanical behavior of soil–rock mixtures based on abstraction idea. Acta Geotechnica, 16(2): 595–611.

Doshi-Velez, F., Kim, B., 2017. *Towards a Rigorous Science of Interpretable Machine Learning*. arXiv preprint arXiv:1702.08608.

Dosovitskiy, A., Beyer, L., Kolesnikov, A., Weissenborn, D., Zhai, X., Unterthiner, T., Dehghani, M., Minderer, M., Heigold, G., Gelly, S., Uszkoreit, J., and Houlsby, N. 2021. *An Image is Worth 16x16 Words: Transformers for Image Recognition at Scale.* arXiv.

Du, L., Lu, Z., and Li, D. 2022. Broodstock breeding behaviour recognition based on Resnet50-LSTM with CBAM attention mechanism. Computers and Electronics in Agriculture, 202: 107404.

Fong, R.C., Vedaldi, A., 2017. Interpretable explanations of black boxes by meaningful perturbation. In: *Proceedings of the IEEE International Conference on Computer Vision.* Venice, Italy, pp. 3429–3437.

Gavin, H.P. 2019. The Levenberg-Marquardt algorithm for nonlinear least squares curve-fitting problems. Department of civil and environmental engineering, Duke University, 19.

Giang, P.H.H., Van Impe, P., Van Impe, W., Menge, P., Cnudde, V., and Haegeman, W. 2017. Effects of particle characteristics on the shear strength of calcareous sand. Acta Geotechnica Slovenica, 14(2): 77–89.

Goodfellow, I., Bengio, Y., Courville, A., 2016. *Deep Learning.* MIT press, Cambridge, MA.

Gómez-Ramírez, J., Ávila-Villanueva, M., and Fernández-Blázquez, M.Á. 2020. Selecting the most important self-assessed features for predicting conversion to mild cognitive impairment with random forest and permutation-based methods. Scientific Reports, 10(1): 20630.

Gu, J., Wang, Z., Kuen, J., Ma, L., Shahroudy, A., Shuai, B., Liu, T., Wang, X., Wang, G., Cai, J., 2018. Recent advances in convolutional neural networks. *Pattern Recognition* 77, 354–377.

Guan, S., Qu, T., Feng, Y.T., Ma, G., and Zhou, W. 2023. A machine learning-based multi-scale computational framework for granular materials. Acta Geotechnica, 18(4): 1699–1720.

Habibagahi, G., and Bamdad, A. 2003. A neural network framework for mechanical behavior of unsaturated soils. Canadian Geotechnical Journal, 40(3): 684–693.

Hao, S., Pabst, T., 2022. Prediction of CBR and resilient modulus of crushed waste rocks using machine learning models. *Acta Geotechnica* 17 (4), 1383–1402.

He, K., Zhang, X., Ren, S., Sun, J., 2016. Deep residual learning for image recognition. In: *Proceedings of the IEEE Conference on Computer Vision and Pattern Recognition.* Las Vegas, NV, pp. 770–778.

Honkanadavar, N.P., Gupta, S.L., 2010. Prediction of shear strength parameters for prototype riverbed rockfill material using index properties. In: *Proceedings of the Indian Geotechnical Conference*—2010, Mumbai, India. pp. 16–18.

Huang, G., Liu, Z., Van Der Maaten, L., Weinberger, K.Q., 2017. Densely connected convolutional networks. In: *Proceedings of the IEEE Conference on Computer Vision and Pattern Recognition.* Venice, Italy, pp. 4700–4708.

Huang, Y., Li, J., Ma, D., Gao, H., Guo, Y., and Ouyang, S. 2019. Triaxial compression behaviour of gangue solid wastes under effects of particle size and confining pressure. Science of The Total Environment, 693: 133607.

Indraratna, B., Ionescu, D., & Christie, H. D. 1998. Shear behavior of railway ballast based on large-scale triaxial tests. *Journal of geotechnical and geoenvironmental Engineering*, 124 (5), 439–449.

Indraratna, B., Wijewardena, L., Balasubramaniam, A.S., 1993. Large-scale triaxial testing of grey wacke rockfill. *Géotechnique* 43 (1), 37–51.

Janiesch, C., Zschech, P., and Heinrich, K. 2021. Machine learning and deep learning. Electronic Markets, 31(3): 685–695.

Javadi, A.A., Faramarzi, A., and Ahangar-Asr, A. 2012. Analysis of behaviour of soils under cyclic loading using EPR-based finite element method. Finite Elements in Analysis and Design, 58: 53–65.

Javadi, A.A., and Rezania, M. 2009. Applications of artificial intelligence and data mining techniques in soil modeling. Geomechanics and Engineering, 1(1): 53–74.

Kaunda, R., 2015. Predicting shear strengths of mine waste rock dumps and rock fill dams using artificial neural networks. *International Journal of Mining and Mineral Engineering* 6 (2), 139–171.

Khan, A., Sohail, A., Zahoora, U., Qureshi, A.S., 2020. A survey of the recent architectures of deep convolutional neural networks. *Artificial Intelligence Review* 53, 5455–5516.

Labuz, J.F., and Zang, A. 2012. Mohr–Coulomb Failure Criterion. Rock Mechanics and Rock Engineering, 45(6): 975–979.

Lazarevska, M., Knezevic, M., Cvetkovska, M., Trombeva-Gavriloska, A., 2014. Application of artificial neural networks in civil engineering. *Tehnički vjesnik* 21 (6), 1353–1359.

Lecun, Y., Bengio, Y., Hinton, G., 2015. Deep learning. *Nature* 521 (7553), 436–444.

Li, J., Monroe, W., Jurafsky, D., 2016. *Understanding neural networks through representation erasure*. arXiv preprint arXiv:1612.08220.

Li, Z., Chow, J.K., and Wang, Y.H. 2017. Applying the Artificial Neural Network to Predict the Soil Responses in the DEM Simulation. IOP Conference Series: Materials Science and Engineering, 216: 012040.

Ling, H., Qian, C., Kang, W., Liang, C., Chen, H., 2019. Combination of Support Vector Machine and K-Fold cross validation to predict compressive strength of concrete in marine environment. *Construction and Building Materials* 206, 355–363.

Liu, D., Chen, H., 2019. Relationship between porosity and the constitutive model parameters of rockfill materials. *Journal of Materials in Civil Engineering* 31 (2), 04018384.

Liu, Z., Shao, J., Xu, W., Wu, Q., 2015. Indirect estimation of unconfined compressive strength of carbonate rocks using extreme learning machine. *Acta Geotechnica* 10 (5), 651–663.

Ly, H., Pham, B.T., Dao, D.V., Le, V.M., Le, L.M., Le, T., 2019. Improvement of ANFIS model for prediction of compressive strength of manufactured sand concrete. *Applied Sciences* 9 (18), 3841.

Mahmoodzadeh, A., Mohammadi, M., Ibrahim, H.H., Abdulhamid, S.N., Salim, S.G., Ali, H.F.H., Majeed, M.K., 2021. Artificial intelligence forecasting models of uniaxial compressive strength. *Transportation Geotechnics* 27, 100499.

Mohammad-Alikhani, A., Nahid-Mobarakeh, B., and Hsieh, M.-F. 2023. One-Dimensional LSTM-Regulated Deep Residual Network for Data-Driven Fault Detection in Electric Machines. IEEE Transactions on Industrial Electronics,: 1–10.

Nassr, A., Esmaeili-Falak, M., Katebi, H., and Javadi, A. 2018. A new approach to modeling the behavior of frozen soils. Engineering Geology, 246: 82–90.

Patil, H., Dwivedi, A., Bidkar, K., 2022. Prediction of upgraded properties of the concrete with the wash sand waste. *Journal of Building Pathology and Rehabilitation* 7 (1), 1–12.

Ovalle, C., Linero, S., Dano, C., Bard, E., Hicher, P.-Y., and Osses, R. 2020. Data Compilation from Large Drained Compression Triaxial Tests on Coarse Crushable Rockfill Materials. Journal of Geotechnical and Geoenvironmental Engineering, 146(9): 06020013.

Peng, X., Wang, Z., Luo, T., Yu, M., and Luo, Y. 2008. An elasto-plastic constitutive model of moderate sandy clay based on BC-RBFNN. Journal of Central South University of Technology, 15(S1): 47–50.

Penumadu, D., and Zhao, R. 1999. Triaxial compression behavior of sand and gravel using artificial neural networks (ANN). Computers and Geotechnics, 24(3): 207–230.

Phoon, K.-K. 2020. The story of statistics in geotechnical engineering. Georisk: Assessment and Management of Risk for Engineered Systems and Geohazards, 14(1): 3–25.

Phoon, K.-K., and Retief, J.V. (Editors). 2016. Reliability of geotechnical structures in ISO2394. CRC Press, Taylor & Francis Group, Boca Raton, FL.

Phoon, K.-K., and Zhang, W. 2023. Future of machine learning in geotechnics. Georisk: Assessment and Management of Risk for Engineered Systems and Geohazards, 17(1): 7–22.

Qu, T., Di, S., T. Feng, Y., Wang, M., Zhao, T., and Wang, M. 2021. Deep Learning Predicts Stress–Strain Relations of Granular Materials Based on Triaxial Testing Data. Computer Modeling in Engineering & Sciences, 128(1): 129–144.

Qu, T., Guan, S., Feng, Y.T., Ma, G., Zhou, W., Zhao, J., 2023. Deep active learning for constitutive modelling of granular materials: From representative volume elements to implicit finite element modelling. *International Journal of Plasticity* 164, 103576.

Rashidian, V., and Hassanlourad, M. 2014. Application of an Artificial Neural Network for Modeling the Mechanical Behavior of Carbonate Soils. International Journal of Geomechanics, 14(1): 142–150.

Reich, Y. 1997. Machine Learning Techniques for Civil Engineering Problems. Computer-Aided Civil and Infrastructure Engineering, 12(4): 295–310.

Russakovsky, O., Deng, J., Su, H., Krause, J., Satheesh, S., Ma, S., Huang, Z., Karpathy, A., Khosla, A., Bernstein, M., 2015. Imagenet large scale visual recognition challenge. *International Journal of Computer Vision* 115 (3), 211–252.

Sadrekarimi, A., Olson, S.M., 2010. Particle damage observed in ring shear tests on sands. *Canadian Geotechnical Journal* 47 (5), 497–515.

Samek, W., Montavon, G., Lapuschkin, S., Anders, C.J., Müller, K., 2021. Explaining deep neural networks and beyond: A review of methods and applications. *Proceedings of the IEEE* 109 (3), 247–278.

Schmidhuber, J., 2015. Deep learning in neural networks: An overview. *Neural Network* 61, 85–117.

Sevi, A., Ge, L., 2012. Cyclic behaviors of railroad ballast within the parallel gradation scaling framework. *Journal of Materials in Civil Engineering* 24 (7), 797–804.

Shang, R., He, J., Wang, J., Xu, K., Jiao, L., Stolkin, R., 2020. Dense connection and depthwise separable convolution based CNN for polarimetric SAR image classification. *Knowledge-Based Systems* 194, 105542.

Shrikumar, A., Greenside, P., Kundaje, A., 2017. Learning important features through propagating activation differences. In: *International Conference on Machine Learning*. PMLR. Sydney, Australia, pp. 3145–3153.

Sun, C., Shrivastava, A., Singh, S., Gupta, A., 2017. Revisiting unreasonable effectiveness of data in deep learning era. In: *Proceedings of the IEEE International Conference on Computer Vision*. Venice, Italy, pp. 843–852.

Szegedy, C., Ioffe, S., Vanhoucke, V., Alemi, A.A., 2017. Inception-v4, inception-resnet and the impact of residual connections on learning. In: *Thirty-First AAAI Conference on Artificial Intelligence*, Honolulu, HI, 4278–4284.

Szegedy, C., Liu, W., Jia, Y., Sermanet, P., Reed, S., Anguelov, D., Erhan, D., Vanhoucke, V., Rabinovich, A., 2015. Going deeper with convolutions. In: *Proceedings of the IEEE Conference on Computer Vision and Pattern Recognition*. Boston, MA, pp. 1–9.

Szegedy, C., Vanhoucke, V., Ioffe, S., Shlens, J., Wojna, Z., 2016. Rethinking the inception architecture for computer vision. In: *Proceedings of the IEEE Conference on Computer Vision and Pattern Recognition*. Las Vegas, NV, pp. 2818–2826.

Tan, M., Le, Q., 2019. Efficientnet: Rethinking model scaling for convolutional neural networks. In: *International Conference on Machine Learning*. PMLR. Lugano, Switzerland, pp. 6105–6114.

Tang, C., and Phoon, K.-K. 2018. Statistics of Model Factors and Consideration in Reliability-Based Design of Axially Loaded Helical Piles. Journal of Geotechnical and Geoenvironmental Engineering, 144(8): 04018050.

Tang, C., and Phoon, K.-K. 2019a. Characterization of model uncertainty in predicting axial resistance of piles driven into clay. Canadian Geotechnical Journal, 56(8): 1098–1118.

Tang, C., and Phoon, K.-K. 2019b. Evaluation of Stress-Dependent Methods for the Punch-Through Capacity of Foundations in Clay with Sand. ASCE-ASME Journal of Risk and Uncertainty in Engineering Systems, Part A: Civil Engineering, 5(3): 04019008.

Vadyala, S.R., Betgeri, S.N., Matthews, J.C., and Matthews, E. 2022. A review of physics-based machine learning in civil engineering. Results in Engineering, 13: 100316.

Varadarajan, A., Sharma, K.G., Abbas, S.M., Dhawan, A.K., 2006. Constitutive model for rockfill materials and determination of material constants. *International Journal of Geomechanics* 6 (4), 226–237.

Varadarajan, A., Sharma, K.G., Venkatachalam, K., Gupta, A.K., 2003. Testing and modeling two rockfill materials. *The Journal of Geotechical and Geoenvironmental Engineering* 129 (3), 206–218.

W. C. Krumbein. 1941. Measurement and Geological Significance of Shape and Roundness of Sedimentary Particles. SEPM Journal of Sedimentary Research, 11(2), 64-72.

Wang, K., and Sun, W. 2019. Meta-modeling game for deriving theory-consistent, microstructure-based traction–separation laws via deep reinforcement learning. Computer Methods in Applied Mechanics and Engineering, 346: 216–241.

Wang, M., Qu, T., Guan, S., Zhao, T., Liu, B., and Feng, Y.T. 2022. Data-driven strain–stress modelling of granular materials via temporal convolution neural network. Computers and Geotechnics, 152: 105049.

Wang, W., Xie, E., Li, X., Fan, D., Song, K., Liang, D., Lu, T., Luo, P., Shao, L., 2021. Pyramid vision transformer: A versatile backbone for dense prediction without convolutions. In: *Proceedings of the IEEE/CVF International Conference on Computer Vision*. Montreal, Quebec, Canada, pp. 568–578.

Wei, Y., Xing, Z., Jian, C., Wang, K., Wu, S., Chiam, K., 2021. Use of tree-based machine learning methods for stratigraphic classification in 3D geological modelling. In: *IOP Conference Series: Earth and Environmental Science*. IOP Publishing, 2021. Bristol, UK, p. 072039.

Wu, C., He, H., Song, R., Zhu, X., Peng, Z., Fu, Q., and Pan, J. 2023a. A hybrid deep learning model for regional O3 and NO2 concentrations prediction based on spatiotemporal dependencies in air quality monitoring network. Environmental Pollution, 320: 121075.

Wu, M., and Wang, J. 2022. Constitutive modelling of natural sands using a deep learning approach accounting for particle shape effects. Powder Technology, 404: 117439.

Wu, M., Xia, Z., and Wang, J. 2023b. Constitutive modelling of idealised granular materials using machine learning method. Journal of Rock Mechanics and Geotechnical Engineering, 15(4): 1038–1051.

Xiao, Y., Liu, H., Chen, Y., Jiang, J., 2014. Strength and deformation of rockfill material based on large-scale triaxial compression tests. II: Influence of particle breakage. *The Journal of Geotechical and Geoenvironmental Engineering* 140 (12), 04014071.

Xiao, Y., Liu, H., Chen, Y., Jiang, J., Zhang, W., 2014. Testing and modeling of the state-dependent behaviors of rockfill material. *Computers and Geotechnics* 61, 153–165.

Xiao, Y., Liu, H., Ding, X., Chen, Y., Jiang, J., and Wengang Zhang. 2016. Influence of Particle Breakage on Critical State Line of Rockfill Material. International Journal of Geomechanics, 16(1): 04015031.

Xu, M., Song, E., Chen, J., 2012. A large triaxial investigation of the stress-path-dependent behavior of compacted rockfill. *Acta Geotechnica* 7 (3), 167–175.

Yamashita, R., Nishio, M., Do, R.K.G., Togashi, K., 2018. Convolutional neural networks: an overview and application in radiology. *Insights Imaging* 9 (4), 611–629.

Zhang, B., Zou, G., Qin, D., Ni, Q., Mao, H., and Li, M. 2022. RCL-Learning: ResNet and convolutional long short-term memory-based spatiotemporal air pollutant concentration prediction model. Expert Systems with Applications, 207: 118017.

Zhang, C., Bengio, S., Hardt, M., Recht, B., Vinyals, O., 2021. Understanding deep learning (still) requires rethinking generalization. *Communication of the ACM* 64 (3), 107–115.

Zhang, N., Shen, S.-L., Zhou, A., and Jin, Y.-F. 2021. Application of LSTM approach for modelling stress–strain behaviour of soil. Applied Soft Computing, 100: 106959.

Zhang, P., Yin, Z., Jin, Y., Liu, X., 2021. Modelling the mechanical behaviour of soils using machine learning algorithms with explicit formulations. *Acta Geotechnica*, 17, 1403–1422.

Zhang, P., Yin, Z., Sheil, B., 2023. A physics-informed data-driven approach for consolidation analysis. *Géotechnique*, 74 (7), 620–631.

Zhang, S., Yao, L., Sun, A., Tay, Y. 2020. Deep Learning based Recommender System: A Survey and New Perspectives. *ACM Computing Surveys*, 52(1): 1–38.
Zhang, W., Wu, C., Zhong, H., Li, Y., Wang, L., 2021. Prediction of undrained shear strength using extreme gradient boosting and random forest based on Bayesian optimization. *Geoscience Frontiers* 12 (1), 469–477.
Zhao, H., Huang, Z., and Zou, Z. 2014. Simulating the Stress-Strain Relationship of Geomaterials by Support Vector Machine. Mathematical Problems in Engineering, 2014: 1–7.
Zhou, J., Li, E., Wei, H., Li, C., Qiao, Q., Armaghani, D.J., 2019. Random forests and cubist algorithms for predicting shear strengths of rockfill materials. *Applied Sciences* 9 (8), 1621.
Zhu, X., Chu, J., Wang, K., Wu, S., Yan, W., Chiam, K., 2021. Prediction of rockhead using a hybrid N-XGBoost machine learning framework. *Journal of Rock Mechanics and Geotechnical Engineering* 13 (6), 1231–1245.

Chapter 9

Evaluation of compressibility properties for soft marine clays

Monica S. Löfman and Leena Korkiala-Tanttu

Low-carbon ground improvement methods such as preloading with surcharge require reliable compressibility properties as input for the settlement prediction. This chapter demonstrates how the database approach can be used to evaluate the needed parameters for the reliability-based design of geo-structures dealing with soft marine clays. The special features related to soft and sensitive Nordic clays are investigated and discussed: inherent variability, defined via coefficient of variation or COV, seems to be smaller than that of COV guidelines based on global summaries since such soft soil deposits are often quite homogeneous. Furthermore, most of the general transformation models used to predict compressibility properties from index properties are biased when applied to soft marine clays; Accordingly, smaller transformation uncertainties may be acquired if a regional clay database is used to calibrate existing transformation models and to derive carefully fitted models with outlier detection. Finally, it is shown how the cross-correlation between compressibility parameters may be effectively modeled via database approach and Monte Carlo simulation with copula.

9.1 INTRODUCTION

Given the climate change mitigation goals, it is essential to prefer low-carbon ground improvement over carbon-intensive foundation methods such as piled slabs whenever possible. According to recent studies (Kivi 2022; Perttu 2023), preloading with surcharge embodies lower carbon emissions than that of other ground improvement methods used at soft soil sites: on average, the carbon dioxide equivalent is only 20–40% of the emissions embodied in the lime–cement column stabilization or the use of lightweight fills (Kivi 2022). Optimized design of preloading (with or without vertical drains) requires reliable compressibility properties as an input for the settlement prediction. However, high-quality undisturbed samples are needed for the oedometer tests, and sampling from soft and sensitive clays has been shown to be challenging (e.g., Mataić 2016). Meanwhile, such soft clays are prevalent, especially in the coastal areas of Finland and other Nordic countries, where the pressure to construct on soft soil sites is increasing as the cities are becoming denser. Hence, there is a growing need to properly characterize the compressibility properties – and the related uncertainties – for soft marine clays.

In a conventional serviceability limit state (SLS) verification according to Eurocode 7, the uncertainty in compressibility properties is considered via the determination of characteristic value. Yet, case studies have demonstrated that this method may lead to overdesigning when applied to ground-supported embankments on soft Finnish clays: López Ramirez, Löfman, and Korkiala-Tanttu (2022) used statistically defined characteristic values to compute settlements and observed that the obtained settlement value was in the

 DOI: 10.1201/9781003441946-9

scale of 3–6 standard deviations (SDs) apart from the mean settlement, thus indicating a safety margin that is unnecessarily large for the SLS design of preloading. Moreover, the safety margin varied between study cases, depending on the amount of non-linearity in the stress–strain response of the subsoil. This non-linearity is a result of both the overconsolidation and the stress–strain behavior of soft marine clay (López Ramirez, Löfman, and Korkiala-Tanttu 2022).

Uncertainty quantification and reliability analysis provide a systematic and robust approach to handling various sources of uncertainty in geotechnical design (Lacasse 2016; López Ramirez, Löfman, and Korkiala-Tanttu 2022; Nadim 2017; Phoon 2017). This chapter describes how the database approach can be harnessed to evaluate the compressibility properties of soft marine clays. A database of laboratory tests is used to characterize the inherent variability and transformation uncertainty in the compressibility properties of Finnish clay soils. Moreover, it demonstrates how the database approach provides insight into the behavior of soft sensitive Nordic clays and their differences compared with other marine clays. Further, the chapter discusses the phenomena of cross-correlation between compressibility parameters and demonstrates its significance to the uncertainty quantification of settlement response in soft marine clays.

9.2 COMPRESSION BEHAVIOR OF SOFT MARINE CLAYS

Soft marine clays are characterized by non-linear stress–strain response defined by a clear distinction between the small and elastic strain in the overconsolidated (OC) region and the larger plastic strain in the normally consolidated (NC) region at stresses beyond the preconsolidation pressure σ_p'. The compression index (C_c) method is one of the most widely used methods for estimating the one-dimensional vertical strain ε_v caused by primary consolidation in clay:

$$\varepsilon_v = \begin{cases} \varepsilon_{NC} = \dfrac{C_c}{1+e_0}\log_{10}\left(\dfrac{\sigma_{v0}' + \Delta\sigma_v}{\sigma_p'}\right) & if\, \sigma_p' \approx \sigma_{v0}' \\ \varepsilon_{OC} = \dfrac{C_s}{1+e_0}\log_{10}\left(\dfrac{\sigma_{v0}' + \Delta\sigma_v}{\sigma_{v0}'}\right) & if\, \sigma_{v0}' + \Delta\sigma_v < \sigma_p' \\ \varepsilon_{OC+NC} = \dfrac{C_s}{1+e_0}\log_{10}\left(\dfrac{\sigma_p'}{\sigma_{v0}'}\right) + \dfrac{C_c}{1+e_0}\log_{10}\left(\dfrac{\sigma_{v0}' + \Delta\sigma_v}{\sigma_p'}\right) & if\, \sigma_{v0}' + \Delta\sigma_v \geq \sigma_p' \end{cases} \tag{9.1}$$

where σ_{v0}' is the effective in-situ stress; $\Delta\sigma_v$ is the vertical external load; C_c is the compression index; C_s is the swelling index; and e_0 is the initial void ratio. The compression ratio is defined as $CR = C_c/(1+e_0)$. Compression indices C_c and C_s are analogous to Cam–Clay parameters λ and κ with the following conversions: $\lambda = C_c/\ln(10)$ and $\kappa = C_s/\ln(10)$.

Many Finnish clays, similar to other soft Nordic clays (e.g. Larsson 1977), are characterized by significant compressibility after σ_p'. Indeed, marine clays that have been deposited in seawater are characterized by a flocculated structure, which leads to a high void ratio e_0 (e.g. Rankka et al. 2004). If these marine clays are later subjected to leaching (reduction in pore water salinity), higher sensitivity S_t and compressibility are induced (Bjerrum 1967; Keinonen 1963; Rosenqvist 1953). Figure 9.1 illustrates a typical

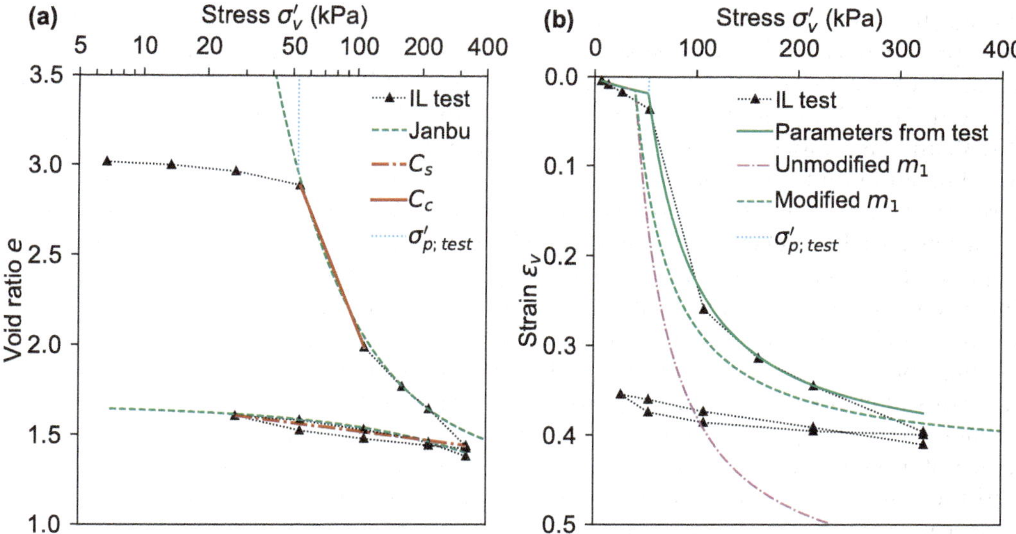

Figure 9.1 Compression behavior of soft Nordic clays: (a) comparison between the C_c method and the Janbu method when stress is in the $\log_{10}$ scale; and (b) modification of Janbu parameter m_1 ($\sigma_{p;calc}{}' = 40$ kPa; $\sigma_{p;test}{}'$=53 kPa; $m_{1;test} = 4.4$; $\beta_1 = -1.1$) (Source: modified from Löfman 2022, with permission).

incrementally loaded (IL) oedometer curve for a Finnish clay specimen: large deformations occur after $\sigma_p{}'$, and the curve is non-linear in semi-logarithmic scale. As a result, the C_c method (which assumes linear stress–strain response in a semi-logarithmic scale) is applicable only for the stress increment, which is used to interpret C_c.

Due to the limitations of the C_c method, Janbu (1963, 1967) proposed a tangent modulus method that would be applicable to all soil types. This method is also known as the “Ohde–Janbu method” as the formulation was first suggested by Ohde (1939). In the Janbu method, the tangential modulus M is defined as:

$$M = m\sigma_{ref}\left(\frac{\sigma'}{\sigma_{ref}}\right)^{1-\beta} \tag{9.2}$$

where m is the modulus number, σ_{ref} is the reference stress (100 kPa), and β is the stress exponent. Strain ε_{NC} is calculated with (Janbu 1963, 1967):

$$\varepsilon_{NC} = \int_{\sigma_p{}'}^{(\sigma_{v0}{}'+\Delta\sigma_v)} \frac{d\sigma'}{M} = \begin{cases} \dfrac{1}{m_1\beta_1}\left[\left(\dfrac{\sigma_{v0}{}' + \Delta\sigma_v)}{\sigma_{ref}}\right)^{\beta_1} - \left(\dfrac{\sigma_p{}'}{\sigma_{ref}}\right)^{\beta_1}\right] & if\ \beta_1 \neq 0 \\ \dfrac{1}{m_1}\ln\left(\dfrac{\sigma_{v0}{}' + \Delta\sigma_v}{\sigma_p{}'}\right) & if\ \beta_1 = 0 \end{cases} \tag{9.3}$$

Janbu parameters m_1 and β_1 are defined from the virgin compression curve using curve-fitting methods: m_1 is the stiffness parameter, while β_1 defines the shape of the curve.

Soft clays often have negative β_1, i.e., concave shape as in Figure 9.1a. Silts often have $\beta_1 > 0$ (i.e., convex oedometer curve, see also Figure 9.3a). If $\beta_1 = 0$, the Janbu method corresponds to the C_c method and the following applies:

$$m_1 = \frac{(1+e_0)\ln(10)}{C_c} = \frac{\ln(10)}{CR} \qquad (\beta_1 = 0) \tag{9.4}$$

Equation (9.3) can also be used to calculate ε_{OC}: OC parameters, referred to as m_2 and β_2, are typically using the recompression (reloading) sequence of the oedometer test (see Figure 9.10a). $\beta_2 = 1$ is often assumed, which corresponds to linear stress–strain dependency.

The Janbu method is flexible but also prone to user errors especially if β_1 is negative: if the value of σ_p' used in the calculation is lower than the one defined while curve-fitting m_1 and β_1, the oedometer curve is extrapolated and hence the settlement is overestimated. The following m_1 modification, suggested by Länsivaara (1995; 2003), may be used to avoid this error:

$$m_{1;calc} = m_{1;test}\left(\frac{\sigma'_{p;test}}{\sigma'_{p;calc}}\right)^{-\beta_1} \tag{9.5}$$

where $m_{1;test}$, $\sigma_{p;test}'$, and β_1 are the parameters defined from the oedometer test, while $\sigma_{p;calc}'$ is the value to be used in settlement calculation. Figure 9.1b illustrates how the m_1 modification preserves the shape of the original oedometer test curve. Meanwhile, much larger strains would be computed if the unmodified m_1 were used with $\sigma_{p;calc}' < \sigma_{p;test}'$ ("unmodified" in Figure 9.1b). Equation (9.5) implies that no modification is needed when $\beta_1 = 0$; that is, the C_c method is not affected by this stress dependency of stiffness parameters. According to the Finnish design guidelines (Finnish Transport Agency 2012), this modulus number modification should be applied, and it has also been incorporated into the Finnish geotechnical software GeoCalc (Civilpoint).

The m_1 modification is typically needed when a slightly overconsolidated clay is assumed to be NC, or when the $\sigma_{p;test}'$ value is from a CRS (constant rate of strain) oedometer test and needs to be rate-corrected to correspond to a lower strain rate (usually IL strain rate). In the case of Finnish clays, the rate correction of σ_p' may be performed using the Sällfors (1975) graphical method or via an empirical method based on Finnish clay data (Finnish Transport Agency 2012):

$$\sigma'_{p;calc} = \frac{\sigma'_{p;test}}{k} \qquad , where\ k = \left(\frac{\dot{\varepsilon}_{test}}{\dot{\varepsilon}_{calc}}\right)^{B} \tag{9.6}$$

where $\dot{\varepsilon}_{test}$ and $\dot{\varepsilon}_{calc}$ are the CRS strain rate and the target strain rate (10^{-7} 1/s for IL test), respectively. B is the empirical factor, for which the value $B = 0.0728$ is recommended (Finnish Transport Agency 2012).

9.3 DATABASE OF FINNISH CLAYS FI-CLAY/14/856

FI-CLAY/14/856 is a partly multivariate database consisting of 856 odometer tests on Finnish clay soils (undisturbed soil specimens). All the data points include specimen-specific

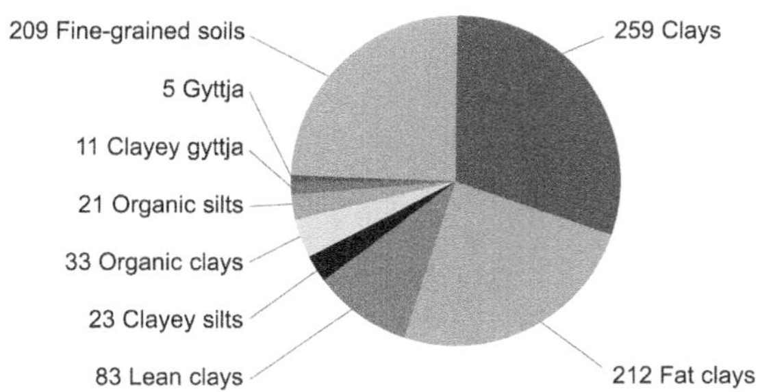

Clay fraction	Soil type
>10...30 %	Clayey silt
>30...50 %	Lean clay
>50 %	Fat clay

Organic content	Soil type
>2...6 %	Organic silt or Organic clay
>6...20 %	Silty gyttja or Clayey gyttja
>20 %	Gyttja

Figure 9.2 Pie chart of soil types included in the FI-CLAY/14/856 database and the Finnish classification system for fine-grained soils (Source: modified from Löfman and Korkiala-Tanttu 2022).

water content, void ratio, compression index, and compression ratio. Some are accompanied by additional laboratory test results for the same or nearby soil specimen (up to 14 parameters in total). The included data originated from various sources, although the majority were from research projects. Most of the specimens represent clays, as illustrated in Figure 9.2 (following the Finnish classification system, i.e., Korhonen, Gardemeister, and Tammirinne 1974). Full description of the sources and procedure used to compile the FI-CLAY/14/856 database can be found from Löfman and Korkiala-Tanttu (2022). The database is included in the TC304 compendium of databases known as 304dB (http://140.112.12.21/issmge/tc304.htm).

Since the Janbu method has been the preferred over C_c method in Finland, some of the data in FI-CLAY/14/856 lacked the C_c interpretation. For those oedometer test results, the C_c was calculated from the Janbu parameters (m_1, β_1, and σ_p') by assuming identical strains at stress increase from σ_1' to σ_2' (Löfman and Korkiala-Tanttu 2019):

$$C_c = \frac{(1+e_0)\dfrac{1}{m_1\beta_1}\left[\left(\dfrac{\sigma_2'}{\sigma_{ref}}\right)^{\beta_1} - \left(\dfrac{\sigma_1'}{\sigma_{ref}}\right)^{\beta_1}\right]}{\log_{10}\left(\dfrac{\sigma_2'}{\sigma_1'}\right)} \tag{9.7}$$

The approximate value for C_c was calculated by setting $\sigma_1' = \sigma_p'$ and $\sigma_2' = 2\sigma_p'$ since usually two load increments after σ_p' are used to interpret C_c. Figure 9.3a illustrates this C_c approximation using Equation (9.7). However, if σ_p' was greater than 120 kPa, the final stress was selected to be $\sigma_2' = \sigma_p' + 120$ kPa in order to better replicate the typical infrastructure geometries in Finland (120 kPa corresponds to a 6-m-high embankment). Figure 9.2a presents the verification of the approximation based on those tests where C_c interpretation was available ("actual C_c"): it can be seen that the scatter around the "no bias" line is moderate.

In a study reported by Löfman (2022), the FI-CLAY/14/856 database was extended: some oedometer test results were complemented with Janbu parameters in addition to compression index parameters. This database is referred to as FI-CLAY/14/856-B, and it has been published in an online repository (Löfman and Korkiala-Tanttu 2021).

Table 9.1 summarizes the statistics for the soil properties included in the FI-CLAY/14/856-B database. The table also includes the mean value from a global

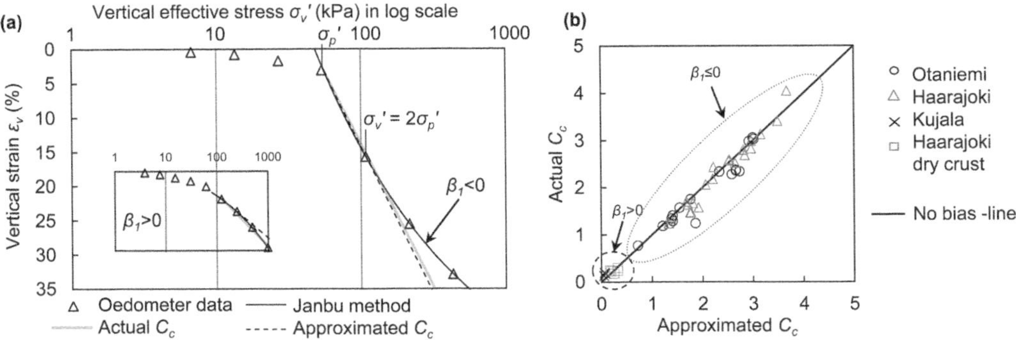

Figure 9.3 Approximation of C_c from Janbu parameters: (a) methodology and (b) verification (Source: modified from Löfman 2022, with permission. Based on Löfman and Korkiala-Tanttu 2019).

Table 9.1 Statistics for the FI-CLAY/14/856-B database and the mean values for the global clay database for comparison

Soil property	**n**	*Min*	*Median*[a]	*Mean*	*Max*	*COV*	*Mean, global database*[b]
Water content, w_n (%)	856	23.0	85.0	84.5	187	0.37	79.1
Void ratio, e_0	856	0.520	2.302	2.309	5.200	0.36	1.992
Liquid limit, w_L (%)	354	19.0	69.0	69.7	203.9	0.39	76.6
Plastic limit, w_P (%)	370	12.0	26.6	27.3	55.6	0.22	N/A
Plasticity index, PI (%)	354	1.0	41.5	42.5	153.7	0.54	41.4
Unit weight, γ (kN/m^3)	799	12.0	15.0	15.3	20.6	0.10	N/A
Organic content, Org (%)	369	0.0	0.8	1.2	7.8	1.26	N/A
Clay content, Cl (%)	263	12.7	54.5	56.6	95.0	0.36	N/A
Undrained shear strength, s_u (kPa)	413	5.2	22.5	30.2	240.0	1.01	N/A
Sensitivity, S_t	281	1.7	17.0	23.2	163.0	0.84	8.0
Preconsolidation pressure, σ_p' (kPa)	843	6.0	49.0	66.7	604.1	0.93	N/A
Overconsolidation ratio, OCR	817	0.3	1.4	2.2	41.3	1.32	2.4
Compression index, C_c	856	0.026	1.287	1.417	9.284	0.70	0.847
Compression ratio, CR	856	0.014	0.378	0.394	2.088	0.56	0.254
Modulus number, m_1	746	0.700	7.60	11.25	339.0	1.43	N/A
Stress exponent, β_1	746	−4.00	−0.31	−0.34	1.23	1.50	N/A
Swelling index, C_s	656	0.002	0.101	0.109	0.611	0.55	N/A
Modulus number, m_2	609	2.25	59.0	77.82	903.3	0.96	N/A
Stress exponent, β_2	609	−1.00	1.00	0.87	3.00	0.48	N/A

Notes: n = number of measurements; COV = coefficient of variation.

[a]50% sample percentile; [b]global database of marine soils compiled by Kootahi and Moradi (2017) for model building.

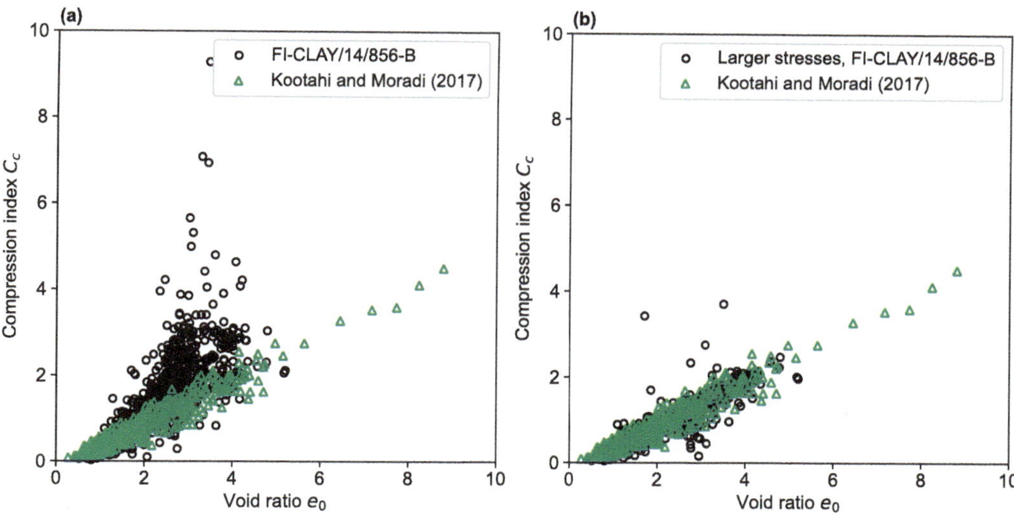

Figure 9.4 Relationship between e_0 and C_c in the global database of marine clays and in (a) Finnish clays (stress range approximately from σ_p' to $2\sigma_p'$) and (b) Finnish clays without structure (stress range from $2\sigma_p'$ to $4\sigma_p'$).

database of marine clays (Kootahi and Moradi 2017) for comparison. On average, the specimens in FI-CLAY/14/856-B have higher values for sensitivity S_t and compressibility (C_c and CR). Indeed, the mean and median values for the stress exponent β_1 are negative, which is typical for soft Nordic clays.

The difference between the Finnish and global clay databases is also visible in the e_0–C_c relationship, as shown in Figure 9.4a. Finnish clays tend to have greater C_c mainly due to their higher sensitivity (Di Buò et al. 2019; Leroueil et al. 1983). However, it is noteworthy that the interpretation of C_c, namely the considered stress range, also contributes to this phenomenon. High C_c values are acquired when the slope is fitted to the stress range right after the preconsolidation pressure σ_p'. On the other hand, if the slope is fitted to the "tail" of the oedometer curve, the acquired C_c values lack the effect of structure on the compressibility (similar to the compression line of reconstituted specimen, see e.g., Mataić 2016; Karstunen et al. 2005). Figure 9.3b shows the C_c values calculated from Janbu parameters using Equation (9.7), when the considered stress range is from $2\sigma_p'$ to $4\sigma_p'$: after removing the effect of structure, the e_0–C_c relationship of soft Nordic clays resembles other marine clays. Further, the data scatter is reduced.

9.4 INHERENT VARIABILITY

Total uncertainty in an estimated soil property is affected by inherent variability, measurement error, statistical uncertainty, and transformation uncertainty (Phoon and Kulhawy 1999). The inherent variability (also known as natural or spatial variability) is a result of geological processes. The magnitude of inherent variability may be estimated from the observed variability within a site or geotechnical layer (while assuming that the measurement error is small by comparison, see Phoon and Kulhawy 1999). The

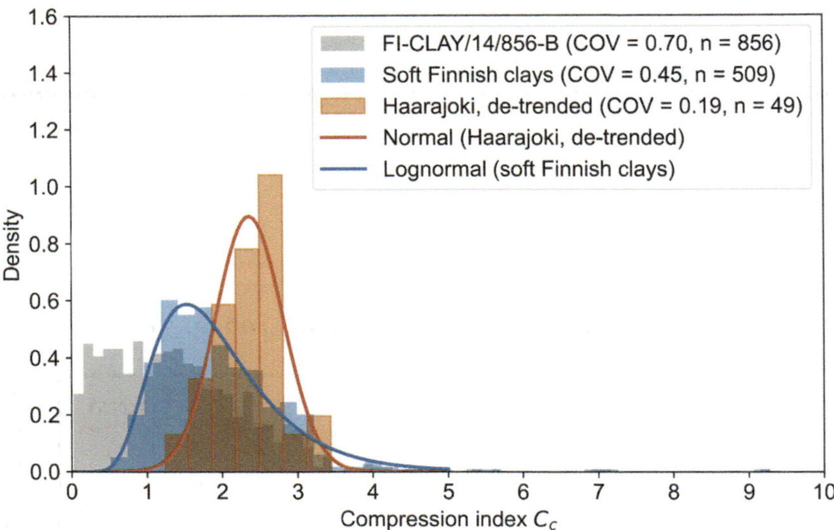

Figure 9.5 Observed variability in C_c: the effect of scale.

magnitude of inherent variability (or any other aleatoric uncertainty) may be characterized via coefficient of variation (COV), which is defined as the standard deviation divided by the absolute value of the mean.

Inherent variability is overestimated if data from different geological layers are mixed (Phoon and Kulhawy 1999). Figure 9.5 compares the histograms for C_c: even though all the considered datasets consist of Finnish clay soils, the variability decreases from COV = 0.70 to COV = 0.45 when the regional database FI-CLAY/14/856-B is filtered to contain soft clays only (for the criteria, refer to Section 9.6). Further, within a homogeneous clay layer at a specific site (Haarajoki), the observed COV is only 0.19. It is noteworthy that also the shape of histogram and hence the probability density function (pdf) changes with the considered scale.

If there is a linear trend with depth, de-trending should be applied before assessing COV of inherent variability (Phoon and Kulhawy 1999). De-trended standard deviation, which measures the scatter around the trend line, is defined by Lacasse et al. (2007):

$$SD_{de-trended} = \sqrt{\frac{1}{n-2}\sum_{i=1}^{n}\left(y_i - \left(b_o + b_1 z_i\right)\right)^2} \tag{9.8}$$

where y_i is the measured soil property value at depth z_i and $(b_0+b_1z_i)$ is the mean soil property value predicted by the linear trend (defined by intercept b_0 and slope b_1).

Knuuti and Löfman (2022) have summarized the observed variability (taken as the estimate of inherent variability) of geotechnical properties within homogeneous clay layers in various sites in Finland. Table 9.2 presents those values (relevant to settlement problems) and the corresponding COV values based on the global clay database or studies. It is noteworthy that some Finnish clay sites contain more than one homogeneous soil layer that was analyzed. The studied sites are also included in the FI-CLAY/14/856 database, but

Table 9.2 Inherent variability (observed variability) for Finnish clays and COV based on global study for comparison

Soil property[a]	*Soil types in the dataset*	***n**, sites*	*Ref.*[b]	***n**, sample size*	*Soil property mean, range*	*Soil property COV, range*[c]	*COV, (approx. guideline)*	*COV, mean or guideline (global)*
w_n (%)	Clay, silt, gyttja	8	A, B, C	15–105	40–146	0.03–0.14	0.1	0.15[e]
e_0	Clay, gyttja	5	A, C	10–53	1.4–3.9	0.02–0.13	0.1	0.07–0.30[f]
γ (kN/m^3)	Clay, silt, gyttja	8	A, B, C	11–80	13–18	0.01–0.04	<0.1	<0.1[f]
Fall cone s_u (kPa)	Dry crust clay	1	A	16	103	0.32–0.58	0.4	0.28[e]
	Clay, silt, gyttja	7	A, B	8–46	12–35	0.10–0.27	0.2	
	Clay/silt with sand layers	1	B[d]	12	35	0.61	0.6	
σ_p' (kPa)	Clay, gyttja	4	A	10–32	26–75	0.11–0.27	0.2	0.10–0.35[f]
OCR	Clay, gyttja	5	A, B	9–34	1.2–2.6	0.12–0.31	0.2	0.18[e]
C_c	Clay, gyttja	5	A, B	10–34	0.49–2.70	0.15–0.21	0.2	0.36[e]
CR	Clay, gyttja	5	A, B	10–34	0.19–0.69	0.15–0.20	0.2	N/A
m_1	Clay, gyttja	4	A[d]	10–34	3.5–12.6	0.09–0.32	0.2	N/A
β_1	Clay, gyttja	3	A[d]	9–34	(−0.27)–(−1.3)	0.33–0.55	0.5	N/A
β_1 (≈ 0)	Clay	1	A[d]	15	−0.09	≈1.00	1.0	N/A
C_s	Clay, gyttja	4	A[d], B	8–15	0.08–0.19	0.08–0.43	0.3	0.42[e]
m_2	Clay, gyttja	4	A[d]	8–10	49–97	0.15–0.52	0.3	N/A
β_2	Clay, gyttja	4	A[d]	8–10	0.49–1.12	0.26–0.60	0.3	N/A
c_v (m^2/a)	Clay	3	A	12–19	0.10–0.50	0.28–0.61	0.5	0.33–0.68[f]
k_1 (10^{-9} m/s)	Clay	3	A	10–13	0.45–1.40	0.29–0.56	0.5	0.68–0.90[f]
$C_{\alpha\varepsilon}$ (%)	Clay	3	A	12–25	0.67–2.47	0.26–0.52	0.4	N/A

[a] c_v = coefficient of consolidation, k_1 = permeability at zero strain, $C_{\alpha\varepsilon}$ = coefficient of secondary compression; [b] A = Löfman and Korkiala-Tanttu (2021b), B = Löfman and Korkiala-Tanttu (2019), C = Löfman (2016); [c] range includes COV values with and without de-trending, [d] based on the same dataset, but not reported in the publication; [e] Guan et al. (2021); and [f] Uzielli et al. (2006).

the analyzed data contains more laboratory test results (mainly index test results) than the FI-CLAY/14/856 database, which focuses on oedometer test results. The datasets used in the inherent variability analysis reported by Löfman and Korkiala-Tanttu (2021a) have been published open access in a data repository (Löfman and Korkiala-Tanttu 2021c).

Table 9.2 shows that the COV values for Finnish clays are mostly within the global guideline ranges or close to the global mean value. However, for properties such as s_u for soft soils, compressibility, and consolidation properties, the observed upper bound for COV is systematically lower than the global value. On the other hand, the global summaries and guidelines are typically based on datasets containing various soils and test types. For example, COV for s_u was between 0.3 and 0.6 in the datasets containing stiff Finnish clay soils (i.e., dry crust clay, and silt–clay mixtures with sand layers), while significantly smaller values were observed in soft Finnish clay or gyttja soil layers (COV = 0.10–0.27). Indeed, soft Finnish clays tend to be rather homogeneous (Gardemeister 1975).

Figure 9.5 further illustrates the relationship between the mean and COV values for selected compressibility-related soil properties (defined using the dataset published by Löfman and Korkiala-Tanttu 2021c). Table 9.2 and Figure 9.6 show that the decrease in observed COV via de-trending can be rather significant even though the analysis aimed to identify pseudo-homogeneous clay layers.

The results align with the hypothesis that COV values of the same physical properties are independent of the geological age of the soil (Uzielli et al. 2006). The analysis included various sites in Finland, and the clay sediments have different geological histories (for more detailed discussion on the geological characterization, see Löfman and Korkiala-Tanttu 2019). The inherent variability analyses also showed that the COV values for the compressibility-related properties of Finnish clays were rather consistent with respect to mean values, as noted in earlier summary studies (Guan et al. 2021; Phoon and Kulhawy 1999).

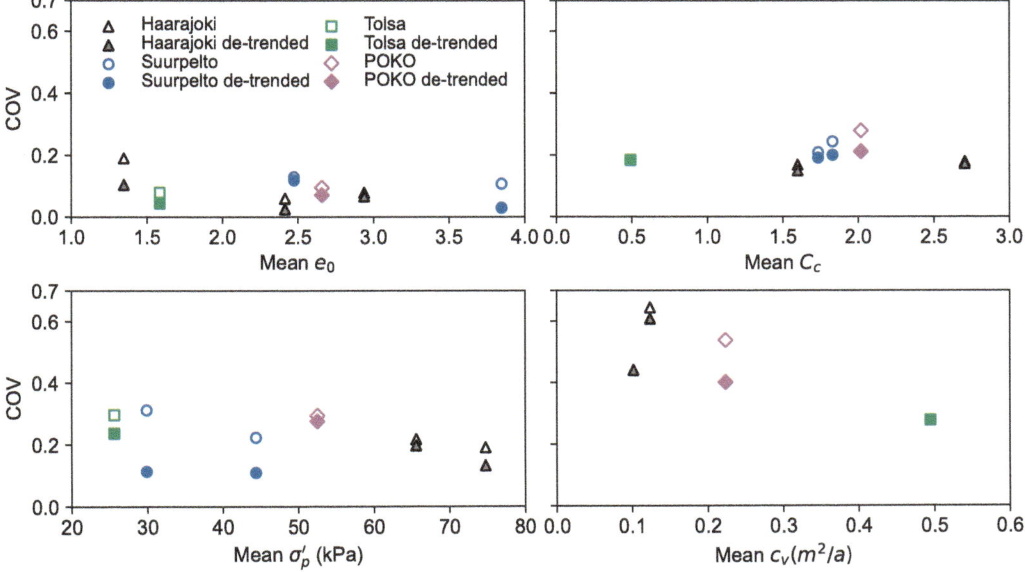

Figure 9.6 Observed variability in compressibility parameters as a function of mean value. The data represent homogeneous clay soil layers (n = 10...42).

9.5 PROBABILITY DENSITY FUNCTIONS

The probability density functions (pdfs) most often used for soil properties have been investigated in a survey by Lacasse and Nadim (1996): normal distribution is often used for index properties such as unit weight or density, while also lognormal distribution is commonly used for strength and overconsolidation. The suitability of different pdfs to represent the observed variability for Haarajoki clay (marine clay deposit from Finland) was investigated in a study by Löfman and Korkiala-Tanttu (2021b). The data was first de-trended (normalized) with depth, so that results from depth z = 2…18 m could be combined to represent one homogeneous clay layer. Figure 9.7 illustrates how the de-trending decreases the COV value for C_c (dataset is from Löfman and Korkiala-Tanttu 2021c). Next, Shapiro–Wilk normality test was used to evaluate the suitability of normal or lognormal distribution. Table 9.3 also shows the results for the homogeneous soft clay layer at depth z = 2…6.5 m, which did not require de-trending with depth (see Figure 9.8).

The results shown in Table 9.3 are consistent with the pdf preference observed in the survey. According to the analysis of Haarajoki clay, pdf suitability seems to be dependent on the considered soil layer and the consideration of trend with depth; For example, normal distribution was suitable for the de-trended C_c, while both normal and lognormal distributions were found appropriate when the homogeneous soft layer was considered (without de-trending).

If the Janbu method is used instead of the C_c method, it is important to apply the necessary corrections prior to the assessment of pdf suitability (or inherent variability). Namely, rate correction is needed when the dataset contains both incremental and CRS oedometer tests. Preconsolidation pressure σ_p' is first rate-corrected with Equation (9.6) to correspond incremental oedometer strain rate, and that lower value is then used to correct m_1 via Equation (9.5) in the case of specimens with negative β_1. It is also

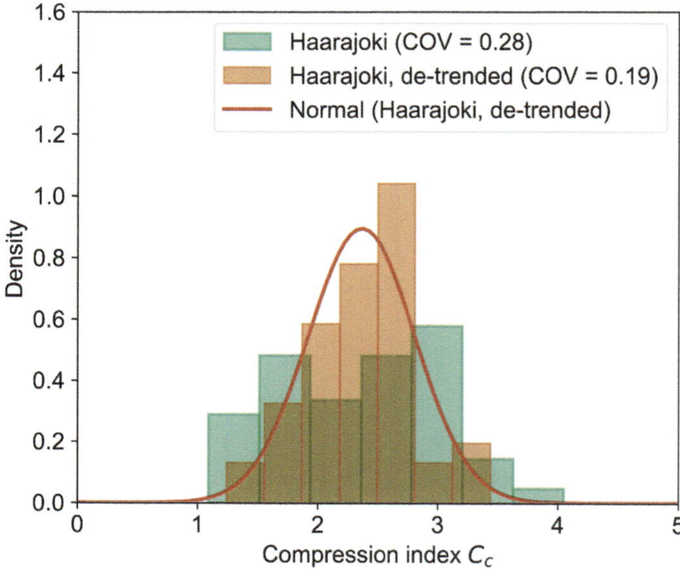

Figure 9.7 Observed variability in C_c: the effect of de-trending (Haarajoki clay layer, depth z = 2…18 m, n = 49).

Table 9.3 Suitable pdfs (*N* = normal; LN = lognormal) for geotechnical properties relevant to settlement calculations in clay subsoil

Soil property	*Haarajoki soft clay, **z** = 2...6.5 m, pdf*[a]	*Haarajoki clay, **z** = 2...18 m, de-trended pdf*[b]	*Survey, pdf*[c]	*Survey, soil type*[c]
Stress-normalized s_u	N/A	N/LN (*n* = 121)	N/LN	Clay
γ	N/A	N (*n* = 121)	N	All soils
e_0	N/LN (*n* = 39)	N/LN (*n* = 59)	N	All soils
OCR	N/LN (*n* = 39)	LN (*n* = 45)	N/LN	Clay
σ_p' (kPa)	LN (*n* = 39)	N/LN (*n* = 53)	N/A	N/A
C_c	N/LN (*n* = 39)	N (*n* = 49)	N/A	N/A
CR	N/LN (*n* = 39)	N/LN (*n* = 49)	N/A	N/A
m_l	N/LN (*n* = 39)	N/A	N/A	N/A
β_l	–[d]	N/A	N/A	N/A
Coefficient of creep $C_{\alpha\varepsilon}$	N/A	N/LN (*n* = 28)	N/A	N/A

[a]Data from homogeneous layer and rate correction of σ_p' and m_l using Equations (9.5) and (9.6); [b]normality test for laboratory test data de-trended with depth (Löfman and Korkiala-Tanttu 2021b). Rate correction of σ_p' based on Sällfors (1975) method; [c]survey on pdf selection (Lacasse and Nadim 1996); and [d]neither normal or lognormal was suitable (note: β_l is negative). Johnson's system-bounded (SB) distribution was found to be suitable (see Löfman 2022).

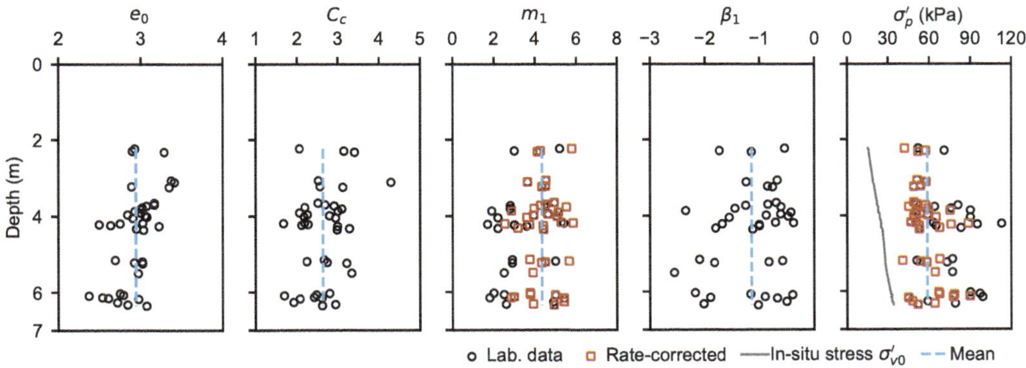

Figure 9.8 Haarajoki clay, homogeneous soft clay layer (z = 2...6.5 m). Preconsolidation pressure σ_p' and modulus number m_l from CRS test results have been rate-corrected.

noteworthy that since β_1 is usually negative, non-negative pdfs such as lognormal cannot be used.

9.6 TRANSFORMATION UNCERTAINTY IN EXISTING AND NEW TRANSFORMATION MODELS

If there are no site-specific oedometer tests available, index properties (e.g., w_n) can be used to indirectly estimate compressibility properties (e.g., C_c). Such transformation models, which are used to predict a target soil property from one or more predictor variables, are known to be marked by some degree of transformation uncertainty. Transformation models that are defined for given site-specific or regional data may be biased if applied to other sites (e.g., Ching and Phoon 2014), whereas "global" models tend to be less biased on average but also less precise when applied to a specific site (Ching and Phoon 2012). Indeed, regional soil databases may be utilized to calibrate existing transformation models with the approach proposed by Ching and Phoon (2014). This approach incorporates transformation error ε_i of ith data point, which is defined as the actual target value (C_c from the oedometer test) divided by the predicted target value (such as C_c predicted from w_n):

$$\varepsilon_i = \frac{\text{actual target value}}{\text{predicted target value}} \tag{9.9}$$

The arithmetic mean of all errors ε_i is the model bias b while the sample COV of ε_i represents the transformation uncertainty δ. For an unbiased transformation model, which has no data scatter around the trend curve, $b = 1$ and $\delta = 0$.

With the calibration study results, the actual (non-biased) target value is acquired from (Ching and Phoon 2014):

$$\text{actual target value} = \text{predicted target value} \times b \times \varepsilon \tag{9.10}$$

where ε is the variability term for the transformation model with a mean value equal to unity and COV = δ. Lognormal distribution may be assumed for the variability term (e.g., Ching 2017).

Löfman and Korkiala-Tanttu (2022) calibrated existing transformation models for predicting the compressibility of soft Finnish clays using the FI-CLAY/14/856 database. Some selected results for compression index C_c, compression ratio CR, and swelling index C_s are presented in Table 9.4. Transformation models for Janbu parameters m_1 and β_1 were not investigated in the study, because these curve-fitting parameters are dependent on the stress state ($\sigma'_{p;test}$) and should not be treated separately (unless the cross-correlation is modeled, see Section 9.6.)

The bias factors confirmed that almost all the existing models, including those most commonly used in Finland (i.e., Helenelund 1951; Janbu 1998), underestimate the value of C_c or CR of soft Finnish clays ($b > 1$). On average, models for C_s were less biased. Transformation uncertainty δ, which represents the amount of scatter around the trend curve, was $\delta = 0.4–0.8$ (up to 1.47 in the case of the transformation model for C_s). In the classification proposed by Phoon and Tang (2019), this corresponds to medium to high variability.

Due to the high transformation uncertainties, Löfman and Korkiala-Tanttu (2022) used the FI-CLAY/14/856 database to derive new transformation models for the compressibility of Finnish clays. To properly quantify the transformation uncertainties, the other sources of uncertainties (mainly measurement errors) should be minimized (Phoon and Kulhawy 1999). Hence, care was taken to remove disturbed specimens and other outliers from the database: the new transformation models were derived using the filtered database FI-CLAY/14/822. Finally, those data points where the residual error ε_{res} was more than three standard deviations ($\sigma_{\varepsilon;res}$) away from the zero-mean were considered outliers and hence removed before performing the final regression fitting, as illustrated Figure 9.9. Natural logarithm transformation was used to ensure that the residual error around the trend (defined as a polynomial regression curve) is normal distributed (see Löfman and Korkiala-Tanttu 2022 for further details).

Table 9.4 Selected results of the calibration study for some existing models for compressibility against FI-CLAY/14/856

Reference	*Applicability*	*Transformation model*	*n*	*b*	δ
Helenelund (1951)	Finnish and Danish clays, silts, and organic soils	$C_c = 0.85\left(\frac{w_n}{100}\right)^{1.5}$	856	1.97	0.44
Di Buò et al. (2019)	Finnish soft sensitive clays (stress range σ_p' + 10…50 kPa)	$C_{c(+10kPa)} = 0.000102\left(w_n\right)^{2.284}$	856	0.54	0.51
		$C_{c(+20kPa)} = 0.000289\left(w_n\right)^{2.031}$	856	0.57	0.46
		$C_{c(+50kPa)} = 0.0011\left(w_n\right)^{1.669}$	856	0.73	0.43
Moran et al. (1958)	Organic soils, peats, organic silts, and clays	$C_c = 0.0115w_n$	856	1.34	0.49
Kootahi and Moradi (2017)	Marine fine-grained soils	$C_c = 0.012\left(w_n - 7.75\right)$	856	1.42	0.47
		$C_c = 0.510\left(e_0 - 0.33\right)$	856	1.30	0.46
		$C_c = 0.012\left(w_L - 8\right)$	354	1.50	0.64
		$C_c = 0.020\left(PI + 0.65\right)$	354	1.43	0.80
		$C_c = 0.374\left(e_0 + 0.01w_L - 0.47\right)$	354	1.22	0.47
		$C_c = -0.117 + 0.009w_n + 0.004w_L$	354	1.26	0.52
Nakase et al. (1988)	Reconstituted Japanese marine clays	$C_c = 0.0104\left(PI + 4.42\right)$	354	2.36	0.67
Nishida (1956)	All clays	$C_c = 0.54\left(e_0 - 0.35\right)$	856	1.24	0.45
Ogawa and Matsumoto (1978)	Japanese marine clays	$C_c = 0.015\left(w_L - 19\right)$	354	1.63	0.89
Andersen (2012)	Danish clays, silts, and organic soils	$CR(\%) = 28.9\log_{10}\left(\frac{w_n}{11.4}\right)$	856	1.56	0.46
Janbu (1998)	All normally consolidated clays	$CR = \frac{\ln 10}{700} \times w_n \left(m_l = \frac{700}{w_n}\right)$	856	1.40	0.42
Elnaggar and Krizek (1970)	Various clays and silts	$CR = 0.156e_0 + 0.0107$	856	1.04	0.42
Di Buò et al. (2019)	Finnish soft sensitive clays	$C_s = 0.0113\left(w_n\right)^{0.67}$	656	0.49	0.44
Kootahi and Moradi (2017)	Marine fine-grained soils	$C_s = 0.0014\left(w_n - 10\right)$	656	1.09	0.43
		$C_s = 0.0540\left(e_0 - 0.3\right)$	656	1.05	0.41
		$C_s = 0.0017\left(w_L - 21\right)$	265	1.68	1.13
		$C_s = 0.0025\left(PI - 3\right)$	264	1.64	1.47
		$C_s = 0.0210\left(e_0 + 0.06w_L - 1.60\right)$	267	1.16	0.52
		$C_s = 0.0330\left(e_0 + 0.04PI - 0.55\right)$	266	1.05	0.44
Kulhawy and Mayne (1990)	All clays	$C_s = 0.0027PI$	267	1.18	0.87

Source: Data from Löfman and Korkiala-Tanttu (2022).

Notes: n = number of data points; b = bias (mean value of errors); and δ = transformation uncertainty (COV of errors).

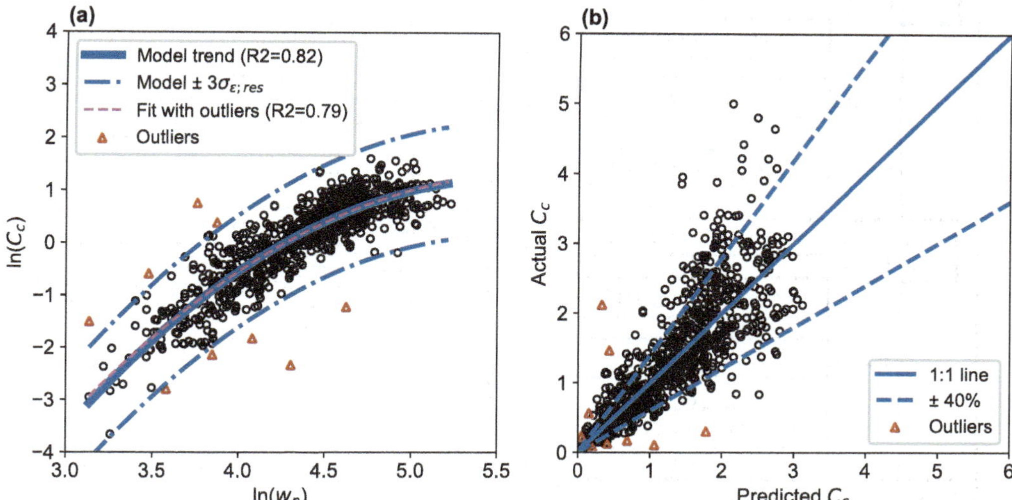

Figure 9.9 Fitting of $C_c(w_n)$-transformation model using the FI-CLAY/14/822 database: (a) polynomial regression curve and outlier detection and (b) predicted versus actual C_c.

Some newly derived transformation models and their statistics are collected in Table 9.5. Standard deviation of residual error $\sigma_{\varepsilon;res}$ quantifies the transformation uncertainty in the model. In addition, the statistics for the transformation error ε_i are provided to allow comparison with existing transformation models. It can be observed that via careful outlier detection and fitting of the transformation model, the transformation uncertainty in predicting C_c and CR for Finnish soft clays can be reduced. Moreover, the transformation uncertainty can be further reduced by incorporating a second predictor; For example, considering plastic limit w_P in addition to e_0 reduced $\sigma_{\varepsilon;res}$ for C_c from 0.346 to 0.317. Further analysis verified that for the same dataset used to derive C_c (e_0, w_P) model, the transformation uncertainty for $C_c(e_0)$ model was indeed greater, and hence not a result of dataset differences (Löfman and Korkiala-Tanttu 2022). In the case of C_s, transformation uncertainty was on average greater, and adding a second predictor did not reduce the performance of the model notably. Löfman and Korkiala-Tanttu (2022) also observed that liquid limit or plasticity index did not perform well as predictors for Finnish clays, contrary to the other studies dealing with marine clays (e.g. Kootahi and Moradi 2017; Kulhawy and Mayne 1990).

Figure 9.10 compares the newly derived and some existing transformation models against the database FI-CLAY/14/856-B. Figure 9.10a illustrates how the newly derived transformation model for C_c properly models the skewed error (shown as ± 2 standard deviations). Similarly, bias statistics in Table 9.5 indicate non-symmetric error distribution (mean ≠ median). This observation aligns with the assumption of lognormal pdf for variability term ε (e.g., Ching 2017). Figure 9.10a also shows how Di Buò et al.'s (2019) transformation models for sensitive Finnish clays form the upper bound for the data, which is due to differences in data and in C_c interpretation: Di Buò et al. (2019) considered a rather small stress range after σ_p' (10–50 kPa), and the data consisted of CRS tests with significant compression after σ_p', thus resulting in very high C_c values.

Figure 9.10b, which compares transformation models for C_s, shows that the general transformation model for marine clays (Kootahi and Moradi 2017) is quite suitable

Table 9.5 Transformation models for compressibility properties derived using a database of Finnish clays FI-CLAY/14/822

Model variables		Regression model: $Y = b_0 + b_1X_1 + b_2X_2 + b_3(X_1)^2 + b_4X_1X_2 + b_5(X_2)^2 + \varepsilon_{res}$									Transformation error statistics		
X_1	X_2	b_0	b_1	b_2	b_3	b_4	b_5	$\sigma_{\varepsilon;res}$	n	R^2_{adj}	Median	b	δ
Models for $Y = \ln(C_c)$:													
$\ln(w_n)$	N/A	−21.57	8.227	N/A	−0.743	N/A	N/A	0.357	813	0.82	1.00	1.07	0.37
$\ln(e_0)$	N/A	−1.586	2.861	N/A	−0.720	N/A	N/A	0.346	816	0.83	1.01	1.06	0.35
$\ln(e_0)$	$\ln(w_P)$	−7.516	1.911	4.429	−0.365	0.248	−0.807	0.317	357	0.88	1.02	1.05	0.32
Models for $Y = \ln(CR)$:													
$\ln(w_n)$	N/A	−20.75	7.986	N/A	−0.791	N/A	N/A	0.350	814	0.67	1.00	1.06	0.36
$\ln(e_0)$	N/A	−2.279	2.353	N/A	−0.831	N/A	N/A	0.343	816	0.69	1.01	1.06	0.35
$\ln(e_0)$	$\ln(w_P)$	−8.193	1.395	4.420	−0.481	0.252	−0.806	0.317	357	0.78	1.02	1.05	0.32
Models for $Y = \ln(C_s)$:													
$\ln(w_n)$	N/A	−12.339	3.618	N/A	−0.302	N/A	N/A	0.440	619	0.48	1.09	1.09	0.40
$\ln(e_0)$	N/A	−3.253	1.480	N/A	−0.317	N/A	N/A	0.435	619	0.49	1.10	1.09	0.38

Source: Data from Löfman and Korkiala-Tanttu (2022).

Notes: $\sigma_{\varepsilon;res}$ = standard deviation of residual errors; n = number of data points; R^2_{adj} = adjusted R-squared (considers the number of predictor variables); b = bias (mean value of errors); and δ = transformation uncertainty.

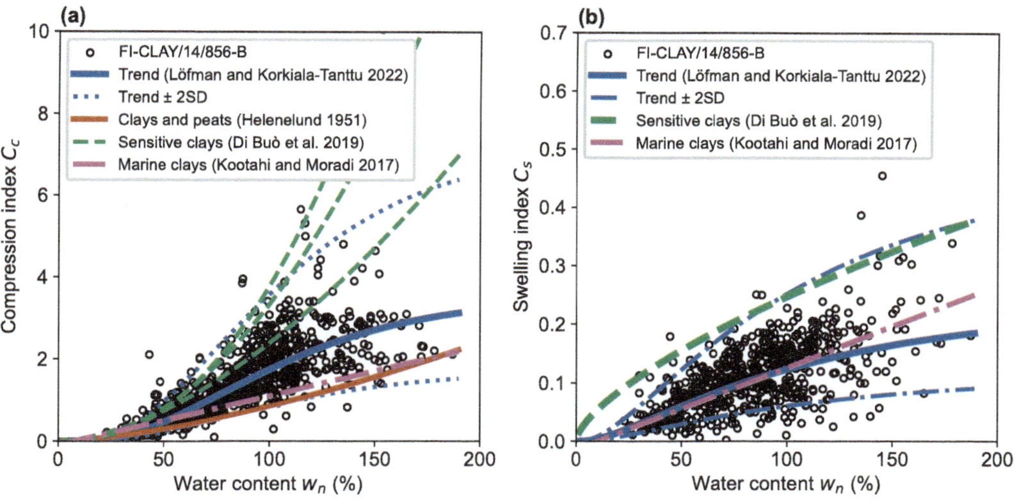

Figure 9.10 Transformation models for the compressibility of Finnish clays: (a) $C_c(w_n)$ models and (b) $C_s(w_n)$ models.

for soft Finnish clays, and the transformation error statistics (Tables 9.4 and 9.5) confirm that there is no notable improvement from the newly derived model. Meanwhile, the model for sensitive clays suggested by Di Buò et al. (2019) overestimates C_s. This bias is due to the differences in the interpretation (Di Buò et al. 2019 used an initial recompression curve, which is affected by sample disturbance) and weak correlation ($R^2 = 0.16$).

9.7 CROSS-CORRELATION AND CORRELATION COEFFICIENTS

Cross-correlation refers to a statistical dependency between two random variables. Ignoring a significant cross-correlation between calculation parameters may result in erroneous estimates of the probability of exceeding the limit state (e.g., Baecher and Christian 2003; Li et al. 2012).

The magnitude of cross-correlation may be measured by means of correlation coefficient ρ . Pearson product–moment correlation coefficient (ρ_{XY}) measures the linear correlation between two random variables, X and Y. Another common measure is Spearman's rank correlation coefficient. To evaluate the significance of cross-correlation, categories suggested by Evans (1996) may be used (Zhou et al. 2021): $|\rho| \geq 0.8$ means very strong, $0.6 \leq |\rho| < 0.8$ means strong, while $0.4 \leq |\rho| < 0.6$ means moderate cross-correlations. Smaller values indicate weak to very weak cross-correlation.

The measured correlation coefficient ρ varies between considered sub-populations. In a global clay database (containing multiple sites), cross-correlation between certain geotechnical parameters may be weak, while for a specific site or soil layer, strong cross-correlation is sometimes observed. This scale effect is noticeable in Table 9.6, which compares Pearson's ρ defined with site-specific, soil type-specific data, and regional or global data. For example, cross-correlation between m_1 and σ_p' is weak on a regional scale (all Finnish sites and data combined), but within a certain site (soft Haarajoki clay or Suurpelto clay), very strong cross-correlation is observed.

Table 9.6 Pearson's correlation coefficient ρ for certain populations

Considered population	$\rho_{e_0C_c}$	$\rho_{e_0\sigma_p'}$	$\rho_{C_c\sigma_p'}$	$\rho_{C_cC_s}$	$\rho_{m_1\sigma_p'}$	$\rho_{m_1\beta_1}$	$\rho_{\beta_1\sigma_p'}$
Site-specific ρ, Haarajoki clay	0.53	−0.67	0.09	0.50[a]	−0.88	0.86	−0.64
Site-specific ρ, Suurpelto clay	0.54	−0.86	−0.30	0.35[a]	−0.87	0.73	−0.73
Site-specific average ρ, soft Finnish clays	0.72	−0.32	0.05	N/A	−0.73	0.34	−0.28
Regional ρ (based on FI-CLAY/14/856-B database)	0.76	−0.44	−0.27	0.59[a]	0.25	0.42	0.11

[a]To compare, for the global clay database, Zhou et al. (2021) found a median value of 0.46 for site-specific ρ.

Indeed, cross-correlation between m_1 and σ_p' is the most prominent in soft Finnish clays with negative stress exponent β_1. A regional clay database may be utilized to define site-specific values for correlation coefficients, and the average or mean value can then be utilized in a reliability analysis if the amount of site-specific data is too limited. Löfman (2022) reported a study, where the FI-CLAY/14/856-B database was used to define the average correlation coefficients for soft Finnish clays. The database was first filtered to be representative of soft Finnish clays ($\beta_1 < 0$, $m_1 < 10$, $w_n > 75\%$). Only the datasets with $n \geq 10$ variable pairs were considered, as in Zhou et al. (2021). Figure 9.11 illustrates how the most consistent ρ values are related to the variable pairs m_1–σ_p' and e_0–C_c, and others are marked with greater variability between sites.

Monte Carlo simulation (MCS) with Gaussian copula was then used to compare the computed stress–strain behavior for Haarajoki clay and Suurpelto clay in different scenarios (see Löfman (2022) for further details):

1. Case Ind.: cross-correlation ignored ($\rho = 0$).
2. Case Corr.: cross-correlation defined using site-specific oedometer test results.
3. Case Datab.: average site-specific ρ for NC parameters (Haarajoki clay and Suurpelto clay not included in the average).

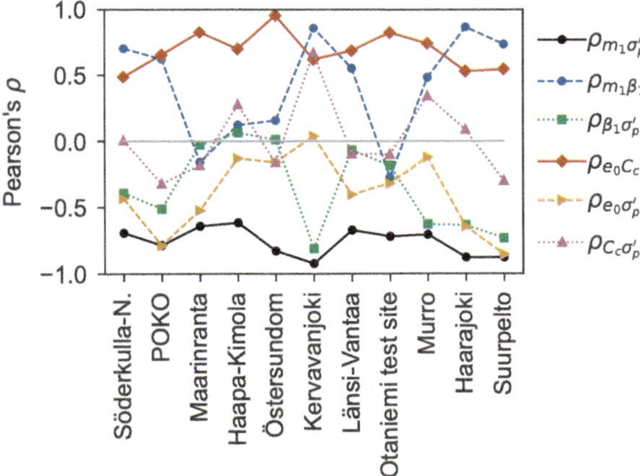

Figure 9.11 Site-specific cross-correlation between compressibility parameters in soft Finnish clays (Source: modified from Löfman 2022, with permission).

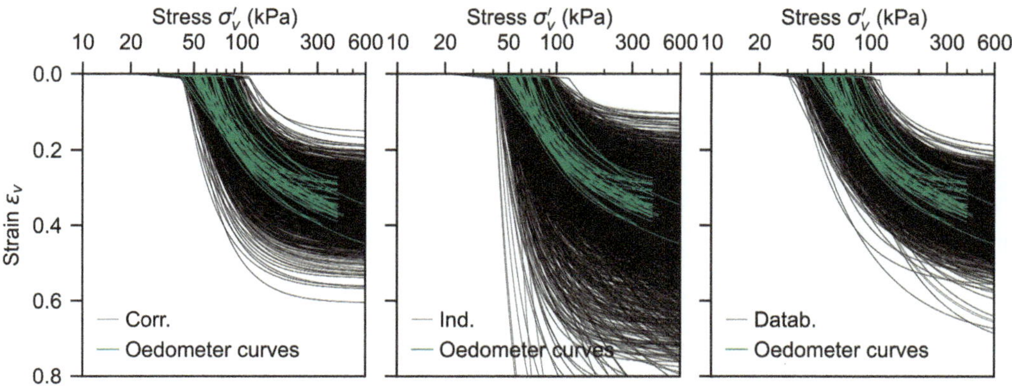

Figure 9.12 Janbu method: simulated oedometer curves for the Haarajoki clay layer in the different calculation scenarios (Source: modified from Löfman 2022, with permission).

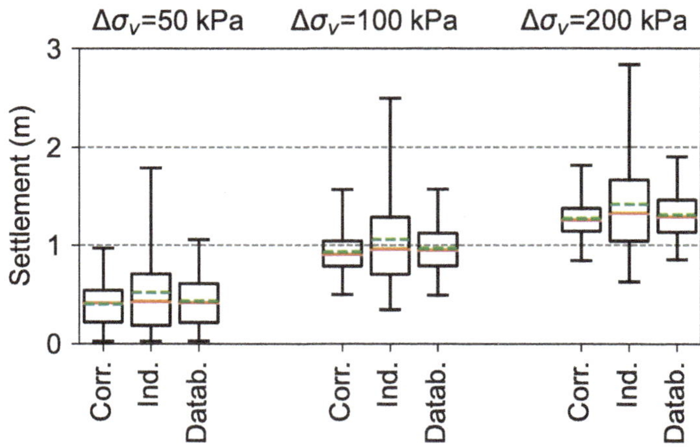

Figure 9.13 Janbu method: settlement boxplots for Haarajoki clay while assuming stress increase of 50, 100, or 200 kPa (Source: modified from Löfman 2022, with permission).

Figure 9.12 illustrates a sub-sample (2000 simulation runs) of stress–strain curves constructed with the simulated Janbu parameter sets. Oedometer curves were reconstructed using the Janbu parameters. Figure 9.13 presents the simulated settlements as boxplots for the Haarajoki clay layer. In the boxplots, the boundaries of the boxes represent the interquartile range, the dashed horizontal line is the mean, and the solid line is the median. The whiskers show the 2nd and 98th percentiles. The number of simulation runs was 10^5 for each calculation depth. It is evident that ignoring the cross-correlation between Janbu parameters (mainly variable pairs m_1–σ_p') will lead to an overestimation of uncertainty in the settlement response in soft clay. On the other hand, the correlation matrix defined via database analysis (Case Datab.) provided a fair replication of the parameter dependence (note that also the COV values were defined using literature values defined for Finnish clays, see Löfman 2022).

Meanwhile, if the C_c method (corresponding to $\beta_1 = 0$) is used to compute the stress–strain response, the influence of cross-correlation is less notable: Figure 9.14 presents the

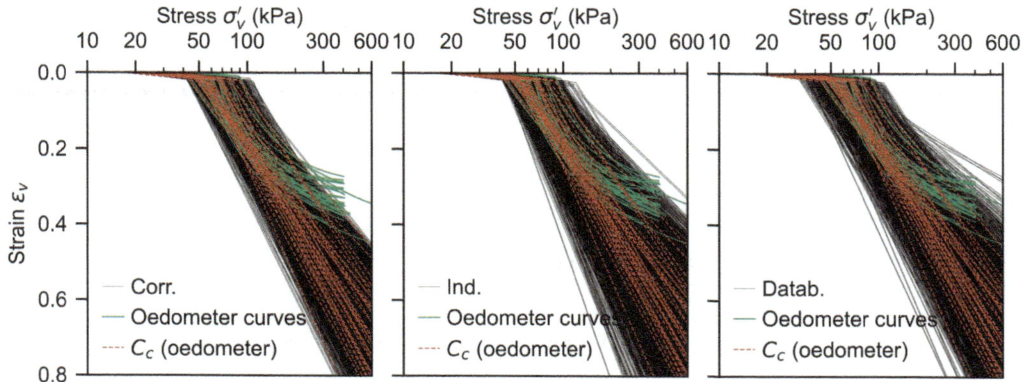

Figure 9.14 C_c method: simulated oedometer curves for the Haarajoki clay layer in the different calculation scenarios (Source: modified from Löfman 2022, with permission).

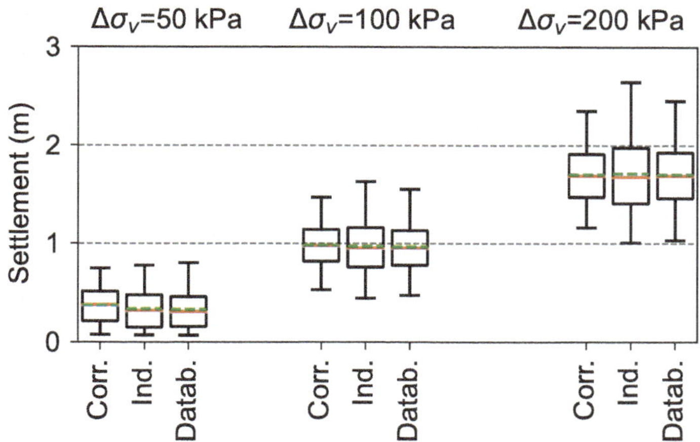

Figure 9.15 C_c method: settlement boxplots for Haarajoki clay while assuming stress increase of 50, 100, or 200 kPa (Source: modified from Löfman 2022, with permission).

oedometer curves constructed using both Janbu and C_c parameters, together with the simulated stress–strain response based on the C_c method. Figure 9.15 shows the simulated settlements as boxplots, and the calculation scenarios are rather similar. Hence, in most cases, the cross-correlation between C_c parameters may be ignored for the sake of simplicity. However, for soft soils with negative β_1 (such as Haarajoki clay), the C_c method will lead to an overestimation of strain (and thus settlement), especially at large load increments as illustrated in Figure 9.14. Hence, if the C_c method is used for soft marine clays, it is crucial that the problem-specific stress range is considered when the value of C_c is interpreted from the oedometer test result.

In a deterministic analysis, Equation (9.5) should be used to calculate the stress-corrected m_1. Using this procedure accounts for the cross-correlation between Janbu parameters. In fact, using Equation (9.5) within MCS allows us to approximate the result of Case Corr., even if the correlation coefficients are set to zero (Löfman 2022; López Ramirez, Löfman, and Korkiala-Tanttu 2022). However, at larger stress increments, such approximation leads to underestimation of the uncertainty in strain response (see Löfman 2022 for further details).

9.8 SUMMARY

This chapter has demonstrated how the database approach can be used to evaluate the needed input for the reliability-based SLS design of geo-structures dealing with soft marine clays. The special features related to soft and sensitive Nordic clays have been investigated and discussed: inherent variability (COV) for soft marine clays was found to be somewhat smaller compared to COV guidelines based on global summaries, since such soft soil deposits are often quite homogeneous. Similarly, smaller transformation uncertainties may be acquired if a regional clay database is used to calibrate existing transformation models and to derive carefully fitted models with outlier detection. Finally, it was shown how the cross-correlation between Janbu parameters may be effectively modeled via a database approach and Monte Carlo simulation with copula.

The COV values describing inherent variability were found to be rather consistent, and the literature values defined for Finnish marine clays may be utilized in the case of too limited data. In a typical situation, the amount of site-specific data is sufficient to determine the mean soil property μ for the considered homogeneous layer. Then, standard deviation (SD) is defined as μ multiplied by COV. Simple pdfs, such as normal or lognormal distribution, may be defined using the first two moments, μ and SD. For example, if C_c from one oedometer test is 2.95 (i.e., $\mu \approx 2.95$), and the specimen represents soft and homogeneous Finnish clay (COV = 0.2, see Table 9.2), SD = 0.2 × 2.95 = 0.59. If needed, the remaining sources of uncertainty are used to assess the total uncertainty COV_{tot} (see Phoon and Kulhawy 1999), and SD is calculated using the acquired COV_{tot} value.

Once the transformation uncertainties have been defined using a regional soil database (see Table 9.5), they can be properly accounted for in the case of indirectly evaluated compressibility properties. For example, if the water content in the studied layer is w_n = 100 %, $C_c = \exp[-21.567 + 8.227 \ln(w_n) - 0.743[\ln(w_n)]^2] = 1.75$. In a simplified approach, transformation uncertainty is modeled via Equation (9.10), meaning that the predicted target value (1.75) is multiplied by bias factor b (= 1.07). Hence, the (non-biased) mean value is qual to μ = 1.07 × 1.75 = 1.87. Next, total uncertainty (here statistical uncertainty, measurement error, and variance reduction from averaging are ignored for the sake of simplicity) may be defined via the sum of squared COV values according to Phoon and Kulhawy (1999):

$$COV_{tot} = \sqrt{COV_{wn}^2 + \delta^2} \tag{9.6}$$

where COV_{wn} is the inherent variability of water content (0.10, Table 9.2) and δ is transformation uncertainty (0.37, see Table 9.5). For the indirectly estimated C_c, we acquire COV_{tot} = 0.383. Hence, SD is 0.383 × 1.87 = 0.717.

For C_c within a homogeneous clay layer, lognormal pdf may be assumed (see Table 9.3).

Figure 9.16 presents the pdfs based on one oedometer test result and transformation model. Further, the figure shows the data from soft Haarajoki clay layer (with parameters μ = 2.65 and SD = 0.531), from which the oedometer test result originated. The chosen value for w_n = 100 % is close to the layers' mean value (107 %). In this example, the actual data falls within the pdf defined from the transformation model, but the clay layer in

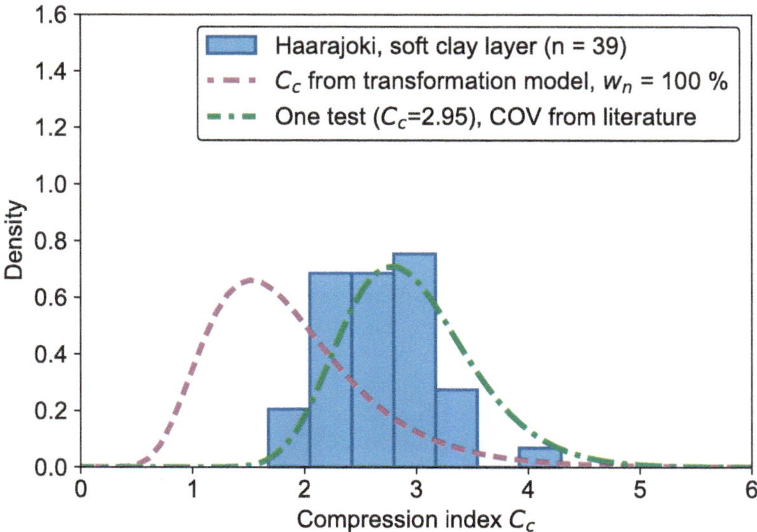

Figure 9.16 Simple example of evaluating the pdf of soft marine clay based on transformation model or an oedometer test and literature COV.

question represents marine clays with higher compressibility; Hence, the settlement prediction will be biased to some extent, even though the pdf of the settlement response will be wider accordingly. The figure also illustrates that even though COV value was lower for the oedometer test case, higher mean C_c will lead to greater SD (and therefore wider pdf).

REFERENCES

Andersen, J. D. 2012. Prediction of compression ratio for clays and organic soils. In *Proceedings of the 16th Nordic Geotechnical Meeting (NGM 2021)*. Copenhagen: Danish Geotechnical Society, 303–310.

Baecher, G. B., and J. T Christian. 2003. *Reliability and Statistics in Geotechnical Engineering*, Chichester: John Wiley & Sons.

Bjerrum, L. 1967. Engineering geology of normally consolidated marine clays as related to settlements of buildings. *Géotechnique* 17(2), 83–119.

Ching, J. 2017. Transformation models and multivariate soil databases. *Joint TC205/TC304 Working Group on "Discussion of statistical/reliability methods for Eurocodes" – Final Report (Sep 2017), ISSMGE.*

Ching, J., and K.-K. Phoon. 2012. Establishment of generic transformations for geotechnical design parameters. *Structural Safety* 35, 52–62.

Ching, J., and K.-K. Phoon. 2014. Transformations and correlations among some clay parameters - The global database. *Can Geotech J* 51(6), 663–685.

Di Buò, B., Länsivaara, T., Franza, A., and P. Mayne. 2019. Compressibility of Finnish sensitive clay. In: *Proceedings of the XVII European Conference on Soil Mechanics and Geotechnical Engineering*. The Icelandic Geotechnical Society.

Elnaggar, H. A., and R. J. Krizek. 1970. Statistical approximation for consolidation settlement. *Highway Res Rec*, 323, 87–96.

Evans, J. D. 1996. *Straightforward Statistics for the Behavioral Sciences*. Pacific Grove, CA: Brooks/Cole Publishing Company.

Finnish Transport Agency. 2012. *Tien geotekninen suunnittelu [Geotechnical design of roads]. Guidelines of Finnish Transport Agency* 10/2012. Helsinki: Finnish Transport Agency.
Gardemeister, R. 1975. *On Engineering-Geological Properties of Fine-Grained Sediments in Finland*. Doctoral thesis. Turku: University of Turku.
Guan, Z., Chang, Y.-C., Wang, Y., Aladejare, A., Zhang, D., and J. Ching. 2021. Site-specific statistics for geotechnical properties. In: *State-of-the-Art Review of Inherent Variability and Uncertainty in Geotechnical Properties and Models*, ed. J. Ching and T. Schweckendiek. TC304, ISSMGE.
Helenelund, K. V. 1951. *Om konsolidering och sättning av belastade marklager. Maa- ja vesiteknillisiä tutkimuksia* 6. Helsinki.
Janbu, N. 1963. Soil compressibility as determined by oedometer and triaxial tests. In: *Proceedings of European Conference on Soil Mechanics and Foundation Engineering*, Wiesbaden, Vol. 1, 19–25.
Janbu, N. 1967. Settlement calculations based on the tangent modulus concept. Three guest lectures at Moscow State University, Bulletin 2, *Soil Mechanics*. Norwegian Institute of Technology.
Janbu, N. 1998. *Sediment Deformations*. Bulletin 35. Trondheim: Department of Geotechnical Engineering, Norwegian University of Science and Technology.
Karstunen, M., Krenn, H., Wheeler, S. J., Koskinen, M., and R. Zentar. 2005. Effect of anisotropy and destructuration on the behavior of Murro test embankment. *Int J Geomech* 5(2), 87–97.
Keinonen, L. 1963. *On the Sensitivity of Water-Laid Sediments in Finland and Fac-Tors inducing sensitivity*. Publication 77. Helsinki: The State Institute for Technical Research.
Kivi, E. 2022. *Pohjanvahvistusmenetelmät Suomessa – käyttömäärät ja hiilijalanjälki [Ground improvement methods in Finland – usage rates and carbon footprint]*. Master thesis. Espoo: Aalto University School of Engineering.
Knuuti, M. and M. S. Löfman. 2022. *Epävarmuudet ja luotettavuus geoteknisessä mitoituksessa – Luotettavuus-workshop 2022 [Uncertainty and reliability in geotechnical design - Reliability workshop 2022]*. Publications of the FTIA 37/2022. Helsinki: Finnish Transport Infrastructure Agency.
Kootahi, K., and G. Moradi. 2017. Evaluation of compression index of marine fine-grained soils by the use of index tests. *Mar Georesources Geotechnol* 35(4), 548–570.
Korhonen, K.-H., Gardemeister, R., and M. Tammirinne, M. 1974. *Geotekninen maaluokitus [Geotechnical soil classification]. Geotekniikan laboratorio, tiedonanto 14*. Otaniemi: VTT Technical Research Centre of Finland.
Kulhawy, F. H., and P. W. Mayne. 1990. *Manual of Estimating Soil Properties for Foundation Design*. Palo Alto, CA: Electric Power Research Institute.
Lacasse, S. 2016. Hazard, Reliability and Risk Assessment - Research and Practice for Increased Safety. In: *Proceedings of the 17th Nordic Geotechnical Meeting (25–28 May 2016, Reykjavik, Iceland)*, 17–42, Reykjavík: The Icelandic Geotechnical Society,
Lacasse, S., and Nadim, F. 1996. Uncertainties in characterising soil properties." In: *Uncertainty in the Geologic Environment: From Theory to Practice*, Geotechnical Special Publication No. 58, C. D. Shackleford, P. P. Nelson, and M. J. S. Roth, eds., 49–75. New York: ASCE.
Lacasse, S., Nadim, F., Rahim, A., and T. R. Guttormsen. 2007. *Statistical Description of Characteristic Soil Properties*. Offshore Technology Conference.
Larsson, R. 1977. *Basic Behaviour of Scandinavian Soft Clays*. Linköping: Swedish Geotechnical Institute.
Leroueil, S., Tavenas, F., and J. P. L. Bihan. 1983. Propriétés caractéristiques des argiles de l'est du Canada. *Can Geotech J* 20(4), 681–705.
Li, D., Tang, X., Zhou, C., and K.-K. Phoon 2012. Uncertainty analysis of correlated non-normal geotechnical parameters using Gaussian copula. *Sci China Tech Sci* 55(11), 3081–3089.
López Ramírez, A., Löfman, M. S., and L. K. Korkiala-Tanttu. 2022. Safety margin from characteristic values in settlement calculations of embankments on clay. *Proc Inst Civ Eng: Geotech Eng* 175(6): 605–617.

Länsivaara, T. 1995. Stress-strain-strain rate relation in oedometer tests. In: *Proceedings, 70 years of Soil Mechanics*, International symposium, Istanbul.

Länsivaara, T. 2003. Problems related Scandinavian clays to settlement calculations on soft Scandinavian clays." In: *Int. Workshop on Geotechnics of Soft Soils -Theory and Practice*, ed. Vermeer, Schweiger, Karstunen, and Cudny. VGE.

Löfman, M. S. 2016. *Perniön saven parametrien luotettavuuden ja saven eri ominaisuuksien välisten korrelaatioiden arviointi* [*Estimation of the Reliability of Perniö Clay Parameters and Correlations between Clay Properties*]. Master thesis. Espoo: Aalto University School of Engineering.

Löfman, M. S. 2022. *Uncertainty Quantification for Compressibility and Settlement Response of Clays*. Doctoral thesis. Espoo: Aalto University School of Engineering.

Löfman, M. S. and L. Korkiala-Tanttu. 2019. Variability and typical value distributions of compressibility properties of fine-grained sediments in Finland. In: *Proceedings of the 7th International Symposium on Geotechnical Safety and Risk (ISGSR)*, ed. J. Ching, D.-Q. Li, and J. Zhang, 182–187. Singapore: Research Publishing.

Löfman, M. S., and L. Korkiala-Tanttu. 2021a. *Data and code for 'Influence of cross-correlation on the modelled uncertainty in stress–strain behavior of soft clays.' Zenodo*, https://doi.org/10.5281/zenodo.5040083

Löfman, M. S. and L. Korkiala-Tanttu. 2021b. Inherent variability of geotechnical properties for Finnish clay soils. In: 18th International Probabilistic Workshop. *IPW 2021*, ed. J. C. Matos *et al.*, Lecture Notes in Civil Engineering, vol 153, 431–443. Cham: Springer.

Löfman, M. S. and L. Korkiala-Tanttu. 2021c. Dataset for 'Inherent variability of geotechnical properties for Finnish clay soils'. https://doi.org/10.5281/zenodo.4709003

Löfman, M. S., and L. Korkiala-Tanttu. 2022. Transformation models for the compressibility properties of Finnish clays using a multivariate database. *Georisk* 16(2), 330–346.

Mataić, I. 2016. *On Structure and Rate Dependence of Perniö Clay*. Doctoral thesis. Espoo: Aalto University School of Engineering.

Moran, Proctor, Mueser, and Rutlege. 1958. *Study of Deep Soil Stabilization by Vertical Sand Drains*. Washington, DC: US Department of Commerse OT.S.

Nakase, A., Kamei, T., and O. Kusakabe. 1988. Constitutive parameters estimated by plasticity index. *J Geotech Eng* 114(7), 844–858.

Nishida, Y. 1956. A brief note on compression index of soil. *J Soil Mech Found Div* 82(3), 1–14.

Ogawa, F., and K. Matsumoto. 1978. The correlation of the mechanical and index properties of soils in harbour districts. *Rep Port Harbour Res Instit* 17(3), 3–89.

Ohde, J. 1939. *Zur Theorie der Druckverteilung im Baugrund*. Berlin: Der Bauingenieur.

Perttu, O. 2023. *Pohjanvahvistuksen optimointi osana vähähiilistä esirakentamista [Optimization of Ground Improvement as a Part of Low-Carbon Preconstruction]*. Master thesis. Espoo: Aalto University School of Engineering.

Phoon, K.-K. 2017. Role of reliability calculations in geotechnical design. *Georisk* 11(1), 4–21.

Phoon, K.-K., and F. H. Kulhawy 1999. Characterization of geotechnical variability. *Can Geotech J*, 36 (4), 612–624.

Phoon, K.-K., and C. Tang. 2019. Characterisation of geotechnical model uncertainty. *Georisk* 13(2), 101–130.

Rankka, K., Andersson-Sköld, Y., Hultén, C., Larsson, R., Leroux, V., and T. Dahlin. 2004. *Quick Clay in Sweden*. Report 65. Linköping: Swedish Geotechnial Institute.

Rosenqvist, I. T. 1953. Considerations on the sensitivity of Norwegian quick-clays. *Géotechnique* 3(5), 195–200.

Sällfors, G. 1975. *Preconsolidation Pressure of Soft, High-Plastic Clays*. Doctoral thesis. Göteborg: Chalmers University of Technology.

Uzielli, M., Lacasse, S., Nadim, F., and K.-K. Phoon. 2006. Soil variability analysis for geotechnical practice. In: *Proceedings of the 2nd International Workshop on Characterisation and Engineering Properties of Natural Soils*. Singapore.

Zhou, Y., Zhang, D., and J. Ching. 2021. Site-specific correlations between soil/rock properties. In: *State-of-the-Art Review of Inherent Variability and Uncertainty in Geotechnical Properties and Models*, ed. J. Ching and T. Schweckendiek. TC304, ISSMGE.

Chapter 10

An engineering geological parameter database of tunnel surrounding rock and its application

Junjie Ma and Tianbin Li

A generic rock mass database consisting of eight parameters is compiled from ten tunnels. The eight surrounding rock engineering geological parameters are the rock harness, rock mass integrity, dicontinuity condition, weathering degree, rock mass structure, groundwater condition, in-situ stress condition, and surrounding rock grade. This generic database, labeled as SurroundingRock/8/286, consists of 286 surrounding rock cases. This chapter aims to build an intelligent and accurate surrounding rock classification model for generic drilling and blasting excavation sites by combining SurroundingRock/8/286 with machine learning. Four methods are used to construct the surrounding rock classification models: convolutional neural network, random forest, Gaussian process, and support vector machine. The 5-fold cross-validation, 10-fold cross-validation, precision, recall, F-measure, and 24 actual tunnel validation results show that the support vector machine classification model is the most effective. Finally, a tunnel information management system was developed with cloud technology and the best surrounding rock classification model, which is now well-used in the Leshan–Xichang Expressway.

10.1 INTRODUCTION

Tunnel excavation will reveal new engineering geological conditions ahead. The different engineering geological conditions of the surrounding rock determine the support measures for the current section and the excavation method for the next section. Therefore, the collection and effective use of new engineering geological information revealed by tunnel excavation is essential for determining the tunnel surrounding rock's stability and the tunnel's dynamic excavation and support.

Many scholars have graded the quality of the surrounding rock based on the engineering geological information revealed by tunnel excavation, used this to determine the surrounding rock's stability, and developed corresponding support measures and excavation methods for different grades of surrounding rock quality. For different engineering geological conditions, the quality of the tunnel surrounding rock is graded, which is called surrounding rock classification or rock mass classification (Ma et al. 2023). Bieniawski (1973) collected and analyzed over 300 tunnel cases and proposed a rock mass classification method for jointed rock mass, called the rock mass rating (RMR) system. The engineering geological parameters considered in the RMR method include rock strength, rock quality designation (RQD), joint spacing, joint state, groundwater, and the relationship between joint attitude and tunnel axis. Barton et al. (1974) created a database of 212 tunnel cases and proposed the Q system for rock mass classification. The calculated parameters of the Q system include RQD, joint set number, joint roughness number, joint alteration number, joint water reduction factor, and stress reduction factor. Barton (2002)

 DOI: 10.1201/9781003441946-10

analyzed the 1050 new tunnel cases and revised the *Q* system. Hoek (1994) proposed the Geological Strength Index (GSI) method and considered the rock mass structure and weathering conditions. In China, the mainly used rock mass classification method is basic quality (BQ), which is determined by a statistical analysis of 460 cases. The BQ method (GB/T 50218–2014) considers the rock strength, rock mass integrity, groundwater condition, in-situ stress condition, and the relationship between weak structural plane occurrence, and tunnel axis direction. Different rock mass classification methods have been widely used in traffic tunnels, mining tunnels, and engineering slopes. However, with the advent of many deep-buried and ultra-long tunnels, traditional classification methods cannot meet the requirements of real-time and accurate acquisition of surrounding rock grades on site.

In the 21st century, the rapid development of artificial intelligence (AI) technology has brought opportunities for geotechnical engineering (Phoon, Ching, and Cao 2022; Phoon, Ching, and Shuku 2022; Phoon and Zhang 2022). Many scholars build a database based on the engineering geological information of the site and combine AI technology to build an intelligent model to realize intelligent prediction of some characteristics of the site. Santos et al. (2021) collected 3216 samples with RMR input parameters in open pit mines, and four rock mass classifiers were established using naive Bayes (NB), random forest (RN), artificial neural networks (ANNs), and support vector machine (SVM). Wang et al. (2021) built a database based on the 299 samples of the drilling parameters of the intelligent rock drilling trolley and established a classification system for surrounding rock. Bo et al. (2022) built a TBM boring parameter database containing 4464 samples, and a rock mass classification model was established using catboost integrated with sequential model-based optimization. In particular, some scholars use AI technology to extract surrounding rock information and use it for surrounding rock classification. Liu et al. (2018) built an image database containing 6000 tunnel face pictures, which was used to extract the surrounding rock information using deep learning technology. Chen et al. (2021) built a rock tunnel fracture database with 3000 tunnel face images, which was used to extract the rock fracture information using deep learning technology. In summary, building a generic and complete database of engineering geological information is exceptionally challenging. And most of the existing intelligent surrounding rock classification methods are limited to theoretical study and have little practical application.

In summary, most current research focuses on TBM-excavated tunnels and little research on intelligent systems for drilling and blasting excavation. Moreover, there are few studies on constructing a generic surrounding rock engineering geological information database for the traditional drilling and blasting tunnel. The above problems are the focus of this chapter. The novelty of this study is twofold:

1. This chapter compiles a new generic database for surrounding rock parameters named SurroundingRock/8/286. This database consists of 286 surrounding rock cases from ten tunnels for eight surrounding rock engineering geological properties, including rock harness, rock mass integrity, discontinuity condition, weathering degree, rock mass structure, groundwater condition, in-situ stress condition, and surrounding rock grade. Figure 10.1 presented in the following summarizes the generic and site-specific statistics for surrounding rock engineering geological parameters in SurroundingRock/8/286. In particular, each sample in SurroundingRock/8/286 contains eight complete surrounding rock engineering geological parameters.

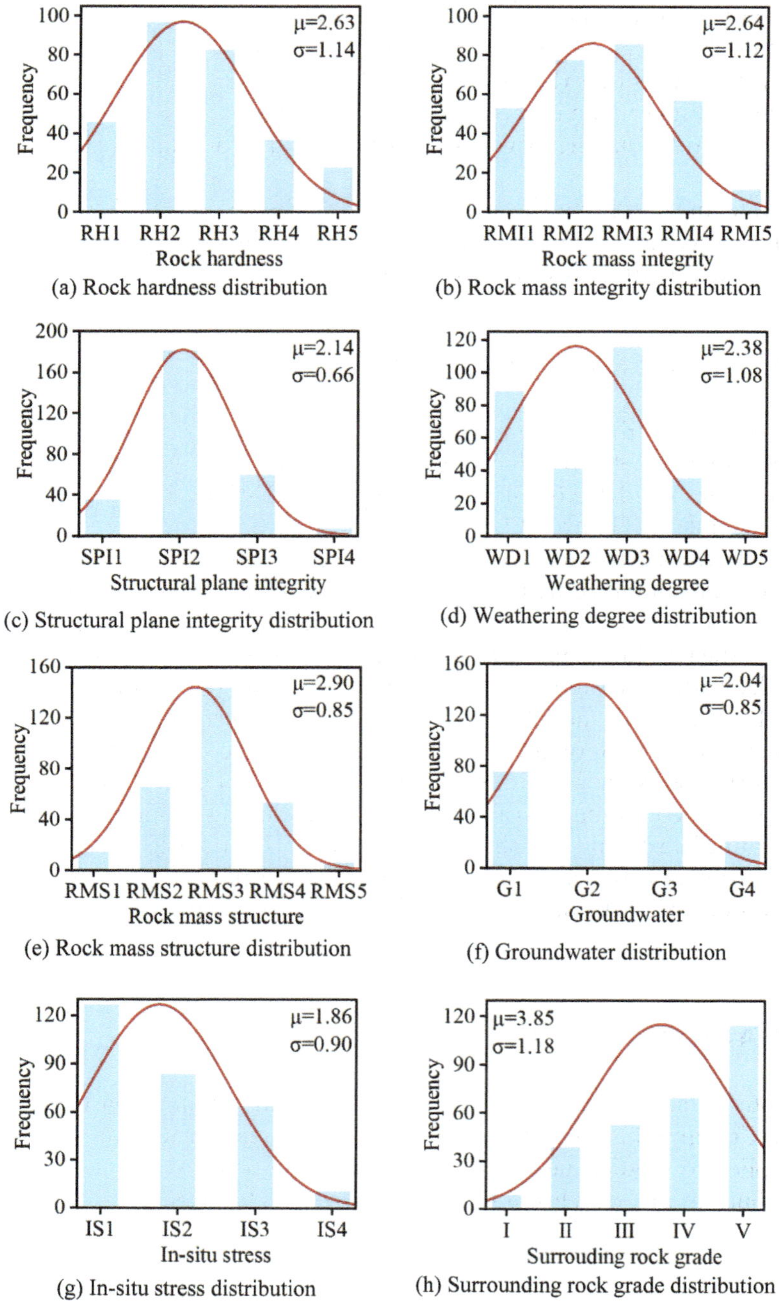

(a) Rock hardness distribution

(b) Rock mass integrity distribution

(c) Structural plane integrity distribution

(d) Weathering degree distribution

(e) Rock mass structure distribution

(f) Groundwater distribution

(g) In-situ stress distribution

(h) Surrounding rock grade distribution

Figure 10.1 Distribution of the engineering geological parameters in SurroundingRock/8/286. Note: The abbreviations for each ***x***-axis in Figure 10.1 correspond to the category abbreviations in Table 10.1.

2. This chapter further investigates the four methods (convolutional neural network (CNN), random forest (RF), Gaussian process (GP), and SVM) of constructing the surrounding rock classification model that SurroundingRock/8/286 supports. Further, accuracy, precision, recall, and *F*-measure are used to evaluate model performance comprehensively. Moreover, 24 actual tunnel cases are used further to test the practicability of the surrounding rock classification (SRC) model. Moreover, combined with cloud technology, a tunnel information management system with the best SVM-based classification model as the core is constructed. The tunnel information management system is now well-used in the Leshan–Xichang (Lexi) Expressway.

10.2 DATABASE SURROUNDINGROCK/8/286

This study complies a generic database (SurroundingRock/8/286) from the tunnel construction site for eight surrounding rock engineering geological parameters. In the tunnel construction site, generic databases have been complied for metamorphic, sedimentary, and igneous rocks. The SurroundingRock/8/286 database consists of 286 different surrounding rock cases from ten tunnels, each case with eight complete surrounding rock engineering geological parameters. The ten tunnels from western China include the Niba mountain tunnel, Erlang mountain tunnel, Wangjiangling tunnel, Zaozilin tunnel, Zhegu mountain tunnel, Longxi tunnel, Futang tunnel, Micang mountain tunnel, Miyaluo #3 tunnel, and Shiziping tunnel.

The eight surrounding rock engineering geological parameters in SurroundingRock/8/286 include rock harness, rock mass integrity, discontinuity condition, weathering degree, rock mass structure, groundwater condition, in-situ stress condition, and surrounding rock grade. Since there are quantitative and qualitative parameters in SurroundingRock/8/286, it is necessary to discretize them. In Table 10.1, the parameters in SurroundingRock/8/286 are discretized according to the parameter division range recommended by "Code for Engineering Geological Investigation of Water Resources and Hydropower" and "Standard for Engineering Classification of Rock Mass." The quantitative and qualitative division bases of the eight surrounding rock engineering geological parameters are shown in Tables S10.1–Table S10.7 in the supplementary material.

Figure 10.1 shows the distribution of the engineering geological parameters in SurroundingRock/8/286. The data in our database covers a wide range of values for these eight surrounding rock engineering geological parameters. The SurroundingRock/8/286 database has the following features:

1. Figure 10.1(a) shows that the rock strength in the database is predominantly slightly hard and slightly soft. The database also has 20–50 hard, soft, and extremely soft rock samples.
2. Figure 10.1(b) shows that the distribution of rock mass integrity in the database is relatively balanced, except for the extremely broken surrounding rock samples, which are only 12.
3. Figure 10.1(c) shows that the unbalanced distribution of discontinutiy condition in the database is mainly ordinary.

Table 10.1 Intervals of the engineering geological parameters in SurroundingRock/8/286.

Parameter		*Intervals (category abbreviation)*
Rock hardness (RH)	Intervals (MPa)	(60, ∞), (30, 60], (15, 30], (5, 15], (0, 5]
	States	Hard (RH1), slightly hard (RH2), slightly soft (RH3), soft (RH4), extremely soft (RH5)
Rock mass integrity (RMI)	Intervals (K_v)	(0.75, 1], (0.55, 0.75], (0.35, 0.55], (0.15, 0.35], (0, 0.15]
	States	Complete (RMI1), slightly complete (RMI2), slightly broken (RMI3), broken (RMI4), extremely broken (RMI5)
Dicontinuity condition (Dc)	States	Good (Dc1), ordinary (Dc 2), bad (Dc 3), very bad (Dc 4)
Weathering degree (WD)	Intervals (k_r)	(0.9, 1], (0.8, 0.9], (0.6, 0.8], (0.4, 0.6], [0.2, 0.4]
	States	Fresh (WD1), slight (WD2), medium (WD3), severe, and extreme (WD4)
Rock mass structure (RMS)	States	Integral/extremely thick-layered (RMS1), block/thick-layered (RMS2), fractured block/medium-thick-layered/interlayer/thin-layered/mosaic (RMS3), fragmented (RMS4), granular (RMS5)
Groundwater condition (G)	States	Dry/wet (G1), moist/dripping (G2), rain-like dripping/linear water (G3), tubular water/gushing water (G4)
In situ stress condition (IS)	Intervals [R_c/σ_{max}]	[9, +∞), [7, 9), [4, 7), (0, 4)
	States	Low (IS1), medium (IS2), high (IS3), extremely high (IS4)
[BQ]	Intervals	(550, 710], (450, 550], (350, 450], (250, 350], (0, 250]
	States	I, II, III, IV, V

4. Figure 10.1(d) shows that the weathering degree of rock mass is mainly fresh and medium weathering in the database. There are only three samples of the extreme weathering rock mass.
5. Figure 10.1(e) shows that the database's normal distribution trend of rock mass structure is apparent. There are fewer granular structures in the database.
6. Figure 10.1(f) shows that the groundwater condition in the database is predominantly no or little water, with few samples for tubular water/gushing water.
7. Figure 10.1(g) shows that the in-situ stress condition in the database is predominantly low, medium, and high, with only 11 samples for extremely high in situ stress.
8. Figure 10.1(h) shows that the database is dominated by classes III, IV, and V surrounding rocks, with a small number of classes I and II surrounding rocks. The small and unbalanced surrounding rock database corresponds to the actual situation on the tunnel project site. However, machine learning algorithms such as SVM can perform well in small and unbalanced sample databases.

10.3 INTELLIGENT METHODS FOR SURROUNDING ROCK CLASSIFICATION

10.3.1 Convolutional neural network

Deep learning is a subfield of machine learning (Ketkar and Moolayil 2021). Convolutional neural network (CNN) is one of the representative algorithms of deep learning, which was first reported in 1989 for machine vision (LeCun et al. 1989). In the past 20 years, a series of classification networks with different architectures have been proposed based on the CNN principle, such as LeNet, AlexNet, GoogleNet, VGGNet, etc. (Khan et al. 2020). In particular, VGGNet has significantly improved network depth, computing speed, and complexity.

Simonyan and Zisserman proposed the VGGNet in 2015, which was mainly developed based on the AlexNet. The VGGNet used a small-sized filter to reduce the number of parameters and the computational complexity. In addition, the VGGNet used random inactivation to mitigate the effect of overfitting. Therefore, the VGGNet has a more significant improvement than the AlexNet regarding recognition accuracy (Khan et al. 2020). In particular, the convolutional and fixed pooling layers of the same size allow the VGGNet to be easily applied to various scenarios.

The VGGNet consists of 3×3 convolution kernel (stride 1 and padding 1) and pooling layer (stride 2). The distinction between the VGGNet and the AlexNet is that a stack of 3×3 convolution kernels is used instead of the large-sized convolution kernels (5×5 and 7×7). For example, the VGGNet16 stacks two identical 3×3 convolution kernels and three identical 3×3 convolution kernels to replace one 5×5 convolution kernel and one 7×7 convolution kernel, respectively. The number of parameters generated by two 3×3 convolution kernels is $2(32D2) = 18D2$, and the number of parameters generated by a 5×5 convolution kernel is $52D2 = 25D2$ (D is the depth of the channel of the feature matrix). Therefore, the number of parameters produced by two 3×3 convolution kernels is significantly less than that produced by one 5×5 convolution kernel. Similarly, the number of parameters generated by three 3×3 convolution kernels is much lower than that generated by one 7×7 convolution kernel. Thus, the VGGNet with relatively few parameters can achieve a more efficient training process. On the other hand, after the 3

$\times$ 3 convolution kernel replaces the 5 $\times$ 5 and 7 $\times$ 7 convolution kernels, the new network will have a deeper structure, which can further improve the recognition performance of the VGGNet. The distinction between the VGGNet16 and the VGGNet19 is the difference in network depth. The VGGNet16 contains 16 hidden layers (thirteen convolutional layers and three fully connected layers). The VGGNet19 contains 19 hidden layers (sixteen convolutional layers and three fully connected layers). More details about the VGGNet can be found in Simonyan and Zisserman (2015).

10.3.2 Random forest

Random forest (RF) is a statistical learning theory that uses the bootstrap resampling method to take multiple samples from the original sample, model the decision tree for each bootstrap sample, and then combine the predictions of multiple decision trees to obtain the final prediction result by voting (Breiman 2001, Genuer and Poggi 2020).

Random forest classification (RFC) is a combinatorial classification model composed of many decision tree classification models $\{h(X, \Theta_k), k=1, 2, \ldots\}$, and the parameter set $\{\Theta_k\}$ is an independent and homogeneous random vector. Given the independent variable X, each decision tree classification model is selected by one vote to select the optimal classification result.

The basic idea of RFC: first, bootstrap sampling is used to extract k samples from the original training set, and each sample has the same sample size as the initial training set; Secondly, k decision tree models were established for k samples to obtain k classification results. Finally, each record is voted on based on the results of the k classifications to determine its final classification.

RF increases the difference between classification models by constructing different training sets, thereby improving the extrapolation prediction ability of combined classification models. Through k round training, a classification model sequence $\{[h_{11}(X), h_{12}(X), \ldots, h_{1n}], [h_{21}(X), h_{22}(X), \ldots, h_{2n}], \ldots, [h_{k1}(X), h_{k2}(X), \ldots, h_{kn}(X)]\}$ is obtained, and then they are used to form a multi-classification model system. In contrast to the original publication (Breiman 2001), the scikit-learn implementation combines classifiers by averaging their probabilistic prediction instead of letting each classifier vote for a single class.

The final classification decision formula is as Equation (10.1).

$$H(x) = \max\left(\begin{array}{l} \dfrac{h_{11}(X)+h_{21}(X)+\cdots+h_{k1}(X)}{k}, \dfrac{h_{12}(X)+h_{22}(X)+\cdots+h_{k2}(X)}{k}, \\ \ldots, \dfrac{h_{1n}(X)+h_{2n}(X)+\cdots+h_{kn}(X)}{k} \end{array}\right) \tag{10.1}$$

where $H(x)$ represents the combinatorial classification model, k is the number of decision tree classifiers, and n is the number of categories that predict labels. $H_{kn}(X)$ refers to the probability that the kth decision tree classifier predicts the label class number. Equation (10.1) illustrates the use of averaging probability decisions to determine the final classification.

10.3.3 Gaussian processes

A Gaussian process is a generalization of the Gaussian probability distribution. A probability distribution describes random variables that are scalars or vectors (for multivariate distributions), whereas a stochastic process governs the properties of functions (Rasmussen and Williams, 2005).

The mean function and the covariance function uniquely determine a Gaussian process. A classification problem based on a Gaussian process can be described as a given training set (T) as Equation (10.2).

$$T=\{(x_i,y_i),i=1,2,\cdots,m\} \tag{10.2}$$

where x_i is the input value and y_i is the category label. When data x^* is the input, the predicted output y^* can be given. For binary classification problems, y is either 1 or –1.

A standard method used in solving binary classification problems in Gaussian processes is to convert the predicted output of a regression model into probability through a response function. Binary classification based on Gaussian processes essentially adds Gaussian processes prior directly to the latent function $f(x)$ and then obtains the prior probability $p(y=1|x)=\sigma(f(x))$ through the logistic function, and finally finds the probability distribution of $p(y^*=1|X, y, x^*)$ (see Equation 10.3). Since the integrable function in Equation (10.3) is not a Gaussian distribution, the Laplace approximation method or the expectation propagation approximation method is usually used to approximate the non-Gaussian joint posterior distribution to obtain an analytical solution. More details about the Laplace approximation can be found in Rasmussen and Williams (2005).

$$p\left(y^*=1\middle|X,y,x^*\right)=\int p\left(f^*\middle|X,x^*,f\right)p\left(f\middle|X,y\right)df \tag{10.3}$$

where X is the input vector, y is the output label, f is the latent function, and the posterior probability of the latent function is shown in Equation (10.4).

$$p\left(f\middle|X,y\right)=\frac{p\left(y\middle|f\right)p\left(f\middle|X\right)}{p\left(y\middle|X\right)} \tag{10.4}$$

The covariance function in the Gaussian process model is equivalent to the kernel function. Commonly used kernel functions include Constant, Radial basis function (RBF), Matern, RationalQuadratic, Exp-Sine-Squared, and DotProduct kernels. It is also possible to add, multiply, and other operations on the primary kernel function to form a new kernel function of the Gaussian process.

Gaussian process classification (GPC) supports multi-class classification by performing either one-versus-rest or one-versus-one-based training and prediction. In one-versus-rest, one binary Gaussian process classifier is fitted for each class, which is trained to separate this class from the rest. In "one_vs_one," one binary Gaussian process classifier is fitted for each pair of classes, which is trained to separate these two classes. The predictions of these binary predictors are combined into multi-class predictions.

10.3.4 Support vector machine

Support vector machine (SVM) can solve regression and classification tasks with relatively few datasets (Drucker et al. 1996), which was proposed by Cortes and Vapnik in 1995. In particular, SVM can be applied to multi-classification problems. The theory of SVM can be summarized as follows.

Suppose there are training sample sets $(\boldsymbol{x}_i, y_i)$. $\boldsymbol{x}_i$ is the sample features of the i-sample set ($i = 1, 2, \cdots, k; i, k \in \mathrm{N}^*$). y_i is the label of the i-sample set ($y_i \in \{+1, -1\}$). The hyperplane and its condition can be defined as Equation (10.5). Obviously, the hyperplane with the largest margin is the optimal hyperplane (Cristianini and Shawe-Taylor 2000).

$$\begin{cases} \boldsymbol{w}^T \boldsymbol{x} + b = 0 \\ y_i\left(\boldsymbol{w}^T \cdot \boldsymbol{x}_i + b\right) - 1 \geq 0 \end{cases} \tag{10.5}$$

where $\boldsymbol{w}^{\mathrm{T}}$ is the weight vector, x is the feature vector, and b represent the bias.

Introducing the Lagrange function, α_i are Lagrange multipliers ($\alpha_i \geq 0$). The linearly separable problem can be rewritten as Equation (10.6).

$$L(w,b,\alpha) = \frac{\|w\|^2}{2} - \sum_{i=1}^{k} \alpha_i \left[y_i \left(\boldsymbol{w}^T \cdot \boldsymbol{x}_i + b \right) - 1 \right] \tag{10.6}$$

Then, the hyperplane of the SVM is determined according to the Karush–Kuhn–Tucker (KKT) conditions (Equation 10.7).

$$\begin{cases} \dfrac{\partial L}{\partial w} = w - \displaystyle\sum_{i=1}^{k} \alpha_i y_i x_i = 0 \\ \dfrac{\partial L}{\partial b} = -\displaystyle\sum_{i=1}^{k} \alpha_i y_i = 0 \end{cases} \tag{10.7}$$

Thus, the primal problem can be transformed into a dual problem. The optimal solution α_i^*, the optimal weight vector $\boldsymbol{w}^*$, and the optimal bias b^* are obtained, where $\boldsymbol{w}^* = \sum_{i=1}^{k} \alpha_i^* y_i \boldsymbol{x}_i$.

Finally, the decision function is shown as in Equation (10.8). The sgn() is a symbolic function.

$$f(x) = \mathrm{sgn}\left[\sum_{i=1}^{k} \alpha_i^* y_i \left(x_i^T \cdot x \right) + b^* \right] \tag{10.8}$$

In particular, the slack variable ξ_i is introduced to seek a balance between the goals of maximum margin and minimum training sample error rate. The condition of Equation (10.5) can be transformed as Equation (10.9).

$$y_i\left(\boldsymbol{w}^T \cdot \boldsymbol{x}_i + b\right) - 1 + \xi_i \geq 0 \tag{10.9}$$

subject to Equation (10.10).

$$\begin{cases} \xi_i \geq 0 \\ \displaystyle\sum_{i=1}^{k} \alpha_i y_i = 0 \\ C \geq \alpha_i \geq 0 \end{cases} \tag{10.10}$$

where C represents the penalty value.

Therefore, the original Lagrange function can be rewritten as Equation (10.11).

$$L(w,b,\alpha,\xi_i) = \frac{\|w\|^2}{2} - \sum_{i=1}^{k} \alpha_i \left[y_i \left(w^T \cdot x_i + b \right) - 1 + \xi_i \right] + \frac{C}{2} \sum_{i=1}^{k} \xi_i^2 \quad (10.11)$$

In order to ensure that the classifier obtained by training has a high generalization ability. The original input space must be mapped into a high-dimensional space when the training samples are non-linearly separable.

In order to avoid complex operations in high-dimensional space, the kernel function $K(\boldsymbol{x}_i^T, \boldsymbol{x})$ is introduced to calculate the inner product of feature vectors. At this point, the Mercer conditions must be satisfied (Mercer and Forsyth 1909). Equation (10.8) can be rewritten as Equation (10.12).

$$f(x) = \mathrm{sgn}\left[\sum_{i=1}^{k} \alpha_i^* y_i K\left(x_i^T \cdot x \right) + b^* \right] \quad (10.12)$$

Commonly used kernel functions include linear kernel, polynomial kernel, RBF kernel, and sigmoid kernel. More details about kernel function can be found in Cervantes et al. (2020).

In practice, the rock mass classification is a typical multi-classification problem. The one-versus-one (OVO) and the one-versus-rest (OVR) are used to transform multi-classification problems into multiple binary classification problems in SVM. Further details about the multi-classification methods in SVM can be found in Cervantes et al. (2020).

10.3.5 Model training and evaluation

This work employs SVM, RF, GPC, and CNN to build surrounding rock classification (SRC) models based on the SurroundingRock/8/286 database. Figure 10.2 depicts the process of building the SRC model. Firstly, the SurroundingRock/8/286 database was established through on-site investigation and literature review. Secondly, the data is discretized by employing OrdinalEncoder or OneHotEncoder methods in scikit-learn. Thirdly, the intelligent SRC model is trained and optimized by adjusting the hyperparameters. Finally, the performance of the SRC model is evaluated by accuracy, precision, recall, and F-measure to select the optimal model.

10.3.5.1 CNN-based SRC model

Four CNN-based SRC models are established by VGGNet16 and VGGNet19, which are the representative algorithm of CNN. In a ratio of 7:3, the SurroundingRock/8/286 database is divided into a training set (200 cases) and a validation set (86 cases).

The four CNN-based SRC models are trained with the Adam optimizer, 16 and 32 batch sizes, categorical_crossentropy loss function, and 200 epochs. And the learning rate is set to 0.0001, which means the model training is terminated when the validation loss value is less than 0.0001, and the model is determined to be optimal. Figure 10.3 depicts the changes in the accuracy of the four established CNN-based SRC models with the number of iterations, respectively.

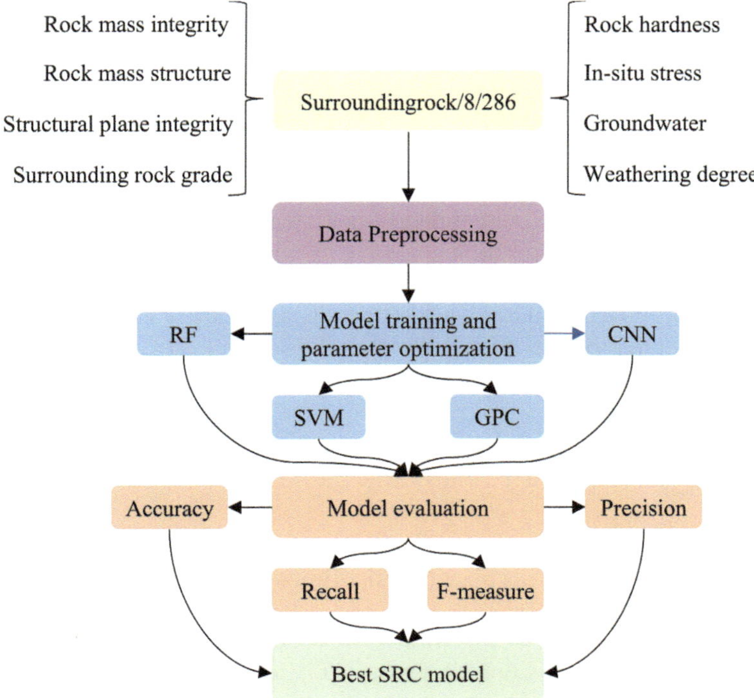

Figure 10.2 The workflow of the intelligent SRC model.

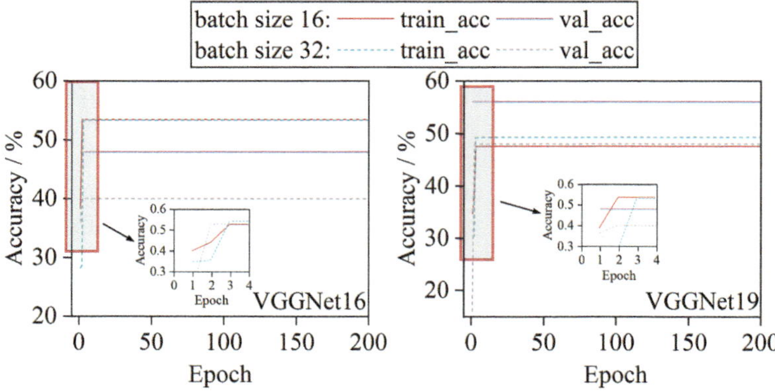

Figure 10.3 Curves of the accuracy of training and validation sets of the VGGNet16-based and VGGNet19-based SRC models.

As is shown in Figure 10.3, the training accuracy and validation accuracy are stabilized after three epochs in VGGNet16-based and VGGNet19-based SRC models. Especially the validation accuracy of the VGGNet19-based SRC model (batch size 16) is stabilized in the first epoch. As a result, the accuracy of the training and validation sets of the four VGGNet16-based SRC models are lower than 60.00% (see Table 10.2), so further model performance evaluation is unnecessary.

Table 10.2 Hyperparameter values and validation set accuracy for intelligent SRC models

Algorithms	*Kernel*	*Gamma*	*Degree*	*Coef0*	C	*Accuracy (%)*	*5-fold (%)*	*10-fold (%)*
SVM	RBF	0.03	–	–	14.09	**93.02**	**93.50**	**94.00**
	Linear	–	–	–	4.30	87.21	90.00	92.50
	Sigmoid	0.01	–	−0.20	18.98	87.21	84.00	84.00
	Poly	10	1	0	0	86.05	88.00	87.00
RF	n_estimators	max_depth	min_samples_split	class_weight		Accuracy (%)	5-fold (%)	10-fold (%)
	25	None	2	None		83.72	84.50	82.00
				Balanced		80.23	85.00	81.50
				balanced_subsample		83.72	85.00	83.00
	100			None		82.56	85.50	83.50
				Balanced		**83.72**	**87.00**	**83.00**
				balanced_subsample		81.40	84.50	84.00
GPC	Kernel	length_scale	length_scale_bounds	multi_class		Accuracy (%)	5-fold (%)	10-fold (%)
	RBF	2.1	(1e-5, 1e5)	one_vs_rest		82.56	81.00	83.00
		1.5	(1e-5, 1e5)	one_vs_one		**86.05**	**87.00**	**91.50**
		[3,3,2,2,1,1,1]	(1e-5, 1e7)	one_vs_rest		82.56	88.00	86.00
		[3,3,2,2,1,1,1]	(9e-9, 9e21)	one_vs_one		86.05	86.00	88.50
	DotProduct	sigma_0	sigma_0_bounds	multi_class		Accuracy (%)	5-fold (%)	10-fold (%)
		1	(1e-5, 1e5)	one_vs_rest		76.74	75.00	74.00
		1	(1e-5, 1e5)	one_vs_one		86.05	82.50	87.00
	Matern	Length_scale	Nu	multi_class		Accuracy (%)	5-fold (%)	10-fold (%)
		2	1.5	one_vs_rest		83.72	84.50	85.50
		1	1.5	one_vs_one		86.05	87.00	88.00
	RationalQuadratic	Length_scale	Alpha	multi_class		Accuracy (%)	5-fold (%)	10-fold (%)
		1	1	one_vs_rest		84.88	84.00	87.00
		1	1	one_vs_one		86.05	90.50	92.50

(Continued)

Table 10.2 (Continued) Hyperparameter values and validation set accuracy for intelligent SRC models

Algorithms	*Kernel*	*Gamma*	*Degree*	*Coef0*	*C*	*Accuracy (%)*	*5-fold (%)*	*10-fold (%)*
CNN	Model	Batch size		Validation set accuracy (%)				Training set accuracy (%)
	VGGNet16	16		53.42				48.00
		32		53.42				40.00
	VGGNet19	16		47.49				56.00
		32		49.32				48.00

Note: Accuracy means the accuracy of a random validation set. 5-fold means the average accuracy of five random validation sets. 10-fold means the average accuracy of ten random validation sets. More details about the cross-validation can be found in Rao et al. (2008).

In summary, the four established VGGNet16-based SRC models cannot meet the high accuracy requirements in tunnel construction. However, as more cases are collected, the VGGNet-based models can be further optimized, or the latest CNN architecture is introduced to build the newer SRC model.

10.3.5.2 RF-based SRC model

The RF-based SRC model is trained and validated in a ratio of 7:3. The hyperparameters of the RF-based SRC model include n_estimators, max_depth, min_samples_split, and class_weight. The n_estimators means the number of trees in the forest. The max_depth means the maximum depth of the tree. The default value of the max_depth is None, which means the nodes are expanded until all leaves are pure or until all leaves contain less than min_samples_split samples. The min_samples_split means the minimum number of samples required to split an internal node. The class_weight is used to assign weight to the class. The 5-fold and 10-fold cross-validation are preliminarily used to evaluate the model's performance.

Let the model parameters in condition 1 be n_estimators = 100, max_depth = None, min_samples_split = 2, and class_weight = balanced. Condition 2: n_estimators = 100, max_depth = None, min_samples_split = 2, and class_weight = balanced_subsample. As shown in Table 10.2, in the case of condition 1, the average accuracy of 5-fold cross-validation of the RF-based SRC model is the highest, 87.00%. In the case of condition 2, the average accuracy of 10-fold cross-validation of the RF-based SRC model is the highest, 84.00%. The average accuracy of the 10-fold cross-validation of the model under condition 1 is only 1% lower than in condition 2. The average accuracy of the 5-fold cross-validation of the model under condition 1 is 2.50% higher than in condition 2. The model's random validation accuracy under condition 1 is 2.32% higher than in condition 2. In summary, the RF-based SRC model's performance is best in condition 1.

10.3.5.3 GPC-based SRC model

The GPC-based SRC model is trained and validated in a ratio of 7:3. The mainly hyperparameters of the GPC-based SRC model are kernel and multi_class. Ten GPC-based SRC models were trained using RBF, DotProduct, Matern, and RationalQuadratic kernel functions. The parameter values for multi_class are "one_vs_rest" and "one_vs_one." In "one_vs_rest," one binary Gaussian process classifier is fitted for each class, which is trained to separate this class from the rest. In "one_vs_one," one binary Gaussian process classifier is fitted for each pair of classes, which is trained to separate these two classes. In addition, different kernel functions have their own unique hyperparameters (see Table 10.2).

As is shown in Table 10.2, the accuracy of the random validation set of GPC-based SRC models is generally greater than 80%. Let the model parameters in condition 3 be kernel = RBF, length_scale = 1.5, length_scale_bounds = (1e−5, 1e5), multi_class = "one_vs_one." Condition 4: kernel = RationalQuadratic, length_scale = 1, alpha = 1, multi_class = "one_vs_one." The GPC-based SRC model's accuracy of the random validation set is equal and the highest in conditions 3 and 4, which is 86.05%. The model's average accuracy of the 5-fold and 10-fold cross-validation under condition 4 is higher than that of condition 3. However, the model training has a convergence warning under condition 4. Therefore, the model trained by condition 4 still needs further debugging in practical applications. In addition, under condition 3, the model's average accuracy of the 10-fold cross-verification reached 91.50%. Therefore, the model in the case of condition 3 is the optimal GPC-based SRC model.

10.3.5.4 SVM-based SRC model

The ratio of training set (200 cases) to validation set (86 cases) is 7:3 for SVM-based SRC model. The mainly hyperparameters of the SVM-based SRC model is kernel function, decision_function_shape, and *C*. The SVM-based SRC model is trained using the default parameter values of decision_function_shape ("ovr"). "ovr" is one_vs_rest, as mentioned in the GPC-based SRC model. *C* is regularization parameter, which must be strictly positive. The strength of the regularization is inversely proportional to *C*. Gamma is the kernel coefficient for RBF, Poly, and sigmoid kernel functions. Degree is the degree of the polynomial kernel function, which must be non-negative and ignored by all other kernels. Coef0 is only significant in "poly" and "sigmoid" kernel functions. Table 10.2 shows the SVM-based SRC model with the highest accuracy under different kernel functions and its corresponding hyperparameters.

As is shown in Table 10.2, the SVM-based SRC model established with the RBF kernel function has the highest accuracy (the average accuracy of 5-fold and 10-fold cross-validation is 93.05% and 94%, respectively). The accuracy of random validation of the SVM-based SRC model established with the linear kernel function did not reach 90%. The average accuracy of cross-validation of the SVM-based SRC model established with the sigmoid and Poly kernel functions reached less than 90%. The SVM-based SRC model established with the RBF kernel function performs better.

10.3.5.5 Performance evaluation of each class of the intelligent SRC model

The accuracy of optimal RF-based, GPC-based, and SVM-based SRC models exceeds 80%. It is necessary to evaluate further the RF-based, GPC-based, and SVM-based SRC models using precision, recall, and *F*-measure. Accuracy measures the model's ability to predict samples correctly. Precision and recall measure the model's ability to correctly predict the majority and minority classes. However, the classification of surrounding rock needs to be as accurate as possible for each type of sample. Therefore, the *F*-measure is a comprehensive indicator considering both precision and recall. See Equations (10.13)–(10.16) for calculating accuracy, precision, recall, and *F*-measure. The meaning of TP, FP, FN, and TN is shown in Table 10.3.

$$\text{Accuray} = \frac{\text{TP} + \text{TN}}{\text{TP} + \text{FN} + \text{FP} + \text{TN}} \tag{10.13}$$

$$\text{Precision} = \frac{\text{TP}}{\text{TP} + \text{FP}} \tag{10.14}$$

$$\text{Recall} = \frac{\text{TP}}{\text{TP} + \text{FN}} \tag{10.15}$$

$$\text{F-measure} = \frac{2\text{Recall} \cdot \text{Precision}}{\text{Recall} + \text{Precision}} \tag{10.16}$$

Prediction of surrounding rock classification is a five-classification problem, and it can be transformed into five binary classification problems. Table 10.3 shows the confusion matrix (taking the class III as an example). Table 10.4 shows the prediction results of the validation set based on the RF-based, GPC-based, and SVM-based SRC models.

As is shown in Table 10.4, the RF-based, GPC-based, and SVM-based SRC models correctly predict class I surrounding rock samples. Both SVM-based and GPC-based SRC models predict class II surrounding rocks correctly. The RF-based and GPC-based SRC model's predictions of class III surrounding rocks have a much higher error than the SVM-based SRC model. The SVM-based SRC model also had the least number of samples to predict class IV surrounding rock errors. The RF-based, GPC-based, and SVM-based SRC models predicted that the class V surrounding rocks were all wrong by only one sample.

According to Table 10.4 and Equations (6.13)–(6.16), the calculation results of precision, recall, and *F*-measure are shown in Figures 10.4–10.6.

Figure 10.4 shows that the precision values of the SVM-based SRC model in class II, class III, class IV, and class V are greater than those of the RF-based and GPC-based SRC models. Figure 10.5 depicts that the recall values of the SVM-based SRC model in class III and class IV are greater than those of the RF-based and GPC-based SRC models. The SVM-based, RF-based, and GPC-based SRC models have equal recall values in class I, class II, and class V. Figure 10.6 shows that the *F*-measure values of the SVM-based SRC model in class II, class III, class IV, and class V are all greater than those of the RF-based and GPC-based SRC models. As a result, the average precision, average recall, and average *F*-measure of the best SVM-based SRC model are 93.87%, 94.41%, and 93.92%, respectively. In summary, the established SVM-based SRC model performs better than RF-based, and GPC-based SRC models.

10.3.6 Validation with new cases

Twenty-four real cases are used to verify the above RF-based, GPC-based, and SVM-based SRC models. The actual verification cases come from nine tunnels of the Leshan–Xichang (Lexi) Expressway, including the Dafengding tunnel, Daliangshan No. 1 tunnel, Daliangshan No. 2 tunnel, Guihua tunnel, Qingheng No. 1 tunnel, Waju No. 2 tunnel, Wuyiwan tunnel, Xiaoliangshan No. 2 tunnel, and Yinchanggou tunnel. In particular, the SurroundingRock/8/286 database does not contain the above 24 actual verification cases. Table 10.5 shows the surrounding rock engineering geological parameters and prediction results of the validation cases.

As is shown in Table 10.5, the SVM-based SRC model predicts correctly in all validation cases, the RF-based SRC model predicts two cases incorrectly, and the GPC-based

Table 10.3 Binary confusion matrix (taking the prediction of class III as an example)

		Predicted	
		III	I/II/IV/V
Actual	III	True positive (TP)	False negative (FN)
	I/II/IV/V	False positive (FP)	True negative (TN)

SRC model predicts one case incorrectly. In summary, the SVM-based SRC model performs better in real verification cases.

10.4 FIELD APPLICATION

Combined with the above-mentioned best SVM-based SRC model and cloud technology, a tunnel information management system is developed. The tunnel information management system can provide data storage, query, and intelligent analysis services for tunnel construction. It is currently applying to a project – the Lexi Expressway project in China.

The Lexi Expressway project is currently using the tunnel information management system (TIMS). The Lexi Expressway starts in Mabian Yi Autonomous County of Leshan City, passes through Leibo County and Meigu County in Liangshan Prefecture, and ends at the south side of Zhaojue County, where it connects with the planned Xichang–Zhaotong Expressway (see Figure 10.7). The total length of the Lexi Expressway is about 152.48 km, and there are 41 tunnels on the whole line. The lithology of the extension line of the project is sandstone, siltstone, mudstone, shale, limestone, basalt, etc. The tunnels are distributed as class III, class IV, and class V surrounding rocks.

The tasks of TIMS can be divided into four categories: management of tunnel design information, real-time intelligent grading of surrounding rock, management of advanced geological forecast information, and dynamic monitoring of surrounding rock deformation. TIMS stores the engineering geological information related to four kinds of tunnel tasks in the server. Users can set filter criteria to query and access the required data. This

Table 10.4 Confusion matrix of SVM-based, RF-based, and GPC-based SRC models

		Predicted (SVM-based)					*Predicted (RF-based)*					*Predicted (GPC-based)*				
		I	II	III	IV	V	I	II	III	IV	V	I	II	III	IV	V
Actual	I	1	0	0	0	0	1	0	0	0	0	1	0	0	0	0
	II	0	15	0	0	0	0	14	1	0	0	0	15	0	0	0
	III	0	3	17	0	0	0	6	13	1	0	0	6	12	2	0
	IV	0	0	1	19	1	0	0	1	16	4	0	0	1	18	2
	V	0	0	0	1	28	0	0	0	1	28	0	0	0	1	28

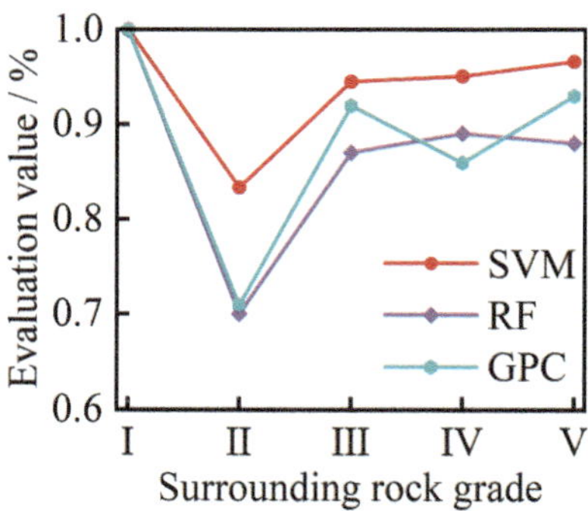

Figure 10.4 The precision of SVM-based, RF-based, and GPC-based SRC models.

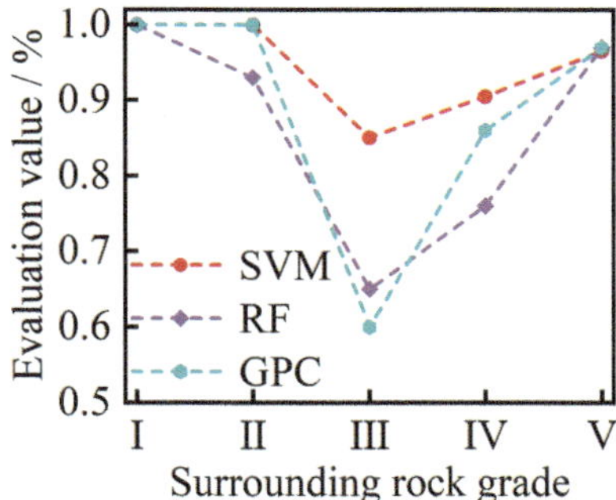

Figure 10.5 The recall of SVM-based, RF-based, and GPC-based SRC models.

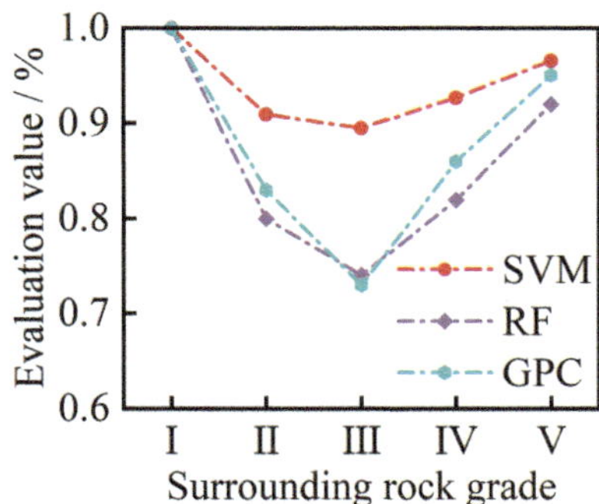

Figure 10.6 The *F*-measure of SVM-based, RF-based, and GPC-based SRC models.

study visualized the existing data of the system and displayed it on the TIMS homepage in the form of tables and graphs (see Figure 10.8). From Figure 10.8, the homepage of TIMS displays the current distribution of surrounding rock quality grades, alterations in design surrounding rock grades, the time distribution of new surrounding rock information, to-do tasks, monitoring and warning of surrounding rock deformation, and system announcements. In particular, user can jump to the details page of the target data by clicking the data on the homepage. In addition, this study have customized standard management procedures for each of the four tunneling tasks in TIMS.

The tunnel information management system was launched in July 2020, and the real-time intelligent surrounding rock classification function is gradually being used in all tunnels of the Lexi project. In particular, more than 5,500 surrounding rock data had been submitted to the system as of December 2022. Figure 10.9 shows that most tunnel excavation sections did not have the design alteration of surrounding rock grade. And 850 excavation sections initiated the online alteration procedure to the designed surrounding rock grade.

More importantly, the online real-time intelligent classification of surrounding rocks dramatically improves the efficiency of on-site office work. (1) The identification of the surrounding rock grade at the construction site is faster. (2) All parties involved in the tunnel construction can view the engineering geological information of the excavation face in real-time in the TIMS system. (3) The online alteration procedure of the surrounding rock grade reduces on-site meetings and increases productivity. Real-time surrounding rock classification is significant for guiding dynamic tunnel design and construction.

Table 10.5 Verification results of the new tunnel cases

Tunnel	*Mileage*	*RH*	*WD*	*RMS*	*Dc*	*RMI*	*IS*	**G**	*Actual*	*SVM*	*RF*	*GPC*
Dafengding	K68 + 586	Slightly hard	Slight	d	Ordinary	Slightly broken	Medium	Rain-like dripping	IV	IV	IV	IV
	K68 + 670	Slightly soft	Medium	e	Good	Slightly complete	Medium	Rain-like dripping	IV	IV	IV	IV
	ZK68 + 536	Slightly hard	Slight	d	Ordinary	Broken	Medium	Dripping	IV	IV	IV	IV
Daliangshan No. 1	ZK82 + 068	Soft	Slight	c	Ordinary	Slightly broken	Low	Dry	V	V	V	IV
	ZK86 + 496	Slightly soft	Medium	e	Ordinary	Broken	Medium	Moist	V	V	V	V
	K75 + 542	Slightly hard	Medium	d	Ordinary	Slightly broken	Low	Rain-like dripping	IV	IV	IV	IV
	PK75 + 417	Soft	Severe	c	Bad	Slightly broken	Low	Dripping	V	V	V	V
Daliangshan No. 2	K107 + 709	Slightly soft	Medium	a	Ordinary	Slightly complete	High	Dripping	IV	IV	IV	IV
	ZK113 + 510	Slightly hard	Fresh	d	Ordinary	Slightly broken	Low	Dripping	IV	IV	IV	IV
	ZK108 + 474	Slightly soft	Medium	a	Ordinary	Slightly complete	High	Moist	IV	IV	IV	IV
	ZK104 + 821	Soft	Medium	c	Very bad	Slightly broken	Medium	Dripping	V	V	V	V
Guihua	K62 + 745	Slightly hard	Medium	e	Bad	Broken	Low	Gushing	V	V	V	V
	K62 + 755	Slightly hard	Medium	e	Bad	Broken	Low	Linear	V	V	IV	V
	K62 + 795	Slightly hard	Medium	f	Bad	Broken	Low	Linear	V	V	IV	V
Qingheng No. 1	K128 + 665	Slightly hard	Slight	b	Good	Slightly complete	Low	Moist	III	III	III	III
	ZK128 + 660	Slightly hard	Slight	b	Good	Slightly complete	Medium	Dry	III	III	III	III

(*Continued*)

Table 10.5 (Continued) Verification results of the new tunnel cases

Tunnel	*Mileage*	*RH*	*WD*	*RMS*	*Dc*	*RMI*	*IS*	*G*	*Actual*	*SVM*	*RF*	*GPC*
Waju No. 2	K23 + 300	Soft	Slight	e	Bad	Broken	Medium	Dripping	V	V	V	V
Wuyiwan	ZK33 + 131	Soft	Medium	c	Ordinary	Broken	Medium	Dry	V	V	V	V
Xiaoliangshan No. 2	K39 + 240	Slightly hard	Slight	a	Ordinary	Slightly complete	Low	Dripping	III	III	III	III
Yinchanggou	K55 + 452	Hard	Slight	c	Ordinary	Slightly broken	Low	Dry	III	III	III	III
	ZK57 + 600	Slightly soft	Medium	g	Very bad	Extremely broken	Low	Moist	V	V	V	V
	K55 + 509	Hard	Slight	d	Ordinary	Slightly broken	Low	Dripping	III	III	III	III
	ZK55 + 596	Hard	Slight	d	Ordinary	Slightly broken	Low	Moist	III	III	III	III
	ZK57 + 540	Soft	Severe	g	Very bad	Extremely broken	Low	Moist	V	V	V	V

Note: (1) The RH, WD, RMS, SPI, RMI, IS, and GW represent rock hardness, weathering degree, rock mass structure, structural plane integrity, rock mass integrity, in situ stress, and groundwater, respectively. (2) The a, b, c, d, e, f, and g represent block structure, thick-layered structure, medium-thick-layered structure, mosaic structure, fractured block structure, fragmented structure, and granular structure, respectively. (3) The "Actual" column in the table is the surrounding rock grade referenced by the actual support at the tunnel construction site.

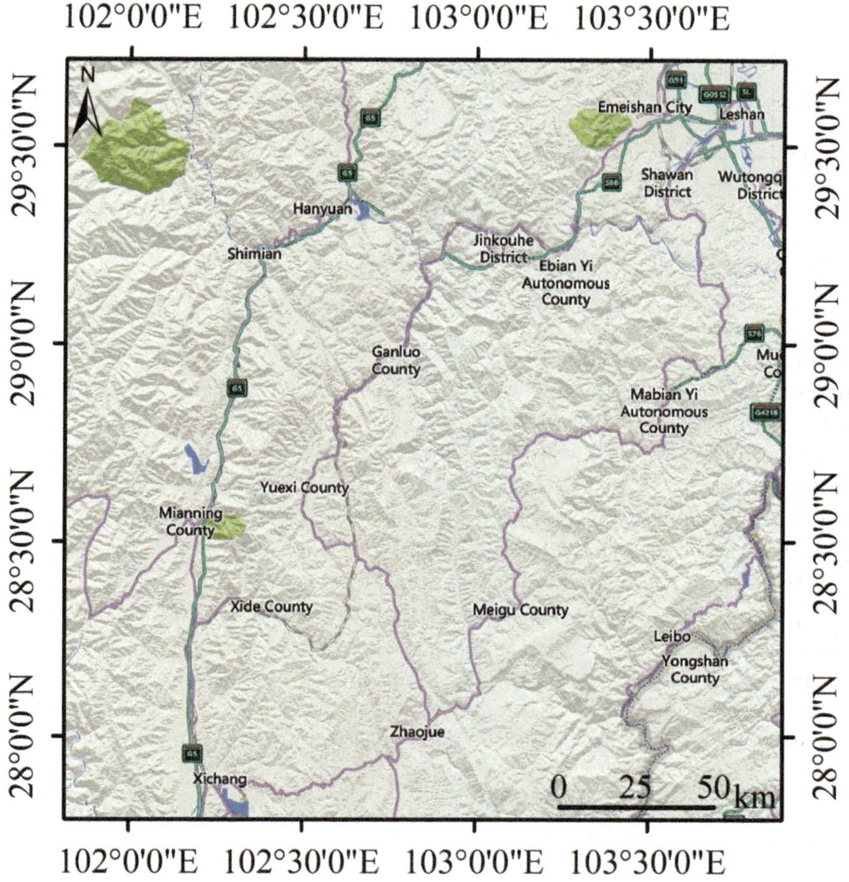

Figure 10.7 Route diagram of the Lexi Expressway.

10.5 CONCLUSIONS

This chapter presents a new generic database for surrounding rock engineering geological parameters named SurroundingRock/8/286. Four types of surrounding rock grade classifiers are established with CNN, RF, GPC, and SVM algorithms. A tunnel information management system is developed with SVM-based SRC model and cloud technology, which is now well-used in the Lexi Expressway project. Based on the results of this study, the following conclusions can be drawn:

1. The SurroundingRock/8/286 database consists of eight parameters (rock harness, rock mass integrity, discontinuity condition, weathering degree, rock mass structure, groundwater condition, in-situ stress condition, and surrounding rock grade), which can well characterize the surrounding rock engineering geology conditions.
2. Based on the SurroundingRock/8/286 database, four intelligent SRC models are built with CNN, RF, GPC, and SVM algorithms, and the SVM-based SRC model has better performance. In particular, the 5-fold cross-validation accuracy, 10-fold cross-validation accuracy, average precision, average recall, and average *F*-measure

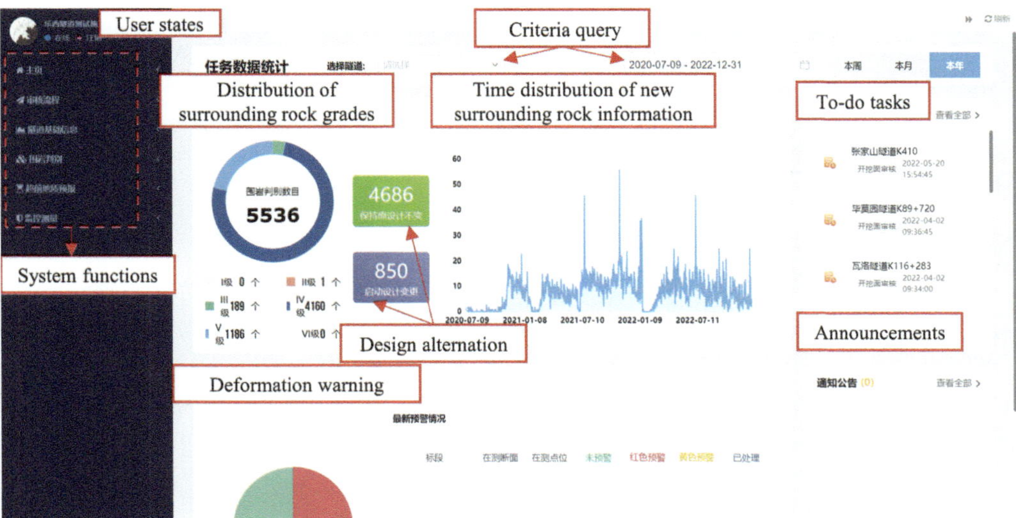

Figure 10.8 Data visualization of the TIMS homepage.

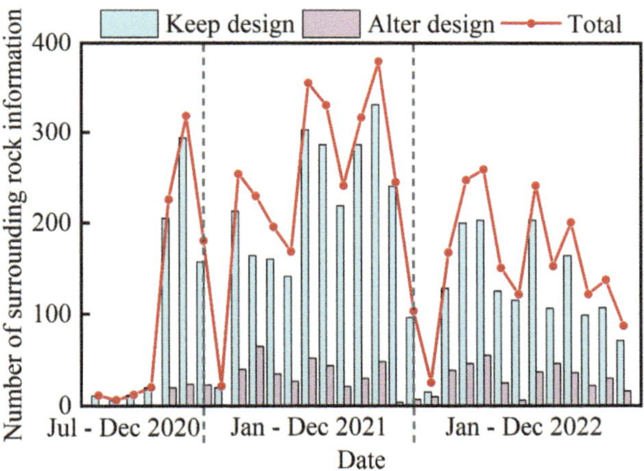

Figure 10.9 Monthly distribution of alteration in the design grade of the surrounding rock.

of the best SVM-based SRC model are 93.50%, 94.00%, 93.87%, 94.41%, and 93.92%, respectively. Further, the SVM-based SRC model predicts all 24 actual tunnel cases correctly.

3. The tunnel information management system can predict the surrounding rock grade accurately and online. And the system used in the Lexi Expressway project has dramatically improved the efficiency of construction and management.

In conclusion, the intelligent SRC models built on the SurroundingRock/8/286 database and machine learning methods have demonstrated excellent performance in other sites. However, the accurate acquisition of surrounding rock engineering geological parameters is an existing problem in the complex environment of the site. Therefore, our study will focus on using deep learning technology to automatically, quickly, and accurately obtain the surrounding rock engineering geological parameters to expand the surrounding rock database further.

DISCLOSURE STATEMENT

No potential conflict of interest was reported by the author(s).

FUNDING

This work was supported by the Science & Technology Department of Sichuan Province, China [Grant No. 2021YFS0317]; and the National Natural Science Foundation of China (NSFC) [Grant No. U19A20111].

REFERENCES

Barton, N. 2002. "Some New Q-Value Correlations to Assist in Site Characterisation and Tunnel Design." *International Journal of Rock Mechanics and Mining Sciences* 39 (2): 185–216. https://doi.org/10.1016/S1365-1609(02)00011-4.

Barton, N., R. Lien, and J. Lunde. 1974. "Engineering Classification of Rock Masses for the Design of Tunnel Support." *Rock Mechanics* 6 (4): 189–236. https://doi.org/10.1007/BF01239496.

Bieniawski, Z. T. 1973. "Engineering Classification of Jointed Rock Masses." *The Civil Engineering in South Africa* 15 (12): 335–343.

Bo, Y., Q. S. Liu, X. Huang, and Y. C. Pan. 2022. "Real-Time Hard-Rock Tunnel Prediction Model for Rock Mass Classification Using CatBoost Integrated with Sequential Model-Based Optimization." *Tunnelling and Underground Space Technology* 124: 104448. https://doi.org/10.1016/j.tust.2022.104448.

Breiman, L. 2001. "Random Forests." *Machine Learning* 45: 5–32. https://doi.org/10.1023/A:1010933404324.

Cervantes, J., F. Garcia-Lamont, L. Rodríguez-Mazahua, and A. Lopez. 2020. "A Comprehensive Survey on Support Vector Machine Classification: Applications, Challenges and Trends." *Neurocomputing* 408: 189–215. https://doi.org/10.1016/j.neucom.2019.10.118.

Chen, J. Y., M. l. Zhou, H. W. Huang, D. M. Zhang, and Z. C. Peng. 2021. "Automated Extraction and Evaluation of Fracture Trace Maps from Rock Tunnel Face Images via Deep Learning." *International Journal of Rock Mechanics & Mining Sciences* 142: 104745. https://doi.org/10.1016/j.ijrmms.2021.104745.

Cortes, C., and V. Vapnik. 1995. "Support-Vector Networks." *Machine Learning* 20 (3): 273–297. https://doi.org/10.1023/A:1022627411411.

Cristianini, N., and J. Shawe-Taylor. 2000. *An Introduction to Support Vector Machines and Other Kernel-Based Learning Methods*. Cambridge: Cambridge University Press. https://doi.org/10.1017/CBO9780511801389.

Drucker, H., C. Burges, L. Kaufman, A. Smola, and V. Vapnik. 1996. "Support Vector Regression Machines." In *Advances in Neural Information Processing Systems*, MIT Press, Cambridge, MA, 779–784.

Genuer, R., and J-M. Poggi. 2020. *Random Forests with R*. Cham, Switzerland: Springer Nature Switzerland AG. https://doi.org/10.1007/978-3-030-56485-8.

Hoek, E. 1994. "Strength of rock and rock masses." *ISRM News* J 2 (2): 4–16.

Ketkar, N., and J. Moolayil. 2021. *Deep Learning with Python: Learn Best Practices of Deep Learning Models with PyTorch.* California: Apress Media. https://doi.org/10.1007/978-1-4842-5364-9.

Khan, A., A. Sohail, U. Zahoora, and A. S. Qureshi. 2020. "A Survey of the Recent Architectures of Deep Convolutional Neural Networks." *Artificial Intelligence Review* 53 (8): 5455–5516. https://doi.org/10.1007/s10462-020-09825-6.

LeCun, Y., B. Boser, J. S. Denker, D. Henderson, R. E. Howard, W. Hubbard, and L. D. Jackel. 1989. "Backpropagation Applied to Handwritten Zip Code Recognition." *Neural Computation* 1 (4): 541–551. https://doi.org/10.1162/neco.1989.1.4.541.

Liu, H. X., W. S. Li, Z. Y. Zha, W. J. Jiang, and T. Xu. 2018. "Method for Surrounding Rock Mass Classification of Highway Tunnels Based on Deep Learning Technology." *Chinese Journal of Geotechnical Engineering* 40 (10): 1809–1817. https://doi.org/10.11779/CJGE201810007 (in Chinese).

Ma, J. J., T. B. Li, G. Yang, K. K. Dai, C. C. Ma, H. Tang, G. W. Wang, J. F. Wang, B. Xiao, and L. B. Meng. 2023. "A real-time intelligent classification model using machine learning for tunnel surrounding rock and its application." *Georisk: Assessment and Management of Risk for Engineered Systems and Geohazards* 17 (1): 148–168. https://doi.org/10.1080/17499518.2023.2182891.

Ministry of Water Resources of the People's Republic of China. 2014. *Standard for Engineering Classification of Rock Mass (GB/T 50218-2014).* Beijing: China Planning Press (in Chinese).

Mercer, J., and A. R. Forsyth. 1909. "Functions of Positive and Negative Type, and Their Connection the Theory of Integral Equations." *Philos Trans R Soc Lond* 209: 415–446. https://doi.org/10.1098/rsta.1909.0016.

Phoon, K. -K., J. Ching, and Z. Cao. 2022. "Unpacking data-centric geotechnics." *Underground Space* 7: 967–989. https://doi.org/10.1016/j.undsp.2022.04.001.

Phoon, K. -K., J. Ching, and T. Shuku. 2022. "Challenges in Data-driven Site Characterization." *Georisk: Assessment and Management of Risk for Engineered Systems and Geohazards* 16 (1): 114–126. https://doi.org/10.1080/17499518.2021.1896005.

Phoon, K. -K, and W. Zhang. 2022. "Future of Machine Learning in Geotechnics." *Georisk: Assessment and Management of Risk for Engineered Systems and Geohazards* 17 (1):7–22. https://doi.org/10.1080/17499518.2022.2087884.

Rao, R. B., G. Fung, and R. Rosales. 2008. "On the Dangers of Cross-Validation. An Experimental Evaluation." *Proceedings of the 2008 SIAM International Conference on Data Mining*, 558–596. https://doi.org/10.1137/1.9781611972788.54.

Rasmussen, C. E., and C. K. I. Williams. 2005. *Gaussian Processes for Machine Learning.* London, England: The MIT Press. https://doi.org/10.7551/mitpress/3206.001.0001.

Santos, A. E. M., M. S. Lana, and T. M. Pereira. 2021. "Evaluation of Machine Learning Methods for Rock Mass Classification." Neural Computing and Applications 34: 4633–4642. https://doi.org/10.1007/s00521-021-06618-y.

Simonyan, K., and A. Zisserman. 2015. "Very Deep Convolutional Networks for Large-Scale Image Recognition." *International Conference on Learning Representations.* https://doi.org/10.48550/arXiv.1409.1556.

Wang, M. N., S. G. Zhao, J. J. Tong, Z. L. Wang, M. Yao, J. W. Li, and W. H. Yi. 2021. "Intelligent Classification Model of Surrounding Rock of Tunnel Using Drilling and Blasting Method." *Underground Space* 6 (5): 539–550. https://doi.org/10.1016/j.undsp.2020.10.001.

SUPPLEMENTARY MATERIAL

Table S10.1 Qualitative and quantitative classification of rock hardness

Hardness	*Qualitative characteristics*	R_c/*MPa*
Hard	The hammering sound is crisp with rebound, shaking hands, harder to break; after soaking in water, usually there is no water absorption reaction.	(60, ∞)
Slightly hard	The hammering sound is slightly crisp with slightly rebound, slightly shaking hands, hard to break; after soaking in water, slight water absorption reaction.	(30, 60]
Slightly soft	The hammering sound is not clear and crisp, no rebound, easy to break; after soaking in water, nails can be imprinted.	(15, 30]
Soft	The hammering sound is dumb, no rebound, dents, easier to break; after soaking in water, the hand can be broken.	(5, 15]
Extremely soft	The hammering sound is dumb, no rebound, deep dents, and the hands can be crushed; after soaking in water, it can be crushed into a ball.	(0, 5]

Source: Modified by the Standard for Engineering Classification of Rock Mass GB/T 50218–2014.

Table S10.2 Qualitative and quantitative classification of rock weathering degree

Degree	*Qualitative characteristics*	k_r	k_f
Fresh	The rock structure does not change, and the rock quality is fresh.	(0.9, 1.0]	(0.9, 1.0]
Slight	The rock structure, mineral composition, and color are basically unchanged, and some fracture surfaces contain iron and manganese or are slightly discolored.	(0.8, 0.9]	(0.8, 0.9]
Medium	The rock structure is partially destroyed; the mineral composition and color are obviously changed, and the fracture surface is severely weathered.	(0.6, 0.8]	(0.4, 0.8]
Severe	Most of the rock structure is destroyed, and the mineral composition and color are obviously changed. The feldspar, mica, and iron–magnesium minerals are weathered and altered.	(0.4, 0.6]	(0, 0.4]
Extreme	The rock structure is completely destroyed, disintegrated, and decomposed into loose soil or sand. All minerals are discolored, and the luster disappears. Most of the minerals except quartz particles are weathered and eroded into secondary minerals.	[0.2, 0.4]	–

Source: After the Standard for Engineering Classification of Rock Mass GB/T 50218–2014.

Table S10.3 Classification of rock mass structure

Type	*Subtype*	*Characteries of rock mass structure*
Block structure	Integral structure	The rock mass is complete with huge block shape and undeveloped structural plane, and the spacing is greater than 1 m.
	Block structure	The rock mass is slightly complete with block shape and slightly developed structural plane, and the spacing is generally 1–0.5 cm.
	Fractured block structure	The rock mass is slightly complete with fractured block shape and moderately developed structural plane, and the spacing is generally 0.5–0.3 m.
Layered structure	Extremely thick-layered structure	The rock mass is complete with extremely thick-layered shape and no developed layers, and the spacing is greater than 1 m.
	Thick-layered structure	The rock mass is slightly complete with thick-layered shape and slightly developed layers, and the spacing is generally 1–0.5 cm.
	Medium-thick-layered structure	The rock mass is slightly complete with medium-thick-layered shape and moderately developed layers, and the spacing is generally 0.5–0.3 cm.
	Interlayer structure	The rock mass is slightly complete or has poor integrity with inter-layered shape and slightly developed or developed layers, and the spacing is generally 0.3–0.1 m.
	Thin-layered structure	The rock mass has poor integrity with thin-layered shape and developed layers, and the spacing is generally less than 0.1 m.
Mosaic structure		The rock mass integrity is poor, the rock blocks are tightly embedded, and the structural plane is slightly developed to developed. The spacing is generally 0.3–0.1 m.
Fragmented structure		The rock mass is broken, the structural plane is well developed, and the spacing is generally less than 0.1 m.
Granular structure		The rock mass is broken. Rocks, cuttings, and mud are mixed together.

Source: After the Code for Engineering Geological Investigation of Water Resources and Hydropower GB 50487–2008.

Table S10.4 Classification and characteristics of the discontinuity condition

Degree	*Characteristics*
Good	1) The opening degree of the structural plane is less than 1 mm, and the filling is siliceous, iron, or calcium cement. Otherwise, the structure plane is rough and there is no filling. 2) The opening degree of the structural plane is 1–3 mm, and the filling is siliceous or iron cement. 3) The structural plane is rough and the opening degree is greater than 3 mm, and the filling is siliceous cement.
Ordinary	1) The structural plane is straight and the opening degree is less than 1 mm, and the filling material is cement with calcareous mud or there is no filling material. 2) The opening degree of the structural plane is 1–3 mm, and the filling is calcium cement. 3) The structural plane is rough and the opening degree is greater than 3 mm, and the filling is iron or calcium cement.
Bad	1) The structural plane is straight and the opening degree is 1–3 mm, and the filling is argillaceous cement or calcium argillaceous cement. 2) The opening degree of the structural plane is greater than 3 mm, and the filling is mostly muddy or rock debris.
Very bad	The filling is muddy or mud with rock cuttings, and the thickness is greater than the undulation difference.

Source: After the Standard for Engineering Classification of Rock Mass GB/T 50218–2014.

Table S10.5 Quantitative classification of rock mass integrity

Rank	*Number of joint sets*	*Joint spacing*	*Degree of major structural plane bonding*	K_v	J_v
Complete	1–2	>1.0	Good or ordinary	(0.75, 1]	[0, 3)
Slightly complete	1–2	>1.0	Bad	(0.55, 0.75]	[3, 10)
	2–3	0.4–1.0	Good or ordinary		
Slightly broken	2–3	0.4–1.0	Bad	(0.35, 0.55]	[10, 20)
	≥3	0.2–0.4	Good or ordinary		
Broken	≥3	0.2–0.4	Bad	(0.15, 0.35]	[20, 35)
		≤0.2	Ordinary or bad		
Extremely broken	Unordered		Very bad	(0, 0.15]	[35, +∞)

Source: After Ma et al. (2023).

Table S10.6 Classification of in-situ stress

Level	*Low*	*Medium*	*High*	*Extremely high*
R_c/σ_{max}	[9, +∞)	[7, 9)	[4, 7)	(0, 4)

Source: After Ma et al. (2023).

Table S10.7 Classification of water outflow from tunnel excavation face

Type	*Qualitative description*	*Quantitative division*
Dry	Excavation face is completely dry.	$p \in [0, 0.01]$ or $q \in [0, 5]$
Damp	Wet marks can be seen in some areas of the excavation face.	
Wet	Water can be seen or leached from the excavation face.	$p \in (0.01, 0.1]$ or $q \in (5, 25]$
Dripping	Water drips from the excavation face.	
Rain-like dripping	There are drops of water resembling rain on the excavation face.	$p \in (0.1, 0.5]$ or $q \in (25, 125]$
Linear-like flowing	The excavation face has similar water pipes, and the water column is smaller (the water volume and water pressure are smaller).	
Tubular-like flowing	The excavation face has similar water pipes, and the water column is relatively thick (the water volume and water pressure are very large).	$p \in (0.5, \infty)$ or $q \in (125, \infty)$
Gushing-like flowing	There is a large water outlet on the excavation face; the flow and velocity are large, which often causes great damage to the project.	

Source: After Ma et al. (2023).

Chapter 11

Exploring challenges via analysis of multivariate geotechnical properties

Insights from large-scale local sampling of Japanese marine clay

Yu Otake, Taiga Saito, Wu Stephen, Ikumasa Yoshida, and Daiki Takano

We are actively engaged in the systematic collection and organization of geotechnical investigation data relevant to civil engineering projects in Japan. This effort is aimed at advancing data-driven methodologies in geotechnical design and construction. Our approach is inspired by the analytical frameworks of Ching and Phoon (2014), Ching et al. (2021), and Wu et al. (2022), with a focus on the detailed analysis of multivariate geotechnical properties. Our study includes a comparative evaluation with key benchmark databases, such as CLAY/10/7490, represented as "Soil Type/Number of Indicator Types/Number of Data Entries." We concentrate on a 10 km^2 area, marked by uniform geological features in its foundational layers. Notably, the top 30 m of this region contains a soft clay type that is commonly found in Japan. Our dataset, comprising 67,760 samples, aligns partially with the parameters of CLAY/10/7490 but includes some data gaps. In contrast to CLAY/10/7490, a Big Indirect Database (global BID) recognized globally and collected from numerous countries, our dataset, named "Tokyo-CLAY/14/67760," is a localized BID, meticulously assembled through intensive sampling within this specific area. In this chapter, we present a thorough comparison between our high-density local BID and the global BID, aiming to categorize and illuminate the distinctive features of our dataset. We also offer visual representations of statistical dispersion across different sites, operating under the premise that each borehole represents a single site. This methodology uncovers patterns of site-specific diversity that correspond with global trends, yielding insights that are both anticipated and unprecedented. Building on the fundamental statistical analysis of our data, we identify and explore the critical challenges that must be addressed to further the field of data-centric geotechnical engineering.

11.1 INTRODUCTION

This chapter conducts a rigorous investigation of the statistical attributes essential to a geotechnical investigation database, with a primary focus on the soft clays typical of the Tokyo lowlands in Japan (Ishii (1985), and Watabe and Tanaka (2004)). The uniqueness of this database lies in its comprehensive compilation of geotechnical data, encompassing 67,760 samples across a 10 km^2 area. Our primary objective is to unravel the statistical complexities of multivariate ground parameters derived from this local sampling. Additionally, by comparing this database with global counterparts, particularly CLAY/10/7490 – as detailed by Ching and Phoon (2014) and represented by "Soil Type/Number of Indicator Types/Number of Data Entries" – we aim to provide insights crucial for addressing current challenges in data-centric geotechnical engineering.

DOI: 10.1201/9781003441946-11

The geotechnical attributes of the Tokyo lowlands, vital for seismic and geological hazard assessments as well as infrastructure development planning, are exhaustively analyzed. The stratigraphic delineations in this region, especially the Nanagō and Yūrakuchō Layers, have evolved due to climatic variations, sea level changes, and the Last Glacial Maximum (LGM), forming key components of the Kanto Plain's geology. The Nanagō Layer, found at depths of 30–50 m below the surface and characterized by its alternating sand and clay sequences, measures 20–30 m in thickness. Its *N*-value, indicative of compaction, varies between 20 and 50, with significant fluctuations due to the layer's stratification, necessitating careful analysis for foundational material applications.

Above the Nanagō Layer is the Yūrakuchō Layer, distinguished by its fine clay composition. This layer, reaching a total depth of around 30 m, is typically divided into upper and lower segments. The upper segment, containing sandy elements and extending 5–10 m, generally has an *N*-value ranging from 2 to 5, which may increase above 10 in some areas. The lower segment, approaching a depth of 20 m, features a significantly lower *N*-value, oscillating between 0 and 2, and is marked by high water retention and substantial porosity. The clays of the Yūrakuchō stratum, known for their high plasticity index, show increased sensitivity to moisture changes.

Globally, sedimentary terrains and their soil properties exhibit remarkable diversity, as evidenced in internationally compiled databases, such as those described by Ching et al. (2021). However, our database is uniquely sourced from a specific geographic area. This discussion aims to spotlight the statistical differences between our local database and global datasets. Through comparative analysis, we seek to shed light on the diverse nature of data in geotechnical assessments .

11.2 OUTLINE OF THE DATABASE

In this chapter, we have conducted a comprehensive examination of various soil parameters, including saturation (S_r), wet density (ρ_t), specific gravity of soil particles (G_s), void ratio (e), liquid limit (LL), plastic limit (PL), water content (W), compression index (C_c), coefficient of consolidation (C_v), deformation modulus (E_{50}), preconsolidation pressure (P_c), uniaxial compressive strength (S_u), and the *N*-value, comprising blow count (N_b) and penetration (N_p). The depths at which these parameters were observed are denoted as "Depth." Our database classifies any collection of data that includes at least one of these indicators as an individual dataset. Following this criterion, the cumulative count of datasets in our study amounts to 67,760. Figure 11.1 graphically represents the database with a scatter plot, where the top-right triangular section displays the pairwise counts between parameters. The frequency of these correlations varies significantly, with some exceeding 10,000 occurrences, while the average count is approximately 5,000.

Tables 11.1 and 11.2 systematically present the basic statistical measures and percentile values for each soil parameter, respectively. Table 11.1 provides the mean, coefficient of variation (COV), minimum, and maximum values of each parameter. We have categorized the parameters into three groups: indicative, mechanical, and sounding – the latter akin to those evaluated in the standard penetration test. Indicative parameters like saturation (S_r), wet density (ρ_t), specific gravity of soil particles (G_s), void ratio (e), liquid limit (LL), plastic limit (PL), and water content (W) essentially describe the inherent properties of soil. Meanwhile, mechanical parameters include compression index (C_c), coefficient of consolidation (C_v), deformation modulus (E_{50}), preconsolidation pressure

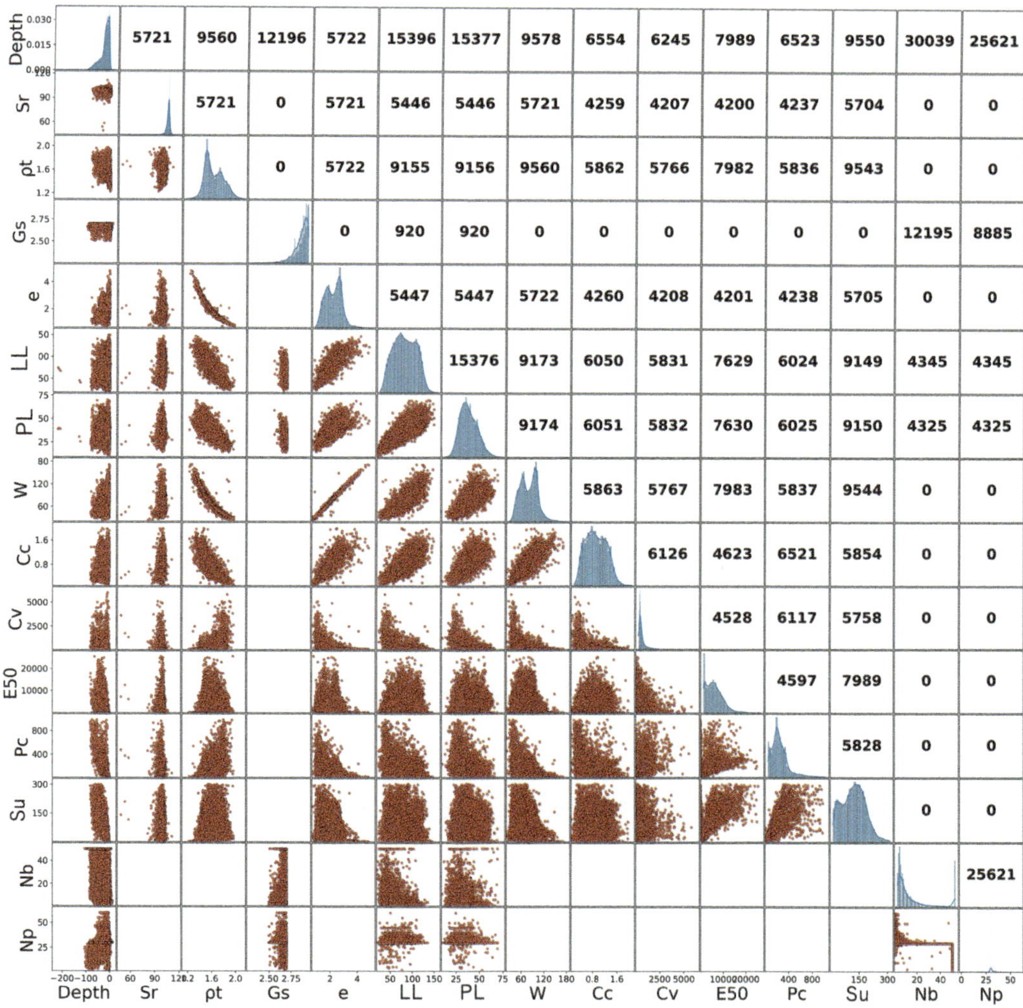

Figure 11.1 Scatterplot matrix of all collected data: pairwise plots between two indicators are shown, as the data includes some missing values.

(P_c), and uniaxial compressive strength (S_u). The N-value, encompassing blow count (N_b) and penetration (N_p), is categorized as a sounding parameter, vital in the Japanese design parameter estimation.

This chapter focuses on analyzing the coefficient of variation across different parameters. Insights from Table 11.1 indicate that certain indicative parameters, particularly saturation (S_r), wet density (ρ_t), and specific gravity of soil particles (G_s), have significantly lower coefficients of variation compared to others. This aligns with established geotechnical views on sub-seafloor clays, supported by research from Phoon (1995), Phoon and Kulhawy (1999a,b), TC304 (2021), and Otake and Honjo (2022). Parameters like LL, PL, and W show coefficients of variation around 0.300, consistent with the higher range observed in the TC304 dataset (LL: 0.034–0.39, PL: 0.029–0.38, W: 0.035–0.46). Mechanical parameters such as C_c, C_v, E_{50}, P_c, and S_u exhibit trends akin to the indicative

Table 11.1 List of basic statistical measures for each parameter

	n	*Mean*	*CV*	*Min*	*Max*
Depth	67,760	−22.69	−0.913	−220.00	24.50
S_r	5721	98.95	0.022	48.60	111.30
ρ_t	9560	1.58	0.075	1.23	1.97
G_s	12,196	2.67	0.010	2.50	2.70
e	5722	1.98	0.285	0.70	4.83
LL	15,396	75.02	0.324	20.30	148.70
PL	15,377	35.34	0.252	10.30	69.90
W	9578	73.15	0.299	22.60	171.50
C_c	6554	0.83	0.397	0.11	1.97
C_v	6245	435.36	1.372	1.68	6000.00
E_{50}	7989	4901.31	0.769	1.02	25,386.96
P_c	6523	205.42	0.740	4.22	983.00
S_u	9550	108.72	0.554	1.32	299.84
N_b	30,039	12.72	1.181	1.00	50.00
N_p	25621	30.15	0.207	1.00	60.00

Table 11.2 List of basic percentile value measures for each parameter

	Percentile value						
	0.025	*0.05*	*0.25*	*0.5*	*0.75*	*0.95*	*0.975*
Depth	−74.30	−65.40	−30.10	−16.50	−8.00	−1.80	−0.90
S_r	94.05	95.67	98.27	99.35	100.00	101.23	101.88
ρ_t	1.40	1.42	1.49	1.56	1.67	1.79	1.83
G_s	2.60	2.62	2.66	2.68	2.69	2.70	2.70
E	1.03	1.12	1.53	2.00	2.41	2.83	3.04
LL	32.80	36.40	55.80	74.40	95.00	113.00	118.00
PL	20.50	22.40	28.70	34.30	41.50	51.00	54.50
W	37.10	40.20	55.66	73.63	89.50	107.51	116.91
C_c	0.29	0.33	0.57	0.81	1.09	1.36	1.45
C_v	66.11	80.02	140.00	220.00	450.00	1550.00	2140.00
E_{50}	5.42	44.42	1971.14	4383.57	7058.34	11,669.72	13,851.89
P_c	13.73	21.57	110.82	174.56	264.78	519.65	646.38
S_u	8.58	14.08	61.34	109.83	150.78	208.39	231.47
N_b	1.00	1.00	3.00	6.00	15.00	50.00	50.00
N_p	11.00	16.00	30.00	30.00	32.00	38.00	42.00

parameters, with their coefficients approaching the upper limits of previous studies (C_c: 0.181–0.473, S_u: 0.159–0.57). The *N*-value stands out, registering at 1.181, significantly exceeding the site-specific variation ranges noted in earlier studies (0.159–0.570).

Figure 11.2 presents a scatter plot highlighting five indices: uniaxial compressive strength (S_u), liquid limit (LL), plastic limit (PL), water content ratio (*W*), and depth of observation. These indices align with those in CLAY/10/7490, a globally recognized database. The figure juxtaposes our database with CLAY/10/7490 using an overlaid

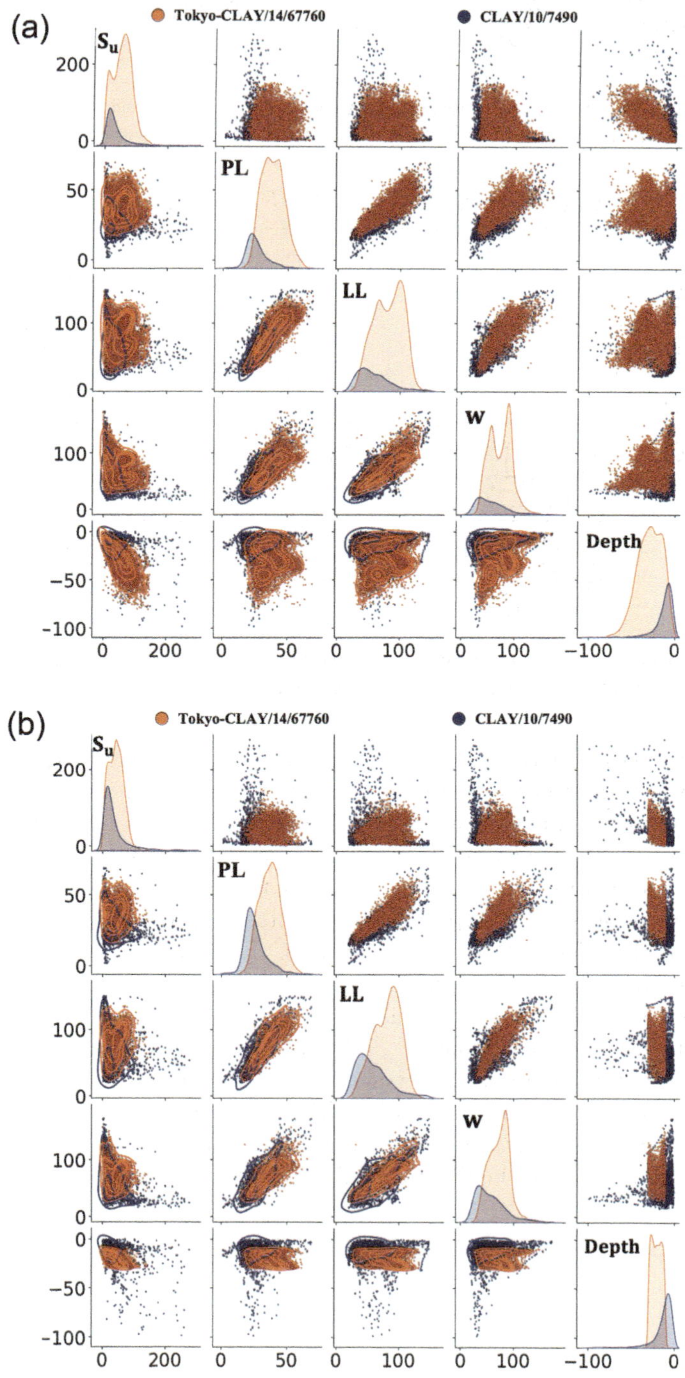

Figure 11.2 Detailed display of scatter plots for indicators competing with CLAY/10/7490.

representation. The diagonal of the figure illustrates the kernel density distribution for each index, with the upper right triangle showing a straightforward scatter plot and the lower left triangle integrating the simultaneous kernel density distribution within the scatter plot. It is important to note that subsets (a) and (b) utilize distinct plotting methods specific to this database: (a) displaying the entire dataset while (b) focusing on depths between 10 and 30 m. The lower section of the Yūrakuchō Layer, located at depths of 10–30 m, is characterized as a pure clay layer with minimal sand content and is distinct from other strata.

While our database is derived from extensive sampling within a specific part of the same soil layer, its variability in ground parameters parallels that in CLAY/10/7490. This similarity is further illustrated in Figure 11.3, which plots data on a plasticity diagram similar to Figure 11.2. However, this representation lacks soil layer distinction in (a) and limits depth to the 10–30 m range in (b). The distribution of our database shows a range that encompasses the intrinsic variation of CLAY/10/7490.

Geotechnical parameter variations fall into two primary categories: intra-subgrade variation (variability within a specific site or soil type) and inter-subgrade variation. The

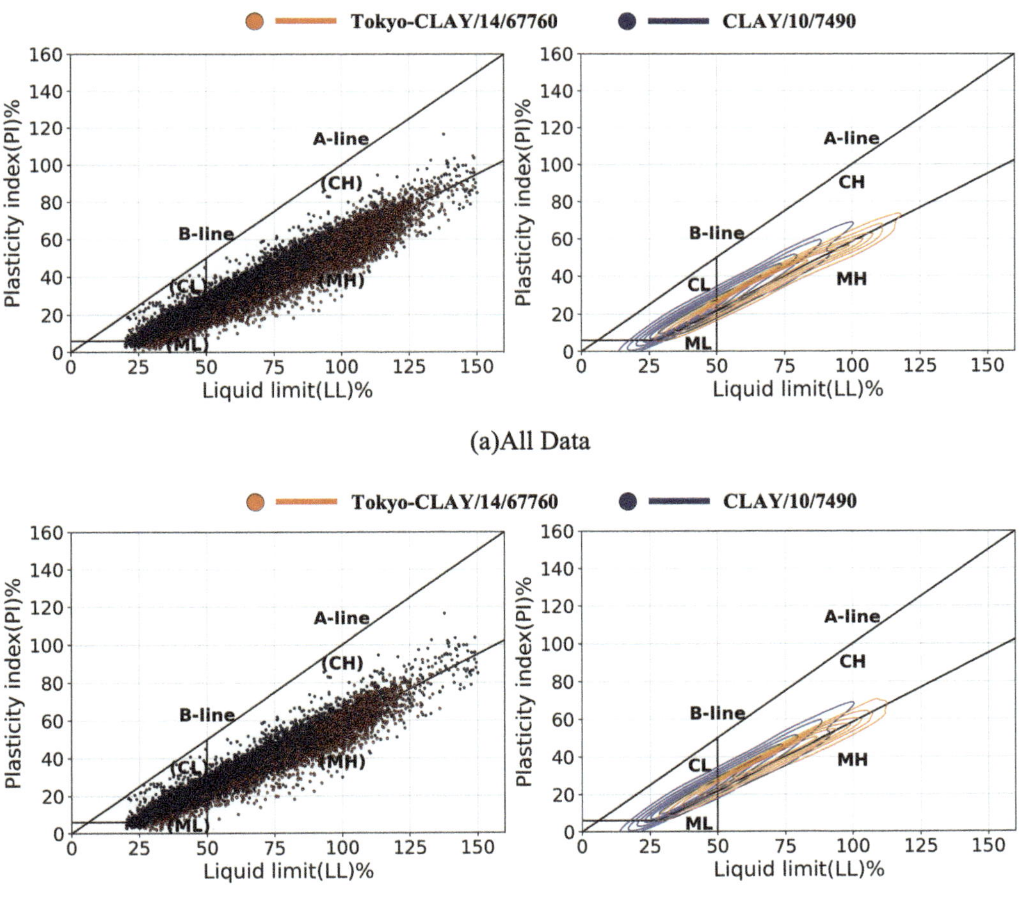

Figure 11.3 Scatter plots on the plasticity diagram.

former encompasses the inherent variability of the soil, representing dispersion specific to a single site. The latter, on the other hand, can be viewed as a bias across different soil types, where unique site characteristics lead to variations in soil parameters once data from multiple locations are aggregated, also known as site-to-site variability.

To explore the effects of inter-subgrade variation, our study focuses on the dispersion of statistics per site, using the hierarchical Bayesian modeling approach proposed by Ching et al. (2021) and Wu et al. (2022). Following a similar trajectory as CLAY/10/7490, we examine the dispersion in mean vectors and covariance matrices computed per site. However, our database, by its nature, limits the discussion on site-to-site variability since it is derived from a single site. Therefore, we adopt a methodology where each borehole is treated as an individual site. Our database includes 2972 boreholes, each containing between 1 and 275 data points, with an average of 29 data points per borehole. For comparison, CLAY/10/7490 (referenced in Ching and Phoon, 2014) encompasses 651 sites, each with 1 to 419 data points and an average of 30 data points per site.

Figure 11.4 displays pairwise scatter plots of the sample means for each site. Following the format of Figure 11.2 and Figure 11.3, these plots distinguish between (a) the comprehensive dataset and (b) data confined to the 10–30 m depth interval. Figures 11.5 and 11.6 further expand on this by presenting scatter plots of the sample means for each site.

A modal difference exists between our database and CLAY/10/7490, suggesting a slight bias. Figures 11.5 and 11.6 show the pairwise scatter plots of the sample standard deviation and sample correlation coefficient, respectively. In these figures, the bias observed in the mean versus scatter plot in Figure 11.4 appears to be reduced. Additionally, the inter-site variability, in terms of standard deviation and correlation coefficient, seems to be consistent between our database and CLAY/10/7490. Notably, the inter-site variability in our database aligns closely with that of the global database, CLAY/10/7490, indicating that the primary differences are likely in the attributes of the sample means (Figure 11.7).

However, these interpretations are fundamentally qualitative. Due to the challenges in visually interpreting correlation coefficients from pairwise scatter plots, we aim to conduct a comprehensive quantitative analysis of these characteristic differences in our future research.

11.3 CONCLUSION

In this chapter, we have examined the statistical properties of multivariate geotechnical parameters derived from extensive sampling within a representative marine clay deposit near Tokyo, Japan. These properties were compared and analyzed in conjunction with CLAY/10/7490, a well-established global database of multivariate geotechnical parameters for clays. The key conclusions from our study are as follows:

1. The overall variability of the geotechnical parameters, gathered from extensive sampling in the localized region, is consistent with the distributions observed in the global database, CLAY/10/7490.
2. We chose to treat each borehole as a single site. This approach revealed primary differences, notably in the fluctuation of mean values between sites. Interestingly, no significant variations were found in the standard deviations or correlation coefficients among the parameters.

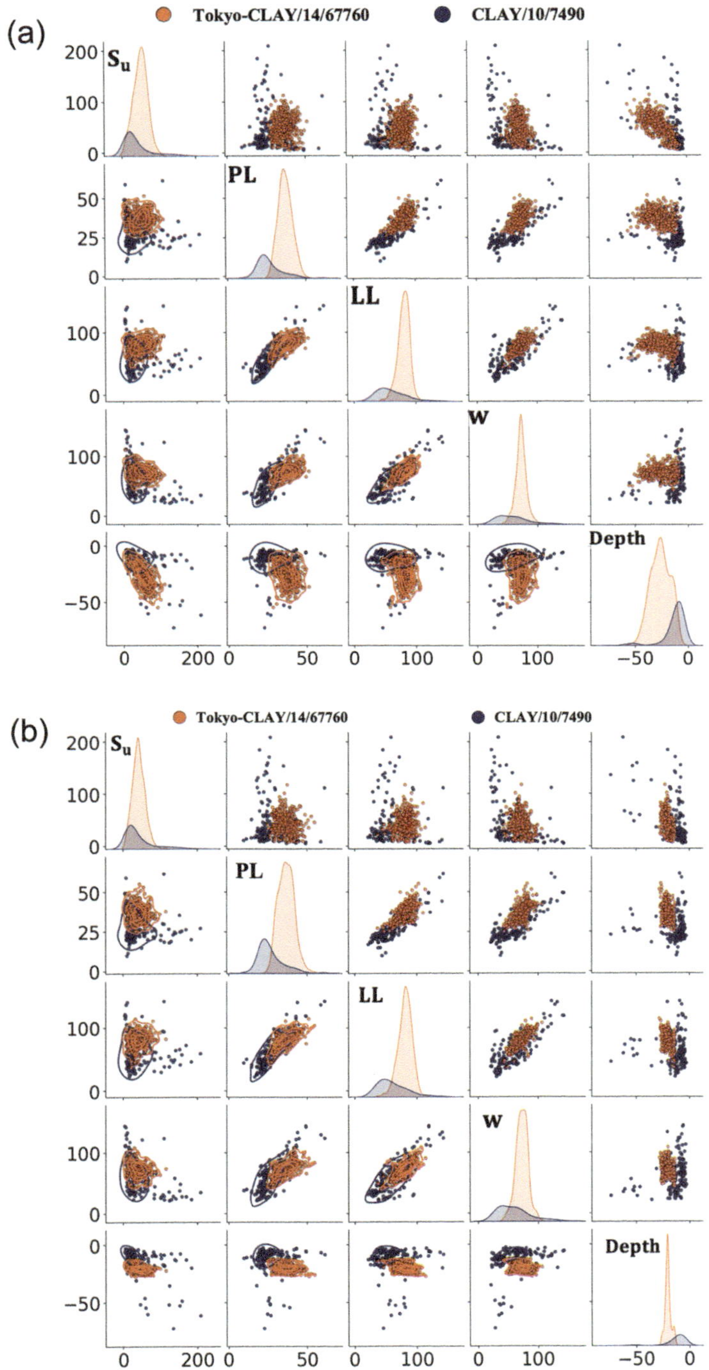

Figure 11.4 Scattering of means per site.

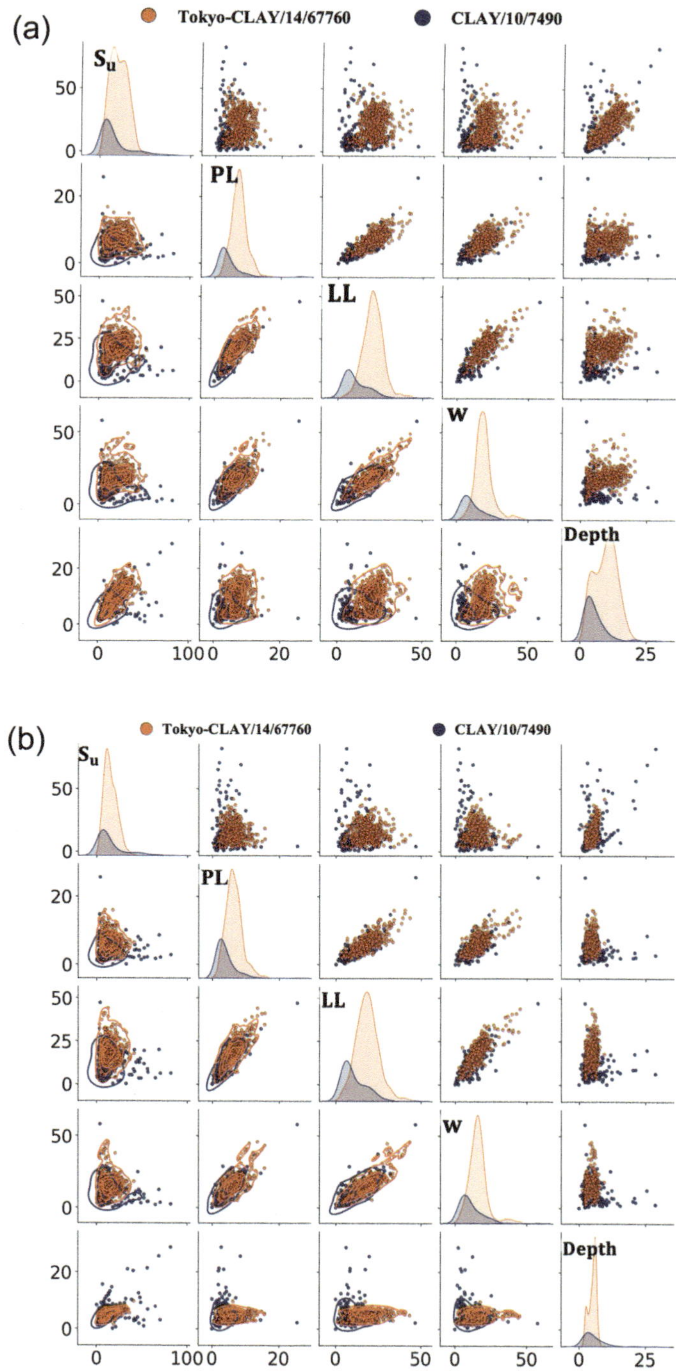

Figure 11.5 Scattering of standard deviations per site.

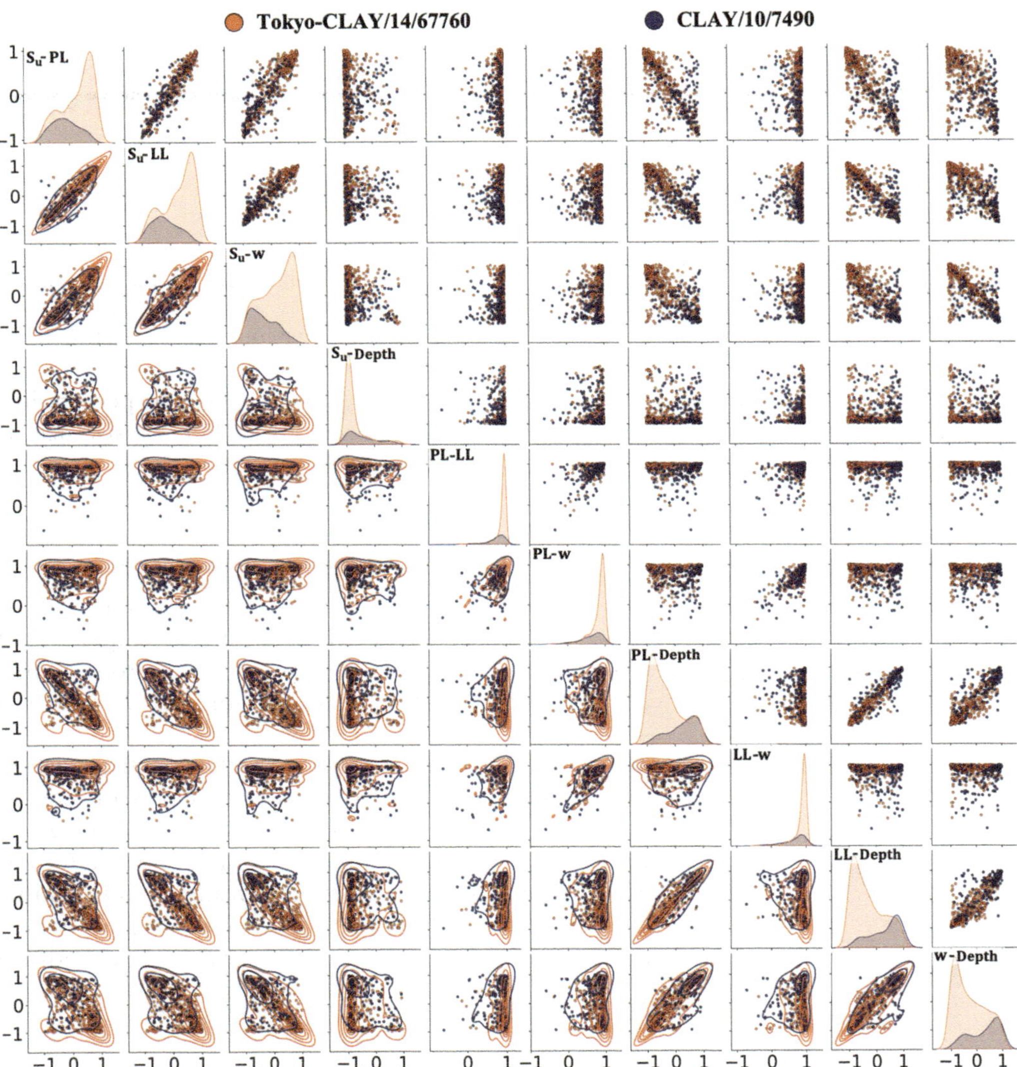

Figure 11.6 Scattering of correlation coefficients per site (all data).

3. Our findings suggest that intensive sampling within a specific area may closely mirror global databases. While the creation of comprehensive databases is vital for advancing data-driven geotechnical engineering, a global scale of data collection may not be essential. Targeted sampling within accessible sites could contribute significantly to building a robust database.
4. Developing methodologies to enhance generic databases is crucial, including techniques like similarity assessment (as referenced in Atma et al., 2022). Identifying hidden patterns in extensive databases is key to effectively filling in missing values.
5. In our analysis, each borehole was treated as an individual site. Standardizing the definition of sites is imperative for constructing a global, generic database. Managing trend components related to soil groupings and depth distributions is also a critical element in developing a universal database.

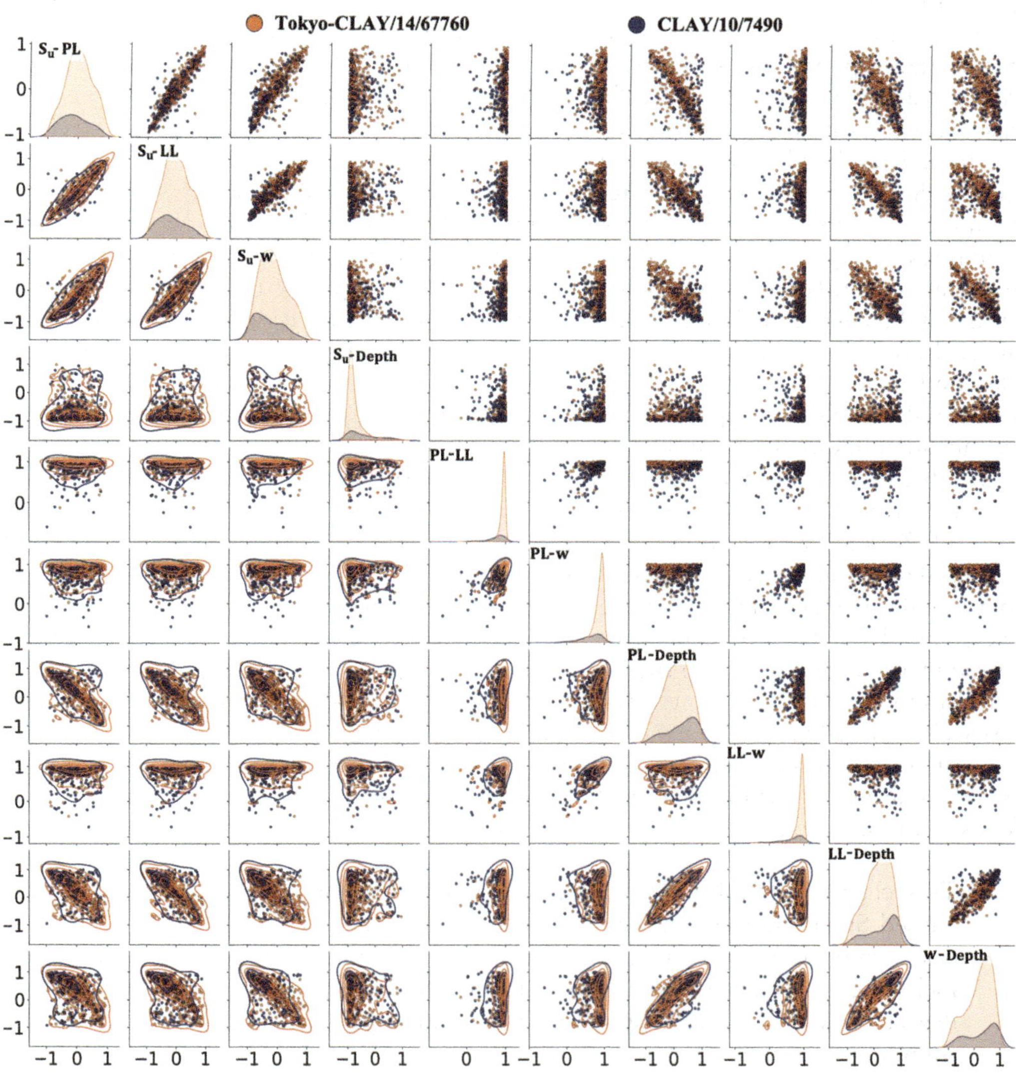

Figure 11.7 Scattering of correlation coefficients per site (Lower Yurakutyo Layer (depth = 10–30 m)).

Through a systematic analysis of results from local sampling, we have arrived at these conclusions and identified directions for future research. Our forthcoming endeavors include meticulous data screening, refined differentiation of soil categories, and reevaluation with consideration of the temporal aspect of surveys. Following comprehensive data management, we aim to develop methods to supplement missing soil parameters. This involves integrating similarity assessments and hierarchical Bayesian modeling to enhance the estimation of multivariate 3D spatial distributions, which will assist in optimally positioning observation points.

REFERENCES

Sharma, A., Ching, J. and Phoon, K. -K. 2022. A hierarchical Bayesian similarity measure for geotechnical site retrieval. *Journal of Engineering Mechanics*, 148(10), 04022062.

Ching, J. and Phoon, K. K. 2014. Transformations and correlations among some clay parameters — the global database. *Canadian Geotechnical Journal*, 51(6): 663–685.

Ching, J., Wu, S. and Phoon, K.-K. 2021. Constructing quasi-site-specific multivariate probability distribution using hierarchical Bayesian model. *Journal of Engineering Mechanics*, 147(10), 04021069.

Ishii, I. 1985. On the development of soil data base system, Technical Note of the Port and Harbour Research Institute, No.515, pp. 1–69.

Otake, Y. and Honjo, Y. 2022. Challenges in geotechnical design revealed by reliability assessment: Review and future perspectives. *Soils and Foundations*, 62(3), 101129.

Phoon, K., Kulhawy, F. and Grigoriu, M. 1995. *Reliability-Based Design of Foundations for Transmisssion Line Structure*. Report TR-105000, Palo Alto, Electric Power Research Institute.

Phoon, K. and Kulhawy, F. 1999a. Characterization of geotechnical variability. *Canadian Geotechnical Journal*, 36, 612–624.

Phoon, K. and Kulhawy, F. 1999b. Evaluation of geotechnical property variability. *Canadian Geotechnical Journal*, 36, 625–639.

Wu, S., Ching, J. and Phoon, K. K. 2022. Quasi-site-specific soil property prediction using a cluster-based hierarchical Bayesian model, *Structural Safety*, 99, 102253.

TC304, S.I., 2021. *State-of-the-Art Review of Inherent Variability and Uncertainty in Geotechnical Properties and Models*. ISSMGE Technical Committee 304.

Watabe, Y. and Tanaka, M. 2004. History of geotechnical database for port and airport construction in Japan. *Proceedings of the International Conference on Engineering Practice and Performance of Soft Deposits*, IS-Osaka 2004, pp. 409–414.

Chapter 12

In situ test-based evaluation of soil effective stress strength properties and stress history

Zhongkun Ouyang, Paul W. Mayne, and Shehab Agaiby

12.1 INTRODUCTION

To investigate the mechanical response of soils due to external loading, one needs to grasp the stress–strain characteristics of the geomaterials, which often could be manifested theoretically via elasticity, plasticity, cavity expansion, and various other models. Such analysis often requires the establishment of a series of parameters to be used in geotechnical problems ranging from stability, foundation-bearing capacity, and deformational investigation (settlement and time rate of consolidation). Additionally, these parameters could be evaluated based on data-driven approaches using regression, statistical, and probabilistic analyses. Comprehensively conducting all available geophysical, in situ, and laboratory testing techniques would be the optimal strategy for obtaining the most accurate and realistic values for these geotechnical parameters. As stated by the first principles of soil mechanics, the effective stress is equal to the total overburden stress minus the porewater pressure, thus soil effective stress strength parameter determination plays a crucial role in geotechnical site investigations involving embankment and slope stability, foundation capacity, and excavations. Also, the stress state analysis of the soils in the ground involves the understanding of soils' unit weight, which is an important soil physical property.

In most geotechnical projects, the degree of preconsolidation of the natural soils is particularly significant as it is a state parameter and represents the current and past stress history. The stress history is commonly represented by the overconsolidation ratio (OCR):

$$OCR = \sigma_p^{'} / \sigma_{v0}^{'} \tag{12.1}$$

where $\sigma_p^{'}$ = preconsolidation stress and $\sigma_{v0}^{'}$ = current effective overburden stress. It is well known that OCR governs the undrained shear strength (s_u), lateral stress coefficient (K_0), pore pressure behavior (A_f), elastic moduli (E' and E_u), and small-strain shear modulus (G_{max}) of soils, and serves as a key parameter in critical-state soil mechanics (CSSM) that has been proven to be a valuable and rational framework in which to organize the mechanical response of soils to loading.

This chapter covers evaluating soils' effective stress strength and stress history parameters from two widely used in situ testing techniques, namely the piezocone penetration testing (CPTu) and flat plate dilatometer tests (DMT). The emphasis in this chapter is on parameters commonly used in geotechnical analysis and design concerned with pilings, shallow foundations, slopes, earth-retaining structures, roadways, and embankments. It is not intended to be exhaustive and mainly focuses on geotechnical engineering parameters of usual concern including the following: (1) unit weight (γ_{total}), (2) preconsolidation

DOI: 10.1201/9781003441946-12

stress or effective yield stress (σ'_p = preconsolidation stress = OCR$\cdot\sigma'_{v0}$), and (c) effective shear strength parameters (ϕ', c'). Additional details and information may be found in Kulhawy and Mayne (1990), Mayne et al. (2002), Loehr et al. (2016), and Ouyang and Mayne (2017a, 2017b, 2018a, 2018b, 2019, 2023). The emphasis in this chapter is on using the results of in situ tests to estimate soil properties via a data-driven method.

12.2 SOIL UNIT WEIGHT

The behavior of soils depends upon the effective stresses in the ground. The total overburden stress is the accumulation of soil unit weight times soil layer thickness and is caused by gravitational forces acting on the soil particles. As such, the unit weight of soils is needed to calculate the total overburden stresses (also later effective stresses) with depth. For other soil parameters involving stresses in soil in general, soil unit weight often time is a required input.

12.2.1 Unit weight from shear wave velocity

For the following, soil unit weight (γ_{total}) is required to evaluate (1) total (σ_{v0}) and effective overburden (σ'_{v0}) stresses, (2) the small-strain shear modulus (G_{max}) from the shear wave velocity (V_s), and (3) DMT horizontal stress index (K_D), normalized cone penetration test (CPT) parameters, and net cone resistance. According to the following, unit weight and mass density (ρ) are related:

$$\gamma_{total} = \rho_{total} g$$

where g = acceleration due to gravity = 9.8 m/s^2.

For direct-push probes and geophysical measurements, unit weights can be calculated in the lab using weight–volume relationships, or they can be inferred using empirical equations. Mayne et al. suggested a correlation between total unit weight and shear wave velocity (V_s). The following are the regression expressions:

$$\gamma_{total}(pcf) = 49.8\log(V_s) - 26.5\, with\, V_s in\ ft/s \quad (12.2)$$

These data come from a range of uncemented soil types, including clays, silts, sands, gravels, and mixed soils, as well as some peats. The measured unit weights came from undisturbed sampling, while the vast majority (92%) of the shear wave velocities were determined using seismic downhole tests. It should be pointed out that the formulas could be applied to soils with a shear wave velocity of less than 3000 m/s.

12.2.2 Unit weight from cone penetration test (CPT)

An empirical association between the total unit weight and the CPT's recorded sleeve friction is proposed by Mayne (2014). The dataset comprises unit weights measured on clays, silts, sands, and mixed soils, and the approach is suitable for uncemented, insensitive, and inorganic soils. This regression analysis's trendline is shown in a dimensionless form as follows:

$$\gamma_{total} = \gamma_w[1.22 + 0.345\log(100 f_s / \sigma_{atm} + 0.01)] \quad (12.3)$$

12.2.3 Unit weight from flat dilatometer test

Flat dilatometer test (DMT) was developed in Italy by Prof. S Marchetti in 1980. As an important in situ testing method, it gained popular acceptance since its introduction due to its simplicity, repeatability, and quick application in geotechnical explorations. By employing nitrogen gas to inflate a flexible steel membrane with 60 mm diameter at each test depth, the DMT generates two pressure readings (A and B). When the membrane is flush with the flat blade face, the corrected A reading or contact pressure, p_0, occurs. The corrected B reading or expansion pressure, p_1, happens when the membrane extends outward 1.1 mm.

For soft clays with a high groundwater level, the corrected contact pressures (p_0) show a linear trend and increase with depth and the expansion pressures (p_1) in various materials exhibit a similar tendency with depth. Regression analyses can be used to define a new parameter (m_{p0}) and produce the best-fit line (intercept = 0):

$$m_{po} \quad = \quad \Delta p_0 \,/\, \Delta z \tag{12.4}$$

For the majority of the soft to firm inorganic clay soils, it is revealed that the total unit weights track with the slope parameter m_{p0}, as shown in Figure 12.1. A few organic-containing clays displayed varying trends and were offset to the right.

An expression can be developed between the total unit weight and slope parameter m_{p0} for soft to firm clays based on the pattern depicted in Figure 12.1, shown by Equation (12.5):

$$\gamma_t = \gamma_w + 0.22 m_{po} \tag{12.5}$$

This empirical expression can therefore be used to forecast the total unit weight of each of these soft clay soils (Ouyang and Mayne 2016). Figure 12.2 shows a comparison of the laboratory-measured unit weights to the predicted unit weights using the empirical equation for a worldwide DMT clay database, and Figure 12.3 compares the lab unit weight profile to the predicted unit weight using DMT data at the Anacostia soft clay site in Washington, DC. It should be noted that in stiff to hard clays with high OCRs or fissured OC clays, further study is required.

12.3 SOIL EFFECTIVE STRESS STRENGTH PARAMETERS

When penetrometer probes (CPTu and DMT) are advanced into clays and fine-grained soils at a standard push rate of 20 mm/s, the interpretation by geotechnical engineers traditionally centers around a total stress analysis, and consequently, the evaluation is focused on the undrained shear strength (s_u). However, it is well recognized that the fundamental behavior of soils belongs to an effective stress framework, as established by several avenues: (1) laboratory stress path response from triaxial tests (e.g., Lambe and Whitman 1979; Mayne et al. 2009), (2) critical-state soil mechanics (e.g., Schofield and Wroth 1968; Holtz et al. 2010), and (3) numerical modeling of soil behavior in triaxial compression, extension, and simple shear modes (Whittle et al. 1994). Therefore, the fundamental strength of clays is represented by effective stress strength parameters, including ϕ' = effective friction angle and c' = effective cohesion intercept. Most commonly, the effective stress strength of soils is represented by a Mohr–Coulomb criterion:

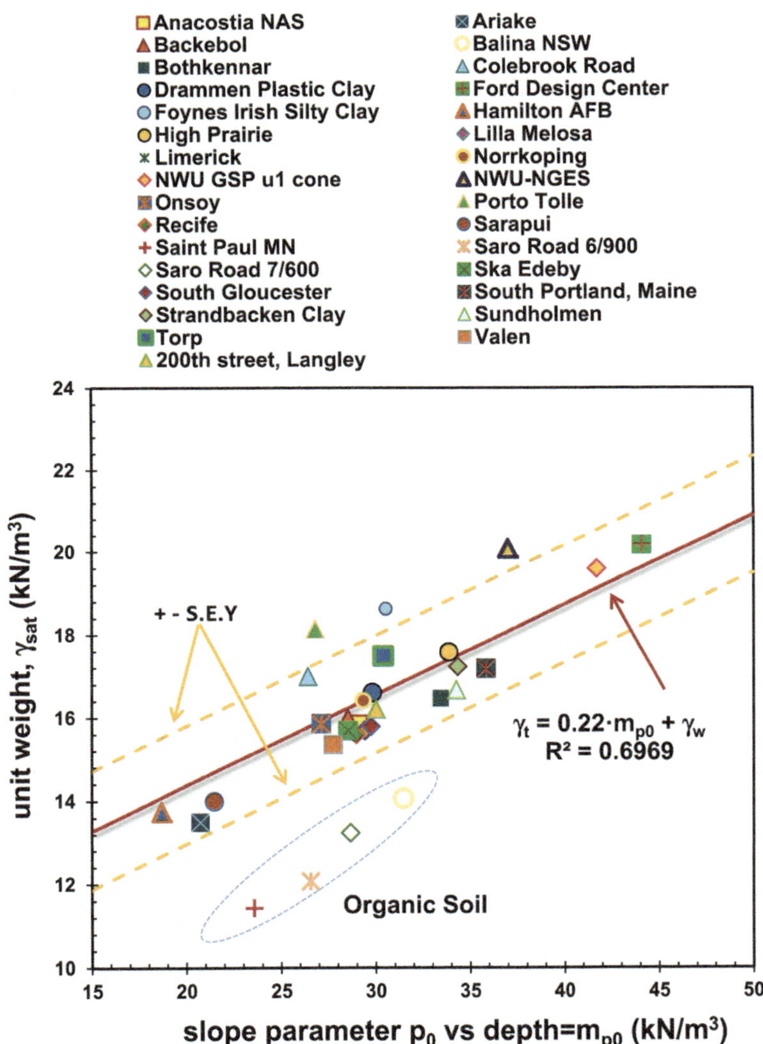

Figure 12.1 Total unit weight versus slope parameter $m_{p0.}$

$$t_{max} = c' + s' \cdot \tan\phi \tag{12.6}$$

where τ_{max} = shear strength (i.e., maximum shear stress), c' = effective cohesion intercept, σ' = effective normal stress, and ϕ' = effective stress friction angle.

A traditional means of evaluating the effective stress strength parameters c' and ϕ' has been via laboratory strength tests, such as the direct shear (DS), isotropically and/or anisotropically consolidated undrained triaxial test with pore pressure measurements (CIU/CAU), and consolidated drained triaxial test (CD). The focus of this section is on evaluating these effective stress strength parameters of clays via in situ tests and examined through comparison with laboratory benchmark values obtained from undrained consolidated anisotropic (CAUC) and undrained compression (CIUC) triaxial tests made on undisturbed samples.

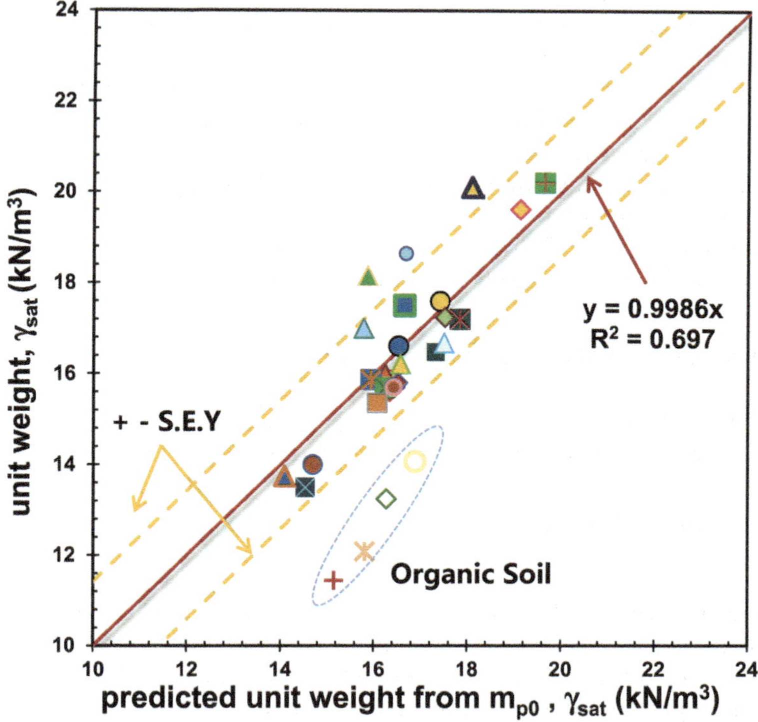

Figure 12.2 Lab soil unit weight versus predicted unit weight (organic clay excluded).

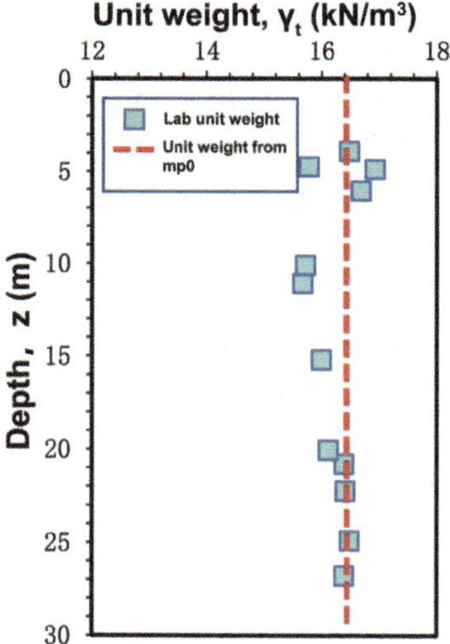

Figure 12.3 Profile of lab-measured unit weight and DMT predicted unit weight for Anacostia soft clay in Washington, DC.

12.3.1 Friction angle criteria

The effective friction angle obtained from triaxial tests can be evaluated on the basis of different criteria, which include: (a) value at maximum deviator stress (q_{max}); (b) value at maximum obliquity $(\sigma_1'/\sigma_3')_{max}$; and (c) value taken at large strains, normally at 15% axial strain, or in some cases, at 20% strain. The latter criterion is sometimes interpreted as the value corresponding to the critical state (ϕ_{cs}'). For CIUC and CAUC tests on soft inorganic clays of low sensitivity and low OCRs, these criteria often occur at similar points on the effective stress path, so only minor differences are notes in their magnitudes (Lade 2016; Ouyang & Mayne 2018a). A study of 19 natural clays generally showed minor differences in values of ϕ' defined at q_{max} and ϕ' at $(\sigma_1'/\sigma_3')_{max}$ (Bjerrum & Simons 1960). For a large series of CK0UC tests on soft Chicago clays, Finno and Chung (1992) showed that $\phi' = 28.3°$ at $(\sigma_1'/\sigma_3')_{max}$; and $\phi' = 27.6°$ at large strains, thus <1° difference. For sensitive and structured clays, however, the values can be of several degrees different (Leroueil & Hight 2003). In most of the database, only a single reported value of ϕ' was provided by the sources of information and the adopted criterion was often not identified. In some cases, an effective stress envelope was provided, yet the criterion used was not noted. In a few cases, a completely effective stress path and/or stress–strain curve was provided, and more than value could be assessed. It is believed that the database primarily contains results from inorganic and insensitive clays where differences in ϕ' from the different criteria are minor.

12.3.1 Effective stress strength parameters for intact clays from piezocone penetration test

Excess pore pressures are observed when the penetrometer is advanced into intact clays and silts at a standard rate of 20 mm/s. In order to assess the effective stress friction angle (ϕ') for a variety of fine-grained soils, ranging from natural lean plastic clays and clayey silts from marine, alluvial, lacustrine, deltaic, and glaciofluvial origins, an effective stress limit plasticity solution for piezocone penetration tests (CPTu) developed at NTH (Norwegian Institute of Technology) by Janbu and Senneset (1974), Senneset et al. (1989), and Sandven (1990) and further modified by Ouyang and Mayne (2018a, 2019, 2023) is illustrated below for the task.

The relationship between soils' effective stress friction angle (ϕ') and the CPTu measurements are given as:

$$N_{mc} = Q' = \frac{\tan^2(45° + \phi'/2) \cdot \exp(\pi \cdot \tan\phi') - 1}{1 + 6 \cdot \tan\phi' \cdot (1 + \tan\phi') \cdot B_q} \tag{12.7}$$

where $N_{mc} = Q'$ = modified resistance number from CPTu= Q/OCR^{Λ}, Q = normalized cone tip resistance = $(q_t - \sigma_{v0})/\sigma'_{vo}$), OCR = overconsolidation ratio of soil, Λ = plastic volumetric strain ratio≈0.7–0.8, and B_q= normalized porewater pressure = $(u_2 - u_0)/(q_t - \sigma_{v0})$.

Typical value of Λ ranges from 0.5 to 0.7 for artificially prepared samples such as kaolin and kaolinitic–sand mixture to around 0.7–0.9 for inorganics and insensitive clays to as high as 1 for structured and sensitive soils.

For intact clays with $18° \leq \phi' \leq 45°$ and $0.05 \leq B_q \leq 1.0$, the above solution for the effective stress friction angle (ϕ') can be numerically approximated by:

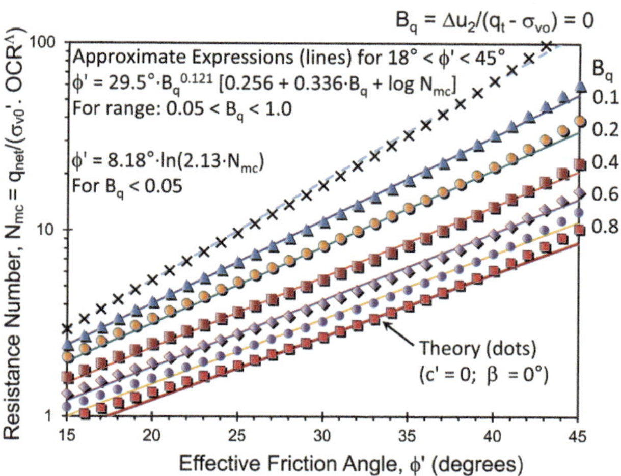

Figure 12.4 NTH method for evaluating ϕ′ from CPTu using analytical and approximate solutions.

$$\phi' = 29.5° \cdot B_q^{\ 0.121} \cdot [0.256 + 0.336 \cdot B_q + \log N_{mc}] \tag{12.8}$$

Figure 12.4 displays both the approximation equation and the closed-form solution. The effective cohesion intercept can be taken as $c'= 0$ for uncemented soft to hard clays and silts, especially those that are normally consolidated (NC) to lightly overconsolidated (LOC) with OCRs < 2 (Mayne and Stewart 1988; Mesri and Abdel-Ghaffar 1993). For overconsolidated (OC) clays, a low value of c' may be reasonable for interpreting the effective stress parameters (φ' and c') (Sellountou et al. 2004; Lade 2016). Since c' really refers to the part of the yield surface that extends above the frictional envelope, many, if not most, c' = 0 also applies to OC clays (Leroueil and Hight 2003; Mayne 2016).

12.3.1.1 Case study: CPTu in soft intact NC clays at Sandpoint, Idaho

A case study from Sandpoint, Idaho, is presented to show the process of interpreting effective friction angle using the NTH method. The Idaho Department of Transportation required soils information for design of approach embankments and a new bridge for State Route 95 highway. Figure 12.5 shows a representative CPTu at the site indicating over 80 m of soft silty clays, with occasional sand layers, lenses, and seams.

For the case of these NC clays having OCR = 1, the resistance number Q' is simply the normalized cone tip resistance Q, and Figure 12.6 shows the procedure for determining Q and B_q. The parameter Q is found as the slope from plotting net cone resistance versus the effective overburden stress. In this example, we force the line through the origin (assuming c'=0) to obtain Q =4.2. By the same token, the porewater parameter B_q is determined as the slope of the Δu versus q_{net}, giving the porewater parameter B_q = 0.75 for the Sandpoint site.

Results from 32 laboratory CIUC triaxial compression tests on undisturbed samples taken from 5 to 80 m depths are used as the reference benchmark values. Figure 12.7 shows the laboratory s–t' space of the triaxial tests that provided a lab friction angle of 33.1°. The NTH CPTu outputs an interpreted friction angle of 32.8° that is in excellent agreement.

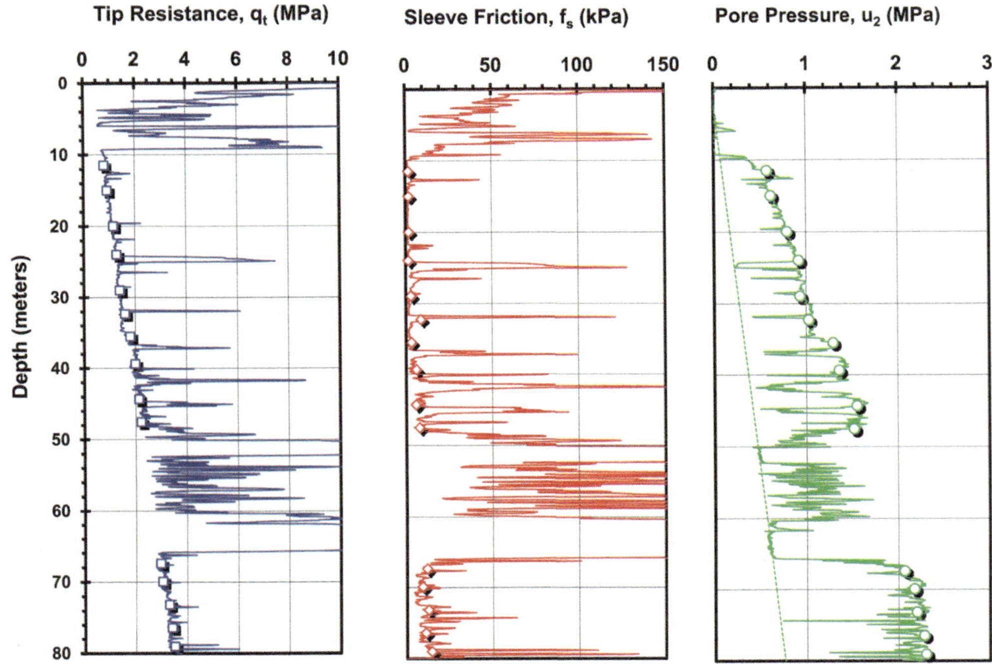

Figure 12.5 Deep CPTu profile at Sandpoint, Idaho with select data points for analysis.

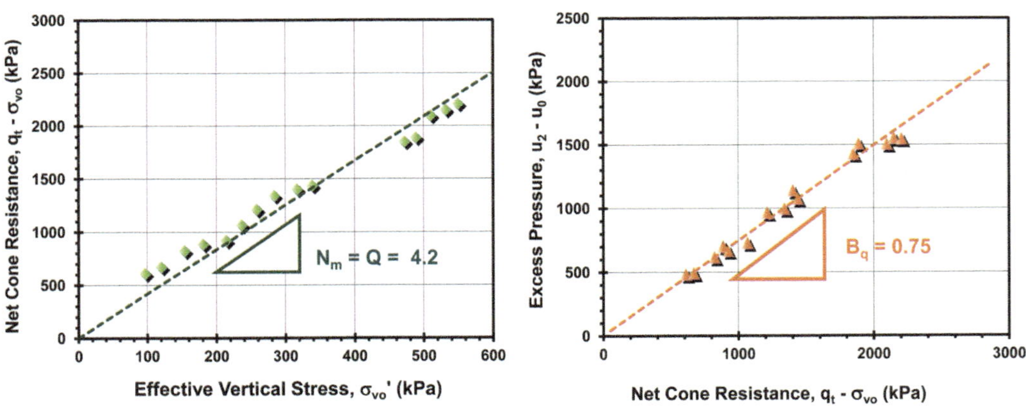

Figure 12.6 NTH postprocessing of CPTu data for Q and B_q at Sandpoint

12.3.1.2 Case study: CPTu-ϕ' profile interpretation for soft intact NC Chicago clays

The national geotechnical experimental site at Northwestern University (NWU) in Evanston, Illinois, is underlain by a 10-m-thick sand layer over 12 m of soft Chicago clay, which overlie other deeper soil layers. A variety of in situ and laboratory tests have been conducted here to investigate the properties of the clay soils (Finno et al. 2000). Figure 12.8 shows a representative portion of a CPTu sounding in the soft clay from 10 to 22 m depths with respective q_t, f_s, and u_2 values obtained using the Georgia Tech cone

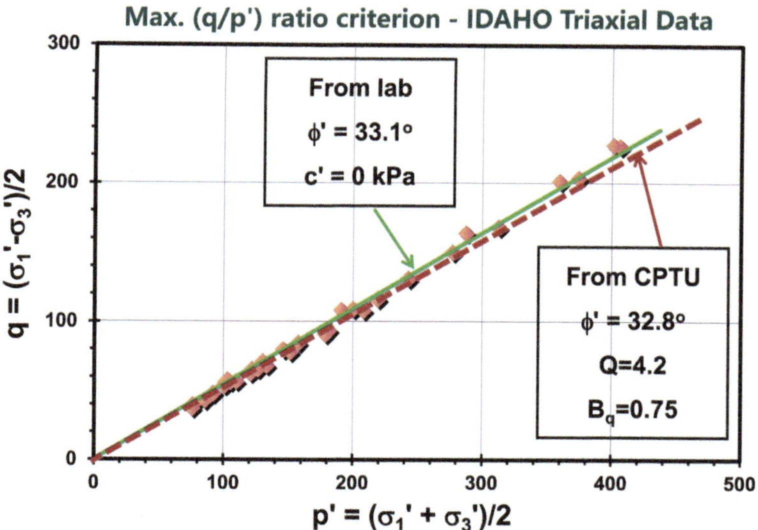

Figure 12.7 Laboratory triaxial effective stress friction angle ϕ' versus interpreted ϕ' from NTH using CPTu data at Sandpoint.

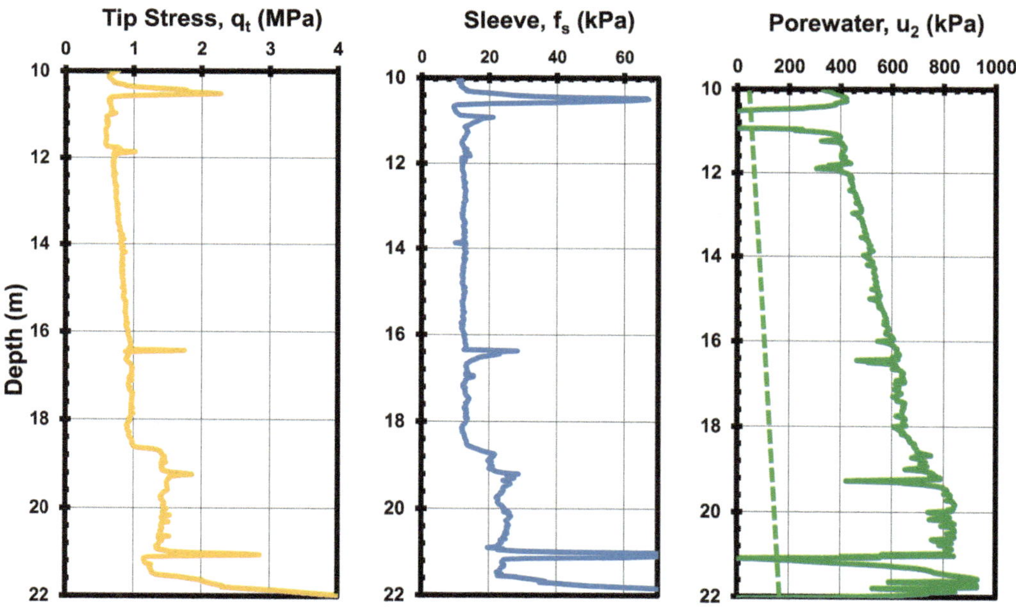

Figure 12.8 CPTu sounding of soft Chicago clays.

truck (Mayne 2007a). The soft clay under investigation has a mean unit weight of 20 kN/m^3, water content of 20%, liquid limit of 38%, and plasticity index (PI) = 12%.

The laboratory value of ϕ' obtained from a series of CAUC triaxial tests conducted on undisturbed samples of these soft clays (Figure 12.9), as reported by Finno and Chung (1992) and Chung and Finno (1992) are adopted to scrutinize the performance of the approximate NTH solution for ϕ' profile determination using Equation (12.8). The

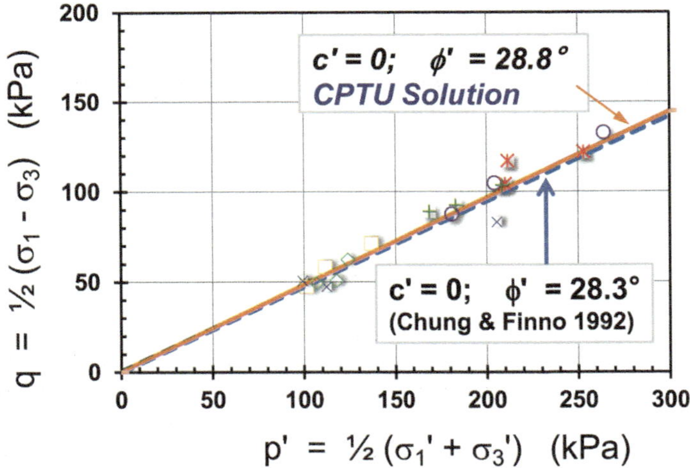

Figure 12.9 Comparison between laboratory triaxial tests showing ϕ′ at q_{max} and NTH friction angle from CPTu at NWU site (Note: triaxial results from Chung and Finno 1992).

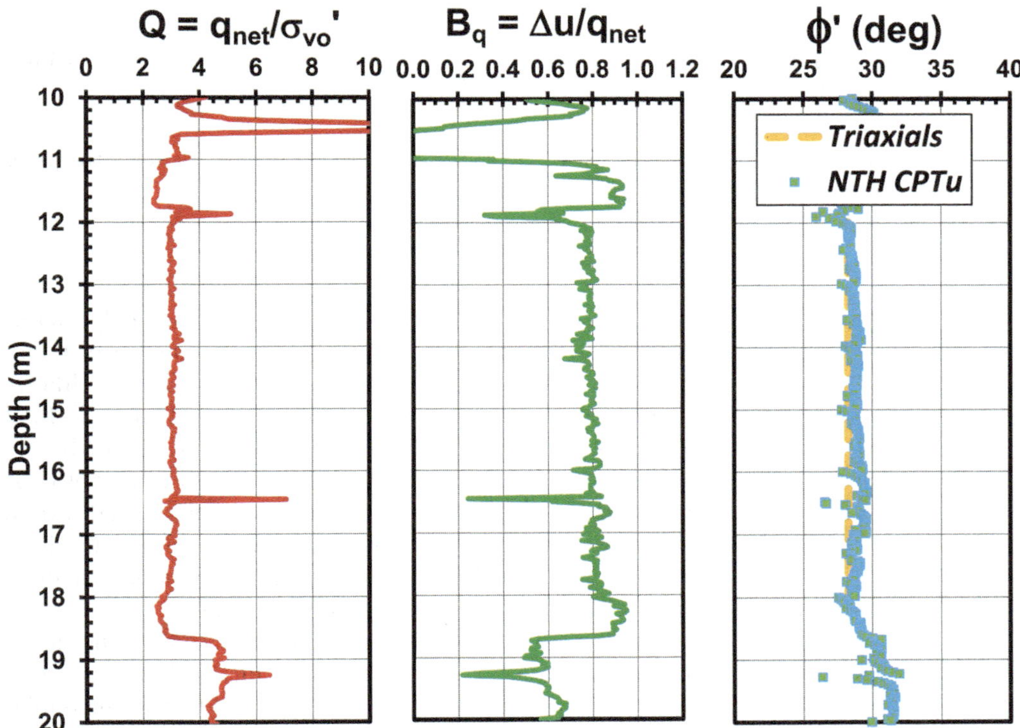

Figure 12.10 Profiles of resistance number (Q), normalized porewater pressure (B_q), and evaluated ϕ′ profile for soft Chicago clays.

profiles of Q and B_q with depth from the CPTu data are presented in Figure 12.10. Here it can be seen that the approximate NTH solution gives results in agreement with the laboratory triaxial tests.

12.3.1.3 Case study: CPTu in overconsolidated Presumpscot clay, ME

For CPTu in overconsolidated intact clays, the same methodology applies, however, the cone resistance number must be adjusted for stress history effect using Equation (12.7). A field case study with CPTu soundings in natural overconsolidated intact clays is presented using data from Martin's Point Bridge, near Portland, Maine (Hardison and Landon 2015). The general stratification of the site consists of an organic silt layer underlain by firm to stiff OC Presumpscot clay that overlies lower strata comprised of glacial outwash sand and bedrock. Shelby tube samples of Presumpscot clay were collected at the site from different elevations and tested for index properties, consolidation parameters, and strength characteristics. The results of laboratory testing on the OC clay layer gave an average total unit weight γ_t = 16.5 kN/m^3, natural water contents ≈ 30%–40%, liquid limits in the range of 20%–45%, and a plasticity index (PI) between 10% and 20%.

Figure 12.11(a) shows a representative CPTu sounding at the site with profiles of q_t and u_2 with depth. Depths of 2–14 m represent the extent of the OC Presumpscot clay of interest at this location with the groundwater table located at about 2 m. The corresponding profile of overconsolidation ratio (OCR = σ'_p/σ'_{vo}) determined by constant-rate-of-strain (CRS) consolidation tests is shown in Figure 12.11(b) (Hardison and Landon 2015).

The profiles of the cone resistance number (N_m and N_{mc}) using Λ = 0.7) with normalized porewater parameter B_q are shown in Figure 12.12(a). These are utilized to produce depth profiles of ϕ' in direct comparison with the CIUC triaxial data that are shown at their respective sampling depths in Figure 12.12(b). Overall, the values from CIUC results are in good agreement with those from CPTu using the modified NTH solution.

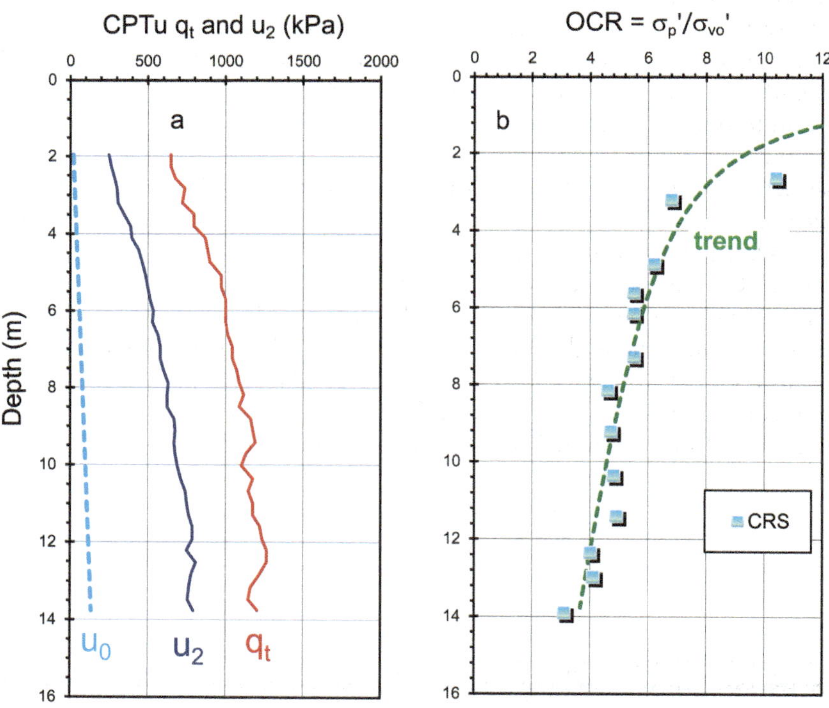

Figure 12.11 Profiles at Martin's Point Bridge, Maine: (a) piezocone readings and (b) OCR from CRS consolidation tests (Source: data from Hardison and Landon 2015).

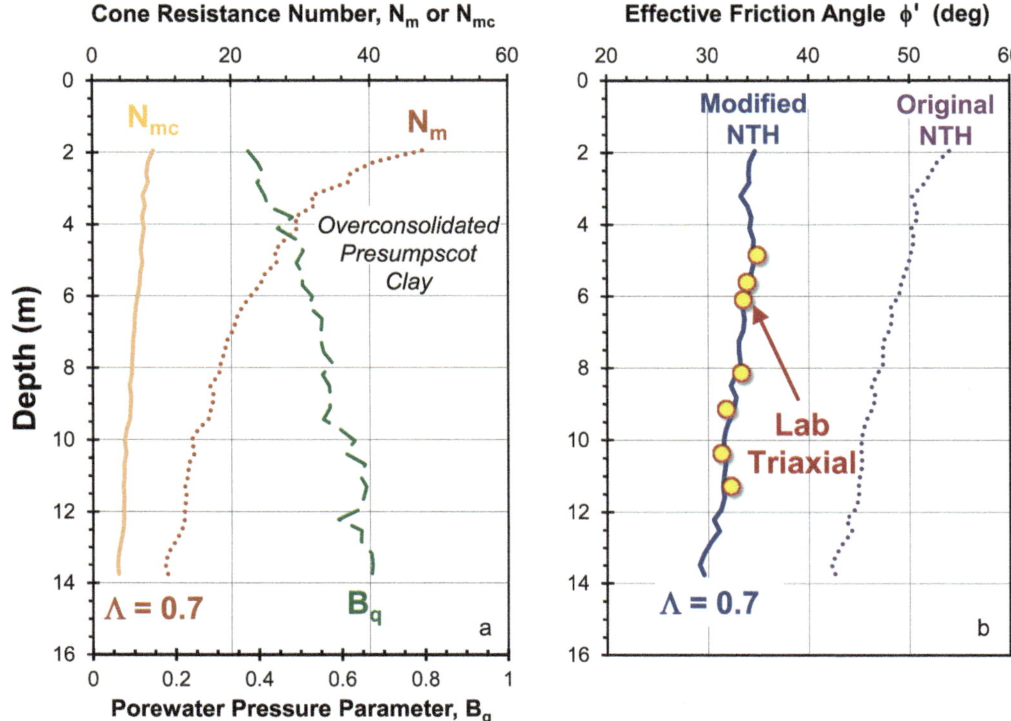

Figure 12.12 Profiles for Presumpscot clay at Martin's Point Bridge: (a) N_m, N_{mc}, and B_q using Λ =0.7; and (b) NTH ϕ', modified NTH ϕ', and CIUC triaxial ϕ'.

12.3.2 Effective stress strength parameters for fissured clays from piezocone penetration test

In fissured overconsolidated fine-grained soils, the porewater pressure u_2 registered by CPTu often demonstrates a near zero magnitude, $u_2 \approx 0$ (Ouyang and Mayne 2019, 2020). To interpret the value of effective stress friction angle ϕ' in OC fissured clays, a practical solution is to take $u_2 \approx 0$, and $B_q < 0.05$ is usually observed, thus ϕ' in OC fissured clays is given by the following equation:

$$\phi' \approx 8.18° \cdot \ln(2.13 N_{mc}) \tag{12.9}$$

12.3.2.1 Case study: Fissured London Clay at Brent Cross, United Kingdom

Brent Cross serves as an experimental test site involving London Clay (Powell and Quarterman 1988). London Clay is an overconsolidated Eocene marine formation that is notable for the presence of discontinuities in the form of fissures and cracks (Lunne et al. 1986). The characteristic effective strength of the fissured clay can be represented by a lower bound envelope given by c'=0 and ϕ'=19.5° (Hight and Jardine 1993). Laboratory testing on the London Clay at Brent Cross has shown a representative total unit weight

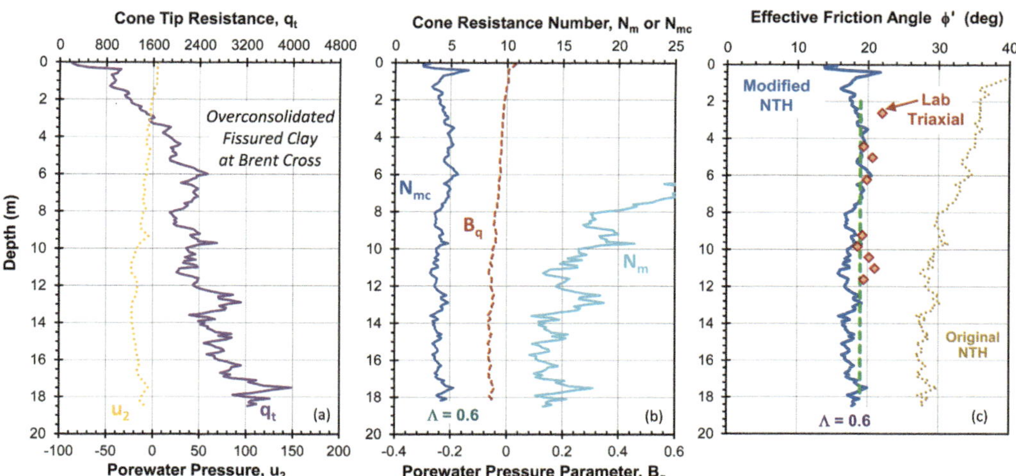

Figure 12.13 Profiles for fissured London Clay at Brent Cross: (a) Piezocone penetration data; (b) N_m, N_{mc}, and B_q using Λ = 0.6; and (c) NTH ϕ' and modified NTH ϕ' and CAUC triaxial ϕ' (Source: data from Lunne et al. 1986; Hight et al. 2003).

γ_t = 20 kN/m^3, natural water contents ≈ 28%–32%, liquid limits in the range of 65%–85%, and plasticity index (PI) between 55% and 65%. A representative exponent Λ = 0.6 has been assigned to the fissured London Clay at Brent Cross. Representative data from piezocone tests at Brent Cross are shown in Figure 12.13(a) with the profiles of q_t and u_2 with depth (Lunne et al. 1986). In Figure 12.13(b), the cone resistance number N_m and modified cone resistance number N_{mc} are presented along with the porewater pressure parameter B_q. It is observed that the value of B_q is essentially zero throughout the entire profile, which again is indicative of a fissured clay. The modified NTH solution using the approximate expression for $B_q = 0$ appears in good agreement with the laboratory triaxial interpretation of the effective friction angle, specifically $\phi' \approx 19.5°$.

12.3.2.2 Triaxial – Piezocone database for NC–OC clays

For the examination of the modified NTH solution for normally consolidated and overconsolidated clays, a master summary plot for the triaxial-measured ϕ' versus the CPTu-determined values via the modified NTH solution is shown in Figure 12.14.

The above statistics generally support that the revised NTH method gives a reasonable evaluation, albeit about 4% conservative, on the effective friction angle when referenced to the benchmark triaxial value.

12.3.2.3 Special considerations for organic soils, sensitive clays, and structured geomaterials

For the data that were considered in these studies, CPTu and triaxial data were obtained from clays and clayey silts that were primarily inorganic soils of low to medium sensitivity. The characteristic porewater pressure response for this group exhibits: 0.5 < Bq < 0.6 (Ouyang and Mayne 2019). However, it is recognized that natural organic clays tend to generate lower normalized porewater pressure (B_q < 0.5) while sensitive and structured

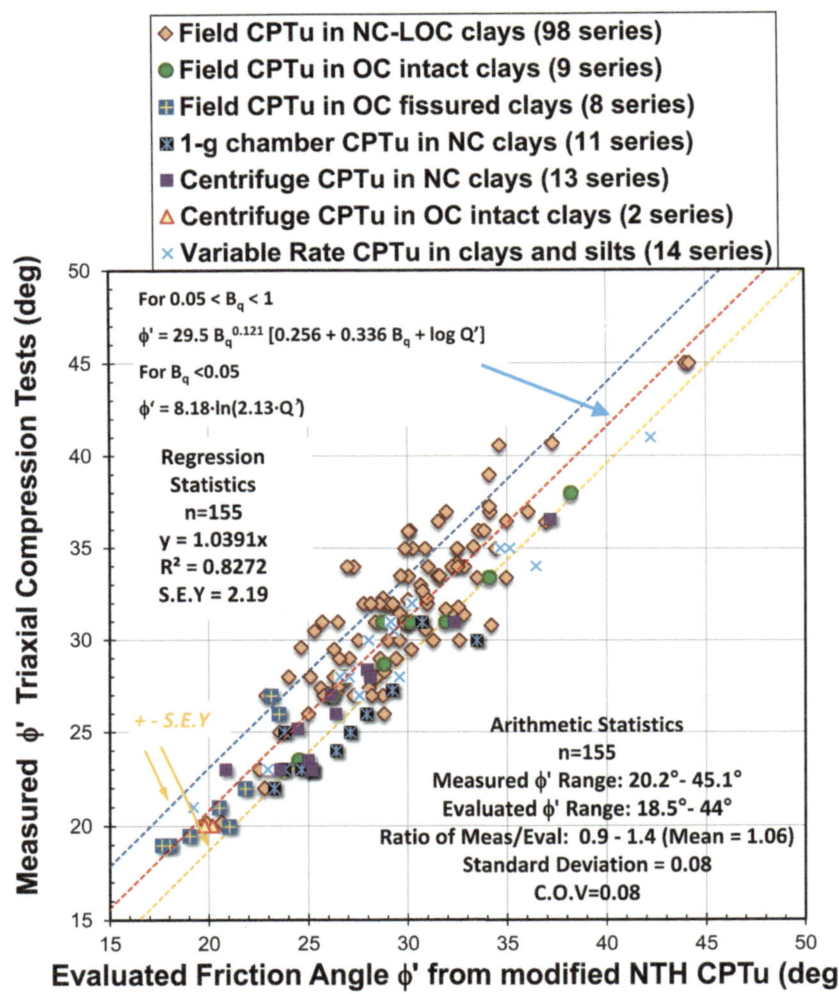

Figure 12.14 Summary plot of laboratory triaxial-measured ϕ' versus CPTu ϕ' from different test series using modified NTH method (Source: modified after Ouyang & Mayne 2023).

clays generally exhibit higher values ($B_q > 0.8$) when compared to normal or well-behaved clays. Therefore, special caution should be taken when evaluating CPTu in organic clays, sensitive or structured soils, or clays of special mineralogy, as the application of the NTH solution should only be undertaken after careful calibration with triaxial tests on high-quality laboratory samples to confirm the appropriate effective stress friction angles are valid in the specific geologic setting of study.

12.3.3 Effective stress strength parameters for intact clays from flat dilatometer test

The flat plate dilatometer test (DMT) was developed in Italy by Silvano Marchetti in the mid-1970s and quickly gained popularity among both practitioners in the geoengineering profession and academicians in geotechnical research who were seeking new and improved methods for site exploration. The DMT is a quick, reliable, and robust in situ test that investigates subsurface characteristics and allows for the interpretation of geoparameters in an efficient, economic, and expedient manner along the soil profile. Details concerning the DMT equipment, field test procedures, and interpretations are given by Marchetti (1980) and ASTM D6635-15, Standard Test Method for Performing the Flat Dilatometer Test (DMT). Updates to the test are discussed by Marchetti et al. (2006) and Marchetti (2015).

For DMT and CPTu soundings in soft to firm clays, a theoretical nexus has been established based on the spherical cavity expansion (SCE) theory by Ouyang and Mayne (2017a, 2017b, 2018). As a result, it is possible to express an equivalent net cone resistance ($q_{net\text{-}DMT}$) and excess porewater pressure (Δu) from DMT soundings as follows:

$$q_{net-DMT} = 2.93p_1 - 1.93p_0 - u_0 \tag{12.10}$$

$$\Delta u_{DMT} = p_0 - u_0 \tag{12.11}$$

where p_0 and p_1 are the contact and expansion pressures from the DMT, respectively.

For soils that are classified as inorganic and insensitive intact clays with OCRs < 2.5, the above relationship is valid. Thus, DMT data can be used to predict the effective friction angle of soft to firm clays using the following equation:

$$\phi' \approx 29.5 \quad (U^*_{DMT}/Q_{DMT})^{0.121}\left[0.256 + 0.336\,(U^*_{DMT}/Q_{DMT}) + \log Q_{DMT}\right] \tag{12.12}$$

where

$$U^*_{DMT} = (p_0 - u_0)/\sigma_{v0}' \tag{12.13}$$

$$Q_{DMT} = (2.93p_1 - 1.93p_0 - u_0)/\sigma_{v0}' \tag{12.14}$$

This equation is valid for the following ranges: $20° \leq \phi' \leq 45°$ and $0 \leq U^*_{DMT} \leq 4$.

12.3.3.1 Case studies: Route 17 Bridge, Chesapeake, Virginia

A roadway improvement project was constructed off Route 17 in Chesapeake, Virginia, in the United States between Interstate 64 and Military Highway where the Norfolk Western Railroad crosses the project site. The Chesapeake area is located within the Atlantic Coastal Plain geologic province and underlain by approximately 800 m of marine and estuarine deposits. Standard soil test borings were performed to depths up to 25 m at the site. Recovered undisturbed tube samples were subjected to consolidation testing for soil settlement characteristics. In addition, isotropically consolidated undrained triaxial compression tests (CIUC) were conducted to determine the total and effective stress strength parameters of the primary clay layer. In situ flat dilatometer tests were also carried out at the project site. Field boring performed for this project encountered a subsurface stratigraphy consisting of three basic strata, where upper strata I and bottom strata III show interlayered loose to dense clean to clayey fine sands. The strata II soils consist predominately of soft to firm high to low plasticity clays about 10 m thick (Ouyang and

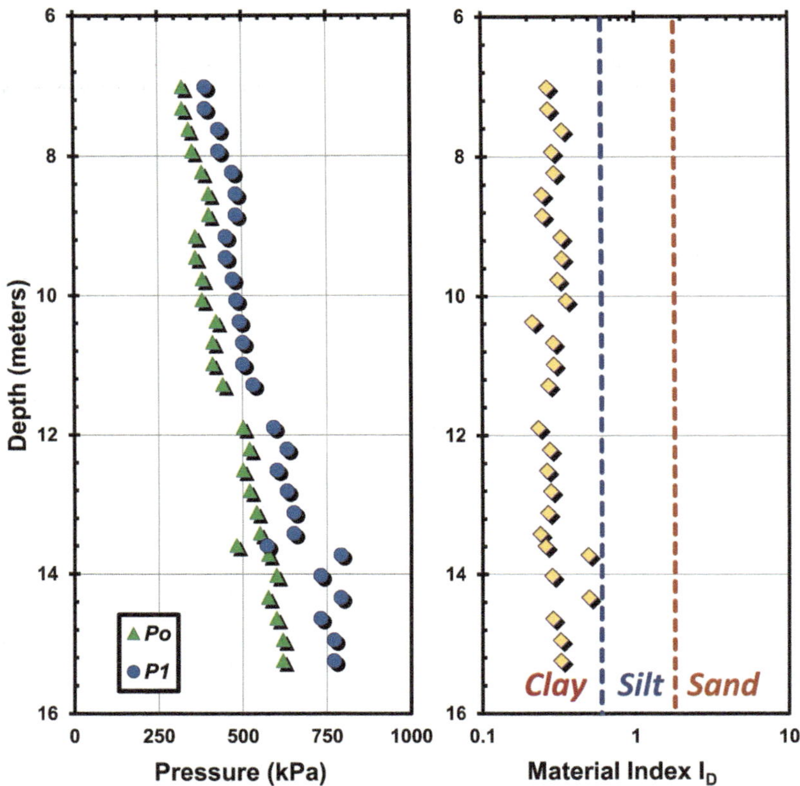

Figure 12.15 Representative DMT sounding at Route 17 bridge site, Chesapeake, VA (Source: after Ouyang and Mayne 2018).

Mayne 2018). This layer created a major problem to approach embankment stability and settlement and characterized by an average unit weight of 15.6 kN/m^3. The natural moisture contents ranged from 53% to 92% and the plasticity index (PI) ranged from 31% to 68%. The ground water level was about 1 m below the surface.

Figure 12.15 shows a representative portion of a DMT sounding in the soft clay layer from 6 to 16 m depths with respective contact pressures (p_0) and expansion pressures (p_1). The procedure for determining the equivalent DMT cone resistance number Q_{DMT} is found as the slope from plotting the equivalent DMT net resistance q_{netDMT} versus the effective overburden stress, as illustrated in Figure 12.16a. In this example, we force the line through the origin (assuming $c'=0$) to obtain Q_{DMT} = 9.73. By the same token, the equivalent DMT porewater parameter $B_{q\text{-}DMT}$ is determined as the slope of measured $\Delta u_{DMT} = (p_0 - u_0)$ versus q_{netDMT}, giving the value of $B_{q\text{-}DMT}$ = 0.52 for the clay layer at this site, as indicated by Figure 12.16b. These values gave an effective friction angle ϕ'=38.6° for the soft Chesapeake Clay. The DMT-NTH interpreted ϕ' is then compared with the laboratory value of the effective stress friction angle from a series of CIUC triaxial tests on the soft clay. The results from the NTH method match reasonably with the laboratory data, as shown by Figure 12.17.

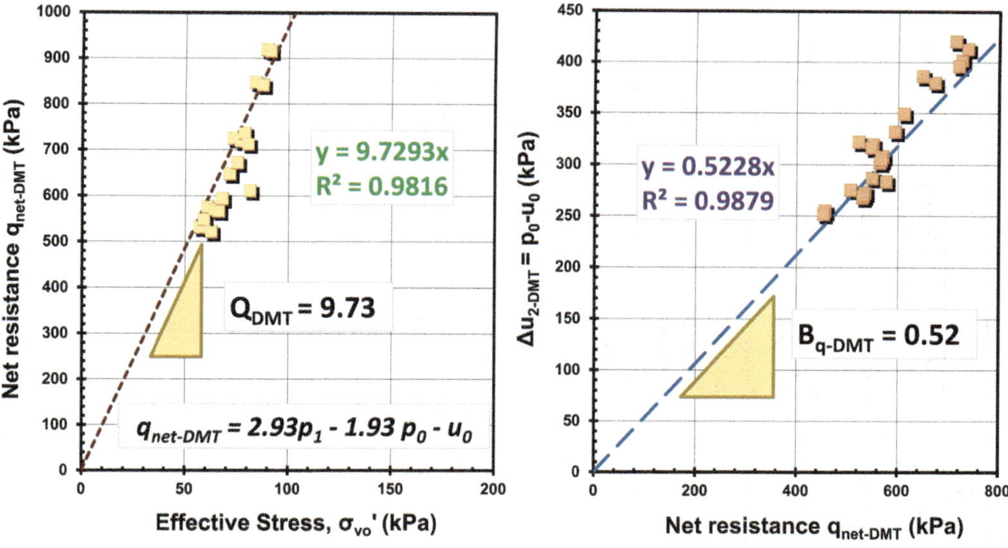

Figure 12.16 Postprocessing of the DMT data in clay layer in Chesapeake, Virginia, for determining (a) DMT-equivalent resistance number, Q_{DMT}; and (b) DMT-equivalent porewater parameter, $B_{q\text{-}DMT}$.

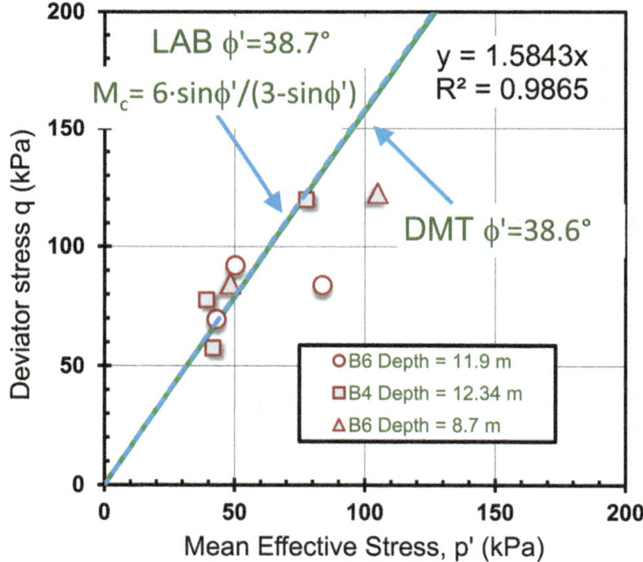

Figure 12.17 Comparison of effective friction angle measured by laboratory triaxial tests and the NTH solution from DMT data in Chesapeake, Virginia.

12.4 DISCUSSION AND APPLICABILITY

12.4.1 Friction angle criteria

For the clays reviewed in the above DMT analysis, the effective stress friction angle ϕ' was obtained from either isotropically and/or anisotropically consolidated undrained laboratory triaxial test (CIUC, CK_0UC, and/or CAUC) on undisturbed samples to establish as the benchmark value. Criteria considering the choice of the laboratory benchmark effective stress friction angle is discussed in Section 12.3. In general, recognized guidelines for the test procedures such as ASTM standards (D4767), Bishop and Henkel (1962), Germaine and Germaine (2009), and Lade (2016) were followed.

12.4.2 Effective cohesion intercept, c′

For this study, the effective cohesion intercept (c') has been taken as zero ($c' = 0$) since the vast majority of the clays considered are NC to LOC with low OCRs < 2.5 (e.g., Finno & Chung 1992; Diaz-Rodriguez et al. 1992). Even for overconsolidated clays, a small value of c' is appropriate (Lade 2016). If a value of c' is necessary, the ratio $c'/\sigma_p' \approx 0.03$ can be used, where σ_p' = effective preconsolidation stress (Larsson et al. 1987; Mayne, 2016).

A recent study by Amundsen and Thakur (2017) investigated the effects of sample disturbance on soft clay parameters. It was found that sample quality, storage time, and testing time between sampling and shearing appreciably affect the magnitude of preconsolidation stress, undrained shear strength, and constrained modulus. However, the magnitude of effective stress friction angle was little affected by these same factors.

12.4.3 Partial drainage effects

The aforementioned approach is valid only for undrained conditions. For a standard penetration rate of 20 mm/s for both CPTu and DMT, it is generally accepted that clay soils with low permeability will experience an undrained penetration process. However, drainage conditions during DMTs in intermediate permeability soils, such as silts, clayey silts, and sandy clays may be different. Schnaid et al. (2016) indicated that the DMT is essentially undrained in soft clay and dominated by penetration pore pressures, whereas the test is drained in clean sands and partially drained in intermediate permeability soils, such as silts.

For the CPTu, special series of tests at variable rates is termed "twitch" testing that allows the demarcation of undrained, partially drained, to fully drained behavior (Randolph 2004). DeJong et al. (2012) give an approximate range of c_v (coefficient of consolidation) that applies to clays, silts, and sands for field operation of CPTu during twitch testing. For clays, the expected range and upper limit for the coefficient of consolidation is approximately $c_v < 0.05$ cm²/s. Thus, for soils with higher values, such as applicable to silts and sands, the SCE nexus in this study would no longer apply.

Also, as noted earlier, CPTu readings q_t and u_2 are taken during penetration at time t = 0, whereas the DMT readings p_0 and p_1 are obtained at approximately t = 15 s and t = 45 s after penetration, respectively. Thus, the method can be invalidated when dissipations occur in soils that do not maintain undrained conditions at constant volume.

Finally, special considerations should also be given when conducting site investigations in more difficult soils, such as highly organic clays, peats, and muskegs, as well as sensitive and structured geomaterials, as the aforementioned relationships may not apply. In highly stratified deposits where clay layers are sandwiched between more permeable silty and sandy layers, then the effects of drainage and partial drainage may also influence the results, especially for thin clay strata. In these cases, caution is warranted.

12.5 SOIL STRESS HISTORY PARAMETERS IN CLAYS

12.5.1 Stress history for intact clays from piezocone penetration test

12.5.1.1 SCE-CSSM solution

A hybrid formulation of spherical cavity expansion (SCE) and critical-state soil mechanics (CSSM) expresses the overconsolidation ratio (OCR) of clays in three separate formulations using net cone tip resistance ($q_{net} = q_t - \sigma_{vo}$), excess pore pressure ($\Delta u = u_2 - u_0$), and effective cone resistance ($q_E = q_t - u_2$). Details of the solution for insensitive inorganic clays are given by Mayne (1991), Chen and Mayne (1994), and Burns and Mayne (1998), whereas the application in sensitive and structured clays is provided by Agaiby and Mayne (2018), Mayne et al. (2018, 2019), DiBuö et al. (2019), Mayne and Benoît (2020), and Agaiby and Mayne (2021).

Three separate algorithms relate the OCR to normalized CPTu parameters: $Q = q_{net}/\sigma_{vo}'$ and $U = \Delta u_2/\sigma_{vo}'$, where $q_{net} = q_t - \sigma_{vo}$ = net cone resistance and $\Delta u_2 = u_2 - u_0$ = excess porewater pressure. These are expressed by the following:

$$OCR = 2 \cdot \left[\frac{Q / M_{c1}}{0.667 \cdot \ln(I_R) + 1.95} \right]^{1/\Lambda} \tag{12.15}$$

$$OCR = 2 \cdot \left[\frac{U^* - 1}{0.667 \cdot M_{c2} \cdot \ln(I_R) - 1} \right]^{1/\Lambda} \tag{12.16}$$

$$OCR = 2 \cdot \left[\frac{Q - \frac{M_{c1}}{M_{c2}} \cdot (U^* - 1)}{1.95 \cdot M_{c1} + \frac{M_{c1}}{M_{c2}}} \right]^{1/\Lambda} \tag{12.17}$$

where $\Lambda = 1 - C_s/C_c$ = plastic volumetric strain potential, C_s = swelling index, C_c = virgin compression index, $I_R = G/s_u$ = undrained rigidity index, $M = 6 \cdot \sin\phi'/(3 - \sin\phi')$ = frictional parameter in q–p' space. The value of M_{c1} is defined at peak strength (i.e., ϕ' at q_{max}), whereas M_{c2} is the value defined at large strains which occurs at maximum obliquity (i.e., ϕ' when the ratio $[\sigma_1'/\sigma_3']$max). For insensitive clays, the value of ϕ' at q_{max} is equal to ϕ' at $(\sigma_1'/\sigma_3')_{max}$, and thus $M_c = M_{c1} = M_{c2}$. For insensitive clays, the value of $\Lambda = 0.80$, whereas for sensitive clays, a value of $\Lambda \approx 1.0$ is more suitable.

While Equations (12.15) and (12.16) both depend on the I_R of the clay, Equation (12.17) is independent of the I_R and is obtained by combination of the first two formulations. The rigidity index is thus given directly from:

$$I_R = \exp\left[\frac{1.5 + 2.925 \cdot M_{c1} \cdot a_q}{M_{c2} - M_{c1} \cdot a_q}\right] \tag{12.18}$$

where $a_q = (U - 1)/Q = (u_2 - \sigma_{vo})/(q_t - \sigma_{vo})$. The parameter a_q can be determined as a single value for any clay layer or uniform clay deposit by taking the slope of a plot of the parameter $(U - 1)$ versus Q, or alternatively taken as the slope of $(u_2 - \sigma_{vo})$ versus $(q_t - \sigma_{vo})$. Using regression analyses, slightly different slope values for a_q are obtained.

12.5.1.2 Simplified approach for insensitive and inorganic clays

A series of simplifications can be made for insensitive inorganic clays, or "regular" and "normal" clays. For one, Equation (12.16) can be approximated by:

$$OCR \approx 2 \cdot \left[\frac{U}{0.667 \cdot M_{c2} \cdot \ln(I_R)}\right]^{1/\Lambda} \tag{12.19}$$

As noted previously for regular clays, the values $M_c = M_{c1} = M_{c2}$, therefore Equation (12.17) reduces to:

$$OCR = 2 \cdot \left[\frac{1}{1.95M_c + 1}\left(\frac{q_t - u_2}{\sigma_{vo}'}\right)\right]^{1/\Lambda} \tag{12.20}$$

For a first-order estimate of σ_p' in regular clays, further simplifications are achieved by (a) setting the exponent $\Lambda = 1$ to reduce the power law format to linear equations; (b) adopting a characteristic effective friction angle of clay $\phi' = 30°$ ($M_c = 1.2$), and (c) using a default value of $I_R = 100$ (Houlsby & Teh 1988; Mayne 2007). The reduced expressions become:

$$\begin{aligned} s_p' &\approx 0.33 q_{net} = 0.33(q_t - s_{vo}) \\ s_p' &\approx 0.53 Du_2 = 0.53(u_2 - u_o) \\ s_p' &\approx 0.60 q_E = 0.60(q_t - u_2) \end{aligned} \tag{12.21}$$

$$\sigma_p' \approx 0.53\Delta u_2 = 0.53\ (u_2 - u_o) \tag{12.22}$$

$$\sigma_p' \approx 0.60\ q_E = 0.60\ (q_t - u_2) \tag{12.23}$$

12.5.2 Stress history for sensitive clays from piezocone penetration test

A review of the CPTu data on different selective sensitive soils as presented in Agaiby and Mayne (2021) show that when the three expressions from Equations (12.21)–(12.23) are used, unmatched OCR profiles among each other and evident disagreement when compared to lab reference values from consolidation tests are present. Moreover, a consistent hierarchy can be observed:

$$0.60\ q_E < 0.33\ q_{net} < 0.53\Delta u_2 \tag{12.24}$$

which gives the signature identification of sensitive clays.

To assess the OCR from CPTu in sensitive clays, Equations (12.15)–(12.17) are utilized with corresponding M_{c1} and M_{c2} that can be obtained from the original NTH expression (Ouyang & Mayne 2018) and modified NTH expression (Ouyang & Mayne 2019), respectively. Results from laboratory triaxial compression tests also confirm and validate these values.

12.5.3 Stress history for organic clays from piezocone penetration test

A review of the CPTu data on different organic soils show that when the three expressions from Equations (12.21)–(12.23) are used, unmatched OCR profiles occur in the following hierarchal order (Mayne & Agaiby 2019):

$$0.53\Delta u_2 < 0.33\ q_{net} < 0.60\ q_E \tag{12.25}$$

Therefore, Equation (12.25) serves as the CPTu signature that is characteristic of organic fine-grained soils.

For evaluating the yield stress of soft organic soils by CPTu, it has been recommended to lower the coefficients of Equations (12.21)–(12.23). This could be justified because organic clays and peats exhibit rather high friction angles ($\phi' > 40°$) when compared to regular clays. By use of a higher friction angle (e.g., $M_c > 1.2$) in Equations (12.15)–(12.17) and 12.20, the resulting set of simplified Equations (12.21)–(12.23) would have smaller coefficients for q_{net}, Δu, and q_E.

For instance, for the soft organic clays of Brazil, modified expressions have been developed (Baroni and Almeida, 2017):

$$\sigma_p' = 0.125\ q_{net} \tag{12.26}$$

$$\sigma_p' = 0.154\ q_E \tag{12.27}$$

In another approach, for CPTu in a variety of soil types, a generalized solution retains the 0.33 coefficient of Equation (12.21) and employs a power law algorithm in the form (Mayne, 2017; Agaiby and Mayne, 2019):

$$\sigma_p' = 0.33 q_{net}^{m'} \left(units\ of\ kPa\right) \tag{12.28a}$$

where the exponent m' depends upon the soil behavioral type (= 1.0 intact inorganic clays; 0.90 organic clays; 0.85 silt mixtures; 0.80 silty sands; and 0.72 clean quartz–silica sands). Figure 12.18 shows the yield stresses σ_p' for various soil types plotted versus the net cone resistance. The corresponding expression in dimensionless terms is given by:

$$\sigma_p' = 0.33\ q_{net}^{m'}\ (s_{atm}\ /\ 100)^{1-m'} \tag{12.28b}$$

where σ_{atm} = reference pressure equal to 1 atm ≈ 1 bar = 100 kPa.

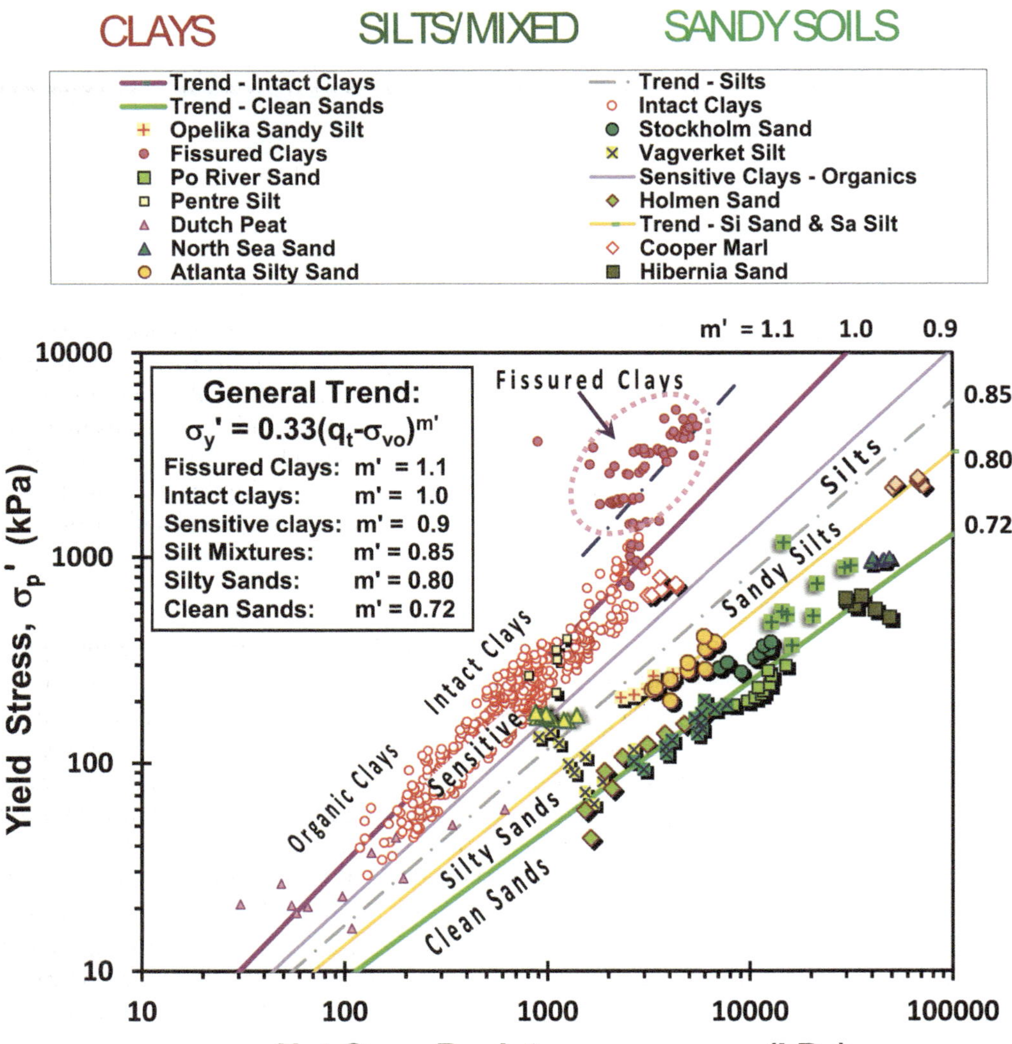

Figure 12.18 General relationship for yield stress of soils and CPT net cone resistance in different geomaterials (Source: modified after Agaiby and Mayne, 2019).

REFERENCES

Agaiby, S. S. and Mayne, P. W. (2018). Evaluating undrained rigidity index of clays from piezocone data. *Proc. 4th Intl. Symposium on Cone Penetration Testing*, TU Delft, Taylor & Francis, CRC Press, 65–72.

Agaiby, S. S. and Mayne, P. W. (2021). CPTU identification of regular, sensitive, and organic clays towards evaluating preconsolidation stress profiles. *AIMS Geosciences*, 7 (4): 553–573.

Amundsen, H. A., Jønland, J., Emdal, A., and Thakur, V., 2017. An attempt to monitor pore pressure changes in a block sample during and after sampling. *Géotechnique Letters*, 7, 2, 119–128. https://doi.org/10.1680/jgele.16.00176

Baroni, M. and Almeida, M. S. S. (2017). Compressibility and stress history of very soft organic clays. *Proceedings of the Institution of Civil Engineers, London—Geotechnical Engineering*, 170 (2): 148–160.

Bishop, A. W. and Henkel, D. J. (1962). *The Measurement of Soil Properties in the Triaxial Test*. 2nd edition, Edward Arnold Publishers Ltd., London. 227 p.

Bjerrum, L. and Simons, N. E. (1960). *Comparison of Shear Strength Characteristics of Normally Consolidated Clays*, presented at the Research Conference on Shear Strength of Cohesive Soils, Boulder, CO. Reston, VA: American Society of Civil Engineers, 711–726.

Burns, S. E. and Mayne, P. W. (1998). Monotonic and dilatory porewater pressures during piezocone dissipation tests in clay. *Canadian Geotechnical Journal*, 35 (6): 1063–1073.

Chen, B. Y. and Mayne, P. W. (1994). *Profiling the Overconsolidation Ratio of Clays by Piezocone Tests*, Report No. GIT-CEE/GEO-94-1 submitted to National Science Foundation by Georgia Institute of Technology, Atlanta, 280pp.

Chung, C. K. and Finno, R. J. (1992). Influence of depositional processes on the geotechnical parameters of Chicago glacial clays. *Engineering Geology*, 32 (4): 225–242.

Di Buò, B., D'Ignazio, M., Selãnpaã, J., Länsivaara, T., and Mayne, P. W. (2019). Yield stress evaluation of Finnish clays based on analytical CPTU models. *Canadian Geotechnical Journal*, 57 (11): 1623–1638.

DeJong, J. T., Jaeger, R. A., Boulanger, R. W., Randolph, M. F., and Wahl, D. A. J. (2012). Variable penetration rate cone testing for characterization of intermediate soils. *Geotechnical and Geophysical Site Characterization 4*. Vol. 1, Taylor and Francis, London: 25–42.

Diaz-Rodriguez, J. A., Leroueil, S., and Aleman, J. D. (1992). Yielding of Mexico City clay and other natural clays. *Journal of Geotechnical Engineering*, 118(7): 981–995. doi:10.1061/(ASCE)0733-9410(1992)118:7(981).

Finno, R. J., Gassman, S. L., and Calvello, M. (2000). *The NGES at Northwestern University. National Geotechnical Experimentation Sites (GSP 93)*. Reston, VA: ASCE, 130–159.

Germaine, J. T., and Germaine, A. V. (2009). *Geotechnical Laboratory Measurements for Engineers*. John Wiley & Sons, New York. 359 p.

Hardison, M. A. and Landon, M. (2015). Correlation of engineering parameters of the Presumpscot Formation to the seismic cone penetration test (SCPTu), Technical Report 15-12, Department of Civil Engineering, University of Maine, submitted to Maine Dept. of Transportation: 393 p.

Hight, D. W. and Jardine, R. J. (1993). Small-strain stiffness and strength characteristics of hard London tertiary clays. *Geotechnical Engineering of Hard Soils & Soft Rocks*, Balkema, Rotterdam: 533–552.

Holtz, R. D., Kovacs, W. D., and Sheahan, T. C. (2010). *An Introduction to Geotechnical Engineering*, 2nd ed. Pearson, NJ: Prentice-Hall.

Houlsby, G. T. and Teh, C. I. (1988). Analysis of the piezocone in clay. *Penetration Testing*, Vol. 2 (Proc. ISOPT-1, Orlando). Balkema, Rotterdam, 777–783.

Janbu, N. and Senneset, K. (1974). Effective stress interpretation of in-situ static penetration tests. *Proceedings of the 1st European Symposium on Penetration Testing*, Vol. 2. Stockholm: Swedish Geotechnical Society, 181–193. Download available at: www.usucger.org

Kulhawy, F. H. and Mayne, P. W. (1990). *Manual on Estimating Soil Properties for Foundation Design*. Report EL-6800, Palo Alto, CA: Electric Power Research Institute.

Lade, P. V. (2016). *Triaxial Testing of Soils*. Hoboken, NJ: John Wiley & Sons, Ltd., 402pp.

Lambe, T. W. and Whitman, R. V. (1979). *Soil Mechanics, SI version*. New York: Wiley.

Larsson, R., Bergdahl, U., and Eriksson, L. (1987). Evaluation of Shear Strength in Cohesive Soils with Special Reference to Swedish Practice and Experience. *Geotech. Test. J.*, Vol. 10, No. 3, pp. 105–112, https://doi.org/10.1520/GTJ10942J

Leroueil, S. and Hight, D. W. (2003). Behaviour and properties of natural soils and soft rocks. *Characterisation and Engineering Properties of Natural Soils Workshop*, Vol 1. (Singapore), Balkema, Rotterdam, the Netherlands, 29–254.

Loehr, J. E., Lutenegger, A. J., Rosenblad, B. and Boeckmann, A. (2016). *Geotechnical Site Characterization (GEC 5)*. Report FHWA NHI-16-072, Washington, DC: National Highway Institute, Federal Highway Administration.

Lunne, T., Eidsmoen, T., Powell, J. and Quarterman, R. (1986). Piezocone testing in overconsolidated clays. *Proceedings, 39th Canadian Geotechnical Conference: In-Situ Testing and Field Behavior*. Ottawa, 209–218.

Marchetti, S. (1980). In-situ tests by flat dilatometer. *Journal of the Geotechnical Engineering Division* ASCE 106 (GT3): 299–321.

Marchetti, S. (2015). Some 2015 updates to the TC16 DMT report 2001. *Proc. 3rd Int. Conf. on the Flat Dilatometer DMT'15*. Rome, Italy. www.dmt15.com

Marchetti, S., Monaco, P., Calabrese, M., and Totani, G. (2006). Comparison of moduli determined by DMT and back figured from local strain measurements under a 40 m diameter circular test load in the Venice area. *Proc. 2nd Int. Conf. on the Flat Dilatometer*, Washington DC: pp. 220–230.

Mayne, P. W. (2016). Evaluating effective stress parameters and undrained shear strengths of soft-firm clays from CPT and DMT. *Aust. Geomech. J.* 51 (4): 27–55.

Mayne, P. W. and Stewart, H. E. (1988). Pore pressure response of K0-consolidated clays. *Journal of Geotechnical Engineering*, 114 (11): 1340–1346.

Mayne, P. W. (1991). Determination of OCR in clays by piezocone tests using cavity expansion and critical state concepts. *Soils and Foundations*, 31(2): 65–76.

Mayne, P. W. (2007). *In-Situ Test Calibrations for Evaluating Soil Parameters, Characterization & Engineering Properties of Natural Soils*, Vol. 3. London: Taylor & Francis, 1602–1652.

Mayne, P. W. (2017). Stress history of soils from cone penetration tests. *34th Manual Rocha Lecture, Soils & Rocks*, Vol. 40 (3). Saö Paulo: Brazilian Society for Soil Mechanics, 203–218. http://www.soilsandrocks.com.br/

Mayne, P. W. (2014). Keynote: interpretation of geotechnical parameters from seismic piezocone tests. *Proceedings of the 3rd International Symposium on Cone Penetration Testing*. Las Vegas, NV, 47–73.

Mayne, P. W. and Agaiby, S. (2019). Profiling yield stresses and identification of soft organic clays using piezocone tests. *Proceedings XVI Pan American Conference on Soil Mechanics & Geotechnical Engineering*, Paper 0149, Cancun, Mexican Society of Geotechnical Engineering (SMIG): www.issmge.org

Mayne, P. W. and Benoît, J. (2020). Analytical CPTU Models Applied to Sensitive Clay at Dover, New Hampshire. *Journal of Geotechnical and Geoenvironmental Engineering*, 146 (12): 04020130.

Mayne, P. W., Christopher, B., Berg, R., and DeJong, J. (2002). Subsurface Investigations - Geotechnical Site Characterization. Publication No. FHWA-NHI-01-031, National Highway Institute, Federal Highway Administration (FHWA). Washington, DC: U.S. Department of Transportation.

Mayne, P. W., Greig, J., and Agaiby, S. (2018). Evaluating CPTu in sensitive Haney clay using a modified SCE-CSSM solution. *Proceedings 71st Canadian Geotechnical Conference: GeoEdmonton*, Paper ID No. 279, Canadian Geotechnical Society: www.cgs.ca

Mayne, P. W., Coop, M. R., Springman, S., Huang, A. -B., and Zornberg, J. (2009). State-of-the-art paper (SOA-1): geomaterial behavior and testing. *Proceedings of the 17th International Conference on Soil Mechanics & Geotechnical Engineering*, Vol. 4 (ICSMGE, Alexandria). Rotterdam: Millpress/IOS Press, 2777–2872.

Mayne, P. W., Paniagua, P., L'heureux, J. -S., Lindgård, A., and Emdal, A. (2019). Analytical CPTu model for sensitive clay at Tiller-Flotten site, Norway. *Proc. XVII ECSMGE: Geotechnical Engineering Foundation of the Future*, Paper 0153, Reykjavik, Icelandic Geotechnical Society: www.issmge.org

Mesri, G. and Abdel-Ghaffar, M. E. M. (1993). Cohesion intercept in effective stress-stability analysis. *Journal of Geotechnical Engineering*, 119 (8): 1229–1249.

Ouyang, Z. and Mayne, P. W. (2018a). Calibration of NTH method for friction angle using centrifuge CPTUs in clays. *Proceedings 4th International Symposium on Cone Penetration Testing (CPT'18, Delft)*, London: Taylor & Francis Group, www.cpt18.org

Ouyang, Z. and Mayne, P. W. (2018b). Effective friction angle of clays and silts from piezocone. *Canadian Geotechnical Journal*, 55(9): 1230–1247.

Ouyang, Z. and Mayne, P. W. (2019). Modified NTH method for assessing effective friction angle of normally consolidated and overconsolidated clays from piezocone tests. *ASCE Journal of Geotechnical & Geoenvironmental Engineering*, 145(10): doi.org/10.1061/ (ASCE) GT.1943-5606.0002112

Ouyang, Z. and Mayne, P. W. 2018c. Effective stress strength parameters of clays from DMT. ASTM *Geotechnical Testing Journal*. 41 (5): 851–867.

Ouyang, Z. and Mayne, P. W. (2023). Evaluating friction angles for clays: piezocone tests versus Atterberg limits. *Geotechnical Engineering, Proc. Institution of Civil Engineers* 177(2), 147–157.

Ouyang, Z., Mayne, P. W. (2019). Evaluating rigidity index, OCR, and su from dilatometer data in soft to firm clays. *Proceedings, 16th Panamerican Conference on Soil Mechanics and Geotechnical Engineering*. Cancun: Mexican Geotechnical Society. www.issmge.org

Ouyang, Z. and Mayne, P. W. (2016). New DMT method for evaluating soil unit weight in soft to firm clays. *Geotechnical and Geophysical Site Characterisation*, Lehane, B., Acosta-Martínez, HE, Kelly, R., *Eds*, 785–790.

Ouyang, Z. and Mayne, P. W. (2017a). Spherical cavity expansion nexus between CPTu and DMT in soft-firm clays. *Proceedings of the 19th International Conference on Soil Mechanics and Geotechnical Engineering*, Vol. 1. Seoul: Korean Geotechnical Society, 631–634. www.issmge.org

Ouyang, Z. and Mayne, P. W. (2017b). Effective friction angle of soft to firm clays from flat dilatometer tests. *Geotechnical Engineering*, Vol. 170 (2). London: Proc. Institution of Civil Engineers, 137–147.

Ouyang, Z. and Mayne, P. W. (2020). Modified NTH solution for overconsolidated fissured clays. *Proc. Sixth International Conference on Geotechnical and Geophysical Site Characterization ISC*-6.isc6.org.

Ouyang, Z. and Mayne, P. W. (2021). Variable rate piezocone data evaluated using NTH limit plasticity solution. *ASTM Geotechnical Testing Journal*. 44. (1), 174–190. https://doi.org/10.1520/GTJ20190183.

Powell, J. J. M. and Quarterman, R. S. T. (1988). The interpretation of cone penetration tests in clays with particular reference to rate effects. *Penetration Testing*, Vol. 2 (Proc. ISOPT, Orlando). Rotterdam: Balkema, 903–910.

Randolph, M. F. (2004). Characterization of soft sediments for offshore applications. *Geotechnical & Geophysical Site Characterization*, Vol. 1 (Proc. ISC-2, Porto), Millpress, Rotterdam: 209–232.

Sandven, R. (1990). *Strength and Deformation Properties Obtained from Piezocone Tests*. PhD thesis, Trondheim: Norwegian University of Science & Technology, 342pp.

Schnaid, F., Odebrecht, E., Sosnoski, J., and Robertson, P. K. (2016). Effects of Test Procedure on Flat Dilatometer Test (DMT) Results in Intermediate Soils. *Can. Geotech. J.*, 53, 8, 1270–1280, https://doi.org/10.1139/cgj-2015-0463

Schofield, A. and Wroth, P. (1968). *Critical State Soil Mechanics*. London: McGraw-Hill.

Sellountou, E. A., Vipulanandan, C., and O'Neill, M. W. (2004). Estimation of drained shear strength parameters of Beaumont overconsolidated clay from piezocone data. *Proc., Center for Innovative Grouting Materials and Technology (CIGMAT)*. Houston: Dept. of Civil Engineering, Univ. Houston.

Senneset, K., Sandven, R., and Janbu, N. (1989). Evaluation of soil parameters from piezocone tests. *Transportation Research Record*, Vol. 1235. Washington DC: National Academies Press, 24–37.

Whittle, A. J., DeGroot, D. J., Ladd, C. C., and Seah, T-H. (1994). Model prediction of anisotropic behavior of Boston Blue Clay. *J. Geotech. Eng.* 120 (1): 199–224.

Chapter 13

Mechanical–statistical evaluation of soil properties

Changhong Wang, Kun Wang, Daofei Tang, and Tian Fang

This chapter delves into the mechanical-statistical evaluation of soil properties, emphasizing the utility of in-situ testing techniques like the Piezocone Penetration Test (CPTU) for assessing physical state and mechanical characteristics of soil layers. Despite fruitful applications of the empirical and regression formulas, their universality is constrained by regional soil property variations. The chapter underscores the significance of appropriate constitutive models and accurate geotechnical parameters in soil mechanics analysis, advocating for a more integrated approach leveraging CPTU data. The text outlines methods for evaluating cohesive and cohesionless soil properties using CPTU data, detailing the estimation of state indexes, strength indexes, and deformation indexes. Further, the chapter explores the application of constitutive models like the Modified Cam Clay (MCC) and Clay and Sand Model (CASM) within the cavity expansion theory to interpret CPTU data accurately. The integration of Bayesian-random field theory in geotechnical parameter evaluation is also discussed, highlighting its potential in accounting for natural variability and uncertainty in geotechnical media. The chapter presents stochastic mechanics methods, including Monte Carlo simulation and the Response Surface Method (RSM) to simulate the stochasticity and spatial variability of geotechnical parameters.

13.1 EVALUATION METHOD OF SOIL PROPERTIES BASED ON CPTU

In-situ testing techniques such as the Piezocone Penetration Test (CPTU) provide high-quality, reproducible data, and can directly evaluate the physical state and mechanical properties of soil layers, which is suitable for engineering exploration activities in deep underground space. The interpretation of CPTU data is a long-term important task. At present, CPTU data has achieved fruitful results in soil layer classification and engineering classification, soil physical and mechanical indexes, soil liquefaction discrimination, and other fields. However, these results are mainly based on empirical and regression formulas established from laboratory experiments and field data. Due to the difference in soil properties of the different areas, the universality of the empirical formula is low. Therefore, CPTU test data have not been fully applied in geotechnical engineering.

Proper constitutive models and accurate geotechnical parameters are the two supports of soil mechanics analysis. CPTU data based on the constitutive model can quickly and accurately obtain the geotechnical parameters required for scientific calculation.

DOI: 10.1201/9781003441946-13

13.1.1 Main mechanical characteristics of cohesive soil

CPTU data can be used to evaluate and estimate the major properties of soil, including state indexes, strength indexes, and deformation indexes. The natural weight-specific density and overconsolidation ratio (OCR) are the main state indexes, the undrained shear strength is the main strength index, and the compression modulus and undrained Young's modulus are the main deformation indexes. This section introduces the evaluation method of the major properties of clay based on the CPTU data.

13.1.1.1 Natural weight-specific density of soil

Based on the analysis of experimental data from Switzerland, Norway, and the United Kingdom, Larsson and Mulabdic[1] proposed a rough estimation of soil density by using net cone-tip resistance $(q_t - \sigma_{v0})$ and pore pressure parameter B_q.

Mayne[2] proposed that the natural weight-specific density of soil is related to shear wave velocity and depth, which is described as follows:

$$\gamma_t = 8.63\lg V_s - 1.18\lg z - 0.53, \tag{13.1}$$

where γ_t is the natural weight (kN/m^3), V_s is the shear wave velocity (m/s), and z is the depth (m). The statistical data of the formula include clay and incohesive soil, and the shear wave velocity measured by the seismic CPTU can be used to estimate the natural weight of the soil and determine the density of the soil.

13.1.1.2 Overconsolidation ratio (OCR)

In general, for the mechanically overconsolidated soil, the OCR is defined as the maximum effective consolidation stress of the soil layer in the history, while for the structural soil layer, OCR may represent the ratio of the yield stress to the current effective stress. The methods for estimating OCR from Cone Penetration Test (CPT) and CPTU data can be roughly divided into three categories: estimating OCR by undrained shear strength S_u, estimating OCR by the range of cone-tip resistance q_t, and estimating OCR directly from CPTU data.

For normally consolidated clays, the normalized cone-tip resistance varies with the plasticity index I_p and generally varies in the following ranges:

$$Q_t = \frac{q_t - \sigma_{v0}}{\sigma'_{v0}} = 2.5 \sim 5.0(\text{depends on } I_P). \tag{13.2}$$

If the soil layer is beyond the range of the above calculation, the soil layer may be overconsolidated.

13.1.1.3 Undrained shear strength

The undrained shear strength of soil (S_u) depends on the failure mode of the soil, the heterogeneity of the soil, and stress history. Hence, for the same soil, different geotechnical test methods will obtain different S_u values. At present, there are two kinds of methods to obtain undrained shear strength by using CPTU: theoretical method and empirical formula method.

13.1.1.4 Compression modulus

The compression modulus E_s of soil based on CPTU data can be expressed as a function of net cone-tip resistance.

For the overconsolidated clay, Senneset et al.[3][4] suggest a linear model:

$$E_{si} = \alpha_i q_n = \alpha_i (q_t - \sigma_{v0}), \tag{13.3}$$

where α_i is the coefficient, which varies from 5 to 15 for most clays.

For normally consolidated clay, a similar relationship exists:

$$E_s = \alpha_n q_n = \alpha_n (q_t - \sigma_{v0}), \tag{13.4}$$

According to Senneset et al.[4], the value α_n ranges from 4 to 8.

Kulhawy and Mayne[5] proposed a more general relationship:

$$E_s = 8.25(q_t - \sigma_{v0}). \tag{13.5}$$

13.1.1.5 Undrained Young's modulus

When using CPT or CPTU data to estimate the undrained Young's modulus E_u, the empirical formula is generally used to estimate indirectly:

$$E_u = nS_u, \tag{13.6}$$

where n is a constant, depending on factors such as shear stress level, overconsolidation ratio, and soil layer sensitivity.

13.1.2 Main mechanical characteristics of cohesionless soil

CPTU probes are usually inserted in the undrained clay and silt. When the CPTU is performed under the undrained conditions, the pore pressure generated near the probe will affect the measured values of cone-tip resistance q_c and sleeve friction f_s. The CPTU in coarse-grained soil such as sandy soil is penetrated under drainage conditions, and none of the excess pore pressure is generated during penetration, that is, only in-situ static pore pressure can be measured. Therefore, only the q_c and f_s can be used for the interpretation of the test data under the condition of complete drainage. However, due to the high compressive stress near the probe, pore pressure can be measured at the probe tip position u_1 [6]. In very dense sand or silt, negative pore pressure may be recorded at the probe shoulder position u_2 due to the soil dilatancy.

13.1.2.1 Relative compactness

Relative compactness is the main state characteristic of cohesionless soil and is usually used as an intermediate soil parameter. Relative compactness D_r or compactness index I_D is defined as:

$$I_D = D_r = \frac{e_{max} - e}{e_{max} - e_{min}}, \tag{13.7}$$

where e_{max} and e_{min} are the maximum and minimum pore ratios, respectively; and e is the in-situ porosity ratio.

The large-scale calibration tank test shows that the cone-tip resistance depends on the density of sand, in-situ vertical stress, horizontal effective stress, and compressibility of sand. According to the sand calibration tank test, different scholars put forward a series of prediction formulas D_r.

13.1.2.2 Effective internal friction angle of sand

There are many methods to evaluate the effective internal friction angle ϕ' of sand based on the cone-tip resistance of CPTU, which can be divided into empirical correlation, bearing capacity theory, and cavity expansion theory. For the pure sand, the effective stress friction angle ϕ' is directly evaluated when the drainage is complete, and the effective cohesion is assumed to be zero (i.e., $c' = 0$). To normalize the stress level of the cone-tip resistance q_t, the following formula[7] is proposed for estimation:

$$\phi' = 17.6° + 11\log q_{t1}, \tag{13.8}$$

where $q_{t1} = (q_t / \sigma_{atm}) / (\sigma'_{v0} / \sigma_{atm})^{0.5}$ is the normalized cone-tip resistance.

13.1.2.3 Young's modulus

The study on the test data of the calibration tank shows that the drained Young's modulus of sand mainly depends on the relative density, overconsolidation ratio, and the current stress level. For the secant Young's modulus (E'_s), when the mean axial strain over the stress history and aging range is 0.1%, the stiffness of the normally consolidated aged sand (i.e., >1000 years) falls between the newly normal consolidated sand and the over-consolidated soil[8].

13.2 CONSTITUTIVE MODEL AND CAVITY EXPANSION THEORY

The dynamic process of CPTU is accompanied by large deformation of the soil around the cone tip, which makes it very difficult to calculate the stress and displacement of the soil. The major approximate theories are bearing capacity theory, moving point dislocation method, pore expansion theory, and strain path method. The pore expansion theory not only considers the elastic deformation, plastic deformation, and initial stress state of soil, but also can describe the mechanical behavior of CPTU more accurately than other theoretical methods.

Cavity expansion theory can be divided into cylindrical cavity expansion and spherical cavity expansion methods. The similarity between cavity expansion theory and CPTU was first proposed by Bishop[9] et al. Gibson and Anderson[10] introduced the theory to geotechnical engineering. The study of cavity expansion theory needs to consider the theoretical knowledge of equilibrium equations, geometric equations, and constitutive models, among which the most important factor is the latter. In recent years, many excellent constitutive models have emerged, from the initial linear elastic constitutive model to the ideal elastic–plastic model, and then to various critical elastic–plastic constitutive models commonly used today, which can more accurately describe the complex stress state of soil. As classical critical state models, the Modified Cam Clay (MCC) model and Clay And Sand Model (CASM) are widely used to describe the mechanical properties of normal consolidated clay and sand, respectively.

13.2.1 Cavity expansion theory of the MCC model

Different from other laboratory tests, CPT does not directly measure the mechanical properties of soil and it should be interpreted. It is required to convert the force on the probe into geotechnical parameters. The penetration process of CPT can be considered as the radial compression of soil under the action of cone indentation, which is highly consistent with the cylindrical cavity expansion theory. Because of the succinct mechanical principle and simple solution form, the cavity expansion theory is widely used in geotechnical engineering, such as in-situ CPTU and pressuremeter tests. According to cylindrical cavity expansion theory, the mathematical relationship between CPTU data and constitutive parameters can be established.

13.2.1.1 Definition of the problem

A cylindrical cavity with an initial radius a_0 in the infinite soil mass is subjected to the radial stress σ_r, the tangential stress σ_θ, and the vertical stress σ_z. In the initial conditions, $\sigma_{r0} = \sigma_{\theta 0} = \sigma_h$, $\sigma_{z0} = \sigma_v$, where σ_h and σ_v are the in-situ horizontal stress and gravitational stress, respectively. When the inner diameter a_0 of the cylindrical cavity expands to a, the expansion pressure correspondingly increases from σ_{r0} to σ_a. Under the action of expansion pressure, the cavity wall firstly yields and enters the plastic region subsequently. As the inner diameter increases, the radius of the plastic area expands from R_{p0} to R_p. At that time, two regions appear around the cavity wall, the internal plastic region and the external elastic region. The schematic diagram of cylindrical cavity expansion is shown in Figure 13.1.

Cylindrical cavity expansion is a simplification of the axisymmetric problem in space, where the radial stress σ_r and tangential stress σ_θ are assumed to be only the functions

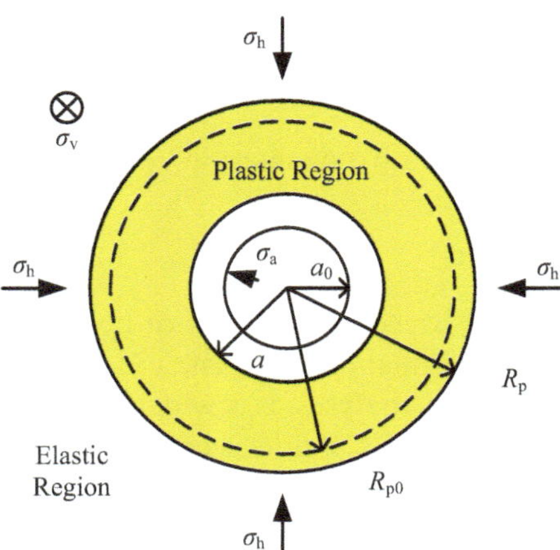

Figure 13.1 Schematic diagram of cylindrical cavity expansion.

of radius r, and the effects of shear stress $\tau_{r\theta}$ are ignored. In both elastic and plastic regions, the equilibrium differential equation scan be written as,

$$\frac{d\sigma_r}{dr}+\frac{\sigma_r-\sigma_\theta}{r}=0. \tag{13.9}$$

Alternatively,

$$\frac{d\sigma'_r}{dr}+\frac{du_w}{dr}+\frac{\sigma'_r-\sigma'_\theta}{r}=0, \tag{13.10}$$

where σ'_r and σ'_θ are the radial and tangential effective stress, respectively, and u_w is the pore pressure.

13.2.1.2 Elastic analysis

Cylindrical cavity expansion theory is subjected to the generalized Hooke's law in the elastic region. According to the elastic/plastic boundary conditions $r = R_p$ and $\sigma_r = \sigma_{rp}$, radial stress σ_r, tangential stress σ_θ, and vertical stress σ_z can be easily obtained in the elastic region, which can be written as[11],

$$\begin{aligned} \sigma_r &= \sigma_h + (\sigma_{rp}-\sigma_h)\left(\frac{R_p}{r}\right)^2 \\ \sigma_\theta &= \sigma_h - (\sigma_{rp}-\sigma_h)\left(\frac{R_p}{r}\right)^2, \\ \sigma_z &= \sigma_v \\ u_r &= \frac{\sigma_{rp}-\sigma_h}{2G_0}\frac{R_p^2}{r} \end{aligned} \tag{13.11}$$

where σ_{rp} is the total radial stress at the boundary of the elastic/plastic region. G_0 is the initial shear modulus. Shear modulus G can be calculated by the mean effective stress p', the specific volume υ, Poisson's ratio v, and the slope of the loading–reloading line κ in the $\upsilon-\ln p'$ plane, which is derived from equation[12] $G=\frac{3(1-2v)}{2(1+v)}\frac{\upsilon p'}{\kappa}$.

13.2.1.3 Elastoplastic analysis

The MCC model is established based on the tests of normal and light overconsolidated clay. The MCC model can explain the elastoplastic deformation characteristics of soil from experimental and theoretical analyses. It is widely used in scientific studies and engineering practices. In addition, the MCC model is easy to extend and a series of representative constitutive models are developed based on it, such as the CASM[13] and the Uniform Hardening (UH) model[14]. For sand–clay mixed soil, when the sand content is lower than 70% ~ 80%, it is considered that the soil still shows the mechanical characteristics of clay[15]. In the layer ⑧$_2$ of silty clay with sand, the sand is mainly silty sand and the content is less than 8%. Thus, the MCC model is adopted in Shanghai, China to perform the elastoplastic analysis in this study. The yield function of the MCC model can be written as,

$$f = \frac{\lambda - \kappa}{1 + e_0} \ln \frac{p'}{p'_0} + \frac{\lambda - \kappa}{1 + e_0} \ln \left(1 + \frac{q^2}{M^2 p'^2} \right) - \varepsilon_{\mathrm{v}}^{\mathrm{p}} = 0, \tag{13.12}$$

where λ is the slope of normal compression in the $\upsilon - \ln p'$ plane, M is the slope of the critical state line, and e_0 is the initial void ratio. In the MCC model, the hardening parameter is a function of the plastic volumetric strain $\varepsilon_{\mathrm{v}}^{\mathrm{p}}$.

Deviatoric stress q and mean effective stress p' are defined as[16],

$$\begin{aligned} p' &= \frac{1}{3}\left(\sigma'_r + \sigma'_\theta + \sigma'_z\right) \\ q &= \sqrt{\frac{1}{2}\left[\left(\sigma'_r - \sigma'_\theta\right)^2 + \left(\sigma'_\theta - \sigma'_z\right)^2 + \left(\sigma'_z - \sigma'_r\right)^2\right]} \end{aligned}. \tag{13.13}$$

Drainage can affect the volumetric strain, which in turn affects the excess pore water pressure and the effective stress components around the cylindrical cavity. For the two cases of undrained and drained conditions, Chen and Abousleiman[16][17] adopted the large deformation theory and the associated flow rule to derive the elastoplastic solution for cylindrical cavity expansion in the plastic region.

For the undrained condition, the elastoplastic solution is expressed as[16],

$$\begin{aligned} \frac{\mathrm{d}\sigma'_r}{\mathrm{d}r} &= \frac{a_{11} - a_{12}}{\Delta r} \\ \frac{\mathrm{d}\sigma'_\theta}{\mathrm{d}r} &= \frac{a_{21} - a_{22}}{\Delta r} \\ \frac{\mathrm{d}\sigma'_z}{\mathrm{d}r} &= \frac{a_{31} - a_{32}}{\Delta r} \end{aligned}. \tag{13.14}$$

Moreover, the pore water pressure u_a at the cavity wall is expressed as

$$u_a = \sigma'_{r\mathrm{p}}(r) - \sigma'_a - \int_{R_\mathrm{p}}^{a} \frac{\sigma'_r(r) - \sigma'_\theta(r)}{r} \mathrm{d}r + u_0, \tag{13.15}$$

where σ'_a is the radical effective stress at the cavity wall.

For the drained condition, an auxiliary independent variable ξ is denoted as the ratio of the particle displacement u_r in the radial direction to its present radius r, which is written as $\xi = \dfrac{u_r}{r}$.

The elastoplastic solution is expressed as[17]

$$N = 1 - \xi - \frac{\upsilon_0}{\upsilon(1-\xi)}$$

$$\frac{d\sigma'_r}{d\xi} = -\frac{\sigma'_r - \sigma'_\theta}{N}$$

$$\frac{d\sigma'_\theta}{d\xi} = -\frac{b_{21}}{b_{11}}\left[\frac{\sigma'_r - \sigma'_\theta}{N} + \frac{b_{11} - b_{12}}{\Delta(1-\xi)}\right] - \frac{b_{22} - b_{21}}{\Delta(1-\xi)} \tag{13.15}$$

$$\frac{d\sigma'_z}{d\xi} = -\frac{b_{31}}{b_{11}}\left[\frac{\sigma'_r - \sigma'_\theta}{N} + \frac{b_{11} - b_{12}}{\Delta(1-\xi)}\right] - \frac{b_{32} - b_{31}}{\Delta(1-\xi)}$$

$$\frac{d\upsilon}{d\xi} = -\frac{\upsilon\Delta}{b_{11}}\left[\frac{\sigma'_r - \sigma'_\theta}{N} + \frac{b_{11} - b_{12}}{\Delta(1-\xi)}\right]$$

where

$$a_r = \frac{p'(M^2 - \eta^2)}{3} + 3(\sigma'_r - p')$$

$$a_\theta = \frac{p'(M^2 - \eta^2)}{3} + 3(\sigma'_\theta - p')$$

$$a_z = \frac{p'(M^2 - \eta^2)}{3} + 3(\sigma'_z - p')$$

$$y = \frac{\lambda - \kappa}{\upsilon p'^3(M^4 - \eta^4)}$$

$$b_{11} = \frac{1}{E^2}(1 - v^2 + Ea_\theta^2 y + 2Eva_\theta a_z y + Ea^2{}_z y)$$

$$b_{22} = \frac{1}{E^2}[1 - v^2 + Ea_r^2 y + 2Eva_r a_z y + Ea_z^2 y] \tag{13.16}$$

$$b_{33} = \frac{1}{E^2}[1 - v^2 + Ea_r^2 y + 2Eva_\theta a_r y + Ea_\theta^2 y]$$

$$b_{12} = b_{21} = \frac{1}{E^2}[-Ea_r(a_\theta + va_z) + v(1 + v - Eva_\theta a_z y + Ea^2{}_z y)]$$

$$b_{13} = b_{31} = \frac{1}{E^2}[-Ea_r(va_\theta + a_z)y + v(1 + v + Eva^2{}_\theta y - Ea_\theta a_z y)]$$

$$b_{23} = b_{32} = \frac{1}{E^2}[v + v^2 + Eva_r^2 y - Ea_\theta a_z y + Eva_r(a_\theta + a_z)y]$$

$$\Delta = -\frac{1+v}{E^3}\begin{bmatrix}\left(-1 + v + 2v^2\right) + E\left(-1+v\right)a_r^2 y + E\left(-1+v\right)a_\theta^2 y \\ -2Eva_\theta a_z y - Ea_z^2 y + Eva_z^2 y - 2Eva_r\left(a_\theta + a_z\right)y\end{bmatrix}$$

At the elastic/plastic boundary, the effective stress components can be written as

$$q(r_p)=Mp'_0\sqrt{OCR\left[1+\left(\frac{q_0}{M+p'_0}\right)^2\right]-1}$$

$$\sigma'_r(r_p)=\sigma'_{r0}+\sqrt{\sigma'^2_{r0}-\frac{1}{3}(4\sigma'^2_{r0}+\sigma'^2_{z0}-2\sigma'_{r0}\sigma'_{z0}-q(r_p)^2)}$$

$$\sigma'_\theta(r_p)=\sigma'_{r0}-\sqrt{\sigma'^2_{r0}-\frac{1}{3}(4\sigma'^2_{r0}+\sigma'^2_{z0}-2\sigma'_{r0}\sigma'_{z0}-q(r_p)^2)} \tag{13.17}$$

$$\sigma'_z(r_p)=\sigma'_{z0}=\frac{3}{1+2K_0}p'_0$$

where q is the initial deviatoric stress, and p'_0 is the initial mean effective stress.

Δ and b_{ij} are the elastoplastic matrix coefficients, which are all explicit functions of the three effective stress components σ'_r, σ'_θ, σ'_z and model parameters of λ,κ,M,E, v, and υ. q_p is the deviatoric stress on the initial yield locus, which can be calculated by q_0, p'_0, M, and OCR. Please refer the Chen and Abousleiman[16][17] for the detailed derivation process.

For the MCC model, the constitutive parameters mainly include the OCR, the slope of the loading–reloading line κ in the $\upsilon-\ln p'$ plane, the slope of normal compression λ in the $\upsilon-\ln p'$ plane, the slope of the critical state line M, the Poisson ratio v, and the initial void ratio e_0. Taking the constitutive parameters into Equation (13.17), the effective stress components at the elastic/plastic boundary can be obtained. The elastic–plastic boundary displacement and expanded cavity radius a/a_0 are taken as the lower and upper limits of the integral. The fourth-order Runge–Kutta method is used to solve a set of first-order simultaneous differential equations of Equation (13.13) or (13.15) in the plastic zone under undrained and drained conditions. By numerical calculation, the stress distribution and expansion stress in the plastic zone can be obtained.

13.2.1.4 CPTU mechanical model

In recent years, due to the developments of multifunctional sensors, CPT technologies have been further updated. However, single/double-bridged probes and CPTU are the most commonly used CPT specifications in the geotechnical engineering practice. Compared with CPTU, the double-bridged probe can only obtain the tip resistance q_c and side friction f_s, but it cannot obtain the excess pore water pressure u. Double-bridged probes and CPTU are widely used in the shallow soil layers. However, for the deep soil layers, which are located in a high-stress state, the probe faces cost difficulties in penetration and extraction. As well as it will produce a large excess pore water pressure in the clay layers. Excess pore water pressure will exceed the measuring range, making it difficult to obtain the full data. According to the above reasons, double-bridged probes and CPTU are limited in deep geotechnical investigation. The single-bridged probe is a CPT specification unique to China, which can obtain the total stress during penetration. It has been widely used in China[18][19][20], and a large number of empirical formulas have been summarized in engineering practice. In addition, a single-bridged probe is cheaper and simpler, which plays an important role in in-situ tests. Hence, the study of the single-bridged probe test mechanism has a practical significance. Moreover, specific penetration resistance p_s, cone resistance q_c, and side friction resistance f_s share a clear statistical law[21].

A single-bridge probe can only measure the total resistance F during penetration. Assuming that A is the cone cross-sectional area, the specific penetration resistance p_s is defined as

$$p_s = \frac{F}{A}. \tag{13.18}$$

The mechanical behavior of CPT is similar to the cylindrical cavity expansion process. Specific penetration resistance p_s has a particular relation to the stress components at the cavity wall. In the process of penetration, the stresses applied to the probe consist of cone resistance p_c, cylindrical friction f_s, and pore water pressure u_w as shown in Figure 13.2.

According to the principle of effective stress, the total stress of saturated soil consists of effective stress and pore pressure. Different from the effective stress, pore pressure cannot resist shear stress. Research[16][17] indicates that when the expansion radius a reaches 3–5 times of the initial radius a_0, the stress at the cavity wall reaches a limit stress state. The penetration process can be considered as a cylindrical cavity expansion from the initial inner diameter of nearly zero to the probe radius R. In that scenario, the soil around the probe approaches the limit stress state, i.e., $\sigma'_{rc} = \sigma'_{rs} = \sigma'_{rlim}$. Thus, the probe cylindrical friction is defined as

$$f_s = \mu\sigma'_{rlim}, \tag{13.19}$$

where μ is the friction coefficient between soil and steel probe.

The interface between the probe and the soil is a conical surface, which is different from the cylindrical cavity expansion. It assumes that there is a rigid slide block that can only move horizontally and the steel probe indirectly interacts with the surrounding soil as shown in Figure 13.2, Δ_{LMN} is the slide block of soil. As an axisymmetric problem,

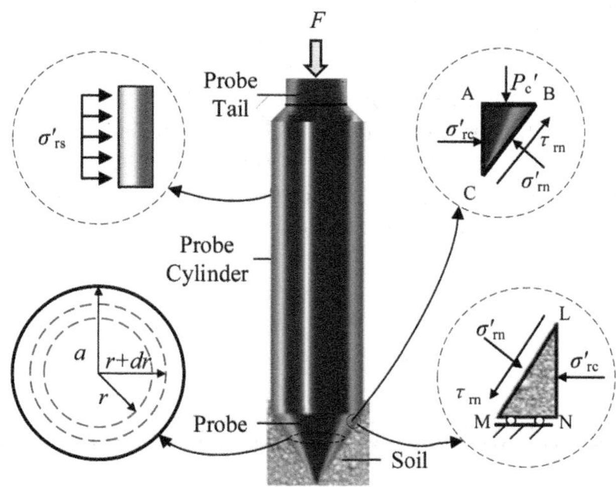

Figure 13.2 Schema of CPT mechanical model.

the forces on the slide block are decomposed vertically and horizontally. Firstly, the stress equilibrium is established on the horizontal plane for slide blocks, which is written as

$$\sigma'_{\rm rc} l_{\rm LN} = \left(\sigma'_{\rm rn}\cos\alpha - \tau_{\rm rn}\sin\alpha\right) l_{\rm LM}, \tag{13.20}$$

where $\tau_{\rm rn}$ and $\sigma'_{\rm rn}$ are the shear and the normal stress of the probe–soil interface, respectively. Shear stress $\tau_{\rm rn}$ can be calculated by $\tau_{\rm rn} = \mu\sigma'_{\rm rn}$. α is the half of cone angle that constantly equals 30°.

According to Equation (13.20), the relationship between expansion stress $\sigma'_{\rm rc}$ and normal stress $\sigma'_{\rm rn}$ of the soil–probe interface can be obtained by

$$\sigma'_{\rm rn} = \frac{l_{\rm LN}}{\left(\cos\alpha - \mu\sin\alpha\right) l_{\rm LM}}\sigma'_{\rm rc}. \tag{13.21}$$

Correspondingly, the stress equilibrium is established on the vertical plane $\Delta_{\rm ABC}$ for the cone probe, which is written as

$$p_{\rm c} l_{\rm AB} = \left(\sigma'_{\rm rn}\sin\alpha + \tau_{\rm rn}\cos\alpha\right) l_{\rm BC}. \tag{13.22}$$

According to Equation (13.22), the relationship between cone resistance $p_{\rm c}$ and normal stress $\sigma'_{\rm rn}$ of the soil–probe interface can be obtained by

$$p_{\rm c} = \frac{l_{\rm BC}}{l_{\rm AB}}\left(\sin\alpha + \mu\cos\alpha\right)\sigma'_{\rm rn}, \tag{13.23}$$

where $l_{\rm BC} = l_{\rm LM}$, substituting Equation (13.21) into Equation (13.23) can get the cone resistance,

$$p_{\rm c} = \frac{1+\sqrt{3}\mu}{1-\sqrt{3}\mu/3}\sigma'_{\rm rc}. \tag{13.24}$$

The drained condition means that the penetration process does not produce excess pore water pressure and the pore pressure $u_{\rm a}$ at the cavity wall always equals the initial pore pressure u_0. The undrained condition means that the penetration process produces excess pore pressure $u_{\rm a}$ without dissipation, which is calculated by Equation (13.16). Under these two drainage conditions, assume that the positions of the probe cone, probe cylinder, and probe tail have equal pore water pressure. The influence of pore water pressure on specific penetration resistance $p_{\rm s}$ is mainly related to the cross-sectional area $A_{\rm t}$ of the probe tail. Hence, the static equilibrium formula along the axis of the probe can be written as

$$p_{\rm s}A = p_{\rm c}A + f_{\rm s}A_{\rm s} + A_{\rm t}u_{\rm a}. \tag{13.25}$$

where $A_{\rm s}$ is the surface area of the probe cylinder.

Taking Equation (13.19) and Equation (13.24) into Equation (13.25), the formula of specific penetration resistance $p_{\rm s}$ can be obtained by

$$p_{\rm s} = \left(\frac{1+\sqrt{3}\mu}{1-\sqrt{3}\mu/3} + \mu\frac{A_{\rm s}}{A}\right)\sigma'_{\rm rlim} + \frac{A_{\rm t}}{A}u_{\rm a}. \tag{13.26}$$

In Equation (13.26), the area of A_t, A_s, and A is determined by the CPT probe size. The friction coefficient μ is determined by back-analysis using the parameters of the drilling test near the CPT curve at a specific depth. Assuming that expansion stress at the cavity wall reaches a limit stress state at an expanded cavity radius of $a/a_0 = 10$. According to Equation (13.13)–(13.15), the excess pore pressure u_a and limit expansion stress σ'_{rlim} can be obtained. To sum up, the specific penetration resistance p_s is calculated.

13.2.2 Cavity expansion theory of the CASM model

Cavity expansion theory can simulate the changes in stress, displacement, and pore water pressure caused by the expansion and contraction of a cylindrical or spherical cavity. The expansion principle is a dynamic process under the internal pressure of a small cavity close to radius zero, which is highly consistent with the mechanism of CPTU. Hence, it is particularly suitable to interpret the in-situ test data, such as cone-tip resistance, q_t, sleeve friction, f_s, and pore water pressure, u. Derivations of spherical cavity expansion theory and mechanical transformation equations for CPTU are provided as follows.

13.2.2.1 Constitutive mode

Roscoe et al. proposed the Original Cam Clay (OCC) model based on London Clay, which successfully described the characteristics of normally consolidated clay as elastic–plastic geomaterial[22]. To predict the mechanical properties of clayey soil under specific loads, the OCC model was modified by many scholars, for example, MCC is a well-known improved model for lightly overconsolidated soil.

Yu proposed the CASM that can uniformly consider the mechanical behaviors of clay and sand[23]. Compared with the OCC and MCC models, CASM adopts the non-associated flow law and introduces two new constitutive model parameters, spacing ratio r^* and stress-state coefficient n. As shown in Figure 13.3, r^* is the ratio of p'_c to p'_x, which is used to define the shape of the yield surface, where p'_c is the preconsolidation stress at the isotropic normal consolidation line (NCL) and p'_x is the intersection point of the swell

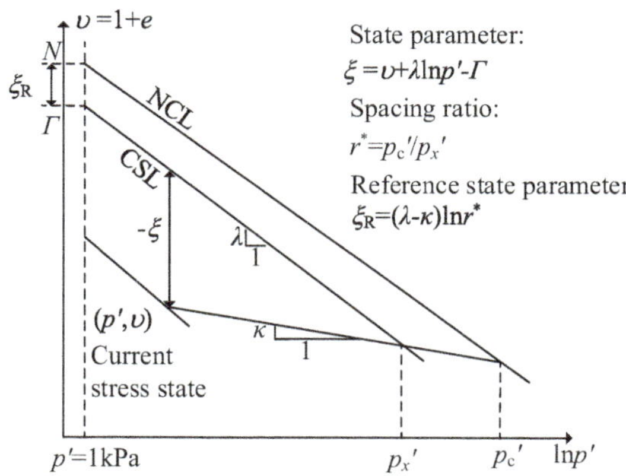

Figure 13.3 Definitions of state parameter, space ratio, and reference state parameter.

line and the critical state line (CSL). n defines the size of the yield surface. r^* is a fixed value for the OCC or MCC models, specifically. It varies with the change of the stress in CASM.

Parameter definitions of CASM are shown in Figure 13.3. υ is the specific volume, $\upsilon = 1 + e$, e is the void ratio. κ is the slope of the swell line, and λ is the slope of CSL in $\upsilon - \ln p'$ space. The state parameter ξ is the difference between the specific volume υ and the counterpart at the CSL under the same effective stress p'. The reference state parameter ξ_R is the difference between the specific volumes at the CSL and the NCL, which describes the loosest state that the soil can reach. The concept of state parameter ξ is proposed by Been and Jefferies based on the critical state soil mechanics, it can describe the sandy soils by combining the effects of void ratio and confining pressure[24]. The yield function can be defined by the state parameter as

$$f(p', q, \xi) = \left(\frac{\eta}{M}\right)^n - 1 + \frac{\xi}{\xi_R} = 0, \tag{13.27}$$

where $\eta = q/p'$ is the stress ratio. M is the critical stress ratio, which is related to the internal friction angle. For the OCC model, $n = 1$, $r^* = 2.718$. For the MCC model, $n = 1.5 \sim 2$, $r^* = 2$. To compare with the MCC model, the initial values of r^* and n adopted in CASM are $n = 1.667$, $r^* = 2$. The plastic potential function can be defined as

$$g(p', q, \beta) = 3M \ln\frac{p'}{\beta} + (3{+}2M)\ln\left(\frac{2q}{p'} + 3\right) - (3 - M)\ln\left(3 - \frac{q}{p'}\right) = 0, \tag{13.28}$$

where the parameter β reflects the size of the plastic potential surface, which can be determined easily for any given stress state (p', q) by solving Equation (13.28). Yield surfaces of the CASM, OCC, and MCC models in the normalized p'-q space are shown in Figure 13.4.

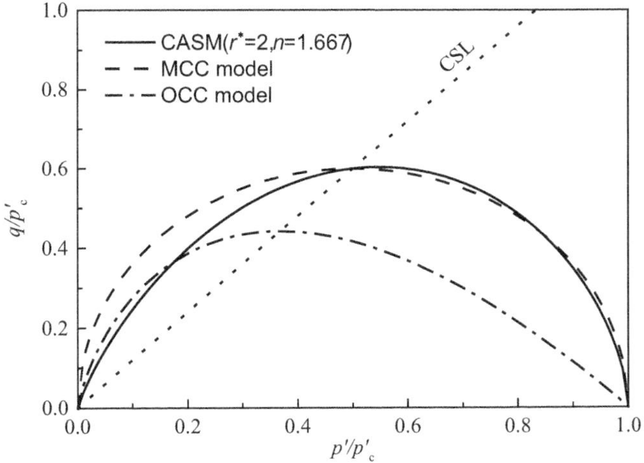

Figure 13.4 Yield surfaces of the CASM, OCC, and MCC models.

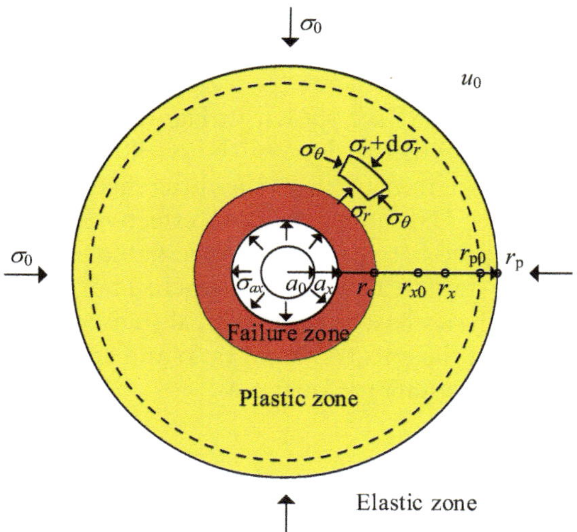

Figure 13.5 Schematic diagram of the spherical cavity expansion.

13.2.2.2 Spherical cavity expansion theory

As shown in Figure 13.5, a schematic process of spherical cavity expansion, which has an initial radius a_0, initial stress σ_0, initial pore water pressure u_0, and initial mean total stress p_0, is illustrated. Positive values should be given for the compressive stress and strain. As the internal cavity pressure increases, the cavity radius will gradually grow. When the pressure of the cavity wall reaches a threshold value, the surrounding soil begins to deform plastically. If the internal pressure continuously increases to a limit value, a failure zone begins to appear. Mean total stress is p, deviatoric stress is q, and the pore water pressure is u at any specific position r_x.

As the internal cavity pressure increases from σ_0 to σ_{ax}, the cavity wall radius gradually expands from a_0 to a_x. While the internal cavity pressure continues to increase, a failure zone is formed in a range of $a_x < r < r_c$, and plastic deformation is occurring in $r_c < r < r_p$, where r_c is the radius of the interface between the failure zone and the plastic zone, r_p is the position of the elastic–plastic boundary, with which the initial position is r_{p0}. Any specific point r_x located in the plastic zone has an initial position of r_{x0}. The margin area outside the plastic zone remains in the elastic state. Due to the spherical symmetry, tangential stress σ_θ is equal to the circumferential stress σ_ϕ. The total mean principal stress and deviatoric stress are represented by p and q, respectively.

$$p = \frac{\sigma_r + 2\sigma_\theta}{3}, \tag{13.29}$$

$$q = \sigma_r - \sigma_\theta, \tag{13.30}$$

where σ_r and σ_θ are the total radial and tangential stresses. The initial radial stress and initial tangential stress $\sigma_{r0} = \sigma_{\theta 0} = \sigma_0 + u_0$, the initial mean stress $p_0 = \sigma_0 + u_0$. In the elastic and plastic zone, the radial and tangential stresses at the position r from the spherical expansion center satisfy the equilibrium differential equation,

$$\frac{\partial \sigma_r}{\partial r} + 2\frac{\sigma_r - \sigma_\theta}{r} = 0, \tag{13.31}$$

or expressed in the form of effective stress

$$\frac{\partial \sigma'_r}{\partial r} + \frac{\partial u}{\partial r} + 2\frac{\sigma'_r - \sigma'_\theta}{r} = 0. \tag{13.32}$$

13.2.2.3 Elastic zone analysis

The small deformation theory is adopted in the elastic region, then the radial strain increment $d\varepsilon_r$ and tangential strain increment $d\varepsilon_\theta$ at the radius r $(r > r_p)$ can be expressed as follows:

$$d\varepsilon_r = -\frac{\partial du_r}{\partial r},\ d\varepsilon_\theta = -\frac{du_r}{r}, \tag{13.33}$$

where du_r is radial displacement increment.

Stress–strain relationship in the elastic zone can be obtained by Hooke's law:

$$\begin{bmatrix} d\varepsilon_r \\ d\varepsilon_\theta \end{bmatrix} = \frac{1}{E}\begin{bmatrix} 1 & -2v \\ -v & 1-v \end{bmatrix}\begin{bmatrix} d\sigma'_r \\ d\sigma'_\theta \end{bmatrix}, \tag{13.34}$$

where v is the Poisson's ratio, E is the Young's modulus, and $d\sigma'_r$ and $d\sigma'_\theta$ are the effective radial and tangential stress increments. Combining equations (13.31), (13.33), and (13.34), the total stresses, radial displacement, and the pore water pressure in the elastic region can be expressed as[17]

$$\begin{cases} \sigma_r = \sigma_0 + (\sigma_{rp} - \sigma_0)(\frac{r_p}{r})^3 \\ \sigma_\theta = \sigma_0 - \frac{1}{2}(\sigma_{rp} - \sigma_0)(\frac{r_p}{r})^3 \\ u_r = \frac{\sigma_{rp} - \sigma_0}{4G_0}(\frac{r_p}{r})^3 r \\ u = u_0 \end{cases}, \tag{13.35}$$

where u_r is the radial displacement at r and σ_{rp} denotes the total radial stress at the elastic–plastic boundary.

13.2.2.4 Plastic zone analysis

Non-associated flow rule is adopted in CASM, and the increment of plastic strain can be expressed as

$$d\varepsilon_r^p = \Lambda\frac{\partial g}{\partial \sigma'_r} = \Lambda\left(\frac{\partial g}{\partial p'}\frac{\partial p'}{\partial \sigma'_r} + \frac{\partial g}{\partial q}\frac{\partial q}{\partial \sigma'_r}\right), \tag{13.36}$$

$$\mathrm{d}\varepsilon_{\theta}^{\mathrm{p}} = \Lambda \frac{\partial g}{\partial \sigma_{\theta}'} = \Lambda \left(\frac{\partial g}{\partial p'} \frac{\partial p'}{\partial \sigma_{\theta}'} + \frac{\partial g}{\partial q} \frac{\partial q}{\partial \sigma_{\theta}'} \right), \tag{13.37}$$

where Λ is a scalar multiplier. Due to the spherical symmetry, $\mathrm{d}\varepsilon_{\phi}^{\mathrm{p}}$ is known to be equal to $\mathrm{d}\varepsilon_{\theta}^{\mathrm{p}}$. The plastic volumetric strain $\mathrm{d}\varepsilon_{\mathrm{v}}^{\mathrm{p}}$ can be expressed as

$$\mathrm{d}\varepsilon_{\mathrm{v}}^{\mathrm{p}} = \mathrm{d}\varepsilon_{r}^{\mathrm{p}} + 2\mathrm{d}\varepsilon_{\theta}^{\mathrm{p}} . \tag{13.38}$$

It can also be expressed by Yu[25]

$$\mathrm{d}\varepsilon_{\mathrm{v}}^{\mathrm{p}} = \frac{\partial g / \partial p}{H} \times \left[\left(\frac{\partial f}{\partial p'} + \frac{\partial f}{\partial \xi} \frac{\partial \xi}{\partial p'} \right) \mathrm{d}p' + \frac{\partial f}{\partial q} \mathrm{d}q \right], \tag{13.39}$$

where H denotes the plastic hardening modulus. The elastic–plastic stiffness matrix is obtained by combining equations (13.28), (13.36)–(13.39) and Hooke's law:

$$\begin{bmatrix} \mathrm{d}\sigma_r' \\ \mathrm{d}\sigma_\theta' \end{bmatrix} = \frac{1}{\Delta} \begin{bmatrix} a_{11} & a_{12} \\ a_{21} & a_{22} \end{bmatrix} \cdot \begin{bmatrix} \mathrm{d}\varepsilon_r \\ \mathrm{d}\varepsilon_\theta \end{bmatrix}, \tag{13.40}$$

where a_{11}, a_{12}, a_{21}, and a_{22} are the matrix coefficients. Soil deformation is subjected to the large deformation theory, thus logarithmic strain is adopted in the plastic zone. The relationship between strain and displacement is

$$\mathrm{d}\varepsilon_r = -\frac{\partial \mathrm{d}r}{\partial r}, \mathrm{d}\varepsilon_\theta = -\frac{\mathrm{d}r}{r}, \tag{13.41}$$

where $\mathrm{d}r$ is the infinitesimal deformation at r. the volume change $\mathrm{d}\varepsilon_v$ is zero under the undrained conditions, hence the radial strain increment can be given

$$\mathrm{d}\varepsilon_r = -2\mathrm{d}\varepsilon_\theta = 2\frac{\mathrm{d}r}{r} . \tag{13.42}$$

Substitute Equation (13.42) into Equation (13.40), the first-order ordinary differential equations are solved to establish the relationships among the effective radial, tangential stresses, and the radius of spherical cavity expansion in the plastic zone,

$$\frac{\mathrm{d}\sigma_r'}{\mathrm{d}r} - \frac{2 \times a_{11} - a_{12}}{\Delta \cdot r} = 0, \tag{13.43}$$

$$\frac{\mathrm{d}\sigma_\theta'}{\mathrm{d}r} - \frac{2 \times a_{21} - a_{22}}{\Delta \cdot r} = 0 . \tag{13.44}$$

Through the integral of Equation (13.32), the pore water pressure at the r_x can be obtained,

$$u(r_x) = -2 \int_{r_{\mathrm{p}}}^{r_x} \frac{\sigma_r' - \sigma_\theta'}{r} \mathrm{d}r + \sigma_{r\mathrm{p}}' - \sigma_r(r_x) + u_0 . \tag{13.45}$$

The stress at any specific point r_x can be calculated by Equations (13.43)–(13.45) in the plastic zone, but the precondition is the radius of the plastic zone r_{p} and the initial stress at the r_x. This question will be discussed below.

13.2.2.5 Elastic–plastic boundary analysis

According to Equations (13.29) and (13.35), the effective radial stress and the effective tangential stress at the elastic–plastic boundary are as follows,

$$\sigma'_{rp} = \sigma'_0 + \frac{2}{3}q_p, \sigma'_{\theta p} = \sigma'_0 - \frac{1}{3}q_p, \tag{13.46}$$

where q_p denotes the deviatoric stress at the elastic–plastic boundary, and $p' = p'_0$. Hence, the deviatoric stress at the elastic–plastic boundary can be defined as

$$q_p = Mp'_0 \sqrt[n]{\frac{\ln OCR}{\ln r^*}}, \tag{13.47}$$

where OCR is the overconsolidation ratio. The plastic zone radius r_p can be obtained by the volume invariable,

$$\frac{r_p}{a} = \left[\left(\left(\frac{a_0}{a} \right)^3 - 1 \right) \Big/ \left(\left(1 - \frac{\sigma'_{rp} - \sigma'_0}{4G_0} \right)^3 - 1 \right) \right]^{-\frac{1}{3}}. \tag{13.48}$$

13.2.2.6 Mechanical transformation model

Pseudo-static equilibrium model between the corrected cone-tip resistance q_t, the limit expansion stress σ_L, and the tangential stress σ_θ is shown in Figure 13.6. While the penetrometer indents into the soil, the radius of the spherical cavity expands from a_0 to a, the soil within the range of r_c is destroyed and reaches to the critical state. Hence, the limit expansion stress σ_L and the shear stress τ_n act on the 1/4 spherical surface BC with the radius r_c, and the tangential stress σ_θ acts on both sides of the cone shoulder. As an axisymmetric problem, the forces on BC are decomposed vertically and horizontally, and the stress equilibrium is established on the vertical plane.

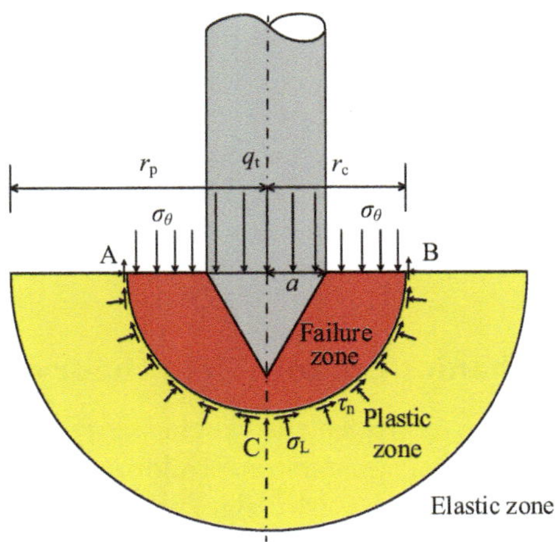

Figure 13.6 Pseudo-static equilibrium model of cone-tip resistance.

$$q_t \cdot a + \sigma_\theta \cdot (r_c - a) = \int_0^{\frac{\pi}{2}} (\sigma_L \sin\theta + \tau_n \cos\theta) r_c d\theta, \tag{13.49}$$

$$q_t = \frac{(\sigma_L + \tau_n) \cdot r_c - \sigma_\theta (r_c - a)}{a}, \tag{13.50}$$

where $\sigma_L = \sigma_a' + u_c$, u_c is the pore water pressure at the radius r_c. $\tau_n = c + \sigma_L' \tan\varphi'$, c is the cohesion which is much smaller than the limit expansion stress, hence it is assumed to be zero. The relationship between the corrected cone-tip resistance q_t and the uncorrected cone-tip resistance q_c is

$$q_t = q_c + (1 - \beta)u \tag{13.51}$$

where β denotes the cone area ratio, which is assumed to be 0.87[26]. Pore water pressure u can be directly determined by the measurement u_a at the radius a,

$$u = u_a \tag{13.52}$$

Cone penetration is regarded as a fast dynamic process in clay; hence, the pore water pressure does not dissipate immediately.

Unknown r_c is the key to derive corrected cone-tip resistance q_t. There are no specific methods to solve r_c. Therefore, the numerical solution of r_c obtained by MATLAB codes may be feasible. As explained above, the effective stress within the radius r_c does not change in the plastic zone. The algorithm can find out the critical point r_c, with which the effective stress alters from constancy to variation.

13.3 BAYESIAN-RANDOM FIELD THEORY

Due to the natural variability of the geotechnical medium, the stochastic nature of geotechnical parameters must be evaluated. Static penetration analysis involving geotechnical parameters, the stochastic description of the geotechnical parameters, and the stochastic simulation of cylindrical cavity expansion theory are anointed stochastic mechanics methods.

Due to the uncertainty of geotechnical parameters, limited site-investigation data and unknown disturbances have a great influence on the acquisition of geotechnical parameters. Therefore, the stochastic mechanics based Bayesian method is an effective tool that helps engineers make a reasonable decision, as it can integrate the prior information with in-situ test data to decrease the uncertainties.

13.3.1 Stochastic mechanics/random field theory

Stochastic mechanics needs to consider the stochasticity of geotechnical parameters, which defines the geotechnical parameters as random variables. Stochastic simulation methods mainly include the Monte Carlo algorithm and Response Surface Method (RSM). In addition, in the study of the statistical characteristics (i.e., mean value and standard deviation) of small samples, Bayesian theory is used to study the stochasticity of geotechnical parameters and the uncertainties of geotechnical parameters are analyzed based on in-situ test data.

13.3.1.1 Monte Carlo simulation

Monte Carlo method is based on probability theory and mathematical statistics, and generally uses the probability density function (PDF) to achieve sampling of random variables. It is necessary to determine the probability distributions of the random variables in advance, such as uniform distribution, normal distribution, and lognormal distribution. The Monte Carlo method randomly takes samples from the prior distributions to simulate the real variables[27][28].

13.3.1.2 Response surface method

For the complicated geotechnical mechanics, it is very hard to directly use the Monte Carlo method for numerical calculation. Therefore, Box and Wilson[29] put forward RSM to determine objective function, which uses Central Composite Design (CCD) to approximate the real state surface. In this research, a quadratic polynomial without cross-terms is established between CPT data and geotechnical parameters instead of numerical analysis. The functional relationship between RSM $g(\psi)$ and the target objective $\mathbf{z}_{\mathrm{b}}$ (i.e., Type B data) is

$$\mathbf{z}_{\mathrm{b}} = g(\psi) + \varepsilon, \tag{13.53}$$

where ψ denote the key geotechnical parameter vector, ε are the model errors.

13.3.1.3 Multivariate spatial random fields

The spatial variability of a geotechnical parameter is simplified as aleatory uncertainty with a random variables assumption. However, the spatial random field theory can not only take the stochasticity into account but also emphasize the spatial correlation. The intrinsic uncertainty of a geotechnical parameter can be depicted by a spatial random field, $\theta(x)$. The letter x defines the Euclidean coordinates. In addition to the random variables (i.e., random variables with only the mean values, μ, and the variance, σ_0^2), spatial random fields use the correlation function[30] to depict the spatial variability,

$$\rho_{i,i}(h) = \frac{E\{[\theta_i(x+h) - \mu_i] \cdot [\theta_i(x) - \mu_i]\}}{\sigma_i^2}, \tag{13.54a}$$

$$\rho_{i,j}(h) = \frac{E\{[\theta_i(x+h) - \mu_i] \cdot [\theta_j(x) - \mu_j]\}}{\sigma_i \sigma_j}, \tag{13.54b}$$

where $\rho_{i,i}(h)$ is the autocorrelation function, $i = 1,\ldots,m$, and m is the number of key geotechnical parameters. The cross-correlation function, $\rho_{i,j}(h)$, $(i,j = 1,\ldots,m, i \neq j)$, depicts the mutually spatial variation of spatial random fields. It may be assumed that $\rho_{i,j}(h) = \rho_{j,i}(h)$. In this chapter, the deterministic drift is a constant. After removing the deterministic drift, the mean value is satisfied $E[\theta(x)] = E[\theta(x+h)] = \mu$.

The exponential model of theoretical function adapts the scenario of a bigger correlation length, while the Gaussian model can depict the reversed curves of a correlation function. The spherical model[31] has the advantages of concision and robustness. Hence, this model is applied widely in the correlation function,

$$\rho(h) = \begin{cases} 1-\left[1.5(h/a_1)-0.5(h/a_1)^3\right] & 0 \le h \le a_1 \\ 0, & h > a_1 \end{cases}, \tag{13.55}$$

where $\rho(h)$ is the correlation coefficient of the lag distance h, and a_1 is the primary correlation length.

The three-dimensional heterogeneity of the spatial random fields can be considered with the weighted lag distance h [32] as,

$$h = \sqrt{\left(\frac{h_1}{\eta_1}\right)^2 + \left(\frac{h_2}{\eta_2}\right)^2 + \left(\frac{h_2}{\eta_3}\right)^2}, \tag{13.56}$$

where η_1 is equal to a constant value of 1.0, and η_2, η_3 are the correlation length ratio of the second correlation length a_2 and the third correlation length a_3 divided by a_1, respectively.

13.3.2 Bayesian inverse modeling method

The spatial variability of key geotechnical parameters cannot be completely described by the initial site investigation. Nevertheless, construction measurements are continuously obtained during excavations. The Bayesian method can dynamically calibrate the statistical characteristics of key geotechnical parameters by assimilating multi-source data.

13.3.2.1 Data classification

Two types of data are defined in the spatial random fields based Bayesian method. The borehole testing data are originally measured for the spatial random fields, $\theta(\boldsymbol{x})$, at spatial positions, $\boldsymbol{x}$, which in turn are defined as the input data, z_a. The tunneling measurements are deductively connected to the spatial random fields, and these measurements are defined as the output data, z_b. The two data types, z_a and z_b, are referred to in the following equations:

$$z_a = f(\boldsymbol{x}_a) + \varepsilon_a, \tag{13.57a}$$

$$z_b = \chi \cdot M(\boldsymbol{\theta}, z_a) + \varepsilon_b, \tag{13.57b}$$

where $f(\cdot)$ is a mathematical operator based on the spatial positions, $\boldsymbol{x}_a$, and the tunnel monitoring data z_b are indirectly derived from the soil–fluid coupling model $M(\theta, z_a)$. ε_a and ε_b are the zero-mean testing errors. The model factors χ are given by the ratios of the real measurements divided by the theoretical predictions, which are defined as independent random variables.

13.3.2.2 Bayesian framework

The unknown objectives, $\Theta = [\theta, \chi]$, include the key geotechnical parameters θ and the model factors χ. The spatial random fields based Bayesian framework is designed to assimilate prior knowledge, $p(\Theta)$, borehole testing data, z_a, and tunnel monitoring data, z_b. Accordingly, the posterior distribution, $p(\Theta \mid z_a, z_b)$, is derived from

$$p(\Theta \mid z_a, z_b) = \frac{L(z_b \mid \Theta, z_a)p(\Theta)}{\int_{\Theta} L(z_b \mid \Theta, z_a)p(\Theta)d\Theta}, \tag{13.58}$$

where the likelihood function $L(z_b \mid \Theta, z_a)$ is a probabilistic distribution of the tunnel monitoring data, z_b, which is also subject to the borehole testing data, z_a, prior distribution, $p(\Theta)$, and soil–fluid coupling model $M(\theta, z_a)$. The posterior distribution $p(\Theta \mid z_a, z_b)$ accounts for the test errors, ε_a and ε_b.

13.3.3 Algorithm implementation

The spatial random fields based Bayesian method is difficult to use to obtain analytical results. Thus, a sampling algorithm is widely used for a sampling solution. The algorithm consists of a numerical analysis, prior distribution, likelihood function, and posterior distribution modules. The analyses are referred to in Figure 13.7.

13.3.3.1 Prior distribution

Before any on-site observations (e.g., z_a and z_b) are used in the Bayesian framework, $p(\Theta)$ summarizes the knowledge that is available on the key geotechnical parameters θ and the model factors χ. Accordingly, the multivariate normal distribution is generalized to depict m geotechnical parameters at n positions, as shown in the following:

$$p(\theta) = |\mathbf{C}|^{-\frac{1}{2}} (2\pi)^{-\frac{m \times n}{2}} \exp\left[-\frac{1}{2}(\theta - \mu)^T \mathbf{C}^{-1}(\theta - \mu)\right], \tag{13.59}$$

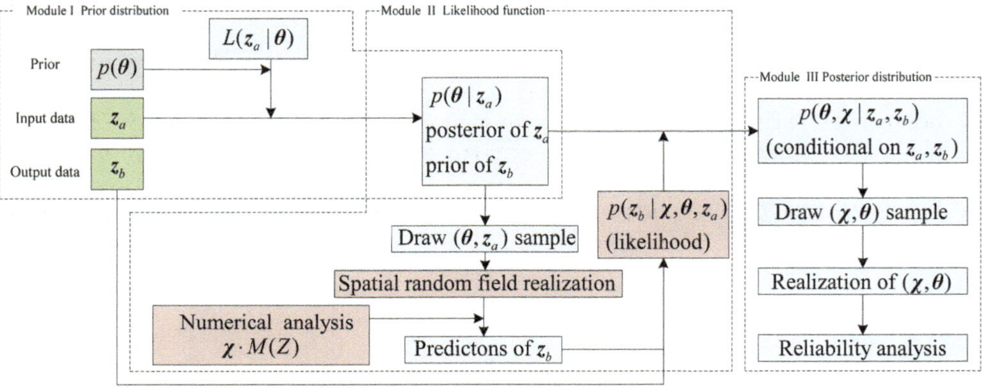

Figure 13.7 Data flow and algorithm schema of spatial random fields-based Bayesian method.

where C is the covariance matrix of the spatial random fields, $C = \rho_n \otimes C_m$, C_m is the covariance matrix of the number m of normal random variables, the matrix ρ_n consists of autocorrelation and cross-correlation functions $\rho_{i,j}(h)$, $(i,j = 1,\ldots,m)$, and the symbol "$\otimes$" denotes the operator of matrix multiplication. In a special scenario, the lag distance $h = 0$, which means that the spatial random fields degenerate into random variables.

The model factors χ are defined as the real measurements divided by the analytical results, which can be assumed to be multiplying independent normal random variables. Examples are given for the case studies. Thus, the prior distribution $p(\Theta)$ is presented as

$$p(\Theta) = p(\theta)\prod_{i=1}^{n_b} p(\chi_i), \tag{13.60}$$

where n_b is the total number of tunneling measurements, and $\prod_{i=1}^{n_b} p(\chi_i)$ denotes the prior joint distribution of the model factors.

13.3.3.2 Likelihood function

The spatial random fields based Bayesian method can assimilate prior knowledge $p(\Theta)$, borehole testing data z_a, and tunnel monitoring data z_b to obtain the posterior distributions of the key geotechnical parameters and model factors, $\Theta = [\theta, \chi]$. The likelihood function calculation includes two key components: the general driver of the soil–fluid coupling model $M(\theta, z_a)$ and the multivariate discretization of spatial random fields.

A general driver is designed to communicate with the Bayesian framework with the software FLAC^{3D} 3.0, as shown in Figure 13.8. This depends on a control file, and several input and output functions are used to configure the stochastic analysis of the soil–fluid coupling mechanics. The command stream of FLAC^{3D} 3.0 can execute numerical analysis and thereby obtain the analytical results for the tunneling measurements.

There are several well-known discretization algorithms of spatial random fields. Examples include the Cholesky decomposition[33], turning band method[34], local average method[35], Karhunen–Loeve decomposition method[36], sequential Gaussian simulation[37], and random harmonic function[38]. These algorithms are effective in accounting for spatial discretization for univariate spatial random fields.

Sequential Gaussian simulation has the advantage of robustness due to the spatial interpolation principle, and it is convenient to apply sequential Gaussian simulation in the conditional multivariate discretization of spatial random fields. As such, the trivariate spatial

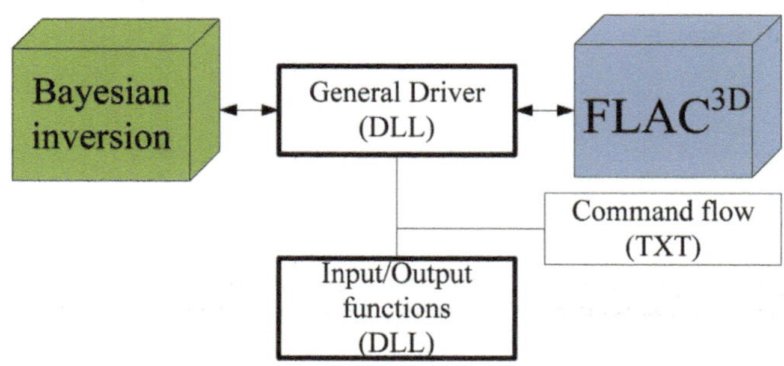

Figure 13.8 General master driver of connecting stochastic discretization and mechanics software.

random fields $\theta_i(x)$, $i=1, 2, 3$, are used for demonstration. In this case, the known quantities include the mean values, μ_i, standard deviation values, σ_i, autocorrelation function, $\rho_{i,i}(h)$, and cross-correlation function, $\rho_{i,j}(h), i=1, 2, 3, i \neq j$. There is n_i conditional data $\theta_i(x_k), k=1,\dots,n_i$ adjacent to point x_0. The mean value $\mu_i(x_0)$ and the standard deviation value $v_i(x_0)$ of the unknown $\theta_i(x_0)$ are calculated with linear co-kriging equations[39].

A sequential Gaussian simulation executes the spatial discretization of the stochastic elements point by point, and individual iteration takes a sample from the joint distribution. The following five key steps involve conditional multivariate discretization.

After defining a random path on the stochastic elements, each midpoint of an element is traversed once in the iteration, and the element edge size is less than the smallest correlation length, which guarantees the depiction of the spatial variation.

The mean values $\mu_1(x_k)$, $\mu_2(x_k)$, $\mu_3(x_k)$ and the standard deviation values $\sigma_1(x_k)$, $\sigma_2(x_k)$, $\sigma_3(x_k)$, at the midpoint, x_k, are predicted. The recognized quantities include the prior mean values, μ_1, μ_2, μ_3, the standard deviation values, σ_1, σ_2, σ_3, autocorrelation functions, $\rho_{1,1}(h)$, $\rho_{2,2}(h)$, $\rho_{3,3}(h)$, and the cross-correlation functions, $\rho_{1,2}(h)$, $\rho_{1,3}(h)$, $\rho_{2,3}(h)$. The conditional data consists of the original borehole testing data z_a (i.e., the errors ε_a are randomly added to the observations), and the recently discretized samples $\theta(x_1),\dots,\theta(x_{i-1})$.

Three normal distributions with $N[\mu_1(x_k),\sigma_1^2(x_k)]$, $N[\mu_2(x_k),\sigma_2^2(x_k)]$, and $N[\mu_3(x_k),\sigma_3^2(x_k)]$ are built for the midpoint x_k, and one sample is taken. The outcomes, $\theta_1(x_k)$, $\theta_2(x_k)$, and $\theta_3(x_k)$, are saved as the new conditional data for the next midpoint x_{k+1}.

Moving to the next midpoint x_{k+1}, Steps 2 and 3 are repeated until the sampling is done for all the stochastic elements.

One multivariate discretization of the spatial random fields is finished.

The stochastic analysis of the soil–fluid coupling model $M(\theta, z_a)$ provides analytical results for the tunnel monitoring data z_b of the ground surface settlements. Accordingly, the likelihood function $L(z_b \mid \theta, z_a)$ can be constructed for $\varepsilon_b = z_b - \chi M(\psi, z_a)$, which can be expressed by a multivariate normal distribution, as shown in Equation (13.61):

$$L(z_b \mid \psi, z_a) = \frac{\exp[-\frac{1}{2}(z_b - \chi M(\psi, z_a))^T C_{\varepsilon_b}{}^{-1}(z_b - \chi M(\psi, z_a))]}{\sqrt{(2\pi)^{n_b} \mid C_{\varepsilon_b} \mid}}, \tag{13.61}$$

where $C_{\varepsilon b}$ is the covariance matrix of the monitoring errors ε_b and $|C_{\varepsilon_b}|$ is the determinant of C_{ε_b}.

Furthermore, the correlation coefficient ρ_{ij} ($i,j=1,\dots,k$) among ε_b is simplified into the function of the mutual distance[40], as shown in Equation (13.62),

$$\rho_{ij} = 1 - \frac{D(x_i, x_j)}{\max D}, \tag{13.62}$$

where $D(x_i, x_j)$ is the Cartesian distance between two monitoring points x_i and x_j, and $\max D$ is the maximum distance of all the monitoring points.

13.3.3.3 Posterior distribution

The stochastic analysis of soil–fluid coupling mechanics $\chi M(\theta, z_a)$ can consume a large amount of computational resources. In the current analysis, a Markov chain Monte Carlo simulation is used to obtain a numerical solution quickly. The Markov chain Monte Carlo analysis process consists of the following major six stepwise procedures.

(1) The prior distribution, $p(\Theta = [\theta, \chi])$, is prepared for spatial random fields, which includes the mean value vector μ and the covariance matrix C.
(2) The ith multivariate discretization $\theta^i(x)$ is subjected to the prior distribution $p(\theta)$ and the original condition data z_a are extracted.
(3) The tunneling measurements z_b are predicted using the stochastic analysis of the soil–fluid coupling model, $M(\theta^i, z_a)$, and the threshold, $\gamma_i = \min\left(\frac{L(z_b \mid \theta^i, z_a)p(\theta^i)}{L(z_b \mid \theta^{i-1}, z_a)p(\theta^{i-1})}, 1\right)$, is calculated.
(4) A random number u in the uniform distribution 0~1 is taken and compared with γ_i. If $\gamma_i > u$, the spatial discretization $\theta^i(x)$ is accepted as a posterior sample. Otherwise, Step (2) is repeated.
(5) The prior mean value vector μ and the covariance matrix C from the accepted posterior samples for acceleration purposes are updated regularly, and then Step (2) is repeated.
(6) The predefined maximum samples, $i \le N$, are reached, and the posterior distribution $p(\Theta \mid z_a, z_b)$ is calculated.

13.4 ENGINEERING APPLICATION

This chapter introduces three engineering application examples of CPTU, two of which are located in Shanghai, China, and another one is located in the Bothkennar site in the United Kingdom.

13.4.1 National Convention and Exhibition Center in Shanghai

The National Convention and Exhibition Center (Shanghai) is the largest single-block building in the world. It is apart 1.5 km from Shanghai Hongqiao Transportation Hub. The primary facilities include exhibition areas A, B, C, and D and a transportation center as shown in Figure 13.9A. In this chapter, we focus on the site characterization of exhibition region A that is leaf shaped, with the largest dimension equals to 350 m in the Figure 13.9B.

The information available from the geotechnical investigation report includes the interfaces separating the various soil layers, groundwater table, and geotechnical parameters such as compression modulus E_s(MPa), consolidated shear cohesion coefficient c_{cq}(kPa), and internal friction angle $\varphi_{cq}(^0)$ as shown in Figure 13.9C. The silty clay soil (i.e., the third soil layer from the ground surface) is the load-bearing soil layer of the shallow structure foundation, the average thickness is about 2.0 m, and the three-dimensional domain is shown in Figure 13.9D. The 109 CPT data curves are sampling each 0.1 m along the drill bar as shown in Figure 13.10A, total $n_a = 1381$ penetration resistance data p_s(MPa) (i.e., the sum of tip resistance and sleeve friction) covers the whole three-dimensional

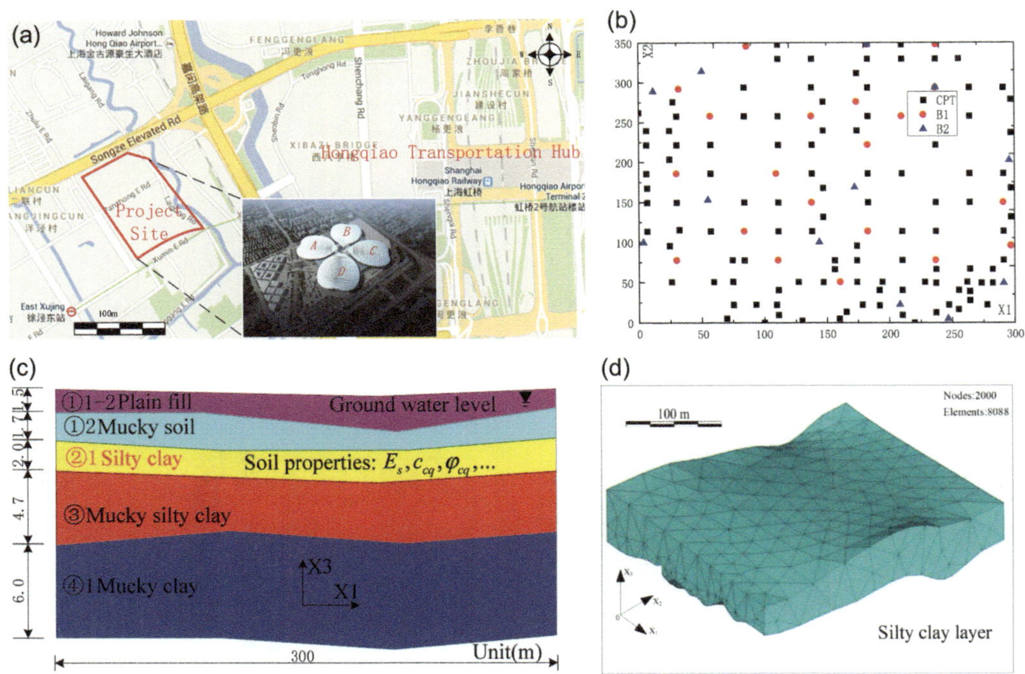

Figure 13.9 National Convention and Exhibition Center (Shanghai), China: (a) the geographic location of the background project, (b) planar geotechnical investigation layout of CPTs and boreholes of the region *A*, B1 provides compression modulus E_s(MPa) samples, B2 provides consolidated shear samples of cohesion coefficient c_{cq}(kPa) and shear internal friction angle $\varphi_{cq}(^\circ)$, (c) typically vertical section of the soil layers of region *A*, and (d) three-dimensional layer of the silty clay soil, the vertical displaying scale has been amplified 30 times.

domain, and histogram plot is shown in Figure 13.10B. To calibrate the transformation model between CPT data and laboratory data, the borehole group "B1" provides $n_{b1} = 21$ laboratory samples for compression modulus E_s(MPa), the histogram plot is shown in Figure 13.10C, and Table 13.1 lists the detailed sampling coordinates; the borehole group "B2" provides $n_{b2} = 12$ laboratory samples for consolidated shear cohesion coefficient c_{cq}(kPa) and internal friction angle $\varphi_{cq}(^0)$, histogram plot is shown in Figure 13.10D, and detailed sampling coordinates are shown in Table 13.2.

CPT and laboratory tests cannot be collocated at the same spatial location since the above geotechnical investigation belongs to destructive testing technologies, and the traditional regression method between CPT data and regular geotechnical parameters ignored the factors of spatial coincidence and uncertainty propagation of geotechnical investigation. The transformation model would be improved within the three-dimensional spatial random field theory. First, the massive CPT data is denoted as a three-dimensional anisotropic spatial random field variable, then the prediction of CPT data will be projected to the borehole sampling location coincidently with the laboratory data; finally, transformation models between CPT data and regular geotechnical parameters would be calibrated by giving spatial variability of CPT data, measurement errors, prior knowledge uncertainty of transformation model, statistical uncertainties of structural parameters, and Bayesian inversely modeling method itself.

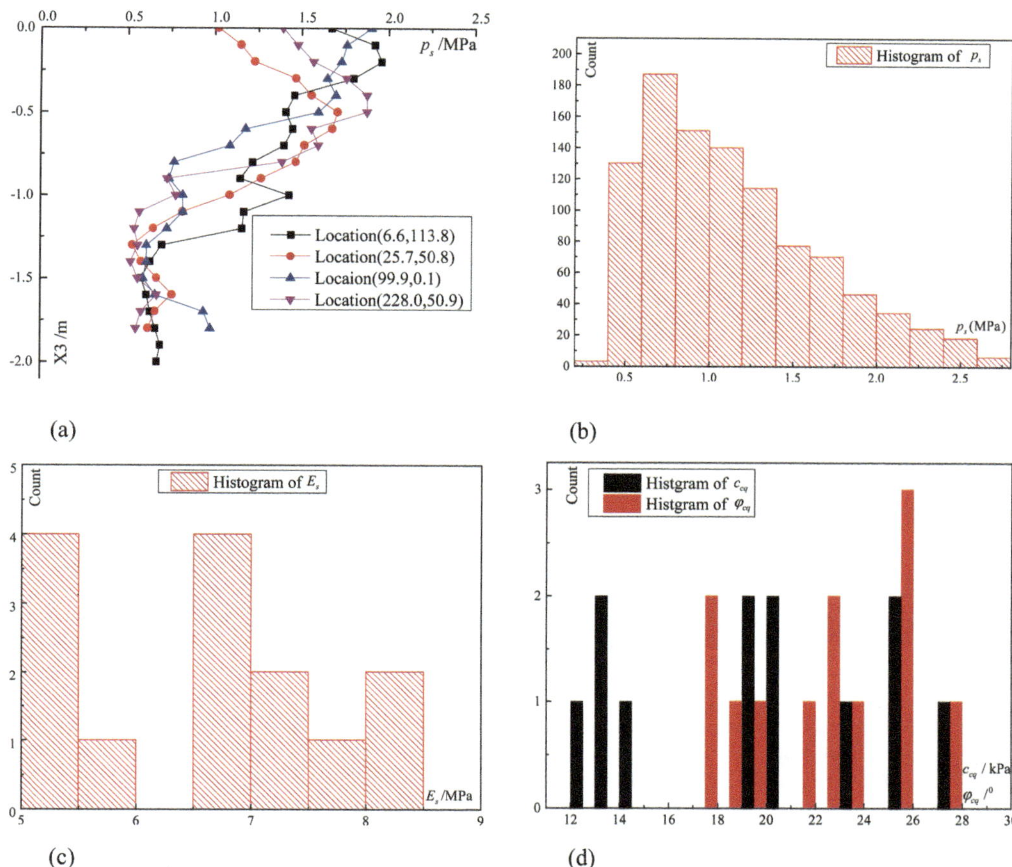

Figure 13.10 CPT data and laboratory data of the silty clay soil in the background project: (a) typical CPT data curves in the in-situ test; (b) histogram plot of whole CPT penetration resistance p_s(MPa); (c) histogram plot of compression modulus E_s(MPa); and (d) histogram plots of consolidated shear cohesion coefficient c_{cq}(kPa) and internal friction angle $\varphi_{cq}(^0)$.

Reversely evaluate transformation models within spatial random field theory and then convert CPT data p_s(MPa) to regular geotechnical parameters E_s(MPa), c_{cq}(kPa), and $\varphi_{cq}(^0)$ of the silty clay soil in the background project. CPT penetration resistance p_s(MPa) is represented by a spatial random variable, and the case studies are divided into two parts: a pseudo-transformation model M between CPT data p_s(MPa) and compression modulus E_s(MPa) is assumed at first, and the impacts of inherent variability and epistemic uncertainty will be taken into account; then, the calibrated transformation models are compared with linear regression models for the geotechnical parameters E_s(MPa), c_{cq}(kPa), and $\varphi_{cq}(^0)$ of the silty clay soil in cross-validation; finally, the calibrated transformation models will be employed in the three-dimensional site characterization.

13.4.1.1 Spatial variability of CPT data

There are 109 CPT data curves with total $n_a = 1381$ penetration resistance p_s(MPa) collected in the silty clay soil, and the classical statistics parameters are the mean value

Table 13.1 Spatial sampling coordinates of compression modulus E_s(MPa) in the laboratory test

Coordinates/m			E_s / MPa	Coordinates/m			E_s / MPa	Coordinates/m			E_s /MPa
X1	X2	X3		X1	X2	X3		X1	X2	X3	
30.00	185.8	2.00	6.94	111.10	77.80	2.64	9.43	183.10	113.80	2.42	4.73
30.10	77.80	2.13	8.09	138.00	257.80	1.29	6.93	210.10	257.90	2.12	4.61
32.10	292.10	2.40	6.72	138.10	149.80	2.61	6.75	237.00	293.80	3.29	4.63
57.10	257.80	2.58	7.00	160.60	50.90	1.95	7.62	237.10	77.90	1.03	5.27
84.10	113.90	1.68	8.10	174.00	275.90	1.48	5.39	237.60	347.90	2.72	4.96
86.40	345.80	2.48	4.51	183.00	347.80	1.92	7.15	291.10	149.90	2.27	5.98
109.60	185.80	1.54	4.76	183.10	221.80	1.68	5.05	297.10	95.90	1.91	5.29

Table 13.2 Spatial sampling coordinates of consolidation shear cohesion coefficient c_{cq}(kPa) and internal friction angle φ_{cq}(^{0}) in the laboratory test

Coordinates/m					Coordinates/m				
X1	*X2*	*X3*	c_{cq} / kPa	φ_{cq} /°	*X1*	*X2*	*X3*	c_{cq} / kPa	φ_{cq} /°
183.00	347.80	1.92	23.50	18.75	237.00	293.80	3.29	27.00	22.50
172.40	168.80	0.67	19.00	17.50	55.10	152.90	1.71	14.00	22.50
291.30	49.60	1.48	20.00	17.50	144.10	100.30	2.43	20.00	27.50
246.80	4.90	1.25	19.00	19.00	50.50	314.00	2.27	13.00	21.50
296.00	202.40	2.68	25.00	25.00	208.40	22.60	1.96	12.00	25.50
3.60	99.80	2.47	25.00	25.00	11.60	289.10	2.08.	23.75	13.00

1.148 MPa without ε_a consideration. According to the research of Tan Phoon and Hight et al.[40], CPT measurement error ε_a is usually depicted by the coefficient of variation (COV), which equals 2~5%, here COV equals to 3% (denoted as ε_a =3%) is adopted in this study, the standard deviation σ_a equals to 3% multiplying 1.148 MPa, which will be put into a zero-mean normal distribution $\varepsilon_a \sim N(0, 0.034^2)$. Moreover, CPT penetration resistance p_s(MPa) is denoted as a spatial random field variable Z, its anisotropic structural parameters $\theta = [\mu, n, s, \lambda, \rho_s, \rho_t]$ can be calculated directly by CPT data $z_a \pm \varepsilon_a$ (there are $n_a = 1381$ samples of ε_a, each corresponding to 1000 times simulation), and the structural parameters θ consist of mean value $\bar{\mu} = 1.148$ MPa, nugget $\bar{n} = 0.040$ MPa^2, partial sill $\bar{s} = 0.211$ MPa^2, major range $\bar{\lambda} = 88.616$m in the $X1$ coordinate, and other two orthogonal range ratio $\bar{\rho}_s = 0.512$ (i.e., range in the $X2$ coordinate) and $\bar{\rho}_t = 0.031$ (i.e., range in the $X3$ coordinate), the theoretical variogram of the spherical model $\gamma(h)$ is fitted by lag distance h as shown in Figure 13.11. The statistical uncertainty of structural parameters θ is listed in Table 13.3, the maximal uncertainty occurs in the secondary range ratio ρ_s, and its uncertainty is less than 5%, for the reason of relieving the Bayesian

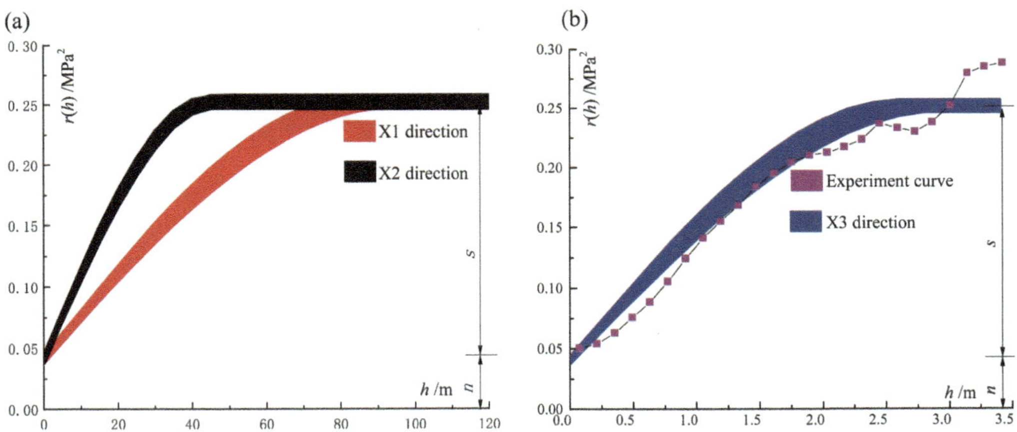

Figure 13.11 Three-dimensional anisotropic variogram of CPT data p_s(MPa), the experimental variogram is simulated with measurement error ε_a consideration. (a) Horizontal variogram of $X1$ $X2$ coordinates and (b) vertical variogram of the $X3$ coordinate.

Table 13.3 Statistical uncertainty of the structural parameters θ of the spatial random field by 1000 times simulation

Structural parameters θ	*Minimum*	*Maximum*	*Mean*	*COV*
μ / MPa	1.152	1.145	1.148	0.001
n / MPa^2	0.043	0.038	0.040	0.022
s / MPa^2	0.216	0.206	0.211	0.008
λ / m	94.310	81.322	88.616	0.030
ρ_s	0.550	0.460	0.512	0.038
ρ_t	0.034	0.029	0.031	0.035

inversely modeling burdens, the structural parameters θ will adopt the mean values to delegate the statistical uncertainty.

Laboratory data z_b consists of $n_{b1} = 21$ compression modulus $E_s(\text{MPa})$ and another $n_{b2} = 12$ consolidated shear cohesion coefficient $c_{cq}(\text{kPa})$, and internal friction angle $\varphi_{cq}(^0)$, respectively. Tan Phoon and Hight et al.[41] suggested that COV equals 10~37% of $E_s(\text{MPa})$, and Thermann Gau and Tiedemann[42] summarized that COV equals 2%–17% of $c_{cq}(\text{kPa})$ and $\varphi_{cq}(^0)$. Therefore, the laboratory measurement error ε_b is uniformly denoted as 10% of $E_s(\text{MPa})$, $c_{cq}(\text{kPa})$, $\varphi_{cq}(^0)$, and then building the zero-mean normal distribution $\varepsilon_b \sim N(0,\sigma_b^2)$, respectively.

13.4.1.2 Transformation model baseline

The spatial random field consists of $n_a = 1381$ CPT data $p_s(\text{MPa})$ and measurement error is $\varepsilon_a \sim N(0, 0.034^2)$, the structural parameters of spatial random field adopt the deterministic mean values $\theta = [\bar{\mu}, \bar{n}, \bar{s}, \bar{\lambda}, \bar{\rho}_s, \bar{\rho}_t]$, and sequential Gaussian simulation algorithm projects CPT data prediction to collocate with laboratory data. The baseline tries to test the effectiveness of applying the Bayesian approach on transformation model calibration within the spatial random field theory, the linear mathematic formula of converting penetration resistance $p_s(\text{MPa})$ to compression modulus $E_s(\text{MPa})$ is taken into demonstration in the silty clay soil,

$$E_s = k \cdot p_s + q \pm \varepsilon_T . \tag{13.63}$$

According to prior knowledge of linear regression between CPT data $p_s(\text{MPa})$ and its spatially closest laboratory data of compression modulus $E_s(\text{MPa})$, the unknown parameters $m = (k, q, \varepsilon_T)$ of the transformation model M is assumed as $k = 4.453$, $q = 0.932$, and $\varepsilon_T \sim N(0, 0.2^2)$. Then, taking out one spatial random field realization given z_a, ε_a, θ, and m will generate the pseudo-laboratory data of $E_s(\text{MPa})$, for convenience, the coordinates of $n_{b1} = 21$ laboratory test of the background project are borrowed in Table 13.4. Proposing the COV of ε_b equals 10% (denoted as $\varepsilon_b = 10\%$) of the pseudo-mean value $6.228\,\text{MPa}$, the measurement error is $\varepsilon_b \sim N(0, 0.623^2)$. The baseline will be used to evaluate the effectiveness of the Bayesian inversely modeling method.

In the Bayesian framework, the prior knowledge of unknown coefficients and uncertainty $m = (k, q, \varepsilon_T)$ are usually assumed as maximal uncertainty when there is a non-informative knowledge assumption: $k \sim U(0,10)$, $q \sim U(-1,5)$, $\sigma_T \sim U(0, 0.7)$, and $\varepsilon_T \sim N(0, \sigma_T^2)$ can cover the potential value intervals of m. The prior joint distribution $p[m = (k, q, \varepsilon_T)]$ is sampled $N_m = 1000$ times from Latin hypercube space (Stein, 1987); furthermore, each sample $m_i, i = 1, \ldots, N_m$ will take $N_a = 200$ spatial random field realizations and $N_b = 100$ calibrations, which is used to calculate the likelihood function value $p(z_b \mid \theta, z_a, m)$. The CPU time for completing the above calculations is 8 h of five threads on a desktop computer that has Intel Xeon (R) dual CPU processors of 2.00 GHz and 4GB RAM.

Figure 13.12 (a–c) gives out the posterior PDF $m = (k, q, \varepsilon_T)$ of the transformation model M baseline, except for the posterior mode of slope k underestimates 4% of the true value, which considers the measurement errors ε_a and ε_b simultaneously as shown in Figure 13.12(d), Bayesian approach captures robustly the true values, the two types of

Table 13.4 Compression modulus E_s(MPa) baseline of spatial random field

Coordinates /m			E_s / MPa	Coordinates /m			E_s /MPa	Coordinates /m			E_s /MPa
X1	X2	X3		X1	X2	X3		X1	X2	X3	
30.00	185.8	2.00	4.75	111.10	77.80	2.64	7.10	183.10	113.80	2.42	4.77
30.10	77.80	2.13	7.72	138.00	257.80	1.29	5.71	210.10	257.90	2.12	4.86
32.10	292.10	2.40	8.09	138.10	149.80	2.61	11.65	237.00	293.80	3.29	6.29
57.10	257.80	2.58	7.53	160.60	50.90	1.95	7.92	237.10	77.90	1.03	3.55
84.10	113.90	1.68	4.24	174.00	275.90	1.48	8.64	237.60	347.90	2.72	3.23
86.40	345.80	2.48	6.97	183.00	347.80	1.92	2.79	291.10	149.90	2.27	6.82
109.60	185.80	1.54	9.27	183.10	221.80	1.68	2.59	297.10	95.90	1.91	6.30

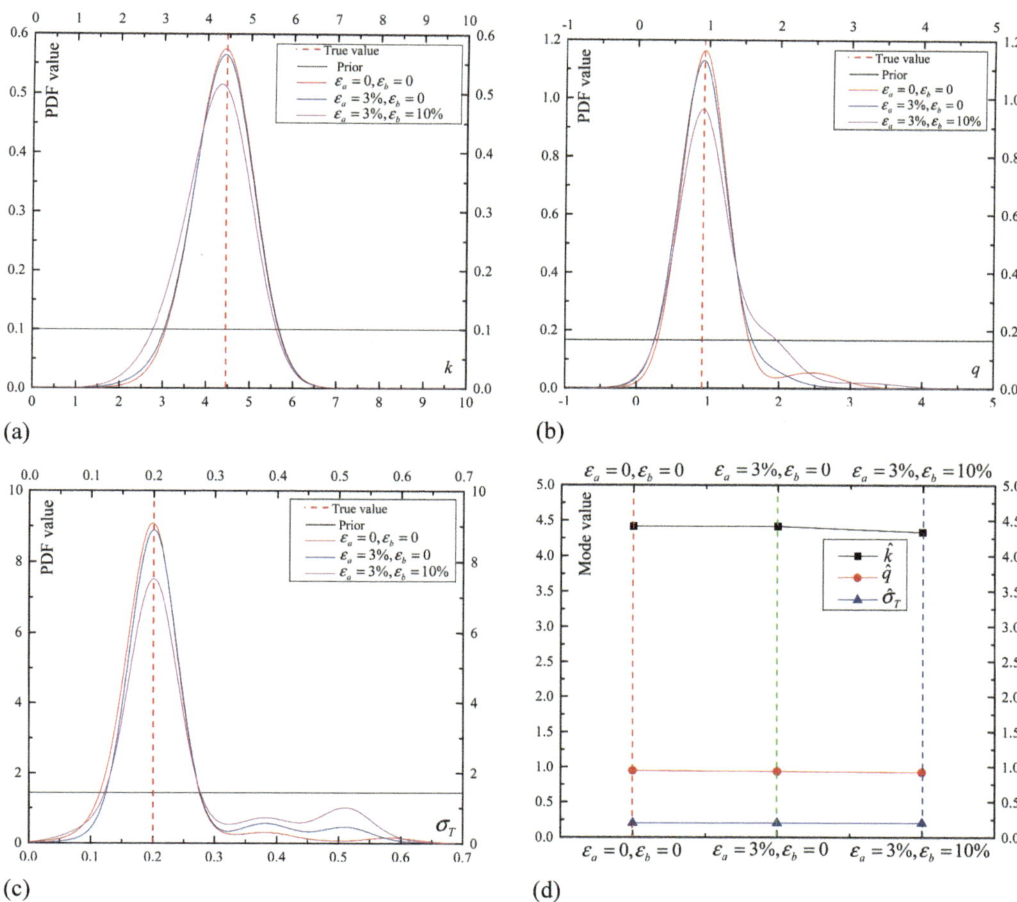

Figure 13.12 Posterior probabilistic characterizations of transformation model baseline within spatial random field theory: (a) posterior PDFs of slope coefficient k, (b) posterior PDFs of intercept coefficient q, (c) posterior PDFs of transformation model uncertainty $\varepsilon_T \sim N(0, \sigma_T^2)$, and (d) posterior mode value comparisons of transformation model baseline.

measurement error impact the posterior variance dramatically, but with subtle influence to the posterior model values, the detailed probabilistic characterizations are listed in Table 13.5.

Sampling-based solutions for the Bayesian framework inevitably bring statistical uncertainty into the system, especially the conditional simulation process of the spatial random field, Latin hypercube sampling algorithm takes $N_m = 1000$ samples, which guarantees to traverse the whole probability space of the joint multivariate distribution $p[m = (k, q, \varepsilon_T)]$. There are $N_a = 200$ spatial random field realizations and $N_b = 100$ calibrations of ε_b corresponding to each sample $m_i, i = 1, ..., N_m$. Spatial random field realizations provide the predictions z_b to calculate the joint multivariate likelihood function $p(z_b \mid \theta, z_a, m)$; therefore, the following discussion will focus on the statistical uncertainty of spatial random field realization to impact the posterior results.

Table 13.5 Posterior probabilistic characterizations of transformation model baseline

Transformation model			*Posterior PDF*			
Coefficient	*True value*	*Measurement errors*	*Mean*	*Mode*	*Variance*	*PDF value*
k	4.453	$\varepsilon_a = 0, \varepsilon_b = 0$	4.392	4.415	0.501	0.574
		$\varepsilon_a = 3\%, \varepsilon_b = 0$	4.356	4.415	0.543	0.564
		$\varepsilon_a = 3\%, \varepsilon_b = 10\%$	4.224	4.251	0.652	0.510
q	0.932	$\varepsilon_a = 0, \varepsilon_b = 0$	1.025	0.941	0.230	1.162
		$\varepsilon_a = 3\%, \varepsilon_b = 0$	0.965	0.928	0.165	1.129
		$\varepsilon_a = 3\%, \varepsilon_b = 10\%$	1.101	0.874	0.343	0.961
$\varepsilon_T \sim N(0, \sigma_T^2)$	0.2	$\varepsilon_a = 0, \varepsilon_b = 0$	0.209	0.199	0.006	9.077
		$\varepsilon_a = 3\%, \varepsilon_b = 0$	0.227	0.200	0.008	8.895
		$\varepsilon_a = 3\%, \varepsilon_b = 10\%$	0.248	0.200	0.014	7.514

In the scenario of taking both measurement errors $\varepsilon_a, \varepsilon_b$ into account, calculating the likelihood function $p(z_b \mid \theta, z_a, m)$ given the different number of spatial random field realizations, $N_a = 50,..,200$, increment of 25 realizations, the results of Figure 13.13 show that the posterior mode value $\hat{k}$ varies from 94% ($N_a = 50$) to 99% ($N_a = 175$) comparing with the final result of $N_a = 200$ realizations, the impacts of spatial random field realization keeps in stable value after $N_a = 175$ realizations for the posterior mode value $\hat{k}$; the variation of the posterior mode value $\hat{q}$ is similar as $\hat{k}$, the value approximates

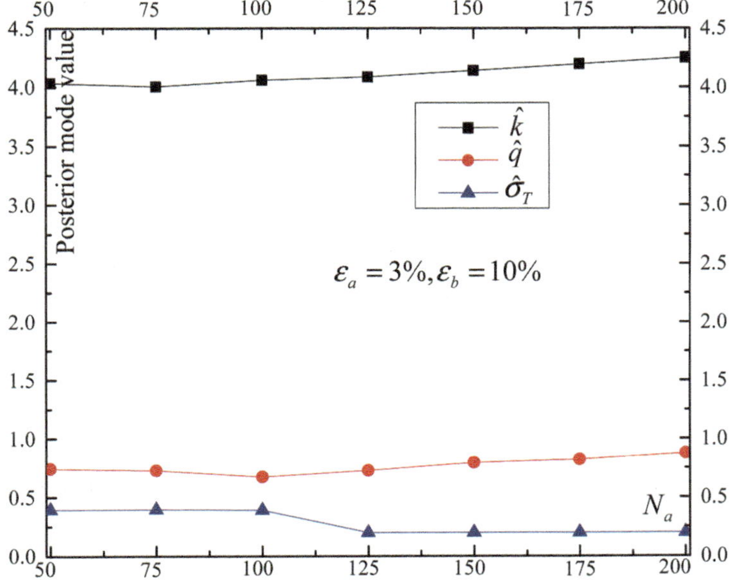

Figure 13.13 Statistical uncertainty of Bayesian reverse modeling method, for example, posterior mode values of transformation model baseline varying with spatial random field realizations.

to 95% ($N_a = 175$) of the final result; but for the transformation model uncertainty, the posterior mode value $\hat{\sigma}_T$ changes significantly from $N_a = 100$ jumping to $N_a = 125$, and then the result keeps in stable value. The statistical uncertainty of the Bayesian inversely modeling method will decrease to 6% when the spatial random field simulations increase $N_a = 175$. Therefore, the realization number $N_a = 200$ would guarantee the Bayesian reversely modeling precision, and the baseline study proves that the Bayesian method is effective for calibrating the transformation model within spatial random field theory.

13.4.1.3 Transformation model calibration

The only difference between the transformation model baseline and engineering application is the latter provides real-life laboratory data, such as geotechnical parameters E_s(MPa), c_{cq}(kPa), φ_{cq}(°) are shown in Tables 13.1 and 13.2, the engineering application is divided into two steps: firstly, the Bayesian inversely modeling method calibrates the transformation model M within spatial random field theory, and comparing with traditional regression models in cross-validation; then applying the calibrated transformation models into three-dimensional site characterization of the background project.

There are $n_a = 1381$ CPT data p_s(MPa), and the COV of measurement error ε_a equals 3%. Laboratory data consists of $n_{b1} = 21$ compression modulus E_s(MPa), $n_{b2} = 12$ consolidation shear parameters c_{cq}(kPa), φ_{cq}(°), and COV of measurement error ε_b equals 10%, respectively. The spatial random field of CPT data still takes the deterministic structural parameters $\theta = [\bar{\mu}, \bar{n}, \bar{s}, \bar{\lambda}, \bar{\rho}_s, \bar{\rho}_t]$, the sampling number of the prior distribution of the transformation model M is $N_m = 1000$, the number of spatial random field realizations is $N_a = 200$, and the number of laboratory data calibrations is $N_b = 100$ corresponding to each sample. The prior knowledge uncertainty of the transformation model M is assumed as non-informative for all the geotechnical parameters in Table 13.6.

All the finally converted geotechnical parameters should be positive values whatever the negative coefficients imported in the transformation models, for example, although the compression modulus E_s(MPa) keeps a significantly negative correlation with CPT data p_s(MPa), the calibrated transformation model should guarantee positive geotechnical parameters; for the internal friction angle, linear-logarithmic transformation model will provide more robust prediction than the linear one, the vertically effective earth pressure σ' of CPT sampling depth is calculated as

$$\sigma' = \sum \gamma_i h_i + \sum \gamma_j' h_j , \tag{13.64}$$

Table 13.6 Non-informative prior knowledge of transformation models between CPT data and regular geotechnical parameters

Geotechnical parameters	*Transformation model*	*Slope k*		*Intercept q*		$\varepsilon_T \sim N(0,\sigma_T^2)$	
		Lower	*Upper*	*Lower*	*Upper*	*Lower*	*Upper*
E_s / MPa	$E_s = k_E \cdot p_s + q_E \pm \varepsilon_{T,E}$	0	10.0	−1.0	5.0	0	0.7
c_{cq} / kPa	$c_{cq} = k_c \cdot p_s + q \pm \varepsilon_{T,c}$	−25.0	0	10.0	60.0	0	7.0
φ_{cq} /°	$\varphi_{cq} = k_\varphi \cdot \ln(p_s / \sigma') + q_\varphi \pm \varepsilon_{T,\varphi}$	0	15.0	−6.0	1.0	0	6.0

where γ_i is the ith soil unit weight with the thickness h_i above the groundwater table and γ'_j is the jth soil unit weight with the thickness h_j beneath the groundwater table.

Figure 13.14 (a–c) provides the posterior PDFs of the unknown parameters $m_E = (k_E, q_E, \varepsilon_{T,E})$ converting CPT data p_s(MPa) into compression modulus E_s(MPa), the posterior mean and mode values are close to each other, the posterior PDF curves show symmetric shapes, and the finally adopted mode value is $\hat{k}_E = 4.442$, $\hat{q}_E = 0.901$, $\hat{\sigma}_{T,E} = 0.2$, and $\hat{\varepsilon}_{T,E} \sim N(0, 0.2^2)$. Scatter data plots between predictions and observations are shown in Figure 13.14 (d), Bayesian result $\hat{m}_E = (\hat{k}_E, \hat{q}_E, \hat{\varepsilon}_{T,E})$ proves that the averaged prediction of compression modulus linearly approximates laboratory data, the gradient equals 0.788, and the R-square value equals 0.713 given $N_m = 1000$ samples $(z_a, \varepsilon_a, \theta, \widehat{m_E})$, each sample corresponding to $N_a = 200$ realizations of a spatial random field, although the linear regression method between CPT data and their closest neighbor laboratory data hold the higher gradient, the R-square value displays that there is the relatively lower linear correlation of traditional regression method comparing with Bayesian method. Figure 13.14 (e) proves that Bayesian predictions always cover the laboratory data within one-time standard deviation (SD), but several laboratory data aren't included in the linear regression prediction with one-time SD (see linear regression method in the Appendix).

Figure 13.15 (a–c) provides the posterior PDFs for unknown parameters $m_c = (k_c, q_c, \varepsilon_{T,c})$ to convert CPT data p_s(MPa) to consolidated shear cohesion coefficient c_{cq}(kPa), the posterior PDF curves show non-normal probabilistic characterizations, and the posterior mean and mode values are in the same magnitude, but the uncertainty is larger than other geotechnical parameters, therefore, the mode values are adopted: $\hat{k}_c = -12.176$, $\hat{q}_c = 33.734$, $\hat{\sigma}_{T,c} = 4.199$, and $\hat{\varepsilon}_{T,c} \sim N(0, 4.199^2)$, which guarantees the biggest posterior PDF values. However, according to scattered data plot of $n_{b2} = 12$ laboratory data and prediction in the cross-validation, Figure 13.15 (d) certificates that averaged prediction of consolidated shear cohesion coefficient approximates diagonally to the laboratory data, the gradient equals to 0.978, it has subtle improvements comparing with linear regression method. In Figure 13.15 (e), Bayesian predictions cover the laboratory data within a one-time SD, and four laboratory data aren't included in the linear regression prediction with a one-time SD.

Figure 13.16 (a–c) provides the posterior PDFs for unknown parameters $m_\varphi = (k_\varphi, q_\varphi, \varepsilon_{T,\varphi})$ to convert CPT data p_s(MPa) into consolidated shear internal friction angle $\varphi_{cq}(^\circ)$, according to former research experiences, there are two subtle changes in the transformation model to improve prediction quality, firstly, the vertically effective earth pressure σ'(MPa) is taken into account; then, the linear-logarithmic formula is adopted. The posterior PDF curves show non-normal probabilistic characterizations, but still in the symmetric shapes, and the posterior mode values approximate to the mean values, the ultimately adopted mode values are $\hat{k}_\varphi = 9.048$, $\hat{q}_\varphi = -2.672$, $\hat{\sigma}_{T,\varphi} = 2.391$, and $\hat{\varepsilon}_{T,\varphi} \sim N(0, 2.391^2)$. According to scattered data plot of $n_{b2} = 12$, laboratory data and prediction in the cross-validation, Figure 13.16(d) proves that the averaged prediction of consolidated shear internal friction angle approximates the laboratory data along with 45° degree straight line, the gradient equals 0.989, and the R-square value equals 0.477, it has been substantially improved comparing with traditional regression method. In Figure 13.16 (e), similar to the other two geotechnical parameters, Bayesian predictions

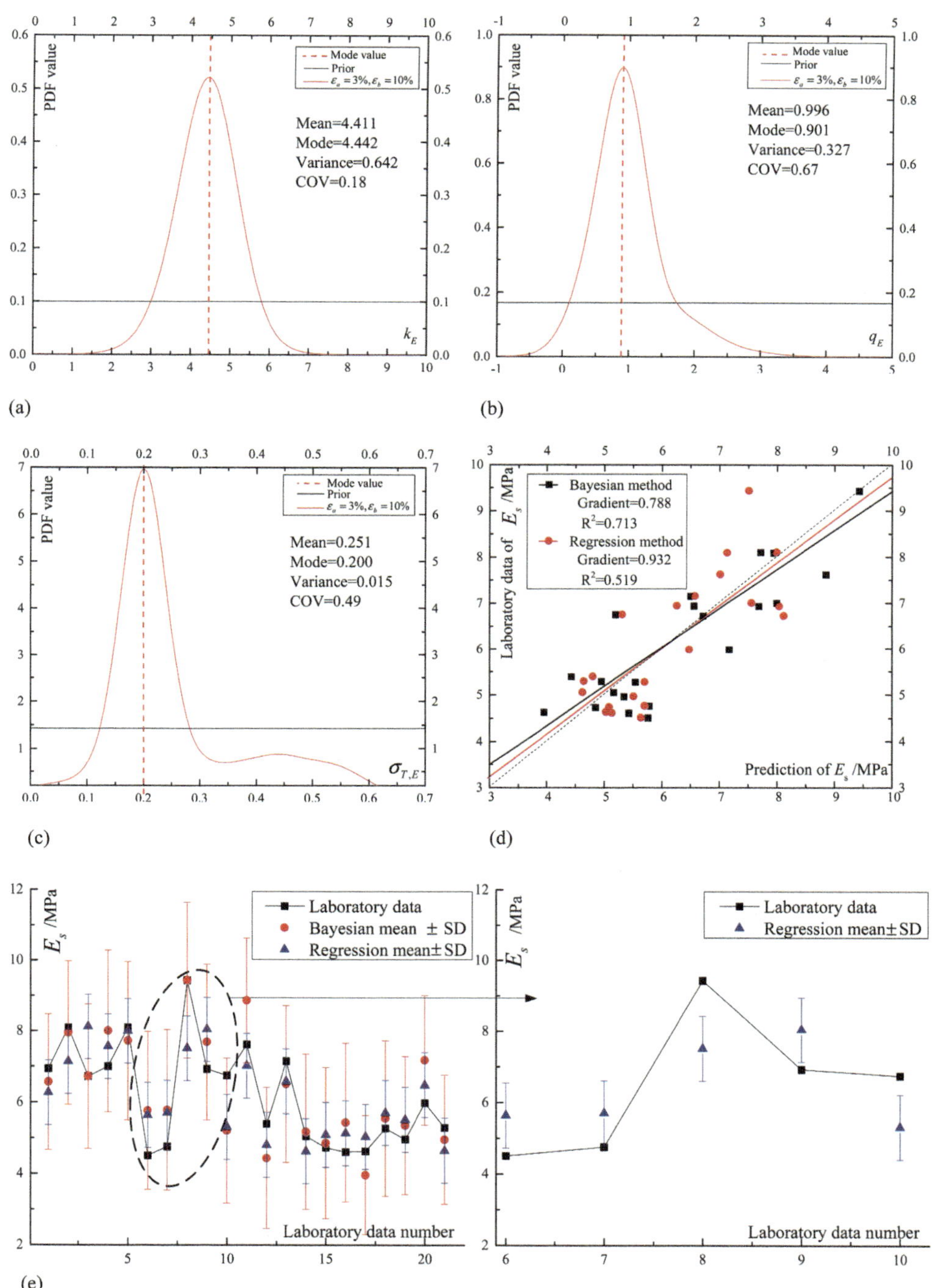

Figure 13.14 Posterior distribution and cross-validation of transformation model M between CPT data p_s(MPa) and compression modulus E_s(MPa). (a) Posterior PDF of coefficient k_E, (b) posterior PDF of coefficient q_E, (c) posterior PDF of transformation model uncertainty $\varepsilon_{T,E} \sim N(0, \sigma_{T,E}^2)$, (d) scattered data plot of compression modulus E_s(MPa) in the cross-validation, and (e) predictive intervals comparison of compression modulus E_s(MPa) in the cross-validation.

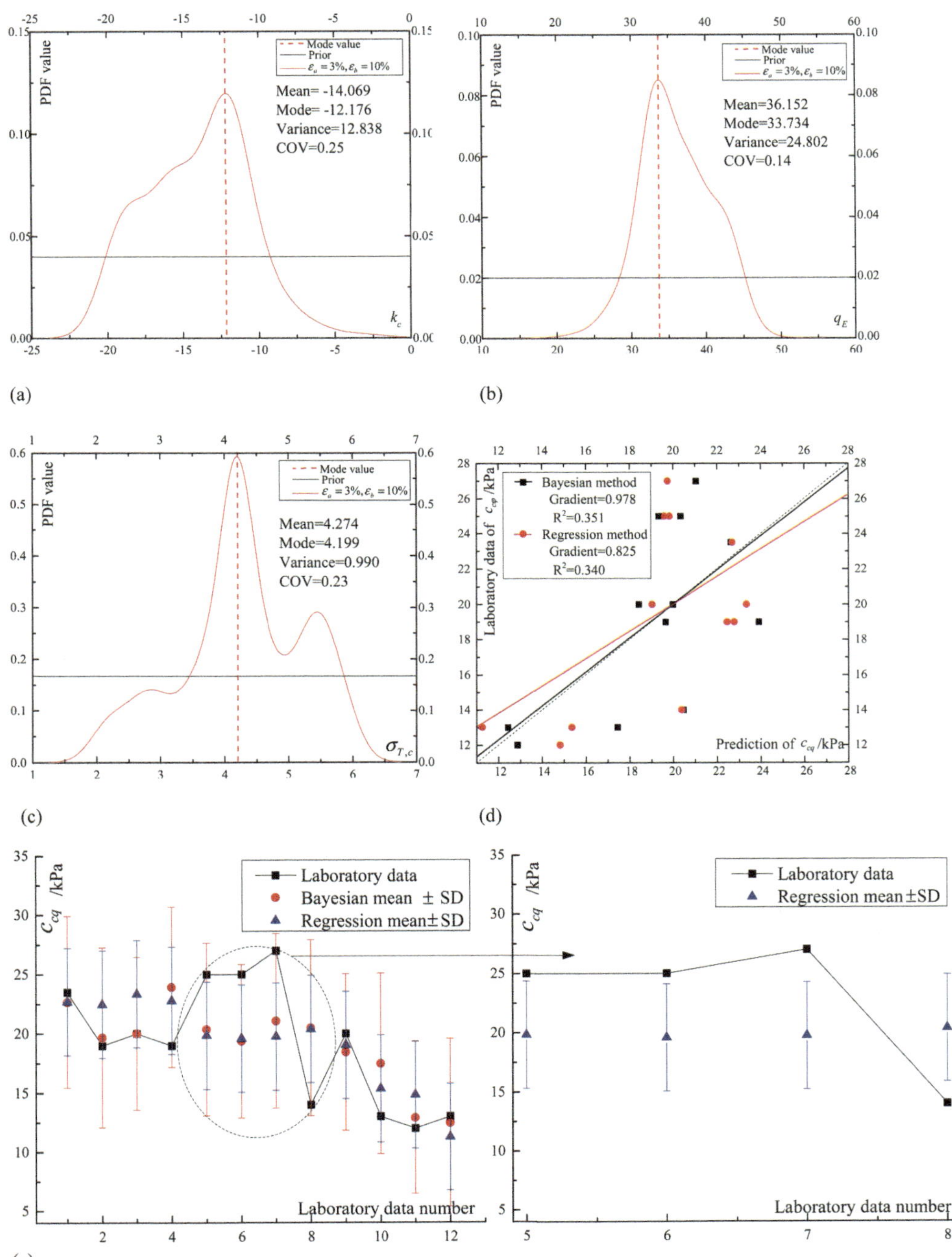

Figure 13.15 Posterior distribution and cross-validation of transformation model M between CPT data p_s(MPa) and consolidated shear cohesion coefficient c_{cq}(kPa). (a) Posterior PDF of coefficient k_c, (b) posterior PDF of coefficient q_c, (c) posterior PDF of transformation model uncertainty $\varepsilon_{T,c} \sim N(0, \sigma_{T,c}^2)$, (d) scattered data plot of consolidated shear cohesion coefficient c_{cq}(kPa)in the cross-validation, and (e) predictive intervals comparison of consolidated shear cohesion coefficient c_{cq}(kPa) in the cross-validation.

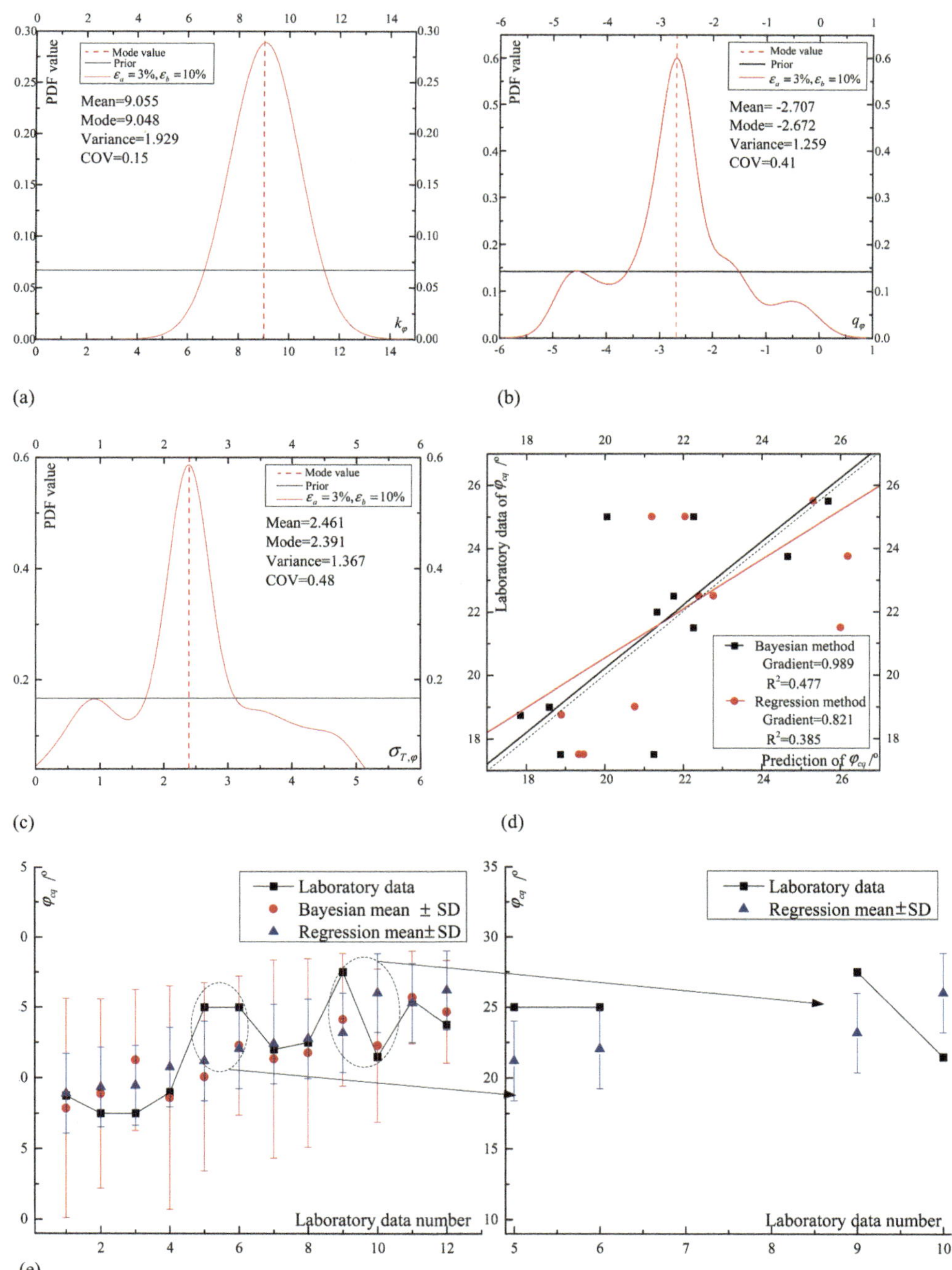

Figure 13.16 Posterior distribution and cross-validation of transformation model M between CPT data p_s(MPa) and consolidated shear internal friction angle φ_{cq}(°). (a) Posterior PDF of coefficient k_φ, (b) posterior PDF of coefficient q_φ, (c) posterior PDF of transformation model uncertainty $\varepsilon_{T,\varphi} \sim N(0, \sigma_{T,\varphi}^2)$, (d) scattered data plot of consolidated shear internal friction angle φ_{cq}(°) in the cross-validation, and (e) predictive intervals comparison of consolidated shear internal friction angle φ_{cq}(°) in the cross-validation.

Table 13.7 Calibrated transformation models between CPT data and regular geotechnical parameters in the silty clay soil

Geotechnical parameters	*Transformation model*	*Slope* $\hat{k}$	*Intercept* $\hat{q}$	$\hat{\varepsilon}_T \sim N(0, \hat{\sigma}_T^2)$
E_s / MPa	$E_s = k_E \cdot p_s + q_E \pm \varepsilon_{T,E}$	4.442	0.901	0.200
c_{cq} / kPa	$c_{cq} = k_c \cdot p_s + q \pm \varepsilon_{T,c}$	-12.176	33.734	4.199
φ_{cq} /°	$\varphi_{cq} = k_\varphi \cdot \ln(p_s / \sigma') + q_\varphi \pm \varepsilon_{T,\varphi}$	9.048	-2.672	2.391

cover the laboratory data within one-time SD, and there are still three laboratory data excluded from the linear regression prediction with one-time SD.

Bayesian reverse modeling method successfully calibrated the transformation models between CPT data p_s(MPa) and geotechnical parameters E_s(MPa), c_{cq}(kPa), and φ_{cq}(°) within spatial random field theory, which takes uncertainty propagation of geotechnical investigation into consideration, the finally adopted parameters of transformation models are listed in Table 13.7, which will be employed into three-dimensional site characterization of the background project.

13.4.1.4 Three-dimensional site characterization

The massive CPT data can support the three-dimensional site characterization of the background project, which provides discrete elements for further soil mechanics analysis. The whole domain of the silty clay layer is meshed into 2000 nodes, and 8088 tetrahedral elements; the maximal element edge size is meshed less than the smallest range of spatial random field, which guarantees the spatial variability of soil would be considered in the mechanical analysis.

The CPT data p_s(MPa), measurement error ε_a, structural parameter θ, and calibrated transformation model M are utilized in the conditional simulation of spatial random field, it needs to be pointed out that the laboratory data is also denoted as conditional observation, Figure 13.17 lists out the three-dimensional discretization (e.g., mean and COV values) given $N_m = 1000$ samples of ($z_a, \varepsilon_a, \theta, m$), here, only measurement error ε_a is in variation, other parameters keep in constant values; each sample corresponds to $N_a = 200$ realizations of spatial random field, and then transforming the penetration resistance p_s(MPa) realization of each element's centroid into geotechnical parameters E_s(MPa), c_{cq}(kPa), and φ_{cq}(°).

Three-dimensional site characterizations show that E_s(MPa) c_{cq}(kPa) versus φ_{cq}(°) keeps a globally negative correlation, mean value $\bar{E}_s$ located in the interval [3.24, 10.76], and the COV is confined in [0.15, 0.47] kPa; the variation of c_{cq} is bigger than E_s and φ_{cq}, the mean value $\bar{c}_{cq}$ varies from 6.55 kPa to 27.61 kPa, and the biggest COV value equals to 1.00; linear-logarithmic transformation model is adaptive to consolidated shear internal friction angle, the mean value $\bar{\varphi}_{cq}$ varies from 13.35° to 28.49°, and the COV of φ_{cq} equals to 0.10 ~ 0.66.

In summary, the three-dimensional site characterization depends on the massive CPT data, spatial random field theory, and transformation models, the conditional simulation method would give out the spatial distribution of the regular geotechnical parameters.

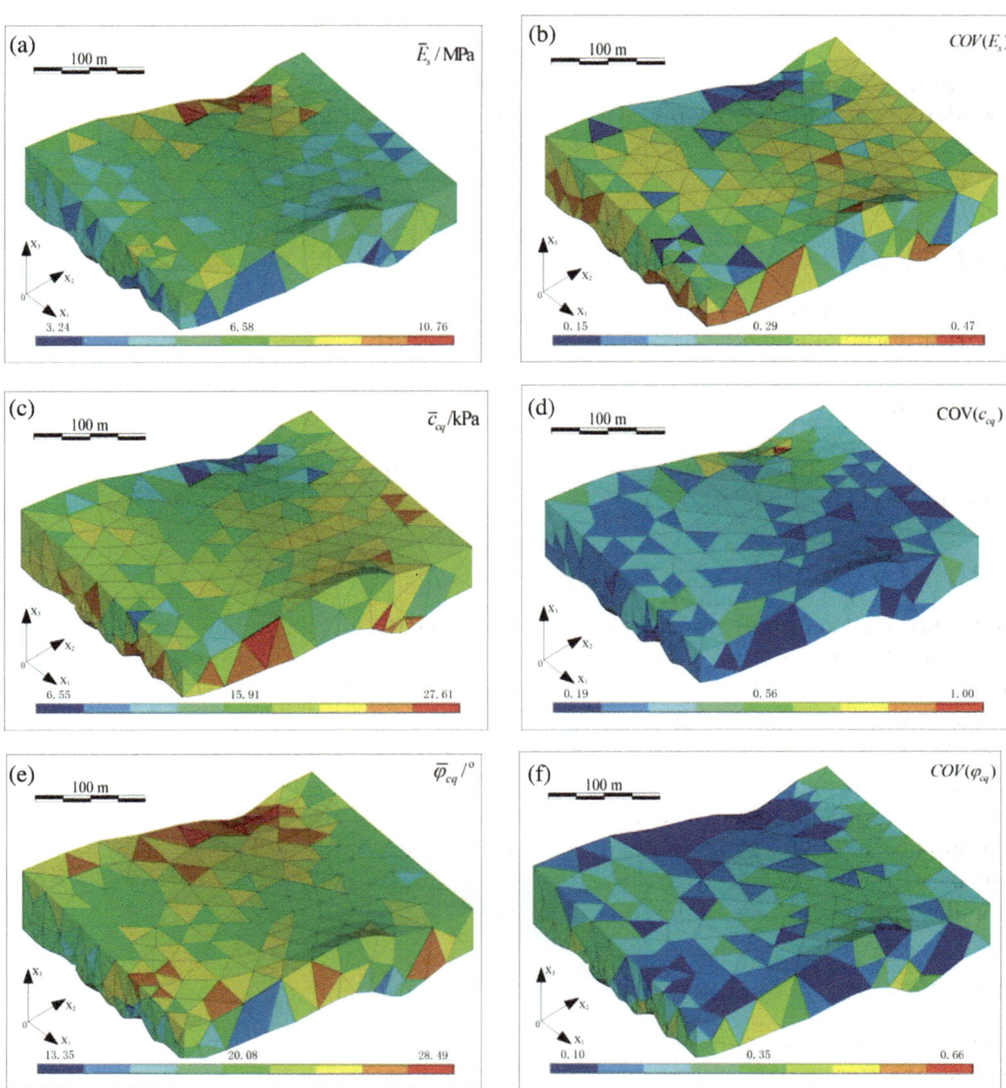

Figure 13.17 Three-dimensional site characterization of geotechnical parameters E_s(MPa), c_{cq}(kPa), and φ_{cq}(°) in the silty clay layer: (a) mean value $\bar{E}_s$(MPa) spatial distribution; (b) COV spatial distribution of E_s(MPa); (c) mean value $\bar{c}_{cq}$(kPa) distribution; (d) COV spatial distribution of c_{cq}(kPa); (e) mean value $\bar{\varphi}_{cq}$(°) distribution; and (f) COV spatial distribution of φ_{cq}(°).

13.4.2 Suzhou river drainage and storage pipeline project

Suzhou River Deep Tunnel Reservoir Project adopts a shield excavating method with a total length of 15 km and a buried depth of 50 m~60 m. This project includes eight extra-deep shafts and complexes. The trial part is about 1.67 km long, which includes Yunling and Miaopu complexes. The depth of the underground diaphragm wall reaches 105 m

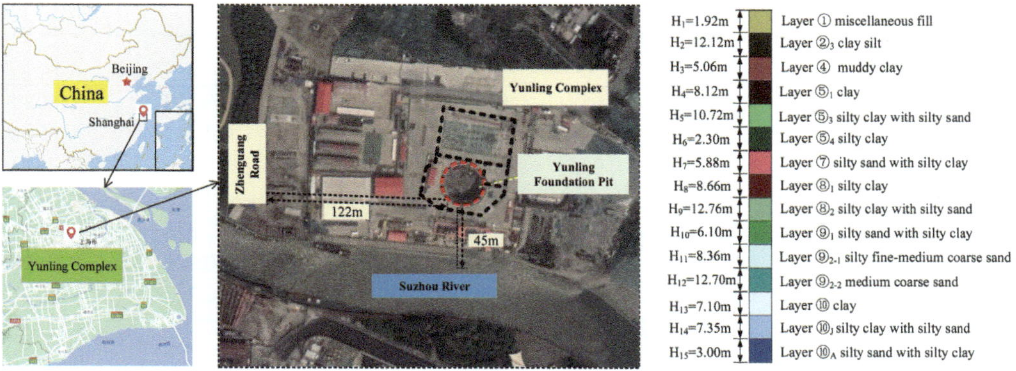

Figure 13.18 Schematic diagram of the surrounding environment of Yunling complex.

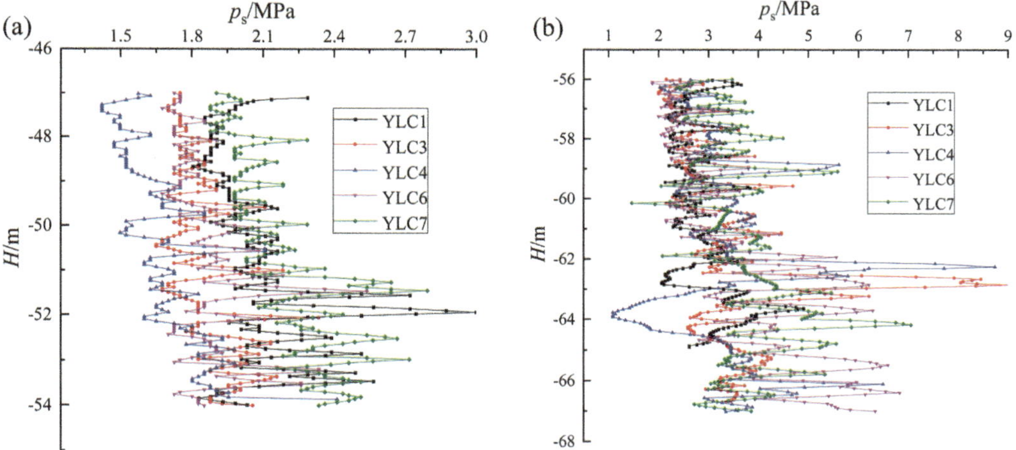

Figure 13.19 The in-situ CPT data in layer ⑧. (a) In-situ specific penetration resistance p_s of layer ⑧$_1$ silty clay; (b) in-situ specific penetration resistance p_s of layer ⑧$_2$ silty clay with sand.

and the excavation depth approaches 57.8 m. The surrounding environment and soil layers are shown in Figure 13.18.

There are five CPT curves in Yunling complexes, as shown in Figure 13.19. In the figure, "YLC" represents the in-situ CPT borehole, which is in Yunling complexes, the numbers "i" (i = 1, 3, 4, 6, 7) represent the borehole number of CPT test.

13.4.2.1 Prior knowledge

Before the stochastic analysis, the key constitutive parameters are necessarily determined by sensitivity analysis. According to the preliminary geotechnical investigation, the constitutive parameters of the soil layer ⑧ are divided into seven parametric levels as shown in Table 13.8. According to the division, the orthogonal table $L_{49}\left(7^8\right)$ is selected for the sensitivity analysis, which has 49 parameter combination. In the calculation, the last two columns of the orthogonal table are regarded as error columns. Substitute each parameter

Table 13.8 Parametric levels of the constitutive parameters

Parametric level	*M*	λ	κ	*OCR*	ν	e_0
1	1.259	0.059	0.007	1.01	0.308	0.602
2	1.276	0.076	0.012	1.26	0.312	0.692
3	1.293	0.093	0.017	1.51	0.316	0.782
4	1.31	0.11	0.022	1.76	0.32	0.872
5	1.327	0.127	0.027	2.01	0.324	0.962
6	1.344	0.144	0.032	2.26	0.328	1.052
7	1.361	0.161	0.037	2.51	0.332	1.142

combination into Equation (13.26) to obtain the corresponding specific penetration resistance p_s. And then, according to Equations (13.30)–(13.32), the critical value F_i is obtained and the key constitutive parameters are determined.

Table 13.9 lists the sensitive orders of constitutive parameters under undrained/drained conditions, which are $OCR > \kappa > \lambda > e_0 > M > \upsilon$ and $OCR > \kappa > e_0 > \lambda > M > \upsilon$, respectively. In general, the overconsolidation ratio OCR and the slope of the loading/reloading line κ have an extreme sensitivity. The sensitivity under drained conditions is greater than that under undrained conditions. Compared with the undrained condition, the slope of normal compression λ and the initial void ratio e_0 have a high sensitivity under drained conditions. In the preliminary geotechnical investigation, a total of 105 initial void ratios e_0 are obtained, which has a strong correlation with the depth. Although the initial void ratio e_0 is an important physical parameter in soil mechanics, sensitivity analysis shows that it has less impact on the specific penetration resistance p_s, as well as it is easy to obtain in engineering practice. Thus, it is not considered the key constitutive parameter. To reflect the difference in the initial void ratio e_0 at a specific depth accurately, the polynomial regression method is used to establish the function between the initial void ratio e_0 and the depth. In addition, as an important constitutive parameter, the slope of the critical state line M should be calibrated. Hence, $\theta = [M, \lambda, \kappa, OCR]$ are denoted as the key constitutive parameters.

Prior distribution is the first step to calibrate constitutive parameters, which is organized based on the former geotechnical investigation. Due to the limitation of cost and time, only a few borehole logs and laboratory test data can be obtained. It is difficult to denote the constitutive parameters completely and accurately. Hence, uniform distribution is regarded as the PDF of prior knowledge. Prior information on the constitutive parameters in soil layer ⑧ is listed in Table 13.10.

13.4.2.2 Posterior distribution

Preliminary data of the geological investigation indicate that silty clay layer $⑧_1$ has a poor permeability. In the process of CPT, pore water pressure could not discharge, which results in a large value. For the silty clay with a sand layer, the amount of sand will affect the permeability. Silty clay with sand layer $⑧_2$ increases the permeability and the excess pore pressure dissipates immediately. At the same time, the CPT data will be affected by the penetration speed. The slower the penetration speed, the greater the excess pore pressure dissipates. According to the permeable condition, soil sub-layers $⑧_1$ and ⑧ adopt the undrained/drained formulas to be calibrated, respectively.

Table 13.9 Variance of sensitivity analysis

Constitutive parameter	The sum of deviation squares S_i^2		Degrees of freedom f_i	Analytic value of F		Critical value of F	Significance	
	Undrained	Drained		Undrained	Drained		Undrained	Drained
M	43.0	216.3	6	1.25	1.51	$F_{0.01}(6,12) = 4.82$	L	L
λ	53.3	441.8	6	1.55	3.09	$F_{0.05}(6,12) = 3.00$	L	H
κ	767.1	4305.0	6	22.36	30.11	$F_{0.10}(6,12) = 2.33$	E	E
OCR	4488.7	4467.2	6	130.87	31.24		E	E
v	34.1	132.5	6	0.99	0.93		L	L
e_0	51.9	470.4	6	1.51	3.29		L	H
ε_e	68.6	286.0	12	-	-		-	-

Table 13.10 Prior information of the constitutive parameters

Soil Sub-layer	***H**/m*	*M*	λ	κ	*OCR*	v	K_0	e_0
⑧$_1$	46–55	1.26–1.33	0.11–0.16	0.015–0.027	1.01–2.51	0.324	0.48	0.904–1.125
⑧$_2$	55–68	1.28–1.36	0.06–0.16	0.007–0.037	1.01–2.51	0.319	0.47	0.619–1.051

Using the MCMC algorithm to analyze every CPT data will require a lot of computational resources. In addition, the extreme value will have a great influence on the data quality. It is reasonable to choose a specific range of CPT curves for calibration. Thus, statistics of all CPT data within 1.0 m are used as the indirect data z_b^i.

The joint prior distribution $p(\theta)$ takes totally $T = 10{,}000$ Markov chain samples. Statistical characteristics of the posterior distribution for the key constitutive parameters $\theta = [M, \lambda, \kappa, OCR]$ are obtained based on the accepted posterior samples. The margin posterior distributions are shown in Figure 13.20 (a) and (b).

Figure 13.20 demonstrates the prior and posterior distributions of the key constitutive parameters for soil layer ⑧. The mean values of the key constitutive parameters $\theta = [M, \lambda, \kappa, OCR]$ in silty clay layer ⑧$_1$ are 1.296, 0.134, 0.024, and 1.438, respectively. The posterior mean values of the corresponding parameters in the silty clay with sand layer ⑧$_2$ are 1.318, 0.110, 0.019, and 1.699. Except for the critical state stress ratio M, the SDs of the other three parameters in the silty clay with sand layer ⑧$_2$ are greater than that of the silty clay layer ⑧$_1$, which indicates that the silty clay with sand layer ⑧$_2$ has greater variability. The overconsolidation ratio (OCR) is less than 2.0 after the calibration. Hence, both the two soil sub-layers are in a light overconsolidation state, and the MCC model can describe the mechanical characteristics.

13.4.2.3 Prediction of deep shaft pit deformation

To provide accurate constitutive parameters by stochastic mechanics based Bayesian method and verifying whether the posterior statistics characteristics can reflect the mechanical properties of the deep soil layers, the above calibrated constitutive parameters are used to carry out numerical analysis for the Yunling shaft pit of Suzhou River Deep Tunnel Reservoir Project in Shanghai, China. Then, the numerical simulation results are compared with the field measurements.

13.4.2.3.1 Numerical model

The shaft pit of the Shanghai deep tunnel project is a complex engineering practice. If a two-dimensional model is adopted, it needs to set up a middle beam as an internal support. At the same time, the supporting element of the annular beam, framework beam, and lining cannot be simulated. In addition, the anti-seepage wall is installed outside of the shaft pit. To simulate the actual engineering situation, according to the design schema of the Yunling shaft pit, a three-dimensional finite element model consisting of width = 160 m, length = 200 m, and height = 150 m is demonstrated in Figure 13.21. The inner diameter of the shaft is 34 m. The bottom of the shaft is located in silty clay with a sand layer ⑧$_2$. The thickness and depth of the diaphragm wall are 1.5 m and 105 m, respectively.

The MCC model can simulate light overconsolidated clay, but it cannot describe the mechanical properties of sand. As a classical plastic model, the Mohr–Coulomb (MC) model can simulate the primary mechanical behavior of clay and sand simultaneously. In addition, the effect of seepage is also considered in the numerical simulation. If the MCC model is used in the clay stratum and the MC model is used in the sand stratum, the mixture of the constitutive models will greatly cause the problem of computational non-convergence. In addition, the preliminary geotechnical investigation only provides limited

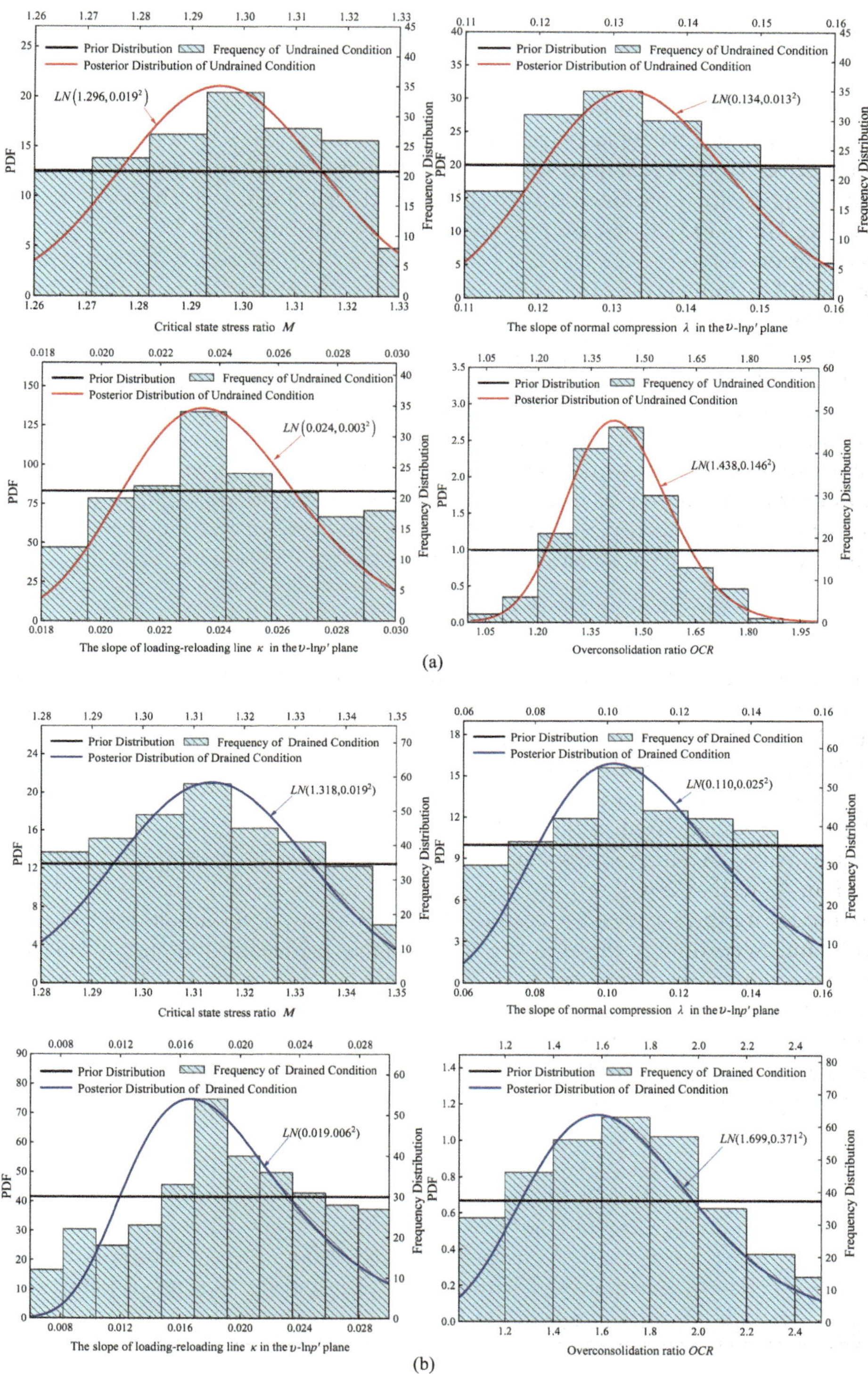

Figure 13.20 Prior and posterior distributions of the key constitutive parameters θ. (a) Silty clay layer ⑧$_1$; and (b) silty clay with sand layer ⑧$_2$.

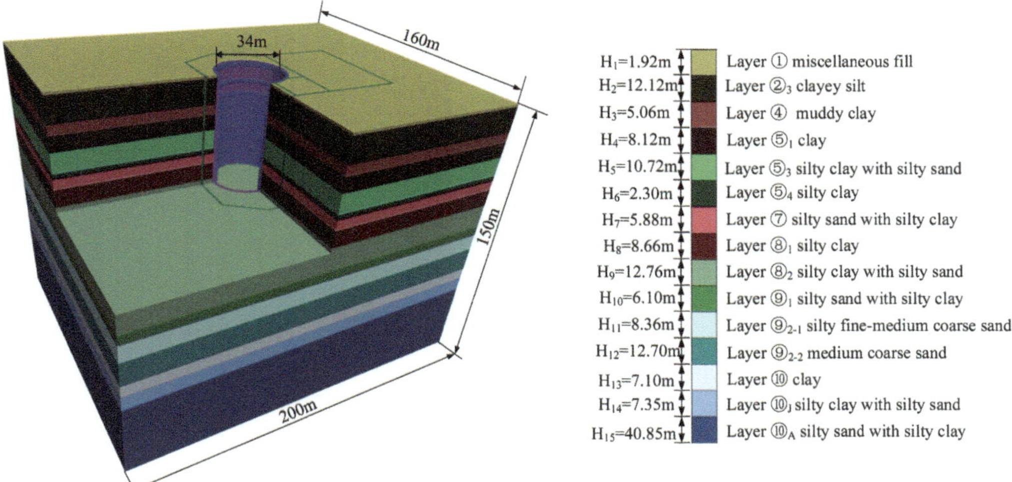

Figure 13.21 Three-dimensional numerical model.

Table 13.11 Basic geotechnical parameters of the MC model

Soil layer	*Density* ρ (kg/m^3)	*Compression modulus* E_s (MPa)	*Poisson's ratio* ν	*Cohesion* c (kPa)	*Internal frictional angle* φ (°)
①	1900	5.14	0.35	10	10
②$_3$	1860	9.38	0.36	6	27
④	1700	2.68	0.42	3	27.9
⑤$_1$	1760	3.75	0.35	4	30.5
⑤$_3$	1790	5.14	0.36	5	31.6
⑤$_4$	1950	6.38	0.36	16	32.4
⑦	1950	11.58	0.34	2	34
⑨$_1$	1940	14.66	0.31	0	35.5
⑨$_{2-1}$	2020	15.69	0.3	0	38.4
⑨$_{2-2}$	1910	17.21	0.31	0	39.8
⑩	1990	11.58	0.34	19	31.6
⑩$_J$	1940	8.56	0.32	16	32
⑩$_A$	1930	12.37	0.32	2	33.5

MCC constitutive parameters. Hence, the MCC model is only used for layer ⑧. For the other soil layers, the MC model is adopted. The basic geotechnical parameters of the MC model are listed in Table 13.11, which are obtained by the preliminary geotechnical investigation. In addition, the dilatancy angle ψ also affects the soil deformation, which adopts the proposed value given by Bolton[43]. Thus, the dilatancy angle ψ of silty clay is set to zero, as well as the dilatancy angle of silty clay with sand is calculated $\psi = \varphi' - 30°$.

For numerical analysis, solid elements are selected for supporting structures such as diaphragm wall, beam, and lining, as well as C30 concrete is used as the elastic material.

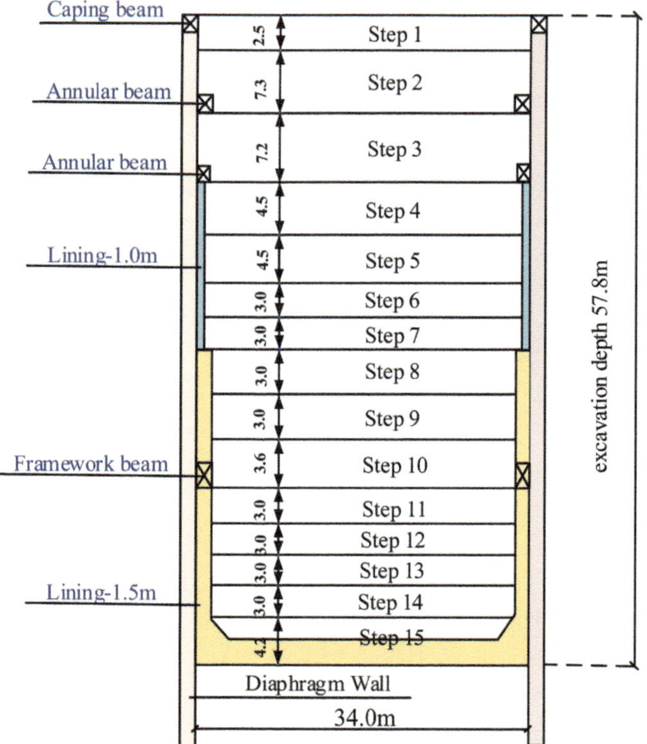

Figure 13.22 Excavation steps of Yunling shaft pit.

The deep shaft is excavated in 15 steps as shown in Figure 13.22. Drainage is carried out according to excavation conditions, and the underground water level is always 3~8 m below the excavation surface.

13.4.2.3.2 Prediction of the pit bottom heave

During foundation pit construction, two monitoring points are set at the depths of 45 m and 50 m, which are arranged at the center of the pit bottom. Figure 13.23 shows the changes in the soil heave at each excavation step, and the "measurement" represents the field measurement. From the perspective of probability, when the normal distribution is followed, the probability of falling within the range of 1.0, 2.0, and 3.0 standard deviations is 68.26%, 95.44%, and 99.74%, respectively. Considering the particularity of geotechnical engineering, 1.0 standard deviation is selected as the value range of geotechnical parameters. In this chapter, the prediction value of the pit bottom heave is calculated by the posterior mean value. Then, according to the range of posterior value, 16 calculated values of the pit bottom heave were obtained after $2^4 = 16$ calculations and the maximum and minimum prediction values were determined. The predicted values are close to the measurements, and the measurements are enclosed in the predicted envelope. It indicates that the stochastic mechanics-based Bayesian method can provide accurate constitutive parameters for the prediction of the pit bottom heave.

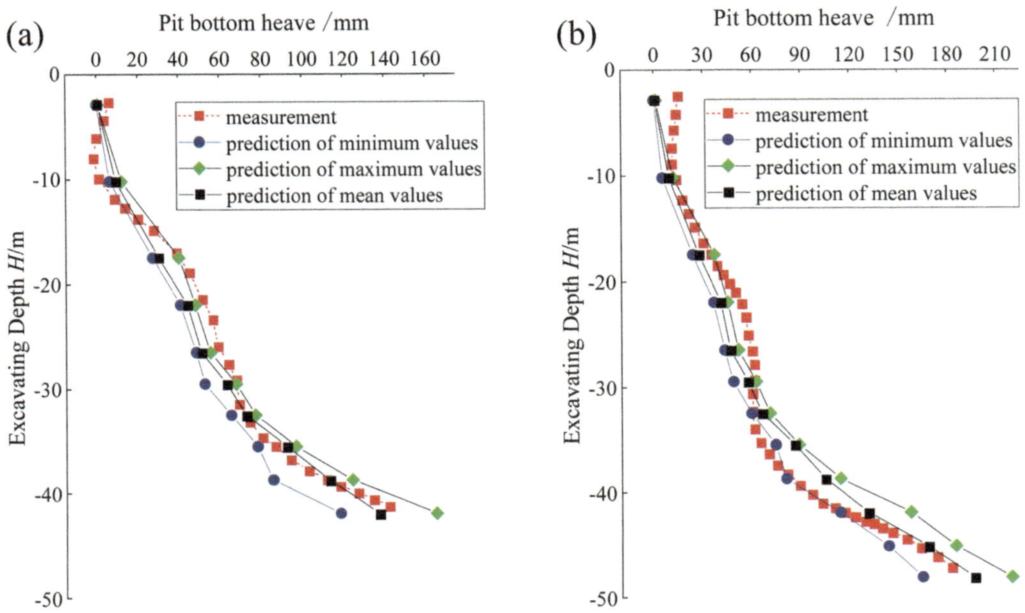

Figure 13.23 Pit bottom heave. (a) Monitoring point at a depth of 45 m; and (b) monitoring point at a depth of 50 m.

13.4.2.3.3 Prediction of the lateral displacement of the diaphragm wall

In addition to soil heave at the center of the pit bottom, the diaphragm wall lateral displacement is another important monitoring item. Figure 13.24 (a, b) shows the lateral displacement of the diaphragm wall when the foundation pit is excavated to 47.6 m and

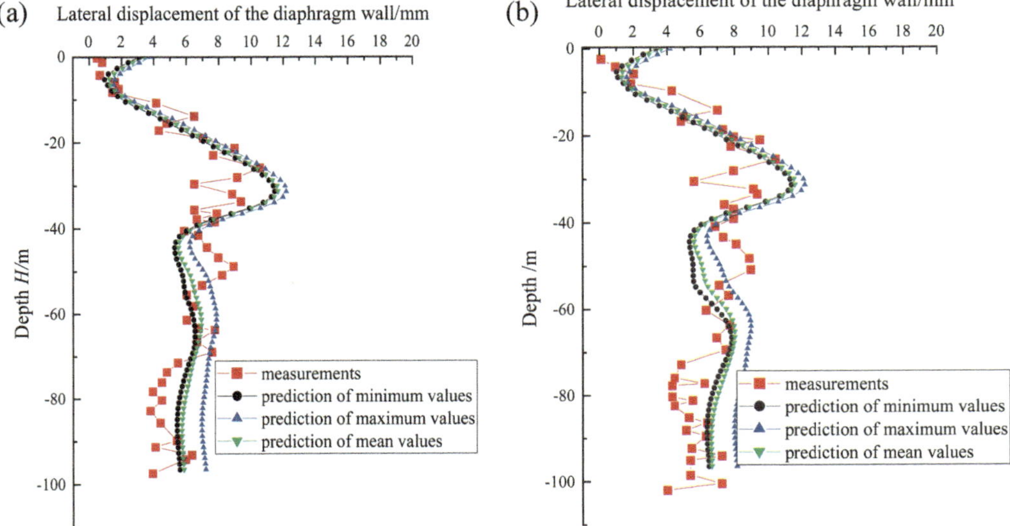

Figure 13.24 Lateral displacements of the diaphragm wall: (a) excavation to 47.6 m; and (b) excavation to 53.6 m.

53.6 m, respectively. The simulated results of the posterior maximum and minimum constitute the upper and lower limits of the lateral displacements for the diaphragm wall. Although the basic geotechnical parameters of other soil layers are not calibrated, the predicted values coincide with the measurements, which further verify the effectiveness of the stochastic mechanics-based Bayesian method.

13.4.3 Bothkennar site in the United Kingdom

Bothkennar Parish is a soft clay testing ground, which is located on the south bank of the Forth River, near the Kincardine Bridge in Scotland, UK. The site is approximately 11 hectares of a low-lying alluvial plain, which is surrounded on three sides by flood dykes. The surface layer is the modern tidal flat sediment with a thickness of 1.5 m. The water table is about 0.75 m below the ground[44]. The geotechnical parameters of the Bothkennar area are listed in Table 13.12. The bottom layer is silty clay with a lower overconsolidation as shown in Figure 13.25.

Many engineers and scholars have carried out geological surveys on the site. Jacobs and Coutts conducted experiments to study the influence of different piezocone types on pore water pressure and found that the filter material has little influence, while the filter location has a greater influence[45]. Two CPTU resistance curves extracted from Hight et al.[44] and Nash et al.[26] are shown in Figure 13.26.

13.4.3.1 Parametric analysis

According to the yield function of CASM, the key constitutive parameters can be selected from M, λ, κ, OCR , r^*, n and e. To determine the influences of constitutive parameters on the CPTU resistances, the single-factor method is utilized to take the parametric sensitivity analysis, and the statistical correlations between the parameters are ignored temporarily. The parametric values at each level are listed in Table 13.13, the mean μ of a single parameter is selected as the baseline of the sensitivity analysis, and the level factors are equally distributed with the baseline, i.e., $\Delta x = (A_{\max} - A_{\min})/6$, where $A_{\max}$ is the maximum, $A_{\min}$ is the minimum of the parameter. Based on the past experimental results and engineering experiences of clayey soil, the value n ranges from 1.0 to 5.0, and the value r^* ranges from 1.5 to 3.0[23]. The values of M, λ, κ, and OCR are listed in Table 13.12. The influences of the constitutive parameters on the corrected cone-tip resistance q_t and pore water pressure u are shown in Figure 13.27 (a, b).

The sensitivity factors of an individual constitutive parameter are listed in Table 13.14. The sensitivity order of a single parameter to q_t is $r^* \approx M > OCR > n > e > \kappa > \lambda$. On the other hand, the sensitivity order to u is $M > OCR > r^* > e \approx \kappa > n > \lambda$. In summary, by comprehensive consideration of parameter acquisition methods and sensitivity results, the key constitutive parameters are selected as r^*, n, M, and OCR

13.4.3.2 Preliminary prediction of CPTU resistances

According to the original geotechnical parameters of the clayey soil as listed in Table 13.13, the CPTU resistances are preliminarily predicted by spherical cavity expansion theory based on CASM. Hence, the theoretical CPTU resistance curve is shown in Figure 13.28 (a, b). The theoretical values of corrected cone-tip resistance q_t and

Table 13.12 Geotechnical parameters of Bothkennar site

Depth **h**(*m*)	**OCR**	λ	κ	s_u (*kPa*)	G_s (*MPa*)	e	*M*	φ' (°)
2.7~17.3	1.4~1.6	0.2~0.3	0.015~0.025	20.1~50.3	2.6~8.6	0.6~1.2	1.374~1.418	34~35

Notes: Effective unit weight $\gamma = 6.5$ kN/m^3, Poisson ratio $\nu = 0.3$.

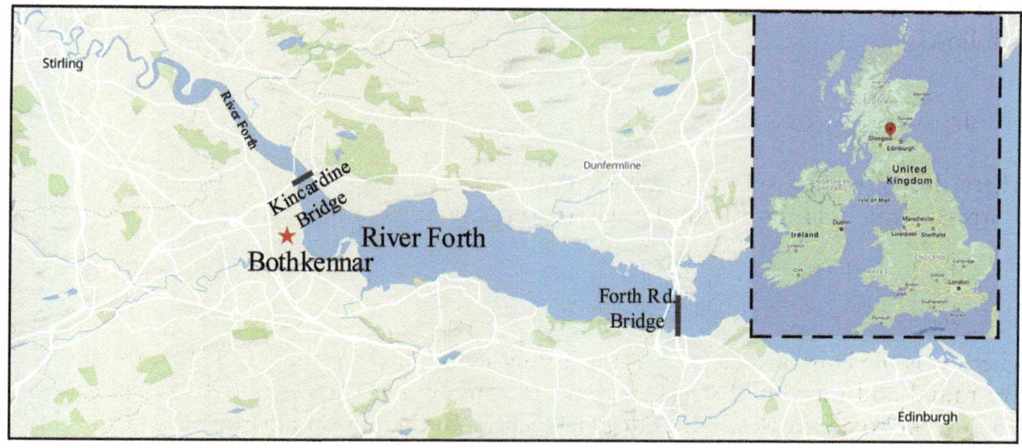

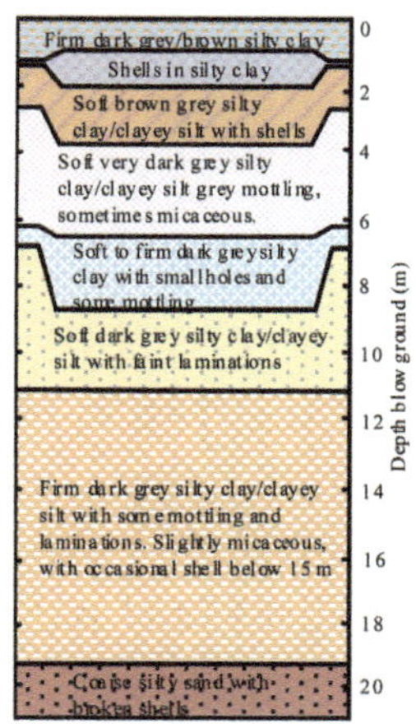

Figure 13.25 Location and soil distribution of the Bothkennar site.

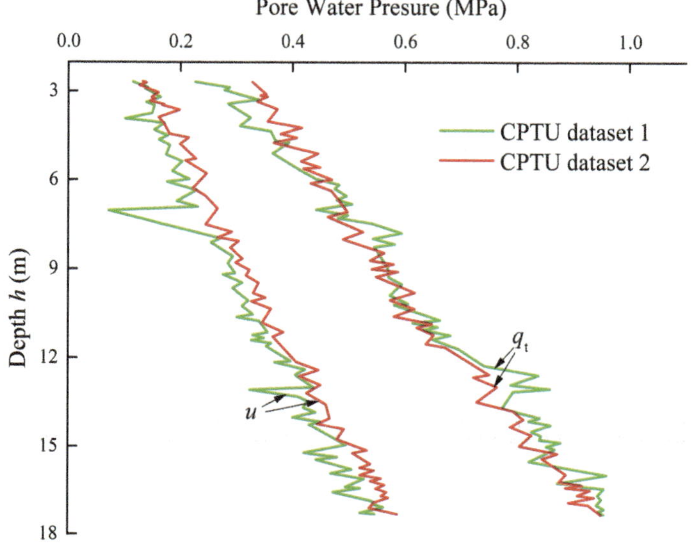

Figure 13.26 In-situ test data of CPTU curves.

Table 13.13 Statistics of parametric sensitivity analysis

Level factor	*M*	λ	κ	r^*	*OCR*	*n*	e
Baseline μ	1.394	0.250	0.020	2.25	1.500	3.00	0.9
$\mu - 3\Delta x$	1.370	0.200	0.014	1.50	1.400	1.00	0.6
$\mu - 2\Delta x$	1.378	0.217	0.016	1.75	1.433	1.67	0.7
$\mu - \Delta x$	1.386	0.233	0.018	2.00	1.467	2.34	0.8
$\mu + \Delta x$	1.402	0.267	0.022	2.50	1.533	3.67	1.0
$\mu + 2\Delta x$	1.410	0.283	0.024	2.75	1.567	4.34	1.1
$\mu + 3\Delta x$	1.418	0.300	0.026	3.00	1.600	5.00	1.2

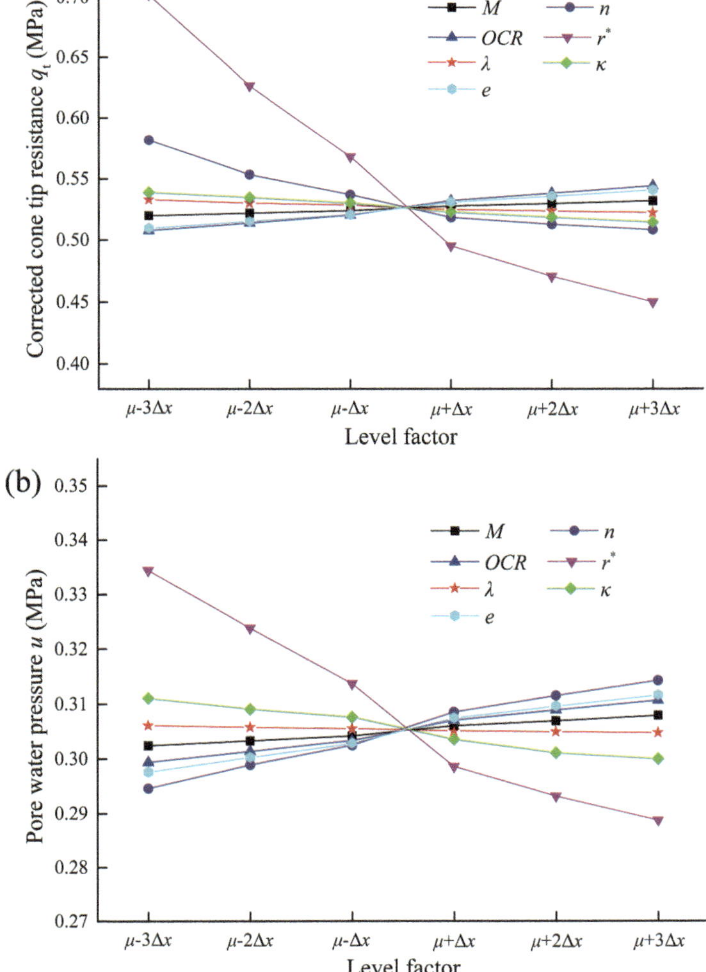

Figure 13.27 Parametric sensitivity of constitutive parameters. (a) Influence on corrected cone-tip resistance q_t, and (b) influence on pore water pressure *u*.

Table 13.14 Sensitivity factors of constitutive parameters

CPTU resistances	*Level factor*	*M*	n	*OCR*	r^*	λ	κ	e
q_t	$\mu \pm 3\Delta x$	0.655	0.103	0.525	0.803	0.049	0.076	0.089
	$\mu \pm 2\Delta x$	0.643	0.086	0.524	0.666	0.045	0.075	0.088
	$\mu \pm \Delta x$	0.602	0.079	0.515	0.621	0.048	0.070	0.086
u	$\mu \pm 3\Delta x$	0.499	0.053	0.277	0.249	0.011	0.060	0.068
	$\mu \pm 2\Delta x$	0.499	0.049	0.276	0.233	0.011	0.065	0.068
	$\mu \pm \Delta x$	0.499	0.047	0.276	0.223	0.011	0070	0.068

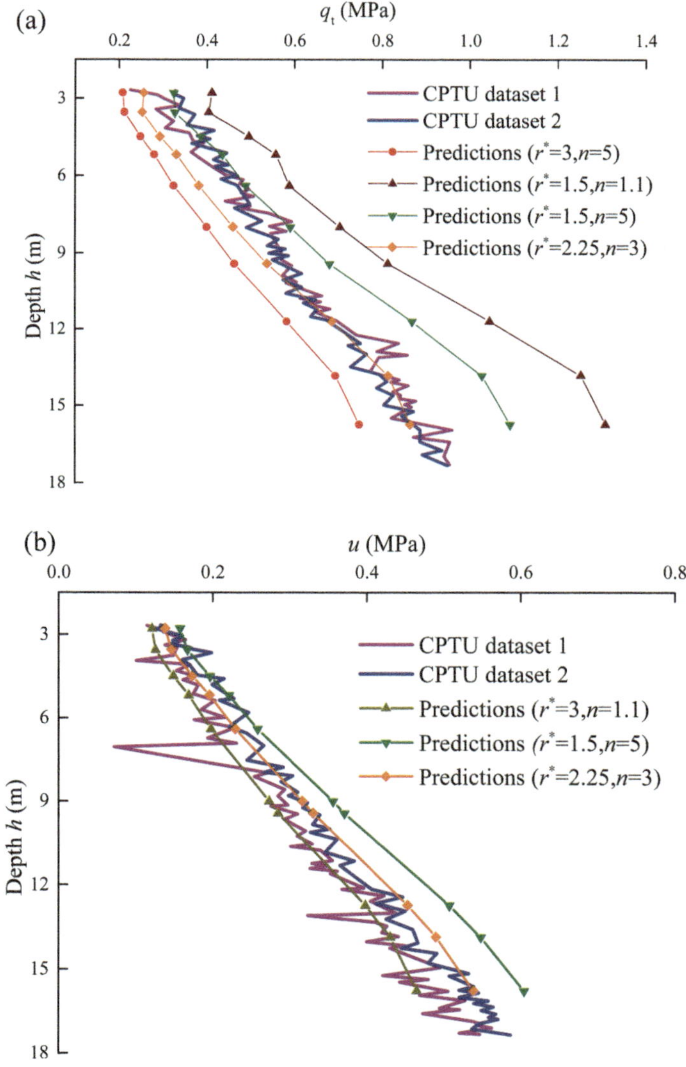

Figure 13.28 Predicted curves of the CPTU resistances by the original geotechnical parameters: (a) cone-tip resistance q_t; and (b) pore water pressure u.

Table 13.15 Prior knowledge of the key constitutive parameters ψ

Prior knowledge	r^*	n	M	OCR
Minimum	1.5	1.0	1.37	1.4
Maximum	3.0	5.0	1.42	1.6

pore water pressure u differ greatly from the in-situ test data because some constitutive parameters are random, fuzzy, and uncertain.

13.4.3.3 Constitutive parameters calibration

To reduce the boundary disturbance of the soil layers, the in-situ CPTU resistances at the depth of 3.5–15.75 m are taken as the z_b . Due to the insufficient test data, the prior knowledge of multiple random variables $\psi = [r^*, n, M, OCR]$ is subjected to uniform distribution at the beginning stage, which is listed in Table 13.15.

The joint prior distribution $p(\psi)$ is randomly sampled 10,000 times. Hence, the posterior results of the key constitutive parameters through the Bayesian calibration by CPTU resistances are shown in Figure 13.29. The posterior distribution of the spacing ratio subjects, $r^* \sim LN(2.447, 0.267^2)$, is shown in Figure 13.29 (a), stress-state coefficient subjects, $n \sim LN(2.952, 1.092^2)$, is shown in Figure 13.29(b), critical stress ratio subjects,

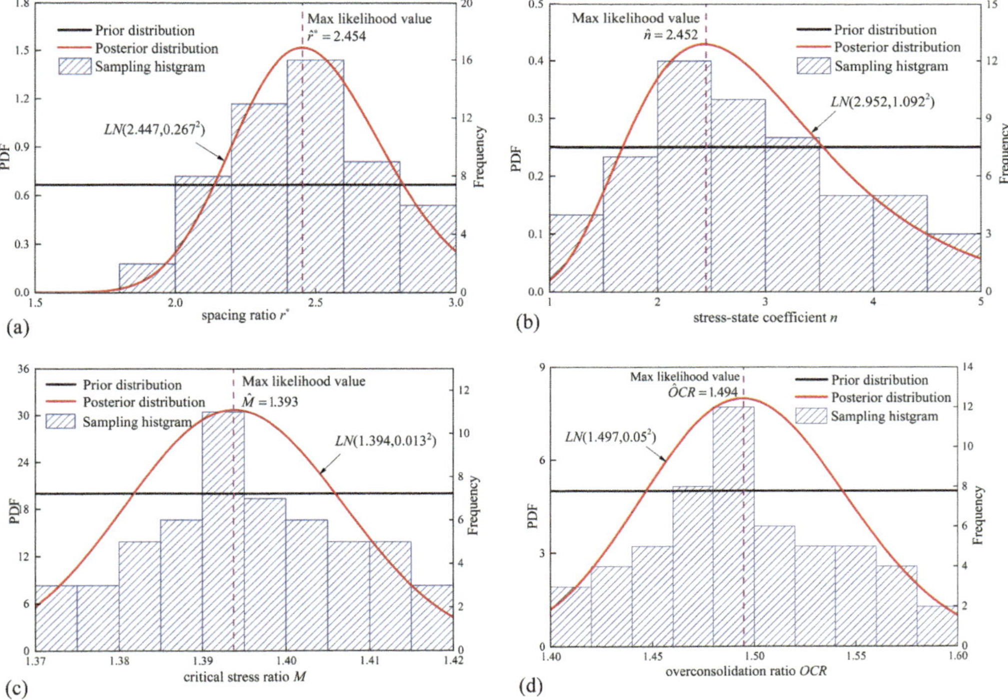

Figure 13.29 Posterior distributions of the key constitutive parameters: (a) spacing ratio r^*; (b) stress-state coefficient n; (c) critical stress ratio M; and (d) overconsolidation ratio OCR .

Table 13.16 Correlation coefficients among the key constitutive parameters

Constitutive parameter	r^*	n	M	OCR
r^*	1.000	−0.162	−0.132	0.266
n		1.000	−0.005	−0.003
M			1.000	0.040
OCR				1.000

$M \sim LN(1.394, 0.013^2)$, is shown in Figure 13.29(c), and overconsolidation ratio subjects, $OCR \sim LN(1.497, 0.05^2)$, is shown in Figure 13.29(d). The coefficient of variation is $\delta_{r^*} = 0.109$, $\delta_n = 0.368$, $\delta_M = 0.009$, and $\delta_{OCR} = 0.033$, respectively. The maximum likelihood value is $\hat{r}^* = 2.454$, $\hat{n} = 2.452$, $\hat{M} = 1.393$, and $\hat{O}CR = 1.494$, individually.

Through the posterior discrete samples of the key constitutive parameters $\psi = [r^*, n, M, OCR]$, the correlation coefficients among them are calculated as listed in Table 13.16.

Table 13.16 indicates that the key constitutive parameters exist cross-correlations between each other. The correlation coefficients between r^* and n, M, OCR are −0.162, −0.132, and 0.266, respectively. There is an unignorable correlation between r^* and OCR. However, there is a subtle correlation between the n and M, OCR, even M and OCR.

ACKNOWLEDGMENT

This work is sponsored by the National Natural Science Foundation of China (51208303), Program for Professor of Special Appointment (Eastern Scholar) at Shanghai Institutions of Higher Learning (TP2018042), and the Shanghai Pujiang Programme (18PJ1403900). Our research is also supported by the Science and Technology Commission of Shanghai Municipality (21DZ1204300). The contributors include Kun Wang, Daofei Tang, Chaoxing Wu, Chengtao Ma, and Tian Fang.

REFERENCES

1. Larsson R, Mulabdic M. Piezocone tests in clay. Report 42. Linköping: Swedish Geotechnical Institute, 1991.
2. Mayne PW. Stress-strain-strength-flow parameters from enhanced in-situ tests. Proceedings International Conference on In-Situ Measurement of Soil Properties and Case Histories, Bali, Indonesia, 2001.
3. Senneset K, Janbu N, Svano G. Strength and deformation parameters from cone penetration tests. Proceedings of the 2nd European Symposium on Penetration Testing, ESOPT-II. Rotterdam: Balkema Publishers, 1982: 863–870.
4. Senneset K, Sandven R, Janbu N. Evaluation of soil parameters from piezocone tests. Journal of Transportation Research, 1989(1): 24–37.
5. Kulhawy FH, Mayne PH. Manual on Estimating Soil Properties for Foundation Design. Electric Power Research Institute, EPRI, Ithaca, 1990.

6. Bruzzi D, Battaglio M. Pore pressure measurements during cone penetration test. ISMES Research Report No. 229. 1987.
7. Kulhawy FH, Mayne PW. Manual on estimating soil properties for foundation design. Electric Power Research Institute, 1990.
8. Baldi G, Bellotti R, Ghionna VN, et al. Modulus of sands from CPTs and DMTs. Proceedings of the 12th International Conference on Soil Mechanics and Foundation Engineering, Rio de Janeiro, 1989.
9. Bishop R, Hill R, Mott NF. Theory of indentation and hardness tests. *Proceedings of the Physical Society* 1945, 57: 147–169.
10. Gibson R, Anderson W. Insitu measurement of soil properties with the pressuremeter. *Civil Engineering and Public Works Reviews*, 1961, 56(658): 615–618.
11. Wood DM. *Soil Behaviour and Critical State Soil Mechanics*. Cambridge University Press, London, 1990.
12. Cao Z, Zheng S, Li D, Phoon KK. Bayesian identification of soil stratigraphy based on soil behaviour type index. *Canadian Geotechnical Journal*, 2019, 56: 570–586.
13. Yu H. CASM: a unified state parameter model for clay and sand. *International Journal for Numerical and Analytical Methods in Geomechanics*, 1998, 22: 621–653.
14. Yao Y. Advanced UH models for soils. *Chinese Journal of Geotechnical Engineering*, 2015, 37: 193–217 (in Chinese).
15. Miller EA, Sowers GF. The strength characteristics of soil-aggregate mixtures & discussion. *Highway Research Board Bulletin*, 1958, 183:16–32.
16. Chen S, Abousleiman YN. Exact undrained elasto-plastic solution for cylindrical cavity expansion in modified Cam Clay Soil. *Géotechnique*, 2012, 62: 447–456.
17. Chen S, Abousleiman YN. Exact drained solution for cylindrical cavity expansion in modified Cam Clay soil. *Géotechnique*, 2013, 63: 510–517.
18. Zhang W, Zou J, Zhang X, Yuan W, Wu W. Interpretation of cone penetration test in clay with smoothed particle finite element method. *Acta Geotechnica*, 2021, 16: 2593–2607.
19. Ma H, Chen Z, Yu S. Correlations of soil shear strength with specific penetration resistance of CPT in Shanghai area. *Rock and Soil Mechanics*, 2014, 35(2): 536–542 (in Chinese).
20. Zheng S, Cao Z, Li D, Au Suikui. Identification of underground stratigraphy based on specific penetration resistance. *Engineering Journal of Wuhan University*, 2018, 51(8): 679–687 (in Chinese).
21. Fan X, Wei G, Ma X. The relationship between single bridge probe and double bridge prob cone penetration test in Shanghai. *Geotechnical Investigation & Surveying*, 2007, 2007(9):10–12.
22. Roscoe KH, Thurairajah A, Schofield AN. Yielding of clays in states wetter than critical. *Géotechnique*, 1963, 13(3): 211–240.
23. Yu H. CASM: a unified state parameter model for clay and sand. *International Journal for Numerical and Analytical Methods in Geomechanics*, 1998, 22(8): 621–653.
24. Been K, Jefferies MG. A state parameter for sands. *Géotechnique*, 1985, 35(2): 99–112.
25. Yu H. *Cavity Expansion Methods in Geomechanics*. Dordrecht: Springer, Netherland, 2000.
26. Nash FT, Powell JJM, Lloyd IM. Initial investigations of the soft clay test site at Bothkennar. *Géotechnique*, 1992, 42(2): 163–181.
27. Gong W, Tien YM, Juang CH, Martin JR, Luo Z. Optimization of site investigation program for improved statistical characterization of geotechnical property based on random field theory. *Bulletin of Engineering Geology and the Environment*, 2017, 76(3): 1021–1035.
28. Qi X, Zhou W. An efficient probabilistic back-analysis method for braced excavations using wall deflection data at multiple points. *Computers and Geotechnics*, 2017, 85: 186–198.
29. Box GEP, Wilson KB. On the experimental attainment of optimum conditions. *Breakthroughs in Statistics*. Springer, New York, 1992: 270–310.
30. Matheron G. Principles of geostatistics. *Economic Geology*, 1963, 58(8):1246–1266.

31. Vanmarcke EH. Probabilistic modelling of soil profiles. *Journal of Geotechnical Engineering Division,* 1977, 103(11): 1227–1246.
32. Wang C, Osorio-Murillo C, Zhu H, Rubin Y. Bayesian approach for calibrating transformation model from spatially varied CPT data to regular geotechnical parameter. *Computers and Geotechnics,* 2017, 85: 262–273.
33. Dereniowski D, Kubale M. Cholesky factorization of matrices in parallel and ranking of graphs. *International Conference on Parallel Processing and Applied Mathematics*, Czestochowa, 2003: 985–992.
34. Mantoglou A, Wilson JL. The turning bands method for simulation of random fields using line generation by a spectral method. *Water Resources Research*, 1982, 18(5): 1379–1394.
35. Zhu H, Griffiths DV, Fenton GA, Zhang L. Undrained failure mechanisms of slopes random soil. *Engineering Geology*, , 2015, 191: 31–35.
36. Huang S, Quek ST, Phoon KK. Convergence study of the truncated Karhunen-Loeve expansion for simulation of stochastic processes. *International Journal for Numerical Methods in Engineering*, 2001, 52(9): 1029–1043.
37. Pebesma EJ. Multivariable geostatistics in S: the gstat package. *Computers and Geosciences*, 2004, 30(7): 683–691.
38. Liang S, Ren X, Li J. A random medium model for simulation of concrete failure. *Science China Technological Sciences*, 2013, 56(5): 1273–1281.
39. Eldeiry AA, Garcia LA. Comparison of ordinary kriging, regression kriging, and cokriging techniques to estimate soil salinity using LANDSAT images. *Journal of Irrigation and Drainage Engineering*, 2010, 136: 355–364.
40. Bhattacharjee A. Distance correlation coefficient: an application with Bayesian approach in clinical data analysis. *Journal of Modern Applied Statistical Methods*, 2014, 13(1):354-366.
41. Tan TS, Phoon KK, Hight DW, Leroueil S. Characterisation and engineering properties of natural soils. *Two Volume Set: Proceedings of the Second International Workshop on Characterisation and Engineering Properties of Natural Soils*, Singapore, 29 November–1 December 2006, Vol. 3. CRC Press, London.
42. Thermann K, Gau C, Tiedemann J. Shear strength parameters from direct shear tests–influencing factors and their significance. *IAEG2006 Paper*, 2006, 484:1–12.
43. Bolton MD. The strength and dilatancy of sands. *Géotechnique*, 1986, 36: 65–78.
44. Hight DW, Bond AJ, Legge JD. Characterization of the Bothkennaar clay: an overview. *Géotechnique*, 1992, 42(2): 303–347.
45. Jacobs PA, Coutts JS. A comparison of electric piezocone tips at the Bothkennar test site. *Géotechnique*, 1992, 42(2): 369–375.

Chapter 14

Data-centric seismic soil liquefaction assessment

Approaches, data, and tools

Chaofeng Wang, Qiushi Chen, and Pedro Arduino

This chapter discusses seismic soil liquefaction and the pressing need for enhanced assessment methods to better evaluate this devastating geological phenomenon. The main focus revolves around the shift from traditional, primarily empirical, assessment methods toward more comprehensive, data-centric approaches. While conventional assessment practices have been useful, they exhibit limitations due to the inherent complexity of soil and seismic dynamics. This chapter explores how contemporary approaches harness large-scale, diverse datasets, cutting-edge computational power, and advanced analytical methodologies, such as artificial intelligence, to provide deeper insights into soil liquefaction and enhance prediction accuracy. The multiscale applicability of the data-centric approach is also underscored, demonstrating its potential from microscale studies to macroscale and regional assessments. Finally, the chapter reports on a case study utilizing extensive datasets and advanced AI tools for high-resolution, city-scale liquefaction hazard estimation, demonstrating the effectiveness of this innovative approach. This contributes to our evolving understanding of seismic soil liquefaction and our efforts to predict and mitigate its damaging effects.

14.1 INTRODUCTION

In fluid-saturated, geologically unconsolidated soil, liquefaction is a phenomenon caused by a sudden change in stress due to a rapid earthquake loading, during which the pore-water pressure increases and the effective stress reduces so that the soil loses much of its stiffness and strength and behaves like a liquid. As summarized in the Kavazanjian et al.'s [28] report, the consequences of liquefaction may include vertically or laterally displaced ground, landslides, slumped embankments, foundation failures, and mixtures of soil and water erupting at the ground surface. In turn, these effects may lead to settlement, distortion, and the collapse of buildings; the disruption of roadways; the failure of earth-retaining structures; the cracking, sliding, and over-topping of dams, highway embankments, and other earth structures; the rupture or severing of sewer, water, fuel, and other lifeline infrastructure; the lateral displacement and shear failure of piles supporting bridges and waterfront structures; and the uplift of underground structures.

Liquefaction is a complex phenomenon – a multiscale, multiphysics problem that can be characterized at the microscale (particle-pore), mesoscale (particle cluster), and macroscale (field). Modern liquefaction engineering was initiated in the wake of two devastating earthquakes: the 1964 Niigata (Japan) and the 1964 Great Alaska Earthquakes. Since then, liquefaction research has been a continuous topic. Over the decades that have followed, significant progress has been made.

DOI: 10.1201/9781003441946-14

In the early stage, researchers focus on the assessment of the likelihood of initiation of liquefaction in clean, sandy soils. As observations continue to increase, there is a growing awareness of the problems associated with silty and gravelly soils in the context of earthquakes. Additionally, the issues of post-liquefaction strength and the stress-deformation behavior of soils have also begun to attract more attention. Quantitative assessment of the "triggering"/initiation of liquefaction and the consequencing deformations of soils are two topics of great importance. The dominant approach in common engineering practice is to correlate observed field behavior with various in-situ "index" tests including the standard penetration test (SPT), the cone penetration test (CPT), shear-wave velocity (V_s), etc. These in-situ test methods have now reached a level of sufficient maturity. A plethora of such empirical and semi-empirical methods have been proposed and utilized in the past (e.g., [24, 44]). Despite their effectiveness to a certain degree, these methods tend to oversimplify the intricate dynamics of soil behavior under seismic loading. This is primarily due to their limitations in handling the complexity and diversity of soil and seismic activity data.

To date, state-of-the-art developments in liquefaction predict it at the macroscopic scale. To enable a more complete understanding of this devastating instability, it is imperative that the micro-mechanical origins of liquefaction should be understood. Numerical simulation tools such as finite-element method (FEM) and discrete element method (DEM) have been used to solve geotechnical problems since they were invented. Numerous efforts have been made to use FEM and DEM or other numerical methods for modeling the liquefaction of soils [15, 36, 43, 49, 66]. However, none of these numerical methods could fully address the cross-scale issue independently. The multiscale issue in the numerical simulation of granular media has attracted more and more attention. However, less attention has been paid to the multiscale simulation of the liquefaction phenomenon.

In recent years, there has been a growing interest in the data-centric approach, which leverages advanced data collection technologies, computational power, and analytical methods to improve the understanding of seismic soil liquefaction. At its core is the pattern recognition enabled by large-scale and diverse datasets collected from different sources such as geological surveys, lab testing, and field observations after earthquakes. These growing datasets provide rich and multidimensional information about soil and seismic characteristics. Analyzed by new methods, such as artificial intelligence algorithms, these datasets can potentially shed light on the complexities of seismic soil liquefaction. By identifying patterns and relationships within the data, we can develop more accurate models for predicting soil liquefaction. This is a significant advancement over traditional methods, which often struggle with handling the complexity of soil and seismic data. Moreover, the data-centric approach is not confined to a single scale. It can be employed at various levels – from the particle-pore scale to the regional assessments. This multiscale applicability further amplifies the potential of the data-centric approach in seismic soil liquefaction assessment. With the growth of data, the data-centric approach shows promise in modeling the cross-scale nature of soil liquefaction within a single unified framework.

This chapter first outlines the state-of-the-art liquefaction assessment in terms of the methodology, data, and tools. Then, we present a case study that leverages a large dataset and advanced artificial intelligence tool to estimate the liquefaction hazard at the city scale.

14.2 DATA

Grasping the data that underpins seismic soil liquefaction assessments is crucial for crafting models that effectively predict hazards. The rich and varied data sources feeding into the extensive datasets now increasingly harnessed for predicting seismic soil liquefaction underline the complexity of the task. Essential input data for evaluating the liquefaction risk include earthquake intensity measures (IMs), geological and technical site condition characteristics, and site topography. A comprehensive data would enable a nuanced understanding and prediction of liquefaction hazards.

14.2.1 Ground motion and intensity measures

Ground motion and intensity measures (IMs) are fundamental to both data-driven and mechanics-based approaches in the assessment of liquefaction potential. These approaches necessitate detailed input data regarding the seismic activity expected or experienced at a site.

1. **Data-driven approaches**: These methods leverage simplified empirical models to predict liquefaction susceptibility. They typically require inputs such as the earthquake moment magnitude, which provides a measure of the total energy released by the earthquake, and various intensity measures, including peak ground acceleration (PGA), which indicates the maximum acceleration of ground motion at a site during an earthquake; spectral acceleration ($Sa(T)$), reflecting the maximum acceleration of a structure responding at its own period T; and Arias intensity (A_I), a measure of the earthquake's energy content or severity as it affects the site. The choice of IM is critical and is dictated by the specific analytical model in use, aiming to capture the response of soil layers to seismic waves accurately.
2. **Mechanics-based simulations**: These approaches are more sophisticated, often requiring full ground motion records for analysis. They simulate the soil's physical and mechanical response to seismic waves, offering a detailed perspective on how different layers and conditions of soil contribute to liquefaction risk.

14.2.2 Site characteristics

Site characteristics include geologic and geotechnical site conditions, water table, and topography.

14.2.2.1 Geologic characteristics

A significant portion of the data is derived from geological surveys. These comprehensive studies generate geological maps that show various soil types, their distribution, and the underlying geological formations in different regions. They include extensive data on soil and rock characteristics, stratigraphy, and structural features like faults and folds, essential for assessing soil liquefaction potential. Regional liquefaction assessment depends on such geological information. Youd and Perkins [64] proposed a method for classifying the liquefaction susceptibility of numerous geologic units. This requires the mapping of geological units, which are generally characterized by their depositional environment, physical characteristics, and age. The depth of the groundwater table is also required.

14.2.2.2 In-site geotechnical characteristics

In-situ geotechnical characterization plays an important role in understanding the ground conditions of a site, offering insights critical for the assessment of liquefaction hazards. This characterization involves a comprehensive examination of soil properties directly at the site, providing data that reflects the nature of the soil layers. Key to in-situ testing is the utilization of methods designed to measure soil response and characteristics. The cone penetration test (CPT) is one such technique, known for its ability to provide a continuous profile of soil resistance, from which soil type, shear strength, and stratification can be inferred. Similarly, the standard penetration test (SPT) measures the resistance of soil to penetration by a standard-sized hammer, offering valuable data on soil density and composition. Shear-wave velocity (V_s) measurements, on the other hand, yield information on the elastic properties of the soil, crucial for seismic response analysis. These in-situ tests are indispensable for the accurate modeling of soil behavior under seismic loading, allowing engineers to predict how different soil layers will react to earthquake-induced forces. The choice of tests and the interpretation of their results require expertise in geotechnical engineering, as the data must be analyzed within the context of the site's overall geologic setting and seismic history.

14.2.2.3 Laboratory geotechnical characteristics

Complementing in-situ analyses, laboratory geotechnical testing offers a controlled setting to delve into the fundamental properties of soil samples. These tests are essential for a granular understanding of soil behavior, especially under conditions that might not be directly measurable in the field. Laboratory tests typically involve detailed examinations of undisturbed soil samples, which are carefully extracted from the site to preserve their natural structure and water content. Through these tests, geotechnical engineers can determine a wide range of soil properties, including particle size distribution, which helps in classifying soil types; specific gravity, vital for understanding the density and buoyancy of soil particles; void ratio, indicating the porosity of the soil; permeability, which affects how water flows through soil layers; and shear strength, a key parameter in evaluating the stability of soil and its resistance to sliding or flowing under stress. Despite the invaluable insights provided by laboratory testing, it is important to acknowledge the challenges in extrapolating these controlled results to predict field behavior accurately. The complex and nonlinear nature of soil response, influenced by factors such as heterogeneity, anisotropy, and the dynamic nature of loading conditions, can make it difficult to fully mimic in-situ conditions. Therefore, a combination of laboratory and in-situ tests, along with advanced modeling techniques, is often employed to achieve a comprehensive understanding of soil behavior under both static and dynamic loads. Together, in-situ and laboratory geotechnical characteristics form the backbone of soil mechanics and geotechnical engineering, offering the detailed data necessary for the effective assessment of liquefaction hazards.

14.2.2.4 Water

The water table is a pivotal element in assessing the risk of liquefaction. Sites where the groundwater level is close to the surface are particularly prone to this hazard. In areas with high groundwater levels, this increase in seismic-induced pore water pressure can be more pronounced and rapid, therefore is more effective in turning the soil into a fluid-like state, drastically reducing its ability to support structures. Moreover, the presence of a shallow water table means that even moderate seismic events can trigger significant increases in pore water pressure, leading to widespread soil liquefaction. This susceptibility is exacerbated in soils composed of loose, fine-grained materials, such as silt and sand, which are inherently more prone to rearrangement under shaking. The interaction between the seismic forces and the inherent properties of the soil and groundwater creates conditions ripe for liquefaction, underscoring the importance of water table considerations in geotechnical and seismic risk assessments. Understanding the dynamics of the groundwater table in relation to seismic activity also involves recognizing how changes in the water level, whether seasonal or due to human activities, can affect the liquefaction potential of a site. For example, irrigation, drainage, and changes in land use can alter the depth of the groundwater table, potentially increasing the risk of liquefaction in previously stable areas or vice versa. Given these factors, accurate mapping of the groundwater table is crucial for identifying areas at risk of liquefaction.

14.2.2.5 Topography

Ground topography, referring to the arrangement and features of the ground surface including elevation, slope, orientation, and other aspects of the terrain, has a significant impact on how seismic waves interact with the soil, affecting the risk and extent of liquefaction. Three fundamental topographic parameters serve as crucial inputs in liquefaction assessments: ground slope, free face height, and the distance to a free face. Ground slope, or gradient, is the degree of steepness or incline of a surface. It is typically quantified as the ratio of the vertical change (height) to the horizontal change (distance) over a specific interval. The ground slope influences the behavior of the soil during an earthquake. Generally, the steeper the slope, the higher the risk of slope instability and subsequent soil liquefaction. Therefore, a comprehensive understanding of the ground slope is necessary to estimate the potential for seismic soil liquefaction accurately. A free face refers to an exposed vertical or near-vertical surface of soil or rock, such as cliffs, bluffs, or steep excavation walls, with no material support beyond it. The height of the free face is another important parameter for determining soil stability. Tall free faces are more susceptible to slumping or collapse during a seismic event if the underlying soils undergo liquefaction. Therefore, it is important to consider the free face height in liquefaction assessments, especially when dealing with sloping terrains or engineered structures like embankments and retaining walls. The distance to a free face from the point of interest also plays a significant role in a liquefaction assessment. If a soil mass close to a free face undergoes liquefaction, the loss of strength can trigger a lateral spread or landslide, especially if the soil mass is on a slope. Thus, the proximity to a free face can significantly influence the soil's behavior during an earthquake and the subsequent hazard from soil liquefaction.

The process of modeling soil liquefaction under seismic loading is inherently complex, and the choice of the modeling approach used can greatly affect the types of outputs produced. These outputs offer insights into the nature and magnitude of the potential liquefaction risks at a specific site. The primary outputs that can be generated include liquefaction indices, induced ground displacement, and liquefaction maps.

14.2.3 Liquefaction indices at a site

The liquefaction indices, such as the liquefaction potential index (LPI) and the liquefaction severity number (LSN), are quantitative measures that reflect the likelihood and potential severity of liquefaction at a given site. These indices are calculated based on a range of input data collected from the site, including geotechnical and geological properties, seismic hazard characteristics, and topographic parameters. The LPI, established by Iwasaki et al. [25, 26], offers an initial method to quantify the likelihood and potential severity of liquefaction within a soil profile, based on the depth and characteristics of soil layers and their susceptibility to liquefaction. The LSN [56], a more recent development, refines this approach by incorporating depth-weighted assessments to produce a dimensionless number that better correlates with observed damage from liquefaction. It accounts for the depth to liquefaction layers, emphasizing the impact of liquefaction closer to the surface and considering the initial soil state, which improves the prediction of liquefaction-induced land damage. Both indices serve as critical methods for evaluating liquefaction risks.

14.2.4 Induced ground displacement at a site

Another important output is the induced ground displacement at the site. Liquefaction can cause significant deformations of the ground surface, including vertical settlement and lateral spreading. They can lead to the tilting or collapse of buildings and other structures, and damage to infrastructure like roads, pipelines, and utilities. Liquefaction temporarily increases the pore water pressure in the soil, which in turn reduces the effective stress between soil particles, causing the soil structure to collapse or densify. When the shaking stops and the soil begins to reconsolidate, the soil particles settle into a denser configuration than before, leading to vertical settlement of the ground surface. This settlement can be uneven across a site, depending on the variability of soil conditions and the presence of non-liquefiable layers, leading to differential settlement. Lateral spreading involves the horizontal displacement of the ground, typically observed on gentle slopes or near water bodies like rivers, where the liquefied layer allows the overlying soil to move toward areas of lower elevation or pressure. This movement can be particularly destructive to structures, tearing apart roads, breaking pipelines, and undermining the foundations of buildings and bridges. Lateral spreading is most hazardous in areas where the water table is high, and loosely packed, sandy soils are prevalent, as these conditions are conducive to significant liquefaction-related movements.

14.2.5 Liquefaction maps

Liquefaction maps offer a comprehensive visual overview of areas at risk of soil liquefaction and the potential extent of liquefaction-induced damage. These maps synthesize data from detailed site- and region-specific analyses, including soil composition, groundwater levels, historical seismic activity, and other relevant geotechnical factors, to model the likelihood and potential severity of liquefaction across varying landscapes. The creation of liquefaction maps involves methodologies that take into account the complex interplay between soil properties, seismic forces, and hydrological conditions. By overlaying this data onto geographic maps, engineers and planners can pinpoint zones of varying liquefaction susceptibility, from low to high risk, thus enabling more informed decision-making processes. These maps are invaluable not only for guiding urban and infrastructure development but also for implementing zoning regulations that ensure buildings and critical infrastructure are designed with adequate safeguards against liquefaction effects. In the context of urban planning, liquefaction maps can direct the allocation of resources toward areas requiring ground stabilization efforts or special construction practices, such as deep foundations, soil compaction, or the use of liquefaction mitigation techniques like dynamic compaction or grouting. For existing structures, these maps help prioritize retrofitting efforts to enhance resilience against future seismic events. Beyond planning and development, liquefaction maps are vital for emergency preparedness and response. In the aftermath of an earthquake, these maps offer rapid insight into areas that may have sustained significant ground deformation and infrastructure damage, allowing for more efficient deployment of emergency services, assessment teams, and reconstruction resources. This facilitates quicker and more targeted recovery efforts, ultimately reducing the economic and social impacts of seismic disasters. As research advances and more comprehensive geotechnical data becomes available, the accuracy and utility of liquefaction maps continue to improve. Incorporating technologies such as remote sensing, machine learning, and Geographic Information Systems (GIS) further enhances their precision and usability, making them an indispensable component of seismic risk assessment and mitigation strategies worldwide.

14.3 ASSESSING APPROACHES

Analysis of liquefaction and its consequences is an active area of research and development in geotechnical engineering. Methods for estimating liquefaction triggering and its consequences vary. They fall into two categories: simplified methods and mechanics-based numerical methods.

14.3.1 Simplified methods

In 1998, a consensus was reached within the geotechnical community on the use of an empirical stress-based approach for liquefaction triggering assessment called the "simplified method," first developed by Seed and Idriss [48]. This method is still commonly used in practice today [28, 63].

In a simplified method, a factor of safety (FS) against liquefaction triggering, defined as the ratio between the seismic loading required to trigger liquefaction (i.e., the liquefaction resistance) and the seismic loading expected from the earthquake (i.e., the seismic demand), is computed. Both the seismic demand and the liquefaction resistance are characterized as cyclic stress ratios, defined as the ratio of the equivalent cyclic shear stress to the initial vertical effective stress. The seismic demand is the earthquake-induced cyclic stress ratio (CSR), and the liquefaction resistance is the cyclic resistance ratio (CRR): that is, the cyclic stress ratio required to trigger liquefaction.

Seed and Idriss [48] proposed a simplified equation, based on Newton's second law, to compute a representative CSR for a given earthquake magnitude. This model was later revised by Boulanger and Idriss, Onder Cetin and Seed, and Idriss and Boulanger, and Idriss [4, 10, 22, 23].

The most common approaches used in practice to compute CRR are based on geotechnical field data, e.g., CPT, SPT, and V_s. The most commonly used relationships to estimate CRR from a CPT profile are those developed by Idriss and Boulanger and Wride,

Moss et al., and Robertson, and Robertson [22, 40, 44, 45]. The most commonly used relationships to estimate CRR from SPT blow count are those proposed by Youd and Boulanger and Idriss, and Onder Cetin et al., Idriss [4, 9, 63]. The most commonly used relationships to estimate CRR from a V_s profile were developed by Andrus and Stokoe and Kayen et al. [1, 29].

14.3.2 Mechanics-based numerical simulation

The development and validation of numerical analysis tools and procedures for estimating the effects of liquefaction on the built environment is identified as an overarching research need [7]. Numerical analysis is critical for several reasons, including obtaining insights on field mechanisms that cannot be discerned empirically, providing a rational basis for developing or constraining practice-oriented engineering models, and providing a tool for evaluating complex structures with unique characteristics that are outside the range of empirical observations.

Finite-element (FE) and finite-difference (FD) procedures are the most common procedures used in engineering practice. As pointed out by Bray et al. [7], there are major challenges in developing robust validated numerical analysis procedures for evaluating the effects of liquefaction on civil infrastructure systems due to the variety of multiscale, multiphysics coupled nonlinear interactions that come to the forefront in different scenarios where analytical capabilities for liquefaction effects have not been validated. Currently, research or commercial software platforms have not incorporated the best available solution techniques/options for these challenging problems, such as the coupled, large-deformation analysis of strain-softening, localizations, cracking, and interfaces in two or three dimensions with complex constitutive models. However, significant progresses have been made [3, 35, 37, 39, 41, 50, 60]. An FE method divides the domain into a set of smaller, interconnected elements, where the material behavior is defined at each point within an element, which allows for a more accurate representation of complex behaviors like soil liquefaction. FEM is particularly useful for modeling nonlinear behavior, making it well suited for simulating seismic soil liquefaction. It can handle complex boundary and initial conditions and heterogeneous soil properties. Furthermore, coupled solid–fluid analyses are possible within the FEM framework, enabling the simulation of pore water pressure build-up and dissipation during an earthquake, which is crucial in liquefaction studies. The FD method represents the continuous domain as a set of discrete points and approximates the differential equations using differences at these points. This method is highly efficient and straightforward, making it suitable for large-scale and long-duration simulations like seismic soil liquefaction. By accurately simulating these wave propagations, the method can estimate the pore pressure changes and consequent changes in soil strength, giving valuable insight into the potential for soil liquefaction.

The discrete element method (DEM) and smoothed particle hydrodynamics (SPH) are other powerful approaches for simulating soil behavior subject to liquefaction [16, 51]. DEM provides an excellent framework for studying the micro-mechanics of liquefaction, as it allows for direct simulation of the grain-scale interactions and rearrangements that lead to liquefaction. By incorporating realistic particle shapes and considering the influence of pore water pressure, DEM can effectively simulate the cyclic loading conditions during an earthquake and track the onset and progression of liquefaction. One of the major advantages of DEM is its ability to model large deformations, such as those that occur during the liquefaction process. This makes it a valuable tool for investigating the effects of liquefaction on structures and infrastructures, such as settlement and lateral spreading. SPH allows for modeling of large deformations and material discontinuities, making it well suited for simulating the nonlinear, dynamic behavior of soil during an earthquake. The fluid-like behavior of liquefied soil can be effectively captured using this method.

14.4 ANALYSIS TOOLS AND SOFTWARE

Systems for liquefaction evaluation are divided into two categories according to the method used: simplified empirical methods and mechanics-based numerical methods.

14.4.1 Tools for simplified methods

Simplified methods have been developed for rapid engineering evaluations of site-specific liquefaction. To date, no open-source software is available: but methods mentioned in References [67–70] are all Windows based. These tools provide a user interface that can let the user input the soil profiles and earthquake loading, and then visualize the liquefaction index for each soil layer.

14.4.2 Tools for numerical methods

For mechanics-based numerical methods, creating a constitutive model that captures the soil's behavior under cyclic loads is crucial. Commercial software such as mentioned in References [17, 42] is widely used by the geotechnical community. Both of them are Windows based. OpenSees is the only open-source software identified for dealing with liquefaction. Several well-known liquefaction-capable constitutive models are PM4Sand, PM4Silt, PDMY02, UBCSAND, and DAFALIAS-MANZARI. Except for UBCSAND, they are all available in OpenSees.

The NHERI SimCenter also develops both research and educational tools to facilitate performing site-specific analysis of soil response to earthquakes. The tools currently available focus on the one-dimensional propagation of ground shaking from the bedrock to the free surface.

14.4.3 QS3HARK

The Site-Specific Seismic Hazard Analysis and Research Kit with Uncertainty Quantification (*QS3HARK*), developed by [60], performs site-specific analysis of ground shaking and liquefaction by simulating wave propagation through soil layers. The simulations use the finite-element method to perform the calculations. Several advanced material models are available in QS3HARK (e.g., PM4Sand, PM4Silt, PDMY, PDMY02, PDMY03, ManzariDafalias, Borja-Amies) to support advanced site-response analysis.

14.5 CASE STUDY: MULTISCALE REGIONAL ASSESSMENT

Traditionally, detailed liquefaction hazard assessments have been focused on site-specific analyses, utilizing both empirical methodologies and numerical simulations such as finite-element method (FEM) and discrete element method (DEM). In recent times, the emphasis has expanded toward encompassing high-resolution regional-scale assessments, a shift driven by the need for a broader understanding of seismic risks. Traditional regional hazard maps, which are often qualitative in nature, classify liquefaction susceptibility into categories such as high, medium, or low, relying largely on the geological characteristics of the area. Innovations in this field suggest enhancing these conventional approaches by integrating detailed spatial information regarding soil properties.

While site-specific assessments accurately determine the liquefaction potential at specific points where field tests have been conducted, they fall short in predicting potential in areas without direct testing, due to the inherent spatial variability of soil properties. To bridge this gap, geostatistical methodologies, which model spatial variations as a stochastic process, have been employed effectively. The following sections delve into two primary methodologies that account for the spatial variability of geotechnical properties. The first methodology, termed the local soil property approach, leverages local field data to infer spatial correlations among soil properties. The second methodology, referred to as the averaged index approach, involves calculating a mean soil property or index based on data from tested sites and employing geostatistical techniques to estimate these indices in untested regions.

Acknowledging that spatial variability often occurs across multiple scales, we introduce a multiscale random field model. This innovative approach harmoniously combines multiscale soil property models and averaged index random fields with established empirical liquefaction prediction models. The objective of this study is to explore the impacts of these methodologies on the assessment of liquefaction potential within spatially variable soil environments. Furthermore, the study aims to demonstrate the practical application of these advanced assessment strategies in a region known for its susceptibility to liquefaction, providing valuable insights for future seismic risk management and urban planning efforts.

14.5.1 General framework

In this work, the CPT-based empirical liquefaction model is integrated with geostatistical tools to account for the spatial variability of geotechnical properties for regional liquefaction evaluation. The flow of the general framework is shown in Figure 14.1. Within a liquefaction-prone region, CPT measurements (e.g., the tip resistance and the side friction) and other geotechnical data of interest (e.g., water table, soil unit weight) are first collected and their geostatistical properties are inferred and characterized (e.g., probabilistic distribution, spatial structure). At a CPT sounding, the empirical liquefaction model described in Section 14.5.2 will be used to evaluate the damage potential of liquefaction, quantified here by the liquefaction potential index (LPI). The aforementioned two approaches as detailed in work by [61], i.e., *the averaged index approach* and *the local soil property approach*, are developed to incorporate soil variability into the evaluation of liquefaction over an extended area. Details of multiscale random field model development and implementation will be discussed in Section 14.5.3. Finally, Monte Carlo simulations will be used to generate realizations of the random fields and the results will be used for the probabilistic and spatial assessment of various quantities of interest for liquefaction evaluation over the region.

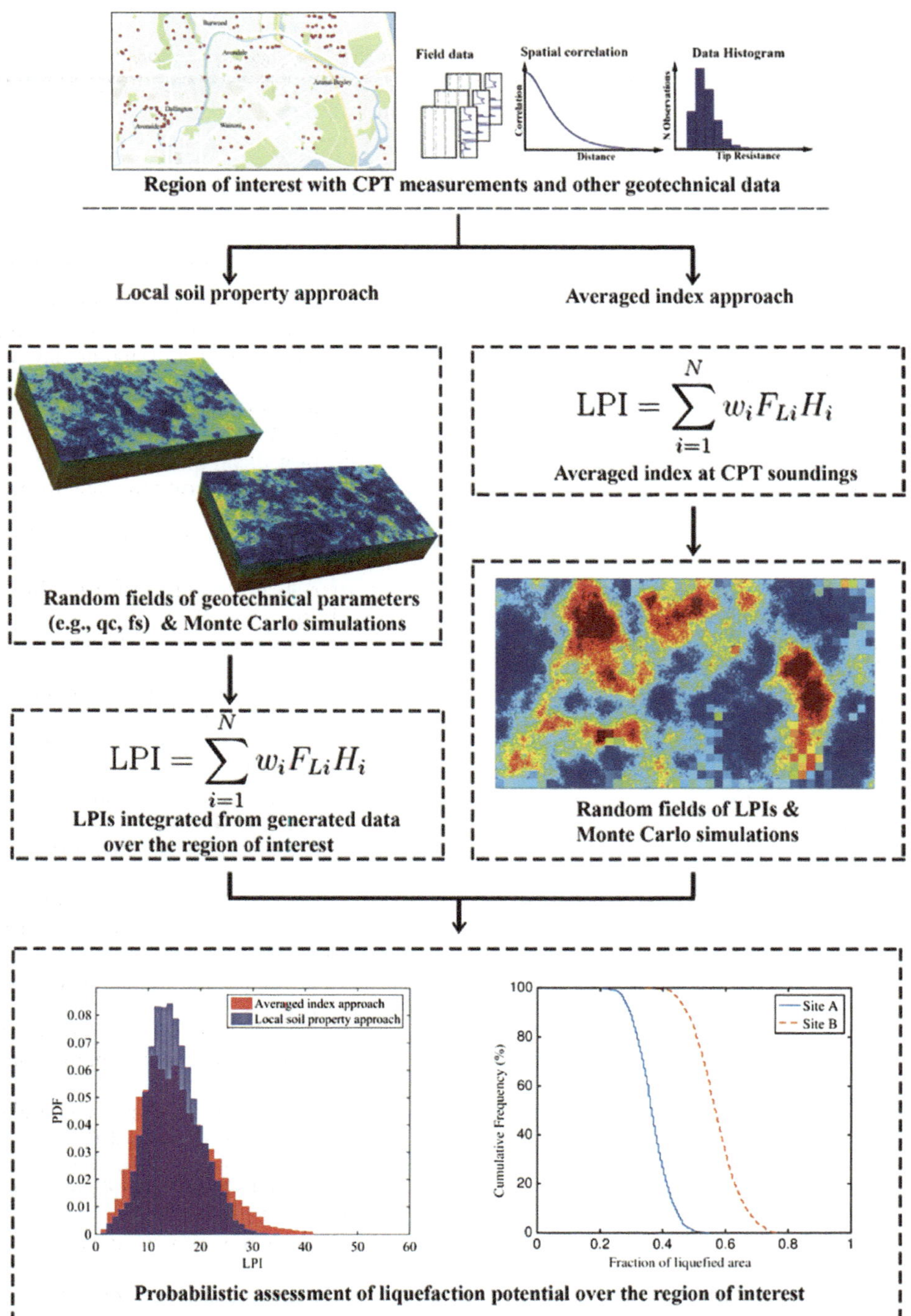

Figure 14.1 General framework of the CPT-based liquefaction potential evaluation over the extended area: the local soil property approach versus the averaged index approach.

14.5.2 Site-specific liquefaction evaluation

Site-specific liquefaction evaluation can be performed using the approach and tool described in Sections 14.3 and 14.4. In this work, we adopt the classical procedure proposed by Robertson and Wride [46] and subsequently updated by Robertson [47] and Ku et al. [32] to evaluate the liquefaction resistance of sandy soils based on CPT data. Herein, the liquefaction potential of a soil layer is evaluated using two variables – the cyclic stress ratio (CSR) and the cyclic resistance ratio (CRR). The detailed equations for CSR and CRR can be found in Wang et al. [61].

Once CSR and CRR are obtained, the factor of safety against liquefaction triggering at a particular depth z can be calculated as

$$FS = \frac{CRR}{CSR} \tag{14.1}$$

which is used to calculate the liquefaction potential index detailed in the following section. In this example, the potential of liquefaction damage will be quantified by an averaged index property, i.e., the liquefaction potential index (LPI), which was originally proposed by Iwasaki et al. [25, 26] and has been subsequently used and calibrated by many investigators, e.g., [12, 20, 21, 27, 33, 52, 53, 55, 57].

Following the definition given in References [25, 26], LPI is usually evaluated for the top 20 m of soil profile as

$$LPI = \int_0^{20} F_L w(z) dz \tag{14.2}$$

where z denotes the depth in meters and $w(z) = 10 - 0.5z$; F_L is a function of FS, the definition of it can be found in work by Sonmez [53].

14.5.3 Multiscale random field characterization

The CPT-based empirical liquefaction model described in Section 14.5.2 evaluates the liquefaction potential at individual locations at the small scale, e.g., the borehole scale. To estimate the extent of liquefaction risk over the entire region of interest, multiscale random field models are introduced and implemented in this section.

14.5.3.1 Spatial correlation

In this study, the spatial correlation of geotechnical parameters is described using the semivariogram, $\gamma(\vec{h})$, which is equal to half the variance of the difference of two random variables separated by a vector distance $\vec{h}$

$$\gamma(\vec{h}) = \frac{1}{2} Var\left[Z(\vec{u}) - Z(\vec{u} + \vec{h}) \right] \tag{14.3}$$

where $Z(\vec{u})$ is a Gaussian random variable at location $\vec{u}$. The vector $\vec{h}$ accounts for both separation distance and orientation and therefore can be used to simulate anisotropic random fields.

In previous studies of liquefaction evaluation, e.g., [33, 34, 58], the correlation $\rho(h)$ is used to describe the spatial dependence of two parameters separated by h and can be related to the semivariogram as

$$\rho(h) = 1 - \gamma(h) \tag{14.4}$$

where $\gamma(h)$ can be a model who's parameter is fit by data. As an example, an exponential model is adopted in this work

$$\gamma(h) = 1 - \exp\left(-\frac{h}{a}\right) \tag{14.5}$$

where a is the range parameter and $3a$ is the practical range, i.e., the distance at which the exponential semivariogram levels off [18].

14.5.3.2 Sequential simulation process

Given a specified probability density function and spatial correlation model, a sequential simulation process [14, 18] is adopted in this work to generate realizations of the random field across the region. Each value is simulated individually conditional upon known information as well as any previously simulated data points.

Denote $\vec{Z}_p$ as a vector of all known and previously simulated points in the random field and Z_n as the next point to be simulated, the sequential simulation process can be illustrated by

$$\begin{bmatrix} Z_n \\ Z_p \end{bmatrix} \sim \mathrm{N}\left(\vec{\mu}, \begin{bmatrix} \sigma_n^2 & \vec{\Sigma}_{np} \\ \vec{\Sigma}_{pn} & \vec{\Sigma}_{pp} \end{bmatrix}\right) \tag{14.6}$$

where $\sim \mathrm{N}(\vec{\mu}, \vec{\Sigma})$ denotes the vector of random variables following a joint normal distribution with mean vector $\vec{\mu}v$ and covariance matrix $\vec{\Sigma}$; σ_n^2 is the prior variance of the next simulated point; $\vec{\Sigma}_{np}$, $\vec{\Sigma}_{pn}$, and $\vec{\Sigma}_{pp}$ are the covariance matrices, where the subscripts "n" and "p" represent "next" (as in next point to be simulated) and "previous" (as in all previously simulated points), respectively. The covariance matrices are obtained by

$$\mathrm{COV}[Z_i, Z_j] = \rho_{ij} \cdot \sigma_i \cdot \sigma_j \tag{14.7}$$

where ρ_{ij} is the correlation between random variables Z_i and Z_j with standard deviations of σ_i and σ_j , respectively.

Using the above model for joint distribution, the distribution of Z_n conditional upon all previously simulated data is given by a univariate normal distribution with updated mean and variance

$$\left(Z_n \mid \vec{Z}_p = \vec{z}\right) \sim \mathrm{N}\left(\vec{\Sigma}_{np} \cdot \vec{\Sigma}_{pp}^{-1} \cdot \vec{z}, \sigma_n^2 - \vec{\Sigma}_{np} \cdot \vec{\Sigma}_{pp}^{-1} \cdot \vec{\Sigma}_{pn}\right) \tag{14.8}$$

It is noted that $\vec{\Sigma}_{np} \cdot \vec{\Sigma}_{pp}^{-1}$ are essentially the weights assigned in the simple Kriging process [18]. For the realization, one value of Z_n is drawn at random from the posterior univariate normal distribution.

Once simulated, Z_n becomes a data point in the vector $\vec{Z}_p$ to be conditioned upon by all subsequent data locations. This process is repeated by following a random path to each unknown location until all the values in the field have been simulated. This sequential simulation process is beneficial for the proposed work because (1) it preserves known information (e.g., field data) precisely at their locations in the simulated random field; and (2) it allows to first simulate the field at only the coarse scale, then add simulation points at the fine-scale probabilistically consistent with the previous coarse-scale realizations, a process detailed in Section 14.5.3.3.

14.5.3.3 Multiscale spatial correlation

The multiscale spatial correlation is based on the notion that material properties at the coarser scales are the arithmetically averaged values of the properties over corresponding areas at the finer scales. This relation allows for the explicit derivation of variances and spatial correlation of quantities of interest between different scales and is visually represented in Figure 14.2. For two scales of interest, we denote the coarser scale as scale "c" and the finer scale as scale "f." The cross-scale spatial correlations can then be calculated as [12]

$$\rho_{Z_I^c, Z_{II}^c} = \frac{\sum_{i=1}^{N}\sum_{k=1}^{N} \rho_{Z_{i(I)}^f, Z_{k(II)}^f}}{\sqrt{\sum_{i=1}^{N}\sum_{j=1}^{N} \rho_{Z_{i(I)}^f, Z_{j(I)}^f}}\sqrt{\sum_{i=1}^{N}\sum_{j=1}^{N} \rho_{Z_{i(II)}^f, Z_{j(II)}^f}}} \tag{14.9}$$

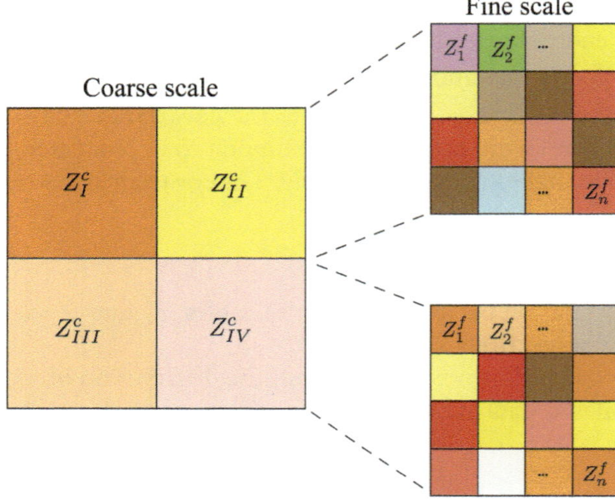

Figure 14.2 Graphic representation of material properties at two scales. The superscripts "c" and "f" refer to "coarse" and "fine" scales, respectively. The subscripts refer to the element number. Roman letters I, II, ... are used for coarse-scale element and Arabic numbers 1, 2, 3, ... are used for fine-scale element.

$$\rho_{z^f, z_I^c} = \frac{\sum_{i=1}^{N} \rho_{z^f, z_{i(I)}^f}}{\sqrt{\sum_{i=1}^{N} \sum_{j=1}^{N} \rho_{z_{i(I)}^f, z_{j(I)}^f}}} \tag{14.10}$$

where $\rho_{z_I^c, z_{II}^c}$ = correlation between two coarse-scale elements I and II; ρ_{z^f, z_I^c} = correlation between a fine-scale element and a coarse-scale element I.

The geotechnical properties will be simulated using random field models at the coarse scale (e.g., the regional scale), and then adaptively refined into smaller scales (e.g., the borehole or structure scales), conditional upon the coarse-scale random field simulations. It is worth noting that the cross-scale correlations (14.9) and (14.10) are applicable to general non-uniform arbitrarily shaped grids, and actual field CPT-based geotechnical measurements will be incorporated into the sequential simulation process.

14.5.4 Numerical examples

In this section, the developed multiscale random field model is applied to evaluate liquefaction potential in the city of Christchurch, New Zealand. The purpose of the numerical examples is twofold: (1) to demonstrate the applicability of multiscale random field models in liquefaction evaluation over extended areas; (2) to assess *the averaged index approach* and *the local soil property approach* in accounting for the spatial variability of CPT-based geotechnical parameters and their implications on the liquefaction evaluation. We assume the stationary of the random field model and focus mainly on the spatial variability of tip resistance, side friction, and the liquefaction potential index. Spatial variability of other geotechnical parameters, e.g., water table and unit weight of soil, are not included in the current study.

14.5.4.1 Analysis region and field data

Christchurch, New Zealand, is a city founded on the boundary of the alluvial Springston Formation and the marine Christchurch Formation [8]. During the period between September 2010 and December 2011, the city of Christchurch was strongly shaken by a sequence of four strong earthquake events known as the Canterbury earthquakes. For the following liquefaction analysis and liquefaction potential mapping, we pick one of the great events, the February 22, 2011, earthquake event. The moment magnitude of the earthquake M_w = 6.2. Typical peak ground surface acceleration $a_{\max}$ in the area of study ranging from 0.34 g to 0.50 g [5]. A median value of $a_{\max} = 0.42$ g is used in the following liquefaction potential calculations. A total of 155 CPT profiles with measured water tables are collected from the New Zealand Geotechnical Database (NZGD). An averaged moist unit weight of $\gamma_m = 18.5$ kN/m^3 is used for soils above water table and an saturated unit weight of $\gamma_{sat} = 19.5$ kN/m^3 is used for soils below the water table following information presented in Reference [31] Figure 14.3.

14.5.4.2 Probabilistic and spatial assessment of liquefaction potentials

In the previous two sections, we have demonstrated typical random field realizations of LPIs by the averaged index approach and the local soil property approach. In this

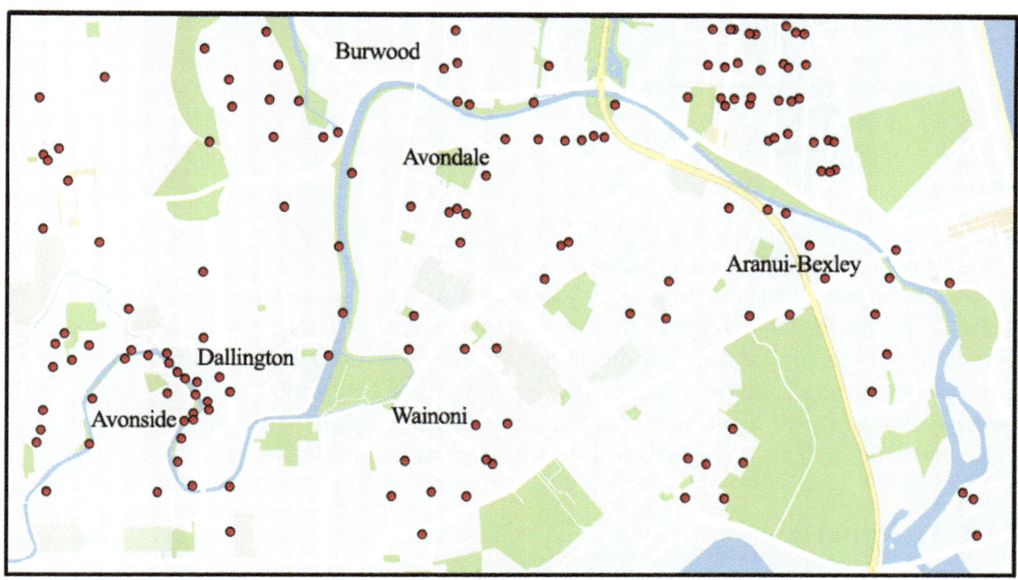

Figure 14.3 Map of the area of study in Christchurch, New Zealand. Red dots represent the locations of the 155 CPT soundings.

section, we will perform Monte Carlo simulations and assess the implications of two approaches on the probabilistic and spatial characteristics of liquefaction potentials in the Christchurch site. In addition, utilizing multiscale random field realizations, liquefaction evaluation will be performed for selected local sites to demonstrate the advantage and applicability of the proposed methodology.

A total of 1000 Monte Carlo simulations are performed to generate realizations of LPIs across the Christchurch site. Figure 14.4a shows the histogram of the averaged LPIs from 1000 Monte Carlo simulations and Figure 14.5 shows the corresponding maps of the averaged LPIs. To attenuate the effect of ergodic fluctuations, empirical semivariograms

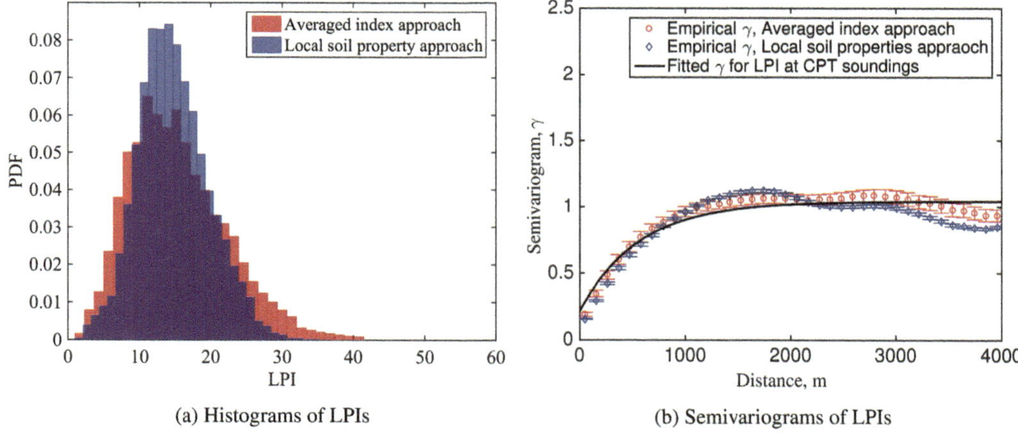

Figure 14.4 Histograms and semivariograms of the averaged LPIs from 1000 Monte Carlo simulations. Error bars in semivariogram plot indicate ± one standard deviation.

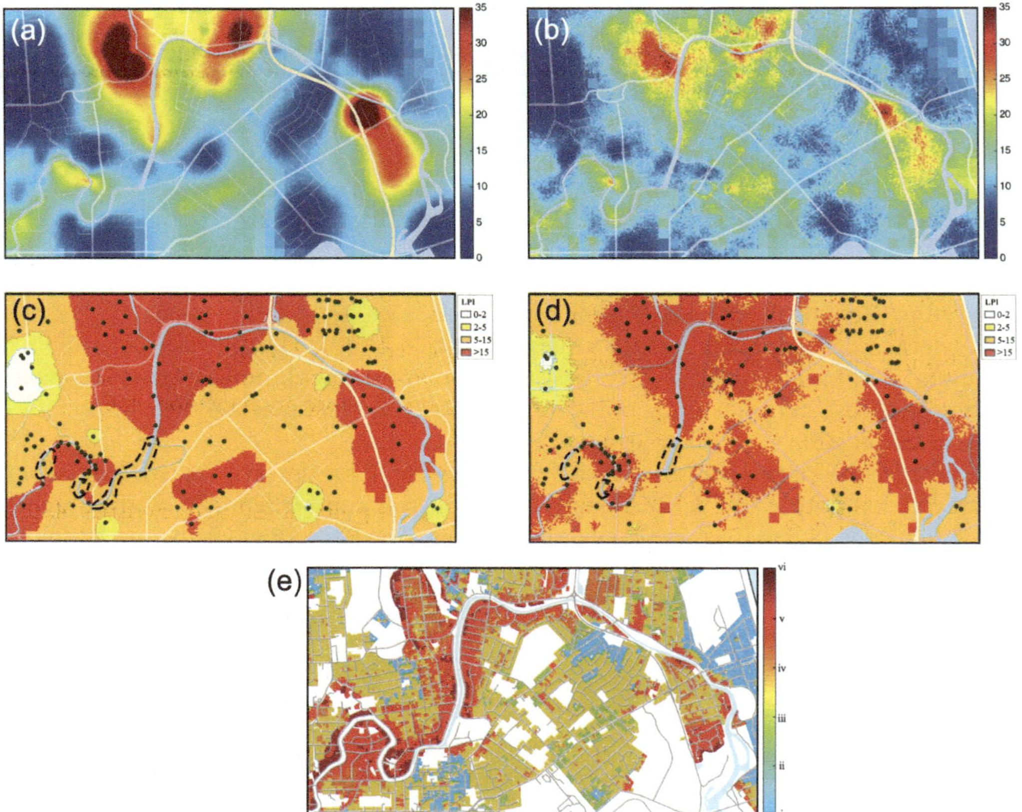

Figure 14.5 Mapping of LPI by two approaches compared with the observed liquefaction phenomena.

are calculated 1000 times, averaged, and plotted in Figure 14.4b. The solid curve is the specified exponential model with parameters fitted using LPIs at CPT soundings. The red dots and blue squares are the empirical semivariograms calculated using Equation (14.3) with LPI values generated by the averaged index approach and the local soil property approach, respectively. The error bars ($\pm$ one standard deviation) show the variations of 1000 Monte Carlo simulations. Fluctuating semivariogram results from weak and unsteady correlation of properties between separated places. This is verified in Figure 14.4b, i.e., great fluctuations (in both the average and standard deviation) of the empirical semivariograms are observed at distances greater than 1000 m. The local soil property approach results in slightly greater fluctuation of semivariogram possibly due to the non-linear transformation from q_c and f_s to LPI.

The generated LPI maps by two different approaches are compared with the map of observed liquefaction phenomena after the February 22, 2011, earthquake event [2], as shown in Figure 14.5. Comparing the maps of predicted LPIs against the observations, it can be seen that both the averaged index approach and the local soil property approach are able to capture the varying severity levels of liquefaction at most locations across the area of study. For instance, the high LPI value area corresponds well with areas with observed moderate to several types of liquefaction. The results of the local soil property approach show more small-scale fluctuations or "noises." This is due to the high nugget effects in the specified semivariograms of q_c and f_s.

Discrepancies between simulated maps of liquefaction potential and observations are also identified. For example, in the dot–circled areas, as shown in Figure 14.5(c, d), the predicted LPI is less than 15, which is not the severest level of liquefaction. In the observation map as shown in Figure 14.5e, however, moderate to severe lateral spreading is observed in these regions and ejected material is also often observed, which usually indicates the highest level of liquefaction. Lateral spreading is a unique form of liquefaction manifestation associated with large lateral permanent ground displacements. The severity of lateral spreading could not be classified by LPI. There are separate criteria for assessing its severity. Therefore, while site profiles with thin liquefiable layers may have low LPI values, these sites may be susceptible to lateral spreading if located near rivers or on sloping ground. Actually, take Christchurch City for an example, there are about 25% of sites with lateral spreading and have LPI values less than 4.5 [38]. This means that while LPI is a useful tool for large-scale hazard assessments, it does not eliminate the need for site-specific analyses or consideration of the influence of local conditions on the manifestation of liquefaction [38].

Results from Monte Carlo simulations can also be used for the probabilistic assessment of liquefaction potentials. Various quantities of interest can be defined. As an example, we evaluate the cumulative frequency distributions of LPIs following the methodology proposed by Holzer et al. [20].

Figure 14.6 plots the cumulative frequencies of LPIs in the area of study obtained with the averaged LPI map from 1000 Monte Carlo simulations. The error bars (± one standard deviation) are also included in the cumulative frequency plots. It can be seen that the cumulative frequency curve by the averaged index approach is very close to that of the known LPI obtained from 155 CPT soundings, while a sharp change in the slope of the cumulative frequency curve for the local soil property approach is observed, which can be explained by its histogram in Figure 14.4a. Most of the LPIs simulated by this approach fall between 10 and 20, resulting in a sharp change in the cumulative frequency curve in this range. For the averaged index approach, the percentages of different LPI values are more "spread-out," which results in smoother cumulative frequency curves.

Comparisons of Figure 14.6(a, b) indicate that multiscale random fields yield consistent results as the single-scale counterparts for cumulative frequencies of LPIs. This is to be expected given the notion that properties of a coarse element are the averaged values of the properties over the corresponding areas at the fine scale.

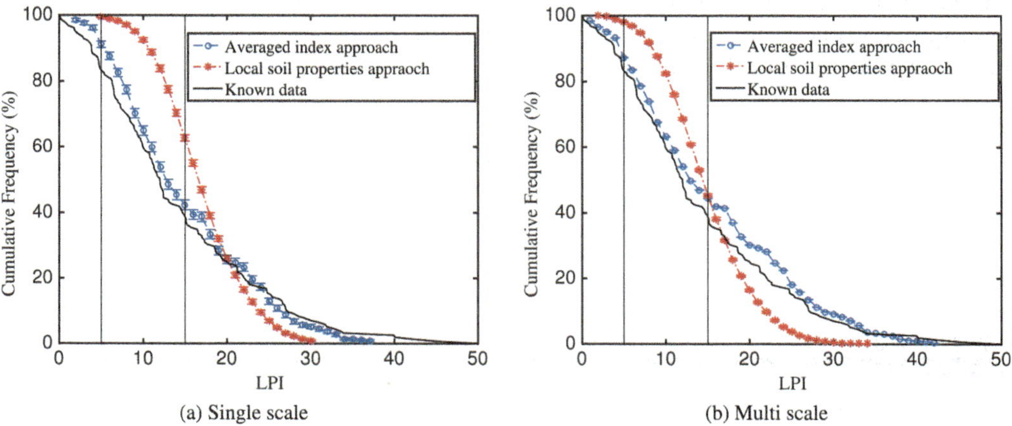

Figure 14.6 Cumulative frequency plot of LPIs for the area of study in Christchurch.

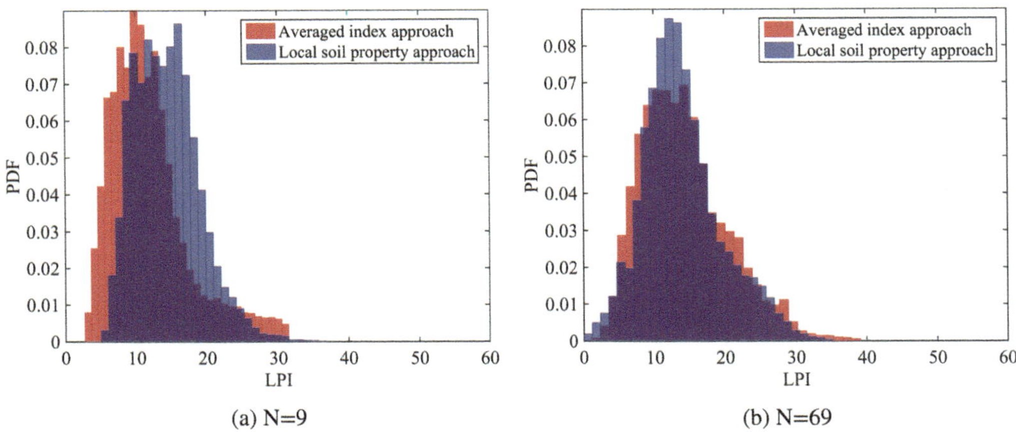

Figure 14.7 Histograms of predicted LPIs in the Christchurch site when a different number of CPT soundings (*N*) is used to infer model parameters.

It should be noted that all previous analyses and comparisons are based on relatively sufficient field data to infer random field model parameters. It is expected that the amount of the field will impact the results of the random field-based liquefaction mapping. While this subject itself deserves a future study, for this Christchurch site, we investigated the effect of data availability on the distribution of predicted LPIs in the area of study. The number of available CPT soundings varied from 9 to 243. It is found that when the number of CPT soundings is small, the local soil property approach yields higher mean LPI values. As the number of CPT soundings increases, the distributions obtained from the averaged index approach and the local soil approach converge. Figure 14.7 plots the histogram of LPIs obtained with 9 CPT and 69 CPT soundings.

14.5.4.3 Fine-scale site-specific liquefaction assessment

The final analysis involves the fine-scale site-specific liquefaction assessment. Compared to single-scale random field, multiscale random field provides higher resolution at local or site-specific scale, e.g., near a critical infrastructure or a particular building of interest. To illustrate this point, we evaluate the liquefaction potential at two local sites in Christchurch. The site locations are shown in Figure 14.8. Site A is a small region that consists of two schools, i.e., the Chisnallwood Intermediate School and the Avondale Primary School. Site B is the location of the Aranui High School. The inset of random fields in Figure 14.8 shows that the multiscale random field could provide much more detailed information than the single-scale counterpart, which makes it possible to perform site-specific liquefaction evaluation while consistently maintaining predictions of liquefaction for the entire region over much larger scales.

Based on the cumulative frequency distribution, Holzer et al. [20] proposed that for a given geologic unit, the percent area predicted to have liquefaction during a given earthquake-shaking scenario that could be estimated from the cumulative frequency distributions at LPI $\geq$ 5. This criterion can be further categorized to indicate the percentage of area to have minor liquefaction ($5 \leq$ LPI < 15) and major liquefaction (LPI ≥ 15).

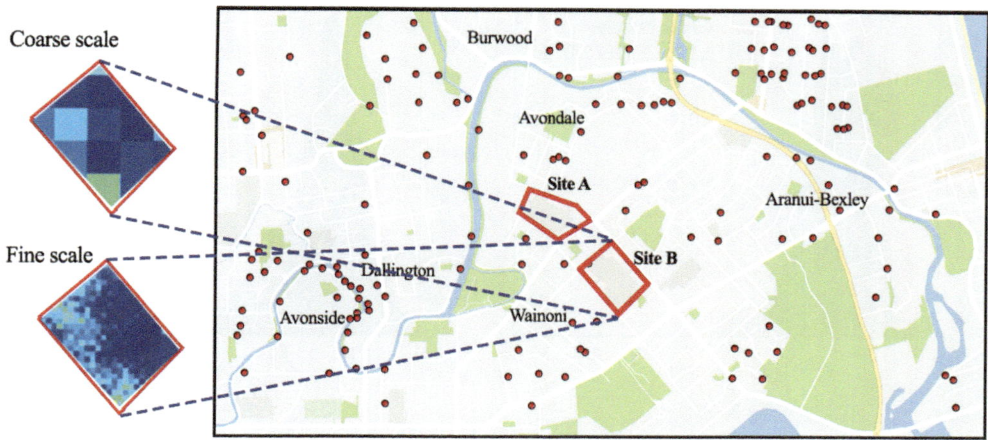

Figure 14.8 Locations of two selected sites for fine-scale liquefaction assessment.

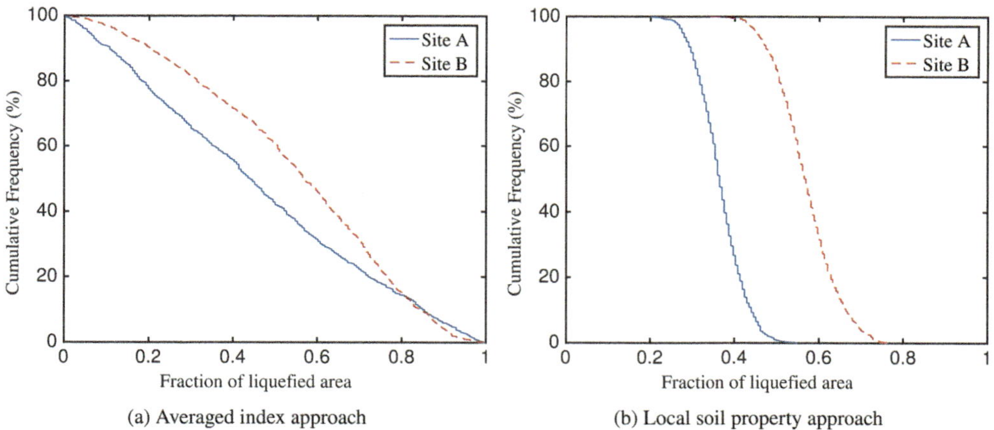

Figure 14.9 Cumulative frequency versus the fraction of liquefied area with LPI ≥ 15 for two local sites A and B selected in the area of study.

Herein, the percentage of areas in the selected sites with LPI ≥ 15 under a given earthquake-shaking scenario ($M_w = 6.2$, $a_{\max} = 0.42$ g) is calculated and plotted against the corresponding cumulative frequency in Figure 14.9.

Figure 14.9 shows cumulative frequencies of LPIs with respect to the fraction of liquefied area. The cumulative frequency curve for site B is consistently "higher" than site A for both the averaged index approach and the local soil property approach, indicating that site B is more likely to liquefy than site A in a given earthquake. Comparisons between two approaches show that the cumulative frequency for the average index approach is more "spread-out" than the local soil property approach. This is consistent with the predictions within the whole site of Christchurch, as shown in Figure 14.6.

The above analysis on the percentage of the liquefiable area underneath a particular site gives an indication of the implied reliability of the site with respect to liquefaction resistance during a given earthquake. These site-specific curves can only be obtained if a higher resolution multiscale random field is generated. Moreover, multiscale random fields are able to obtain high-resolution information at local sites while consistently maintaining low-resolution information at much larger scales, leading to a computationally efficient process.

Finally, it should be noted that the methodologies discussed in this study are sufficiently general to be applied to other earthquake-prone areas and to evaluate quantities of interest such as liquefaction-induced settlement and lateral spreads. This will be explored in future studies.

14.6 RESEARCH GAPS AND OPPORTUNITIES

While significant strides have been made in liquefaction risk analysis, primarily focusing on the prediction and mapping of liquefaction susceptibility, a comprehensive understanding of the consequent ground damage remains elusive. Current methodologies largely concentrate on assessing the likelihood of liquefaction triggering but fall short in quantitatively evaluating the aftermath – specifically, the extent of damage inflicted on the ground and underlying structures. The development of universally recognized and applied techniques for estimating liquefaction-induced damage is an ongoing challenge within the geotechnical community. The HAZUS methodology, despite its widespread application, underscores this gap. It offers a procedure rooted in historical engineering observations for estimating liquefaction-induced permanent ground deformation (PGD) based on peak ground acceleration (PGA) and site-specific susceptibility. Nonetheless, this framework is predicated on analyses of comparatively older seismic events, highlighting the need for updated and more comprehensive approaches.

Recent advancements have illuminated this research path. An increasing number of observed liquefaction instances [7, 13], coupled with the establishment of extensive databases [6], has catalyzed the development of new regression-based ground damage models [30, 54]. These models, which are being rigorously tested across various geographical contexts [11], represent a significant leap forward in our ability to predict and understand the ground damage resulting from liquefaction. Concurrently, emerging research on the vulnerability of overlying structures to liquefaction offers fresh insights into the fragility of built environments in the face of such phenomena.

Moreover, the traditional methodologies for regional liquefaction assessment – primarily reliant on geological datasets – have been superseded by more sophisticated approaches [19]. Recent propositions advocate for the use of random field models for regional-scale liquefaction hazard modeling. These models, by incorporating both geotechnical and geological data, offer a nuanced perspective that accommodates spatial uncertainties, representing a critical advancement in our quest to refine liquefaction risk assessment [59, 62, 65].

As we move forward, the geotechnical engineering field is poised at the cusp of a new era in liquefaction research. The integration of advanced modeling techniques, extensive empirical databases, and a deeper understanding of structural vulnerabilities opens up unprecedented opportunities. The challenge now lies in harnessing these insights to develop methodologies that not only predict liquefaction susceptibility but also quantify the potential damage, thereby enhancing our resilience to future seismic events.

14.7 CONCLUSION

In this chapter, the authors have delved into the complex, multifaceted nature of seismic soil liquefaction and underlined the pressing necessity for improved assessment methodologies to better forecast this catastrophic geological occurrence. Through an examination of traditional empirical assessment methods and their inherent limitations, we underscored the urgent need for a more holistic, data-centric approach.

Our exploration encompassed familiar practices like the standard penetration test (SPT), cone penetration test (CPT), and shear-wave velocity (V_s), acknowledging their contributions while highlighting their challenges due to the intricacies of soil and seismic dynamics. These techniques, while beneficial to a degree, often oversimplify the complexity of soil behavior under seismic loading, leading to less accurate assessments of liquefaction risk.

This chapter emphasized the transformative potential of contemporary approaches that harness large-scale, diverse datasets and cutting-edge computational capabilities to enhance our understanding of soil liquefaction. By leveraging advanced analytical methodologies, such as artificial intelligence, we can discern patterns and relationships within data, facilitating more precise predictions of soil liquefaction. These strides constitute significant advancements over traditional methods, revolutionizing our capacity to predict and mitigate the destructive impacts of seismic soil liquefaction.

Additionally, we touched upon the multiscale applicability of the data-centric approach, demonstrating its potential from microscale laboratory studies to macroscale and regional assessments. This ability to scale seamlessly enhances its potential as a robust assessment tool. A practical illustration of this innovative approach's effectiveness was the city-scale case study, which utilized a large dataset and sophisticated AI tools, such as the random field, to estimate liquefaction hazards. This case study was a testament to the efficacy of data-centric methodologies in understanding and predicting seismic soil liquefaction.

This chapter stands as a step in the ongoing evolution of our understanding of seismic soil liquefaction and our capabilities in predicting and mitigating its ruinous impacts. By persistently pushing the boundaries of traditional methodologies and embracing the potential of data-centric approaches, we are better equipped to address the challenges posed by soil liquefaction. As we continue to deepen our knowledge and refine our tools, we move closer to a future where we can effectively manage and mitigate the risks of seismic soil liquefaction.

REFERENCES

1. R.D. Andrus and K.H. Stokoe II. Liquefaction resistance of soils from shear-wave velocity. *Journal of Geotechnical and Geoenvironmental Engineering*, 126(11):1015–1025.
2. J. Beavan, S. Levick, J. Lee, and K. Jones. Ground displacements and dilatational strains caused by the 2010–2011 Canterbury earthquakes. *GNS Science Consultancy Report*, 67:59, 2012.
3. R.W. Boulanger and K. Ziotopoulou. Pm4sand (version 3): A sand plasticity model for earthquake engineering applications. In *Center for Geotechnical Modeling Report No. UCD/CGM-15/01, Department of Civil and Environmental Engineering*, University of California, Davis, California, 2015.
4. R.W. Boulanger and I.M. Idriss. CPT and SPT based liquefaction triggering procedures. Report No. UCD/CGM.-14, 1, 134.
5. B.A. Bradley and M. Hughes. Conditional peak ground accelerations in the Canterbury earthquakes for conventional liquefaction assessment. In *Technical Report for the Ministry of Business, Innovation and Employment*, New Zealand, 2012.

6. S.J. Brandenberg, P. Zimmaro, J.P. Stewart, D.Y. Kwak, K.W. Franke, R.E.S. Moss, K.Ö. Çetin, G. Can, M. Ilgac, J. Stamatakos, *et al.* Next-generation liquefaction database. *Earthquake Spectra*, 36(2), 939–959, 2020.
7. J.D. Bray, R.W. Boulanger, M. Cubrinovski, K. Tokimatsu, S.L. Kramer, T. O'Rourke, E. Rathje, R.A. Green, P.K. Robertson, and C.Z. Beyzaei. U.S. – New Zealand – Japan International Workshop on Liquefaction-Induced Ground Movements Effects, University of California, Berkeley, California 2–4 November 2016. Report No. 2017/02.
8. L.J. Brown and J.H. Weeber. Geology of the Christchurch urban area, Institute of Geological and Nuclear Sciences 1:25,000 Geological Map 1, 1992. https://natlib.govt.nz/records/21363240.
9. K. Onder Cetin, R.B. Seed, R.E. Kayen, R.E.S. Moss, H.T. Bilge, M. Ilgac, and K. Chowdhury. SPT-based probabilistic and deterministic assessment of seismic soil liquefaction triggering hazard. 115:698–709.
10. K. Onder Cetin and R.B. Seed. Nonlinear shear mass participation factor (rd) for cyclic shear stress ratio evaluation. 24(2):103–113.
11. Qiushi Chen, Chaofeng Wang, and C Hsein Juang. Probabilistic and spatial assessment of liquefaction-induced settlements through multiscale random field models. 211:135–149.
12. Q. Chen, C. Wang, and C.H. Juang. CPT-based evaluation of liquefaction potential accounting for soil spatial variability at multiple scales. *Journal of Geotechnical and Geoenvironmental Engineering*, 142, 04015077, 2015.
13. M. Cubrinovski, J. Bray, C. de la Torre, M. Olsen, B. Bradley, G. Chiaro, E. Stock, L. and Wotherspoon. Liquefaction effects and associated damages observed at the Wellington Centreport from the 2016 Kaikoura earthquake. 50(2):152–173.
14. C.V. Deutsch and A.G. Journel. *GSLIB: Geostatistical Software Library and User's Guide.* Oxford University Press, New York, 1992.
15. A. Elgamal, Z. Yang, and E. Parra. Computational modeling of cyclic mobility and post-liquefaction site response. *Soil Dynamics and Earthquake Engineering*, 22(4):259–271, 2002.
16. U. El Shamy and S. Farzi Sizkow. Coupled sph-dem simulations of liquefaction-induced flow failure. *Soil Dynamics and Earthquake Engineering*, 144:106683, 2021.
17. FLAC.
18. P. Goovaerts. *Geostatistics for Natural Resources Evaluation.* New York, Oxford University Press, 1997.
19. T.L. Holzer, M.J. Bennett, T.E. Noce, A.C. Padovani, and J.C. Tinsley III. Liquefaction hazard mapping with LPI in the greater Oakland, California area. 22(3):693–708.
20. T.L. Holzer, M.J. Bennett, T.E. Noce, A.C. Padovani, and J.C. Tinsley III. Liquefaction hazard mapping with LPI in the greater Oakland, California, area. *Earthquake Spectra*, 22(3):693–708, 2006.
21. T.L. Holzer, J.L. Blair, T.E. Noce, and M.J. Bennett. Predicted liquefaction of East Bay fills during a repeat of the 1906 San Francisco earthquake. *Earthquake Spectra*, 22(S2):261–277, 2006.
22. I.M. Idriss and R.W. Boulanger. *Soil Liquefaction during Earthquakes.* Berkeley, CA, Earthquake Engineering Research Institute.
23. I. Idriss. An update to the Seed-Idriss simplified procedure for evaluating liquefaction potential. In *Proceedings of the TRB Workshop on New Approaches to Liquefaction*, pp. 99–165, Federal Highway Administration.
24. K. Ishihara. Liquefaction and flow failure during earthquakes. *Geotechnique*, 43(3):351–451, 1993.
25. T. Iwasaki, F. Tatsuoka, K. Tokida, and S. Yasuda. A practical method for assessing soil liquefaction potential based on case studies at various sites in Japan. In *Proceedings 2nd International Conference on Microzonation*, pp. 885–896, 1978.
26. T. Iwasaki, K. Tokida, F. Tatsuoka, S. Watanabe, S. Yasuda, and H. Sato. Microzonation for soil liquefaction potential using simplified methods. In *Proceedings of the 3rd International Conference on Microzonation*, Seattle, WA, volume 3, pp. 1310–1330, 1982.

27. C.H. Juang, C.N. Liu, C.H. Chen, J.H. Hwang, and C.C. Lu. Calibration of liquefaction potential index: A re-visit focusing on a new CPTU model. *Engineering Geology*, 102(1):19–30, 2008.
28. E. Kavazanjian, J. E. Andrade, K. Arulmoli, B. F. Atwater, J. T. Christian, E. Green, S.L. Kramer, L. Mejia, J.K. Mitchell, E. Rathje, J.R. Rice, and Y. Wang. *State of the Art and Practice in the Assessment of Earthquake-Induced Soil Liquefaction and Its Consequences.* Washington, National Academies Press.
29. R. Kayen, R. Moss, E. Thompson, R. Seed, K. Cetin, A.D. Kiureghian, Y. Tanaka, and K. Tokimatsu. Shear-wave velocity-based probabilistic and deterministic assessment of seismic soil liquefaction potential. 139(3):407–419.
30. S. Khoshnevisan, H. Juang, Y.-G. Zhou, and W. Gong. Probabilistic assessment of liquefaction-induced lateral spreads using CPT – focusing on the 2010–2011 Canterbury earthquake sequence. 192:113–128.
31. S. Khoshnevisan, H. Juang, Y.G. Zhou, and W. Gong. Probabilistic assessment of liquefaction-induced lateral spreads using CPT – focusing on the 2010–2011 Canterbury earthquake sequence. *Engineering Geology*, 192:113–128, 2015.
32. C.S. Ku, C.H. Juang, C.W. Chang, and J. Ching. Probabilistic version of the Robertson and Wride method for liquefaction evaluation: development and application. *Canadian Geotechnical Journal*, 49(1):27–44, 2011.
33. J.A. Lenz and L.G. Baise. Spatial variability of liquefaction potential in regional mapping using CPT and SPT data. *Soil Dynamics and Earthquake Engineering*, 27(7):690–702, 2007.
34. C.N. Liu and C.H. Chen. Mapping liquefaction potential considering spatial correlations of CPT measurements. *Journal of Geotechnical and Geoenvironmental Engineering*, 132(9):1178–1187, 2006.
35. R. Luque and J.D. Bray. Dynamic analyses of two buildings founded on liquefiable soils during the Canterbury earthquake sequence. *Journal of Geotechnical and Geoenvironmental Engineering*, 143(9):04017067, 2017.
36. J. Lu, A. Elgamal, L. Yan, K.H. Law, and J.P. Conte. Large-scale numerical modeling in geotechnical earthquake engineering. *International Journal of Geomechanics*, 11(6):490–503, 2011.
37. M.T. Manzari, M. El Ghoraiby, M. Zeghal, B.L. Kutter, P. Arduino, A.R. Barrero, E. Bilotta, L. Chen, R. Chen, A. Chiaradonna, *et al.* Leap-2017 simulation exercise: Calibration of constitutive models and simulation of the element tests. In *Model Tests and Numerical Simulations of Liquefaction and Lateral Spreading: LEAP-UCD-2017*, pp. 165–185. Springer, 2020.
38. B.W. Maurer, R.A. Green, M. Cubrinovski, and B.A. Bradley. Evaluation of the liquefaction potential index for assessing liquefaction hazard in Christchurch, New Zealand. *Journal of Geotechnical and Geoenvironmental Engineering*, 140(7):04014032, 2014.
39. C.R. McGann, P. Arduino, and P. Mackenzie-Helnwein. A stabilized single-point finite element formulation for three-dimensional dynamic analysis of saturated soils. *Computers and Geotechnics*, 66:126–141, 2015.
40. R.E. Moss, R.B. Seed, R.E. Kayen, J.P. Stewart, A.Der Kiureghian, and K. Onder Cetin. CPT-based probabilistic and deterministic assessment of in situ seismic soil liquefaction potential. 132(8):1032–1051.
41. F. Oka, A. Yashima, T. Shibata, M. Kato, and R. Uzuoka. FEM-FDM coupled liquefaction analysis of a porous soil using an elasto-plastic model. *Applied Scientific Research*, 52:209–245, 1994.
42. PLAXIS.
43. R. Popescu. Finite element assessment of the effects of seismic loading rate on soil liquefaction. *Canadian Geotechnical Journal*, 39(2):331–344, 2002.
44. P.K. Robertson. Interpretation of cone penetration tests—a unified approach. 46(11):1337–1355.
45. P.K. Robertson and C.E. Wride. Evaluating cyclic liquefaction potential using the cone penetration test. 35:442–459.

46. P.K. Robertson and C.E. Wride. Evaluating cyclic liquefaction potential using the cone penetration test. *Canadian Geotechnical Journal*, 35(3):442–459, 1998.
47. P.K. Robertson. Performance based earthquake design using the CPT. In *Proceedings of the IS-Tokyo*, pp. 3–20, 2009.
48. H.B. Seed and I.M. Idriss. Simplified procedure for evaluating soil liquefaction potential.
49. H.B. Seed, I. Arango, and C.K. Chan. Evaluation of soil liquefaction potential for level ground during earthquakes. A summary report. Technical report, Shannon and Wilson, Inc., Seattle, WA; Agbabian Associates, El Segundo, CA, 1975.
50. R.B. Seed, K. Onder Cetin, R.E.S. Moss, A.M. Kammerer, J. Wu, J.M. Pestana, M.F. Riemer, R.B. Sancio, J.D. Bray, R.E. Kayen, *et al.* Recent advances in soil liquefaction engineering: a unified and consistent framework. In *Proceedings of the 26th Annual ASCE Los Angeles Geotechnical Spring Seminar*, Long Beach, CA, 2003.
51. S.F. Sizkow and U. El Shamy. Sph-dem simulations of saturated granular soils liquefaction incorporating particles of irregular shape. *Computers and Geotechnics*, 134:104060, 2021.
52. H. Sonmez and C. Gokceoglu. A liquefaction severity index suggested for engineering practice. *Environmental Geology*, 48(1):81–91, 2005.
53. H. Sonmez. Modification of the liquefaction potential index and liquefaction susceptibility mapping for a liquefaction-prone area (Inegol, Turkey). *Environmental Geology*, 44(7):862–871, 2003.
54. J.P. Stewart, S.L. Kramer, D.Y. Kwak, M.W. Greenfield, R.E. Kayen, K. Tokimatsu, J.D. Bray, C.Z. Beyzaei, M. Cubrinovski, T. Sekiguchi, *et al.* PEER-NGL project: Open source global database and model development for the next-generation of liquefaction assessment procedures. 91:317–328.
55. S. Toprak and T.L. Holzer. Liquefaction potential index: field assessment. *Journal of Geotechnical and Geoenvironmental Engineering*, 129(4):315–322, 2003.
56. S. Van Ballegooy, V. Lacrosse, M. Jacka, and P. Malan. LSN – a new methodology for characterising the effects of liquefaction in terms of relative land damage severity. In *Proceedings of the 19th NZGS Geotechnical Symposium*. Ed. CY Chin, Queenstown, 2013.
57. S. van Ballegooy, F. Wentz, and R. W. Boulanger. Evaluation of CPT-based liquefaction procedures at regional scale. *Soil Dynamics and Earthquake Engineering*, 79:315–334, 2015.
58. B. Vivek and P. Raychowdhury. Probabilistic and spatial liquefaction analysis using CPT data: a case study for Alameda County site. *Natural Hazards*, 71(3):1715–1732, 2014.
59. C. Wang, Q. Chen, M Shen, and C. Hsein Juang. On the spatial variability of cpt-based geotechnical parameters for regional liquefaction evaluation. *Soil Dynamics and Earthquake Engineering*, 95:153–166, 2017.
60. C. Wang, F. McKenna, P. Mackenzie-Helnwein, W. Elhaddad, A. Zsarnoczay, and M. Gardner. Nheri-simcenter/s3hark: Release v1.5, July 2020.
61. C. Wang, Q. Chen, M. Shen, and C.H. Juang. On the spatial variability of cpt-based geotechnical parameters for liquefaction potential evaluation. *Soil Dynamics and Earthquake Engineering*, 95:153–166, 2017.
62. C. Wang and Q. Chen. A hybrid geotechnical and geological data-based framework for multiscale regional liquefaction hazard mapping. *Géotechnique*, 68(7):614–625, 2018.
63. T.L. Youd and I.M. Idriss. Liquefaction resistance of soils: Summary report from the 1996 NCEER and 1998 NCEER/NSF workshops on evaluation of liquefaction resistance of soils. 127(4):297–313.
64. T. Leslie Youd and D.M. Perkins. Mapping of liquefaction severity index. 113(11):1374–1392.
65. J. Zhu, L.G. Baise, and E.M. Thompson. An updated geospatial liquefaction model for global application. 107(3):1365–1385.
66. O.C. Zienkiewicz, A.H.C. Chan, M. Pastor, B.A. Schrefler, and T. Shiomi. *Computational Geomechanics*. Chichester, Wiley Chichester, 1999.
67. Cliq.
68. LiqIT.
69. Liquefy-Pro.
70. NovoLIQ.

Chapter 15

Prediction for soil design properties based on a multivariate database for Shanghai soft clay

Yelu Zhou, Dongming Zhang, and Hongwei Huang

This study compiles a database for Shanghai soft clay labeled as SH-CLAY/12/4191 based on data collected from 51 investigation reports of geotechnical engineering in Shanghai. The database contains 4191 data points, each of which is composed of multivariate parameters measured simultaneously in close proximity at the same depth. There are five physical parameters and seven mechanical parameters. Then, a 12-dimensional multivariate probability distribution is constructed based on the compiled database. To make it more useful for engineering practice, the transformation models between the design parameters (i.e., K_0, E_s, c, and φ) and the index parameters as well as the in-situ test parameters are derived. The transformation models are explicit nonlinear formulas of multiple variables without cross-term. The transformation models are validated by the data from the new site, showing the generally reliable performance in predicting the design parameters.

15.1 INTRODUCTION

Determination of soil parameters for design is critical for geotechnical engineering. For a desired project, the number of boreholes and soil tests are very limited, which leads to the fact that the mechanical parameters provided in the geotechnical investigation report cannot fully reflect the characteristics of the mechanical properties of the soil at the site, making them an unreliable basis for geotechnical design. On the other hand, the geotechnical investigation report provides a relatively rich set of physic index parameters and field measurements for the soil. In view of this fashion of geotechnical site investigation, transformation models are typically adopted to estimate the value or reduce the uncertainty of the design parameters, i.e., mechanical properties, via these abundant index or field properties (Phoon and Kulhawy, 1999).

In the literature, many transformation models have been constructed based on soil databases (Kulhawy and Mayne 1990; Mayne et al. 2001). However, most of the transformation models are pairwise correlations. It is generally known that design parameters are correlated with multiple index parameters and field measurements. Annex D of ISO2394 (2015) illustrated the advantage of the multivariate probability distribution model considering the correlations between different soil parameters so that more accurate probabilistic characterization of soil design parameters could be obtained based on the multiple information of other soil parameters. Before the construction of the multivariate probability distribution, the compilation of the target multivariate soil database consisting of laboratory and field data is required. It should be noted that incomplete multivariate databases such as those consisting of multiple sets of bivariate data obtained from different locations can be used to construct the multivariate distribution, that is to say, not all parameters are measured simultaneously in close proximity. This is more

 DOI: 10.1201/9781003441946-15

practical because simultaneous measurements of two parameters are more commonly encountered than those of three or more parameters, which is precisely the reason why traditional multivariate regression assumptions are often challenging to meet in geotechnical exploration.

So far, several global or regional multivariate soil databases have been compiled, such as CLAY/10/7490 (Ching and Phoon, 2014a), SAND/7/2794 (Ching et al., 2017), ROCK/10/4025 (Muzamhindo and Ferentinou, 2023), and ROCKMass/9/5876 (Ching et al., 2021) at the global scale and F-CLAY/7/216 (D'Ignazio et al., 2016), S-CLAY/7/168 (D'Ignazio et al., 2016), J-CLAY/5/124 (Liu et al., 2016), FI-CLAY/14/856 (Löfman and Korkiala-Tanttu, 2021), and SH-CLAY/11/4051 (Zhang et al., 2020) at the regional scale, and some multivariate probability distribution models have been derived successfully from these databases. The practical applications of the multivariate probability distribution may include the Bayesian updating and the establishment of the transformation models involving multivariate correlations. However, whether the database is regional or global, the transformation model developed based on it is generic for a specific site. When a generic model is applied to a specific site, the transformation uncertainty can be significantly high. This is because the generic model is designed to encompass a broad range of soil types and site conditions. Meanwhile, data collected from one single site may be too sparse to develop an imprecise site-specific model due to the statistical uncertainty. Hybridization methods for constructing a site-specific multivariate distribution by considering the generic multivariate distribution model and the multivariate site-specific data together have been proposed to reduce the transformation uncertainty (Ching and Phoon, 2019; Ching et al., 2022). In situations where ample data is available at a specific site, the methodologies will predominantly draw upon local data to formulate the transformation model. Conversely, when site-specific data are scant and local experience is limited, the transformation model should place a greater emphasis on the generic data. Nevertheless, the greater the congruity between the soil properties in the database employed to develop a generic model and those of the soil at the specific site, the more advantageous it becomes in the establishment of a site-specific model (Ching et al., 2020). The database for Shanghai's soft clay is thus compiled to facilitate the development of the transformation models for the region, because it is more relevant to Shanghai than the other global databases that cover sites from all around the world.

This chapter introduces the municipal database for the properties of Shanghai Clay labeled SH-CLAY/12/4191. Note this database is an extension of SH-CLAY/11/4051 (Zhang et al., 2020) with additional data points collected from a new site as well as differences in parameters. Transformation models to predict the soil design parameters are derived based on this database. The derived transformation models might still be generic but are practical for the cases where engineers are compelled to use the generic models. These models are also readily adaptable for constructing the site-specific models by incorporating the site-specific data, which, however, is beyond the scope of this study.

The rest of this study is organized as follows. Firstly, some basic information on the location, test condition, and type of clay parameters as well as the basic statistics and correlations pertaining to the database SH-CLAY/12/4191 is presented. Secondly, a multivariate probability distribution is constructed based on this database to characterize the multivariate correlations among the 12 parameters. The transformation models between the design soil parameters and the index parameters as well as the field measurements are then derived and subsequently validated using data from a new site that is independent of the database.

15.2 SHANGHAI CLAY DATABASE

15.2.1 Description of SH-CLAY/12/4191 database

Shanghai is situated in the eastern part of China at the mouth of the Yangtze River. It is a fairly typical soft soil area, characterized by flat terrain formed by the deposition of sediment carried by the Yangtze River over millions of years.

This study compiles a database for Shanghai's soft clay, including soil parameter information for gray muddy silty clay, gray muddy clay, gray clay, and gray silty clay, which are the primary constituents of Shanghai's soft soils. The maximum depth from the ground surface of the collected soil layers is 30 m.

The data points in SH-CLAY/12/4191 are collected from 51 site investigation reports for Shanghai. These sites are distributed in nine administrative areas of Shanghai, including Pudong, Yangpu, Huangpu, Jing'an, Putuo, Changning, Xuhui, Jiading, and Qingpu, covering an approximate area of 145 km^2, about 1/43 the size of Shanghai. There are 574 boreholes, 49 vane shear tests, and 201 cone penetration tests in total. Figure 15.1(a, b), respectively, illustrates the distribution of the 51 sites and the layout for the three types of tests at a specific site.

The collected soil parameters consist of 12 variables, which include: (1) five index parameters – the water content (ω), liquid limit (LL), plastic index (PI), liquid index (LI), and void ratio (e); and (2) seven mechanical property parameters – vertical effective stress (σ'_v), lateral pressure coefficient (K_0), modulus of compressibility (E_s), consolidated quick shear cohesion (c), consolidated quick shear angle of internal friction (φ), undrained shear strength ($S_{u(VST)}$) obtained from the vane shear test (VST), and cone-tip resistance (p_s)

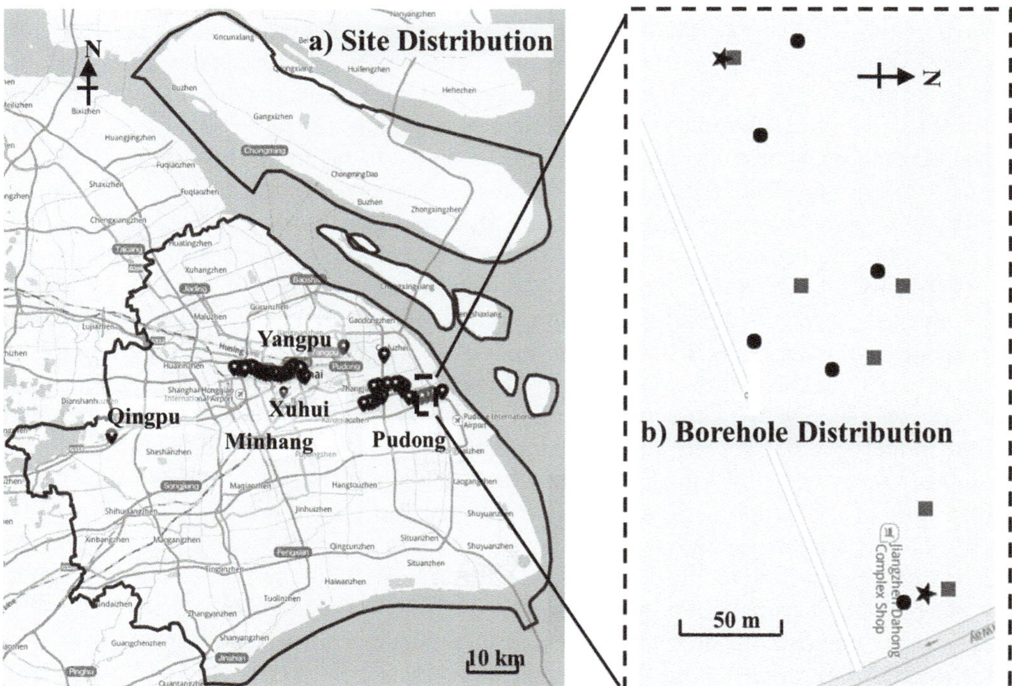

Figure 15.1 Schematic graph of site and borehole distribution.

from the cone penetration test (CPT). Note that p_s is different from the cone-tip resistance q_t adopted in international practice, due to the difference in the CPT test equipment and procedure between China and other countries or regions

There are 4191 records in the database, and each record represents the values of the 12 parameters at the same depth, which are obtained from the different types of boreholes located within a specific proximity to each other. This study constrains this distance to 35 m, within which the parameters are strongly correlated referring to the average value of the correlation distance of Shanghai (Xiao, 2018). However, it is common to find empty cells in each record because some indoor mechanical property tests are not conducted for soil samples at all depths, and data from the boreholes is not readily paired with that from the VSTs due to the less frequent execution of the VST in the proximity of boreholes.

15.2.2 Basic statistics of SH-CLAY/12/4191 database

Table 15.1 provides the statistical values of the 12 parameters in the database, including the sample size, mean value, coefficient of variation (COV), maximum value (Max), and minimum value (Min). It shows the characteristics of Shanghai soft clay with high water content, high compressibility, and low strength.

Gao and Wei (1983) collected the geotechnical test data from 17 sites in Shanghai and calculated the mean value and COV of the index and mechanical parameters for different soil layers at each site. It was found that the mean values of the collected parameters did not vary significantly among different sites, suggesting that Shanghai soft clay is relatively homogeneous. Therefore, statistical analysis of soil parameters at the municipal level is feasible. In addition, the Shanghai local code for geotechnical site investigation conducted statistical analyses on over 100 geotechnical investigation reports in Shanghai, encompassing data from tens of thousands of soil samples. It provided ranges and COV of 22 parameters for different soil layers. By comparing the statistics of parameters shared by the SH-CLAY/12/4191 database with those derived from the investigation code, it is evident that the mean values of all the parameters in SH-CLAY/12/4191 fall within the ranges determined by the investigation code, verifying the rationality of the data collected in this study. Details pertaining to statistics from Gao and Wei (1983) and the investigation code can be found in Appendix A.

Table 15.1 Basic statistics of the soil parameters of Shanghai soft clay

Parameters	*n*	*Mean*	*COV*	*Min*	*Max*
ω (%)	4011	43.29	0.15	23.30	63.60
LL (%)	2368	40.29	0.12	22.80	58.70
PI	4183	18.24	0.18	10.30	30.90
LI	2201	1.14	0.22	0.49	2.19
e	4014	1.23	0.14	0.67	1.79
σ'_v (kPa)	2801	132.61	0.36	28.69	274.38
K_0	300	0.51	0.10	0.43	0.65
E_s (Mpa)	2066	3.08	0.32	1.46	12.91
c (kPa)	1026	14.35	0.22	7.00	28.00
φ (°)	1026	13.66	0.33	5.50	28.50
S_u (kPa)	410	40.07	0.24	18.60	76.60
p_s (MPa)	1357	0.75	0.39	0.30	4.20

15.2.3 Correlation of the parameters

Pearson product–moment correlation coefficient ρ is adopted to evaluate the degree of dependency between the parameter pairs. In total, there are $12 \times (12-1)/2 = 66$ parameter pairs in SH-CLAY/12/4191. Table 15.2 presents the correlation coefficients for the 66 parameter pairs. Several inferences can be drawn from the table as follows. Firstly, there are distinct correlations between the index parameters, with the exception being that the correlation between LI and other index parameters is relatively weak. Secondly, σ'_v exhibits a substantial negative correlation with LI and marked positive correlations with c, S_u, and p_s. Since σ'_v approximates a linear increase with depth, it can be deduced that LI diminishes gradually with the increase of the soil depth, whereas c, S_u, and p_s increase with the soil depth. Other parameters, however, appear to remain unaffected by variations in soil depth. Thirdly, mechanical parameters K_0, E_s, and φ display relatively strong correlations with the index parameters, rendering the estimation of these mechanical properties using the index parameters feasible. Lastly, p_s demonstrates a certain degree of correlation with the strength index c, φ, and S_u. Hence, it is legitimate to estimate the strength index through the more readily measured in-situ parameter, p_s.

15.3 MULTIVARIATE PROBABILITY DISTRIBUTION FOR SH-CLAY/12/4191

The multivariate normal distribution is employed in this chapter due to its computational simplicity and practical applicability. In accordance with the established methodology (Ching et al., 2016), this study constructs the multivariate normal distribution using the following steps: (1) transforming non-normally distributed soil parameters into variables following a standard normal distribution and (2) calculating the correlation coefficients between the transformed variables to constitute a correlation matrix in the normal space, with which the multivariate normal distribution can be defined. Details of the two steps are described in the subsequent subsections.

15.3.1 Normal transformation for soil parameters

Soil parameters typically do not follow the normal distribution in the marginal sense. In addition to its successful applications in many soil databases, the Johnson transformation method was suggested by Zhou et al. (2021) as the optimal approach for constructing the multivariate normal distribution for real soil databases. The recommendation was based on their comparative assessment, which found the Johnson transformation method to be more robust and superior in terms of normal transformation compared to the other two methods.

The Johnson distribution system (Johnson, 1949) comprises three types of distributions: SU (for the unbounded system), SB (for the bounded system), and SL (for the lognormal system), with the commonly known lognormal distribution falling under the SL category. The formula that transforms the non-normal variable Y_i to its corresponding standard normal variable X_i is as follows:

Table 15.2 Correlation coefficients between soil parameters

	ω	LL	PI	LI	**e**	σ'_v	K_0	E_s	**c**	φ	S_u	p_s
ω	1.00	0.75	0.72	0.58	0.99	−0.28	0.70	−0.78	−0.39	−0.63	−0.27	−0.31
LL		1.00	0.95	−0.09	0.77	0.05	0.73	−0.68	0.03	−0.70	0.08	−0.27
PI			1.00	−0.11	0.74	0.17	0.69	−0.66	0.11	−0.72	0.13	−0.15
LI				1.00	0.54	−0.70	0.15	−0.41	−0.64	−0.06	−0.48	−0.39
e					1.00	−0.25	0.72	−0.79	−0.36	−0.64	−0.23	−0.30
σ'_v						1.00	−0.06	0.23	0.61	0.11	0.73	0.44
K_0							1.00	−0.66	−0.08	−0.58	−0.36	−0.36
E_s		Symmetry							1.00	0.23	0.72	0.19
c								1.00	−0.11	0.52	0.33	0.35
φ										1.00	0.06	0.30
S_u											1.00	0.62
p_s												1.00

Table 15.3 Distribution type and distribution parameters for $(Y_1, Y_2, \ldots, Y_{12})$

Random variable	*Soil parameters*	*Distribution type*	*Distribution parameters*			
			a_X	b_X	a_Y	b_Y
Y_1	$\ln(\omega)$	SB	1.60	−0.93	1.11	3.06
Y_2	ln(LL)	SB	1.33	−0.75	0.77	3.21
Y_3	ln(PI)	SB	1.22	−0.63	1.04	2.25
Y_4	ln(LI)	SB	2.11	−0.72	2.01	−1.06
Y_5	ln(e)	SB	1.62	−0.92	1.07	−0.47
Y_6	$\ln(\sigma'_v/\text{Pa})$	SB	1.96	−3.80	6.89	−5.77
Y_7	$\ln(K_0)$	SB	1.09	0.72	0.49	−0.86
Y_8	$\ln(E_s)$	SB	1.23	0.16	1.56	0.34
Y_9	ln(c)	SU	6.05	39.77	0.003	3.72
Y_{10}	$\ln(\varphi)$	SB	1.67	0.50	2.29	1.57
Y_{11}	$\ln(S_u/\sigma'_v)$	SU	1.10	−1.24	0.11	−1.35
Y_{12}	$\ln(p_s/\sigma'_v)$	SU	1.01	−0.82	0.16	1.52

$$X = \begin{cases} b_X + a_X \sinh^{-1}\left(\dfrac{Y - b_Y}{a_Y}\right) & SU \\ b_X + a_X \ln\left(\dfrac{Y - b_Y}{a_Y + b_Y - Y}\right) & SB\ (b_Y \le Y \le a_Y + b_Y) \\ b_X + a_X \ln\left(\dfrac{Y - b_Y}{a_Y}\right) & SL\ (Y \ge b_Y) \end{cases} \quad (15.1)$$

It is a closed-form equation, where the distribution type and the corresponding distribution model parameters (a_X, b_X, a_Y, b_Y) can be determined by using a non-parametric method based on the four percentiles of Y_i proposed by Slifker and Shapiro (1980).

The natural logarithm of the 12 parameters included in SH-CLAY/12/4191 are denoted as $(Y_1, Y_2, \ldots, Y_{12})$, with some of the parameters having undergone dimensionless standardization (as detailed in Table 15.3). Following the abovementioned approach, the distribution types and corresponding parameters for $(Y_1, Y_2, \ldots, Y_{12})$ are derived and presented in Table 15.3. Standard normal variables $(X_1, X_2, \ldots, X_{12})$ can subsequently be generated using the parameters provided in Table 15.3 and Equation (15.1).

Even if each transformed variable X_i follows a normal distribution, the multivariate set $(X_1, X_2, \ldots, X_{12})$ may not necessarily follow a multivariate normal distribution, unless the linear combination of any of the components follows a normal distribution. Nevertheless, for several soil databases, the assumption of a multivariate normal distribution continues to yield satisfactory practical outcomes, despite the transformed parameters not strictly following the multivariate normal distributions. Hence, this study forgoes testing for multivariate normality.

15.3.2 Correlation matrix of multivariate distribution

The probability density function (PDF) of the multivariate joint distribution is given below:

$$f(\boldsymbol{x}) = |\mathrm{C}|^{-1/2} (2\pi)^{-m/2} \exp\left\{-\frac{1}{2}(\boldsymbol{x}-\boldsymbol{\mu})^T \times (\mathrm{C})^{-1} \times (\boldsymbol{x}-\boldsymbol{\mu})\right\} \tag{15.2}$$

where $\boldsymbol{x} = (x_1, x_2, \ldots, x_m)^{\mathrm{T}}$ is a vector composed of m random variables following standard normal distributions, with m being the number of variables (m is taken as 12 in this study), "T" refers to the transpose of the matrix, $\boldsymbol{\mu}$ is the mean vector of $\boldsymbol{x}$, C denotes the covariance matrix, and |C| stands for the determinant of the covariance matrix. As each component of $\boldsymbol{x}$ is a standard normal variable, $\boldsymbol{\mu}$ simplifies to a zero vector. C corresponds to the correlation matrix, which is composed of the Pearson product–moment correlation coefficients (denoted as δ_{ij}) between X_i and X_j.

Figure 15.2 shows the scatter plots for each parameter pair (X_i, X_j). To make the best use of the collected data, δ_{ij} is estimated using independent bivariate datasets, i.e., data points in the corresponding scatter plot in Figure 15.2, which involve two parameters measured in proximity, without the requirement for simultaneous knowledge of other parameters. However, this method has the potential to generate a non-positive definite correlation matrix, rendering it mathematically invalid.

This study follows the approach proposed by Ching and Phoon (2014b) to tackle the problem. To ensure statistical robustness, the process involves obtaining 1000 samples for

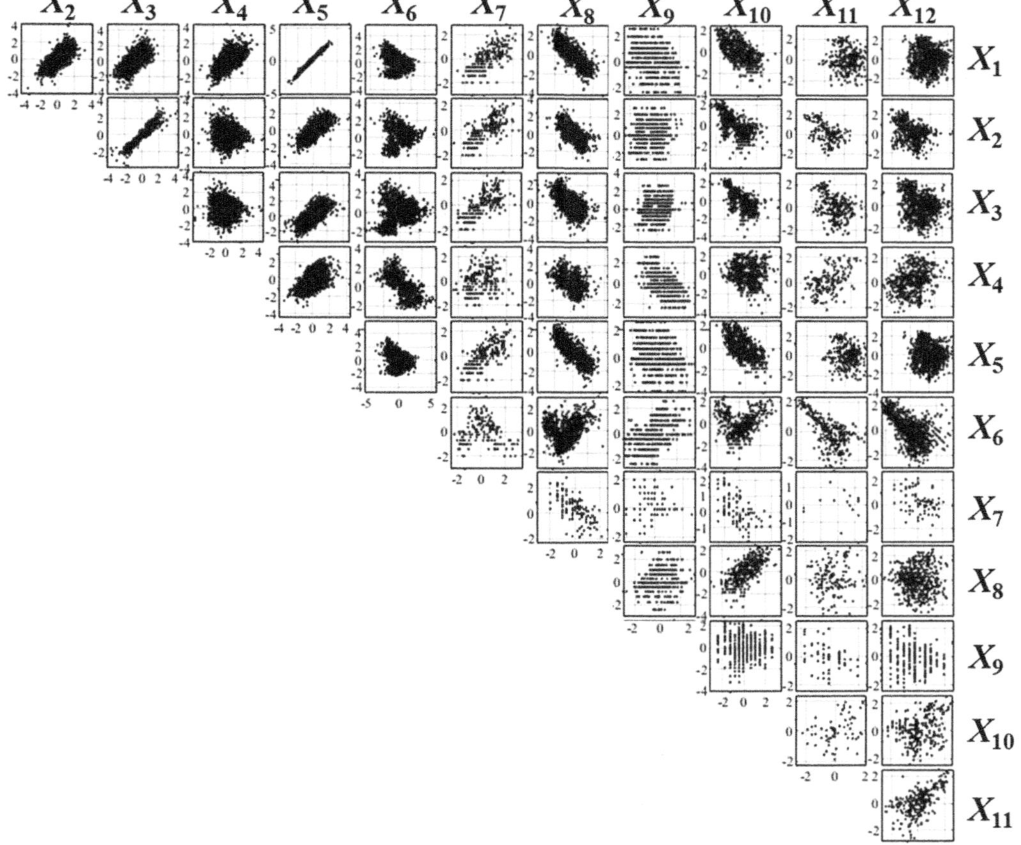

Figure 15.2 Scatter plots between X_i and X_j.

each δ_{ij} through bootstrapping technique. From this pool of samples, one δ_{ij} value is randomly selected from the 90% confidence interval of the 1000 samples, which is repeated 66 times for each parameter pair to form a new correlation matrix **C**. While only retaining the positive definite matrices, 1000 valid correlation matrices **C** are generated by this approach. The final correlation matrix **C** is derived by calculating the average value of the 1000 accepted matrices, as presented in Table 15.4.

15.4 TRANSFORMATION MODEL FOR DESIGN PARAMETERS

The established multivariate normal distribution can serve as a prior distribution for Bayesian updating, and the posterior distribution can be obtained based on the new measurements. The median and 95% confidence interval of the posterior distribution of the target parameter can be considered as predictions for the parameter given other observed data. Consequently, this allows the development of the multivariate transformation model between the target parameter and the other parameters.

15.4.1 Bayesian updating

Let the target parameters to be updated be represented as vector $\mathbf{X}^{u}$, which is an m×1 vector. The observed parameter is denoted as vector $\mathbf{X}^{o}$, an n×1 vector. Their corresponding mean vectors are designated as $\boldsymbol{\mu}^{u}$ (an m×1 zero vector) and $\boldsymbol{\mu}^{o}$ (an n×1 zero vector), and the covariance matrices are referred to as $\mathbf{C}^{u}$ (an m×m matrix) and $\mathbf{C}^{o}$ (an n×n matrix), respectively.

$\mathbf{C}^{uo}$ (an m×n matrix) and $\mathbf{C}^{ou}$ (an n×m matrix) denote the covariance matrices between the target parameters and the observed parameters, with each being the transpose of the other. The updated multivariate distribution continues to follow a multivariate normal distribution due to the conjugacy of the prior distribution, with its mean vector and covariance matrix denoted as $\boldsymbol{\mu}'^{u}$ and $\mathbf{C}'^{u}$, which can be computed using the following equation:

$$\begin{aligned}\mu'^{u} &= \mu^{u} + \mathbf{C}^{uo}(\mathbf{C}^{o})^{-1}(\mathbf{x}^{o} - \mu^{o}) \\ \mathbf{C}'^{u} &= \mathbf{C}^{u} - \mathbf{C}^{uo}(\mathbf{C}^{o})^{-1}\mathbf{C}^{ou}\end{aligned} \tag{15.3}$$

One attractive feature of the Johnson distribution is that the functional form of the marginal distribution for Y_i remains unaltered after conditioning. Additionally, the type of the Johnson distribution, as well as its corresponding a_Y and b_Y parameters, remains constant. Only the parameters a_X and b_X are updated. The expressions for the updated parameters a'_X and b'_X are

$$\begin{aligned}a'_X &= a_X / \sigma'_X \\ b'_X &= (b_X - \mu'_X) / \sigma'_X\end{aligned} \tag{15.4}$$

where $\boldsymbol{\mu}'_x$ is the updated mean value of X_i and σ'_x is the updated standard deviation, which corresponds to the values in $\boldsymbol{\mu}'^{u}$ and $\mathbf{C}'^{u}$ for the component X_i, respectively.

In this study, the design parameters K_0, E_s, c, and φ are selected as the target parameters. Tables 15.5–15.8 present the updated Johnson parameters (a'_X, b'_X) for these four parameters conditioned on other parameters.

Table 15.4 C matrix composed of the average of the 1000 accepted δ_{ij}

X	X_1	X_2	X_3	X_4	X_5	X_6	X_7	X_8	X_9	X_{10}	X_{11}	X_{12}
X_1	1.00	0.74	0.70	0.60	0.99	−0.28	0.69	−0.81	−0.29	−0.63	−0.07	−0.04
X_2		1.00	0.94	−0.03	0.76	0.10	0.72	−0.66	0.09	−0.67	−0.45	−0.32
X_3			1.00	−0.07	0.73	0.18	0.67	−0.64	0.19	−0.65	−0.40	−0.26
X_4				1.00	0.56	−0.69	0.17	−0.44	−0.61	−0.17	0.49	0.33
X_5					1.00	−0.24	0.70	−0.82	−0.26	−0.65	−0.10	−0.07
X_6						1.00	−0.03	0.27	0.55	0.11	−0.64	−0.57
X_7							1.00	−0.63	0.03	−0.64	−0.26	−0.36
X_8		Symmetry						1.00	0.22	0.67	0.08	0.02
X_9									1.00	−0.07	−0.31	−0.17
X_{10}										1.00	0.27	0.26
X_{11}											1.00	0.62
X_{12}												1.00

Table 15.5 Updated Johnson parameters for K_0 under various combinations of information

	Updated Johnson parameter	
Information	a'_X	b'_X
ω	1.50	$-0.95X_1+1.00$
LL	1.57	$-1.04X_2+1.04$
PI	1.47	$-0.90X_3+0.97$
LI	1.11	$-0.17X_4+1.73$
e	1.53	$-0.98X_5+1.01$
σ'_v/P_a	1.09	$0.03X_6+0.72$
S_u/σ'_v	1.13	$0.27X_{12}+0.75$
p_s/σ'_v	1.17	$0.39X_{13}+0.77$
ω, LL, PI, LI, e	1.68	$-0.60X_1-0.60X_2+0.30X_3+0.35X_4-0.44X_5+1.11$
ω, LL, PI, LI, e, σ'_v/P_a	1.68	$-0.62X_1-0.56X_2+0.29X_3+0.41X_4-0.47X_5+0.04X_6+1.12$
ω, LL, PI, LI, e, σ'_v/P_a, S_u/σ'_v	1.68	$-0.62X_1-0.55X_2+0.28X_3+0.40X_4-0.46X_5+0.05X_6+0.01X_{11}+1.11$
ω, LL, PI, LI, e, σ'_v/P_a, p_s/σ'_v	1.84	$-1.30X_1+0.19X_2-0.16X_3+0.72X_4-0.17X_5+0.51X_6+0.61X_{12}+1.22$

Table 15.6 Updated Johnson parameters for E_S under various combinations of information

	Updated Johnson parameter	
Information	a'_X	b'_X
ω	2.09	$1.38X_1+0.28$
LL	1.63	$0.88X_2+0.22$
PI	1.59	$0.83X_3+0.21$
LI	1.36	$0.49X_4+0.18$
e	2.14	$1.43X_5+0.28$
σ'_v/P_a	1.27	$-0.28X_6+0.17$
S_u/σ'_v	1.23	$-0.08X_{12}+0.16$
p_s/σ'_v	1.23	$-0.02X_{13}+0.16$
ω, LL, PI, LI, e	2.17	$-0.49X_1+0.28X_2+0.26X_3+0.35X_4+1.34X_5+0.29$
ω, LL, PI, LI, e, σ'_v/P_a	2.21	$-0.34X_1+0.01X_2+0.32X_3-0.04X_4+1.52X_5-0.30X_6+0.29$
ω, LL, PI, LI, e, σ'_v/P_a, S_u/σ'_v	2.22	$-0.35X_1-0.12X_2+0.43X_3-0.005X_4+1.50X_5-0.39X_6-0.15X_{11}+0.29$
ω, LL, PI, LI, e, σ'_v/P_a, p_s/σ'_v	2.21	$-0.31X_1-0.03X_2+0.35X_3-0.06X_4+1.51X_5-0.33X_6-0.03X_{12}+0.29$

15.4.2 Transformation formula

The transformations from X to Y are as follows:

$$Y = \begin{cases} b_Y + a_Y \sinh\left(\dfrac{X - b_X}{a_X}\right) & SU \\ b_Y + a_Y\left[1 + \exp\left(-\dfrac{X - b_X}{a_X}\right)\right]^{-1} & SB \\ b_Y + a_Y \ln\left(\dfrac{X - b_X}{a_X}\right) & SL \end{cases} \tag{15.5}$$

Table 15.7 Updated Johnson parameters for c under various combinations of information

Information	Updated Johnson parameter	
	a'_X	b'_X
ω	6.32	$0.30X_1+41.55$
LL	6.07	$-0.09X_2+39.93$
PI	6.16	$-0.19X_3+40.50$
LI	7.63	$0.76X_4+50.18$
e	6.26	$0.27X_5+41.18$
σ'_v/P_a	7.24	$-0.66X_6+47.61$
S_u/σ'_v	6.36	$0.33X_{12}+41.83$
p_s/σ'_v	6.14	$0.17X_{13}+40.35$
ω, LL, PI, LI, e	8.03	$-0.18X_1+0.78X_2-0.94X_3+0.77X_4+0.18X_5+52.83$
ω, LL, PI, LI, e, σ'_v/P_a	8.16	$-0.04X_1+0.56X_2-0.91X_3+0.43X_4+0.33X_5-0.27X_6+53.63$
ω, LL, PI, LI, e, σ'_v/P_a, S_u/σ'_v	8.18	$-0.05X_1+0.44X_2-0.82X_3+0.47X_4+0.31X_5-0.34X_6-0.13X_{11}+53.80$
ω, LL, PI, LI, e, σ'_v/P_a, p_s/σ'_v	8.25	$0.16\ X_1+0.29X_2-0.76X_3+0.34X_4+0.22X_5-0.43X_6-0.21X_{12}+54.20$

Table 15.8 Updated Johnson parameters for φ under various combinations of information

Information	Updated Johnson parameter	
	a'_X	b'_X
ω	2.15	$0.81X_1+0.64$
LL	2.25	$0.90X_2+0.67$
PI	2.19	$0.85X_3+0.65$
LI	1.69	$0.17X_4+0.50$
e	2.19	$0.85X_5+0.65$
σ'_v/P_a	1.68	$-0.11X_6+0.50$
S_u/σ'_v	1.73	$-0.28X_{12}+0.52$
p_s/σ'_v	1.73	$-0.27X_{13}+0.52$
ω, LL, PI, LI, e	2.36	$-0.75X_1+0.56X_2+0.19X_3+0.16X_4+1.01X_5+0.71$
ω, LL, PI, LI, e, σ'_v/P_a	2.39	$-0.62X_1+0.34X_2+0.24X_3-0.18X_4+1.15X_5-0.26X_6+0.71$
ω, LL, PI, LI, e, σ'_v/P_a, S_u/σ'_v	2.46	$-0.66X_1-0.001X_2+0.53X_3-0.08X_4+1.12X_5-0.48X_6-0.39X_{11}+0.74$
ω, LL, PI, LI, e, σ'_v/P_a, p_s/σ'_v	2.53	$-0.17\ X_1-0.28X_2+0.63X_3-0.41X_4+0.95X_5-0.63X_6-0.48X_{12}+0.76$

The conditioned Y' can be obtained by substituting the original (a_X, b_X) with the updated (a'_X, b'_X) in Equation (15.5). It is thus easily derived that the p-quantile for the distribution of Y', denoted as y'_p, can be obtained from the corresponding p-quantile for the standard normal distribution, denoted as x_p. Using the Johnson SB distribution as an illustration, the median $(y'_{0.5})$ and the upper $(y'_{0.025})$ and lower $(y'_{0.975})$ boundaries of the 95% confidence interval (CI) for the updated Johnson SB distribution are as follows:

$$y'_{0.5} = b_Y + a_Y\left[1+\exp\left(\frac{b'_X}{a'_X}\right)\right]^{-1}$$

$$y'_{0.025} = b_Y + a_Y\left[1+\exp\left(-\frac{-1.96-b'_X}{a'_X}\right)\right]^{-1} \tag{15.6}$$

$$y'_{0.975} = b_Y + a_Y\left[1+\exp\left(-\frac{1.96-b'_X}{a'_X}\right)\right]^{-1}$$

The median of the conditioned distribution can serve as the point estimate for the target parameter based on other observed information, with the 95% CI indicating the precision of the estimate.

From Tables 15.5–15.8, it is evident that b'_X can be expressed as a linear combination of X_i. Therefore, Equation (15.6) is also an expression involving X_i and can be regarded as a transformation formula between the target parameter and the observed parameters.

Let's consider parameter pairs e (X_5) and K_0 (X_7) as an example to elaborate on the process of establishing the transformation formula between parameters. According to Table 15.5, the updated Johnson parameters for X_7 conditioned on X_5 are $a'_X = 1.53$ and $b'_X = 1.01–0.98\ X_5$. Notably, the parameters a_Y and b_Y remain invariant at 0.49 and −0.86, as recorded in Table 15.3. By substituting these values into Equation (15.6), the expression for Y_7 with respect to X_5 can be derived. Using Equation (15.1) and the corresponding Johnson parameters from Table 15.3, X_5 can be transformed to Y_5. Finally, after taking the exponential transformation, the functional relationship between e and K_0 can be established:

$$K_{0_{0.5}} = \exp\left\{-0.86+0.49\times\left[1+\exp\left(\frac{1.91-1.58\times\ln\left(\frac{\ln(e)+0.47}{0.60-\ln(e)}\right)}{1.53}\right)\right]^{-1}\right\}$$

$$K_{0_{0.025}} = \exp\left\{-0.86+0.49\times\left[1+\exp\left(\frac{3.87-1.58\times\ln\left(\frac{\ln(e)+0.47}{0.60-\ln(e)}\right)}{1.53}\right)\right]^{-1}\right\} \tag{15.7}$$

$$K_{0_{0.075}} = \exp\left\{-0.86+0.49\times\left[1+\exp\left(\frac{-0.05-1.58\times\ln\left(\frac{\ln(e)+0.47}{0.60-\ln(e)}\right)}{1.53}\right)\right]^{-1}\right\}$$

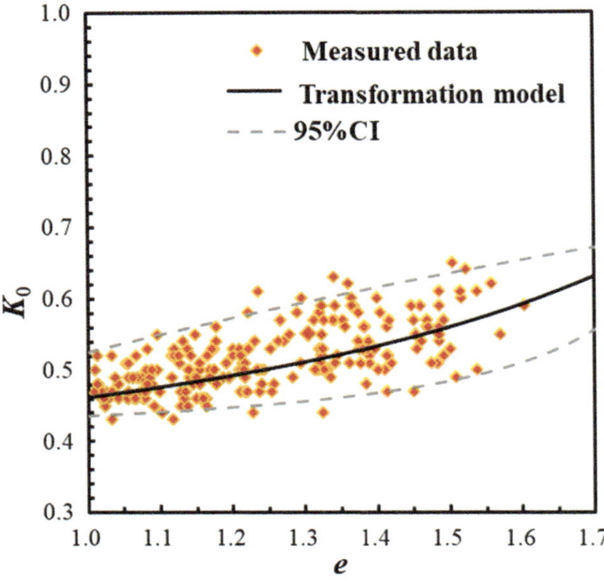

Figure 15.3 Transformation model and observed data between e and K_0.

Figure 15.3 illustrates the derived relationship between (e, K_0) based on Equation (15.7) and the corresponding data points from the database. As can be seen from the picture that the collected data align well with the transformation formula, with the majority of data points falling within the 95% confidence interval. Functional relationship between the target parameters and other univariate or multivariate parameters can also be similarly established through the same approach.

In Tables 15.5–15.8, the coefficient for X_i quantifies the sensitivity of the target parameters with respect to information of X_i. Considering the case where there is only univariate information, the coefficients for X_i exhibit good consistency with the correlation matrix presented in Table 15.4. Taking the example of the target parameter K_0: if the coefficient for X_i is negative, the target parameter increases as X_i increases, while conversely, if the coefficient is positive, the target parameter decreases with increasing X_i. For X_1 to X_5, they are positively correlated to X_7, as evidenced by their negative coefficients in b'_X. Conversely, the other parameters are negatively related to X_7, corresponding to the positive coefficients in b'_X. Additionally, the absolute ratio of the coefficient for X_i to a'_X reflects the degree of influence that X_i has on the target parameter. The ranking of the ratio values for each X_i corresponds to the ranking of the absolute value of the correlation coefficient between each X_i and X_7. The same patterns can be observed for E_s, c, and φ.

Furthermore, according to Equation (15.4), a'_X is inversely proportional to σ'_x. A higher value of a'_X indicates a smaller standard deviation for the updated variable, implying its decreased uncertainty. As can be seen from Tables 15.5–15.8, as more information is incorporated, the value of a'_X gradually increases, signifying a reduction in the uncertainty of the target parameters. This observation is consistent with our common understanding.

However, it is worth noting that for the target parameter K_0, the updated Johnson parameters obtained based on (ω, LL, PI, LI, e), (ω, LL, PI, LI, e, σ'_v/P_a), and (ω, LL, PI, LI, e, σ'_v/P_a, S_u/σ'_v) yield very similar results, with the coefficients for X_6 and X_{11} close to zero. This implies that the inclusion of σ'_v and S_u information has minimal impact on the update and prediction for the parameter K_0. Similarly, the information related to p_s has little effect on the update and prediction for the target parameter E_s.

Moreover, it is noticeable from Tables 15.5–15.8 that when multisources of information are simultaneously known, the coefficients for X_i in b'_X no longer present the same pattern as they did with univariate information. For instance, in the prediction of K_0 using (ω, LL, PI, LI, e), the coefficients for X_3 and X_4 become positive, which contradicts their positive correlation with X_7. Therefore, it is imperative to assess the validity of the transformation model.

15.4.3 Validation of transformation models

Validation of the transformation models is conducted using 156 data points from a new site in Shanghai, located in the Xuhui distinct. The actual values of the target parameters are assumed to be unknown during the prediction process. After the prediction, they are compared with the predicted medians and 95% CIs to verify the performance of the transformation model.

Figure 15.4 illustrates the median values (represented by open dots) and the 95% CI (indicated by black vertical dashed lines) for K_0 conditioning on various combinations of information. Parameters σ'_v/P_a and S_u are not included considering their little impact on the prediction of K_0. The 1:1 line represents the scenario where actual values match the predicted values. If the vertical line representing the 95% CI intersects with the 1:1 line, it indicates that the actual value falls within the predicted 95% CI. The red dashed line is used to annotate the absence of a match if there is no intersection. Note that Figure 15.4(a) signifies the statistical results of the raw data from this new site, without incorporating any information for the parameter updates.

From Figure 15.4, it is clear that the median points gradually converge toward the 1:1 line when K_0 is conditioned on other available information, indicating a closer alignment between the predicted values and actual values. However, it can be observed that the conditioned 95% CIs do not cover the 1:1 line when K_0 equals 0.4, 0.43, 0.44, 0.62, or 0.65. This is because for K_0 = 0.4, 0.43, and 0.65, these data points inherently fall outside the 95% CI of the data used to construct the model, with 0.4 even being lower than the minimum value (=0.43) in the database. Furthermore, the two data points (K_0 = 0.44 and 0.66) are very close to the 2.5% and 97.5% percentiles of the K_0 values in the database.

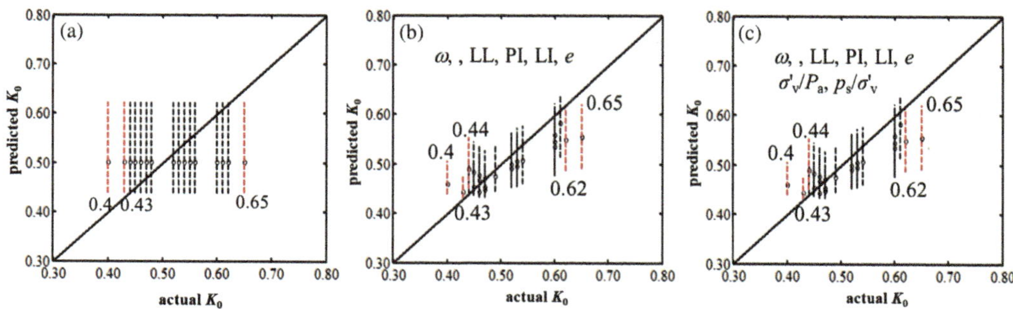

Figure 15.4 Updating results for K_0 by conditioning on: (a) no information; (b) ω, LL, PI, LI, e; (c) ω, LL, PI, LI, e, σ'_v/P_a, p_s/σ'_v.

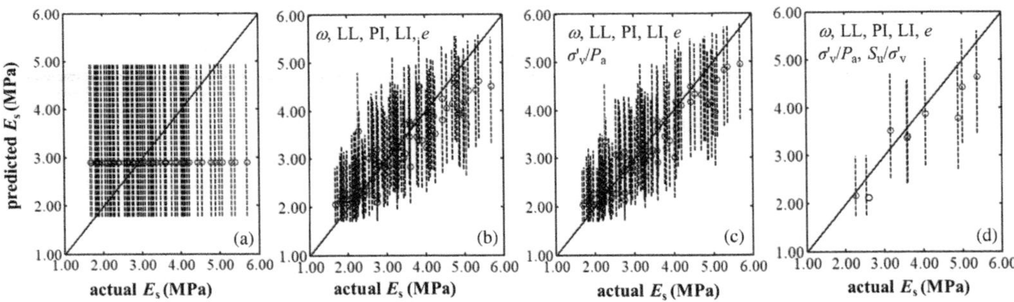

Figure 15.5 Updating results for E_s by conditioning on: (a) no information;(b) ω , LL, PI, LI, e;(c) ω , LL, PI, LI, σ'_v/P_a; (d) ω, LL, PI, LI, e, σ'_v/P_a, S_u/σ'_v.

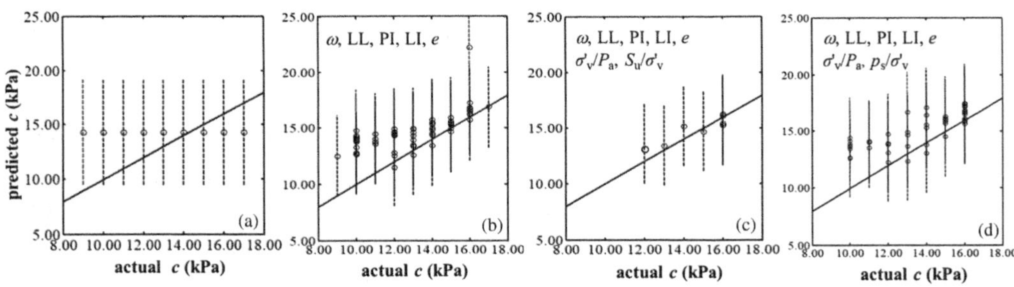

Figure 15.6 Updating results for c by conditioning on: (a) no information;(b) ω , LL, PI, LI, e;(c) ω , LL, PI, LI, σ'_v/P_a, S_u/σ'_v; (d) ω, LL, PI, LI, e, σ'_v/P_a, p_s/σ'_v

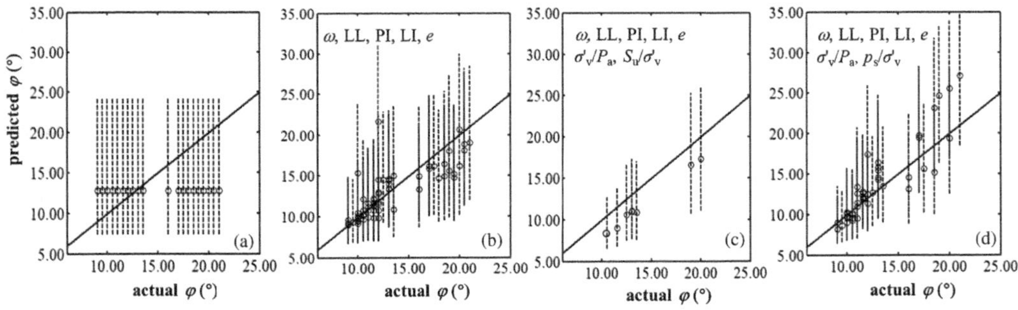

Figure 15.7 Updating results for φ by conditioning on: (a) no information; (b) ω, LL, PI, LI, e; (c) ω, LL, PI, LI, σ'_v/P_a, S_u/σ'_v; (d) ω, LL, PI, LI, e, σ'_v/P_a, p_s/σ'_v.

This implies that if any data point lies outside or close to the 95% CI of the database, extrapolation issues might arise.

Figures 15.5–15.7 show the updating results for E_s, c, and φ conditioning on various combinations of information. The results indicate that the majority of the actual values fall within the 95% CIs of the predicted values. Moreover, there is a noticeable decrease of the width of the confidence interval with the increasing information as well as the closer alignment between the predicted values and actual values.

15.5 CONCLUSIONS

This chapter presents a database for Shanghai soft clay labeled as SH-CLAY/12/4191, whose data points are collected from 51 site investigation reports in Shanghai covering approximately 145 km^2 in area, about 1/43 the size of Shanghai. General information including the geology of Shanghai, the borehole locations, the types of clay parameters, and the basic statistics of these parameters is given. By comparing the marginal statistics of SH-CLAY/12/4191 with those presented in the related lecture, it can be concluded that the quality of the data in SH-CLAY/12/4191 is generally satisfactory. A multivariate probability distribution for the 12 clay parameters is then constructed based on the SH-CLAY/12/4191 database. Finally, transformation models for soil design parameters K_0, E_s, c, and φ in relation to various index and in-situ test parameters are established. The established models are verified using data from a new site, and the results demonstrate the generally reliable performance of the established transformation models.

REFERENCES

Ching, J. & Phoon, K. K. (2014a). Transformations and correlations among some clay parameters—the global database. *Canadian Geotechnical Journal*, 51(6), 663–685.

Ching, J. & Phoon, K. K. (2014b). Correlations among some clay parameters–the multivariate distribution. *Canadian Geotechnical Journal*, 51(6), 686–704.

Ching, J., Li, D. Q., & Phoon, K. K. (2016). *Statistical Characterization of Multivariate Geotechnical Data*. Chapter 4, Reliability of Geotechnical Structures in ISO2394, CRC Press, Balkema, 89–126.

Ching, J., Lin, G. H., Chen, J. R., & Phoon, K. K. (2017). Transformation models for effective friction angle and relative density calibrated based on generic database of coarse-grained soils. *Canadian Geotechnical Journal*, 54(4), 481–501.

Ching, J. & Phoon, K. K. (2019). Constructing site-specific multivariate probability distribution model using Bayesian machine learning. *Journal of Engineering Mechanics*, 145(1), 04018126.

Ching, J., Phoon, K. K., Khan, Z., Zhang, D., & Huang, H. (2020). Role of municipal database in constructing site-specific multivariate probability distribution. *Computers and Geotechnics*, 124, 103623.

Ching, J., Phoon, K. K., Ho, Y. H., & Weng, M. C. (2021). Quasi-site-specific prediction for deformation modulus of rock mass. *Canadian Geotechnical Journal*, 58(7), 936–951.

Ching, J., Phoon, K. K., Yang, Z., & Stuedlein, A. W. (2022). Quasi-site-specific multivariate probability distribution model for sparse, incomplete, and three-dimensional spatially varying soil data. *Georisk: Assessment and Management of Risk for Engineered Systems and Geohazards*, 16(1), 53–76.

D'Ignazio, M., Phoon, K. K., Tan, S. A., & Lansivaara, T. (2016). Correlations for undrained shear strength of Finnish soft clays. *Canadian Geotechnical Journal*, 53(10), 1628–1645.

Gao, D. Z & Wei, D. D. (1983). Probability and statistical characteristics of soil engineering properties in shanghai. In *Proceedings of the 4th Conference of the Chinese Civil Engineering Society on Soil Mechanics and Foundation Engineering, Wuhan, China*. (in Chinese)

ISO2394 (2015). *General Principles on Reliability for Structures*. International Organization for Standardization, Geneva, Switzerland.

Johnson, N. L. (1949). Systems of frequency curves generated by methods of translation. *Biometrika*, 36, 149–176.

Kulhawy, F. H. & Mayne, P. W. (1990). *Manual on Estimating Soil Properties for Foundation Design*. Report EL-6800, Electric Power Research Institute, Palo Alto, CA.

Liu, S., Zou, H., Cai, G., Bheemasetti, B. V., Puppala, A. J., & Lin, J. (2016). Multivariate correlation among resilient modulus and cone penetration test parameters of cohesive subgrade soils. *Engineering Geology*, 209, 128–142.

Löfman, M. S. & Korkiala-Tanttu, L. K. (2022). Transformation models for the compressibility properties of Finnish clays using a multivariate database. *Georisk: Assessment and Management of Risk for Engineered Systems and Geohazards*, 16(2), 330–346.

Mayne, P. W., Christopher, B. R., & DeJong, J. (2001). *Manual on Subsurface Investigations*. Nat. Highway Inst. Sp. Pub. FHWA NHI-01–031. Fed. Highway Administ, Washington, DC.

Muzamhindo, H. & Ferentinou, M. (2023). Generic compressive strength prediction model applicable to multiple lithologies based on a broad global database. *Probabilistic Engineering Mechanics*, 71, 103400.

Phoon, K. K. & Kulhawy, F. H. (1999). Characterization of geotechnical variability. *Canadian Geotechnical Journal*, 36(4), 612–624.

Slifker, J. F. & Shapiro, S. S. (1980). The Johnson system: selection and parameter estimation. *Technometrics*, 22(2), 239–246.

Xiao, L. (2018). *Probabilistic Analysis of Shield Tunnel Convergence by Random Field of Multilayer Soil*. M.S. thesis, Tongji University, Shanghai. (in Chinese)

Zhang, D., Zhou, Y., Phoon, K. K., & Huang, H. (2020). Multivariate probability distribution of Shanghai clay properties. *Engineering Geology*, 273, 105675.

Zhou, Y., Zhang, D., Huang, H., & Xue, Y. (2022). Effect of normal transformation methods on performance of multivariate normal distribution. *ASCE-ASME Journal of Risk and Uncertainty in Engineering Systems, Part A: Civil Engineering*, 8(1), 04021074.

APPENDIX A

Table A1 Statistics of the muddy clay in Shanghai

Site No.	*Sample size*	*Index*	ω (%)	*LL* (%)	*PI*	**e**	φ (°)
1	14	Mean	49.8	45.7	21	1.43	9.5
		COV	0.16	0.20	0.54	–	–
2	16	Mean	51.0	43.9	21.8	1.44	7.4
		COV	0.25	0.23	0.97	–	–
4	30	Mean	50.5	43.8	20.6	1.45	8.0
		COV	0.22	0.23	0.72	–	–
7	9	Mean	48.0	44.1	19.9	–	8.1
		COV	0.14	0.16	0.79	–	–
8	33	Mean	49.0	45.9	22.0	1.41	8.4
		COV	0.26	0.30	0.79	–	–
9	13	Mean	51.2	44.9	20.7	1.47	7.8
		COV	0.17	0.22	0.77	–	–
10	15	Mean	50.0	40.1	18.6	1.45	–
		COV	0.30	0.25	1.04	–	–
11	90	Mean	48.0	–	20.0	1.73	–
		COV	0.26	–	0.88	–	–
Summary	220	Mean	49.1	44.3	20.5	1.6	8.2

Source: Data from Gao and Wei (1983).

Table A2 Statistics of the muddy clay, muddy silty clay, and clay in Shanghai

Soil Layer	*Index*	ω (%)	*LL (%)*	*PI*	e	E_s *(MPa)*	c *(kPa)*	φ (°)
muddy silty clay	Min–Max	36.0~49.7	29.6~40.1	10.3~17.0	1.00~1.36	2.20~5.97	8.5~14.2	12.1~28.0
	COV	0.110	0.066	0.145	0.104	0.292	0.240	0.250
muddy clay	Min–Max	40.0~59.6	34.4~50.2	17.0~25.1	1.12~1.67	1.32~3.58	11.5~15.7	8.5~16.9
	COV	0.080	0.078	0.112	0.075	0.179	0.037	0.162
clay	Min–Max	29.8~42.5	28.3~42.9	10.2~20.0	0.85~1.22	3.00~6.77	11.5~20.0	12.7~27.4
	COV	0.082	0.096	0.152	0.079	0.200	0.223	0.217
Summary	Min–Max	29.8~59.6	28.3~50.2	10.2~25.1	0.85~1.67	1.32~6.77	8.5~20.0	8.5~28.0

Source: Data from Code for Investigation of Geotechnical Engineering in Shanghai, DGJ08-37-201.

Index

F

G

H

I

J

L

M

N

O

P

Q

R

T

U

V

W

Y

For Product Safety Concerns and Information please contact our EU representative GPSR@taylorandfrancis.com Taylor & Francis Verlag GmbH, Kaufingerstraße 24, 80331 München, Germany

Batch number: 10399588

Printed by Printforce, the Netherlands